THE CUTTING EDGE: CONSERVING WILDLIFE IN LOGGED TROPICAL FOREST

Robert A. Fimbel, Alejandro Grajal, and John G. Robinson, editors

Columbia University Press
New York

Columbia University Press
Publishers Since 1893
New York   Chichester, West Sussex

Library of Congress Cataloging-in-Publication Data

The cutting edge: conserving wildlife in logged tropical forests /
edited by Robert A. Fimbel, Alejandro Grajal, and John G. Robinson.
p.   cm. — (Biology and resource management)
Includes bibliographical references (p.).
ISBN 0–231–11454–0 (cloth) — ISBN 0–231–11455–9 (pbk.)
1. Wildlife conservation—Tropics. 2. Rain forests—Management.
I. Fimbel, Robert A. II. Grajal, Alejandro. III. Robinson, John G. IV. Biology and
resource management in the tropics series
QL109.C66   2001
333.95'416'0913—dc21   00–031782

Casebound editions of Columbia University Press books
are printed on permanent and durable acid-free paper.

Printed in the United States of America
c 10 9 8 7 6 5 4 3 2 1

Contents

Foreword   ix

Preface   xiii

Contributors   xvii

Foreword

In recent decades, there has been unprecedented growth in the scale and intensity of tropical forest management. Industrial forestry, both directly and indirectly, has wreaked havoc on the wildlife and the ecological integrity of tropical forests. While the mark of human intervention in modern times is quite apparent, forests have in fact been subject to some form of management since the emergence of human beings. The wildlife of forests that we value so highly today is the result of varying intensities of human management over an extraordinarily long period. And the more we study even the most remote forests, the more surprised we are by the sophistication and extent of its management by traditional peoples.

Just as human populations have exploded over the past century, so our ability to manipulate, modify and replace tropical forests has changed considerably. In only a few decades, we have moved from low-intensity local management driven largely by efforts to meet local needs, to highly disruptive management driven by the desire to satisfy distant industrial needs. It is easy for us to forget that only since World War II have the chain saws and the heavy-tracked vehicles needed to clear and log tropical forests on a large scale become widely available. The unprecedented economic and technological growth of the second half of the twentieth century has totally changed the paradigm under which tropical forests exist and are managed. Most notably, our ability to manipulate forests—in ways that can cause major environmental disturbance—has evolved far faster than the laws, institutions and traditions that previously served to safeguard the broader values of forests.

In all this increased forest disturbance of recent years, the highly visible destruction resulting from industrial "first-cut" logging of old growth tropical forests has, more than any other activity, provoked a groundswell of public and political concern, and a plethora of plans, strategies and projects aimed at mitigating the damage. Most of these conservation responses, however, have had only limited success. For the most part, they have attacked the visible symptoms of bad forestry rather than the underlying causes. This has occurred because much of the response has been rooted in incorrect assumptions about the pre-logging history of our forests and the real nature of the threats they face. Many conservation activists and political decision-makers still hold to the myth that loggers are entering areas of pristine wilderness, never before marked by human presence. They fail to recognize that what we are witnessing is, in large part, simply an increase in the intensity of our manipulation of the forest, albeit a major increase. But, more important, they fail to recognize that the heart of the current problem is that these significantly increased impacts on our forests are occurring at a time when the demand for land to be used for local agriculture or for cultivation of industrial crops is also growing exponentially.

The international debate on forest conservation that is occurring under the auspices of the Convention on Biological Diversity and the Inter-Governmental Forum on Forests has been important in helping us to change our perceptions about the nature of the threats to tropical forests and to think more clearly about possible solutions. In particular, this wide-ranging debate is helping to build a scientific and popular consensus that forestry alone may not be the cause of the problems, and that good forestry, in fact, may be part of the solution. From these inter-governmental processes is emerging a growing acknowledgement that solutions to forest problems must reconcile the needs and wishes of multiple users and beneficiaries of the forest, and that such solutions will entail management for a wide range of forest products and services. Many of the problems we are wrestling with today in forest management have resulted from the appropriation of huge tracts of forest to single operators for extraction of a single product: timber. In contrast, forest management in the future will have to achieve a balance in access to all forest goods and services, reflecting the needs and demands of multiple stakeholders. Furthermore, those engaged in forest management will have to focus less on activities at the stand scale and more on achieving a balance in forest use across the broader landscape. Hopefully, we are moving toward an era in which "ecosystem management" will be the guiding principle of field managers rather than just the rhetoric of planners.

The management of wildlife is central to any new era of tropical forestry. Wildlife as defined in this book encompasses the full range of vertebrate and invertebrate animals that make up the fauna of the forest. Tropical forests harbor the majority of the world's animal species, perhaps 70 to 80 percent of all species. This wildlife is immensely valuable to humankind for a wide variety of goods and services. Wildlife has significant and diverse—and yet still little understood—values in the overall functioning of a forest ecosystem. Wildlife contributes to the livelihood of forest-dwelling people, providing an important source of food, and it has a range of aesthetic and ethical values—even for people who will never visit or encounter these species firsthand.

Reconciling conservation of the full spectrum of forest wildlife values with competing pressures to manage forests for production of industrial products as well as other land uses will require a quantum leap in our understanding about the complexity of forest systems. It will require an increase in our understanding of forest wildlife and how people in different sectors of society value its conservation, for economic, spiritual and other reasons. Ultimately, we must capture these values in our economic decision-making about forest management and land use.

For all of the above reasons, this book is timely and significant. It brings together the accumulated knowledge of many of the leading conservation scientists in wildlife and forest management. Collectively, the chapters make a case for the need for new approaches to integrating wildlife into the management of tropical forests. This book leaves us with no doubt that better forest management is pivotal to the maintenance of much of the valuable biodiversity of the tropics, and suggests how much of what is already known about forestry practices could be modified to achieve better outcomes for forest wildlife.

The book also makes a strong case against an argument that still persists in some quarters: that industrial forestry and wildlife conservation are fundamentally irreconcilable. This view is rooted in the misconception that there is a simple choice between forest use by humans and virgin wilderness; that the alternative to managed forestry is designation of the forest as a nature reserve or national park. This ignores the reality that the biggest threat to forest wildlife lies in agricultural conversion. Experience has shown repeatedly that solutions that focus exclusively on the values of distant stakeholders and ignore the legitimate needs and concerns of local people will almost never succeed. Total protection of a forest is rarely a good option for a poor person dwelling on the edges of it.

Inevitably, multiple-use management of forests must be included in the portfolio of measures needed to meet the challenge of conserving the world's forest wildlife. At the same time, forestry practice will have to be based on more sophisticated application of ecological knowledge. Although there will be some measures that could be applied across the full spectrum of forest types, conditions and uses, most wildlife management will require that forest practitioners have the ecological skills and training to make sound decisions on the basis of local information and needs. Clearly, we will be faced with much uncertainty about the impact of different management interventions on forest wildlife, as well as with changes in related human expectations and demands over time and from place to place. Therefore, good forest management will have to be adaptive, marked by continuing negotiations among all users of the forest and careful monitoring of key measures or indicators of wildlife health. The insights provided by this book will be invaluable in helping managers to respond to these challenges.

Forest managers really have no choice but to proceed along this route. The knowledge provided by this book is essential for countries and forest managers who are dedicated to meeting the commitments being made by their countries under the Convention on Biological Diversity and other mechanisms that are working to define criteria and indicators for assessing sustainability of tropical forests. The maintenance of wildlife populations at certain levels is central to all criteria for recognition of forest management as sustainable. And increasingly, only products that come from forests that are certified as being sustainably managed will qualify for the highest value markets.

The most dramatic damage to forests over the past decades has occurred in countries still at an early stage of their economic and social development. It has often occurred in remote frontier areas where the rule of law was yet to be felt. Much of this damage has been linked to "crony capitalism"—big business riding roughshod over the interests of society at large. Forest misuse has often been a focus of social unrest and of calls for greater justice and equity. Experience suggests that forests are more likely to be well managed in societies that have sound democratic processes and are subject to the rule of law. Thus, the wave of social reform that is now sweeping through many of the less developed countries of the tropics should provide a context in which better forest management can work. The rise of national conservation lobbies as part of emerging civil societies should further increase the demand for good forest management. At this time, the social and economic context is right for the adoption of the technologies discussed in this book.

Many challenges still remain. We are at an early stage of acquiring the knowledge and skills needed to better understand the implications of forest management practices on wildlife. Much more research will be needed, including a great deal of empirical observation of changes in wildlife populations. This book presents the state of the art and makes a convincing case for what is possible. Future generations of scientists and practitioners will need to build on this body of knowledge to establish new forestry techniques that adequately address wildlife conservation needs.

Professor Jeffrey A. Sayer
Director General, Centre for International Forestry Research (CIFOR)

Preface

This book presents a timely treatment of wildlife-forestry interactions, synthesizing current knowledge regarding the direct and indirect impacts of tropical timber-harvesting practices on native forest fauna. More importantly, the authors highlight opportunities for mitigating these environmental impacts in the short-term, and present directions for developing forest management techniques that protect the biological integrity of forest systems. Its origin began with the growing awareness of the conservation potential of tropical production forests, and the current threats that logging activities present to these natural communities. Early in the decade, a number of field scientists and forest resource managers met informally and identified the need to summarize all available knowledge on logging-wildlife interactions. This information was deemed critical to the development of short-term management strategies for reducing harmful impacts on native fauna and their habitats, and the identification of longer-term research priorities to address gaps in our understanding of these interactions and their impacts. These initial discussions led to a November 1996 workshop in Bolivia, hosted by BOLFOR (Bolivia Sustainable Forestry Management Project) and the Wildlife Conservation Society, where short- and longer-term recommendations on how to conserve wildlife in production forests were developed. The findings of this mixed forum, which included foresters, wildlife biologists, resource mangers, and policy makers, served as a springboard for the 30 chapters presented in this volume.

In brief, the book aims to: a) synthesize current knowledge of the impacts of forest management practices, particularly logging, on wildlife in tropical forests; b) present guidelines for timber-harvesting systems that balance economic with ecological considerations; and, c) set directions for future research, in an effort to move sustainable forest management-wildlife conservation from a theoretical entity to an applied state in production forests. Contributors were selected for their expertise in tropical forestry-wildlife interactions, and have provided relatively exhaustive reviews of their topics. In an effort to strengthen the continuity within this volume, the editors have subdivided the contributions into six sections, each with its own brief synopsis. These section overviews present a short introduction to the main theme of the section, condensed abstracts of each chapter, and a brief summary of the major issues discussed in the section. An introduction (chapter 1) and final synopsis chapter (chapter 30) also frame the key issues, present syntheses of the contributor's key points, and outline prospects for wildlife conservation within managed tropical forest landscapes.

We have designed the book to appeal to a wide audience of scientists and nonscientists specifically interested in wildlife-timber harvesting issues, and sustainable forestry in general. In addition to foresters who should find this volume an invaluable tool in assisting their efforts to develop sustainable forest management practices, we hope that this book will interest the academic conservation community; including conservation biologists, forest and wildlife ecologists, social scientists and their students; professional decision-makers (including ecological timber certification organizations, resource managers, and conscientious managers of commercial timber lands interested in practicing good stewardship), nonprofit research and policy institutions (Smithsonian, TROPENBOS, CIFOR, WWF, FAO, BIONET members, etc.), and donor agencies (USAID, GEF, SNV, GTZ, World Bank, etc.).

Finally, and most importantly, we wish to acknowledge the numerous individuals that helped to make this volume a reality. Foremost were the contributors, who expended an enormous amount of time and energy researching their topics, reviewing related chapters, and applying patience during our editorial reviews. Participants in the Bolivia workshop were also instrumental in the creation of the book, serving as a catalyst for its development: E. A. Alvarez, E. Bennett, R. B. Wallace, R. Boot, G. Blate, J. F. Contreras Sanjines, J. C. Cornejo, R. Z. Donovan, J. Falck, R. A. Fimbel, B. Flores, E. Flores, P. Frumhoff, G. Garson, A. Grajal, A. Grimes, D. Guinart Sureda, R. E. Gullison, J. C. Herrera, H. H. Hurtado Balcazar, J. Johnson, I. Kraljevic, W. F. Laurance, H.-F. Maitre, D. Mason, C. E. Murillo Salazar, J. B. Nittler, H. Ormachea, R. L. Painter, L. F. Perez Obando, A. Plumptre, F. E. Putz, L. Ouevedo, J. G. Robinson, R. J. Rivero Bonifaz, S. Rosholt, D. I. Rumiz, E. Sandoval Hurtado, P. Saravia Patón, D. S. Hammond, A. Stocks, A. Taber, G. Taylor, W. R. Townsend, A. R. Vargas Pereira, A. Whitman, M. Yates, and P. Zuidema.

A special thanks is due to Jeff Sayer, who took time from his extremely busy schedule at the helm of the young Centre for International Forestry Research (CIFOR) to write the volume's Foreword.

We especially appreciate the efforts of numerous scientists and resource managers who donated their time as peer reviewers for the chapters (most anonymously), including Cheryl Fimbel who reviewed all chapters (some several times!). The comments provided by these individuals helped enormously to improve the content and clarity of the chapters.

Lastly, we wish to thank the Wildlife Conservation Society for its support throughout the development of the volume, and Tracy van Holt and Rosa Fernando in particular, for their tireless efforts and good humor in typing and formatting the various iterations of the book.

Robert A. Fimbel
Alejandro Grajal
John G. Robinson

Contributors

Chapters

Aguilar, Fernando
Consejo Boliviano para la Certificación Forestal Voluntaria
Av. Cañoto 580, entre México y Centenario, Edif. Vázquez Of. M1
Casilla No 7175
Santa Cruz, Bolivia

Auzel, Philippe
DEA Environnement (ESEM/ERMES)
Rue Léonard de Vinci
45067 Orleans Cedex
France

Barborak, James R.
Conservationist
Mesoamerican and Caribbean Program
Wildlife Conservation Society
4424 N.W. 13th Street, Suite A-2
Gainesville, FL 32609 U.S.A.

Bennett, Elizabeth L., Ph.D.
Wildlife Conservation Society
7 Jalan Ridgeway, 93200 Kuching
Sarawak, Malaysia

Benstead, Jonathan P., Ph.D.
Institute of Ecology
University of Georgia
Athens, GA 30602 U.S.A.

Blake, Stephen
The Nouabalé-Ndoki Congo Forest Conservation Project
Wildlife Conservation Society
185th Street and Southern Blvd.
Bronx, NY 10460 U.S.A.

Boundzanga, Georges C.
Ministère des Eaux et Fôrets
B.P. 98
Brazzaville, Congo

Byron, Neil, Ph.D.
Senior Associate
Center for International Forestry Research
P.O. Box 6596 JKPWB
Jakarta 10065 Indonesia

Caldwell, Janalee P., Ph.D.
Associate Curator of Amphibians and Associate Professor of Zoology
Oklahoma Museum of Natural History and Department of Zoology,
University of Oklahoma, Norman, OK 73019 U.S.A.

Camilo, Gerardo, R., Ph.D.
Department of Biology
Saint Louis University
3507 Laclede Ave.
St. Louis, MO 63103-2010 U.S.A.

Chapman, Colin A., Ph.D.
Department of Zoology
University of Florida
Gainesville, FL 32611 U.S.A.

Davies, Glyn, Ph.D.
Biodiversity in Development (EC/DFID/IUCN)
Avenue de Arts, 50
1000 Brussels, Belgium

Donovan, Richard Z.
Director, SmartWood Program
Rainforest Alliance
61 Millet St.
Richmond, VT 05477 U.S.A.

Dranzoa, Christine, Ph.D.
Makerere Institute of Environment and Natural Resources
PO Box 10066
Kampala, Uganda

Fimbel, Robert A., Ph.D.
Washington Parks and Recreation Commission
Resource Stewardship Program
7150 Cleanwater Lane
Olympia, WA 98504 U.S.A.

Francis, Charles M., Ph.D.
Bird Studies Canada
P.O. Box 160, Port Rowan
N0E 1M0 Canada

Frumhoff, Peter, Ph.D.
Director, Global Resources Program
Union of Concerned Scientists
Two Brattle Square
Cambridge, MA 02238-9105 U.S.A.

Ghazoul, Jaboury, Ph.D.
Huxley School for Environment
Earth Sciences and Engineering
Imperial College, Silwood Park
Ascot, Berkshire, SL5 7PY, UK

Grajal, Alejandro, Ph.D.
Director
Latin America and Caribbean Program
National Audubon Society
444 Brickell Avenue, Suite 850
Miami, FL 33131 U.S.A.

Grieser Johns, Andrew, Ph.D.
Fountain Renewable Resources Limited
The Bell Tower
12 High Street
Brackley
Northants NN12 7DT, UK

Guinart, Daniel S., Ph.D.
Pso. Jaume Gordi 28
08921 Barcelona, Spain

Gullison, R. E., Ph.D.
13810 Long Lake Road
R.R. #3
Ladysmith, B.C. VOR 2E0
Canada

Gumal, Melvin T.
National Parks and Wildlife Division
Sarawak Forest Department, 93660 Kuching
Sarawak, Malaysia

Henderson, Max
Pacific Heritage Foundation
P.O. Box 546
Rabaul, East New Britain,
Papua New Guinea

Herrera F., José Carlos
BOLFOR, Casilla #6204
Santa Cruz
Bolivia

Heydon, Matthew J., Ph.D.
Farming and Rural Conservation Agency
Oxford Spires Business Park, Kidlington, Oxford, OX5 1FR, UK

Hill, Jane, Ph.D.
Dept. of Biological Sciences
University of Durham
Durham DH1 3LE, UK

Jansen, Patrick A.
Wageningen Agricultural University
Department of Environmental Studies
Silviculture and Forest Ecology Group
P.O. Box 342
6700 AH Wageningen
The Netherlands

Kremen, Claire, Ph.D.
Wildlife Conservation Society/Center for Conservation Biology
Dept. of Biological Sciences
Stanford University
Stanford, CA 94305 U.S.A.

Laurance, William F., Ph.D.
Biological Dynamics of Forest Fragments Project
National Institute for Research in the Amazon (INPA)
C.P. 478
Manaus, AM 69022-970
Brazil

Leader-Williams, Nigel, Ph.D.
Durrell Institute of Conservation and Ecology
Department of Anthropology
University of Kent at Canterbury
Canterbury CT2 7NS, UK

Leighton, Mark, Ph.D.
Peabody Museum
Harvard University
Cambridge, MA 02138 U.S.A.

Losos, Elizabeth, Ph.D.
Center for Tropical Forest Science
Smithsonian Tropical Research Institute
900 Jefferson Drive, Suite 2207
Washington, DC 20560 U.S.A.

MacKinnon, John, Ph.D.
88 Wincheap
Canterbury CT1 3RS, U.K.

Marcot, Bruce G., Ph.D.
USDA Forest Service, Pacific Northwest Research Station
Portland Forestry Sciences Lab
1221 SW Yamhill, Suite 200
POB 3890
Portland OR 97208-3890 U.S.A.

Mason, Douglas J., Ph.D.
USAID/Bolivia, Unit 3914
APO AA 34032 (USA only)
Biodiversity and Forestry Advisor
Agencia de los Estados Unidos
Para el Desarrollo Internacional
Casilla Correo 4530
La Paz, Bolivia

Newing, Helen, Ph.D.
4 Blenheim Road
Horspath, Oxford OX33 1RY, UK

Ochoa, José G., Ph.D.
Asociación Venezolana para la Conservación de Areas Naturales (ACOANA)
Apartado 69520
Caracas, 1063-A, Venezuela

Owiunji, Isaiah
Makerere Institute of Environment and Natural Resources
PO Box 10066
Kampala, Uganda

Pinard, Michelle A., Ph.D.
Department of Forestry
University of Aberdeen
Aberdeen, AB24 5UA
Scotland, UK

Plumptre, Andrew J., Ph.D.
Wildlife Conservation Society
International Programs
185th Street and Southern Boulevard
Bronx, NY 10460 U.S.A.

Pringle, Catherine M., Ph.D.
Institute of Ecology
University of Georgia
Athens, GA 30602 U.S.A.

Putz, Francis E., Ph.D.
Department of Botany
University of Florida
P.O. Box 11 8526
Gainesville, FL 32611 U.S.A.

Robinson, John G., Ph.D.
Wildlife Conservation Society
International Programs
185th Street and Southern Boulevard
Bronx, NY 10460 U.S.A.

Rumiz, Damián I., Ph.D.
Museo Noel Kempff Mercado
UAGRM
Santa Cruz, Bolivia

Salafsky, Nick
MacArthur Foundation
140 S. Dearborn Street, Suite 1100
Chicago, IL 60603-5285 U.S.A.

Sidle, John G.
Great Plains National Grasslands
USDA Forest Service
125 N Main St.
Chadron, NE 69337 U.S.A.

Sirot, Laura K.
Department of Zoology
University of Florida
Gainesville, FL 32611 U.S.A.

Solar R., Luciano
Calle Ingavi casi 2do Anillo
Santa Cruz, Bolivia

Soriano, Pascual J.
Departamento de Biología
Facultad de Ciencias
Universidad de Los Andes
Mérida 5101, Venezuela

Thiollay, Jean-Marc, Ph.D.
Laboratoire d'Ecologie, E.N.S.
46, rue d'Ulm 75230 Paris Cedex 05
France

Vitt, Laurie, Ph.D.
Curator of Reptiles and Professor of Zoology
Oklahoma Museum of Natural History and Department of Zoology
University of Oklahoma, Norman, OK 73019 U.S.A.

Wilkie, David S., Ph.D.
Associates in Forest Research and Development
18 Clark Lane
Waltham, MA 02154 U.S.A.

Zakaria, Mohamed Bin Hussin, Ph.D.
Faculty of Forestry
Universiti Pertanian Malaysia
43400 Serdang, Malaysia

Zou, Xiaoming, Ph.D.
Institute of Tropical Ecosystems Studies
University of Puerto Rico
PO Box 363682
San Juan, PR 00936-3682 U.S.A.

Zuidema, Pieter A.
University of Utrecht
Department of Plant Ecology and Evolutionary Biology
PO Box 800.84
3508 TB Utrecht
The Netherlands

Text Boxes

Cherry, Ronnie
Harvard University Laboratory of Tropical Forest Ecology
Peabody Museum
Cambridge, MA 02138 U.S.A.

Gascon, Claude, Ph.D.
Deputy Director, Center for Applied Biodiversity Science
Conservation International
2501 M St. NW, Suite 200
Washington, DC 20037 U.S.A.

Guillery, Phil
Forest Project Director
Institute for Agriculture and Trade Policy
2105 First Avenue South
Minneapolis, MN 55409 U.S.A.

Hammond, David S., Ph.D.
Tropenbos-Guyana
12 Garnett Street
Campbellville, Georgetown, Guyana

Hartshorn, Gary, Ph.D.
Organization for Tropical Studies
North American Headquarters
Box 90630
Durham, NC 27708-0630 U.S.A.

Malcolm, Jay R., Ph.D.
Assistant Professor, Faculty of Forestry University of Toronto
33 Willcocks St., Toronto, Ontario
ON M5S 3B3, Canada

Miller, Bruce, Ph.D
Wildlife Conservation Society
PO. Box 37
Belize City, Belize

Nummelin, Matti, Ph.D.
Environmental Biology
Department of Ecology and Systematics
P.O. Box 7
FIN-00014 University of Helsinki
Finland

Thomas, Raquel
Tropenbos-Guyana
12 Garnett Street
Campbellville, Georgetown, Guyana

Part I

AN INTRODUCTION TO FORESTRY-WILDLIFE INTERACTIONS IN TROPICAL FORESTS

Wildlife serves a crucial role in maintaining the health of natural forests, where up to 90 percent of the plant species in tropical rain forests are dependent on animals for their pollination or dispersal—including many important timber species. Forest animals, in addition, are integral components of all ecosystem processes, such as predator-prey relationships that keep pest species in-check; vectors for mycorrhizal fungi-tree symbiotic relationships; and decomposers-nutrient recyclers that maintain forest productivity.

Timber management of tropical forests has both direct and indirect effects on wildlife populations and their habitats. Part I briefly reviews the role of wildlife in tropical forests, the common silvicultural practices used in these forests, and potential short- and long-term consequences for wildlife, forest health, and productivity as a result of timber management activities. General recommendations for mitigating these impacts in the future are provided by the authors. These first three chapters provide an overview of the physical, biological, and social factors shaping forestry-wildlife dynamics today—setting the stage for more in-depth discussions of these topics in later sections of the book.

Forestry-Wildlife Interactions
in Managed Tropical Forests

In chapter 1, Robert Fimbel, Alejandro Grajal, and John Robinson open the book with a brief introduction to logging and wildlife issues in the tropics. This short chapter provides an overview of:

- Why tropical production forests are important to wildlife conservation
- The ways in which current logging practices are directly and indirectly affecting wildlife and their habitat
- The options for promoting sustainable forest management and the conservation of wildlife in tropical forests

In chapter 2, Jack Putz, Laura Sirot, and Michelle Pinard describe the most common pre- and postfelling silvicultural treatments used in tropical forests, how these activities affect forest structure, and the potential consequences of forest domestication practices (particularly logging) on arboreal animals. Nonflying arboreal mammals are faced with the challenge of moving from tree to tree to meet a variety of needs, such as the search for food and water, mating, and defending their living space. The locomotory capacities of this faunal group vary greatly—from pottos and sloths that can span gaps no wider than an "arm's length," to monkeys capable of leaping several meters across openings. The effective gap size for a nonflying arboreal animal, therefore, is a function of the individual animal (body size, locomotion capacity, etc.) and the distribution of usable supports. Any silvicultural treatment that minimizes the recovery of forest structure and composition (such as harvesting, vine cutting, thinnings, or short reentry periods by loggers) can greatly restrict the movements of non-flying arboreal animals. These conditions increase the energy requirements and predation risks associated with travel. Factors to be considered when predicting the effects of forest management activities on arboreal mammals, and the silvicultural practices capable of minimizing these impacts, are discussed.

In chapter 3, Patrick Jansen and Pieter Zuidema consider the possible consequences of selective logging on vertebrate-mediated seed dispersal, and the potential of cascading effects on the long-term composition and productivity of the forest. Most forestry harvest-regeneration systems in the tropics depend on natural regeneration, yet few management prescriptions include guidelines to ensure seed dispersal. In the Neotropics, where over 70 percent of the exploited timber species have vertebrate-dispersed seeds, unregulated hunting of animals and the reduced availability of food resources threaten the recruitment of some commercial tree species. To what extent compensating mechanisms (such as redundancies among species that disperse seeds) actually occur in logged-over areas remains largely unknown. Management guidelines for minimizing the disruption of vertebrate seed-dispersing activities and research priorities for refining these silvicultural prescriptions are presented in this chapter.

Wildlife Conservation in Managed Tropical Forests

Protected tropical forests are currently inadequate for conserving the wildlife of the region because of their limited size, number, distribution, and composition. In many countries, the large size and varied habitats within production forests can complement existing protected area systems. Taken as part of the landscape, they can make significant contributions to wildlife conservation. This conservation role is strongest where production forests are sustainably managed for both timber and non-timber resources. At present, however, only a fraction of the world's tropical forests are being well managed.

Contributors to this section note several opportunities for conserving wildlife populations in production forests, while improving commercial timber production. Proposed measures include:

- Designing timber harvests that promote the recruitment of commercial species (chapters 2 and 3). Where drastic disturbances are required to regenerate shade-intolerant timber species (resulting in the fragmentation of forest structure), cuts should be positioned within the landscape to allow reserve areas of intact forest. These reserves provide critical habitat areas for wildlife species sensitive to early successional conditions, and maintain a gene pool of commercial tree species in the event of regeneration failure within the logged area

- Planning and initiating practices that minimize disturbance of forest structure at the actual sites of felling and transport. These include creating inventories of harvestable trees and species meriting protection (chapter 3), cutting vines only on trees to be harvested, directional felling of these stems, and constructing narrow, linear roads (chapter 2)

- Strictly regulating hunting practices to protect populations of seed dispersers and other game or commercial trade species at risk of over-exploitation (chapter 3)

The planning and implementation of the reduced-impact silvicultural prescriptions noted above are discussed in greater detail within part V of this volume.

Much can be done today to reduce the negative environmental impacts associated with timber management programs, through the application of reduced-impact logging procedures and strengthened regulations governing management practice. These are just the first steps, however, toward sustainable forest management and conservation of wildlife in tropical forests. We currently have limited knowledge about the ecology of tropical forests and the response of wildlife to forestry practices (part II discusses the interactions between wildlife and logging). Our understanding of hunting (a major indirect pressure on wildlife as a result of forest management practices), has improved (see part III). There is a great need to focus future research on an assessment of the impacts associated with wildlife-logging interactions, and then apply this information to improve conservation of forest resources for the future (part IV) by refining silvicultural practices. Several topics requiring urgent attention include:

- Understanding the ecology of important timber and nontimber species (chapter 2)
- Identifying the role that pollinators and dispersers play in maintaining the biological integrity and productivity of the forest (chapter 3)
- Defining ways to integrate research, management, industry, and government resources to yield the greatest gains for conservation (chapter 1)

The final section of this volume, part VI, considers incentives for implementation of the research and management recommendations proposed in this first section and throughout the book.

LOGGING-WILDLIFE ISSUES IN THE TROPICS

An Overview

Robert A. Fimbel, Alejandro Grajal, and John G. Robinson

Protected areas are currently inadequate to conserve the biological diversity found within tropical forests, because of their limited size, number, distribution, composition, and protection status. In 1990, approximately nine percent of the world's major tropical rain forests were legally preserved in national parks or equivalent reserves (Grieser Johns 1997:4). The area physically protected against intrusions by hunters, settlers, miners, illegal loggers, or road and hydroelectric projects, however, is actually much less. The distribution of preserves also does not represent all forest types or areas of high biodiversity (Frumhoff and Losos 1998).

In many tropical countries, the large size of timber production forests represents an opportunity to complement existing protected area systems, providing critical habitat for wildlife (vertebrate and invertebrate fauna) and native plant species. Although production forests are not a substitute for nature preserves, they provide a complementary role when *sustainably* managed for both timber and non-timber resources. At present, few (if any) forests are successfully managed in a sustainable manner (Poore et al. 1989; R. Donovan, personal communication). For the purposes of this book, *sustainable forest management* is defined as:

> "the stewardship and use of forests and forest lands in a way and at a rate that maintains their biodiversity, productivity, regeneration capacity, vitality and their potential to fulfill, now and in the future, a role of ecological, economic and social functions, at local, national and global levels, and that does not cause [long-term] damage to the ecosystem."
> — (the Ministerial Conference on the Protection of Forests in Europe 1993, as quoted in Myers 1996).

Timber Harvesting and Wildlife in the Tropics

Numerous harvest-regeneration silvicultural systems have been developed for the sustained yield of timber resources in the tropics (reviewed in Nwoboshi 1982; Armitage and Kuswanda 1989; Gómez-Pompa and Burley 1991; Catinot 1997; see chapters 2, 21, and 24). While most forestry departments recommend one or more of these silvicultural treatments for logging operations in the forests they manage, their efficacy is generally limited by a weak matching of prescriptions to forest conditions, and limited supervision during treatment implementation.

The most common logging practice in the tropics today is a variation of the *diameter-limit* or *selective* cut, where most merchantable stems of a species above a specified diameter are harvested in a specific area. For this logging system to work, forests must contain ample commercial species in the small to medium-size diameter classes. These stems must also contribute seed and recruit rapidly into the merchantable size class following cutting. Harvests are generally scheduled on 20 to 40 year intervals, depending on the projected annual incremental growth of the species being harvested (Smith et al. 1997). This simple-to-prescribe harvesting practice seldom reflects the ecological conditions of the forest, however, and often leads to the *high grading* or *creaming* of the logged area (i.e., *mining* the forest in a nonsustainable manner).

All timber-harvesting practices impact forest wildlife and their habitat, with the impacts increasing as the intensity of logging increases, and the planning, supervision, and period of recovery between cutting events decreases. Logging directly impacts forest-dependent wildlife through the destruction or degradation of habitat, disruption of faunal movements, and interruption of ecological interactions between organisms (discussed in chapters 2–14). Logging can also impact wildlife indirectly by increasing accessibility to the forest, which frequently leads to hunting and land conversion activities (see chapters 15–17). While much more research is needed to quantify the long-term responses of wildlife and the various habitats to logging (explained in chapters 18 and 19), we currently know that species dependent on mature, closed-canopy forest conditions decline or are locally extirpated in the wake of logging activities (see chapters 4–14). Even under relatively low-intensity cutting (< 3 trees harvested per hectare), the impact can be high where logging practices are poorly planned and supervised (Grieser Johns 1997).

Wildlife Conservation in Production Forests

What Options Exist for Conserving Wildlife and Habitats in Production Forest Landscapes?

Logging attracts strong political sponsorship in most developing nations, as the ratio of revenue and infrastructure development to investments in forest departments to oversee the management process is high. As a result, there is rarely a choice between *to cut or not to cut*. Arguments for restricting or limiting logging in select areas of high biological diversity (see chapter 20) may attract some support (especially where alternative low-impact options like tourism exist). Wildlife conservation in most tropical forest landscapes, however, is going to depend on diminishing the environmental impact associated with logging in production forests.

The basic technologies currently exist for developing silvicultural systems that regenerate commercial species while reducing some impacts of logging practices on forest structure and composition (outlined in chapters 21, 22, 24, and 25). Guidelines also exist for selecting habitat elements of high conservation value within the management unit (cutting area) that should be reserved from cutting impacts (see chapter 23). While there is a great need to adapt these techniques to local conditions (especially with regard to timber species ecology—see chapters 18 and 19), the application of existing conservation measures can have a considerable (positive) impact on many interior-forest plant and animal species.

What Options Exist to Promote the Application of Existing and Future Conservation Technologies Within Forestry Operations?

The failure to apply technologies that conserve the structure and integrity of production forest landscapes (i.e., practices promoting sustainable forest management) is rooted in a suite of technical, social, political, and economic factors that advance the liquidation of tropical forest resources (Myers 1996). Many aspects of these factors are outside the forestry sector, and most beyond the control of foresters. A number of incentives for advancing better forest stewardship, however, have recently developed: certification, donor support, cost-benefit analyses, and tax incentives (see chapters 26–29). These incentives are starting to catalyze broader conservation reforms related to logging in the tropics.

Efforts to achieve true sustainable forest management are still in their infancy and face complex technical, biological, social, and political hurdles. One step towards breaking the old paradigms of forest exploitation and domestication would be to provide foresters, planners, donors, and politicians with empirically derived information showing:

- The impacts of present logging practices on the environment
- Cost-effective options for balancing timber and biodiversity conservation issues

In the absence of this information, policy debates will continue to be shaped more by rhetoric and suppositions than facts—polarizing opposing perspectives and stalling (even undermining) progress toward sound forestry practices. It is our hope that the information presented in this volume will help fill crucial gaps, and advance the debate surrounding logging-wildlife interactions in tropical forests and the search for management tools to conserve wildlife while sustaining timber yields.

Chapter 2

TROPICAL FOREST MANAGEMENT AND WILDLIFE

Silvicultural Effects on Forest Structure, Fruit Production, and Locomotion of Arboreal Animals

Francis E. Putz, Laura K. Sirot, and Michelle A. Pinard

Silvicultural activities—beginning with timber inventories to estimate exploitable wood volumes and continuing on to include road construction, timber harvesting, treatments to increase stocking of commercial timber trees, and treatments to increase timber yields—all affect wildlife. Given the huge range of logging intensities, the diversity of possible silvicultural treatments, and the incredible diversity of animal species in tropical forests, we cannot expect any single predictive model of the effects of forest management activities on wildlife to be very useful. Silvicultural treatments affect wildlife through a myriad of direct (see chapters 4–14) and indirect (see chapters 15–17) mechanisms, at spatial scales ranging from individual trees to entire landscapes, and over time periods that span centuries.

In this chapter, we concentrate on the impact of stand-level treatments on the locomotion of nonvolant (nonflying) arboreal animals (e.g., sloths, tree kangaroos, and leaf monkeys), and on food availability for frugivores (animals whose diet is comprised primarily of fruits). We begin by providing a general overview of tropical silviculture and the environmental impacts of logging—the most severe of all silvicultural treatments and generally the only one applied in tropical forests. We then consider the specific impacts of logging on forest structure from the perspective of nonvolant canopy animals (see box 2-1). This is followed by a brief discussion of the effects of logging and other silvicultural treatments on fleshy fruit production and frugivores. Last, we assess some of the likely effects of other silvicultural treatments (e.g., vine cutting) on forest structure and wildlife, management options to mitigate these impacts, and research priorities for conserving wildlife in logged forest landscapes.

Box 2-1 *Locomotion of arboreal animals.*

Nonvolant arboreal animals are faced with the challenge of moving from tree-to-tree to meet a variety of needs, including food and water acquisition, reproduction, and territorial defense. Life in the canopy is exceedingly risky, even for animals as agile as gibbons (*Hylobates* spp.) and other primates—as indicated by the substantial proportion of their skeletons with healed fractures (20 to 30 percent; Schultz 1956). Crossing gaps between trees is the most effective means for arboreal animals to reduce the distances they have to travel (Cant 1992; Cannon and Leighton 1994). Canopy gap creation is probably among the most important effects silvicultural treatments can have on nonvolant arboreal animals. The creation of large gaps (by logging and other silvicultural treatments such as liberation thinning), however, renders some trees inaccessible.

When considering the locomotion of nonvolant arboreal animals, it is essential to remember that gaps are three-dimensional and dynamic, and that different species of arboreal animals tend to move at different heights above the forest floor (Fleagle and Mittemeier 1980). In fact, many "arboreal" animals descend to ground levels to feed or cross gaps. This method of crossing gaps, however, increases traveling distances and probably also increases the risks of predation. Whether or not they go all the way down to the ground, the most common deviation from the horizontal pathway results from substantial descents during gap crossing (Gebo and Chapman 1995a). A second important consideration in assessing the effects of silvicultural treatments on arboreal locomotion is that different species are affected in different ways by the same changes in forest structure. In Kibale National Park in Uganda, for example, the maximum leaping distances of nonvolant arboreal animals range from 0 m for pottos (*Perodicticus potto*) to 4 m for red colobus (*Colobus badius*) (Gebo and Chapman 1995a). Some species are able to modify their modes of locomotion when they encounter new habitats, but others are more restricted (Dagasto 1992; Garber and Pruetz 1995; Gebo and Chapman 1995a).

The effective gap size for a nonvolant arboreal animal is a function of 1) the characteristics of the individual animal and 2) the distribution of usable supports. First, the modes of locomotion used by an animal limit the size of the gap it is able to cross. Different modes include bridging, leaping, jumping, brachiating (swinging arm over arm), gliding, and descending to the ground (Emmons 1995). Specialized modes of locomotion are often enabled by morphological characteristics, such as claws (vertical clinging), long forelimbs (brachiation), and prehensile tails (bridging and suspension). Second, body mass determines the characteristics of branches that are usable as supports (e.g., diameter, angle, strength, and elasticity) (Grand 1984). Light animals can generally move from one tree to another using thinner, terminal branches, while heavier animals have to use larger supports and, as a result, may be faced with larger effective gaps between trees (Fleagle 1985). Mass also determines the risks and consequences of falling, as heavy animals are less likely to survive long falls than light animals (Cartmill and Milton 1977; Vogel 1988).

These individual biomechanical characteristics influence the size gap an individual animal is able to cross and the types of usable supports it needs. Other characteristics, such as reproductive state, presence of predators or conspecifics, and distance to the ground also influence the size of gaps an animal is willing to cross, and the supports it is willing to use.

Tropical Silviculture and Its Influence on Wildlife: A General Overview

Silvicultural Treatments to Maintain or Increase Stocking of Timber Trees

Forest managers can employ a wide variety of techniques to enhance seed production, promote seedling establishment, and stimulate growth of commercially valuable tree species. This section concentrates on seed production and seedling establishment, but we acknowledge that these same silvicultural treatments influence post-establishment growth.

Felling regimes (if they are to play a positive role in forest management for timber) should be selected on the basis of the predominant mode of target species regeneration in the stands to be harvested. Non-silvicultural factors, unfortunately, often decide how timber harvesting is to proceed (e.g., market demand, season, concession duration, maintenance of biodiversity or ecosystem functions, processing capacity, and income expectations). For our purposes, however, we focus on methods for enhancing regeneration of commercial timber tree species. To further simplify this vast topic, we divide all tree species into two classes: those well-represented in the understory before logging as seedlings, saplings, and pole-sized trees (i.e., advanced regeneration), and species that mostly regenerate from seed (dormant or freshly dispersed) in large clearings. We use this dichotomy with trepidation due to the unquestionable importance of species with intermediate regeneration requirements (Clark et al. 1993)—many of which are important commercial timber species.

Harvest-Regeneration Systems Favoring Shade-Tolerant to Moderate Shade-Tolerant Tree Species

Where stands are managed for tree species having abundant advanced regeneration in the understory, it is unlikely that these species will flourish in the high temperature, high light intensity, and high vapor-pressure deficit conditions characteristic of moderate-large clearcuts (e.g., Turner and Newton 1990). To promote growth of these shade-tolerant or moderately shade-tolerant species, polycyclic (uneven-aged) methods such as *single-tree* or *group selection* harvesting approaches are generally appropriate. Selective logging, per se, is not explicitly defined, and as currently employed, is difficult to distinguish from *high grading* or *creaming* of the forest (i.e., harvesting only the most valuable trees without regard to future stand production) (Palmer and Synnott 1992; Bruenig 1996; but see Wadsworth 1997). Nevertheless, the environmental conditions resulting from removal of a small portion of the canopy should be appropriate for the survival and growth of many species represented by advanced regeneration (Smith 1986). Where advanced regeneration is especially abundant and responds well to canopy openings, and where vines and other weeds do not threaten timber stand development, selective logging can be followed with further canopy opening techniques such as *frill girdling* (cutting a band around the trunk to interrupt internal flows) or arboricide application. One particularly intensive method of promoting the growth of advanced regeneration—the Malayan Uniform System—was used successfully in some lowland forests in Peninsular Malaysia (Wyatt-Smith 1987). The same approach was disastrous in hill forests in Malaysia, where radical opening of the canopy stimulated proliferation of vines, bamboo, understory palms, and other weeds (Burgess 1975).

Where advanced regeneration is plentiful, the first principle of good forest management is to protect it from damage during logging. Failure to protect advanced regeneration is a major and very common tragedy in many forests, necessitating expensive and problematic modes of artificial regeneration, such as enrichment planting (Johnson and Cabarle 1993). Protection of advanced regeneration during selective logging has the additional advantages of maintaining pre-harvest forest structure and genetic structure. Where reduced-impact logging methods are used (see chapter 21), the deleterious impacts on forest interior animals are likely to be modest, but need to be further investigated.

Where selectively logged stands are not sufficiently stocked with natural regeneration of commercial species (often due to uncontrolled logging), enrichment planting of nursery-grown seedlings or *wildlings* in logging gaps or along cleared lines is often prescribed (Weaver 1987). Failures with enrichment planting are commonplace (Dawkins and Philip 1998), usually due to lack of seedling tending (e.g., vine removal) and canopy closure over light-demanding seedlings (F. E. Putz, personal observation). In spite of high costs and frequent failures, enrichment planting is widely and frequently invoked throughout the tropics (Bruenig 1996). Where successful, the effects of enrichment planting on wildlife will depend on 1) the species selected for planting, 2) planting densities, and 3) whether natural regeneration of other tree species is tolerated or encouraged. Intensive enrichment planting can be tantamount to converting forests into timber plantations. One enrichment planting project in Malaysia is trying to mitigate the likely deleterious effects of such a conversion on wildlife by planting mixtures of seedlings of fleshy-fruited and timber tree species (Moura-Costa 1996).

Silvicultural Systems Favoring Shade-Intolerant Species

Regenerating light-demanding species requires substantial disruption of forest structure. Depending on local conditions such as erosion-proneness, the availability of markets for small-dimension logs of a great variety of species, and the likelihood of postfelling weed infestations, silviculturalists can select from a wide range of felling regimes. These can include strip and patch clearcuts (Hartshorn 1989), clearcuts with seed tree retention (often 5 to 10 mature trees retained per hectare; Lamprecht 1989), and shelterwood methods (Baur 1964; Wadsworth 1997). Most of these approaches are self-explanatory, with the exception of shelterwood harvests.

Shelterwood management calls for two phases of felling. During the first (or regeneration) phase, a substantial portion of the canopy is removed to stimulate reproduction of the remaining trees and enhance seedling establishment and growth (Smith 1986). Once the regenerating cohort is well developed—which takes only a small portion of the full rotation—the remaining canopy trees are removed.

Shelterwood cuts, clearcuts and other *monocyclic* (even-aged) methods, vary in treated stand size from a fraction of a hectare to many hectares—a variable that greatly influences wildlife habitat. Rotation length (time from seedling establishment until the trees are harvested) is also an important factor shaping habitat recovery. Pulpwood rotations of 5 to 15 years are common in well-managed plantations (Evans 1982), whereas sawtimber rotations tend to be much longer. In considering the effects of monocyclic management on wildlife, it is perhaps more appropriate to focus at the landscape scale than at the level of individual cutting units (e.g., should harvesting areas be clustered or dispersed?). A long-term perspective also seems justified.

In monocyclic management, complete or near complete canopy opening is often accompanied by treatments designed to enhance seed germination and seedling establishment, and reduce competition with established seedlings. Broadcast burning of logging debris, mechanical scarification of the soil surface, mechanical or chemical control of

competitors, and direct seeding are frequently prescribed site preparation treatments (Smith 1986). If these treatments are successful in
increasing the stocking of commercial tree species that do not produce
fleshy fruits, the long-term effects of silvicultural management will
not be favorable for frugivorous animals. A recipe for biodiversity
degradation would include short rotations and high planting densities, accompanied by intensive site preparation and stand tending
operations in extensive areas removed from native species seed
sources. This is exactly what is happening, however, in many tropical
and temperate countries.

Silvicultural Treatments Designed to Increase Timber Volume Increments

Most silvicultural techniques used to promote growth of commercial trees are based on competition reduction. Soil fertilization, while
commonplace in plantations, is seldom used in natural forest management. Instead, both soil resources and light are made more available to
potential crop trees through the killing of nearby trees and vines.
Competitors may be girdled, herbicide treated, or felled. Where the
wood is extracted and sold, foresters refer to the operation as *commercial thinning* (Smith 1986). In areas where potential crop trees are
selected from the advanced growth, and competing trees and vines are
killed, the treatment is often referred to as *selection thinning* (Nyland
1996), or by tropical foresters as *liberation thinning* (Hutchinson
1988). A stand is *thinned from below* (Smith 1986) when trees of
shorter stature than the potential crop trees are thinned. When
extremely large remnant trees are killed, the treatment can be called
relict removal (Hutchinson 1988). All of these thinning treatments
have numerous (but little studied) effects on forest structure and composition. The effects on wildlife are even less understood.

A goal of timber stand management operations is to concentrate a
stand's volume increments in trees of future commercial value. When
timber stand improvement is successful, populations of frugivorous
animals may decline because few of the most commercially important
timber-producing species in the tropics produce fleshy fruits (see table
2-1; also see Crome 1991). Changes in stand structure associated with
timber stand improvement are also likely to affect nonvolant arboreal
animals. By providing each potential crop tree with sufficient growing
space all around its crown and removing all vines, a diligent and effective timber stand manager creates a three-dimensional forest structure

that challenges the locomotory abilities of even the most agile nonvolant arboreal animals. Finally, relict tree removal (i.e., removal of large trees with badly formed or hollow stems)—a timber stand improvement activity of dubious silvicultural value) (Korsgaard 1992), can have serious negative impacts on cavity nesting birds and mammals (e.g., Conner 1978). In the Neotropics, one positive offshoot of this otherwise disastrous treatment is that it discourages forest invasion by nest-parasitizing cowbirds (Robbins 1979).

TABLE 2-1 *The Most Important Internationally Traded Tropical Hardwood Timbers Ranked by Volumes Exported by ITTO Countries in 1995 (ITTO 1996)*

Rank	Species	Family	Fleshy Fruits	Continent
1[a]	*Shorea* spp.	Dipterocarpaceae	No	As
2[a]	*Dryobalanops* spp.	Dipterocarpaceae	No	As
3[a]	*Dipterocarpus* spp.	Dipterocarpaceae	No	As
4	*Triplochiton scleroxylon*	Sterculiaceae	No	Af
5	*Terminalia* spp.	Combretaceae	No	Af, LA
6	*Entandrophragma* spp.	Meliaceae	No	Af
7	*Ceiba pentandra*	Bombacaceae	No	LA, Af
8	Koompassia malaccensis	Leguminosae	No	As
9	Antiaris africana	Moraceae	Yes	Af
10	*Calophyllum* spp.	Guttiferae	Yes	As, LA
11	*Dactylocladus* spp.	Crypteroniaceae	No	As
12	*Swietenia macrophylla*	Meliaceae	No	LA
13	*Parashorea* spp.	Dipterocarpaceae	No	As
14	*Chlorophora excelsa*	Moraceae	Yes	Af
15	*Tabebuia* spp.	Bignoniaceae	No	LA
16	*Dyera costulata*	Apocynaceae	No	As
17	*Tectona grandis*	Verbenaceae	No	As
18	*Cedrela* spp.	Meliaceae	No	La
19	*Aucoumea klaineana_*	Burseraceae	No	Af
20	Hymenaea courbaril	Leguminosae	Yes	LA

[a] The three top-ranked taxa contributed more than 50 percent of the reported volumes of exported timber. (Af = Africa, As = Asia, LA = Latin America).

Environmental Impacts Associated with Logging

Although some forests are noteworthy for the abundance of marketable, nontimber forest products (e.g., Plotkin and Famolare 1992), timber is the principal commercial attraction in most forests. Timber harvesting unavoidably changes forest structure, modifies the microclimate, and affects wildlife populations. One obvious impediment to developing generalizations about the effects of logging on wildlife is the wide range of logging intensities and logging methods employed in the tropics.

Logging intensities in the tropics range over several orders of magnitude (expressed as timber volumes or the number of trees harvested per hectare). Low logging intensities characteristic of western Amazonia (e.g., 0.12 trees/ha and < 1 m³/ha harvested in Bolivia; Gullison and Hardner 1993), result from the interplay of various factors, including low stocking of commercial species, poor stand access, long hauling distances, and limited market development. High logging intensities typify well-stocked and accessible areas in regions with well-developed markets, such as in the dipterocarp forests of Southeast Asia (e.g., 8–15 trees/ha and 80–160 m³/ha harvested; Sabah Forest Department 1989) or eastern Amazonia (4–6 trees/ha and 37–52 m³/ha harvested; Johns et al. 1996).

The diversity of the logging methods employed, particularly for timber *yarding* (i.e., different methods for delivering logs to roadsides), further complicates the predicting of environmental impacts of different logging intensities. The least damaging yarding methods employ aerial extraction of logs, either with helicopters or with skyline (cable crane) systems (not to be confused with *high lead* cable yarding in which the logs are skidded along the ground; Conway 1982). Forest damage during aerial yarding is mostly restricted to felling damage, but by using these methods, timber can be yarded from steep slopes that would otherwise not be logged. Clearing narrow (3–5 m wide) cableway corridors for skyline yarding causes slightly more damage than helicopter yarding, but neither method results in any appreciable soil damage, except along roads and in log landings.

Ground-based timber yarding, the predominant approach in the tropics, causes unavoidable and extremely varied amounts of damage to residual stands and soils. A wide range of machinery is used for skidding, including farm tractors, articulated skidders with rubber tires, and bulldozers (crawler tractors; Conway 1982). When widely separated trees are felled, logs are sawn into boards in the forest and the lumber is hauled out manually—there is relatively little damage to the residual forest. Skidding entire or quarter-sawn logs with draft animals is also environmentally benign, at least compared to mechanized extraction (Cordero 1995). Heavier machines are inherently more damaging, especially those designed for other uses such as road building (i.e., bulldozers). With proper training, however, in conjunction with supervision and appropriate incentives, even bulldozers need not cause excessive damage (Marn and Jonkers 1981a; Hendrison 1990; Pinard et al. 1995; Johns et al. 1996; see chapter 21).

It is commonplace in the tropics for untrained and unsupervised logging crews to be paid on the basis of timber volumes delivered to log landings (Johnson and Cabarle 1993). Without proper incentives and training to minimize their impacts, and lacking stock maps (maps of trees to be harvested) and pre-planned skid trails, logging crews cause unnecessary damage to both residual stands and soil (Dykstra and Heinrich 1996). Conventional ground-based logging with bulldozers in Paragominas, Brazil, for example, resulted in 51 trees >10 cm dbh damaged for each tree harvested (Johns et al. 1996). In Sabah, Malaysia, where 10–15 trees are harvested per hectare, 40–70 percent of the residual trees are damaged and as much as 30 to 40 percent of the soil is scraped or compacted (Sabah Forest Department 1989; Jusoff 1991; see photo 2-1).

Logging damage in Paragominas and Sabah, however, was reduced by 50 percent or more with the implementation of fairly straightforward guidelines (e.g., stock-mapping, prefelling vine cutting, directional felling, and skid trail planning) (Pinard and Putz 1996; Johns et al. 1996; see also chapters 21 and 28; see photo 2-2). Given the wide range of logging intensities and the equally wide range in the amount of damage that can result from such logging intensity, predictions of the effects of logging on wildlife will, to some extent, require substantial site-specific information about logging operations (see chapters 18 and 19). Nevertheless, employment of reduced-impact logging (RIL) techniques will likely solve many of the problems faced by wildlife that directly result from silvicultural operations (see chapters 21 and 24).

Depending on how logging is conducted, it can be a purely exploitative process or an important silvicultural treatment. Of all silvicultural treatments (with the exception of forest conversion to plantations), logging is the most intrusive. Even when properly implemented, harvesting canopy trees has unavoidable and wide-ranging environmental effects. Silviculturally appropriate cutting regimes range from single tree selection to clearcuts (see box 2-2), because can-

PHOTO 2-1 *Malaysian hill-dipterocarp forest logged without
supervision or planning. (D. Kennard)*

opy tree species vary in their regeneration requirements—some regenerate in small gaps and others require large clearings (Everham and Brokaw 1996; Whigham et al. 1999). In the next section, we consider different cutting regimes, and other treatments prescribed to promote regeneration of commercial tree species. For each of these treatments, we discuss how logging affects forest structure, locomotion of nonvolant canopy animals, and fleshy fruit production.

PHOTO 2-2 *Malaysian hill-dipterocarp forest logged following reduced-impact logging guidelines. (C. Marsh)*

Box 2-2 *The impacts of silvicultural treatments on a Bolivian seasonally dry tropical forest.*

The seasonally dry forests of the Lomerío region of Bolivia are transitional—between the humid Amazon to the north and the dry chaco and cerrado to the south and east (Killeen et al. 1990). Semideciduous forest and savanna cover the undulating hills characteristic of Lomerío, with more humid, evergreen, gallery forests in the valleys. Gallery forests cover only about 10 percent of the area, but are critical for wildlife as sources of water and fruit during dry periods. The diversity of mammals in the region is high, but population densities are low—probably due to substantial hunting pressures and low resource availability (Guinart 1997). Arboreal mammals rely on a mixed diet of fruits, leaves, nectar, and small animals (including insects). The 16 species of nonvolant, arboreal mammals include five primates, two anteaters, two carnivores, four rodents, and three marsupials.

The canopy is dominated by trees in the Leguminosae, Anacardiaceae, and Apocynaceae families. About 55 percent of the tree species and 75 percent of the tree basal area (trees >10 cm dbh) produce dry fruits. The majority of plants that produce fleshy fruits are riverine, understory, or subcanopy species.

Silvicultural Approach

Lomerío's forests are managed by indigenous community foresters with the goal of sustainable timber production, while minimizing the negative impacts of management activities on the forest's biological and physical resources. A mixed harvesting system that combines single tree selection with strip-shelterwoods seems silviculturally appropriate for the forest. This mixed approach encourages regeneration of light-demanding canopy tree species (43 percent) in the strip-shelterwoods, while maintaining the regeneration of species with more shade-tolerant regeneration (57 percent) in the remaining forest mosaic where isolated or small groups of trees will be harvested (Pinard et al. 1999a). Implementation of a single system—monocyclic or polycyclic—is not recommended because it would be inconsistent with the goal of maintaining current levels of tree diversity and species composition. Streamside habitats are protected from harvesting by buffer strips (see chapter 23), and only a minimum number of road and skid trail crossings are allowed—providing protection to several plant species considered important for frugivores (e.g., *Ficus* spp., *Scheelea* sp., *Capparis* sp.). Though hunting in the management area is not legally permitted, it continues to be an important source of protein for local people (Guinart 1997; and chapter 15). Inadvertent facilitation of hunting in managed forests in Lomerío, as in most other areas in the tropics, is likely to be of much more consequence to wildlife than the silvicultural treatments themselves (see chapters 15–17).

Many of the commercially harvested tree species in Lomerío currently have relatively low market values in Bolivia. Alhough it is likely that the international market for these hardwood species will develop within the next decade, present values preclude substantial investment in silvicultural treatments. Broadcast seeding with the seeds of commercial tree species in felling gaps is presently being tested. Vine cutting may also be justifiable as a postlogging treatment for improving stem form of future crop trees, but it is unlikely to be implemented as a blanket preharvest treatment because few tree crowns are connected by vines. In this analysis, we consider the likely environmental impacts of a silvicultural system that includes timber harvesting, broadcast seeding in felling gaps, and postharvest vine cutting.

Environmental Impacts of Silviculture in Lomerío

The principal impact of silvicultural treatments on forest structure is disruption of canopy continuity. Following a harvest of about 10 m^3/ha, felling gap density (Brokaw 1982) is 5–8 gaps/ha. Eighty percent of the gaps are small (40–70 m^2) and very few are relatively large (300–500 m^2). The road network used for the harvest is minimal (covering <4 percent of the area), and principal roads and skid trails average 5 and 3 m wide, respectively (Camacho 1996). Corridors of vegetation with continuous canopy remain intact along streams. It is hoped that broadcast seeding of commercial tree species in felling gaps will accelerate the recovery of tree cover, but the seed bed (i.e., the soil surface) may require more severe treatment (e.g., controlled burns of logging debris or mechanical scarification) to foster tree seedling establishment.

Changes in plant species composition following implementation of the recommended silvicultural treatments are likely to include at least a temporary increase in the area dominated by early successional species. Herbs and pioneer trees colonize the larger felling gaps and log landings within the first year after logging. Resprouting vegetation generally dominates small gaps. If vine-cutting treatments are effective, vine density will decrease. Because the treatment targets only vines on commercial trees, however, vine diversity is expected to remain unchanged. Efforts to protect the managed areas from wildfires will be important in maintaining the diversity of woody species in the understory and avoiding an increase in grasses and herbaceous vines (Pinard et al. 1999b). The diversity of tree species should be maintained, but the spatial distribution of species may change.

Fleshy fruit production in gallery forests (width varies from 10–20 m) is expected to remain unchanged because gallery forests are not logged or silviculturally treated. Elsewhere in the forest, increased fruit production is expected in (and adjacent to) felling gaps—at least in the short-term. Increased light availability and decreased root competition are expected to stimulate residual trees (e.g., *Casearia* sp. and *Eugenia* sp.) to fruit. The early successional vegetation that is common to the area includes a mix of species that produce fleshy fruits (*Urera* sp., Cucurbitaceae spp.) and species that produce dry, wind-dispersed fruits (e.g., *Eupatorium* sp. and *Mikania* sp.). In the long-term, if broadcast seeding is successful at regenerating the shade-intolerant tree species, the area in gaps (10–15 percent) will be dominated by commercial tree species that produce dry fruits. The reduction in vine density will probably not directly influence fleshy fruit production, as the majority of vines are Bignoniaceae and Hippocrateaceae.

The increase in gaps following logging activities is likely to increase the costs of locomotion for nonvolant arboreal animals. The large felling gaps, in particular, may force animals to take circuitous routes or travel on the ground. Changes in forest structure related to management, however, may be less likely to affect arboreal animals in Lomerío than in other, more humid, tropical forests. Many of the mammals in Lomerío's forests that are considered arboreal are not strictly so—many forage on the ground when the canopy is leafless. The gaps between tree crowns in the upper canopy of mature upland forests in the area also often exceed the threshold for leaping. Because lianas seldom connect tree crowns in this forest, vine cutting may not remove a significant number of arboreal pathways.

The silvicultural treatments prescribed for the forests of Lomerío were selected for their apparent compatibility with the multiple goals of management. Further research is needed on the effectiveness of the mixed harvest system at maintaining both tree species diversity and viable wildlife populations. It would also be useful to strengthen the scientific basis for demarcation of protected areas (see chapter 23). Environmentally concerned foresters will also need to know what sort of management activity, if any, these reserves will require to maintain their intended functions.

Logging, Forest Structure, and Arboreal Mammals

Logging operations in tropical forests tend to focus on large trees of only a few species. Large trees create relatively large canopy openings when they are felled, especially if they are attached to neighbors by woody vines (reviewed in Putz 1991). Even more damage to the residual forest occurs during yarding operations, especially if bulldozer or skidder operators do not employ their winches to draw logs to preplanned skid trails.

Canopy Openings and Gaps in Logged and Natural Forest

Canopy openings caused by logging differ from natural disturbances in several ways. Because only large, living trees are harvested for timber, felling gaps tend to be bigger than canopy gaps opened naturally by trees that die standing and fall apart piecemeal—a mode of death that typifies the natural dynamics of some forests (Appanah and Putz 1984; Kasenene and Murphy 1991). Unnaturally large gaps are also created when harvestable trees occur in clusters—a common pattern in tropical forests. Unless felling rules restrict harvesting all trees in a cluster, information about mean felling intensities often belies the existence of pockets of severe forest structure disruption. Felling clusters of trees can be silviculturally justified (Troup 1928; Smith 1986) in areas where the target tree species require moderate-sized canopy openings for regeneration. Foresters refer to this treatment as *group selection*.

One difference between logging and natural disturbances is that ground-based timber yarding often results in substantial soil damage, whereas naturally uprooted trees disrupt little soil. Logging gaps also are invariably connected to one another and to extraction roads by skid trails. These corridors of disturbed forest are used by wildlife and may provide entryways for invasive species of both plants and animals. Many well constructed and carefully used skid trails in a reduced-impact logging study area in Sabah, Malaysia (Pinard and Putz 1996), for example, still had substantial exposed mineral soil three years after logging due to heavy use by elephants, deer, and bearded pigs (F.E. Putz, personal observation). Depending on the type of yarding equipment used, and the care with which it is operated, skid trails may be open to the sky over little or much of their length. Canopy openings over skid trails and areas where trees have been felled may enlarge over time due to the death of adjacent trees that have been damaged during logging operations—or decrease by lateral crown expansion of adjacent trees. These processes have not been investigated.

While it is tempting to depict logged forests with gaps resembling Swiss cheese (i.e., holes in a solid matrix) (Liebermann et al. 1989), the reality is more complicated. Canopy gaps penetrate different distances down towards the ground (Weldon et al. 1991) and perhaps below ground as well (Wilczynski and Pickett 1993; Ostertag 1998). In addition, the microclimatic effects of canopy openings penetrate into the surrounding forests well beyond the gap *edge* (Popma et al. 1988; Canham et al. 1990; Lawton 1990; Brown 1996; Laurance and Bierregaard 1997). These conditions promote the establishment and growth of both advanced regeneration (Connell et al. 1997) and new seedlings. In recently opened gaps in a montane forest in Costa Rica, for example, more than 90 percent of the gap area was occupied by living plants and the mean leaf area index (m^2 of leaves per m^2 of ground) was 1.6 m^2/ m^2 (Lawton and Putz 1988). Regenerating gaps are also often dominated by sprouts of broken trees that contributed to gap formation (Putz and Brokaw 1989). Finally, where vines are not cut prior to logging, gaps and the adjacent canopy trees often become infested with dense tangles of coppiced vines and those that sprout from fallen stems (Appanah and Putz 1984; Perez 1998). Where vines, bamboo, heliconias, or other aggressive, light-demanding species abound, gap opening can result in weed infestations that may delay canopy tree regeneration for many decades (Chaplin 1985; Putz 1991).

Impacts of Logging on Nonvolant Arboreal Animals

Where logging gaps are large or closely spaced (especially where uncontrolled bulldozer operators yard the timber), some canopy and understory animals suffer (e.g., Thiollay 1992; Datta and Goyal 1996; Laurance 1996; Mason 1996). The large gaps that result from group selection (often 100–1000 m^2), for example, represent obstacles for nonvolant arboreal animals—at least until the trees in the gaps are again 5 to 10 m tall (which may require 3 to 10 years of growth, depending on local conditions and species composition) (e.g., Brokaw 1987). While some arboreal animals descend into short trees or to the ground to cross canopy gaps (Malcolm 1991) or to feed in the crowns of gap-trees, the movements of other species are impeded until canopy structure is more fully restored. Descents to the ground are likely to be particularly problematic for slow moving animals, like sloths and pangolins.

Evaluating the effects of different silvicultural treatments on the locomotion of arboreal animals is clearly very complicated (see box 2-3). The size, quantity, and distribution of the gaps created varies greatly within areas receiving silvicultural treatments. This variation is further exacerbated by the difficulty in predicting the sizes of gaps created by silvicultural treatments and how these gaps will change over time. The multitude of potential direct and indirect impacts of silvicultural treatments on wildlife is also vast, including effects on food availability, changes in social structure, and increased susceptibility to predation.

Box 2-3 *Gibbons, macaques, and silvicultural treatments.*

Here we examine the effect of selective logging on the arboreal locomotion of gibbons and macaques. The two important components of the proposed evaluation are 1) understanding the locomotory abilities of gibbons and macaques, and 2) understanding how selective logging and liberation thinning affect the distribution of noncrossable canopy gaps.

We use agile and lar gibbons (*Hylobates agilis* and *H. lar,* referred to as gibbons) and long-tailed macaques (*Macaca fascicularis*) for our comparison because much is known about their locomotion (Fleagle 1980; Grand 1984; Cant 1988, 1992; Cannon and Leighton 1994; see table 2-2). Gibbons and macaques also have similar body weights (gibbons about 5 kg, macaques about 4 kg) and live sympatrically in Sumatra, Borneo, and Peninsular Malaysia They do, however, have quite different anatomies and modes of locomotion.

Gibbons are characterized by elongated forelimbs and are able to reach around themselves with their forelimbs through almost a complete hemisphere with a radius of their arm length (Grand 1984). Brachiation is their primary mode of locomotion, suspending themselves beneath branches. They also can move on top of branches and use leaping, climbing, bipedalism, and quadrapedalism.

Macaques have approximately equal-length fore and hind limbs and use quadrapedalism most frequently—balancing themselves on top of branches (Cannon and Leighton 1994). They do not use suspensory movement, but climb, leap, and occasionally walk bipedally (Fleagle 1980). Macaques tend to use continuous pathways to move between trees (Grand 1984), but use bridging and leaping to cross gaps (Cannon and Leighton 1994).

The movement of both species often leads to the downward deformation of support branches, changing the effective gap sizes. Gibbons take advantage of this downward deformation by leaping from the depressed branch. Macaques, in contrast, do not generally move from one branch to another while the support branch is swaying because they use both their hands and feet to hold on (Grand 1984). Because macaques travel on top of branches, downward deformation of branches usually increases effective gap size.

In the dipterocarp forests inhabited by gibbons and macaques, a typical recommended series of silvicultural treatments would begin with vine cutting to reduce logging damage (e.g., Wyatt-Smith 1987). Several months later, approximately 12 trees/ha would be extracted (these trees might be 60–100 cm dbh and have crowns 6–12 m in diameter). Felling would be followed by long-term timber stand improvement treatments, in which 20–40 trees/ha would be released from competition by clearing the area within 2–5 m on all sides of the crowns of these potential crop trees, and removing all trees with taller crowns. After 35 years, the crop trees would be harvested.

Assuming that these trees were evenly spaced with 8 m diameter crowns, and that felling each tree results in a gap of about 80 m^2, the first phase would create 12 gaps that could not be crossed by either species. The second stage, liberation thinning, would result in 40 doughnut-shaped gaps of 2–5 m width at the crown level, but of varying widths lower down. Liberation gaps <5 m wide could be crossed by gibbons using brachiation, bridging, and leaping, whereas this treatment would present more of a challenge for macaques. Macaques would not be likely to cross to a selected tree at crown level. Instead, they would need to descend to a level that had usable supports adjacent trees in closer proximity, or (in their absence) move to the ground. The effects of the silvicultural treatments are expected to diminish with time as the crowns of the remaining trees expand to fill the space created.

The third stage of removing 20–30 crop trees/ha would have the most dramatic impacts on the locomotion of both gibbons and macaques. Removal of so many trees would result in a patchwork of gaps that are not crossable at canopy level by either species. Cumulatively, these gaps would cover half of the managed area. As a result, the gaps would present frequent locomotory challenges to both species that require high-energy or high-risk locomotion.

To reduce silvicultural impacts on locomotion of nonvolant arboreal animals, managers should attempt to minimize the number of discontinuities in the canopy. To make recommendations for reducing the total impact of these treatments on nonvolant arboreal animals requires synthesizing or collecting data on changes in other variables—such as food distribution and abundance, and vulnerability to predation and pathogens. Certainly the maintenance of gibbon populations in heavily logged forest in Malaysia (Grieser Johns 1997) for more than a decade suggests that factors other than just forest structure deserve attention.

Where the forest is purposefully cleared down to the ground to enhance tree regeneration from seed, the consequences for forest wildlife may be particularly severe. Cutting all trees in narrow strips through the forest is a classic silvicultural technique (Troup 1929) that has recently received attention from researchers in the tropics (Hartshorn 1989; Gorchov et al. 1993). Like roads through the forest, these so-called *strip clearcuts* or *strip shelterwoods* (even if only 20–30 m wide) create difficult-to-circumvent (although temporary) impediments to the movement of nonvolant animals. Even some species of birds seem to be adversely affected by clearing strips through the forest—at least during the early phases of forest regeneration (Gorchov et al. 1993; Mason 1996; see chapter 8). Research on the effects of strip clearcuts that are of different widths, lengths, and spatial distributions on animals that maintain territories may be particularly instructive. Predation risks and other factors in cleared strips should also be assessed over time, since even young second growth provides suitable canopy pathways for some animals (Malcolm 1991).

Woody Vines, Timber Stand Management, Fruit Availability, and Intercrown Pathways

Vines, including climbing bamboos, pose a very common silvicultural problem in the tropics (e.g., Fox 1968b; Chaplin 1985). Where

TABLE 2-2 *Median Gap Crossed and Median Diameter of Support Used in Different Modes of Locomotion by Gibbons and Macaques (based on Cannon and Leighton 1994)*

	Gibbons	Macaques
Quadrapedal Walking	-	0.1 m
Bipedal Walking	1 m	-
Brachiation	2.5 m	-
Bridge	2 m	0.5 m
Leaping	3 m	1.5 m
Travel Supports	6 cm	female 3 cm male 5 cm
Take-off Supports	4 cm	female 2 cm male 2 cm
Landing Supports	4 cm	female 2 cm male 2 cm

timber production is the primary forest management objective, foresters typically prescribe cutting vines: a) before logging to reduce felling damage, b) during stand regeneration to increase tree seedling establishment, and c) during stand maturation to increase tree growth rates and to decrease the proportion of poorly formed trees (Putz 1991). Vine treatment prescriptions generally call for cutting all vines larger than a particular stem diameter (e.g., 2 cm dbh), or cutting only vines growing in the crowns of trees to be harvested or on future crop trees. In either case, many vines resprout after being cut (Appanah and Putz 1984; Vidal et al. 1997), minimizing the risk of their extirpation from managed forests.

Woody vines increase the accessibility of tree crowns to nonvolant arboreal animals. They also can substantially diminish tree growth rates and increase tree mortality. Given the abundance of vines in many unlogged lowland tropical forests (100–2,500 vines greater than 2 cm dbh/ha; Gentry 1991; Perez 1998), and the large leaf areas supported by even narrow stemmed vines (Putz 1984), these competitive effects on trees are not surprising. Often 20 to 60 percent of trees greater than 20 cm dbh have vines in their crowns, and about 50 percent of the vines that reach the canopy grow in the crowns of more than one tree—often spanning intercrown gaps of 2–3 m (Caballé 1977; Putz 1984; Campbell and Newberry 1993). Climbing plants often increase in abundance after logging, due to decreased forest stature and increased canopy openness.

Vines serve a unique role in nonvolant arboreal animal locomotion because they provide continuous connections between tree crowns (see photo 2-3). For many species of both vertebrates and invertebrates, vines serve an invaluable function in reducing the energetic expenditures and locomotory risks of moving between tree crowns. Long-tailed macaques, for example, use lianas for 18 percent of their total locomotion and 23 percent of their crossings between trees (Cant 1988). Two- and three-toed sloths (*Bradypus variegatus* and *Choloepus hoffmanni*, respectively) in Panama were significantly more common in liana-laden trees than in liana-free trees (Montgomery and Sunquist 1978). Since vines may span intercrown gaps of up to 3–4 m in some areas (Putz 1984), their removal is likely to greatly increase the number of impassable intercrown discontinuities encountered by nonvolant arboreal animals. Some gap leaping species also use tangles of vines as landing platforms, with unique combinations of flexibility and strength.

In addition to providing ready access to trees isolated by crown shyness (spaces around the perimeters of tree crowns opened by mechanical abrasion) (Jacobs 1955; Putz et al. 1984) and larger diameter canopy gaps, some woody vines produce fleshy fruits. Although many vine species have wind-dispersed propagules (e.g., most climbing species of Leguminosae, Bignoniaceae, Sapindaceae, Hippocrateaceae, Apocynaceae, Malphigiaeae, Ancistrocladaceae, and Polygonaceae; Morellato and Leitão-Filho 1996), others produce fleshy fruits that may be important to frugivores (e.g., Annonaceae and Piperaceae). Little is known about the quantities of fruits produced by vines, but climbing plant leaves, flowers, and fruits have been reported to supply up to 40 percent of the food consumed by howler (*Alouatta fusca*) and capuchin (*Cebus apella*) monkeys in southeastern Brazil (Galletti and Pedroni 1994; Galletti et al. 1994). Unfortunately, vine-cutting prescriptions seldom distinguish between vine species, and where fleshy-fruited vines are cut to favor wind-dispersed commercial timber trees, the long-term consequences to frugivores are probably deleterious. Finally, canopy vines also form tangles that trap leaf litter and provide refuge for arthropods (Wolda 1979). Such tangles are frequently visited foraging sites for insectivorous birds (Greenburg 1981) and mammals, and are used as nesting and predator avoidance sites for various animals (F. E. Putz, personal observation).

PHOTO 2-3 *Vines are important structural and compositional components of wildlife habitat in tropical forests. (M. Pinard)*

Logging and Fleshy Fruit Availability

Most commercially important tropical tree species do not produce fleshy fruits (see table 2-1), but many potential timber species do (see chapter 3). There are, however, local exceptions to this pattern. Tree species with mammal- or bird-dispersed seeds, for example, dominate the timber trade in the Guianas (Hammond et al. 1996). Furthermore, as stocks of high-value, wind-dispersed timbers are depleted, more fleshy-fruited species will enter the market. Because of the relatively high-market values of their timbers, however, wind-dispersed trees are likely to remain the main focus of forest management in the short term. Dry-fruited tree species undoubtedly provide critical resources for a diversity of animals, but generally do not depend on frugivores for primary seed dispersal.

Harvesting of trees that produce wind-dispersed seeds should not result in any substantial reductions in fleshy fruit availability (but may reduce availability of seeds and other foods on which wildlife depend). In fact, fleshy fruit production in the residual forest may be higher soon after logging because of increases in both light and soil resource availability. In a comparative study of a selectively logged forest (80 m^3/ha harvested) and an unlogged dipterocarp-dominated forest in Sabah, Zakariah (1994), it was reported that in the year following logging more seeds from fleshy fruits fell in traps in the logged area even though tree density was substantially lowered by destructive harvesting practices. It was also noted that after several decades of logging of *Cynometra*—an explosive-gravity-dispersed legume in Uganda—populations of most of the monitored primate species either increased or were maintained (Plumptre and Reynolds 1994). Several other researchers have reported that reproduction of fleshy-fruited understory plants is enhanced in canopy gaps created by natural treefalls (Wong 1983; Blake and Hoppes 1986; Levey 1988a). Similar increases in plant reproductive activity are reported on forest edges bordering pastures (Restrepo et al. in review) and adjacent to treefall gaps (Augspurger and Franson 1988). Where pioneer species flourish after logging, total fleshy fruit production may be enhanced because pioneer trees typically begin reproduction early and produce large numbers of fruits at frequent intervals (Levey 1988b). Increased fruit production on a mass per area basis, however, does not necessarily herald good conditions for all frugivore species. Fruits of many pioneer species are small, sugary, nutrient poor, and may not be used by many primary forest frugivores (Grieser Johns 1997). Furthermore, if logging interrupts fruiting for even a single season (while the trees are recovering from damage and adjusting to more open conditions), the period of scarcity might result in fauna population crashes—from which it takes decades to recover.

Conserving Wildlife Habitat in Logged Forest

The effects of silvicultural treatments on nonvolant arboreal animal locomotion are determined by how the treatments influence the distribution of usable supports. Many of the silvicultural treatments described in this chapter result in large discontinuities in arboreal pathways. Treatments designed to open the canopy, liberate selected trees, and regenerate light-demanding species all aim to isolate individual trees—a goal that is incompatible with maintaining continuous arboreal pathways. Harvesting clusters of trees and clear cutting lead to large gaps that penetrate to ground level. Even relatively small gaps may enlarge with time, due to incidental death and adjacent tree damage (Lawton and Putz 1988; Young and Hubbell 1991). By eliminating the only truly continuous connections between trees, vine removal also presents a special problem to nonvolant arboreal animal locomotion. This problem is especially severe for small animals that rely on quadrapedalism (e.g., pottos, sloths, and pangolins). The effects of these treatments on canopy animal locomotion can, in part, be assessed and possibly predicted by understanding how the treatments change the distribution of usable supports for the animal of interest.

The silvicultural *improvement* treatments discussed in this chapter—including reduced-impact logging—are applied in a disturbing small proportion of tropical forests (<0.01 percent in 1985, accordingly to Poore et al. 1989). Forest management activities are intensifying, however, as the remaining primary forests in the tropics are diminishing and the certification of products from well-managed forests is developing as a potent market force (e.g., Viana et al. 1996; see chapter 26). The financial costs of improved forest management are relatively minor compared to the long-term benefits for the environment and forest industries (see chapter 28). Furthermore, while logging primary forest may be excessively profitable for concession holders, forest workers (e.g., chainsaw operators) suffer injury and death rates that should not be tolerated and that could be substantially reduced by training and supervision (Tanner 1996). By reviewing some of the possible approaches to tropical silviculture, we hope that this chapter will serve to stimulate wildlife biologists and foresters to jointly address some of the problems that will face us as more forests are intensively *managed*. More studies with titles like *The effects of logging on species* X may be less than effective in promoting the transformation from purely exploitative logging to sustainable forest management if they do not address silviculturally and financially viable alternatives.

The long-overdue transition from timber *mining* to responsible forest *stewardship* will be encouraged by research that reveals methods for mitigating the deleterious impacts of silvicultural treatments on wildlife. To predict the effects of forest management activities on wildlife—particularly nonvolant arboreal animals—the following needs to be known about each animal species of interest:

- How old does a regenerating stand have to be before it again represents suitable habitat for each of the species of concern?
- What is the maximum width of inappropriate habitat patches that the species will cross?
- What is the minimum area of forest that needs to be protected to maintain viable wildlife populations?
- What is the optimum spatial distribution of silviculturally managed and protected areas?

If, in addition to logging, site preparation, weed control, or other *stand improvement* treatments are applied, it is also critical to know their combined effects on forest structure and composition for predicting the local fates of different animal species. Answers to some of these questions are available, yet they represent only a few animal species in even fewer forests (Grieser Johns 1997). Obviously more research is needed, but enough is already known to strongly recommend that all timber-harvesting operations should include reduced-impact logging measures within their design and implementation (see part V).

We focused this chapter on mitigating the effects of forest management for timber on wildlife—not just on describing the effects of logging—because "to log or not to log" is seldom the question that researchers are asked to answer. We hope, instead, that through better understanding of forestry, conservationists will have more opportunities to influence decisions regarding how, when, and where to log, and what silvicultural treatments to apply before and after logging. As we tried to show, many compelling and researchable issues face biologists concerned about how to mitigate the impacts of forest management operations on wildlife.

ACKNOWLEDGMENTS

We appreciate the suggestions about this topic we received from colleagues at BOLFOR, CIFOR, the University of Florida, and several anonymous reviewers. Comments on earlier versions of this manuscript by R. Ostertag, S. Bigelow, C. Chapman, E. Zweede, J. Lambert, S. Boinski, and D. Gorchov were particularly helpful.

Chapter 3

LOGGING, SEED DISPERSAL BY VERTEBRATES, AND NATURAL REGENERATION OF TROPICAL TIMBER TREES

Patrick A. Jansen and Pieter A. Zuidema

Tropical forestry is increasingly focused on finding systems for sustainable timber harvesting from natural forests. So-called *natural regeneration systems* are typically polycyclic, selective logging systems that rely on natural regeneration to produce the next crop of timber (Gómez-Pompa and Burley 1991; see chapters 2 and 21). These systems seek to maintain commercial productivity of logged forests by minimizing damage to residual trees and by conserving the natural regeneration potential of the forest (Grieser Johns 1997). Some, such as the CELOS silvicultural system (Centrum voor Landbouwkundig Onderzoek in Suriname; De Graaf 1986; De Graaf and Poels 1990; De Graaf et al. 1999), also include procedures to enhance regrowth of high value species by applying *liberation* thinnings and other *intermediate* practices (see chapter 2).

In natural regeneration systems it is assumed that regeneration will occur naturally (Grieser Johns 1997), beginning with an acceptable production of seedlings—some of which eventually become harvestable adult trees. The guidelines to ensure natural regeneration are usually designed to conserve the soil and seedling pool and to spare seed sources and pole-sized juveniles of the desirable species (Bruenig 1993). Rarely, however, do these prescriptions explicitly consider conservation of the basic ecological processes involved in natural regeneration—such as pollination, seed dispersal, seed predation, and seedling establishment. Since animals play important roles in these processes, it is conceivable that impacts of logging on the native fauna (see chapters 4–17) can directly and indirectly influence natural regeneration and long-term commercial productivity.

A vast number of studies have investigated the role that seed dispersal plays in the regeneration of tropical trees in natural forests (Gilbert 1980; Howe 1984; Terborgh 1986a, b, 1990; Bawa and Krugman 1991; Janzen and Vázquez-Yanes 1991; Chambers and MacMahon 1994; Hartshorn 1995; Ter Steege et al. 1996). Seed dispersal promotes offspring survival by linking seed sources to suitable establishment sites, making this process a priority for consideration when designing forest management systems. Sustained production of timber requires adequate recruitment of timber species seedlings from continuous seed input. Appropriate seed sources, however, are scarcer in logged forests because many reproductive individuals have been harvested. Management must plan carefully, therefore, to protect not only an adequate number and distribution of seed trees, but also the dispersal agents critical to moving seeds to suitable establishment sites.

This chapter deals with the possible consequences of selective logging and silvicultural procedures on seed dispersal by vertebrates. Studies that examine how logging actually influences vertebrate-mediated dispersal do not exist, but there is abundant literature on how logging affects fauna, on processes of seed dispersal and seedling establishment, and on natural regeneration. We searched for patterns in this literature, but the statements we make often lack direct empirical support. After a brief introduction to the field of seed dispersal, we will address four questions:

- How important are vertebrates for seed dispersal of tropical timber species?
- How might selective logging influence vertebrate seed disperser fauna?
- How can changes in vertebrate disperser fauna influence seed dispersal?
- What may be the ultimate consequences of logging-disperser interactions for the regeneration of timber species?

We conclude with some recommendations for forest management and research.

The Importance of Seed Dispersal

Seed dispersal—the transport of seeds away from a parent plant—is an important process in the regeneration of most higher plants. Its evolutionary importance is illustrated by the mechanisms and structures of plants that promote seed dispersal. Dispersal modes can often be recognized by morphological characteristics of fruits and seeds, such as:

- Wings or panicles for wind dispersal
- Release mechanisms for explosive dispersal
- Sweet or nutritive fruit pulp for dispersal by frugivorous animals
- Nutritive nuts for dispersal by granivorous animals
- Adhesive structures for dispersal in furs
- Airy tissues for dispersal by water (Van der Pijl 1982)

Plant species often have more than one dispersal mode.

Seed dispersal is advantageous to plants when it enhances seed survival and increases reproductive success, and it may do so in various ways. First, dispersal reduces the risk of distance- or density-dependent mortality. Janzen (1970) and Connell (1971) introduced the idea that tropical forest tree seed survival increases with distance from the parent plant, because their seeds and seedlings suffer disproportional mortality due to predation and disease where they occur in high concentrations. The presence of stem canker in juveniles of the bird-dispersed canopy tree *Ocotea whitei* (Lauraceae), for example, is more likely to occur close to conspecific adult trees (Gilbert et al. 1994). Hammond and Brown (1998) reviewed experiments that test whether seed predation conforms to the Janzen-Connell model, and concluded that invertebrate attack generally did, while vertebrate attack generally did not. In a study by Fragoso (1997), bruchid beetle larvae killed 77 percent of *Maximiliana* palm seeds remaining near parent trees, but less than one percent of seeds dispersed away from parent trees by tapirs (*Tapirus terrestris*). Species that suffer heavy seed predation by insects are very common in tropical forests, and include timber species such as *Peltogyne* spp., *Hymenea courbaril* (both Caesalpiniaceae) and *Aspidosperma* spp. (Apocynaceae) in Guyana (Ter Steege et al. 1996), *Virola nobilis* (Myristicaceae) in Panama (Howe et al. 1985), and *Mimusops bagshawei* (Sapotaceae) in Uganda (Chapman and Chapman 1995). Seed dispersal appears to be an essential survival mechanism for tree species that are heavily attacked by invertebrate seed predators.

Second, seed dispersal theoretically enhances a plant's chance to place seeds in suitable establishment sites (Hamilton and May 1977). Any increase in the average distance over which seeds are transported in random directions results in an exponential increase of the area over which seeds are distributed, with a larger area likely to include more suitable establishment sites. Dispersal can also be directional towards suitable sites, for instance, by *Rhinodemmys* tortoises that move between gap areas (Moll and Jansen 1995). At least some distance-related gain in seed and seedling survival may be attributed to a higher probability of encountering high light environments where invertebrate attacks are likely to be less severe and seedling vigor greater (Hammond and Brown 1998; Hammond et al. 1999).

Third, dispersal may improve germination when it involves passage through the gut of animals (Traveset 1998). The activity of chimpanzees (*Pan troglodytes*) (Wrangham et al. 1994) and forest elephants (*Loxodonta africana*) (Chapman et al. 1992) has suggested such enhancement. In some cases, gut passage may even be obligate. Seeds from *Calvia major* (Sapotaceae), for instance, require passage through the gut of large birds (Temple 1977; Janzen 1983). Other plant species require handling by animals to release seeds from the fruits, or to get seeds into suitable condition for germination. Seeds of *Chrysophyllum lucentifolium* (Sapotaceae), for example, suffer heavy predation by frugivorous insect larvae when they are not removed from fruit pulp (P. A. Jansen, personal observation), while *Bertholettia excelsa* (Lecythidaceae) needs agoutis to gnaw open the extremely hard fruit (Peres et al. 1997). Other species need to be buried by scatterhoarding rodents (Leigh et al. 1993), such as the Guianan timber species *Vouacapoua americana* (Caesalpiniaceae; Forget 1990). Burial by rodents may also be accompanied with the inoculation of mycorrhizae (Janos et al. 1995).

Other possible advantages of seed dispersal are colonization, area extension, and gene flow. Even if plants are able to regenerate in the same locality time and again, taxa that move or extend their populations may do better because they avoid inbreeding and are better able to track long-term climatic changes. The avoidance of interactions with siblings is also of possible importance (e.g., Willson 1992). For general reviews of seed dispersal and the advantages it brings to plants, see Ridley (1930); Van der Pijl (1982); Howe and Smallwood (1982); Willson (1992); and Venable and Brown (1993). Reviews specific to animal-mediated seed dispersal include McKey (1975); Janzen (1983); Howe (1986); Jordano (1992); and Stiles (1992).

The evolutionary importance of seed dispersal remains poorly understood because the fitness consequences of seed dispersal have rarely been measured. Indirect evidence, however, argues for the importance of seed dispersal. One can conclude that dispersal enhances the chances of seed survival, as well as the probability of suitable site colonization. Both aspects are of interest to forest managers where timber species are concerned.

Vertebrates as Dispersers of Tropical Trees

The high number and generic richness of frugivorous (species with more than 50 percent of their diets composed of fleshy fruits) birds, bats, and primates in all three tropical regions (e.g., Snow 1981; Fleming et al. 1987) illustrates the importance of vertebrates as seed dispersers in tropical forests. Neotropical examples of seed-dispersing frugivores include toucans (Howe 1977; Howe and Vande Kerckhove 1981), guans (Howe and Vande Kerckhove 1981), cotingas (Snow 1982), fruit bats (Fleming and Heithaus 1981; Fleming 1986), howler monkeys (Estrada and Coates-Estrada 1984; Julliot 1996) and spider monkeys (Van Roosmalen 1985). African examples include hornbills (Whitney et al. 1998), bulbuls (Graham et al. 1995), tauracos (Gautier-Hion et al. 1985; Sun et al. 1997), gorillas (Tutin et al. 1991), chimpanzees (Wrangham et al. 1994), baboons (Lieberman et al. 1979), and cercopithecine monkeys (Gautier-Hion et al. 1985). Australian-Asian examples include hornbills (Grieser Johns 1997), fruit pigeons (Crome 1975; Snow 1981), birds of paradise (Beehler 1983), cassowaries (Stocker and Irvine 1983; Willson et al. 1989), flying foxes (Fujita and Tuttle 1991; Richards 1995), gibbons (Whitington and Tresucon 1991), and macaques (Balasubramanian and Bole 1993).

Tropical dispersers are also found among other vertebrate groups. Ungulates implicated in seed dispersal include African duikers (Gautier-Hion et al. 1985; Feer 1995) and elephants (Alexandre 1978; Chapman et al. 1992; White et al. 1993; Parren and De Graaf 1995; Yumoto et al. 1995), Neotropical tapirs (Fragroso 1997), Asian rhinoceros (Dinerstein and Wemmer 1988), and deer (Balasubramanian and Bole 1993). Examples of seed-dispersing reptiles include tortoises (Rick and Bowman 1961; Moll and Jansen 1995), while seed-dispersing rodents include spiny rats (see photo 3-1; Gautier-Hion et al. 1985; Hoch and Adler 1997) and agoutis (Forget 1990; Hallwachs 1993). Even tropical fish can disperse seeds (e.g., Gottsberger 1978; Horn 1997).

A large percentage of tropical forest plant species produce fleshy fruits or arillate seeds, which are associated with seed dispersal by animals (see table 3-1; Howe and Smallwood 1982; Willson et al. 1989). Approximately 70 percent of woody plants in wet forests— and 35 to 70 percent in dry forests—produce such seeds (Gentry 1982, Willson et al. 1989). In the Neotropics, species with fleshy fruits may account for up to 90 percent of all woody plant species (Howe and Smallwood 1982). Vertebrates might even disperse species

PHOTO 3-1 The spiny rat (*Proechimys sp.*) is a seed-dispersing rodent. (P. Jansen)

whose fruits are not fleshy, as many nut-bearing plant species are dispersed by granivorous birds and mammals.

TABLE 3-1 *Proportions of Tree Species with Fleshy Fruits in Different Tropical Forests of the World (based on Howe and Smallwood 1982 and Willson et al. 1989)*

Location	Forest Type	Species Considered (#)	Fleshy-Fruited (%)	Source
America				
Ecuador (Rio Palenque)	Wet	241	92	Gentry 1982
Costa Rica (La Selva)	Wet	161	91	Frankie et al. 1974
Colombia (Alto Yunda)	Wet	133	89	Hilty 1980
Costa Rica (La Selva)	Wet	320	65	Hartshorn 1978
Panama (Barro Colorado Island)	Moist	422	81	Gentry 1982
Costa Rica (Santa Rosa)	Dry	198	68	Gentry 1982
Costa Rica (Guanacaste)	Dry	104	51	Frankie et al. 1974

TABLE 3-1 *Proportions of Tree Species with Fleshy Fruits in Different Tropical Forests of the World (based on Howe and Smallwood 1982 and Willson et al. 1989) (continued)*

Location	Forest Type	Species Considered (#)	Fleshy-Fruited (%)	Source
Africa				
Gabon (Makakou)	Wet	136	71	Hladik and Miquel 1990
Cameroon (Dja)	Moist	372	83	Sonke 1998
Ghana (Kade)	Moist	115	60–78	Hall and Swaine 1981
Nigeria (Okomu)	Dry	180	46–80	Jones 1955, 1956 in Howe and Smallwood 1982
Asia				
Borneo (Mount Kinabalu)	Wet	360[a]	35–40	Stapf 1894 in Howe and Smallwood 1982
Malaysia	Wet	?	>67	Raemakers et al. 1980 in Willson et al. 1989
Australia				
Australia (Queensland)	Wet	774	84	Willson et al. 1989
Australia (Queensland)	Wet	?	76–100	Willson et al. 1989
Australia (Queensland)	Dry	?	18–63	Willson et al. 1989

[a] Includes other plant species, not just trees

Scientific information on the reproductive biology of tropical timber species is generally scarce (Hartshorn 1995; Hammond et al. 1996), and little is known about the importance of seed dispersal by animals for most of these species. Vertebrate-mediated dispersal is probably as common among timber species as among tree species in general. In the Guianas, for example, 72 percent of all 95 timber species are primarily vertebrate-dispersed (Hammond et al. 1996), while 74 percent of all 46 timber species in Bolivia appear to be animal-dispersed (P. A. Zuidema, unpublished data). Yet, vertebrate-mediated dispersal is not predominant among the tropical timber trees that dominate the current world market (see table 3-2 and chapter 2). Roundwood volumes of species primarily dispersed by vertebrates account for only 18, 33, and 47 percent of the production from Australia-Asia, America, and Africa, respectively. This is due to the fact that the timber market (and consequently forest exploitation) is still focusing on a very limited subset of all potential timber species (see chapter 18). In Peninsular Malaysia, for example, 16 percent of the 2,500 tree species have commercial potential, but only one percent plays a significant role at the international market (Grieser Johns 1997). The marketed subset over-represents long-boled emergent species, which are typically wind-dispersed. Proportions of animal-dispersed species are expected to increase as the market for lesser-known timber species increases. One example of an area where this diversification has begun is the Fazenda 2 Mil in Brazil, where the CELOS silvicultural system is practiced and 74 percent of 46 exploited species are vertebrate-dispersed (De Camino 1998; N. R. De Graaf, personal communication). Asia has lower proportions of vertebrate-dispersed timber species than Africa and the Neotropics. This is because a large proportion of Asian timber species are Dipterocarpaceae, which are generally known as (poorly) wind-dispersed or self-dispersed (Ashton 1988). Dipterocarp seeds, however, are often nutritious and commonly harvested by rodents that store them in spatially-scattered caches (Ashton 1988). Such cached seeds may be in better environments for germinating and developing than unharvested seeds (Jansen and Forget, in press). Dispersal of winged pine seeds by seed-hoarding *Tamias* squirrels in Nevada has been shown to be superior to wind dispersal (Vander Wall 1994), even though the trees appear morphologically adapted for the latter. Aspects of the fruiting strategy of Dipterocarps (large, nutritious seeds, mast fruiting) conform well to rodent dispersal (cf. Vander Wall 1990). Although rodents do not carry Dipterocarp seeds very far, and will consume part of their food reserves later, these seedeaters may have a positive net effect on the survival of Dipterocarp seeds. Animals may thus be important agents for secondary seed dispersal in Asian timber species.

TABLE 3-2 Role of Vertebrates in Seed Dispersal of the Most Important Timber Species for Three Tropical Regions in Terms of Volumes Exported by ITTO-Member Countries During 1994–1996

America			Africa			Asia		
Species	Volume[a] (1,000 m³)	Role of animals[b]	Species	Volume[a] (1,000 m³)	Role of animals[b]	Species	Volume[a] (1,000 m³)	Role of animals[b]
Cedrela spp. (Meliaceae)	1,767.1	-	Aucoumea klaineana (Burseraceae)	6,620.8	P	Shorea spp. (Dipterocarpaceae)	14,559.5	S
Swietenia spp. (Meliaceae)	1,217.5	S	Guibourtia arnoldiana (Leguminosae)	3,087.2	S	Dipterocarpus spp. (Dipterocarpaceae)	4,531.1	S
Hymenaea courbaril (Leguminosae)	926.3	P	Triplochiton spp. (Sterculiaceae)	2,936.8	-	Dryobalanops spp. (Dipterocarpaceae)	3,511.9	S
Tabebuia spp. (Bignonaceae)	502.9	-	Mitragyna ciliata (Rubiaceae)	2,167.2	-	Anisoptera spp. (Dipterocarpaceae)	1,966.6	S
Araucaria angustifolia (Araucariaceae)	240.0	P	Entandophragma spp. (Meliaceae)	2,050.8	-	Homalium foetidum (Samydaceae)	863.0	-
Dinizia excelsa (Leguminosae)	171.4	S	Tieghemelia spp. (Sapotaceae)	1,314.3	P	Koompassia spp. (Leguminosae)	815.9	-
Catostemma commune (Bombacaceae)	164.0	P	Dacryodes spp. (Burseraceae)	1,255.0	P	Tectona grandis (Verbenaceae)	698.0	P
Ocotea spp. (Lauraceae)	119.8	P	Ceiba pentandra (Bombacaceae)	1,201.7	S	Calophyllum spp. (Guttiferae)	674.0	-
Virola spp. (Myristicaceae)	107.7	P	Chlorophora excelsa (Moraceae)	1,181.1	P	Dactylocladus stenostachys (Crypteroniaceae)	594.8	P
Bagassa guianensis (Moraceae)	82.9	P	Khaya spp. (Meliaceae)	805.9	S	Intsia spp. (Leguminosae)	573.4	P

America

Africa

Asia

TABLE 3-2 Role of Vertebrates in Seed Dispersal of the Most Important Timber Species for Three Tropical Regions in Terms of Volumes Exported by ITTO-Member Countries During 1994–1996 (continued)

Species	Volume[a] (1,000 m³)	Role of animals[b]	Species	Volume[a] (1,000 m³)	Role of animals[b]	Species	Volume[a] (1,000 m³)	Role of animals[b]
Brosimum utile (Moraceae)	75.7	P	Baillonella toxisperma (Sapotaceae)	789.1	P	Pouteria spp. (Sapotaceae)	504.0	P
Carapa guianensis (Meliaceae)	71.7	P	Terminalia spp. (Combretaceae)	759.1	S	Parashorea spp. (Dipterocarpaceae)	439.1	P
Ochroma lagopus (Bombacaceae)	64.0	S	Lophira alata (Ochnaceae)	483.2	S	Copaifera palustris (Leguminosae)	388.6	P
Amburana spp. (Leguminosae)	58.3	S	Antiaris africana (Moraceae)	435.8	P	Syzygium spp. (Myristicaceae)	358.4	P
Trattinickia spp. (Burseraceae)	28.0	P	Pycnanthus angolensis (Myristicaceae)	302.5	P	Terminalia spp. (Combretaceae)	332.3	P
Peltogyne pubescens (Leguminosae)	25.4	P	Aningeria spp. (Sapotaceae)	227.8	P	Celtis spp. (Ulmaceae)	310.9	P
Platymiscium pinnatum (Leguminosae)	22.9	S	Erythropleum ivorense (Leguminosae)	153.0	P	Palaquium spp. (Sapotaceae)	302.4	P
Vochysia spp. (Vochysiaceae)	16.0	-	Pterygota macrocarpa (Sterculiaceae)	131.8	S	Buchanania spp. (Anacardiaceae)	277.4	P
Bowdichia nitida (Leguminosae)	11.4	-	Herietia utilis (Sterculiaceae)	111.8	S	Canarium spp. (Burseraceae)	244.3	P
Eperua falcata (Leguminosae)	6.9	P	Nauclea diderrichii (Rubiaceae)	97.5	P	Calophyllum spp. (Guttiferae)	234.4	P
Mora excelsa (Leguminosae)	6.4	S	Distemonanthus benthamianus (Leguminosae)	94.1	-	Gonystylus bancanus (Thymelaeaceae)	231.3	P

TABLE 3-2 Role of Vertebrates in Seed Dispersal of the Most Important Timber Species for Three Tropical Regions in Terms of Volumes Exported by ITTO-Member Countries During 1994–1996 (continued)

America			Africa			Asia		
Species	Volume[a] (1,000 m^3)	Role of animals[b]	Species	Volume[a] (1,000 m^3)	Role of animals[b]	Species	Volume[a] (1,000 m^3)	Role of animals[b]
Cordia goeldiana (Boraginaceae)	5.7	P	*Piptadeniastrum africanum* (Leguminosae)	87.8	S	*Pterocymbium beccarii* (Sterculiaceae)	213.9	-
Aspidosperma spp. (Apocynaceae)	3.4	-	*Afzelia* spp. (Leguminosae)	68.2	P	*Heritiera simplicifolia* (Sterculiaceae)	208.9	S
Pouteria spp. (Sapotaceae)	2.9	P	*Gossweilerodendron balsamiferum* (Leguminosae)	60.6	S	*Dillenia* spp. (Dilleniaceae)	200.9	P
Hyeronima spp. (Euphorbiaceae)	2.0	P	*Lovoa trichilioides* (Meliaceae)	54.9	S	*Pometia pinnata* (Sapindaceae)	196.7	P
Goupia glabra (Goupiaceae)	0.3	P	*Pericopsis elata* (Leguminosae)	49.7	S	*Hopea* spp. (Dipterocarpaceae)	182.7	S
Humiria balsamifera (Humiriaceae)	0.1	P	*Canarium schweinfurthii* (Burseraceae)	49.5	P	*Xylia xylocarpa* (Leguminosae)	179.5	?
			Pterocarpus soyauxii (Leguminosae)	46.1	-	*Masticiodendron* (Rubiaceae)	172.0	P
			Distemonanthus spp. (Leguminosae)	42.0	-	*Agathis* spp. (Araucariaceae)	171.4	S
			Guarea cedrata (Meliaceae)	38.8	P	*Dyera costulata* (Apocynaceae)	170.0	-
Other species	4,171.4		Other species	567.1		Other species	14,802.1	
Total	9,872.0		Total	27,271.2		Total	48,909.2	

TABLE 3-2 Role of Vertebrates in Seed Dispersal of the Most Important Timber Species for Three Tropical Regions in Terms of Volumes Exported by ITTO-Member Countries During 1994–1996 (continued)

Species	America		Africa		Asia	
	Volume[a] (1,000 m³)	Role of animals[b]	Volume[a] (1,000 m³)	Role of animals[b]	Volume[a] (1,000 m³)	Role of animals[b]
Percentage of volume						
Primary dispersal by animals		33		47		18
Only (potential) secondary dispersal by animals		23		26		74
No dispersal by animals		42		27		8
Percentage of genera						
Primary dispersal by animals		59		43		57
Only (potential) secondary dispersal by animals		11		37		23
No dispersal by animals		26		20		17

[a] Roundwood equivalents, calculated with 50% conversion efficiency for plywood and veneer, and 35% for sawnwood.
[b] Dispersal modes; - = no dispersal by vertebrates, P = primary dispersal by vertebrates, and S = only (potential) secondary dispersal by vertebrates.
SOURCES Van Roosmalen 1985, and Hammond et al. 1996 for America; Hall and Swaine 1981, and Hawthorne 1985 for Africa; Soerianegara and Lemmens 1993, Lemmens et al. 1995, and Sosef et al. 1998 for Asia.

Effects of Logging on the Disperser Fauna

Logging operations and silvicultural treatments induce changes in the vertebrate fauna of a forest, both directly (see chapters 4–11, 14, and 24) and indirectly (see chapters 15–17). Populations of some species decline, while others become more common. Changes in the population of frugivorous and granivorous birds and mammals can be brought about by:

- Physical alterations to the habitat
- Alterations of food availability
- Increased hunting pressure

How each of these mechanisms affect seed disperser fauna is not clear because it is difficult to distinguish the effects of physical alterations from those altering food availability. Studies examining these processes are also not always comparable because they differ in methodologies, length of evaluation periods, and degree of control for edge effects and indirect effects such as hunting (see chapters 18 and 19).

Physical Alterations to the Habitat

Physical alterations to the habitat due to logging and silvicultural measures often deviate from natural levels of disturbance (see chapter 22). For example, logging creates new gaps that are usually larger than natural ones (Struhsaker 1997; see chapter 2). As a result, gap environments increase from one percent to 10–20 percent at low extraction levels (Crome et al. 1992; Johns 1992a; White 1994b), to almost 50 percent at higher levels (Uhl and Vieira 1989). Logging usually kills many more trees than those extracted (Grieser Johns 1997). Johns (1988), for example, found that harvesting as few as 3.3 percent of the trees of dbh ≥ 9.5 cm in a Malaysian forest resulted in an overall loss of more than 50 percent of individuals in this category. Other physical effects of logging are habitat destruction and fragmentation through access roads (Ter Steege et al. 1996) and clearings. Consequently, interior forest environments may become scarce in logged forests (Struhsaker 1997).

Frugivorous and granivorous animals are sensitive to these habitat alterations (Janzen and Vázquez-Yanes 1991), but vary in the way they respond. Moderate disturbance can favor certain dispersers, while negatively impacting others (Janzen and Vázquez-Yanes 1991). Bird diversity strongly declined with logging in French Guyana (Thiollay 1992), and significant declines in primate populations have been found several years after logging in many forests (Skorupa 1988, cited in Grieser Johns 1997; Struhsaker 1997). Animals that benefit from the denser undergrowth following habitat disturbance include the tapir in Brazil (Fragoso 1990), duikers in Sierra Leone (G. Davies cited in Grieser Johns 1997), forest elephants in Uganda (Struhsaker 1997), opportunistic frugivorous birds in Indonesia (Lambert 1992; Danielsen and Heegaard 1995), and rodents (Struhsaker 1997). Some species, such as the red-rumped agouti (*Dasyprocta agouti*), seem unaffected (Ter Steege et al. 1996). The next section of this volume (chapters 4–14) examines in detail the direct effects of logging on forest wildlife habitat.

Alteration of the Availability of Food

Logging and silvicultural treatments can also alter the availability of food. Logging may not necessarily affect the overall *quantity* of food. Plants remaining after logging may use the greater resource availability in logged-over forest to increase their production of leaves, flowers and fruits (Johns 1988; cf. Levey 1988a, 1990; Janzen and Vázquez-Yanes 1991). This increased production can buffer the loss of food plants due to logging, although the overall amount of fleshy fruit still tends to be lower in logged forests (Johns 1991b; Grieser Johns 1997). Many frugivorous mammals and birds are able to partially shift their diets towards folivory or insectivory or adapt their foraging behavior. Obligate frugivores such as spider monkeys, diana monkeys (*Cercopithecus diana*) and fruit bats, however, may be scarcer in logged-over forest (Grieser Johns 1997).

Logging and silvicultural operations, however, are likely to affect the *quality* of food resources, particularly when certain species of trees and climbers are selectively removed. Logging can result in a severe decline or even elimination of reproductive individuals of particular timber species (Lambert 1991; Grieser Johns 1997), especially those that are rare or that start fruiting above the minimum diameter for logging (Plumptre 1995). The impact of this on the frugivore fauna might be severe, as many species depend on a very limited set of food plant species during periods of low fruit availability. Such periods occur in most tropical forests (Hilty 1980; Howe and Smallwood 1982; Leighton and Leighton 1983; Gautier-Hion et al. 1985; Sabatier 1985; Heideman

1989; Van Schaik et al. 1993; White 1994a). Some timber species may even be *keystone* food resources that determine frugivore population levels (Gilbert 1980; Terborgh and Winter 1980; Foster 1982; Leighton and Leighton 1983; Howe 1984; Terborgh 1986a; Lambert 1991; Peres 2000a). Harvesting and liberation thinnings might theoretically reduce the availability of keystone food resources to a level at which the disperser community is severely affected. No study, however, has convincingly demonstrated the existence of such a cascading effect.

Some potential keystone food plants have been recognized among timber species. Examples are *Sterculia pruriens* (Sterculiaceae) and *Clathrotropis brachypetala* (Papilionaceae) in Guyana (Ter Steege et al. 1996), and various *Ficus* species in Bolivia (Fredericksen et al. 1999b). Most potential keystone food plants, however, have no direct commercial value—such as the following trees:

Casearia corymbosa (Flacourtiaceae)	Costa Rica (Howe 1977)
Musanga cecropioides (Cecropiaceae)	Guinea (Yamakoshi 1998)
Myristicacaeae and Meliaceae	Borneo (Leighton and Leighton 1983)
Palms	Peru (Terborgh 1986b) Guyanas (Ter Steege et al. 1996) Guinea (Yamakoshi 1998)
Figs	Costa Rica (Wheelwright et al. 1984) Peru (Terborgh 1986a) Malaysia (Lambert and Marshall 1991) Borneo (Leighton and Leighton 1983; Lambert and Marshall 1991; Harrison 2000) Colombia (D. Rumiz, personal communication) Not in Africa (Gautier-Hion and Michaloud 1989)
lianas (e.g., Anonaceae in Asia)	(Leighton and Leighton 1983)

Ironically, many of these keystone plants are categorized as *undesirable species* in silvicultural systems, and are commonly suppressed. Figs in Borneo (and probably elsewhere), for example, are greatly affected by logging because they often grow on trees of commercial size and value (Leighton and Leighton 1983).

Increased Hunting and Live Animal Capture

Increased hunting and live animal capture, either for subsistence or commerce, are often a consequence of making forests accessible for logging operations. Regulation of hunting during and after logging is not commonly practiced. Hunting in tropical forests is well studied (Redford 1992; see chapters 15–17), and harvest data (Ayres et al. 1990; Vickers 1990; Fa et al. 1995; Fitzgibbon et al. 1995; Peres 2000b) indicate that large mammals and birds (primates, ungulates, large rodents, large bats, hornbills, toucans, cracids) are commonly taken. These are the animals that account for most of the seed-disperser biomass, as well as the biomass of handled seeds. Large-bodied frugivores are particularly susceptible to hunting (Chapman and Chapman 1995; see chapters 4 and 5). Unregulated hunting may cause drastic declines in the population sizes—notably of primates and large mammals (Fragroso 1990; Glanz 1990; Mittermeier 1990; Ráez-Luna 1995) and large birds (Robinson et al. 1990; Silva and Strahl 1990)— and is therefore a severe threat to disperser fauna.

Consequences of Logging for Seed Dispersal

Vertebrate-dispersed plants are more susceptible to dispersal failure than wind-dispersed plants (Willson 1992). This raises a question. How great is the risk that changes in the disperser fauna after logging will lead to reduced dispersal success or outright dispersal failure?

Dispersal success, or *effectiveness,* is the product-sum of the quantity of seeds and the quality of dispersal. Dispersal quantity—the number of seeds that are dispersed—depends on the abundance of dispersers and the number of seeds each disperser takes. Dispersal quality reflects the probability that a given seed will produce a new adult, as a function of seed treatment and seed deposition. Low-quality dispersers kill seeds, or deposit seeds in unsuitable conditions, while high-quality dispersers enhance seed germination and/or deposit seeds in favorable environments (Schupp 1993).

Quantities of Seeds Dispersed

Quantities of seeds dispersed seem to be buffered against loss of dispersers by *disperser redundancy* in most vertebrate-dispersed species.

Tropical plants are typically dispersed by a variety of animal species that belong to different taxonomic groups (see box 3-1). Their disperser assemblages are usually *loose* and highly variable in space and time (Howe 1983; Jordano 1993, 1994; Fuentes 1995). Weak interactions between animals and plants are characteristic of species-rich tropical systems (Jordano 1987). It is likely, therefore, that reduced seed dispersal by one species can be compensated for by increased dispersal by other species (Janzen and Vázquez-Yanes 1991).

Box 3-1 *A neotropical seed disperser assemblage in the Sapotaceae tree family.*

Disperser assemblages of tree species may include an array of animal species, even when these tree species appear to be "adapted" to seed dispersal by particular agents. For instance, the Guianan Sapotaceae family includes many species with a fruiting syndrome that appears to be monkey-directed—large, yellow to reddish fruits with sweet, juicy, odorous pulp and large seeds. In fact, spider monkeys (*Ateles paniscus paniscus*) and howler monkeys (*Alouatta seniculus*) are the main—if not only primary—dispersers of many species in this family (Van Roosmalen 1982; Julliot 1996). Fruits and seeds that drop below the parent tree because the monkeys neglect or spoil them, however, are often eaten by an array of frugivorous and granivorous vertebrates that may act as effective secondary dispersers. While feeding on fruit pulp below parent trees of a smaller-seeded genus such as *Manilkara,* large frugivorous birds, such as trumpeters (*Psophia crepitans*), may also ingest and disperse seeds. In addition, many other animals are secondary dispersers for the Sapotaceae family, such as coatis (*Nasua nasua*), tayras (*Eira barbara*), tapirs (*Tapirus terrestris*), and reptiles such as tortoises (*Geochelone denticulata*)(P. A. Jansen, personal observation). Secondary dispersal is also provided by agoutis (*Dasyprocta agouti*) and acouchis (*Myoprocta acouchi*), which harvest seeds larger than 1 g (approximately) from the forest floor and bury them one by one in shallow, widely spaced caches (Forget 1990, 1993; Hallwachs 1993; P. A. Jansen unpublished data). Most of these seeds are relocated and eaten (*i.e.,* killed), but those that escape predation may be in a very good position to establish.

Finally, most of the seeds that are ingested by monkeys pass the digestive tracts of these animals unharmed and are defecated in small clumps. An array of dung beetle species transport and bury these seeds while harvesting dung (Feer 1999). Tortoises ingest seeds while eating dung (P. A. Jansen, personal observation), and rodents harvest seeds from dung piles and bury them (Janzen 1982a; Hallwachs 1993; F. Feer and P. M. Forget, personal communications). These animals provide (mostly short-distance) secondary dispersal.

Substitution of one disperser species by another is possible and likely because dietary overlap between vertebrate dispersers is often great—even between different taxonomic groups (Janzen and Vázquez-Yanes 1991). On islands in the South Pacific, for instance, frugivorous pteropid bats are the most important dispersers of many plant species. In fact, these bats substitute for the large frugivorous birds that have been eliminated in recent historical time (Cox et al. 1991; Rainey et al. 1995). In the Neotropics, Janzen and Martin (1982) argue that certain plant species have persisted after the extinction of their original megafauna dispersers some 10,000 years ago, and Hallwachs (1993) states that agoutis (*Dasyprocta punctata*) substitute for the long-extinct dispersers of the timber tree *Hymenaea courbaril*. Janzen and Vázquez-Yanes (1991) propose that free-ranging cattle and horses also substitute for the original ungulate dispersers of the Central American tree *Crescentia alata* (Bignonaceae).

Plants having one or very few disperser species are at risk of dispersal failure, and certain animals might be *keystone dispersers* for these plants (Futuyma 1973; Howe 1977; Gilbert 1980). Species with few dispersers, however, appear to be uncommon. Well-known examples include the timber species *Cola lizae* (Sterculiaceae) in Gabon, where the only known disperser is the lowland gorilla (*Gorilla gorilla gorilla*; Tutin et al. 1991), and *Hymenaea courbaril* (Caesalpiniaceae) in Costa Rica that is principally dispersed by agoutis (Hallwachs 1993; Asquith et al. 1999). Bond (1995), in his investigation of plant extinction due to pollinator or disperser failure, found no clear case of plant extinction as a result of the loss of its dispersers. One possible example, however, is the regeneration failure of the tree *Calvaria major* as a result of the absence of obligatory seed processing by its historical disperser, the dodo (*Raphus cucullatus*; Temple 1977).

While there is redundancy in vertebrate-mediated dispersal in most plants, this does not imply that dispersal processes are not sensitive to logging. Large-seeded tree species may depend on disperser species that are particularly susceptible to logging and associated hunting practices. Moreover, the loss of any given disperser species may affect a range of plant species. Hornbills, for example, disperse *Guarea* sp., *Celtis mildbraedii*, *Maesopsis eminii*, *Canarium schweinfurtii* and at least 30 other tree species that are exploited for timber (Whitney et al. 1998). Red howler monkeys disperse at least 86 species of plants in forests of French Guiana (see photo 3-2; Julliot 1996). Many plant species on islands in the South Pacific risk losing their principal dispersers because flying foxes are an endangered group (Cox et al. 1991; Fujita and Tuttle 1991). The redundancy capacity of vertebrate-mediated dispersers in logged forests remains largely unknown, however, as no studies have compared relative quantities of seeds dispersed between logged and unlogged forests (but see Guarigata et al. 2000).

PHOTO 3-2 Howler monkeys (*Alouatta palliata*) are important dispersers for many plant species, including at least 86 species in the forests of French Guiana. (P. Jansen)

The Quality of Dispersal

The quality of dispersal may also be affected by changes in the disperser fauna because disperser species differ in the way they treat and deposit seeds (Levey 1986; Loiselle and Blake 1999). Passage through the digestive tracts of one species might cue germination, while passage in another species might kill the majority of seeds (Traveset 1998). One disperser might carry seeds further and deposit seeds in more favorable conditions than another (Schupp 1993). Of the many animals that forage in fruiting trees, very few may actually be good dispersers (Howe 1977).

Any change in the disperser assemblage is likely to influence dispersal quality, even if the total quantity of seeds dispersed remains constant (Janzen and Vázquez-Yanes 1991). Many species that are vulnerable to logging (see chapters 4–11, 14, and 24) are considered high-quality dispersers, even though their dispersal qualities have not actually been measured. This is particularly true for large animals, as they are capable of transporting seeds further than most small animals. Lower dispersal quality leads to fewer seeds surviving disperser handling, fewer seeds leaving the vicinity of the parent tree, and more seeds being deposited in clumps rather than being scattered (Howe 1989).

We can conclude that changes of the vertebrate disperser fauna due to logging may be at the cost of both quantity and quality of seed dispersal. The risk of poor dispersal holds particularly true for species commonly dispersed by highly frugivorous vertebrates and large mammals. Such species commonly are relatively large-seeded. In the following section, we attempt to evaluate the consequences of poorer dispersal for the natural regeneration of these *risk species*.

Possible Consequences of Poor Dispersal for Natural Regeneration

Several researchers have claimed that it is critical for the preservation of natural tropical forests to maintain the animals that facilitate seed dispersal (e.g., Howe 1984; Pannell 1989; Chapman et al. 1992). Poor dispersal should lead to decreased seedling recruitment because fewer seeds reach suitable sites. Those that do still may suffer distance- and density-dependent mortality. This would threaten the long-term persistence of tree species that are unable to recruit under conspecific adults (Chapman and Chapman 1995). While logging can reduce the quantity and quality of seed dispersal (see above), changes in microenvironment after logging may offset lower dispersal rates through enhanced germination and survivability of seeds (Oliver and Larson 1990). This raises the question of whether enhanced seed and seedling survival in logged-over forests counterbalances poor dispersal of risk-species?

Logged-over forest often has higher individual seed and early seedling survival because more gap environments are available. Risk species tend to be large-seeded and shade-tolerant, and these do well in gap edges and small gaps (but not in the core area of large gaps) (Kasenene and Murphy 1991; Hammond and Brown 1995; Zagt and Werger 1998). Their poor performance in gap centers may be linked to their vulnerability to high temperatures and water stress (Whitmore 1991; Brown and Whitmore 1992; Hammond and Brown 1998). Reduced dispersal in logged forests may be counterbalanced if low impact logging measures are applied (see chapters 21 and 24)—as minimum disturbance to forest structure helps to create small gap sizes that are of benefit to many risk species (Veenendaal et al. 1996; Van der Meer et al. 1998). Seed and seedling mortality due to biotic sources may also be lower in logged-over forest. The drier conditions under increased canopy openness tend to reduce fungal attack on seedlings. Greater light availability also enhances the growth of seedlings, making them less susceptible to diseases and predation than seedlings in understory conditions (Hammond and Brown 1998; Hammond et al. 1999).

How logging affects levels of post-dispersal seed predation is not well known (Janzen and Vázquez-Yanes 1991). Rodents tend to become more common in logged-over forest, and the dense vegetation may also attract browsers (e.g., deer and elephants) that hamper recruitment (Struhsaker 1996). Asquith et al. (1997) found that the removal of larger mammals resulted in increased seed predation and seedling hervibory by rats on small islands. Notwithstanding, it is not clear what the net effect of logging is on seedling recruitment (Guarigata and Pinard 1998). So far, there is no conclusive evidence of the long-term sustainability of existing tropical natural regeneration systems (Whitmore 1990; Grieser Johns 1997; N. R. de Graaf, personal communication). Not only may forest growth be over-estimated (Bossel and Krieger 1991; Grieser Johns 1997), but it is also uncertain whether natural regeneration and the recruitment of timber species is sufficient (Frederickson and Mostacedo 2000). Stimulating the survival and growth of *potential crop trees* does not always increase short-term timber production (De Graaf et al. 1999), nor does it contribute to the population growth required for long-term production. While seedling recruitment seems to be the bottleneck, very little is known about the reproductive ecology and establishment requirements of most timber species (Bongers 1998).

Whichever way one looks at it, logging appears to give timber species a comparative disadvantage due to the harvesting of seed sources (and the vulnerability of some of their dispersal systems), which may result in low rates of seedling recruitment relative to nontimber species (Bongers 1998). These disadvantages can be mitigated by offering juveniles a selective advantage through silvicultural measures (refinement and liberation, see chapter 2), but these measures are expensive and might negatively affect food availability for the disperser community.

Given our current knowledge, guidelines to ensure recruitment are limited. Sparing the reproductive individuals of timber trees and protecting soils are essential requirements for providing available seed material and establishment sites, but do not guarantee recruitment. Trees may produce increased seed yields through higher availability of resources (light, nutrients) and less intense predispersal seed predation—but only if they survive habitat alteration due to logging and if there is sufficient pollination (Janzen and Vázquez-Yanes 1991). Seeds of timber species, therefore, should be viewed as an invaluable resource. Any measures that promote seed dispersal may contribute to a better use of this resource by enhancing net seed survival, and will be in management's interest.

Conserving Seed Dispersal Processes in Logged Forest

Management Options

The availability of vertebrate dispersers in a logged-over forest depends on the availability of critical food resources, suitability of the habitat, and the hunting pressure. Forest management guidelines to foster seed dispersers should therefore be aimed at:

- Ensuring the year-round availability of the food resources required by dispersers
- Minimizing habitat alteration
- Restricting the hunting of known dispersers

These approaches are being tested and evaluated at a number of sites in the tropics (see chapters 18, 19, and 26)—including the Fazenda 2 Mil, Manaus Brazil, where an extended form of the CELOS silvicultural system is being applied (De Graaf 1986; De Camino 1998).

Protecting Food Availability

Protecting the most important food plants—especially potential *keystone* food plants—during all silvicultural operations can probably ensure food availability most effectively for seed dispersers. One way to minimize the reduction of overall food availability to dispersers is reduced-impact logging (see chapters 21 and 24). At Fazenda 2 Mil in Brazil, a 100 percent inventory of harvestable trees (and some other woody plants) precedes every harvesting operation. The function of trees and other woody plants as food resources (and seed sources) are taken into account when planning harvesting, liberation, or refinement. Students conducted a literature and field survey to identify which food plants are locally important for monkeys (N. R. De Graaf, personal communication). The goal of these efforts is to use careful planning to avoid endangering the food supply of seed dispersers.

Minimizing Habitat Alteration

Habitat alteration due to logging is unavoidable. It can, however, be limited by reduced-impact logging procedures (Hendrison 1990; see chapters 21, 24, and 25). Setting aside reserve areas can further mitigate the effects on fauna and corridors that are left undisturbed (see chapter 23). Specific requirements of particular dispersers, if known, can also be incorporated into management planning. At Fazenda 2 Mil, careful planning and preparation of felling (including vine cutting and directional felling) and skidding are standard procedures designed to minimize damage and habitat alteration. The concession area also includes 10 percent of its area in reserves (N. R. De Graaf, personal communication).

Controlling Hunting

Hunting seed dispersers should be avoided or at least not exceed the sustainable harvest (e.g., Robinson and Redford 1991). To achieve this, effective regulation is indispensable. At the Fazenda 2 Mil concession, hunting and gathering animals that disperse seeds is strictly forbidden. Employees (many of whom are hunters) are informed about the reasons, and the company sees to it that alternatives for wild meat are available (N. R. De Graaf, personal communication).

Applying these criteria requires a basic knowledge about the reproductive ecology of timber tree species, including an understanding of which vertebrates are important seed dispersers. Habitat and food requirements (or at least which plants are their most important food resources) of these animals must also be known. For most forest areas, such information still needs to be collected. Because the same plant species may have different dispersers in different forests, and the same animals may have different sets of important food species, it would be safest to have such information at the management (or concession) level.

Research Priorities

Very little is known about how silvicultural procedures influence seed dispersal and seedling recruitment. Understanding where and how logging influences disperser fauna—and at what point natural regeneration might be jeopardized due to negative environmental impacts on these animals—is basically a *trial and error* matter (Janzen and Vázquez-Yanes 1991). To address this paucity of information it is recommended that research focus on the following areas:

Fitness Consequences of Seed Dispersal

First, there is a need for research that actually measures the fitness consequences of seed dispersal. Such a study could, for instance, compare the fates of entire cohorts of seeds in areas with and without disperser(s).

Habitat Suitability for Seed Dispersers

Second, there is a clear need for field studies of how silvicultural procedures affect habitat suitability and year-round availability of food for seed dispersers. The identification of keystone food plants for dispersers would be indispensable. This information is testable by comparing frugivore population levels between large forest patches (from which only keystone foods are experimentally removed), untreated forest patches, and forest patches with random experimental removal of food plants up to a comparable level of reduction (see chapter 19).

Dispersers and Seed Dispersal Effectiveness

Third, we need to know how changes in disperser fauna influence the effectiveness of seed dispersal. Seed/seedling distribution patterns, and seed/seedling survival rates between forest patches with and without (or with fewer) individual dispersers could be compared. Comparable locations with different levels of hunting might be suitable sites for such studies. Quantifying dispersal effectiveness of different species, determining what proportions and absolute quantities of seed removed by different species are influenced by the relative abundance of these species, and evaluating the effects of changes in disperser fauna on dispersal effectiveness by modeling are other approaches needing evaluation. With such studies, we could also try to answer the questions raised by Schupp (1993):

- Are there dispersers that maximize both the quantity and quality of dispersal?
- Is quantity or quality more important for effective dispersal?
- How variable is the dispersal quality of a disperser for different plant species?

Logging and Recruitment Sites

Fourth, we still need to know much more about how silvicultural procedures influence the availability of recruitment sites—especially for timber species. We must first collect basic information on the reproductive ecology and establishment requirements of these species and others of commercial, ecological, and social importance. More theoretical and empirical studies are also needed on how the environment influences plant survival and growth in order to determine what levels of seedling stocking are required for reaching the sustained yields of timber that are commonly envisioned.

ACKNOWLEDGMENTS

We are greatly indebted to Frans Bongers, Robert Fimbel, Reitze de Graaf, Jan van Groenendael, Douglas Levey, Jan den Ouden, Marc Parren, Susan Parren, Lourens Poorter, Frank Sterck, Chin Sun, and four anonymous reviewers for the useful information and stimulating comments they provided. While writing this paper, the authors were supported by the Netherlands Foundation for the Advancement of Tropical Research with grant W 84-407 (P. A. Jansen) and by the Dutch Ministry for International Development Cooperation with grant BO 009701 (P. A. Zuidema).

Part II

WILDLIFE AND CHAINSAWS: DIRECT IMPACTS OF LOGGING ON WILDLIFE

Logging practices directly impact wildlife by altering habitat, dispersing populations, and (in rare instances) maiming or killing individual animals. In turn, changes in wildlife numbers and behaviors are believed to affect ecological processes ranging from pollination to decomposition. The following eleven chapters review what is known about the direct impacts of logging on tropical wildlife (indirect impacts are addressed in part III). Taxonomic groups evaluated in this section include mammals, birds, herpetofauna, invertebrates, and aquatic fauna. Responses of individual species and guild assemblages to timber harvesting—both positive and negative—are reviewed. Management options to reduce the impacts of logging on various taxa are also presented. The contributors close their reviews with specific recommendations for future research efforts that can fill the gaps in our understanding of direct impacts associated with logging-wildlife interactions in tropical forests.

Direct Impacts of Logging on Wildlife in Managed Tropical Forests

In chapter 4, Andrew Plumptre and Andrew Grieser-Johns review the effects of selective logging on primates in Africa, Asia and the Neotropics. Primates are one of the more conspicuous animal groups in tropical forests, and the ecology of many species is fairly well known. Monitoring programs at three sites—Tekam, Malaysia; Bundongo, Uganda; and Kirindy, Malaysia—provide valuable long-term, pre- and post-logging responses of primate communities in production forests. Some response patterns include the following:

- Small frugivorous primates appear to benefit from logging in Africa, while folivorous species demonstrate variable responses
- Most primates in southeast Asia are generalists, and their population densities in exploited dipterocarp forests several years after logging often exceed those of unlogged forests
- South American primates appear to respond strongly to microhabitat changes, with different species demonstrating very different responses to habitat modification from logging

In general, primates appear fairly adaptable to selective logging, changing ranging patterns, and diet to accommodate differences in forest structure and composition in response to timber management activities. The greatest threat to these animals from selective logging is the increased hunting activities that often accompany commercial logging operations (see part III) as a result of the improved forest access created by logging roads.

In chapter 5, Glyn Davies, Matt Heydon, Nigel Leader-Williams, John MacKinnon, and Helen Newing examine the effects of timber-harvesting operations on medium- to large-bodied terrestrial mammals. Production forest studies for this group are sparse, often anecdotal, widely scattered in the literature, and supported by very few long-term investigations. The response of most medium to large size mammals to forestry practices, therefore, must be extrapolated from projected impacts on their diet and habitat. As a group, large-bodied forest browsers appear to adapt well to changes in forest habitat brought about by logging, provided the hunting intensity remains low and the land is not converted to other uses. Interior-dependent species, however—such as the bearded pig and the yellow muntjac of southeast Asia—tend to decline in disturbed habitats as generalist competitors (e.g., the ubiquitous wild boar and common muntjac) gain a competitive advantage over their more specialized counterparts. Small carnivores (including civets), whose diets are restricted to vertebrate and invertebrate prey, also seem more sensitive to logging than omnivorous browsers and grazers. In general, many large mammal species appear well adapted to disturbed forest conditions (and thus logging), provided secondary impacts associated with logging do not increase appreciably (e.g., hunting, fires, and forest conversion). Management recommendations to limit the risk of local species extinction, and future areas of research to assess medium-to large-bodied mammal responses to logging, are reviewed in this chapter.

In chapter 6, José Ochoa and Pascual Soriano describe the effects of timber harvesting on small mammal communities in the Neotropics. Information related to the ecology of rodents and marsupials, and their responses to timber extraction and related activities (e.g., hunting), remain largely unknown. Where investigated, changes in forest structure and composition as a result of logging (especially in the lowland rain forests of the Amazonas and the Guiana Shield) have drastically altered select small mammal communities. Within these managed landscapes, rodent populations demonstrate:

- Higher species richness
- Increased abundance of species that exploit early successional habitats (semiarboreal predator-omnivores and terrestrial frugivore-omnivores)

— Reductions in taxa associated with the canopy strata (arboreal frugi-vore-granivores)

Alterations in the availability of primary forest resources (fruits, seeds, roots, and locomotory strata), increases in gap sizes and frequencies, and the modification of micro-climatic patterns at the understory level are thought to be the primary mechanisms that account for these population changes. The consequences of these community changes on forest dynamics, and the potential for local extirpation of some taxa, are discussed in this chapter.

In chapter 7, Pascual Soriano and José Ochoa discuss the response of bat communities to timber exploitation activities, and the capacity of these mammals to serve as indicators of environmental change due to forest management prescriptions. Bat communities represent 50 percent of the mammal fauna in lowland Neotropical forests, and are estimated to include between 90–110 sympatric species in the Amazon Basin alone. Insectivorous and vertebrate predator species (mainly the subfamily *Phyllostominae*) are the most adversely affected by timber exploitation due to the loss of prey, shelter, and changes in microclimate. Species whose populations appear to be favored by logging practices include many of the frugivorous megachiroptera that feed on, and disperse, the seeds of pioneer plant species. The introduction of cattle in areas near forestry concessions may also increase the food resources of hematophagous (vampire) bats, and increase their populations. The ecological consequences of these population changes are reviewed in this chapter, along with management activities that can help to minimize them.

Of all the studies involving wildlife-timber harvesting interactions, the responses of bird communities to forest management practices have been the most thoroughly studied. As a group, birds are useful in evaluating the effects of logging on fauna, owing to:

— Their well-established taxonomy and capacity to be identified in the field

— The availability of biological and ecological information on most bird families and many species

— Their sensitivity to changes in forest structure, micro-climate, and composition

— The key ecological roles they play in pollination, seed dispersal, and seed predation

This portion of the book, including contributions by Douglas Mason and Jean-Marc Thiollay (chapter 8), Mohamed Zakaria Bin

Hussin and Charles M. Francis (chapter 9), and Andrew Plumptre, Christine Dranzoa and Isaiah Owiunji (chapter 10), review the ways in which bird communities react to logging disturbances in Latin America, the Indo-Australia region, and Africa, respectively.

Across the three continents, bird feeding guilds tend to respond in similar manners to timber-harvesting disturbances. Nectarivores and generalist frugivore/insectivores of the canopy demonstrate increases in their populations following logging. These wide-ranging generalists appear well adapted to a broad range of successional conditions. The most seriously-affected birds are those taxa of the primary forest interior. Terrestrial insectivores and low- to mid-understory flycatchers are consistently vulnerable to logging practices. Species normally found in mixed-species flocks—which are among the most abundant and widespread birds of the primary forest—and large terrestrial species are also some of the most negatively impacted. The frequency of gaps and dense regrowth may disrupt traveling or foraging behavior through the understory, and the hotter/drier conditions in these openings may cause a drop in insect prey abundance. Large gaps also pose barriers to some species that will not fly across these open spaces. Finally, several guilds—including predators, arboreal foliage gleaners, and large specialized frugivores—exhibit variable responses to logging practices. These varied observations are attributed to differences in species biology, sampling procedures, intensity of logging, and level of forest recovery since logging. Recommendations are provided for reducing the risk of perturbing populations of sensitive bird species, and identifying areas of urgently needed research to clarify the short-term responses and long-term recovery of vulnerable species and guilds.

In chapter 11, Laurie Vitt and Janalee Caldwell investigate the effects of logging and deforestation on populations of amphibians and reptiles that inhabit tropical forests. Herpetofauna may reach their highest diversity in tropical forests, yet few studies have quantified these populations and even fewer have quantified their responses to timber harvesting operations. Where the forest structure is drastically altered (e.g., during monocylcic silvicultural treatments such as clear cutting), most herpetofauna of the primary forest exhibit severe population declines—especially those species that live their entire life cycle in the leaf litter. When harvests are less intensive, such as during single tree selection cuts, most species of reptiles and amphibians appear capable of persisting within the managed forest matrix (at least in the short term). The invasion of post-harvest secondary forests by large-bodied predaceous heliothermic lizards, however, may lead to the local extinction of some small-bodied frogs and lizards. Finally, not all herpetofauna respond negatively to logging disturbances. Some species of frog, for example, are capable of exploiting the breeding habitat created by stream-crossing impoundments and skidder ruts. A great many unknowns characterize the interaction between herpetofauna and logging. In this chapter the authors outline steps to address this paucity of information in managed tropical forests.

In chapter 12, Jaboury Ghazoul and Jane Hill explore the impacts of selective logging on tropical forest invertebrates. Invertebrates dominate the animal diversity, abundance, and biomass in tropical forest habitats, yet little is known about their trophic associations, habitat requirements, interactions with other community members, or responses to disturbances such as logging. Nevertheless, some projections about invertebrate-logging interactions are possible. Invertebrates are incapable of physiological temperature regulation. Changing climatic regimes in logged forests, therefore, may cause a decline in the populations of invertebrate taxa occurring in microhabitats that normally have restricted climatic variation (e.g., understory, leaf litter, and soil). Furthermore, taxonomic groups such as moths (which are specialist herbivores) appear to be very sensitive to logging disturbances, with significant losses of diversity following some harvests. The above responses can have a negative cascading effect on insect-mediated ecological processes, such as pollination, decomposition, and nutrient cycling. On a positive note, the short generation times and high-potential fecundity provide most invertebrate populations with high resilience to disturbance. In general, the extent to which invertebrate populations recover from the impacts of logging is believed to depend on how closely logging mimics natural disturbance and regeneration processes. Management and research recommendations relevant to the conservation of invertebrates and invertebrate-mediated processes are presented in this chapter.

In chapter 13, Gerardo Camilo and Xiaoming Zou discuss the functional diversity of soil fauna in relation to logging. The diversity of soil animal taxa is poorly known, especially in the tropics, and few studies have examined how logging practices affect the diversity and functionality of soil fauna in forested systems. The authors propose that the removal of timber, coupled with the physical changes to soil properties following logging, directly influences soil fauna diversity. In the case of endogeic and anecic earthworm communities, however, recent data from long-term studies of tree communities in the Luquillo Experimental Forest of Puerto Rico suggest that both small scale clear cutting and low-frequency, large-scale clear cutting (more than 50-year intervals) have limited impact on these groups. The response of other soil organisms to timber harvesting practices, where litter composition and quality are altered over long periods of time, may not demonstrate the same responses as those exhibited by earthworms. Potential long-term impacts of logging practices on soil food webs are also discussed in this chapter.

Catherine Pringle and Jonathan Benstead close the section with a review of the effects of logging on aquatic systems in chapter 14. Timber-harvesting practices perturb tropical watercourses in several ways:

- Destruction and/or fragmentation of critical riverine habitat, including downstream floodplains and estuarine mangroves
- Alteration of stream sediment loads, turbidity, and solute chemistry
- Changes in hydrology and fluvial geomorphology
- Shifts in in-stream incident light and temperature regimes

Physical and chemical alterations of aquatic systems have ecological consequences, including the loss of in-stream and riparian zone biodiversity, decreases in the abundance of fish and benthic invertebrates that are intolerant of high sedimentation, shifts in the ratio of photosynthesis to respiration, shifts in the abundance of different feeding guilds along stream continua (high turbidity limits light to algal primary producers, with cascading effects for other members of aquatic food webs), and facilitation of invasion by exotic species. Two case studies depicting these conditions are presented: 1) the effects of logging on rivers of the Pantanal in Brazil, and 2) the facilitation of invasion by exotic fish species through removal of riparian forests in Madagascar. Management and policy recommendations to mitigate these conditions are also discussed in this chapter.

Steps Toward Mitigating the Direct Impacts of Timber Harvesting on Forest Wildlife

Forest animals respond in a wide variety of ways to timber-harvesting practices. Changes observed in the population of a species after logging depend on:

- The intensity of logging and its impacts on the wildlife habitat (food resources, shelter, microclimate conditions, competitors/predators, etc.)
- The recovery period between silvicultural interventions

⁓ The sampling procedures used to evaluate a species' response to forest management activities (see chapter 4)

Fauna at highest risk of population declines within logged forests are those species associated with interior forest conditions, such as specialized browsers and canopy rodents (chapters 5 and 6), insectivorous bats and birds (chapters 7–10), animals inhabiting the soil litter (chapters 11 and 13), and those hunted for food, trade, or sport (part III focuses on this interaction).

The physical attributes of post-harvested sites that most negatively impact forest interior species are large canopy gaps and excessive soil disturbance. As a rule, logging tends to create more and larger gaps than natural disturbance events, which increases the sunlight reaching the forest floor. As a result, the understory becomes hotter and drier and eventually much denser with plants responding to these new conditions. Many species of wildlife avoid these artificial openings because of the poor quality of plant cover, the low availability of specific food resources, and the hotter climatic conditions that characterize these gaps (see chapters 6–13). At the soil level, many litter-inhabiting organisms are poorly adapted to drastic swings in temperature and moisture regimes, and their populations can plummet in the wake of cover removal and compaction (see chapters 11–13). Finally, the disturbed soils along roads and within large canopy gaps serve as point sources for erosion. Severe soil loss may slow forest recovery through lost productivity, and degrade aquatic habitats through siltation and nutrient loading (see chapter 14).

All contributors to this section believe that the application of good forest management practices today can greatly reduce the impact of timber harvesting on most wildlife species. Several management recommendations received universal support by the authors. To begin, planning at the landscape level was identified as an essential step towards ensuring a sustainable flow of wood, adequate recovery periods between cuts, and the development of a protected area network to conserve the wildlife, water, and plant communities of the forest. This protected network should be designed to protect sensitive (steep and riparian zones) and unique (biologically or culturally) areas throughout the forest, and be designated before concessions are awarded to minimize confusion about their location. Within the managed portion of the forest matrix, silvicultural interventions should be designed to minimize the amount of disturbance to nontarget trees. Beyond conserving large fruiting trees and the habitat in general, these practices also reduce the recovery time of the cut area following the cessation of activities. Numerous Reduced Impact Logging (RIL) procedures have been developed in recent years to achieve this goal. Part V of the book discusses these practices in detail.

Long-term forest protection was another area where management was deemed to have great potential for conserving wildlife within the managed landscape. Every contributor believed that enforcing existing laws that restrict hunting within the timber concession would help enormously to protect wildlife in managed forests while incurring minimal costs to the concession holder. Protection of the permanent forest estate in general, from fire and land conversion activities, was also seen as a critical component of long-term conservation efforts in managed forests—benefiting the forest itself, wildlife, and the nation as a whole by helping to ensure a long-term supply of forest resources. Finally, empowering foresters to implement these procedures, through improved training in management and conservation practices, was identified as essential to implementing the practices noted above. Part VI discusses in greater detail the available options, incentives, and policy reforms for bringing about these conservation initiatives.

While the application of *best* forest management practices today are capable of reducing the environmental impacts associated with timber harvesting in tropical forests, all contributors to this section identified a paucity of quantifiable information that specifically identifies those species being conserved or threatened by timber harvesting practices. In an effort to truly achieve sustainable forest management in tropical forests, including the conservation of wildlife populations, the authors strongly advocate additional site-specific research in the following areas:

- Understanding the basic ecology of forest species, including their demographics, habitat requirements, degree of specialization, and reproductive requirements (priorities should be given to rare, threatened, endangered, and commercial species

- Clarifying ecological processes that are critical to the health of tropical forests, including wildlife-plant regeneration interactions and identifying indicator species that serve as barometers of these processes

- Identifying criteria for the best allocation of protected area networks within a managed landscape (dimensions for riparian zones and small refugia in the logging matrix are especially important topics)

- Refining silvicultural treatments in light of the above points, to maximize the yield of commercial products while minimizing the recovery time of the forest from these exploitation activities

Gathering this information will require carefully-planned experimental designs to overcome many of the shortcomings associated with historical research in this area, and mechanisms to ensure their application in management practices. Part IV of this book addresses these points.

Chapter 4

CHANGES IN PRIMATE COMMUNITIES FOLLOWING LOGGING DISTURBANCE

Andrew J. Plumptre and Andrew Grieser Johns

Primates are a conspicuous part of tropical forest ecosystems. They are important seed dispersers, especially in terms of the volume of seeds dispersed (Wrangham et al. 1994; Chapman 1995), and often form a large part of the animal biomass in these forests (White 1992). Consequently, the effects of logging have been studied better in this group of animals than most others.

Johns and Skorupa (1987) reviewed the effects of forest disturbance on primates and estimated survival ability following disturbance. They argued that survival ability is negatively correlated with the degree of frugivory, and that large-bodied primates are more at risk from forest disturbance than small species. They admit, however, that their analyses were based upon data from different habitats, and pointed out that their dataset combined results from areas of timber harvesting alone and areas where it was combined with shifting cultivation. Furthermore, many primates are highly adaptable and their diets can be very flexible, making correlation calculations less robust. For example, Johns and Skorupa (1987) used a value of about 19 percent fruits, seeds and flowers to describe the diet of black-and-white colobus (*Colobus guereza*), based on data from the Kibale Forest of Uganda. In the Budongo Forest of Uganda, however, these items form up to 60 percent of the diet for this species (A. Plumptre, unpublished data). Similarly, the level of frugivory in gorillas (*Gorilla gorilla*) was based on the data from mountain gorillas (*G. g. beringei*), which have little access to fruit. More recent data from the analysis of dung samples of the western lowland gorilla (*G. g. gorilla*) in Gabon suggests that it consumes large quantities of fruit (Williamson 1988; Williamson et al. 1990).

It is common for research on the effects of logging on biodiversity to exhibit inaccuracies arising from poor comparability of results between different study sites. Almost all the primate studies reviewed by Johns and Skorupa (1987)—and most since—compare population densities (or relative densities) in unlogged and logged areas and assume that the difference is a function of logging. Given the spatial heterogeneity found in tropical forests (Terborgh 1992b), this assumption is erroneous and rarely tested (Johns 1992a, 1992b; see chapter 18).

Rather than revisit the information reviewed by Johns and Skorupa (1987), this chapter summarizes comparisons of primate populations in logged and unlogged forests of the tropics (Africa, southeast Asia, and South and Central America). These comparisons are contrasted against three case studies where pre- and postharvest assessments of primates are made at the same site. The benefits and limitations of these studies are discussed, along with recommendations to conserve primates in production forest landscapes.

Short-Term Comparative Studies of Logging and Primates

All the studies quoted in this chapter employed line-transect censuses to estimate relative or absolute densities of primates (Buckland et al. 1993). Some have fitted mathematical functions to perpendicular distance data to calculate density estimates (e.g., Plumptre and Reynolds 1994), while others have only compared encounter rates (White 1992). Within each study, however, the same method is used to compare results between data from logged and unlogged areas.

It is not possible to provide site descriptions for all the studies referred to in this chapter—the original references should be consulted for details. Additional information, however, will be presented on three *model* sites described in the third section of this chapter.

Africa

The effects of selective logging disturbance on primates have been studied in four main sites in Africa: Kibale, Uganda (Skorupa 1986, 1988; Howard 1991; Struhsaker 1997), Budongo, Uganda (Plumptre and Reynolds 1994, 1996; Fairgrieve 1995), Kalinzu, Uganda (Howard 1991; Hashimoto 1995), and Lopé, Gabon (White 1992,

1994b). A summary of primate responses to logging at each site shows that many species do not change in density in logged sites, but where they do, responses can vary depending on the study (see table 4-1). For example, data averaged from two unlogged and three logged sites in Gabon suggest no change in densities, yet one of the logged sites (logged 3–5 years previously), had significantly lower encounter rates for white-nosed monkeys (*C. nictitans*), moustached monkeys (*C. cephus*), grey-cheeked mangabeys (*Cercocebus albigena*), gorillas (*G. gorilla*), and chimpanzees (*P. troglodytes*) than the unlogged sites. Whether this variation is due to logging history or initial differences between sites is not known. Data given for the Kibale forest (see table 4-1) draws from several logged sites, and shows that responses can vary depending on the level of damage that occurred during logging. Where damage was relatively heavy (as in compartment K15), and forest regeneration was smothered by a matrix of climber tangles and dense shrubs, primate densities declined for most species.

TABLE 4-1 *The Responses of African Primates Following Disturbance from Selective Logging at Four Sites Based on Intersite Comparisons of Unlogged and Logged Areas*

Species	Lopé, Gabon	Kibale, Uganda	Budongo, Uganda	Kalinzu, Uganda
Frugivores				
Cercopithecus mitis		nc	+	nc
Cercopithecus ascanius		+/-	+	+
Cercopithecus cephus	nc			
Cercopithecus pogonias	nc			
Cercopithecus nictitans	nc			
Cercopithecus lhoesti		-		nc
Folivores				
Colobus badius		nc/-		
Colobus guereza		+	+	nc
Colobus satanas	nc			
Seed Predator				
Cercocebus albigena	nc	nc/-		
Apes				
Pan troglodytes	-	nc/-	nc	+
Gorilla gorilla	nc			

TABLE 4-1 *The Responses of African Primates Following Disturbance from Selective Logging at Four Sites Based on Intersite Comparisons of Unlogged and Logged Areas (continued)*

Species	Lopé, Gabon	Kibale, Uganda	Budongo, Uganda	Kalinzu, Uganda
Others				
Mandrillus sphinx	nc			
Galago spp.		-		
Perodicticus potto		-		
Effort (km transect walked)	753	610	1,921	152
Logging Intensity (trees/ha)	1	5–7	6–25	?
Damage Levels (% tree loss)	11	?<50	?>50	?
Years since logging	1–15	12–14	1–50	4–20

+ = increase in density (no./km^2) following logging; nc = no change; - = decrease in density (P < 0.05). At Kibale results are available for lightly and heavily damaged forest: light/heavy. ?=unknown values.

In all sites human hunting pressure was low and the logging concessions surrounded by mature forest. Data sources are as follows: Lopé (White 1992, 1994b), Kibale (Skorupa 1986, 1988; Weisenseel et al. 1993), Budongo (Plumptre and Reynolds 1994), Kalinzu (Howard 1991; Hashimoto 1995).

Most African forests are characterized by moderate or high tree species diversity. Monodominance, however, occurs in several regions (Connell and Lowman 1989; Hart 1985). In these monodominant forests, early successional stages tend to be more diverse than the climax monodominant condition in which a single species comprises 90 percent of the large trees. *Cynometra* monodominance occurs in both Budongo and Kalinzu forests (see table 4-1). In Budongo, primate densities have been estimated in mature *Cynometra* monodominant forest, and a variety of mixed and secondary forests arising from logging activity. In addition, primates have been censused in *Gilbertiodendron* monodominant and mixed forests in the Ituri Forest of eastern Zaire (presently the Democratic Republic of Congo; Thomas 1991a). These studies show that most primates occur at low densities in monodominant forests, although in some areas the dominant tree can form an important part of their diet (Maisels and Gautier-Hion 1994). Results from Budongo suggest that only chimpanzees use the monodominant forests as much as mixed forests, but this estimate is strongly influenced by a single site where chimpanzee nest density was particularly high. At Budongo, forestry interventions (particularly arboricide treatment of non-timber trees) designed to change the forest from mon-odominance to more mixed have been successful. In response to the increased vegetation diversity in these compartments, primate densities have increased (see table 4-1; Plumptre and Reynolds 1994; Plumptre 1996). Detailed studies of *Cercopithecus mitis*, *C. ascanius*, and *Colobus guereza* show that range size is smaller and diets more diverse in logged monodominant forest (C. Fairgrieve 1995; Plumptre unpublished data). Forest management prescriptions that reduce monodominance in forests appear to benefit monkeys.

Frugivores and Seed Predators

In general, the frugivorous guenons (*Cercopithecus* spp.) are adaptable to forest disturbance. Data from the Kibale, Lopé and Budongo forests agree in this respect (see table 4-1). There are exceptions, however. White (1992) suggests that the inclusion of data from other sites around Lopé indicates a decrease in the abundance of some frugivore/seed predator species in response to logging. Many seed predators (Grieser Johns 1997), such as the grey-cheeked mangabey (*Cercocebus albigena*), occur at lower densities or show no change in disturbed forests. Data from two sites indicate that l'Hoest's monkey (*Cercopithecus lhoesti*), may decline in abundance in logged forests, perhaps because it forages to a large extent on the ground and may become more vulnerable to predation. The fact that predation may become a significant risk in logged forests is indicated by the folivorous red colobus (*Colobus badius*) in the Kibale Forest—adult males become highly aggressive in areas where they are forced to travel and forage on the ground due to canopy loss following logging (Skorupa 1988). These animals are at a significant risk of predation by chimpanzees, and the increased aggression is thought to deter the chimpanzees from attacking (Skorupa 1988).

Chimpanzees occur at lower densities in logged mixed forests in both Uganda (Skorupa 1986) and Gabon (White 1992; see photo 4-1), but show no density change and can even increase in logged monodominant forests (Plumptre and Reynolds 1994; Hashimoto 1995). It has been suggested that a drop in chimpanzee density in recently-logged forests at Lopé may be due to inter-group conflict (White and Tutin 2001). Another interpretation, however, is that they avoid areas where humans are present (see Budongo Forest data in next section).

PHOTO 4-1 Male chimpanzee (*Pan troglodytes schweinfurthii*) and figs in the Budongo Forest, Uganda, where logging does not seem to have affected this species greatly. (A. Plumptre)

Folivores

Many folivores, such as black-and-white colobus (*C. guereza*), can benefit from forest disturbance because the increase in light leads to a flush of leaf growth (Ganzhorn 1995). *C. guereza* has a dietary preference for the leaves of certain colonizing tree species, and are often found in very disturbed forests (Oates 1977). Red colobus, however, occur at lower density in most disturbed forests and it is not clear why this is so because food availability is generally higher in the disturbed forests. Red colobus were studied in the Kibale Forest in Uganda and were found to suffer high chimpanzee predation (Skorupa 1988). Consequently, they may avoid the more open disturbed forests because there are fewer escape routes. Gorilla numbers do not appear to respond to logging at low intensity in Gabon (White 1994b) because they can feed off the dense terrestrial herbaceous vegetation (Marantaceae and Zingiberaceae) found in secondary forest.

Relatively few data have been collected on the responses of Malagasy lemurs to logging (see photos 4-2 and 4-3). In general, the abundance of frugivorous lemurs correlates with tree diversity (Hawkins et al. 1990). Tree diversity has been shown to decrease in logged, dry deciduous forests (Ganzhorn et al. 1990), however fruit and leaf production has been shown to increase (Ganzhorn 1995). If this is also the case with other forest types in Madagascar, then positive effects on frugivorous and folivorous lemur populations might be expected following logging. Some incidental observations, however, suggest that the populations of some lemur species may be reduced in degraded tropical forests. More research is required in this area (see photo 4-2). It has been suggested, for example, that the fork-marked dwarf lemur (*Phaner furcifer*) is reduced in numbers in degraded forest (Harcourt and Thornback 1990), but some research indicates that this species doesn't change or may only occasionally increase (Ganzhorn 1995).

In general, it appears that small, frugivorous African primates can tolerate logging disturbance, provided the levels of damage do not lead to climber tangles and a smothering of forest regeneration. Responses of folivores to logging are variable, but predictable, for most species (see table 4-1).

PHOTO 4-2 Madagascar mouse lemur (*Microcebus rufus*) is an insectivorous species of lemur. (A. Plumptre)

Southeast Asia

Southeast Asia covers several biogeographical zones, with different primate species assemblages. The effects of logging on these species have been studied in detail only in the Tekam Forest Reserve in Peninsular Malaysia, the Ulu Segama Forest Reserve in Sabah, and the Nanga Gaat Reserve in Sarawak (Marsh and Wilson 1981; Davies and Payne 1982; Laidlaw 1994; see table 4-2). Additional studies are underway at Barito Ulu and Gunung Palung in East Kalimantan, Indonesia.

Results of these studies are influenced both by spatial heterogeneity and different hunting levels. Hunting was absent at Tekam and Ulu Segama, for example, but occurred at other sites. Primates are particularly targeted as a food source in Sarawak, but this practice may be changing in timber concessions (see chapter 16).

Apes and Macaques

Detailed studies at Tekam have shown adaptability to logging by lar gibbons (*Hylobates lar*), where they increased in density. Adaptability has also been indicated in less detailed studies of *H. muelleri* in Sabah,

PHOTO 4-3 Brown lemur (*Petterus fulus*), a primarily frugivorous species, whose response to logging will depend on how fruit availability is affected by logging. (A. Plumptre)

TABLE 4-2 *The Responses of Southeast Asian Primates to Selective Logging, Based on Intersite Comparisons of Unlogged and Logged Area*

Species	Ulu Segama, Sabah	Nanga Gaat, Sarawak	Tekam, Peninsular Malaysia	Sungai Lalang, Peninsular Malaysia	Kemasul, Peninsular Malaysia
Apes					
Pongo pygmaeus	+				
Hylobates muelleri	nc	-			
H. lar			+	-	-
Leaf Monkeys					
Presbytis rubicunda	-	-			
P. hosei	+	-			
P. melalophos			+	-	-
P. obscura			+	-	-
Macaques					
Macaca fascicularis	nc	-	+	-	-
M. nemestrina	nc	?	nc	+	nc
Effort (km transect walked)	271	304	500	120	120
Logging Intensity (trees/ha)	c.20	10	18	>20	>20[a]
Damage Levels (% tree loss)	32–58	54	51	38	55
Years Since Logging	6–12	0–4	0–12	20–30	20–30

+ = increase in density (no./km^2) following logging; nc = no change; - = decrease in density (P < 0.05).

Data sources are as follows: Ulu Segama (Johns 1989b), Nanga Gaat (Dahaban 1996), Tekam (Johns 1989b), Sungai Lalang and Kemasul (Laidlaw 1994).

[a] Three separate logging events recorded.

where no change in the population was detected. In Sungai Lalang and Kemasul, however, gibbons declined in density following logging (see table 4-2). Unfortunately, there are not enough data on the changes in forest composition and structure to tease out why these differences occur. At Tekam, lar gibbons sang considerably less in recently logged forests. As singing is energetically expensive (Johns 1986a), this might reflect the lesser availability of high-energy fruits. Lar gibbons and banded leaf monkeys (*Presbytis melalophos*) were observed to switch from largely frugivorous diets to largely folivorous diets following a logging event. This ability to survive on a varied diet is probably what enables primate species to survive in logged areas where other less adaptable species are lost.

The orangutan (*Pongo pygmaeus*) exists in a largely nomadic social system (Payne 1988). It has been reported to congregate in patches of unlogged forest surrounded by recently logged forest (Davies 1986), perhaps using such *refuges* as points from which to recolonize logged forest at a later stage of regeneration. Studies in Ulu Segama show that densities are higher in logged forest close to unlogged forest, compared with logged forest distant from unlogged forest (see table 4-3), providing some support for the refuge theory. Factors other than forest structure and fruit availability—such as soil type—have been considered to limit the distribution of orangutans, and perhaps other sympatric species (Payne 1988).

Macaques (*Macaca*) tend to be highly generalist in diet and also range widely. Small macaques often become commensals of humans where their forest habitat is degraded (e.g. Lim and Sasekumar 1979). Long-tailed macaques (*M. fascicularis*) and proboscis monkeys (*Nasalis larvatus*) are typically riverine species, and congregate in these areas in logged forests because of the practice of retaining unlogged riparian corridors in production forest reserves (Johns 1989b). Long-tailed

TABLE 4-3 *Primate Densities (individuals/km²) in Logged Forests Adjacent to and Distant (20–30km) from Blocks of Unlogged Forest (Ulu Segama Forest Reserve, Sabah)*

Species	Unlogged	6-year-old Logged (adjacent)	6-year-old Logged (distant)	12-year-old Logged (adjacent)	12-year-old Logged (distant)
Pongo pygmaeus	0.3	0.9	0.5	2.1	present
Hylobates muelleri	5.3	11.1	4.5	3.9	4.2
Presbytis rubicunda	13.5	12.6	7.0	13.0	4.6
P. hosei	2.1	8.0	0	5.7	2.0
Macaca fascicularis	present	0	present	0	11.5
M. nemestrina	16.0	16.0	4.0	20.0	28.0
Effort (km transect walked)	113	58	39	31	30

Data are from Johns (1989b).

macaques were at lower densities on logged areas in most sites, however—probably due to differences in the availability of these riverine corridors—while pig-tailed macaques (*M. nemestrina*) showed little change or even increases in numbers.

Leaf Monkeys

Leaf monkeys, despite their name, are highly frugivorous and so cannot be compared with the colobines of Africa. This makes the distinction of frugivore/folivore much less clear in Asia when compared with Africa. Studies at Tekam have shown adaptability to logging among banded-leaf monkeys (*P. melalophos*) and dusky-leaf monkeys (*P. obscura*), and Hose's leaf monkey (*P. hosei*) also exhibited increases at Ulu Segama (see table 4-2). Elsewhere in peninsular Malaysia, however, they have shown a decline in numbers where logging has occurred. Laidlaw (1994) observed leaf monkeys ranging outward from unlogged refuges into logged forest on an occasional basis. This observation is supported when densities are compared in logged forest adjacent to unlogged forest, and in logged forest about 30 km away (see table 4-3). Older regeneration stages may be more likely to support similar population densities of these primates, but results are often contradictory (see table 4-3). Changes in primate territories can occur following logging. At Tekam, banded-leaf monkeys switched from territoriality to mutual avoidance in logged forests, and this may be one way that densities were able to increase.

In general, results from dipterocarp forests suggests that primates are able to persist in logged forests where hunting is low, however many species exhibit declines in their populations. The degree of isolation of the logged forests from unlogged forests (see table 4-3), and the extent of isolated patches of unlogged forests within the logged forest matrix (Johns 1989b), is important in determining primate responses.

South and Central America

There have been no detailed studies of the effects of logging on primates in the Neotropics. Most available data are from a 12-month survey at one site in Brazilian Amazonia—Ponta da Castanha (Johns 1991b)—and several short surveys elsewhere in Brazil (Johns 1986b) and Venezuela (Ochoa 1988).

Seed Predators

Surveys suggest that pithecine seed predators, particularly the bearded sakis (*Chiropotes*) and uakaris (*Cacajao*) are adversely affected by intensive logging (see figure 4-1). In the case of the southern bearded saki (*Chiropotes satanas*), some of the main timber trees extracted within its fairly small geographical range are known to be important food resource trees (Johns and Ayres 1987). Uakaris possess certain behavioral adaptations to cope with seasonal food shortages, such as terrestrial foraging for old fallen seeds or exploitation of caterpillars (Ayres 1986)—which may explain its persistence in logged forests compared to other surveyed primates (see figure 4-1). Their habitat, which is flooded during the wet season, is particularly vulnerable to damage through logging activities (Grieser Johns 1997).

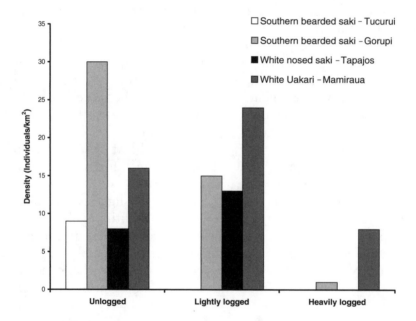

FIGURE 4-1 *Densities of some pithecine seed predators (number of individuals/ km²) in unlogged, lightly logged (<15 percent of trees destroyed) and heavily logged (>50 percent of trees destroyed) forests of Brazilian Amazonia. Forests were logged between 2 and 18 years previously. Data from Grieser Johns (1997). X = the primate species was not detected.*

Other Species

The species-rich, Amazonian primate communities are segregated through sub-habitat specialization to a greater extent than primates elsewhere (Peres 1993). Small-bodied species—tamarins (*Saguinus mystax*), squirrel monkeys (*Saimiri* sp.), titi monkeys (*Callicebus* spp.), and capuchins (*Cebus* spp.)—typically exploit early successional or edge habitats, and have highly variable diets. These species are most likely to use degraded habitats (see table 4-4). The metabolic demands faced by large-bodied species are too great to enable specialization in patchy sub-habitats, and these species are more typical of mature forest habitat. The large-bodied woolly monkeys (*Lagothrix lagotricha*) and spider monkeys (*Ateles paniscus*) have frequently been identified as surviving poorly in degraded habitats, but the effect of habitat degradation is often masked by hunting activity (these species are preferred for meat; see chapter 15).

In general, the larger-bodied seed predators—woolly and spider monkeys—appear to decline in density following logging, while the smaller species appear to be more adaptable to habitat change.

Long-Term/Longitudinal Studies of Primate Responses to Logging

Before-and-after studies at the same site are considered to provide better data on the effects of logging because spatial heterogeneity is controlled for. With most of the studies quoted above, we can never be sure whether differences in densities/encounter rates are due to the effects of logging or inherent site differences. Consequently, the studies presented below represent a more appropriate research method and provide a better indication of the response of primates to logging (see chapters 18 and 19 for a further discussion of logging-wildlife study design).

Budongo Forest Reserve, Uganda

The Budongo Forest Reserve in western Uganda is a medium altitude (900–1200 m elevation), moist, semideciduous tropical forest and covers a total area of 793 km². It has been harvested for timber, using

TABLE 4-4 Densities of Primates (individuals/km²) in Unlogged and Logged Forests at Ponta da Castanha, Brazilian Amazonia

Species	Unlogged Forest	Logged Forest[a]	Unlogged Forest "Island" (35ha)[b]
Saguinus mystax	78	88	16
Callicebus moloch	0.3	1.2	5.6
C. torquatus	2.5	3.7	1.8
Saimiri sp.	32	81	42
Cebus apella	11.5	32	61
C. albifrons	14	31	4.8
Pithecia albicans	9.0	18	2.4
Alouatta seniculus	0.9	0.2	3.8
Ateles paniscus	1.3	1.1	0
Lagothrix lagotricha	6.0	present	0
Effort (km transect walked)	554	170	97

Data from Johns (1991b).

[a]60 percent of trees destroyed 11 years previously

[b]Islands were surrounded by logged forest and shifting cultivation

a selective logging system, since the 1920s. Plumptre (1996) describes the history of the management of this forest.

Eight compartments in the forest (8–11 km² each) were studied from 1991 to 1996. Several compartments were affected by illegal pit-sawing of trees during the four-year interval between surveys. Others were allocated as concessions to registered pitsawyers or sawmills. The history of the eight sites, and the densities of large trees (> 70 cm dbh) removed, are summarized in table 4-5. Only mahoganies (*Khaya* spp. and *Entandrophragma* spp.) were harvested in pit-sawed areas, while in the sawmill concession other species—particularly *Cynometra alexandri*—were also cut. During the period of study, extraction levels were very low (0.5–2.3 trees/ha) as compared with extraction levels from previous logging events (Plumptre 1996).

Plumptre and Reynolds (1994) censused four primate species—blue monkeys (*Cercopithecus mitis*), redtail monkeys (*C. ascanius*), black-and-white colobus (*Colobus guereza*), and chimpanzees (*Pan troglodytes*) in these eight areas during 1992. Censuses were repeated in 1996 along the same transects. Chimpanzees were censused using nest counts because encounters with individuals were rare (Plumptre and Reynolds 1996). Most primates, with the exception of chimpanzees, showed no significant change following the low levels of logging carried out by the illegal pitsawing—mainly because only mahoganies were removed and these do not form an important part of the primates' diets (see table 4-5). In compartment K4, however, where a sawmill has been felling Cynometra since 1992 (with much heavier damage levels since this is the most common tree in the forest), both blue monkeys (*Cercopithecus mitis*) and black-and-white colobus (*Colobus guereza*) have declined in density.

Chimpanzee abundance, as determined by nest counts, did not increase in any site and appear to have decreased in some of the pitsawn areas. This may be because people use the census trails for cutting and carrying timber—and chimpanzees generally avoid areas commonly used by people. This could also explain why there was a decrease in nest density in compartment N3, where the main research

TABLE 4-5 *The Logging History of the Eight Study Compartments Surveyed in the Budongo Forest Between 1992–96*

				Primates			
Compartment	Dates Logged Before 1992	Dates Logged After 1992[a]	Number of Trees Removed /ha Since 1992	*C. mitis*	*C. ascanius*	*C. guereza*	*P. troglodytes*
N15	Unlogged	Unlogged	-	nc	+	nc	nc
K11–13	Unlogged	Unlogged	-	nc		nc	nc
N3	1947–52	Unlogged	-	nc	+	nc	-
B1	1931–36 1982–86	1992–96	0.84	nc	nc	nc	nc
B4	1940–42	1992–96	0.68	nc	nc	nc	nc
N11	1960	1996	0.54	nc	nc	nc	-
W21	1965	1995–96	0.94	nc	+	nc	-
K4	1992–96	1992–96	2.32	nc	nc	-	nc

[a]Compartments B1, B4, and N11 were illegally pitsawn between 1992 and 1996; W21 was a legal pitsawing concession; and K4 was a legal concession to a sawmill.

Differences in densities (individuals/km^2) between a census in 1992 and 1996 for four species of primates are shown (P < 0.05): + = increase in 1996; nc = no change relative to 1992; - = decrease in 1996 relative to 1992. Data from A. Plumptre (unpublished).

site was established, and where many researchers used the intensive network of trails (see table 4-5). Measurement of the distances from transects to 266 nests show a greater than expected frequency of nests away from these census trails. These transects were being used more by people at the time of the second census, which may explain this drop in density (Plumptre and Reynolds 1997). No change was found in the site used by the sawmill, which may be because chimpanzees had moved away from this site prior to 1992 (logging had been going on at a low level before this date).

These findings show that, like the comparative studies presented earlier, low levels of logging did not affect these primate species greatly in most sites. There were, however, declines in some species where mechanized logging occurred. These monkey species tend to occur in both disturbed and undisturbed forest, and therefore may be fairly adaptable. Chimpanzees can be affected more than monkeys, but this is probably a response to the presence of people and machinery rather than the change in forest structure. Once the people leave, we predict the density of chimpanzees would increase again. Whether this movement affects neighboring chimpanzee communities needs further investigation.

Forêt de Kirindy, Madagascar

The Kirindy Forest is a deciduous, dry forest of low stature (trees reaching only 20 m in height) in western Madagascar. Vegetation is tallest on humid soils with a high proportion of clay, and is probably determined by rainfall levels (Ganzhorn and Sorg 1996). The forest was logged up to 1993, and the study referred to here took place in three, 1–1.5 km^2 plots within a single concession. Logging intensity was light, with less than 10 percent loss of forest cover and removal of about 10 m^3/ha of timber (Ganzhorn et al. 1990). On a large scale, this level of forest disturbance was actually indistinguishable from the effects of disturbance through natural treefall and gap formation (Ganzhorn et al. 1990).

Ganzhorn (1995) censused seven species of lemur in three sites during March 1990. At that time, two sites were unlogged and one had been logged in 1985/86 (the logging intensity is unknown). Censuses were repeated along the same transects in March 1992. One of the unlogged sites was logged in late 1990—between the first and second census.

Changes in the density of lemurs following logging in the Kirindy Forest are outlined in table 4-6. There were few changes in the encounter rate before and after the felling event in compartment N5.

TABLE 4-6 *Changes in Lemur Encounter Rates at Three Sites in the Kirindy Forest Between 1990 and 1992*

Species	Compartment S7 (unlogged plot)	Compartment N5 (logged in 1985–86)	Compartment N5 (logged in 1990)
Petterus fulvus	nc	nc	nc
Propithecus verreauxi	nc	nc	nc
Cheirogaleus medius	nc	+	+
Lepilemur mustelinus	nc	nc	nc
Microcebus spp.	nc	nc	+
Mirza coquereli	nc	nc	nc
Phaner furcifer	nc	nc	+

Significant differences in encounter rate (P < 0.05) are shown: + = increase in encounter rate in 1992, nc = no change. Data from Ganzhorn (1995).

Where changes are recorded, they demonstrate an increase in numbers. Ganzhorn (1995) showed that fruit production and protein concentration in leaves increased following logging in N5, and concluded that low-level disturbances may increase the potential carrying capacity of the forest for folivorous and frugivorous lemurs (see photo 4-3).

Tekam Forest Reserve, Peninsular Malaysia

The Tekam Forest Reserve of Peninsular Malaysia is composed primarily of hill, dipterocarp forest (for a description of the forest type see Whitmore 1984), covering an area of 315 km^2 that abuts on three sides, with more extensive areas of production forest and with mono-cultural plantations to the south. Logging began in the Reserve in 1975.

An intensive study was carried out in one compartment of the Reserve (site 13C—an area of about 4 km^2) during 1979–81. This compartment was logged halfway through the initial 26-month study, providing detailed baseline data on the initial effects of logging on primates. The responses of two primates were intensively studied—the lar gibbon (*Hylobates lar*) and banded-leaf monkey (*Presbytis melalophos*)—and other species in less detail (Johns 1986a). During the later stages of this initial study, surveys were undertaken in three other compartments that were logged 0–6 years previously. The four study areas were resurveyed in 1987 and again in 1993, using the original transects (Johns 1989b; Grieser Johns and Grieser Johns 1995).

Logging at Tekam removed mainly *Shorea* and *Dipterocarpus* spp. (Dipterocarpaceae), *Koompassia malaccensis* and *Intsia palembanica*

(Leguminosae), and *Dyera costulata* (Apocynaceae). Less common commercial species were also cut. In compartment 13C, the felling operation removed 18.3 trees/ha—a total of 3.3 percent of the total trees present, but resulted in a total loss of 51 percent of trees through associated logging damage (Johns 1988). Similar extraction and damage levels occurred at the other sites.

Felling operations in compartment 13C resulted in a great deal of infant mortality in the lar gibbon and the two leaf monkey species, although precise mechanisms for this phenomenon could not be determined. Subsequent surveys have demonstrated improvement in the numbers of infant lar gibbons, but continuing low numbers of infant leaf monkeys (see figure 4-2). This might indicate a decreasing population of leaf monkeys, but there is no evidence for this from survey data.

There are no significant trends in primate abundance resulting from surveys undertaken between 1979 and 1993 and encompassing forests logged up to 18 years previously (see figure 4-3). The only consistent feature appears to be a peaking of the abundance of long-tailed macaques (*M. fascicularis*) in forests logged around six years previously.

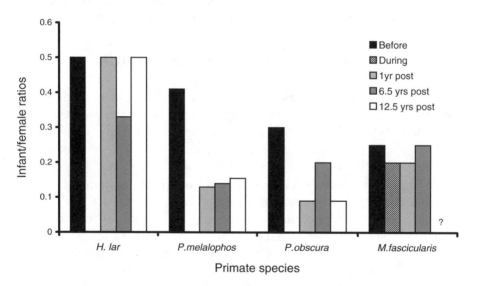

FIGURE 4-2 *Estimated infant/female ratios among primates at compartment 13C, Tekam Forest Reserve, before, during, and after logging. Times elapsed are from the start of tree felling in the study area; the cutting and removal of timber was completed in five months. Data from Grieser Johns and Grieser Johns (1995). ? = no data exist.*

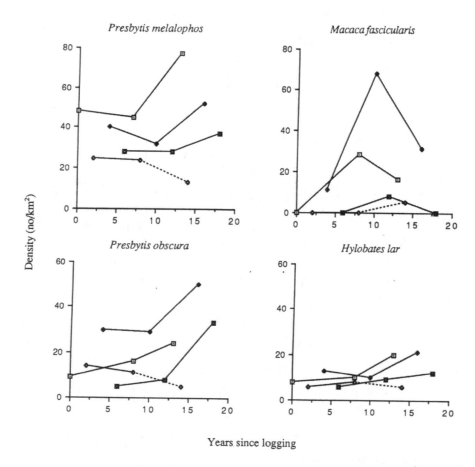

FIGURE 4-3 *Estimated primate densities (number of individuals/km²) up to 18 years postlogging at the Tekam Forest Reserve. Lines link data points from the same study site censused during the period 1989–1993. Four study sites were censused (see Johns 1989). Dotted lines indicate the presence of subsistence hunting at this site, which probably affects results. Data from Grieser Johns and Grieser Johns (1995).*

This species is generally associated with riparian habitats in unlogged forests at Tekam, and appears to be expanding and then contracting its range into logged forests—perhaps to take advantage of resource abundance at a particular stage of the regeneration cycle. Banded-leaf and dusky-leaf monkey densities show little change initially, but then seem to increase (although a couple of populations decreased).

What Are the Causes of Population Change Following Logging?

Methods of Study

Are there any patterns that can be drawn from these primate responses to disturbance from logging? Johns and Skorupa (1987) originally treated this question, and pointed to certain socioecological traits that may assist or prevent survival in logged forest. They showed that populations of large frugivorous primates tended to occur at lower densities in disturbed forest. Their comparisons, however, were unable to separate several causes of disturbance. The changes, therefore, could not be ascribed to selective logging alone. Even with studies designed to measure the effects of selective logging, there are so many factors that influence primate responses that results tend to be inconsistent. These factors include:

- Densities and types of trees harvested
- Damage levels created by the harvesting
- Spatial heterogeneity in tropical forest
- Time taken to harvest an area (human disturbance period)
- Time elapsed since logging
- Instability of tropical forests (unlogged areas naturally change over time)
- Areas of logging concessions
- Levels of hunting by humans

The three detailed studies reported here have tried to control for spatial heterogeneity by following the same sites over time. Two of them follow an unlogged area to monitor changes that occur in undisturbed forest. They do not, however, control for all of these factors. Chapters 18 and 19 of this volume attempt to address study designs that control for most of the factors noted above.

Given these potential confounding factors, are there any general conclusions that can be drawn from all the results presented here? If there are, then given the provisos above it is likely that these conclusions will be fairly robust over a variety of disturbance levels caused by selective logging.

Geographical Areas

In African forests, frugivorous primates appear to increase in abundance or remain the same following logging, except where logging intensity is high. Logging or arboricide treatments of monodominant forests result in mixed forests with higher tree diversity, which in turn benefits many primate species. Folivorous primates show conflicting trends—lemurs and *C. guereza* appear to benefit from light logging, while other colobines occur at lower densities or show no change in disturbed forest. These other colobines, however, are often seed predators where they occur in West Africa. In general, seed predators do not appear to do well in logged forests. Why frugivores do better than seed predators is unclear, but may be due to colonizing tree species that tend to produce *fleshy*—rather than *beanlike*—fruit.

In southeast Asian forests, where the main timber trees (dipterocarps) are not a food resource for primates, the general trend is for primates to be adaptable to logging. In some ways, the dipterocarp forest is similar to monodominant forests in Africa, and the generation of a more diverse, mixed forest can encourage the regeneration of more primate food species. Most primates in Asia are generalists, able to adapt their diets to changing availabilities of food in different types of logged forest. There are as yet no data from the northern tropical forests of Indochina or Myanmar, where dipterocarps are less abundant and primates may respond differently to logging.

Neotropical primates segregate primarily by subhabitat (Peres 1993), and responses will be tied to how these subhabitats are affected. Predictions can be made that edge species, such as the dusky titi monkey (*Callicebus moloch*) will generally benefit from disturbance, while the uakari (*Cacajao* spp.) will be adversely affected if logging occurs in seasonally flooded riparian forest. Seed predators, as in Africa, appear to be adversely affected by logging.

Other Factors Affecting Primate Responses

Some of the results presented here show that presence of patches of unlogged forest, or the proximity of unlogged forest, can be important in determining the response of primates in Malaysia (see table 4-3). More research should investigate the roles of islands of mature forest and corridors of riverine forest within logged forests (see chapter 23), and studies should report whether these exist in the logged areas under investigation.

Hunting has been briefly mentioned in the text of this chapter, but we have tried to concentrate on studies conducted where the hunting of primates is minimal so that the effects of the change in forest structure and composition can be determined. In many areas of the tropics, however, hunting of primates increases following logging operations (Wilkie et al. 1992; Oates 1996; see chapters 15–17)—with larger and slower-moving primates (such as colobines) being particularly targeted (Lahm 1994b). This factor may be the most important in determining primate densities following logging.

General Conclusions

Primates are well known for their ability to adapt to changing situations. It consequently comes as no great surprise to find they can adapt to low-level habitat disturbance. Both frugivores and folivores vary in their responses depending on the species concerned and the continental region where they occur. This is probably a result of floristic differences between regions. Most species, however, have been found to survive in logged forests even if their densities are lower. Care must be taken, however, when interpreting these results. Certain factors that are often associated with logging are more likely to affect primate densities than the disturbance to the forest itself. In particular, the influx of humans into areas that previously had no human contact, and the hunting or farming that often follows, is more likely to cause declines in primate densities than anything else (see chapters 15–17).

ACKNOWLEDGMENTS

We would like to thank Fountain Forestry and the Institute of Biological Anthropology at the University of Oxford for providing office support. Andrew Plumptre is grateful to ODA's Forestry Research Programme and the Wildlife Conservation Society for providing financial support for this work. Andrew Grieser Johns thanks the numerous organizations that have funded him in the past for the research studies included here.

Chapter 5

THE EFFECTS OF LOGGING ON TROPICAL FOREST UNGULATES

Glyn Davies, Matt Heydon, Nigel Leader-Williams,
John MacKinnon, and Helen Newing

Compared to the primates that have just been discussed, the information on the ecology of most forest dwelling, terrestrial mammals is sparse, often anecdotal, widely scattered in the literature, and supported by very few long-term studies. This chapter focuses on ungulates—a group of medium to large-sized terrestrial mammals, which includes elephants, odd-toed hoofed mammals (such as rhinos and tapirs), and even-toed hoofed mammals (such as cattle and antelopes). These mammals are:

- Important sources of subsistence animal protein for many peoples
- Central to the wild meat (bushmeat) trade in many parts of the world
- A source of opportunity for forest ecotourism
- Integral parts of forest ecosystem processes (e.g., seed dispersal)
- Major determinants of forest vegetation structure and composition

It is important, therefore, to review the effects of logging tropical forests on ungulate populations.

We begin with a review of the feeding habits and general ecology of ungulates to assess their reliance on relatively undisturbed closed-canopy forests, and their adaptability to the changes in forest structure and plant species composition that result from logging. Second, we examine reports of the effects of logging on these animals, or where these are unavailable, reports of the species' ability to live in human-altered forest habitats. Third, we describe two case studies that provide recent details on habitat change and its effects on ungulate populations in the forests of West Africa and northern Borneo. We close the chapter with a discussion of conservation, forest management, and research issues.

Although our discussions are restricted to the impacts of habitat change, it is important to consider that the primary impacts of logging are largely inseparable from the secondary consequences of opening up the forest to hunters, trappers, rattan collectors, and farmers (see chapters 15–17, and 20). Opening the canopy also allows desiccation of the forest floor—including the drying of any debris remaining after timber operations—which exacerbates the risk of forest fires (see chapter 21). All of these secondary factors can lead to ungulate population declines, even if species survive the initial habitat changes caused by logging operations.

Logging and Its Influence on Ungulate Food Resources: A General Overview

Variation in Logging Intensity

Logging is a generic term that refers to a process that ranges from clear cutting of forests (usually prior to agricultural development) to extraction of a few timber trees per hectare. In this chapter, we restrict ourselves to selective removal of timber species (see chapter 2). The level of extraction (number of trees taken per ha) and the forest management practices applied during the harvest (e.g., girth limits for felling, reduced impact logging measures, etc.) influences the impact of selective logging (see chapter 21). Damage also tends to be greater as a result of large-scale, mechanized timber extraction than during small-scale community logging (see chapter 25).

Commercial timber harvests range from 70–130 m^3/ha in some southeast Asian forests (Marsh and Greer 1992), where tall, straight-boled dipterocarp trees predominate, to yields of 10–40 m^3/ha in many African forests and even lower volumes in Amazonian forests. The results of logging in southeast Asian forests are predictably more severe, with as much as 62–80 percent damage to the canopy (Johns 1989a), whereas light logging for predominantly one commercial species in areas of West Africa results in damage of only 5–10 percent (Wilks 1990).

Logging and Food Supply for Terrestrial Mammals

Tropical forests produce an abundance of year-round browse, in the form of young and mature leaves. Despite this abundance of potential food, relatively few large mammals found in tropical forests are pure folivores. Frugivores, which rely on seasonally and annually variable supplies of fruits and seeds, tend to be the more diverse group (Terborgh 1986a). Logging impacts on both vegetative and fruit parts affect ungulate populations.

The initial opening of the canopy causes a surge of new growth on the forest floor, including herbs, grasses and shrubs, and a rapid growth of creepers and lianas. Colonizing plants quickly fill canopy gaps (see photo 5-1). Canopy gaps often close after a few years, giving way to a *growth phase* forest patch that develops into a mature/climax phase, closed-canopy forest (Whitmore 1998). These changes mimic successional cycles that occur when old trees die and fall over, but variations in logging intensity mean that changes in the forest age, structure, and tree species may be evident for several decades.

Young leaves are generally more palatable than their mature leaf counterparts (e.g., Waterman and Kool 1994), and the flush of leaf growth that occurs beneath canopy gaps is a key food resource for terrestrial forest browsers. Studies of trees in both the Malay Peninsula and Brazil have shown an increase in young leaf production by upper and middle canopy trees following logging (Johns 1992a), and a similar pattern would be expected for understory plants. Furthermore, colonizing plants tend to invest less in structural molecules and plant defense compounds to protect their leaves from browsers than do climax species (e.g., Janzen 1975; McKey 1978).

Good quality browse for ungulates should therefore increase substantially following logging for two reasons: 1) colonizing plants multiply in the new openings, and 2) there is a surge of young leaf production. One caveat to this generalization, however, is that many factors influence tree leaf chemistry (reviewed by Oates et al. 1990). Young leaves and/or foliage of colonizing plants can have low overall digestibility.

Changes in fruit availability following logging are little studied, yet are probably more important than foliage for the smaller ungulates. Colonizing plants frequently have small-seeded fruits, which may be either fleshy (to attract animal seed dispersers) or dry (for wind dispersal). In contrast, climax species tend to have larger-seeded fruits that are produced less frequently (Swain and Whitmore 1988). For those animals that depend on large fruits and seeds, logging will tend to reduce food supply. But even trees producing small fruits can decline following logging. One example is the strangling figs *(Ficus* spp.)—which are an important keystone species in Asian forests— whose abundance has great influence on the food supplies of both

PHOTO 5-1 *Musanga* tree growing up with dense carpet of creepers in one-year-old logged forest, Gola Forest Reserve, Sierra Leone. (G. Davies)

arboreal (primates, birds, squirrels) and terrestrial (ungulates, civets, pheasants) frugivores (Leighton and Leighton 1983).

As well as changes in fruit tree numbers, patterns of fruit production can change following logging. At Tekam (West Malaysia), the 50 percent of remaining trees after logging showed an immediate increase in fruit production (Johns 1992a)—possibly producing a fruit crop equal to that of previously unlogged forests. By contrast, fruit production in Danum (East Malaysia) declined by 35 to 45 percent up to twelve years after logging (Heydon 1994), and eleven-year-old logged forests in Brazil, the fruit crop was substantially reduced (Johns 1992a). Such declines in production are expected to have a negative impact on terrestrial frugivore numbers.

For grazers, there is a clear increase in grass availability immediately following logging, which persists until the grasses are shaded out by taller plants. In summary, logging often increases grass and browse supplies, but may reduce fruit abundance.

Feeding Habitats, General Ecology, and Effects of Logging on Forest Ungulates

Tropical forest ungulates show marked differences in their digestive systems and diets (Janis 1976; Demment and van Soest 1985). They are all characterized by one of two fermentation digestion systems: midgut (or caeco-colonic fermenters), and foregut ferments, which are often ruminants (i.e., possessing a rumen). The latter group has the option to ferment foods in either the fore- or midsection of the intestinal tract (Kay and Davies 1994).

Ungulates are well adapted to herbivorous diets, but few are obligate folivores (Bodmer 1990a). These mammals can be classified as:

- **Frugivores:** feed on high-energy, low-fiber foods that often contain high levels of readily available energy in the form of simple sugars and fats
- **Browsers:** consume foods with intermediate levels of nutrients and fiber such as dicotyledonous leaves
- **Grazers:** consume low-energy, high-fiber items, typically monocotyledonous leaves such as grasses

Intermediate categories (frugivore/browsers and browser/grazers) switch diets during the year according to variation in the availability of preferred items.

Bodmer (1990a) has shown that frugivorous ungulates are restricted to tropical forests, where round-the-year fruit availability is greatest, and have consistently small body sizes. Browsers—the most common group—show a wide range of body sizes in all habitat types, although they are often most abundant in forest edge and savannas. Pure grazers are absent from tropical forests, being most common in grasslands and savannas.

Medium- and Large-Sized Terrestrial Mammals (150–4,000 kg)

Elephants and Rhinos

African (*Loxodonta africana*) and Asian (*Elephas maximus*) elephants (see photo 5-2) are the largest mammals to occupy tropical forests. The African elephant in particular has occupied a considerably wider range of habitats throughout sub-Saharan Africa than simply tropical forests. African elephants occupying tropical forests in the Congo Basin and West Africa show considerable differences in comparison to savanna elephants, including a smaller body size (some 2000–4000 kg), straighter tusks, and smaller group sizes (Merz 1986; Turkalo and Fay 1995). Asian elephants also live in habitats outside tropical forests, including bamboo forest, flood plain, and grassland habitats. They are of similar body size to the forest-dwelling African elephant, however, and live in small groups—but only the males have tusks (Sukumar 1989).

Both species of elephants have a simple stomach with microbial fermentation occurring in the caecum, which, together with their large body size, limits elephants to eating a large volume of low quality fibrous food that passes rapidly through the gut. African forest elephants are browser-frugivores, with a diet consisting almost entirely of woody plant tissues (bark, leaf and roots), fruits, and herbs (Alexandre 1978; Merz 1981; Short 1981; Lieberman et al. 1987; Dudley et al. 1992; White et al. 1993). In contrast, Asian elephants (and savanna elephants in Africa) are grazer-browsers, with a marked preference for (and nutritional dependence upon) grass in their diet—particularly in the wet season—with a switch to more browse in the dry season (Owen-Smith 1988; Sukumar 1989; Menon et al. 1997). Asian ele-

PHOTO 5-2 *Forest elephant* (Loxodonta africanan cyclotis) *at a saleine (PAN element 1, chapter 23). (L. White)*

phants do well in swampy forests that are rich in palms, and in monsoon forests that have grass in the understory (Santiapillai amd Jackson 1990). They are scarce in dense, evergreen primary forests unless they have access to open grassy areas, forest edge, and riverbanks.

The dietary differences between these two give rise to important predictions about the likely responses of elephants living in African and Asian tropical forests to different degrees of logging. In Africa, these browser-frugivores can both tolerate—and indeed prosper—in secondary forest (Barnes et al. 1991; White 1994c), including logged sites (Merz 1986; Struhsaker 1997). More elephants, therefore, are found in heavily logged forests (tree stem basal area harvested: 21 m^3/ha) than in lightly logged (14 m^3/ha) or unlogged forests in Kibale. Elephants are indeed an important factor in suppressing forest regeneration (Strusaker 1997). The Asian elephant, a grazer noted as widespread outside forest habitats, showed an increase in population density in logged forests at one study area in Peninsular Malyasia (Olivier 1978). These large mammals tend to avoid current villages and roads, however (Barnes et al. 1991).

Two species of rhinoceros occupy tropical forests. Sumatran (*Dicerorhinus sumatrensis*) and Javan (*Rhinoceros sondaicus*) rhinos—weighing up to 800 kg and 1,400 kg, respectively—once occurred widely throughout the tropical forests of southeast Asia, and were largely restricted to forest habitats. Where the two species lived sympatrically (throughout Sumatra and on the Malay Peninsula), there appears to have been a clear ecological separation between them. The Sumatran rhino occurred on higher and steeper ground and in undisturbed primary forests, while the Javan rhino occurred on lower, flatter ground and in disturbed secondary forests, forest edges, and transitional areas (Groves 1967; Borner 1979).

All rhinos, like elephants, are mid-gut fermenters with an enlarged caecua. The Sumatran rhino appears to be almost exclusively a browser—eating foliage and stems from plants in the understory of primary forest (Van Strien 1985)—while the Javan rhino also eats grasses and herbs in forest gaps. There are records, however, of rhinos eating fruits (Strickland 1967; Schenkel and Schenkel-Hulliger 1969; Borner 1979). Again, these dietary differences suggest differences in the response of these two species to logging.

Sumatran rhinos are relatively selective browsers that frequently use tree-fall gaps (Borner 1979). This species was abundant in regenerating thickets in abandoned transmigration schemes near the Way Kambas Reserve in Sumatra (J. MacKinnon, personal observation), and also relatively abundant in logged forests of the Dent Peninsula in Sabah—congregating in areas with natural salt sources (Davies and Payne 1982). Direct studies of its distribution in logged and unlogged areas inside and outside protected areas in Peninsular Malaysia, however, indicate that the species avoids areas where the primary forest habitat has been modified by commercial logging (Flynn 1978; Flynn and Abdullah 1984). In general, the Sumatran rhino appears to tolerate little human disturbance, despite their subsistence on forest gap vegetation in a variety of disturbed habitats. This once widespread species has suffered severe declines during the course of this century, both from wholesale loss of undisturbed habitats, and from over-hunting (Foose and van Strien 1997).

Observations on Javan rhinos (Amman 1985) indicate that they may be more adaptable to disturbed habitats. The only viable populations in Java and Vietnam are found in secondary forests following the Krakatau eruptions (Hoogerwerf 1970; Hommel 1987), and aerial herbicide spraying (Schaller et al. 1990), respectively. Their direct response to logging is unknown.

Tapirs

The South American and Mesoamerican tropical forests have not supported any mega-herbivores since humans caused the demise of the Pleistocene large mammal fauna (Diamond 1989; Burney 1993). Three species of tapir are now the continent's largest extant mammals living in tropical forests. The Brazilian or lowland tapirs (*Tapirus terrestris*), mountain tapirs (*T. pinchaque*) and Baird's tapirs (*T. bairdi*), attain body masses of 225–300 kg, 150 kg, and 150–300 kg, respectively. One slightly larger species of tapir—the Malayan (*T. indicus*), which weighs from 250–450 kg—occurs in southeast Asia. Like their closest living relatives—the rhinos—all tapirs are midgut fermenters.

The Brazilian or lowland tapir lives in forested and grassy habitats with permanent water, and is widespread over lowland areas. The mountain tapir lives at higher altitudes and can occur above the tree line. Baird's tapir lives in swampy or hill forests, while the Malayan tapir lives in a wide variety of habitat types, including primary rain forest. In these different habitats, tapirs feed on a diverse array of leaves, undergrowth, and forest-edge herbs, shrubs, and saplings. Tapirs also seek out a wide variety of fruits in different habitats— which can make up to 30 percent of their diet in some areas—and are very significant seed dispersers for different tree species (Matola 1974; Terwilliger 1978; Williams and Petrides 1980; Janzen 1982b; Bodmer 1990b; Rodriques et al. 1993; Naranjo 1995; Downer 1996).

The generalist nature of most tapir diets suggests that these species will respond well to direct changes in their habitat following logging operations in tropical forests. Browser-frugivores like Baird's tapir, for example, appear to do well in secondary forests, including areas selectively logged (Fragoso 1991b). The Malay tapir has also been reported in logged forests, and has even been recorded as *common* in the logged and burned forests of the Way Kambas Reserve in Sumatra (FAO 1979).

Okapi

Among the large ruminants that occur in tropical forests, folivorous browsers are very unusual. The best example is the okapi (*Okapia johnstoni*), which occurs in the dense rain forest of the Democratic Republic of Congo (formerly Zaire). This giraffid reaches some 210– 250 kg in body weight and almost exclusively selects high quality browse from species of shrubs and trees growing in the forest understory (Hart and Hart 1989).

Studies of the okapi—a specialist browser—have been carried out inside a protected area that offered a contrast between primary forest and naturally occurring tree-fall gaps, but not between logged and unlogged areas. The okapi used leaf more extensively in forest gaps than under undisturbed canopy, probably because the high productivity in gaps provided a greater density of quality forage that was within reach (Hart and Hart 1989). These results suggest that the okapi might tolerate low-impact logging that mimics the effects of tree-fall gaps.

Forest Grazers

Many of the other large ruminants that occur in tropical forests are grazers with more localized distributions in forest edges or clearings. In Africa, the buffalo (*Synceros caffer*) reaches 480–680 kg and, like the elephant, is distributed throughout a very wide variety of habitats. The flush of leafy growth found beneath forest canopy gaps is a key source of food for buffalo living in forests (Hoppe-Dominik 1989). In contrast, the bongo (*Tragelaphus euryceros*) is an antelope that reaches 210–235 kg, and is largely restricted to lowland and montane forest habitats (Emmons et al. 1983). It also occurs, however, in large herds in some grassy/wetland areas (Leuthold 1977).

Several species of wild cattle occur within southeast Asian forests, including the gaur (*Bos gaurus*) reaching 700–940 kg, the kouprey (*B. sauveli*) reaching 700–900 kg, the banteng (*B. javanicus*) reaching 600–800 kg, the much smaller mountain and lowland anoas (*Bubalus quarlesi* and *B. depressicornis*) reaching 100–150 kg and 200–300 kg, respectively, and the recently discovered saola (*Pseudoryx nghetinhensis*) in Vietnam and Laos, which weighs 80–100 kg (J. MacKinnon, personal observation). Most of these wild cattle species have localized or restricted distributions.

The large, primarily grazing ruminants use differing forms of grass-dominated habitats within tropical forests, and are likely to be very differently affected by logging than those species which use browse and fruit. Wharton (1968) has argued that it was largely the spread of humans, and the associated burning of forests, that led to the spread of wild cattle through southeast Asia (also see Gaio et al. 1998). Where logging produces suitable grassy clearings, therefore, favorable conditions are produced for gaur and banteng. The African buffalo has also been reported to penetrate far into the forest along grassy logging roads (Prins and Reitsma 1989), and are common in forest edge/savanna habitats (White 1994c). By contrast, anoas in Sulawesi eat more browse and fruit than the larger wild cattle (*Bos* spp.) and are more dependent on primary forest, but use secondary forest and can survive logging (J. MacKinnon, personal observation). Finally, the saola (a primitive bovid) is a specialist browser that appears to be dependent on closed-canopy forest, feeding mostly on a common herb-plant (*Homalomena aromatica*) that grows only in dark shady river gorges (Dung et al. 1994). This species seems to be much more dependent on closed-canopy forest than the okapi and other forest grazers. Gaps in its geographical distribution may result from historical forest disturbances as well as more recent hunting and trapping (Schaller and Rabinowitz 1995).

These studies suggest that logged forest provides as much, if not more, food for these medium- to large-size mammal generalists than unlogged or undisturbed forest. There is a limit, however, to the amount of habitat disturbance that these species can tolerate, since certain species of fruiting trees—which account for a significant amount of their diet—may be lost as logging intensities increase.

Peccaries and Pigs (25–275 kg)

Peccaries

Peccaries, bush pigs, and bearded pigs are concentrate feeders—eating many things but specializing on fruits and animal matter. Bodmer (1990a) characterized this group as frugivore/omnivores that occupy a variety of habitats (see table 5-1). Collared peccary *(Tayassu tajacu)* are largely frugivorous, with 59–71 percent of their diets consisting of fruit and seeds (Kiltie 1981; Kiltie and Terborgh 1983; Bodmer 1989). They eat many vegetative plant parts as well, including roots and tubers (Kiltie and Terborgh 1984), which act as a resource during times of fruit scarcity (reviewed by Kiltie 1982). White-lipped peccary *(Tayassu pecari)* are also highly frugivorous (61–66 percent fruit and seeds—Kiltie 1981; Bodmer 1989), but the two species have very different ecological strategies. The white-lipped peccary is the larger and migrates over huge areas, at times forming herds of over one hundred individuals. Collared peccaries have smaller group sizes and ranges.

Johns (1986a) noted sightings of collared peccaries during general surveys, and recorded more sightings in logged forest (0.8 individuals/10 km surveyed) than in unlogged forest (0.3 individuals/10 km surveyed). No other data are available comparing peccaries in logged and

TABLE 5-1 *Habitat Preferences of Peccaries and Pigs*

Species	Body Weight	Habitat Preferences
Peccaries:		
White-lipped peccary (*Tayassu pecari*)	35 kg	Herds concentrate near water sources. Declining throughout its range due to loss of suitable habitat[6]
Collared peccary (*Tayassu tajacu*)	25 kg	Closed canopy forest; lower densities in secondary forests and agricultural areas (but could be due to hunting)[2,3]
Pigs:		
Bush pig (*Potamochoerus porcus*)	75–130 kg	Mature forest 15–20-year-old regrowth forest, farmbush-fallow, crop pest[1,4,9]
Giant forest hog (*Hylochoerus meinertzhageni*)	150–275 kg	Primarily mature forest, but also recorded in 15–20-year-old regrowth forest[1]; grasslands and mixed habitats[7,8]
Bearded pig (*Sus barbatus*)	male: 60–145 female: 40–125	Migratory between forest zones according to food availability.[5]

[1]Wilkie and Finn, 1990; [2]Bodmer et al, 1988; [3]Johns, 1986; [4]Tutin et al, 1997; [5]Caldecott, 1991b; [6]Mayer and Wetzel, 1987; [7]d'Huart, 1978 ; [8]d'Huart, 1993; [9]White, 1994c

unlogged forests, but secondary effects are likely to be more important than direct changes to the environment. Peccaries are the major source of wild meat in much of the Amazon (Hvidberg-Hansen 1970), and hunting increases during timber operations to 1) feed timber crews, and subsequently, 2) as a result of the opening of the forest to commercial hunters and settlers. There appears to be a direct relationship between the number of trees extracted and the number of animals hunted (Bodmer et al, 1988). White-lipped peccaries are particularly susceptible to this problem because they live in large herds that attract hunters or local villagers.

There is also evidence that peccary populations fluctuate with the availability of palm fruits (Bodmer et al. 1999). Fruits of palms such as *Astrocaryum* spp., *Jessenia* spp., and *Mauritia flexuosa* are important in the diets of both species of peccaries. Unfortunately, palm fruit is one of the most important nontimber plant resources in the Peruvian Amazon (Bodmer 1994), and two palm species that are important in peccary diets are being reduced at an alarming rate because the fruits are frequently harvested by felling trees.

Pigs

The bearded pig (*Sus barbatus*) has an analogous foraging strategy to the white-lipped peccary. The species formerly occurred throughout Peninsular Malaysia and the islands of Borneo and Sumatra, with related sub-species in Indochina, Java, and the Philippines (Caldecott 1991a). It has since been eliminated over much of its range. In Borneo, its diet is varied, with fruits being very important (Caldecott 1991b). Pig population explosions have been recorded when dipterocarp mast fruiting occurs in two consecutive years (Pfeffer and Caldecott 1986). The pigs follow complex migrations, over tens or hundreds of kilometers, in search of food sources ranging from bamboos (Davies and Payne 1982), to the fallen fruit of camphorwood (*Dryobalanops aromatica*) trees in the Malay Peninsula, and other dipterocarps and oaks in Borneo (Caldecott 1991b). Any disruption to the abundance of these fruiting trees, or to migration routes between fruiting areas (such as may occur with logging), is expected to have a negative impact on bearded pig numbers.

Bearded pigs appear to respond to disturbance in a similar manner to that of the white-lipped peccary. The pigs' dependence on fruits from dipterocarp trees in the lowlands, which are the main timber tree, has resulted in their complete elimination from Peninsular Malaysia as a result of the felling of the last of the camphorwood forests (Caldecott 1991b). Furthermore, their population eruptions attract the attention of scores of hunters, who slaughter thousands of animals—far more than they can consume at the time (Caldecott 1991b; see chapter 16).

In Africa, the giant forest hog (*Hylochoerus meinerzhageni*) is found in bamboo and montane forests of the Rift Valley Mountains, and in scattered and sparse populations in Guineo-Congolian forests (d'Huart 1993). It is largely a grazer that also feeds on many dicotyledonous plants in forest habitats (d'Huart 1978), and prefers landscapes of mixed open and bush/forested areas (d'Huart 1993). Despite being a grazer-browser, it appears to depend on forested patches more than the other pig species. It was observed in Bwindi forest (Uganda) to leave a logged area while bushpigs (*Potamochoerus larvatus*) remained (d'Huart, personal communication). The red river hog (*P. porcus*) of west and central Africa, like its congener, adapts to a wide variety of vegetation types (Davies 1987; White 1994c; Tutin et al. 1997). Logging is not expected, therefore, to have an adverse impact on the two *Potamochoerus* species, but may negatively impact the shyer giant forest hogs.

Small Forest Ungulates (<100 kg)

Small forest ungulates include the duikers (*Cephalophus* spp.), royal antelopes (*Neotragus pygmaeus* and *N. batesi*) and the water chevrotain (*Hyemoschus aquaticus*) in Africa; brocket deer species (*Mazama* spp.) in South America; and the mousedeer (*Tragulus* spp.), muntjacs or barking deer (*Muntiacus* spp.), and sambar deer (*Cervus* spp.) in Asia. All are ruminants except the tragulids (chevrotain and mousedeer), which have sacculated stomachs that allow some fore-stomach fermentation (Nordin 1978). All are generally regarded as frugivores and browser-frugivores, except the *Neotragus* antelopes, which are selective browsers.

Africa

Duikers in Africa represent a diverse radiation of forest mammals. Traditionally lumped together as primitive forest frugivores, stomach content analyses of different species indicate that there is remarkable similarity in the overall fruit/seed intake (between 72–97 percent of stomach contents—Dubost 1984; Hart 1985; Feer 1988; Newing 1994). Hart (1985) also recorded no significant differences in percentage of frugivory between seasons. Only the black-fronted duiker (*Cephalophus nigrifrons*) seems to have a predominantly folivorous annual diet (Plumptre 1990).

Many studies indicate that duikers—other than the savanna-dwelling, red-flanked (*C. rufilatus*), and grey duikers (*Sylvicapra grimmia*)—are most common in the forest, but it is clear that species' habitat preferences differ, ranging from forest to secondary and savanna habitats (see table 5-2). Only zebra (*C. zebra*), Ogilby's (*C. ogilbyi*), and white-bellied (*C. leucogaster*) duikers have not been recorded outside closed forests. This suggests that many duiker species should have a wide tolerance to logging.

One study in a mixture of forest types in Lope Reserve (Gabon)—including areas that had been lightly logged 3 to 25 years previously—indicated that overall duiker densities were low, but not adversely

TABLE 5-2 *Habitat Preferences of African Duikers* (Cephalophus *spp.*)

Duiker Species	Wt (kg)	Habitat Preferences
Blue (*C. monticola*)	3–7	Open parts of forest[1, 4]; logged and unlogged forests[2, 35] —sometimes with lower population densities in the latter[36, 37]; avoid big clearings and thick vegetation near villages[3, 4]; closed-canopy forest; higher densities in forest fragments[30]
Maxwell's (*C. maxwelli*)	6–10	Closed-canopy forest and secondary vegetation[26, 31, 32, 33] Some preference for secondary vegetation/forest edge[8, 11, 12]
Zebra (*C. zebra*)	9–15	Closed-canopy forest[22, 26, 31, 32]
Red (*C.weynsii/ C. harveyi*)	15–19	Logged and unlogged forest[18]—with lower densities in the latter[36, 37]
Red (*C. natalensis*)	10–12	Closed forest and adjacent open areas[13]
White-bellied (*C. leucogaster*)	12–15	Forest[26]
Black-fronted (*C. nigrifrons*)	13–16	Riparian[3] high forest, especially waterlogged areas[26] Associated with swampy habitat, often in boggy meadows[34]
Ogilby's (*C. ogilbyi*)	14–20	Closed-canopy forest only[32]
Black (*C. niger*)	15–20	Closed-canopy forest[11, 26]; actively prefers secondary vegetation; also present in closed canopy forest[32, 31]
Peter's (*C. callipygus*)	15–24	All forest types and edges[2]; prefer dense undergrowth[1] High forest; sometimes secondary forest[26]
Bay (*C. dorsalis*)	19–25	All forest types and edges[2]; plantations and secondary vegetation[1, 32] Dense forest[11, 26]
Jentink's (*C. jentinki*)	70	Near water and swamps[21]; Forest edge and farmlands in rainy season[23] Reported at some sites in closed-canopy forest only[26, 32]
Yellow-backed (*C. silvicultor*)	60–80	Closed-canopy forest and secondary vegetation[18,24,27,30,31] Reported at some sites in secondary vegetation only[17, 32, 38]

[1]Feer (1988); [2]Hart (1985); [3]Dubost (1979); [4]Dubost (1980); [5]Dubost (1983); [6]Dubost and Feer (1988); [7]Crawford (1984); [8]Ralls (1973); [11]Baudenon (1958); [12]Aeschlimann (1963); [13]Perrin et al, 1992 [17]Ansell (1950); [18]Lumpkin and Kranz (1984); [21]Kuhn (1968); [22]Kuhn (1966); [23]Davies and Birkhager (1990); [24]Dekeyser and Villiers (1955); [26]Happold (1973); [27]Wilkie (1987); [28]Nummelin (1990); [29]Feer (1979); [30]Tutin et al, 1997; [31]Davies, 1987; [32]Newing (1994); [33]Fimbel, (1994); [34]Plumptre (1990); [35]Plumptre (1994); [36]McFoy (1995); [37]Struhsaker (1997); [38]White (1994c).
NOTE There is much confusion in relation to red duiker taxonomy.

affected by logging (White 1994c). In western Uganda, studies in the Budongo and Kibale semideciduous/evergreen forests come to different conclusions about the effect of logging on red (*C. weynsi*) and blue (*C. monticola*) duikers. A comparison of eight blocks of forest in Budongo—six logged (at ten-year intervals prior to the study) and two unlogged blocks—indicated that there was no significant difference in blue or red duiker populations in logged and unlogged forests. There were also no differences due to forest formations, although there was a trend towards more duikers in more mature forests (Plumptre 1994; Plumptre and Reynolds 1994). In Kibale, duiker populations in seven blocks of logged forests were much lower than in two unlogged blocks (Struhsaker 1997), even 20 years after logging. This has been attributed to poor forest regeneration, which prevented the development of suitable duiker habitat in the logged forests. One heavily logged Kibale forest block, however, had high duiker densities (McFoy 1995), drawing attention to the probable effects of hunting and trapping which confound interpretation of the impact of vegetation changes following logging (Struhsaker 1997).

Royal antelope prefer secondary growth and plantations (Feer 1979; Davies 1987). Bate's pigmy antelope (*N. batesi*) have been recorded in mature forest, but only near tree falls and along riversides where a more open canopy ensures a supply of young leaves within reach of the forest floor (Feer 1979). The water chevrotain, another diminutive African antelope whose ecology is poorly understood (Dubost 1984; Hart 1985), appears to be commonly associated with watercourses. This species has been recorded more frequently in regrowth forest (15–20 years old) than in mature forest in the Democratic Republic of Congo (Wilkie and Finn 1990). No data are available, but it is plausible that selective logging increases the food available to these terrestrial browsers.

South America and Meso-America

In South America, brocket deer are generally classified as frugivorous, with stomach contents of the gray brocket deer (*Mazama gouazoubira*, 11–18 kg) and the red brocket deer (*Mazama americana*, 24–48 kg) comprising 87 and 81 percent fruit, respectively (Bodmer 1989). Brocket deer stranded on small refuge islands during floods, however, have switched their diets to a greater proportion of leaf/fiber material. Red brocket deer rumen samples in flooded forests contained primarily Gramineae leaves, which were virtually absent in the low-water season. Similarly, during periods of fruit scarcity in the dry season in Surinam forests, brocket deer shift from a fruit to a browse diet (Branan et al. 1985 in Bodmer 1989).

Both species of brocket deer are known to inhabit disturbed habitats (Emmons 1990). During general surveys in Brazilian forests, Johns (1986a) found identical encounter rates (0.3 individuals/10 km surveyed) for red brocket deer in logged and unlogged forests, which concurs with expectations that logging will not affect this adaptable species.

South and Southeast Asia

There are three distinct groups of small forest ungulates in south and southeast Asian forests: mousedeer, muntjacs and sambar deer. Two species of mousedeer are common in southeast Asia, with a diet dependent on fallen fruits—especially those of strangling figs (*Ficus* spp.). Both species appear to show population reduction associated with logging (Heydon and Bulloh 1997—see case study in next paragraph).

The common muntjac eats a high proportion of browse (77.8 percent in a study in Sri Lanka) compared to other muntjac species (Barrette 1977), and may have a competitive advantage over other muntjac species following logging. Consistent with this, Heydon (1994) reports that the numerical advantage of the endemic yellow muntjac (*Muntiacus atherodes)* over the common muntjac in primary forests in Sabah declined after logging. Duff et al. (1984) report higher sighting frequencies of common muntjac in nine-year-old selectively logged forests than in primary forests. The sensitivity to logging of some muntjacs is supported by Giao et al. (1998), who attributes the survival of local endemic muntjacs (such as *Megamuntiacus vuquangensis* and *Muntiacus truongsonensis* in competition with *M. muntjak*) to the persistence of closed forest conditions.

Sambar deer (*Cervus unicolor)* occur throughout the forest of Asia and prefer forest edge, riverbanks, grassy clearings, secondary scrub, and open farmlands (Schaller 1967), but are scarce in dense forests. Sambar are also more common in logged and secondary forests than in primary forests (Heydon 1994). The same probably holds true for C. *timorensis* and *C. elaphus* within their Asian distributions. *Axis* species (A. *axis, A. procinus,* and *A. kuhlii*) occur in large herds in grasslands, swamps and open wooded habitats, and sometimes invade severely logged and fragmented forests and seasonal monsoon forests with grassy undergrowth. Many of the generalist browser/frugivores, and grazer/browsers of south and southeast Asian forests have very wide geographical distributions and may be favored by habitat changes arising from logging operations (Giao et al. 1998). Opposite responses characterize closed-canopy specialists.

Case Studies of Logging-Ungulate Interactions

To provide a fuller picture of the difficulties associated with logging-forest ungulate studies, we include detailed results from two case studies where logged and unlogged forests have been compared: a) lowland Upper Guinean forests in coastal West Africa and b) lowland dipterocarp forests in northern Borneo.

Maxwell's Duikers in Upper Guinean Forests of West Africa

The forest formation in the Gola Forest of Sierra Leone is the *Tarrieta/Lophira* type, characterized by a predominance of *Tarrieta utilis* and *Cryptosepalum tetraphyllum*, with lesser numbers of *Erythrophleum ivorense* and *Lophira alata*, and occasional concentrations of *Brachystegia leonensis* and *Didelotia idea* (Fox 1968a; Davies 1987). This composition is very similar to the tree flora of the Tai area in the Ivory Coast (Newing 1994). Within these forests, Maxwell's duiker (*C. maxwelli*) is the most common. Comparisons of this species' populations in closed-canopy, unlogged forest, logged forest, and farm-fallow habitats—together with an assessment of food availability—has helped to clarify the ability of these small ungulates to adapt to different levels of habitat disturbance.

In 1989–1990, two sites were examined in the Gola North Forest Reserve: one unlogged forest block and one logged block within 10 km of one another (Davies and Richards 1991). Both were at similar altitudes (200–300 m a.s.l.). Timber inventories of the reserve showed a timber stocking of over 40 m³/ha. It is estimated that 20 to 30 m³/ha were extracted from the logged study site 5 to 6 years before the duiker surveys were carried out. Despite this moderate level of timber extraction (as indicated by clear logging tracks, felled trees, and cut stumps still on the ground), tree stand density was similar to that in the unlogged site.

In Tai, the two study sites were also 10 km apart. One area supported a closed-canopy forest that had been lightly logged 20 to 30 years prior to the study (it had a very similar composition to the Gola unlogged site, following this long regeneration phase). The other site was a mosaic of farmed-over land, with a mixture of one-year-old monospecific thickets of *Chromolaena* (22 percent), mixed thickets less than seven years old (37 percent); mature forest fragments (29 percent), other aged secondary growth (13 percent), and bamboo (1 percent). This latter area provided a comparative perspective of extreme habitat alteration, approximating intensive logging impacts.

In Gola, the results of line transects show that more duikers were encountered in the unlogged forest, and that the most common duiker at both sites was Maxwell's duiker (see table 5-3; for discussion of methodology see Davies 1991). These findings suggest that logging leads to a decrease in the population densities of Maxwell's duikers. To interpret these findings as showing that logging leads to a decrease in population densities of duikers, however, is premature, as there were probably differences in hunting and trapping pressures between the two sites. Primates and duikers were hunted in Gola (Davies and Richards 1991), and some hunters confirmed that they had been active in the logged survey site area. While no recent traps were found during the survey period (and very few shotgun cartridges)—suggesting that hunting and trapping activities were moderate to low—the relative ease of access along the logging road (1–2 hours walk from a nearby village) and presence of old hunting camps within 300 m of the survey site indicated a greater influence of hunting and trapping in the logged forest than in the unlogged site.

Different methods of surveying duiker densities can also contribute to differences in duiker population estimates. Duiker drives with crews of men and dogs resulted in Maxwell's duiker estimates of 15 to 30 individuals/km^2 in both logged and unlogged sites (Davies 1991). The twofold

TABLE 5-3 *Encounter Rates of Maxwell's Duikers Using Line Transects in the Gola Forest Reserve, West Africa*

Survey Intensity	Mogbai (unlogged) 60.2 km; 25 days	Koyema (logged) 71.7 km; 26 days
Maxwell's duiker		
sighting rate	0.45/km	0.36/km
standard error	+/- 0.20	+/- 0.19
transect width	40 m	50 m
abundance	11.2 indivs/km^2	7.2 indivs/km^2
Other duiker species sighting rate	0.26/km	0.17/km

increase in densities with this method compared to the line transects noted above concur with similar observations in other studies where duiker line transect encounter rates underestimated population densities by 50 percent compared to home-range mapping (Gabon: Dubost, 1980; Ivory Coast: Newing 1994). Hunting and sampling techniques, therefore, appear to account for more of the variability in duiker populations between logged and unlogged sites than the environmental impacts associated with the harvesting operation.

In Tai, these methodological shortcomings were overcome by radio tracking and mapping the home range of Maxwell's duikers and by determining group sizes based on direct observations (Newing 1994). Calculated population density in closed forest was 63 individuals/km^2, while degraded habitats appeared to support 79 individuals/m^2. These calculations suggest that Maxwell's duikers are capable of tolerating significant habitat conversion—much greater than those generally associated with timber extraction.

Food used by Maxwell's duikers was identified from the stomach contents of animals killed in forest and degraded habitats. Fruit and seed parts accounted for at least 85 percent of the total dry weight in nine of eleven stomachs. During the season of greatest fruit abundance this value rose to over 90 percent for all samples. During seasonal periods of low fruit availability, a greater weight of leaves and flowers was found, and in one case the stomach contained no fruit or seeds at all. It is, therefore, likely that in periods of fruit scarcity, duikers adapt their diets and become more folivorous.

The monthly availability of suitable duiker fruits and seeds (i.e., those under 3 cm in diameter), including fruits and seeds that have been lying on the ground for several months, showed that potential foods were significantly more abundant in closed-canopy forest than in the degraded habitats. Despite a fivefold decrease in the availability of potentially edible fruits and seeds in degraded habitats in comparison to that of closed-canopy forest—including a decrease in timber trees (Ebenaceae), which are important sources of duiker food (Newing 1994)—Maxwell's duikers occurred at similar population densities in degraded *bush* as they did in closed-canopy forest in Tai. The direct effect of habitat change on Maxwell's duikers following logging, therefore, is believed to be negligible compared to losses from indirect impacts such as increased hunting and trapping in the wake of improved forest access.

Ungulates and Small Carnivores in Ulu Segama, Sabah, East Malaysia

Southeast Asian mammals were surveyed during a sixteen-month period (1968–71), using indirect sign and direct sightings in an area of primary dipterocarp forest in Ulu Segama, Sabah, Borneo (MacKinnon 1982). A further survey in 1982, nine years after the original study area had been logged, detected a major increase in the numbers of sambar deer, some increase in banteng, and an increase in elephants. Sumatran rhinoceros were still present and the muntjacs were at similar density levels. Only the bearded pigs, formerly the most numerous large mammal in the area, showed a major drop in densities. A similar pattern of distribution and abundance was noted during other surveys in the region (Davies and Payne 1982).

More detailed, quantitative studies of these species were carried out in the Danum Valley region of Ulu Segama in 1992, and directly related density changes to the abundance of food following logging (Heydon 1994; Heydon and Bulloh 1996, 1997). Timber extraction levels in Danum (averaging 78 m^3/ha) were typical of extraction levels in lowland dipterocarp forests in Sabah (Marsh and Greer 1992). The resulting forest landscape was a mosaic of different aged logged compartments interspersed with small fragments of unlogged forest (see photo 5-3) and adjacent to large areas of remaining primary forest, including the Danum Valley Conservation Area (438 km^2). Hunting was prohibited in the Ulu Segama and little, if any, poaching was apparent during the study (except for the Sumatran rhinos).

To examine the effects of logging on terrestrial mammals, ten straight-line transects (1 m wide and 2–3 km in length) were established: four in unlogged forest, and two each within two-, five-, and twelve-year postlogged forest blocks (for details see Heydon and Bulloh 1997). In addition, one botanical plot (1000 x 20 m) was established in primary forest, and another in twelve-year old logged forest. All trees (> 12 cm dbh) were enumerated and identified, and the number of fruiting trees were recorded monthly between June 1992 and October 1993. Furthermore, the structure of the forest was assessed at all sites to show differences in foliage abundance at six strata, from ground level to upper canopy (see table 5-4).

PHOTO 5-3 *Log landing in dipterocarp forest one year after logging, Danam Valley, Sabah, East Malaysia. (G. Davies)*

TABLE 5-4 *Relative Proportions of Major Vegetation Types (%) Before and After Logging (Timber Extracted at 78 +/- 7.0 m³/ha) in Danum Valley, Sabah (after Heydon and Bulloh 1997)*

Vegetation Type	Unlogged Forest	Logged Forest (years postlogging)		
		2 Years	5 Years	12 Years
Tall trees	59	7	13	13
Intermediate	34.5	30	45	43
Pioneer trees	0	13	17	26
Vine tangle	4	30	13	7
Grass/herbaceous growth	0.5	12	9	3
Damaged vegetation	2	1	3	7
Exposed mineral soil	0	7	0	1
No. trees/ha	466			488

Food Resources

In terms of food resources, the most significant implication of logging for terrestrial herbivores and frugivores is the increase in ground level vegetation. Disruption of the forest canopy dramatically increases the availability of browse within the reach of ungulates (i.e., < 2 m). The benefits, however, were transient. Twelve years after logging, the area of grass/herbaceous vegetation had declined to only one quarter of that available two years after logging (12 percent to 3 percent; see table 5-4), due to pioneer trees establishing new canopies in disturbed areas and shading out the ground plants.

Fruit availability was reduced by logging, and remained so twelve years after timber extraction took place (see table 5-5). This change in fruiting occurred even though tree densities in logged forest (466 vs. 488/ha in unlogged forest), numbers of families (47 vs. 43), genera (95 vs. 105), and species (183 vs. 218), were roughly similar to undisturbed forests. Logged areas were dominated by a relatively small number of pioneer tree genera (e.g., *Macaranga, Octomeles, Neolamarkia, Duabangga* and small, free-standing *Ficus*). In the sampled area, *Macaranga* made up 29 percent of all trees (> 12 cm dbh), 86 percent of which was *M. hypoleuca*. By comparison, the most common tree species in the unlogged forest—*Mallotus wrayi*—made up only six percent of trees, and *Macaranga* less than 0.1 percent.

These changes in species composition are critical for mammalian frugivores as the pioneer species produce either wind- or bird-dis-

TABLE 5-5 *Declines (% change) in Fruit Availability and Strangling Fig Densities (n/km²) Following Logging in Danum Valley, Sabah (after Heydon and Bulloh 1997)*

	Unlogged Forest	Logged Forest (years postlogging)		
		2 Years	5 Years	12 Years
Fruit-on-trail counts (n/km)[a]	4.0	- 45%	- 47%	- 35%
Fruiting phenology (n/km²)[b]	800			- 47%
Strangling fig density (n/km²)	251	- 31%	- 54%	- 30%

[a]median number of independent ripe fruiting sources along transects (5 monthly measurements)

[b]median number of fruiting trees in 2 ha enumerated plot (17 monthly measurements)

persed fruits (with the exception of *Ficus*). Most have not been recorded in the diet of any ungulate. While overall tree density twe;ve years after logging is similar to unlogged forest, both the total number of fruiting trees and the density of those trees producing fruit eaten by mammals (i.e., in the 1–5 gm fruit size range) is considerably reduced: declines of 59 percent for mousedeer, 53 percent for muntjac, and 28 percent for bearded pig (Heydon 1994, and unpublished data). This situation was exacerbated by the seasonal nature of fruit production, which was more pronounced in the logged forest. It varied twice as much between peak and nadir months, leaving a more prolonged period when fruit supply was very scarce (Heydon and Bulloh 1997).

Terrestrial Mammal Abundance

The abundance of mammals in logged and unlogged forest was determined by night time transect surveying. Each of the ten transects was surveyed at monthly intervals between August and December of 1993 (total distances surveyed: 72 km in logged forest and 56 km in unlogged forest). The DISTANCE program (Laake et al. 1993) analyzed sighting data. Densities of sambar deer and the common muntjac were higher in logged forest (see table 5-6). Yellow muntjac and mouse deer densities were lower. For two species, the yellow muntjac *(M. atherodes)* and the mousedeer *(T. napu)*, home ranges were also measured. In mousedeer, a three-fold difference in density between primary and 12-year-old logged forest was mirrored by a 2.7-fold increase in home range area (Ahmad 1994). For the yellow muntjac, home range mapping of this species in the same unlogged forest blocks generated a density of 4.6–6.1 animals per km^2 (Heydon 1994), similar to values from the transect surveys.

The large (i.e., elephant, banteng, and rhino) and nomadic (i.e., bearded pig) ungulate species were observed too infrequently (or not at all in the case of rhinos) to calculate densities using distance-sampling techniques (Buckland et al. 1993). Nevertheless, differences in encounter frequencies (including indirect signs) indicate higher densities of elephant (+380 percent) and banteng (only observed in logged forest) in logged areas, but lower densities of bearded pigs (-16 percent).

Finally, in conjunction with the above surveys, civet densities were also recorded in logged and unlogged forests. While civets are not ungulates, the response of these carnivore/omnivore mammals to log-

TABLE 5-6 *Changes in Ungulate Densities (Individuals/km^2) After Selective Logging in Danum Valley (after Heydon 1994; Heydon and Bulloh 1997)*

Species	Unlogged Forest	Logged Forest	% Change
Cervus unicolor (Sambar deer)	0.2	2.2 (2-5 yr) 0.3 (12 yr)	+1000 +50
Muntiacus atherodes (Yellow muntjac)	4.9	2.2 (12 yr)	-55
Muntiacus muntjac (Common muntjac)	1.0	1.9 (12 yr)	+90
Tragulus napu (Greater mousedeer)	54.5	5.5 (2-5 yr) 16 (12 yr)	-90 -71
Tragulus javanicus (Lesser mousedeer)	30	10.3 (2-5 yr) 15 (12 yr)	-66 -50

ging is briefly noted in box 5-1, as there is virtually no information available on carnivore-logging interactions in the tropics.

Box 5-1 *Civet densities in logged and unlogged forests of the Danum Valey in Sabah, Malaysia.*

All civet species persisted in logged forests, as did the five felid species found in the Ulu Segama (M. Heydon, unpublished data). There were far fewer sightings of civets than ungulates, so data for all civet species were pooled during the Distance sampling analysis (see box table 5-1). The analysis showed that logging was associated with an 80 percent decline in the overall density of civets. Our results suggest that exclusively carnivorous species (e.g., *Hemigalus* and *Prionodon*) were disproportionately affected by logging, as compared to the omnivorous species (*Paradoxu-*

BOX TABLE 5-1 *Densities of Civets (individuals/km^2) in Unlogged and Logged Forest in Danum Valley (after Heydon and Bulloh 1996)*

Species	Unlogged Forest	Logged Forest	% Change
Paradoxurus hermaphroditus	5.6	1.4	-75
Arctogalidia trivirgata	1.1	P[a]	
Arctictis binturong	0.8	1.1	+38
Paguma larvata	P	P	
Hemigalus derbyanus	9.8	1.0	-90
Cynogale bennettii	P	P	

BOX TABLE 5-1 *Densities of Civets (individuals/km²) in Unlogged and Logged Forest in Danum Valley (after Heydon and Bulloh 1996)*

Viverra tangalunga	12.2	2.8	-77
Prionodon linsang	2.1	P	

[a]P = presence confirmed, but not observed during survey sessions.

rus and *Arctictis)*. These findings are at variance with the increased encounter rates of civets reported at Tekam in west Malaysia (Johns, 1983b), which may be a consequence of logging intensity—timber extraction levels at Tekam were 40 percent lower than in the Ulu Segama. The discrepancy highlights the need for studies on carnivores, which remain one of the least-examined vertebrate groups in logged forests.

Ungulate Responses to Selective Logging in Sabah

It appears that larger sambar deer invade newly logged forests in response to logging, and as regeneration proceeds, their numbers decline. Their abundance was positively associated with the area of severely degraded forest (e.g., grass/herbs $r_s = 0.86$, $P < 0.01$, and liana tangle $r_s = 0.91$, $P < 0.001$, Spearman's two-tailed test, table 5-6; see photo 5-3) and negatively with the area of climax stage forest ($r_s = 0.76$, $P < 0.05$). For muntjac, density differences in logged and unlogged forest were not as dramatic, but the yellow muntjac declined in density while the more ubiquitous common muntjac increased. Differences in diet preferences, particularly in relation to browse and fruit, provide a plausible explanation for these changes. Finally, both species of mousedeer showed consistent and prolonged population decline following logging. Mousedeer densities were positively correlated with fruit abundance ($r_s = 0.81$, $P < 0.01$), and in the case of *T. napu*, particularly with the abundance of large strangling fig trees ($r_s = 0.89$, $P < 0.01$). The small size of mousedeer (1.5 and 5 kg; Ahmad 1994), combined with their non-ruminant digestive system, may limit their ability to switch from a fruit to a browse-based diet (Kay 1987)—especially since mature leaf chemistry can be of very low quality in eastern Sabah (Davies 1994) (see photo 5-4).

Timber extraction rates in Sabah are among the highest in the tropics (Sundberg 1983). It is likely, therefore, that the post-logging changes in mammal abundance in Ulu Segama represent the most extreme consequences associated with logging impacts on wildlife habitat. Despite extensive disruption to the forest (Johns 1989b), however, all the ungulate taxa that were studied persisted after logging—although the response of individual taxa varied. Species normally

PHOTO 5-4 *Creepers festoon remaining trees in heavily logged-over forest in Sabah, East Malaysia. (G. Davies)*

showing a preference, or even dependence for forest edge habitats, benefit from the logging disturbance to the forest canopy (e.g., sambar deer and banteng). Species able to adapt their diet to the change in food resources are relatively unaffected (e.g., muntjac), while species dependent on resources that decline during logging, such as fruit, decline sharply in abundance (e.g., mousedeer). Logging in Ulu Segama, however, was not accompanied by human settlement or increased hunting pressure, which was an unusual circumstance in logged forests worldwide. Where hunting occurs in Borneo, densities of large mammals are inversely related to hunting pressure (Bennett et al. 1999; see chapter 16).

Conservation of Forest Ungulates

Ungulate Responses to Logging

Grazers and grazer/browsers, such as the Asian elephant, African buffalo, gaur, banteng and sambar deer, are not dependent on disturbed forests and generally not negatively affected by logging activities (provided hunting is restricted). These species make use of logging roads to colonize forest areas, feeding on the roadside grasses, and their populations often increase after logging. Only the giant forest hog is an anomaly in this group, as it is apparently sensitive to logging.

Generalist browsers and browser-frugivores, such as African elephants, Asian rhinos, tapirs, and even the diminutive royal antelopes are similarly little affected by logging. These species make use of shrubs and climbers growing in canopy gaps, often benefiting from logging that encourages the growth of forest gap vegetation. Many of these species are migratory—or can travel large distances to find suitable habitats—so population increases that occur after logging often arise through immigration rather than birth peaks, and their population dynamics need to be considered on landscape scales.

The intensity of logging that browsers can tolerate will depend on the extent to which they need a mix of forest and non-forest plants. The saola in Indochina depends on plants found only in closed-canopy forests in shady river gorges, while the okapi may need a mix of colonizing and mature-phase forest species. These appear to be the exceptions, however, rather than the rule. In the absence of hunting, browsers adapt well to logging.

The smaller-bodied browser-frugivores are more negatively affected by alterations to their habitat caused by logging. This is epitomized in southeast Asian forests, where forest-specialists are displaced (after logging) by closely related generalists—i.e., the bearded pig losing out to the wild boar, and the yellow muntjac to the common muntjac.

Bearded pigs and white-lipped peccaries travel over large areas of forest to consume tree fruit crops that are either produced in vast quantities every 8–12 years (i.e., dipterocarp mast fruiting), or accumulate in huge piles on the forest floor (e.g., palm nuts). This is a suitable strategy in large tracts of forest, and may have evolved at sites where there are few forest elephants (White 1994c). Logging for timber and tree felling for palm fruits, however, removes the super-abundant food sources that underpin this migratory browser-omnivore strategy—leading to population crashes for both species.

Nonmigratory browser-frugivores, including brocket deer in South America and duikers in Africa, can switch diets in response to seasonal or habitat changes and adapt to opening of the forest that follows logging. Many duikers spend much of their time in tree-fall gaps (Dubost 1980; Feer 1988; Newing 1994), possibly because they are preferred food supply areas rather than just good places to avoid predators. Exceptions to this rule, however, may include the zebra, and Ogilby's and white-bellied duikers, which have not been reported outside closed-canopy forests.

The browser-frugivore guild in southeast Asia has fewer species than the duiker radiation, and does have species which specialize in closed-canopy forest living Yellow muntjac and both mousedeer species show heavy losses following logging. The large-scale and protracted decline of both species of mousedeer suggests that they are obligate frugiovores—unable to increase dietary browse in response to fruit declines following logging—in contrast to the highly folivorous diets of similar-sized (< 4 kg) African antelopes, royal antelopes (*Neotragus* spp), and blue duikers in Natal (Faurie and Perrin 1993).

Conservation Implications for Ungulates in Logged Forest Landscapes

How do these results fit into the broader context of global deforestation, and how do they relate to estimates that one in four mammals is threatened with extinction (Baillie 1996)? A review of forest mammals found at continental sites (i.e., not on islands) showed that 99 species were threatened, but that only five of these (all primates) were threatened solely by habitat change following forestry activities (Grieser Johns 1997). In addition, there was no field evidence of any mammal going extinct locally as a result of simply logging, although some populations were severely depleted. This chapter also reports that the majority of forest ungulates are expected to survive habitat changes that result from logging, although some closed-canopy/old growth forest specialists are likely to suffer major population declines.

Grieser Johns' 1997 review also noted that of the same 99 threatened species, 10 are threatened with hunting (and trapping), and forty-four with hunting/trapping and habitat loss. Logging commonly accelerates hunting and trapping (see chapters 15–17). Terrestrial mammals particularly vulnerable to hunting/trapping following logging are:

- Those that can be shot at night using lamps, and particularly those which feed on roadside vegetation (e.g., forest buffalo and cattle)
- Those eaten and traded for food, irrespective of whether they use roadside habitats (e.g., duikers, deer, pigs, and peccaries)
- Species of special trade interest, such as elephants for ivory and rhinos for traditional Chinese medicine

Even more extreme local population declines occur when logging leads to increased rates of clear felling for agriculture, or desiccation that increases the risks of forest fires. Furthermore, it is very common for a second timber harvest to take place before a full, sustainable yield recovery period has passed following the first cut. Little information exists for areas that have been felled more than once, whether on a sustainable or unsustainable logging cycle, and ungulate populations could be more severely affected by multiple felling than single harvests.

Forest Management

In light of these conclusions, logging operations designed to reduce the risks of global extinction of medium- to large-mammal species need to include a number of measures:

1. Implementation of management practices which minimize the general impact of selective logging on the forest ecosystem; as noted in discussions of minimal damage models (Johns 1992 and b; see chapters 21 and 24). Key measures include minimizing damage to non-harvested trees; reducing soil compaction and avoiding damage to watersheds during extraction; leaving sufficient mature trees of exploited species to re-seed depleted areas, and leaving adequate gaps between logging

cycles for forest recovery. Many of these important practices have long been known, but as yet are still infrequently implemented (eg. ITTO 1993; FAO 1997).

2. Prevention of an increase in hunting and trapping for trade, or to feed timber camp labor (see chapters 15–17). This should be discouraged during logging operations. Most crucially, access to logged forests following timber operations should be restricted by blocking old logging roads and removing bridges. This can be effective in reducing the activities of commercial operators coming in from outside the area, but will only be effective with local communities if they are given a stake in the long-term management of the area. This is the most important aspect of logging management for large mammal conservation.

3. Protection of unlogged forest blocks, ideally in areas that are difficult to access and serve ecosystem functions (e.g., watersheds) or cultural needs (e.g., sacred forests) (see chapter 23). Given the incomplete state of our knowledge, this is necessary both as a safeguard for those species that prove intolerant of the direct effects of logging—yellow munjac, mousedeer, and possibly zebra and Ogilby's duikers—and as a refuge for mobile species that are displaced or eliminated from logged areas. These protected blocks need to be linked by *mammal-friendly* corridors, and logging activities need to be planned to reduce the risk of fragmenting mammal populations.

Conservation Research

Research to inform improved management is most urgently needed on measures to reduce negative impacts of timber extraction on forest ecosystems (see chapters 18 and 19). Such research will only be applicable, however, if it is integrated with economic and social research which supports commercial timber companies in their implementation of such measures, and socioeconomic research that takes account of local communities' wider needs and impacts—particularly in terms of hunting and trapping (Noss 1997; Frumhoff and Losos 1998).

In terms of ecology, very little information is available on what factors limit populations of forest ungulates, including short- and long-term changes in food supplies following logging. There is a need to investigate what foods are found on the forest floor (Newing 1994), as well as patterns of plant part production and phytochemistry. A comparison of differences between forest-dependent and non-forest-dependent relatives may help reveal what factors limit populations following logging.

Interpretation of field results on the effects of logging are consistently confounded by hunting and trapping impacts, and thus more attention needs to be focused on getting detailed measures of past and present hunting/trapping pressures. The line transect methods commonly used to study small ungulates are also very unreliable, and more attention needs to be given to radio tracking, home range mapping, drive counts, and duiker-focused studies—avoiding pellet and track count methods wherever possible (Nummelin 1990; McFoy 1995).

ACKNOWLEDGMENTS

Glyn Davies acknowledges the assistance and support of Mohamed Bakarr in studying the duikers of Gola, and other short-term assistants: D. Edwards, S. Bashir, K. Garcia-Alarcon, and H. Newing. Funding for the research was provided through ESCOR (DFID) to University College-London, and the fieldwork was conducted under the tutelage of Sierra Leone Forestry Department and Njala University College. Helen Newing acknowledges that the Taï case study was supported by SERC, Stirling University, the Centre Suisse de Recherche Scientifique in Abidjan, and the Tropenbos Foundation. Thanks also to M. Denis Gle for his work on evaluation of population census methods. Matt Heydon acknowledges the support of the Economic Planning Unit of the Malaysian Federal Government, the Danum Valley Management Committee, the Sabah Wildlife Department and the Sabah Foundation. Thanks also to C. Marsh, N. Ghaffar, S. Ambi, P. Bulloh and A. H. Ahmad. The project was funded by grants to A. G. Marshall, from Royal Society (UK), under ODA/NERC (contract F3CR26/G1/05). Thanks to Julian Caldecott and Jean-Pierre d'Huart for comments on pigs. The editors' support and contributions to improving the chapter are also much appreciated.

Chapter 6

THE EFFECTS OF LOGGING ON NONVOLANT SMALL MAMMAL COMMUNITIES IN NEOTROPICAL RAIN FORESTS

José Ochoa G. and Pascual J. Soriano

Recent literature on the study of mammal communities in Neotropical rain forests identifies the growing threat to many species (Patterson 1991; Terborgh 1992a, 1992b; Woodman et al. 1995, 1996; Voss and Emmons 1996) as a consequence of increased deforestation rates or degradation of primary forests by unsustainable resource exploitation (Uhl and Vieira 1989; Whitmore and Sayer 1992; Frumhoff 1995; Pinard and Putz 1996; Bryant et al. 1997; Miranda et al. 1998). Among the high levels of diversity that characterize the mammalian fauna in these ecosystems (Emmons and Feer 1990; Chesser and Hackett 1992; Voss and Emmons 1996), some taxonomic groups show a complex adaptive radiation and represent important regulatory elements in forest dynamics (Fleming et al. 1987; Smythe 1987; Emmons and Feer 1990; Terborgh 1992a). Marsupials and small rodents, after bats, represent the dominant fraction of mammals in Neotropical forested areas (Handley 1976; Emmons and Feer 1990; Ochoa et al. 1993; Voss and Emmons 1996), and constitute a major component of ecological processes and animal biomass (Emmons 1984; Robinson and Redford 1989; Janson and Emmons 1990). Very few projects, however, have assessed the ecological impacts of extractive activities on small tropical marsupials and rodents—in contrast to studies of other vertebrate groups (e.g., Wilson and Johns 1982; Johns 1986b, 1988, 1992a; Fenton et al. 1992; Thiollay 1992; Plumptre and Reynolds 1994; Mason 1996; Greiser Johns 1997; see chapters 4, 5, 7–14, and 24). Of those that have occurred on small mammals, most are from Old World localities (e.g., Kemper and Bell 1985; Isabirye-Basuta and Kasenene 1987; Pahl et al. 1988; Laurance and Laurance 1996).

In the Neotropics, there are a limited number of field inventories of small mammal communities (Patterson 1991; Voss and Emmons 1996), and few come from forests influenced by logging (Ochoa 2000). This condition is, in part, due to few researchers in Latin American countries having the taxonomic expertise to examine this topic, as well as few institutions considering this kind of work among their priorities. Even though knowledge of biological resources is fundamental to conservation strategies in this region (Blockhus et al. 1992; WRI et al. 1995), efforts to accomplish this goal lag far behind the wave of economically driven resource exploitation schemes impacting most forested areas (ITTO 1991a; Whitmore and Sayer 1992; Miranda et al. 1998).

This chapter reviews the response of Neotropical small mammals (excluding bats) to timber extraction, based on the available ecological information for this vertebrate group in forested areas and their importance in forest conservation (e.g., Julien-Laferreire and Atramen-tovitz 1990; Johns 1992a; González-M and Alberico 1993; Ochoa 1997a, 1997b, 2000). Research and management priorities are highlighted for addressing the negative impacts on small mammal communities that are associated with logging practices.

The Geographic, Taxonomic, and Ecological Context of This Chapter

In this chapter, Neotropical small mammals refer to marsupials and small rodents ubiquitously distributed across Central and South America, including their continental islands and the oceanic islands of the West Indies, Galapagos, and Falklands (Hershkovitz 1972). This review, however, is restricted to lowland rain forests (< 600 m) located in the following subregional categories (Voss and Emmons 1996): trans-Andean rain forests, coastal Venezuelan rain forests, Amazonian rain forests, and Atlantic rain forests (see figure 6-1). Included in this group of mammals are the Didelphimorphia (Didelphidae family) and Rodentia (Sciuridae, Geomyidae, Heteromyidae, Muridae, and Echimyidae families) orders (see table 6-1). These families comprise approximately 25 percent of the mammalian species recorded in Neotropical lowland rain forests (Handley 1976; Ochoa et al. 1993; Ochoa 1995; Voss and Emmons 1996). The taxonomic nomenclature follows Wilson and Reeder (1993), with several exceptions drawn from Voss and Emmons (1996).

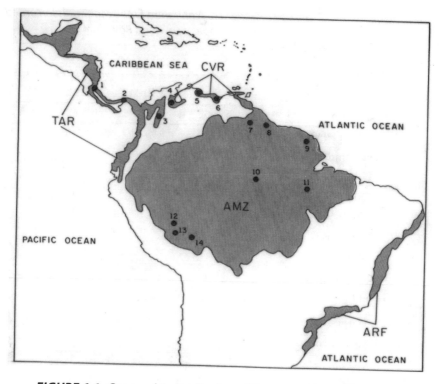

FIGURE 6-1 *Geographic distribution of Neotropical lowland rain forests. Nomenclature for the subregions is presented in Table 6-1. Localities are: 1—La Selva and vicinity, Costa Rica; 2—Barro Colorado Island, Panama; 3—Foothills on Western Llanos of Venezuela; 4—Maracaibo Basin, Venezuela; 5—Foothill in the border of Falcon and Yaracuy States, Venezuela; 6—Guatopo National Park, Venezuela; 7—Imataca Forest Reserve, Venezuela; 8—Kartabo, Guyana; 9—Arataye, French Guiana; 10—Minimal Critical Size Ecosystems Reserves, Brazil; 11—Xingu, Brazil; 12—Balta, Peru; 13—Cocha Cashu, Peru; 14—Amazonian Cuzco, Peru.*

To describe the structural patterns of small mammal communities in Neotropical rain forests and their responses to the ecological impacts of logging, each species was distinguished by its trophic strategy and principal foraging strata within a forest (Handley 1976; Charles-Dominique et al. 1981; Alho 1982; Robinson and Redford 1986; Ochoa et al. 1988; Emmons and Feer 1990; Janson and Emmons 1990; Nowak 1991; Mabberley 1992; Ochoa, 2000). Eight guilds were identified (see table 6-1), including four based on diet (fruigivores, herbivores, predators, and omnivores) and four based on forest strata use (arboreal, semiarboreal, terrestrial, semiaquatic). For some

TABLE 6-1 *Genera of Nonvolant Small Mammals Recorded in Neotropical Lowland Rain Forests*

	Subregion										Ecological Guild
	TAR			CVR			AMZ			ARF	
	a	b	c	d	e	f	g	h	i	j	
DIDELPHIMORPHIA											
Didelphidae											
Caluromys	1	1	1	1	1	1	1	2	1	1	FO/Ar
Caluromysiops									1		FO/Ar
Chironectes	1	1			1	1	1		1	1	P/Sa
Didelphis	1	1	1	1	1	1	2	1	1	1	PO/Sc
Glironia								1	1		FO/Ar[e]
Gracilinanus		1		1					1	2	FO/Sc
Marmosa	1	1	1	1	2	2	1	1	1	1	FO/Sc
Marmosops[a]				2	1		1	1	2	1	FO/Sc
Metachirus		1	1	1			1	1	1	1	PO/Te
Micoureus[a]				1	1		1	1	1	1	FO/Sc
Monodelphis		1		1	1	1	1	1	2	9	PO/Te
Philander	1	1	1	1			1	1	2	1	PO/Sc
RODENTIA											
Sciuridae											
Microsciurus	1	1						1			FO/Sc
Sciurillus							1				FO/Ar
Sciurus	2	1	1	1	1	1	1	1	2	1	FO/Ar
Geomyidae											
Orthogeomys	1										HO/Te
Heteromyidae											
Heteromys	1	1	1	1	1	1					FO/Te
Muridae											
Abrawayaomys										1	PO/Te[e]
Akodon					1					3	PO/Te
Blarinomys										1	PO/Te[e]
Delomys										2	FO/Te[e]
Ichthyomys[a]											P/Sa

TABLE 6-1 *Genera of Nonvolant Small Mammals Recorded in Neotropical Lowland Rain Forests (continued)*

	Subregion										Ecological Guild
	TAR			CVR			AMZ			ARF	
	a	b	c	d	e	f	g	h	i	j	
Isthmomys[a]											HO/Te
Melanomys[b]	1			1							FO/Te
Neacomys[a]					1	1	1	1	2		PO/Te
Nectomys			1	1			1	1	1	1	PO/Sa
Neusticomys[c]							1		1		P/Sa
Nyctomys	1										FO/Ar
Oecomys		2	2	1	2	3	6	5	3	1	FO/Sc
Oligoryzomys	1	1	1		1		1		1	2	HO/Te
Oryzomys	1	1	1	1	1	2	3	3	4	4	FO/Te
Ototylomys[a]											FO/Sc
Oxymycterus								1	1	4	PO/Te
Peromyscus[a]											FO/Te
Phaenomys										1	FO/Ar[e]
Reithrodontomys[a]											HO/Sc
Rhagomys										1	FO/Ar[e]
Rheomys[a]											P/Sa
Rhipidomys			2	1	1	1	3	2	1	1	FO/Ar
Scolomys[d]											PO/Te[e]
Sigmodontomys	1			1							FO/Te
Tylomys	1	1									HO/Sc
Echimyidae											
Dactylomys								1	1		H/Ar
Diplomys		1									FO/Ar
Echimys			1		1	1	2	2	1	5	FO/Ar
Hoplomys	1										FO/Te
Isothrix								1	1		FO/Ar[e]
Kannabateomys										1	H/Ar
Lonchothrix										1	FO/Ar[e]

TABLE 6-1 *Genera of Nonvolant Small Mammals Recorded in Neotropical Lowland Rain Forests (continued)*

	Subregion									Ecological Guild
	TAR			CVR			AMZ			ARF	
	a	b	c	d	e	f	g	h	i	j	
Mesomys							1	1	1		FO/Ar
Nelomys										6	FO/Ar
Proechimys	1	1	1	1	1	1	3	4	4	5	FO/Te

The nomenclature for subregions (figure 6-1) is defined according to Voss and Emmons (1996):

TAR = Trans-Andean Rain forests; **CVR** = Coastal Venezuelan Rain forests; **AMZ** = Amazonian Rain forests; and **ARF** = Atlantic Rain Forests.

Numbers represent the species richness in the following localities (see figure 6-1): **a**—La Selva and vicinity, Costa Rica (Wilson 1990; Voss and Emmons 1996); **b**—Barro Colorado Island, Panama (Glanz 1990; Voss and Emmons 1996); **c**—Foothills on Western Llanos of Venezuela (Ochoa et al. 1988; Utrera 1996; Ochoa unpublished data); **d**—Maracaibo Basin, Venezuela (Handley 1976; Voss and Emmons 1996); **e**—Foothills on the border of Falcon and Yaracuy States, Venezuela (Handley 1976; Ochoa unpublished data); **f**—Guatopo National Park, Venezuela (Ochoa et al. 1995); **g**—Imataca-Venezuela, Kartabo-Guyana and Arataye-French Guiana (Ochoa 1995; Voss and Emmons 1996); **h**—Minimal Critical Size Ecosystems Reserves and Xingu, Brazil (Malcolm 1990; Voss and Emmons 1996); **i**—Balta, Cocha Cashu and Cuzco Amazónico, Peru (Emmons and Feer 1990; Janson and Emmons 1990; Woodman et al. 1995, 1996; Voss and Emmons 1996); **j**—Coastal forests of eastern Brazil (Emmons and Feer 1990; Wilson and Reeder 1993).

Ecological guilds include the following:

Frugivores whose diets are based on fruits or seeds (**F**); Herbivores (**H**); Predators, including insectivores (**P**); Omnivores (**O**); Terrestrials (**Te**); Arboreal (**Ar**); Semiarboreals, corresponding to species using ground and tree strata (**Sa**); and Semiaquatics, integrated by species using terrestrial and aquatic habitats (**Sq**).

[a]Recorded in the trans-Andean region (Emmons and Feer 1990; Voss and Emmons 1996).
[b]Known from the Andean Piedmont (Amazonia) of eastern Ecuador (Voss and Emmons 1996).
[c]Recorded in the Coastal Venezuelan Range from the type locality of Neusticomys venezuelae (Anthony 1929).
[d]Recorded in the Amazonia (Emmons and Feer 1990; Voss and Emmons 1996).
[e]Diet unknown; trophic strategy estimated by morphology and ecological habits (Alho 1982; Emmons and Feer 1990; Nowak 1991).

marsupial and rodents, however, the guild position can change according to the ecological gradients within their geographical range (Charles-Dominique et al. 1981), or as a result of adaptations to variations in the availability of certain resources following anthropogenic disturbances (e.g., decrease in fruit production caused by elimination or reduction of tree strata).

Small Mammal Communities in Neotropical Rain Forests

Species Richness and Taxonomic Composition

Voss and Emmons (1996) assessed the diversity of small mammal communities across Neotropical rain forest types (see figure 6-1). Their inventory includes fifty-two genera of nonvolant small mammals: twelve in the Didelphimorphia order and forty in the Rodentia order (see table 6-1), with the highest levels of richness occurring in the Amazon and the Atlantic forests of Brazil (see table 6-2). Among these taxa, nineteen (37 percent) are broadly distributed (present in more than two subregions), including ten marsupials and nine rodents (83 percent and 23 percent of the genera comprising these orders, respectively).

The trans-Andean forests have the greatest number of endemic or geographically restricted genera (see table 6-2), followed by the Atlantic and Amazon forests. Forest ecosystems located in the coastal range of Venezuela exhibit the lowest richness and are composed of communities whose members have been recorded in two or more subregions (see table 6-1). Among the twenty-seven genera whose known distributions are restricted to one sub-region (52 percent of those genera inhabiting Neotropical rain forests), only two are in the Didelphimorphia (*Caluromysiops* and *Glironia*) order, and both are confined to the Amazon Basin. The amount of diversification in these forests is determined by the evolutionary histories of each phylogenetic group, the colonization patterns that have influenced each biogeographical region, and the responses of small mammals to geographic variations in environmental characteristics such as soil quality and primary productivity (Hershkovitz 1972; Emmons 1984; Voss and Emmons 1996; Ochoa 1997b).

In light of the above patterns, small mammal communities from the rain forests of the Amazon, the Atlantic coast of southwestern Brazil, and the trans-Andean lowlands should be considered a high priority for protection. This recommendation gains particular relevance when the increasing rates of deforestation and primary forest degradation in these subregions are considered (Voss and Emmons 1996; WRI et al. 1996; Bryant et al 1997; Miranda et al. 1998).

TABLE 6-2 *Some Characteristics of Nonvolant Small Mammal Communities in Selected Lowland Rain Forests of the Neotropical Region*

	Subregions			
	TAR	CVR	AMZ	ARF
Species Richness (range)	16–18	15–20	22–33	60
Didelphimorphia	5–7	6–10	7–12	19
Rodentia	10–13	9–12	15–21	41
Total Number of Genera	34	24	30	28
Didelphimorphia	10	10	12	10
Rodentia	24	14	18	18
Endemic or Restricted Genera[a]	11	--	7	8
Didelphimorphia	--	--	2	--
Rodentia	11	--	5	8

[a]Includes endemic genera or those whose geographical distributions in the Neotropical Region embrace only one subregion (e. g., *Glironia* and *Peromyscus*).
NOTE Bibliographic references and geographical context of subregions are described in table 6-1

Community Structure and Its Importance in Regulating Adaptability to Forest Disturbance

The guild structure of small mammal communities in several forested areas of the Neotropics is presented in figure 6-2. The following trends can be observed:

- Most species are semiarboreal or terrestrial. In all subregions, arboreal or semiaquatic species constitute the minority
- Most species, excluding those in aquatic habitats, have omnivore diets. Trophically specialized guilds, including strict predators and herbivores, contain the lowest number of species

In tropical forests, many nonvolant small mammals appear to be resilient to disturbances (Wilson and Johns 1982; Ochoa et al. 1988; Johns 1992a; Ochoa 1993, 2000; Laurance and Laurance 1996). This may be a function of their generalist habits and their adaptability to the ecological conditions found in secondary environments (e.g., scar-

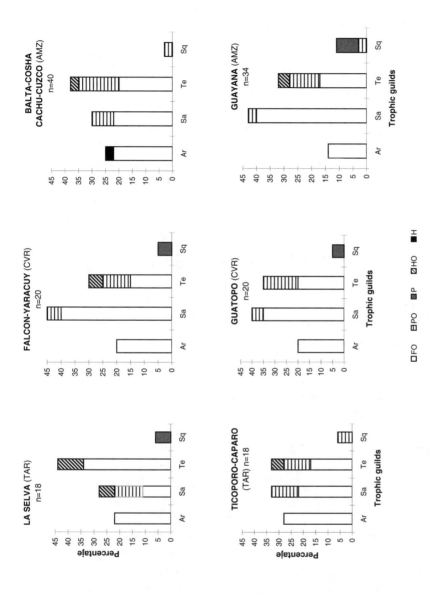

FIGURE 6-2 *Estimated structures of small mammal communities in forested localities of the Neotropical Region. Nomenclature for the subregions and ecological guilds are described in table 6-1. The Atlantic rain forest subregion was excluded due to a lack of detailed inventories from specific localities.*

city of some trophic resources, lower relative humidity, and higher temperature at the understory level) (Ochoa 1997b, 1998, 2000). Communities with the highest proportions of folivores, predators (mainly insectivores), and omnivores—in addition to a low abundance of strictly arboreal species—have the maximum probability of maintaining their original components after timber extraction. Those communities dominated by specialists (e.g., strict frugivores, granivores, and canopy species) are likely to be more susceptible to drastic changes in forest composition and structure (Kikkawa and Dwyer 1992; Laurance and Laurance 1996; Ochoa, 2000).

In general, the elevated proportion of omnivores comprising small mammal communities in Neotropical rain forests (see figure 6-2), together with the ability of semiarboreal and terrestrial taxa to exploit those resources available at the lower strata of these ecosystems, suggest that small mammal communities may be resilient to selective timber extraction.

Response of Small Mammals to Logging and Other Disturbance Events

Predicting the Sensitivity of Small Mammal Communities to Forest Disturbances

To assess the sensitivity of small mammals to forest disturbances, it is necessary to understand their ecological characteristics and requirements at both the population and community levels (e.g., demographic tendencies, competitive and predator-prey relationships, trophic strategies, and ecophysiological patterns). While many of these have not been explored for most small mammals, the use of ecological predictions supported by local extinction probabilities allows a preliminary assessment of how select taxonomic groups are likely to respond to changes in primary forest conditions (Ochoa et al. 1993, 1995; Utrera 1996).

An extinction probability model was developed for 140 species of mammals (grouped in nine orders and twenty-eight families) in a timber concession of the Venezuelan Guyana region (Ochoa 1997a). In this analysis, the extirpation potential of different members of the mammal community was related to:

⁓ Species-specific affinity to primary forest environments

⁓ Dependence on arboreal (mainly canopy) resources

⁓ Geographic distribution pattern and population densities

⁓ Estimated ecophysiological requirements

⁓ Vulnerability to the hunting activities of both local people and loggers (Arita et al. 1990; Emmons and Feer 1990; Ochoa et al. 1993; Wilson and Reeder 1993; Ochoa 1995; Voss and Emmons 1996)

In logged forests, members of the Sciuridae and Echimyidae families exhibited the greatest extirpation probabilities among nonvolant small mammals. A high proportion of the species in these families depend on resources in the canopy (e.g., arboreal/frugivore-granivores like *Sciurus* spp., *Echimys* spp., and *Isotryx bistriata*; and terrestrial/frugivore-granivores of the genus *Proechimys*). Some species in these families also occur at low densities or have restricted geographic distributions (e.g., *Isotryx bistriata*, *Dactylomys dactylinus*, and *Proechimys cuvieri*). In contrast, members of the Didelphidae and Muridae families appear less sensitive to logging. Most species in this group have widespread distributions, or they use resources found within the understory strata of secondary forests. Predictions of this type facilitate efforts to conserve small forest-dwelling mammals within areas to be logged, especially when rapid assessments are needed to identify those species at risk (Ochoa et al. 1993, 1995).

Forest Fragmentation

In logged areas it is common to find fragments of primary forests surrounded by secondary vegetation (sometimes highly degraded), plantations, or areas deforested for agriculture and infrastructure development (Johns 1988; Uhl and Vieira 1989; Ochoa 1993). Even though specific studies on small mammal communities in these kinds of fragments have not been published, the theoretical and experimental work on forest fragments (e.g., Bennett 1987; Friend 1987; Johns 1988; Terborgh 1992a; Andren 1994; see box 6-1) allows general predictions of the effect of fragmentation on such communities.

Box 6-1 *Unifying the study of fragmentation: external vs. internal forest fragmentation. (Jay R. Malcolm)*

The extent of deforestation in the tropics varies greatly from one region and development activity to another. Perhaps equally important, the spatial grain of deforestation also exhibits variability. At one extreme, large-scale industrial agriculture can result in coarse-grained, externally fragmented landscapes, wherein large forest fragments are interspersed with large clearings. At the other extreme, in what is termed *internal fragmentation* or *forest perforation* (Forman 1995), selective logging results in a fine-grained mosaic of small clearings, treefall gaps, and roads. Although both configurations are known to have important implications for the flora and fauna of the remaining forest, possible parallels between the two have only recently begun to be investigated. The identification of these links has great potential for conservation. In the light of accelerating loss and disturbance of tropical forests, and urgent needs for effective conservation guidelines and management techniques, the ability to apply knowledge and understanding from one type of landscape transformation to another is needed.

Recent research suggests that the ability to link and apply results from one type of forest fragmentation to another can be made through a common feature of human disturbances—the creation of canopy openings. Small gaps resulting from treefalls have long been known to play an important role in tropical forests because of the change in the physical environment that they bring and the resultant implications for plant growth and regeneration (Richards 1996). The importance of the large and/or abundant gaps created during human activities, however, has only recently been appreciated. In coarse-grained landscapes, where forest fragments abut large clearings, enormous changes occur in the forest close to clearings (Laurance and Bierregaard 1997). A suite of pervasive ecological changes quickly follows the creation of edges. Understory vegetation growth accelerates in response to increased light levels. Trees along the edge are exposed to the ravages of wind turbulence, and as they blow over, further possibilities for the growth of understory plants are provided. As a result of these changes, obvious edge-induced changes extend 50–100 m into fragments within a few years after edge creation (Malcolm 1994) and subtle effects may extend much further (Laurance and Bierregaard 1997). Several vertebrate communities, including both small mammals and birds (see chapters 8–10), respond quickly to the changes (Bierregaard and Lovejoy 1989, Malcolm 1997). Pronounced changes in species diversity and community composition are observed, especially in small fragments where perimeter-to-area ratios are high and edge impacts are strong.

Although canopy openings created during selective logging are much smaller that those arising from modern agricultural development, the density of clearings may be high—especially in areas where many trees are being harvested. Under such conditions, the relatively weak edge effects of individual clearings may begin to overlap and interact and result in edge effects that are strong in an overall sense (see chapter 2). Computer simulations that incorporate this overlap highlight the potential for widespread and powerful edge effects (Malcolm 1998, 2001). Perhaps most seriously, they suggest that even slight increases in harvest rates can lead to disproportionate edge-induced habitat change.

Edge effects may thus provide a common currency for understanding some of the changes that diverse types of development activities bring. In some cases, because of pervasive edge-induced habitat change, it appears that very different patterns of forest clearing can result in similar wildlife responses. As increasingly strong edge effects in forest fragments led to increased small mammal abundance and diversity in the central Amazon (see box figure 6-1A), increasingly strong edge effects along 12- and 19-year-old log extraction routes in the Central African Republic also led to increased small mammal abundance and diversity (Malcolm and Ray 2000) (box figure 6-1B). The increases in abundance, which were true of nearly all rodent species found in undisturbed forest, suggest that the forest changes that accompany edge formation may be positive for at least some rain forest taxa.

For many other rain forest groups (including birds), however, edge effects led to a strong decline in diversity—both in coarse-grained fragmented landscapes in the central Amazon (box figure 6-1C; Bierregaard and Lovejoy 1989) and in fine-grained selectively logged landscapes in French Guiana (box figure 6-1D; Thiollay 1992). In these last two studies, individual bird species responded in similar ways despite the differences in the grain of the forest clearing. Of the twenty-four species affected by the landscape transformation in both studies (a species was judged to be affected by the landscape transformation if its abundance was consistently highly or lower in disturbed forests than in undisturbed forest), the sign of the impact (either positive or negative) was the same for 17 of them (P = 0.06, two-tailed Binomial Test). Evidently, many of the bird species that are adapted to the conditions of undisturbed forest are intolerant of edge-induced habitat change, both in the coarse- and fine-grained landscapes. Even the increase in small mammal abundance and diversity along edges may, unfortunately, have negative implications for the rest of the ecosystem. Small mammals are important predators of seeds and seedlings, and a superabundant small mammal fauna may alter patterns of plant regeneration. Struhsaker (1997) suggested that seed and seedling predation by small mammals was a key causative agent in suppressing tree regeneration in heavily logged forests in Africa—resulting in transformation of the original rain forest to a more-or-less stable herbaceous and semi-woody plant community. From the standpoint of conserving intact rain forest ecosystems, therefore, edge creation appears to have strong negative impacts.

These results provide a clear message for forest managers in the tropics: reduce canopy openings (and hence edge effects) wherever and whenever possible. This can be accomplished in fragmented landscapes by setting aside large reserves and wide corridors that provide the interior forest conditions required by so many tropical forest organisms (see chapter 23). In logged landscapes, the frequency and size of openings should be minimized through low harvest rates and careful logging techniques (see chapter 21, 24). An important task of future research will be to measure the responses (and tolerance thresholds) of tropical species to these edge effects, and to determine the exact manner in which forest openings interact in determining overall levels of edge-induced habitat change.

Preliminary results of field inventories in the Western Llanos of Venezuela and the Guayana region immediately after logging (Ochoa unpublished data), where small forest fragments (< 50 ha) surrounded

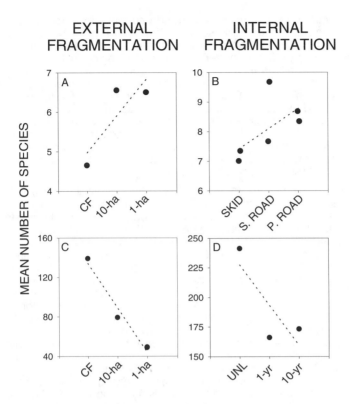

EXTERNAL FRAGMENTATION

INTERNAL FRAGMENTATION

MEAN NUMBER OF SPECIES

BOX FIGURE 6-1 *Small mammal (part A) and avian (part C) species richness in continuous forest (CF) and 10- and 1-ha forest fragments in the central Amazon (Bierregaard and Lovejoy 1989; Malcolm 1997) compared with small mammal richness in unlogged forest and along timber extraction routes in the Central African Republic (part B; UNL = unlogged forest, SKID = skidder trails, S.ROAD = secondary access roads; P. ROAD = primary access roads; Malcolm and Ray unpublished data) and avian richness in French Guiana (part D; UNL = unlogged, 1-yr = 1 year postlogging, 10-yr = 10 years postlogging; Thiollay 1992). In each plot, edge-modified sites are to the right of undisturbed forest sites.*

by selectively logged areas (8–10 m³/ha) and deforested sectors were examined, showed these fragments to contain a high proportion of small mammal species found in undisturbed forest communities. Some species even increased in their relative abundance. Three years later, population densities of the most specialized taxa declined (e.g., arboreal/frugivore-granivores), and in some areas, forest-dwelling species were found inter-mixed with some typical small mammals of non-forest environments (e.g., *Zygodontomys brevicauda* and *Sigmodon alstoni*;

Soriano and C. Lulow 1988). The latter penetrated into newly logged areas using forest openings as colonization routes. Survival of typical, forest-dwelling small mammal communities in these fragments depends on the range of available resources (partially influenced by habitat complexity and heterogeneity in the isolated patch of forest), and the likelihood of gaining access to complementary food sources within the habitat matrix found in surrounding areas (Uhl and Vieira 1989; Johns 1992a; Terborgh 1992a; Frumhoff 1995; Malcolm 1995: see box 6-1). These trends, however, suggest that some species remaining in fragmented forests after logging could be a risk, at least in the medium or long term—which, in turn, may influence certain ecological processes within the forest (Terborgh 1992a; Ochoa 1997a, 2000).

Studies by Fonseca (1989) and Fonseca and Robinson (1990) in the Brazilian Atlantic coast provide a preliminary evaluation of the combined effect of fragmentation processes and the level of perturbation on the isolated forest patch as a consequence of tree extraction. Their results suggest a *fragment size/small mammal diversity* relationship that is partially explained by habitat structure and specific interactive mechanisms, such as predation. These authors found that in large secondary forest fragments (35,973 ha), mammal communities included the top predators (e.g., *Leopardus* spp.), which regulate population levels of generalist species with high trophic plasticity and competitive abilities (e.g., *Didelphis marsupialis*). When unregulated, these generalists have the potential to exclude other ecologically related faunistic elements (*Metachirus nudicaudatus*, *Marmosa spp.*, and *Oryzomys trinitatis*) (see photo 6-1), such as found in small secondary fragments (60 ha). Some of these excluded mammals play a crucial role in forest dynamics and regeneration, mainly through the dispersal of seeds and the pollination of many tree species (Charles-Dominique et al. 1981; Emmons and Feer 1990).

Changes in Availability of Food and Denning Resources After Logging

The composition of vertebrate communities reflects the ecological conditions at a given locality (Smythe 1986; Terborgh 1986; Fleming et al. 1987; Johns 1988; Kikkawa and Dwyer 1992). These conditions influence a species' access to often scarce and highly dispersed resources such as roosts in hollows of mature trees, fruits, and nectar produced by some trees. These scarce resources are expected to decline in exploited areas (Johns 1986b, 1988, 1992a, 1992b; Uhl and Vieira

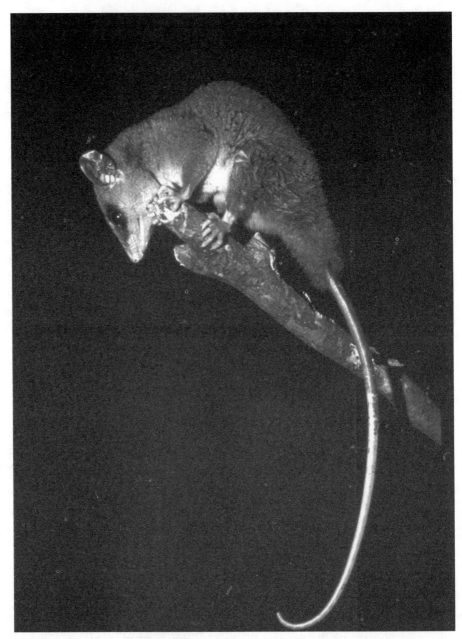

PHOTO 6-1 The marsupial *Metachirus nudicaudatus* is one of the most common terrestrial small mammals in some logged forests of the New World. It is an omnivore species which prefers invertebrate preys (e.g., worms and beetles). (J. Ochoa G., Imataca Forest Reserve, Venezuelan Guayana Region)

1989; Kikkawa and Dwyer 1992; Frumhoff 1995) where logging and other related activities lead to:

- A reduction in the absolute availability of canopy-linked resources (e.g., large den trees and canopy epiphytes, nectar and large-seeded fruits)
- A reduction in the density of keystone plants (e.g., some species of Chrysobalanaceae, Lecythidaceae, and Hippocrateaceae)
- Modifications to phenological patterns of trees, with a predominance of vegetative growth phases over reproductive ones
- An alteration in microclimatic conditions in the understory, with consequences for some species with ecophysiological restrictions (e.g., *Erisma uncinatum* and *Anacardium rhinocarpus*)

In spite of these predictions, several studies have shown increases in small mammal species' richness and density in secondary forests (Isabirye-Basuta and Kasenene 1987; Fonseca 1989; Julien-Laferreire and Atramentowitz 1990; Frumhoff 1995; Malcolm 1995). This is sometimes the result of the arrival of savanna/forest ecotone species in disturbed areas (e.g., *Zygodontomys brevicauda* and *Sigmodon alstoni*), and an increase in densities of some interior-forest mammals, which can use the resources present in exploited areas (e.g., *Didelphis* spp. and *Proechimys* spp.). The greater vegetative cover at the understory level of secondary forests (as a result of the high density of heliophitic plants and fallen trunks), and the higher structural complexity of the midlevel forest strata, also increase feeding and shelter opportunities for many terrestrial or semiarboreal species (Fonseca 1989; Ochoa 1997a).

Studies in the lowland rain forests of southern Venezuela (locality 7 in the Amazon rain forests subregion of figure 6-1) indicate that after four years of selective timber extraction (trees > 40 cm dbh, with an average volume removal of 5.8 m^3/ha), small mammal communities undergo significant changes in their original structure and composition (Ochoa 1997b, 2000). These changes lead to a predominance of semiarboreal/omnivore-predators in secondary habitats (e.g., *Didelphis* spp. and *Philander opossum*). Terrestrial/frugivore-omnivores (e.g., *Proechimys* spp.), the most common foraging guild in undisturbed forests, also continue to be important in logged forests. Both guilds are able to use the abundant resources after logging disturbances: e.g., fruits of colonizer plants, invertebrates from the soil, and denning sites under fallen trunks (Charles-Dominique 1986; Fonseca 1989; Emmons and Feer 1990) (see photo 6-2). In contrast, the more arboreal frugivores and granivores (e.g., *Rhipidomys* spp. and *Micoureus demerarae*) (see photo 6-3) were not found in intensively

PHOTO 6-2 Interior view of a logged forest in the Venezuelan Guayana (Imataca Forest Reserve). In this area the understory is characterized by a high density of colonizer plants and fallen trunks, where communities of nonvolant small mammals are dominated by semiarboreal/predator-omnivore marsupials (e.g., *Philander opossum*) and terrestrial/frugivore-omnivore rodents (e.g., *Proechimys* spp.). (J. Ochoa G.)

PHOTO 6-3 The semiarboreal marsupial *Micoureus demerarae* is among the group of small mammals showing reductions of population levels in some intensively logged forests of the Venezuelan Guayana Region. This species has preferences by food resources located at the canopy level.
(J. Ochoa G., Imataca Forest Reserve, Bolivar State)

disturbed areas and did not show significant lower abundances (Ochoa 1997a, 2000).

Edaphic conditions that affect forest productivity appear to have an effect on small mammal responses to logging (Emmons 1984). In southern Venezuela, sandy soils with low water and nutrient retention capabilities are the predominant edaphic conditions, primarily in the lowlands of the Guayana Shield (Mogollón and Comerma 1995). This results in low productivity in the understory of closed-canopy forests, with limited availability of some resources for primary terrestrial consumers (Emmons 1984; Voss and Emmons 1996). Following logging in these areas, the semiarboreal/omnivore-predator, the common gray four-eyed opossum (*Philander opossum*), occurred with a relative abundance that was 4.3 times higher than the value found in undisturbed areas (see figure 6-3), while primary consumers like spiny rats (*Proechimys* spp.) showed increases approaching 2.1 times (Ochoa 1997b, 2000). These same species occur in forests of the Venezuelan Andean foothills (characterized by alluvial soils with high fertility). On these more productive sites, however, the genus *Proechimys* was the dominant component in primary and secondary forests (see figure 6-3). While site productivity may influence the carrying capacity in terms of animal density and biomass, opening of the forest canopy (and subsequent increases in understory growth following logging) appears to have a greater effect on small mammal populations than the site's inherent productivity potential. Finally, it is important to consider the implications of logging on trees with flowering cycles that provide key resources for nectar-feeding small mammals (Lumer and Schoer 1986; Terborgh 1986; Emmons and Feer 1990; Julien-Laferreire and Atramentowitz 1990). In many cases, marsupials and other small mammals depend on the nectar produced by valuable timber trees (e.g., *Ceiba pentandra, Eriotheca spp., and Pachira quinata*), especially during seasonal periods of low food availability (Julien-Laferreire and Atramentovitz 1990; Kikkawa and Dwyer 1992). Harvesting practices that remove a large percentage of these *keystone* resources may promote the local extinction (or reduction in population levels) of some nectar-feeding species that are highly dependent on these resources. Where these animals play critical roles in the reproduction of timber species (principally through pollination and secondary dispersal), overexploitation of trees may lead to a cascading collapse in their reproductive success through the loss of ecological services obtained by mutualistic associations with small mammals (Lumer and Schoer 1986; Gribel 1988).

Effects on Reproductive Behavior

The reproductive strategies of some Neotropical marsupials can be influenced by changes in patterns of food availability after forest disturbance (Julien-Laferreire and Atramentowitz 1990). Primary forests tend to vary widely in their food abundance for small mammals, with longer scarcity periods, lower densities of fruiting trees, and reduced reproductive effort for many plants compared to secondary forests (Charles-Dominique 1986). In response to these conditions, Julien-Laferreire and Atramentowitz (1990) found that the bare-tailed woolly opossum and the four-eyed opossum show more continuous reproductive patterns and higher juvenile survival rates in primary forests. These characteristics are thought to be a function of: 1) their ability to focus foraging efforts on a limited number of resources that are relatively abundant during periods of fruit scarcity (mainly nectar); and 2) their lower population densities compared to disturbed areas (see figure 6-3). When the arboreal strata is drastically reduced, fragmented, or even eliminated, the most specialized species (and those

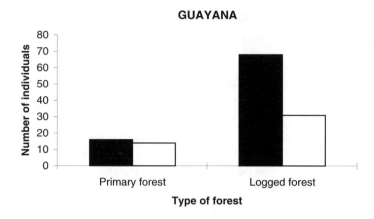

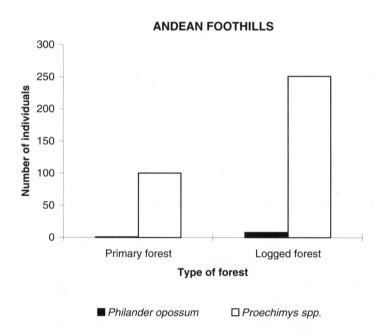

■ *Philander opossum*   □ *Proechimys spp.*

FIGURE 6-3 Relative abundance of *Philander opossum* (Didelphimorphia: Didelphidae) and *Proechimys* spp. (Rodentia: Echimyidae) in primary and logged forests in two biogeographic regions of Venezuela: Andean Foothills and Guiana Shield (based on 1,500 and 5,000 trap nights respectively; Ochoa et al. 1988; Ochoa 1997b).

generalists with a high level of resilience to forest disturbance) will probably be affected by significant trophic constraints and a drastic change in demographic patterns (Johns 1992b; Malcolm 1997).

Hunting Pressures

Logging activities are often associated with increases in hunting pressure by both local people and timber company personnel, who augment their economic income through trade or consumption of wildlife products—including an important number of mammals (Johns 1986b; Bisbal 1994; Frumhoff 1995; Ochoa 1997a; see chapters 15–17). In most Neotropical areas, mammal extraction is focused on medium to large-sized species (Redford and Robinson 1987, 1991). In some areas where large species have been overexploited, however, hunters target meat from marsupials and small rodents (including members of the families Didelphidae, Heteromyidae, Muridae and Echimyidae families: e.g., rice rats [*Oryzomys* spp.], spiny rats [*Proechimys* spp.], and the armored rat [*Hoplomys gymnurus*]; Suárez et al. 1995; chapters 15 and 16).

Depending on the locality, hunting patterns can influence the population status of certain taxa, leading to a significant reduction in the density and relative abundance of some species—especially where human settlements have exerted hunting pressures for a long time (Suárez et al. 1995) and wildlife use is translated into benefits for logging companies (Ochoa 1997a). These population changes may modify the patterns of community assemblages and some ecological processes within the forest (J. Ochoa unpublished data).

The Ecological Role of Marsupials and Small Rodents in Neotropical Forest Dynamic and Regeneration

An extensive area of Neotropical rain forest has been allocated for timber production, under management schemes that scarcely consider the dynamics of these ecosystems (Uhl and Vieira 1989; Johns 1992a; Whitmore and Sayer 1992). During the past two decades, field studies have demonstrated the importance of small mammals—including bats (see chapter 7)—as connective elements in a complex web of mutualistic interactions, which help to maintain the taxonomic and structural diversity of forests (Uhl et al. 1981; Charles-Dominique 1986).

Seed dispersal and predation represent two of the most studied roles of marsupials and rodents in rain forests from the New World (Charles-Dominique et al. 1981; Gautier-Hion et al. 1985; Smythe 1986, 1987; Adler and Kestell 1998; see box 6-2). Many small mammals have trophic strategies closely associated with the consumption of fruits or their constituent parts. At the same time, an important number of plants depend on small mammals as seed dispersers or pollinators (Charles-Dominique 1986; see box 6-2). These mutualistic interactions are often associated with coevolutionary patterns and are strongly related to the population dynamics of many commercial trees (e.g., *Spondias mombin*, *Inga* spp., *Carapa guianensis*, *Erisma uncinatum* and *Protium* spp.). Features such as phenology and the reproductive patterns of plants, and morphology and the nutritional quality of fruits, maximize the efficiency of dispersing processes (Janson 1983; Gautier-Hion et al. 1985; Herrera 1984). These mammals have also developed a set of strategies to access scarce food resources and avoid competition with ecologically related species through strategies such as the consumption of size-differentiated fruits or the hoarding of fruits and seeds (Smythe 1986, 1987; Terborgh 1986; Fleming et al. 1987; Kikkawa and Dwyer 1992; see box 6-2).

> **Box 6-2** *Implications of logging on seed dispersal and predation by rodents in the lowland rain forests of central Guyana.*
> *(David S. Hammond and Raquel Thomas)*
>
> Rodents regulate the recruitment of trees by selectively predating and/or dispersing seeds. By selecting the seeds of certain species over others, they can actively influence the competitive hierarchy which establishes the relative recruitment success of co-occurring canopy tree species.
>
> Timber tree species in the Guianas tend to have relatively large seeds. More than 50 percent of the eighty-seven timber species currently harvested in Guyana are dispersed by mammals—an estimated 28 percent being solely dispersed by rodents, especially by red-rumped agouti (*Dasyprocta agouti*), red acouchy (*Myoprocta acouchi*), Cuvier's spiny rat (*Proechimys cuvieri*), and the Guyanian spiny rat (*P. guyannensis*) (Hammond et al. 1996). The dispersal or burial of seeds by rodents often results in enhanced recruitment. One of Guyana's most important timber tree species, Greenheart (*Chlorocardium rodiei*: Lauraceae) has very large seeds that fall beneath dense aggregations of adult stems during the main fruiting period from January to April. The seeds (and later seedlings) are attacked by a bark beetle (Scolytidae), which reduces per capita survivorship significantly beneath parent trees in comparison to sites at greater distances (Hammond et al. in press). Though most Greenheart

stems are found in aggregations, the minority of offspring recruiting away from these clumps sustain the regeneration of this species—especially given that most harvesting is directed to these densely stocked areas (i.e., high advanced regeneration mortality due to incremental damage during logging). At the same time, the extremely prolonged germination of Greenheart seeds makes them an ideal resource for seed-eating rodents during periods between high fruit production. Most Greenheart seeds were found to be consumed during the period when overall resource availability was lowest, suggesting that this species is not a preferred resource, but one that is used during periods of scarcity by at least five species of rodents. Seed losses incurred through rodent foraging during periods of scarcity is a minor trade-off for the tree, however, in light of the important role rodents play in establishing new Greenheart regeneration when cached seeds are later forgotten.

The healthiest stems of Greenheart are the individuals typically harvested. This also means that the most reproductively active adult trees are also removed from the population—an activity that can lead to a precipitous decline in the number of seeds of these species that are available each year. Many rodents are also territorial, and this spatial segregation can mean that some intensively logged areas can no longer support the pre-harvest inhabitants. After logging at one 15 ha study site, the estimated abundance of acouchys (*Myoprocta* spp.).—accounting for changes in visibility—was lower than in a same-sized, unlogged plot with nearly identical abundances of the most common tree species (Hammond et al. 1992). A conservative approach to Greenheart seed tree retention would maximize the seed stock available for maintaining seed-dispersing rodents during periods of resource scarcity.

Size plays an important role in determining which animals will attempt to consume a fruit or seed. Generally, a greater number of species can consume the smaller seeds. The fact that many larger-bodied rodent species specialize in consuming large seeds is likely a product of the reduced competition for these resources. Small rodent species (such as those in the Muridae) are able to efficiently consume much smaller-seeded species, many of which profit from heavy disturbance. Seed dispersal by these small rodents can lead to a rise in the abundance of small-seeded colonizing plant species (Forget and Hammond in press).

Few applied studies have been carried out to determine the role of seed-consuming rodents in the regulation of timber tree recruitment after logging and the impact that logging intensity can have on the habitability of the site by these animals.Rodents, along with bats, represent the greatest diversity of tropical forest dwelling mammals. Attempts to preserve this diversity need to consider the conservation and role of these species in timber management.

Marsupials are commonly frugivorous and include species that exploit the understory and arboreal strata in Neotropical forests (Emmons and Feer 1990). They consume fleshy fruits of different sizes (Charles-Dominique et al. 1981), particularly those belonging to colonizer plants (*Cecropia* spp. and *Solanum* spp.) that produce numerous small seeds (Charles-Dominique 1986). In most cases, passage of seeds through the digestive tracts of the disperser agent represents an essential step in increasing the germination capacity and reproductive success of some pioneer plants (Fleming and Heithaus 1981; Fleming 1988).

In contrast to marsupials, small, forest-dwelling rodents prefer seeds of unripe and hard fruits (usually small or medium-sized), which are carried to feeding sites (Charles-Dominique 1981; Gautier-Hion et al. 1985). In this case, the successful contribution of these mammals as disperser agents depends on the proportion of transported seeds that find suitable microhabitats for germination, or whose viability is not affected by predation (Gautier-Hion et al. 1985; Smythe 1986).

Seed predation has been recorded for several small Neotropical rodents belonging to the Sciuridae, Heteromyidae, Muridae and Echimyidae families (Emmons and Feer 1990; Nowak 1991). These mammals may act as important regulators of tree composition and density (Terborgh 1992a), and represent key elements for the management of forests (see box 6-2). Among them, the genus *Proechimys* has been considered one of the main seed predators-dispersers in Neotropical rain forests (Smythe 1986, 1987; Adler and Kestell 1998) due to its high population densities across a broad geographic range (Handley 1976; Emmons 1982; Janson and Emmons 1990; Malcolm 1990) (see photo 6-4).

Even species that are primarily seed predators can positively influence the regeneration potential of secondary forests. Spiny rats, for instance, disperse the mycorrhizal sporocarpus (e.g., *Sclerocystis coremioides*) and spores (e.g., *Glomus* spp.) (Emmons 1984; Janos et al 1995). This role has been studied for other small rodents (including the Neotropical genera, *Oryzomys* and *Mesomys*), demonstrating their contributions to forest conservation (Janos et al 1995; Johnson 1996).

Recommendations for Conserving Small Mammals in Logged Landscapes

Although most small mammals appear relatively adaptable to logging disturbance, strategies for forest management and conservation must consider that some species may be at risk—especially the arboreal frugivores-granivores and those taxa with limited range distributions. The loss of these mammals, and the additions of some savanna-secondary forest colonizing species in intensively disturbed forests, will cause changes in ecological processes that could negatively influence the regeneration success of important timber trees.

PHOTO 6-4 The terrestrial rodents of the genus *Proechimys* are considered among those mammals with a high influence on Neotropical rain forest regeneration. They are seed predators, but at the same time they contribute with dispersion of micorrhizaes and seeds produced by many timber tree species. (P. J. Soriano)

A prudent approach would be to tread as lightly as possible during logging, to avoid significant impacts on mammal communities and other biodiversity components typically inhabiting primary forests (see chapters 21 and 24). This issue is especially important until we know more about how these communities respond to logging activities, mainly in those regions with the highest diversity.

Technical criteria in forest policies adopted by governments and logging industries must be focused on the protection of small mammal communities, if the long-term productivity, health, and economic value of production forests are to persist. Possible approaches include the creation of primary forest corridors in association with areas managed for timber (Ochoa 1993; MARNR 1995; see chapters 20 and 23). Several authors have explored reducing the impacts of logging on forest biodiversity (Johns 1986b; Thiollay 1992; Whitmore and Sayer 1992; Hernández et al 1994; Frumhoff 1995; Mason 1996; Greiser Johns 1997; Ochoa 1997b, 1998, 2000; Miranda et al. 1998; see chapters 21 and 24). These approaches include:

- Designing plans for selective tree cutting and log removal that minimize the level of disturbance to the residual vegetation structure and composition

- Taking into account the ecological role of commercial trees in providing key resources for marsupials and rodents (e.g., seeds and roosts in mature trees), thereby ensuring an adequate number of residual stems for both regeneration and mutualistic interaction purposes
- Using silvicultural methods based on an evaluation of their implications on primary forest regeneration and biodiversity conservation (see chapter 22)
- Controlling and monitoring hunting activities

Success of these initiatives will depend on technical and scientific cooperation from all stakeholder groups, including governmental agencies, research institutions, the logging industry, and local people. The integration of these sectors represents a basic step towards implementation of management models based on the conservation and sustainable use of Neotropical rain forests.

Long-term field inventories and monitoring efforts are also needed to understand the composition and structure of small mammal communities inhabiting Neotropical rain forests—especially those areas with high extractive pressures and/or low available biological information (Malcolm 1990, 1991; Patterson 1991; Woodman et al. 1995, 1996; Voss and Emmons 1996). These inventories serve as the foundation for additional ecological research and conservation planning. Such projects are likely to be difficult to implement in Latin American countries—at least in the short term—where financial constraints, an absence of institutional policies that promote forest evaluations, and a limited number of local taxonomic experts exist.

The analysis of plant-animal interactions as regulatory mechanisms and their importance in the regeneration and maintenance of primary forests (especially timber species), represent a crucial aspect for success of forestry management activities and the recovery of areas with high levels of disturbance (Frumhoff 1995). It is necessary to identify and assess those small mammal groups with a major influence on key ecological processes—including seed dispersion and predation—and pollination of those tree species with a high commercial value. Based on long-term monitoring programs, these evaluations must identify the levels of disturbance beyond which some marsupials and rodents are irreversibly affected (Blockhus et al. 1992; Johns 1992b; see chapter 19).

ACKNOWLEDGMENTS

We are very grateful to Robert Fimbel for his editorial revision and important suggestions during the elaboration of this manuscript. Lionel Hernández, Rosa

Fernández, Alejandro Grajal, and John Robinson provided indispensable assistance at different stages of our work. Edjuly Marquez, Mariapia Bevilacqua, and Elizandra Delgado helped us with computerized figures and bibliographic recompilation.

Chapter 7

THE CONSEQUENCES OF TIMBER EXPLOITATION FOR BAT COMMUNITIES IN TROPICAL AMERICA

Pascual J. Soriano and José Ochoa G.

The wealth of biological diversity in tropical forest ecosystems owes its existence to mutual relationships between species (Gilbert 1980). Some Neotropical forest bats (*Chiroptera*) are the mobile links in these interrelationships, benefiting plant species that depend on their ecological *services* for successful reproduction and seed dispersal. The loss of such links may have cascading effects on the health and productivity of tropical forest systems (Terborgh 1986b; Johns 1992a). In Venezuela, for example, approximately 1.3 million hectares of forest were cut in the western lowlands between 1950 and 1975, leading to the severe degradation of the three most important forest reserves north of the Orinoco River (Veillon 1976; Catalán 1993). The long-term ecological ramifications of these actions are virtually unknown, and the capacity of these reserves to recover to a state where they can once again provide a wide range of ecological and social services remains in doubt. Understanding this complex set of interrelationships is a major challenge confronting forest ecologists and resource managers, making management planning for the sustainable use of resources a difficult process (Whitmore 1980; WRI 1990; Thiollay 1992; Mason 1996; Ochoa 1998).

Planning and careful implementation of timber-harvesting practices, in an effort to minimize the impacts of logging on forest biodiversity and ecological processes, seldom occur in tropical forests. This failing is, in part, a function of our limited understanding of wildlife and their habitat requirements and the costs involved in obtaining such information (Mondolfi 1976; Ochoa 1992; see chapter 6). There are currently few examples of reasonably complete lists of vertebrates that inhabit these exploited areas (Voss and Emmons 1996), and even fewer studies of the effects of timber extraction on these animals (see other chapters in this section).

In this chapter, we explain the taxonomic and functional structure of bat communities in Neotropical lowland forests. We provide a synthesis of the available information regarding the effects of timber exploitation on these bat communities, and propose some management and research priorities that will help to safeguard the biological diversity of Neotropical forest bats in production forest landscapes.

Composition and Structure of the Bat Communities in Neotropical Rain Forests

Bat species represent approximately 50 percent of the mammal fauna associated with lowland Neotropical humid forest ecosystems, where between 35 and 65 species are commonly found per major forest type (i.e., Handley 1976; Ibáñez 1981; Mok et al. 1982; Webster and Jones 1984; Jones et al. 1988; Brosset and Charles-Dominique 1990; Ochoa et al. 1988, 1993; Ascorra et al. 1991, 1993; Medellín 1993, 1994, Pacheco et al. 1993; Voss and Emmons 1996). A recent survey carried out in Venezuelan Guayana found over 78 species coexisting in the Imataca Forest Reserve (Ochoa 1995; Ochoa 1997a). It is estimated that between 90 and 110 sympatric species exist in the lowland forests of the Amazon Basin (Voss and Emmons 1996). These communities show a similar taxonomic composition at supraspecific (families and genera) levels (see table 7-1). Inherent limitations in the survey methodologies, however, and a lack of taxonomic specialists mean that most surveys are incomplete and biased in their listing of the actual assemblages (Handley 1967; Voss and Emmons 1996). Only bat species occupying the lower strata of the forest are well represented in samples, while the majority of species in the Molossidae family—which are found above the canopy—are absent (see box 7-1).

> **Box 7-1** *Beyond mist-nets: what the rest of the bats can tell us about forests. (Bruce Miller)*
>
> Determining which species are truly rare is a necessary first step before an assessment can be done of the impact of logging on Neotropic bat communities. Until recently, bat surveys in the Neotropics have been biased by traditional sampling methods—primarily mist-netting—towards the readily captured leaf-nosed bats (*Phyllostomidae*). Consequently, more is known of the phyllostomids than other families of bats in the Neotropics (Kalko 1997; see this chapter).

The important ecological role that leaf-nosed bats (*Phyllostomidae*) play in seed dispersal and pollination in the Neotropics is widely acknowledged (Findley 1993; see this chapter). The insectivorous bats—primarily the non-phyllostomids—play an equally important role by controlling leaf herbivores, stem and shoot borers, and seed predators. As this important component of Neotropical bat communities has not previously been adequately sampled, we are uncertain what species compositions or population levels exist in most forests.

Double-frame harp traps can augment mist-netting surveys by capturing bats that use echolocation to avoid nets (LaVal and Fitch 1977; Voss and Emmons 1996). These traps are double-framed arrays of vertical monofilament lines spaced approximately 2.5 cm apart (with the lines offset). The echolocating bats do detect these lines, but because they are adept at flying through clutter, they simply tilt and fly between the lines of the first panel where they strike the second set of offset lines and fall unharmed into a collection bag under the trap. Although harp traps provide a better means of sampling non-phyllostomids bats than mist-nets, they still collect too few samples of many species (Miller 1998).

The identification of free-flying bats using their unique vocal signatures (Kalko 1995; O'Farrell 1997; O'Farrell and Gannon 1999; O'Farrell and Miller 1997, 1999; O'Farrell et al. 1999a, b) has allowed the greatest insight into the non-phyllostomid community (Kalko 1995, 1997; Kalko et al. 1996; Miller 1998; H. Schnitzler and E. Kalko, unpublished data). To obtain the highest quality, least-biased inventories, acoustic identification should be combined with other sampling methods such as harp traps and mist-nets (Kalko 1997; Miller 1998; O'Farrell and Gannon 1999).

From October 1995 through October 1997, new acoustic sampling techniques were developed during a study of the nonphyllostomid bat community in northwestern Belize. The tropical hardwood forests of the study site reflect more than 150 years of light, selective timber extraction. This extraction ended in the mid-1960s, and most trees in the canopy today are old-growth species with substantial regeneration in the undisturbed understory (Brokaw and Mallory 1993). Nine vegetation habitats (Wright et al. 1959) were surveyed using harp traps and new acoustic methods to evaluate habitat preferences of many species for which little ecological information previously existed. As a result, two species considered rare—Thomas's bat, or shaggy bat (*Centronycteris centralis*) (Gardener et al. 1970; Timm et al. 1989) and Van Gelder's bat (*Bauerus dubiaquercus*) (Engstrom et al. 1987)—were found to be among the most common species in many forest habitats. This implies that rarity, as defined from studies using mist-nets, may be an artifact of biased sampling and not of actual low population numbers (Gaston 1994; Kalko et al. 1996).

In the same study noted above, it was also discovered that *C. centralis, Natalus stramineus* (Mexican funnel-eared bat), *B. dubiaquercus*, and *Myotis elegans* (elegant myotis) appeared restricted to forested areas, suggesting these species may serve as indicators of mature, closed-canopy forests. At a finer level, *C. centralis* was abundant and *M. elegans* common in one forest type (Sapote-Silion-Mahogany). In another forest type (Sapote-Ramon-Spice), the reverse pattern was found. As both of these species are suspected tree roosters (Timm et al. 1989; LaVal and Fitch 1977), it may be that a yet unknown tree species association exists. This suggests that these bats may serve as indicators of habitat at a much finer level of detail than previously known in two very similar forest types.

In most production forests, timber extraction has profound effects on forest structure and bat roost sites (Crampton and Barclay 1996; Grindal 1996; Perdue and Steventon 1996; see chapter 2). In Australia, logging was found to have its greatest impact at the microhabitat level for many bat species (Crome and Richards 1988; Crome et al. 1996). In northwestern Belize, bat species that prefer medium to dense understory disappeared from survey sites after brushing (<2 m width) was carried out along trails. Although these species returned to the sites after understory vegetation regenerated, the apparent sensitivity to very localized structural changes suggests that larger scale changes that result from logging may have management implications. Finally, the impact herbivorous insects can have on tropical forests is significant (Lowman 1984, 1991, 1992a,b; Aide and Zimmerman 1990; Gerhardt 1993; Dial and Roughgarden 1995). The extent of damage from herbivorous insects can be related to the presence or absence of insectivores (Marquis and Whelan 1994). It follows then, that if habitat modification decreases the aerial insectivore component of the bat community, the resulting increase in deleterious insects could have adverse effects on forest resources for years or even decades.

Using multiple survey methods, a more complete picture emerges of bat species composition in tropical forests, and a better understanding is gained of the often subtle differences in species-habitat associations. As these new methods are implemented in previously surveyed areas, species considered rare or yet undocumented are likely to be found. Once we understand the *where* and *why* of species occurrence, we will be in a better position to evaluate the impacts of anthropogenic habitat change on bat communities and the ecological functions they perform. The richness of coexisting species may be partially explained by the *guild* or functional structure of bat communities. Bat communities are organized in large, independently functional categories (guilds) related to principal food sources (Bonaccorso 1979; Fleming and Heithaus 1981b): insectivores, frugivores, nectarivores, carnivores, piscivores, and sanguivores (vampires) (McNab 1971; Wilson 1973; LaVal and Fitch 1977; Bonaccorso 1979). Variations also exist between species found in the same guild, in terms of methods of obtaining food, size differences of certain anatomical features, and the use of flying space. These characteristics allow greater resolution—or niche partitioning—of each assemblage (e.g., Handley 1967; Smith and Genoways 1974; Carranza et al. 1982; Humphrey and Bonaccorso 1979; Humphrey et al. 1983; Soriano 1983, 1985; Ascorra and Wilson 1992).

The richness of coexisting species may be partially explained by the *guild* or functional structure of bat communities. Bat communities are organized in large, independently functional categories (guilds) related to principal food sources (Bonaccorso 1979; Fleming and Heithaus 1981b): insectivores, frugivores, nectarivores, carnivores, piscivores, and sanguivores (vampires) (McNab 1971; Wilson 1973; LaVal and Fitch 1977; Bonaccorso 1979). Variations also exist between species found in the same guild, in terms of methods of obtaining food, size differences of certain anatomical features, and the use of flying space.

TABLE 7-1 *Taxonomic Composition of Several Neotropical Rain Forest Bat Communities*

| Locality | Emb | Noc | Mor | Phyllostomidae | | | | | Nat | Fur | Thy | Ves | Mol | Total Species |
				Phy	Glo	Car	Ste	Des						
La Selva (Costa Rica)[a]	8	2	2	19	6	3	13	1	0	1	1	7	2	65
BCI (Panama)[a]	7	2	2	16	3	3	16	1	0	0	2	6	6	64
Primary forest (French Guiana)[b]	7	0	3	27	6	3	18	1	0	1	2	4	6	78
Sn. Juan Manapiare (Vzla)[c]	10	1	1	19	6	2	16	2	0	0	0	5	6	68
Imataca (Venezuela)[d]	9	2	1	18	8	3	20	1	0	0	1	6	9	78

Abbreviations: Emb. = Emballonuridae, Noc. = Noctilionidae, Mor. = Mormoopidae, Phy. = Phyllostominae, Glo. = Glossophaginae, Car. = Carolliinae, Ste. = Stenodermatinae, Des. = Desmodontinae, Nat. = Natalidae, Fur. = Furipteridae, Thy. = Thyropteridae, Ves. = Vespertilionidae, Mol. = Molossidae

SOURCE [a]Voss and Emmons 1996, [b]Brosset et al. 1996, [c]Handley 1976, [d]Ochoa 1996.

These characteristics allow greater resolution—or niche partitioning—of each assemblage (e.g., Handley 1967; Smith and Genoways 1974; Carranza et al. 1982; Humphrey and Bonaccorso 1979; Humphrey et al. 1983; Soriano 1983, 1985; Ascorra and Wilson 1992).

At least two general foraging strategies can be clearly differentiated in the most well represented trophic groups—the insectivores and frugivores. Insectivorous bats may be classified as either aerial insectivores, which capture their prey in the air while flying, or foliage gleaners, which pluck their prey from surfaces. Representatives of the former can be stratified vertically in the forest according to their preferred area of activity, including the area above the forest canopy. High-flying species may show a marked reduction in the width of the wings—a characteristic found in members of the Molossidae family (which show a flying behavior similar to that of swifts and swallows)—making them more efficient in the capture of prey from open spaces above the forest canopy (Fenton 1972, 1990; Norberg and Rayner 1987). In contrast, foliage gleaners detect and capture invertebrates that are sitting on the ground or on leaves. Bats using this strategy usually have elongated ear pinnae, which allow more efficient detection of sounds made by prey (Norberg and Rayner 1987; Fenton 1990).

Amongst fruit eaters, nomadic frugivores consume fruits from plant species with relatively wide-spaced individuals and fairly high fruit production during short periods of time. Variations in fruit patterns may cause the movement of nomadic frugivore populations to new areas of the forest when resources become scarce (Soriano 1983). Morrison (1979) showed that the common large fruit-eating bat (*Artibeus jamaicensis*) can fly up to ten kilometers in one night between its refuge and its foraging area. This temporal variation often requires the near-daily identification of new roost sites, which are usually found in the foliage of large trees. In contrast to these nomads, sedentary frugivorous bats have fairly fixed search itineraries every night, and prefer fruits of plant species that produce continuously but offer only a few ripe fruits per plant each night. Species with more sedentary strategies inhabit proportionally smaller fixed areas and their refuges are more stable, such as hollow tree trunks and caverns (see photo 7-1). Members of the Carolliinae and Phyllostominae subfamilies, and the genus *Sturnira* (Stenodermatinae) represent this group (Soriano 1983, 1985). These different feeding habits and strategies make each group sensitive to the environmental changes related to its particular requirements. The consequences of the effect of timber exploitation on each functional component of the bat community may be very different..

Effects of Timber Exploitation on Bat Communities

The activities involved in selective exploitation of timber species, such as the use of heavy machinery to construct roads, affect the structure and climate within forests (Uhl and Guimaraes 1989; Johns 1992a; Thiollay 1992; Mason 1996; Ochoa 1998). These environmental changes include:

- Complete deforestation of certain areas
- Fragmentation and isolation of intact forest
- Breaks in the continuity of the forest canopy
- Changes in the microclimate beneath the canopy (i.e., luminosity, temperature, humidity)

In extreme cases, the decrease in relative humidity may facilitate forest fires, which in turn causes severe modifications in the composition and structure of the understory and availability of prey populations.

PHOTO 7-1 *Small cavern that serves as refugia for bat species with sedentary strategies. (WCS)*

The wide range of differences between bat guilds, resulting from the different kinds of resources used by each group, makes Chiropterans ideal indicators of environmental disturbance (Fenton et al. 1992). The available information on the effects of timber exploitation on bat communities in Neotropical localities is very scarce, however, with two general trends recognized to date: 1) some species show a negative response to changes in the structure and composition of the forest, and 2) other species in the community show population increases due to a reduction in competition or an increase in preferred resources.

Species Whose Populations Are Adversely Affected by Timber Exploitation

Studies by Brosset et al. (1996) in French Guiana and Ochoa (2000) in the Imataca Forest Reserve (Venezuelan Guayana Region) have demonstrated the local extinction of certain species—and the reduction in guild complexity of bat communities—where trees larger than 40 cm dbh were logged (approximate 5.8 m³/ha harvested). Guild reduction was greater where silvicultural treatments such as *enrichment strips* were established after logging (see table 7-2). The most affected species were the specialized insectivorous and vertebrate predator species (Kikkawa and Dwyer 1992; Gardner 1977) found in the Phyllostominae subfamily, which seems to have the highest sensitivity to timber exploitation (Fenton et al. 1992; Brosset et al. 1996; Utrera 1996; Ochoa 2000). The abundance of some species appreciably decreases (e.g., the black large fruit-eating bat [*Artibeus obscurus*], small yellow-eared bat [*Vampyressa bidens*], spear-nosed bat [*Phyllostomus elongatus*], and the round-eared bat [*Tonatia saurophila*],) or even disappears (e.g., the false vampire bat [*Vampyrum spectrum*], round-eared bat [*Tonatia sylvicola*], and the woolly false vampire bat [*Chrotopterus auritus*]) in disturbed forests (see photo 7-2). Some Emballonuridae such as the common doglike sac-winged bat (*Peropterix kappleri*) also show similar patterns (Ochoa 2000). These data are in agreement with those of Brosset et al. (1996), who reported the loss of rare species in disturbed areas of French Guiana.

Trees of greatest size are preferentially selected during timber exploitation, which may cause drastic reductions in roost availability for Chiropteran species (i.e., round-eared bats [*Tonatia* spp.], the woolly false vampire bat, the false vampire bat, and the free-tailed bat family Molossidae). This effect may be especially pronounced in regions were the geomorphology does not favor the formation of caverns that offer alternative roost sites. Causes for the reductions of carnivorous and insectivorous Phyllostomid bats have not been completely established. We postulate the joint action of several factors:

- Reduction in the food supply resulting from a numeric decrease or disappearance of prey (mainly invertebrates), which may be affected in their various larval stages by microclimatic changes related to the loss of canopy continuity (Fenton et al. 1992)

- Breaks in the canopy that cause a reduction in specialized fruit and flower availability (Charles-Dominique 1986; Johns et al. 1985, Johns 1988; Uhl and Guimaraes 1989). In some instances these conditions sever interdependent plant-animal links, negatively impacting keystone species of commercial value (i.e., *Inga* spp., *Spondias mombin* and *Ceiba pentandra*) and bat communities in the season of lowest fruit productivity (Fleming et al. 1987; Johns et al. 1985; Johns 1988; Terborgh 1986b; see chapter 6)

- Reduction or elimination of refuges such as holes in trunks or branches due to the felling of trees with diameters greater than 40 cm

- Loss of termite nests (usually constructed in large trees) which are consumed by certain species of Phyllostominae bats

Finally, some very specialized frugivores such as the visored bat (*Sphaeronycteris toxophyllum*), and the wrinkle-faced bat (*Centurio*

TABLE 7-2 Differences in Taxonomic Composition Between Primary and Exploited Forests at Unit V of Imataca Forest Reserve, Venezuela

| Imataca Forest Reserve | Bat Families and Subfamilies | | | | | | | | | | Total Species |
| | Emb | Noc | Mor | Phyllostomidae | | | | | Ves | Mol | |
				Phy	Glo	Car	Ste	Des			
Primary forest	6	0	1	13	3	3	13	1	3	2	45
Exploited forest	6	2	1	10	3	3	13	1	6	5	50
Enrichment stripes	2	0	1	5	2	3	13	0	0	1	27

Abbreviations: Emb. = Emaballonuridae, Noc. = Noctilionidae, Mor. = Mormoopidae, Phy. = Phyllostominae, Glo. = Glossophaginae, Car. = Carolliinae, Ste. = Stenodermatinae, Des. = Desmodontinae, Ves. = Vespertilionidae, Mol. = Molossidae. (Source: Ochoa 2000).

cenex)—as well as some small-sized species of Honduran white bats (*Ectophylla* spp.), and small yellow-eared bats (*Vampyressa* spp.)—are probably affected by disturbances associated with forest management practices such as elimination of understory palms required for tent roosts and/or loss of certain critical foods. Limited information about the natural histories of these species does not allow us to make any definitive conclusions.

Species Whose Populations Are Favored by Timber Exploitation

Bats that appear to be favored by logging include many of the seed-dispersing frugivorous genera (i.e., short-tailed fruit bats [*Carollia*], fruit-eating bats [*Artibeus*], and the yellow-shouldered bats [*Sturnira*]) (see photo 7-3a) that are mutualistically associated with pioneer plants (i.e., *Piper, Cecropia, Solanum,* and *Vismia genera*). This relationship generates a positive feedback effect: the more these plants prosper in an area, the greater the carrying capacity for their associated Chiropteran species (Charles-Dominique 1986; Fleming 1988; Fenton et al. 1992; Kikkawa and Dwyer 1992; Brosset et al. 1996; Utrera 1996; Ochoa 2000). These bats may quadruple their population densities in disturbed sites, as reported by Brosset et al. (1996) in French Guiana. Some of the large species of fruit-eating bats (*Artibeus* spp.) in clearings seem to have a greater influence on seed dispersal of *Cecropia* spp. than most frugivorous birds. These bats typically fly long distances, favoring the colonization of pioneer plant species (Charles-Dominique 1986), and thereby creating a positive feedback mechanism that favors the

PHOTO 7-2 Photographs of select insectivores or predator bat species whose populations are adversely affected by timber exploitation: (top) *Tonatia saurophila*; (bottom) *Phyllostomus elongatus*. (P. J. Soriano)

population growth of these bat species. Similarly, the introduction of cattle to areas of exploited forests that are converted into pastures may serve to supply large additional feeding sources for the vampire bats (see photo 7-3b), leading to increases in their population levels (Johns et al. 1985; Wilkinson 1985; Johns 1988, 1992a; Fenton et al. 1992).

Finally, the loss of continuity in the forest canopy caused by clearing large areas for logging roads gives certain species of flying insectivores—which restrict their activity to the upper layer of the canopy in natural conditions—access to lower forest strata. An increase in prey availability, combined with access to additional water sources, partially explains the positive response exhibited by this group. Some structures, such as logging camps, may also provide alternative refuges for species with more flexible requirements. This is the case for some Emballonuridae and Molossidae, which may roost in or on man-made structures and forage in camp clearings (Brosset et al. 1996; Ochoa 2000).

Protection of Chiropteran Communities in Logged Landscapes

Timber exploitation affects Chiropteran communities by altering the composition and structure of vegetation, availability of food resources, number of refuges, and microclimate (Johns 1992a; Macedo and Anderson 1993; Frumhoff 1995; Ochoa 1997a; see chapter 6). Measures taken to protect bat communities, therefore, should focus on mitigating these disturbance factors.

Of the possible measures for protecting bats in timber production forests, the most important is protecting old-growth areas in forest landscapes destined for timber exploitation (Thiollay 1992; Ochoa 1993; see chapter 23). When selecting these sites, the natural composition and structure of the forest should be considered in an effort to maintain representative areas of preharvest habitat across the managed forest (Lovejoy and Bierregaard 1990; Frumhoff 1995). Corridors of uncut vegetation, such as gallery forests, should connect these *within* landscape refugia, thereby allowing movement between the different protected areas (Johns 1992a; Thiollay 1992; Mason 1996; Ochoa 1993, 1997a; see chapters 21 and 23).

Within areas slated for exploitation, measures must be taken to reduce the impacts of harvesting practices. Selective logging is often very *unselective*, impacting more than 50 percent of the trees in

PHOTO 7-3 Photographs of select bat species whose populations are favored by timber exploitation: (top) the short-tailed fruit bat, *Carollia perspicillata*; and (bottom) the vampire bat, *Desmodus rotundus*. (P. J. Soriano)

exploited stands (Johns et al. 1985; Johns 1992a; Frumhoff 1995; Ochoa 1998). It is, therefore, very important to implement reduced impact-logging procedures (see chapter 21), and to strengthen forest planning that favors a mix of exploited and reserve sites (Frumhoff 1995; see chapter 20), to preserve the integrity of production forest habitat. Such measures should be accompanied by sufficient financial support to guarantee the implementation of basic and applied scientific research programs, designed to:

- Assess the ecological roles (pollination and dispersal) and requirements of individual bat species
- Monitor their activities in recovering stands
- Train technical personnel to translate research findings into improved forest management practices (Johns et al. 1985; Ochoa 1992; see chapters 18 and 19)

Research Priorities for Conserving Forest Bat Communities

Baseline information of bat communities is lacking for most forests. The most complete inventories possible need to be compiled, combining the use of different capture and detection methodologies for sympatric species. With sufficient information, geographical distribution of the Chiropteran species could be generated. These inventories should begin in the regions under greatest forestry use, so that rare, threatened, and sensitive species may be detected and conservation strategies developed.

Surveys of bat communities cannot be made adequately and reliably without resolving the inherent taxonomic problems associated with certain genera (i.e., *Choeroniscus, Platyrrhynus,* and *Molossus*), which are urgently in need of revision (Voss and Emmons 1996). This task cannot be guaranteed without an adequate series of specimens and the support of collections where the voucher specimens are housed. It is of strategic importance to put an aggressive program into practice to stimulate the development of already existing zoological collections and museums in different countries within the region, to create new collections where they are lacking, and to employ appropriate personnel to oversee their administration and conservation.

Although important progress has been made in the knowledge of the diet and habitat requirements of many Chiropteran species found in Neotropical forests, the available information is far from complete (Gardner 1977). Obtaining this information remains a priority—particularly for Phyllostominae species that are sensitive to logging practices, and Glossophaginae and Stenoderminae species that are likely to be potential pollinators or seed dispersers of commercially important timber and non-timber plants. Research based on radio telemetry will allow better knowledge of the movement patterns and home ranges of sensitive species, and the establishment of minimum area requirements for protected areas (Frumhoff 1995). The goal of such an effort is to better understand the links between bats and forest tree species, thereby helping to design restoration plans for degraded areas and sustainable forest management plans for natural forests managed for timber resources.

ACKNOWLEDGMENTS

We thank R. Fimbel and A. Grajal for their invitation to participate in this book and their assistance and encouragement in elaborating the first draft of this chapter. We are also grateful for the comments of two anonymous reviewers who greatly improved the quality of this manuscript.

Chapter 8

TROPICAL FORESTRY AND THE CONSERVATION OF NEOTROPICAL BIRDS

Douglas J. Mason and Jean-Marc Thiollay

Latin America is the largest remaining frontier for tropical forestry. While southeast Asia continues to produce most of the world's tropical timber, over half of the world's remaining tropical forests occur in Latin America. Nearly 5,000,000 ha of tropical forests are logged in Latin America each year—a total that is expected to increase as Asian stocks are rapidly depleted and Latin America becomes the primary producer of tropical hardwoods (Grainger 1987; WRI 1994). Even where intact forest remains, huge areas of forest have already been allocated as concessions.

These forests have tremendous value, not only for their wood, but also for the biodiversity they conserve. Even by tropical standards, Latin American forests are spectacularly diverse. Some sites in western Amazonia harbor from 450 to 550 bird species, almost twice the totals found at similar sites in tropical Africa and Asia (Ridgely and Tudor 1989).

Conserving this diversity will depend on more than just strictly protected areas. While an essential tool for conservation, most protected areas are too small to maintain the region's biodiversity by themselves. The best hope for tropical biodiversity may be to develop economically viable land uses, such as forestry, that depend on large areas of tropical forest (Brokaw et al., in press). Current forestry practice, unfortunately, is careless and unsustainable—often leading to severe forest degradation and loss—particularly because logging roads provide access to areas that were previously inaccessible (Dudley et al. 1995). Colonists clearing land for agriculture often eliminate the trees remaining after selective logging, and hunting threatens many species where forest remains (see chapters 15–17). Forestry, therefore, represents both an opportunity and a problem. The challenge for the region is to make the transition from unsustainable, short-term extraction to long-term forest management and conservation.

Birds are a useful group for evaluating some of the effects of logging on fauna. Their taxonomy is well established—most species can be identified in the field, and at least some information is available about the biology of most bird families, if not species (see photo 8-1). More importantly, bird species are sensitive to changes in forest structure, microclimate, and composition. Birds are also important because of their roles in key forest processes, such as the pollination, seed dispersal, and seed predation of plants (also see chapters 3, 9, and 10). Despite their value as indicators, and the importance of their ecological roles to forest health and productivity, few studies have evaluated the effects of logging on Neotropical birds.

This chapter summarizes what is known about the relationship between timber management and birds in Latin America. The first section describes how timber management affects birds, both directly and indirectly. The following section describes how avian diet plays an important role in rendering some species more vulnerable than others. Both of these sections draw heavily from fieldwork in northern South America (the Guianan Shield) and parts of Amazonia. The third sec-

PHOTO 8-1 Latin America's tropical forests are extremely diverse, fostering some unique feeding adaptations, such as the unusual bill of this red-billed scythebill, *Campylorhamphus trochilirostris*. (D. J. Mason)

tion outlines some potential long-term impacts of forest management on birds. The last section highlights some steps that can reduce both the short- and long-term impacts of timber management on birds.

Direct Effects of Timber Management: Habitat Alteration

The direct effects of timber management result from logging and silvicultural operations themselves. These operations often alter the structure, microclimate, and composition of the forest, while also fragmenting forests into smaller and more isolated blocks. All of these changes can affect bird communities (see box 8-1 for a case study from French Guiana).

Box 8-1 *Ecological characteristics of bird species sensitive to logging in French Guiana.*

Thiollay (1992) surveyed the composition of forest bird communities in French Guiana (northeastern Amazonia) in one- and 10-year intervals after selective logging of 1 to 3 trees (7–15 m^3) per hectare, comparing them with the bird communities of an undisturbed primary forest. A point count method was used, in which 937 0.25 ha plots were each censused for 20 minutes. To compare the responses of each group to logging, the data are reanalyzed below by species groups. Abundances are reported as the number of birds per 100 0.25 ha plots. The 328 primary forest plots were selected to match the proportion of samples in each habitat type in logged forest areas. The number of species in logged forest was corrected for differences in sample sizes with primary forest by rarefaction (Heck et al. 1975). Significant differences (positive or negative) in abundance within logged forests, when compared to abundance in primary forest, were assessed using Kruskall-Wallis Analyses of Variance (Zar 1984).

Bird species richness and abundance both declined following logging (see box table 8-1). While overall abundances were lower in both logged forests, abundance was more depressed 10 years after logging than after one year. Many of the changes in the forest (and subsequently bird populations) occur years after actual extraction. Although most species were able to persist in logged forests, many were greatly reduced in abundance and some were extirpated. Insectivores almost always tended to be more affected than nectarivores, omnivores, and frugivores. The large species in most categories were also more extinction-prone after logging than were smaller species. Hunting was probably a factor in the vulnerability of these large species

Habitat selection was the major determinant of vulnerability to logging. Among small frugivorous manakins, for example, only species closely associated with open, dark, low-level understory in high mature forest actually decreased after logging (the white-fronted manakin [*Pipra serena*] and the white-throated manakin [*Corapipo gutturalis*]). Other Pipridae—that remained stable or increased—were naturally associated with gaps, subcanopy levels, dense vegetation in swampy areas, or low second growth, microhabitats most similar to those likely to occur in logged areas.

Obligate flocking behavior of understory species, and their resulting extensive foraging movements, may increase the sensitivity of such species to forest gaps and disturbance. This may explain the striking decline of mixed-species flocks and army-ant followers, even though their unspecialized insect prey (and even the army ants) had not obviously decreased.

BOX TABLE 8-1 *Responses of Guilds of Birds in French Guiana to Selective Logging*

Species Group (guild)	Forest Strata	Primary Forest (328 plots)		Logged 1 year (335 plots)		Logged 10 years (273 plots)	
		Species	Abundance	Species	Abundance	Species	Abundance
Large terrestrial game birds	U	6	18.9	3	0.6^2	2	0.4^2
Small terrestrial insectivores	U	11	25.2	3	1.5^2	3	2.3^2
Small low-level solitary insectivores	U	6	27.7	4	6.6^2	3	4.7^2
Army ant followers	U	4	14.1	4	5.4^1	1	0.4^2
Regular mixed flock insectivores	U	21	131.3	11	28.8^2	10	16.3^2
Large mid-upper level solitary insectivores	U	14	17.5	6	5.4^1	6	4.6^1
Small mid-upper level solitary insectivores	U	13	55.8	9	24.6^1	9	20.4^1
Frugivores and allies	U	19	100.8	11	58.4^1	13	64.0^1
Hummingbirds	U	7	30.2	4	20.5^1	5	19.6^1
Small gap insectivores	U	8	45.5	8	40.9	8	29.4^1
Low second growth specialists	U	12	20.1	12	36.0^1	11	43.4^1
Large to medium frugivores	C	25	53.0	18	44.3	19	30.4^1
Small insectivores and omnivores	C	60	160.5	51	226.7^1	44	161.9
Woodpeckers	All	14	8.8	12	10.4	10	9.8
Vine tangle insectivores	All	10	27.8	7	12.8^1	10	29.3
Edge or large gap species	All	9	3.9	10	3.3	20	17.3^2
TOTAL		239	741.1	172	526.2^2	174	454.2^2

U = Mature forest understory; C = Mature forest upper canopy; All = species present in all forest strata. Species seen outside sample plots in a given forest are included in the species total, but do not appear in the abundance index (number of birds/100 one-quarter ha plot).
Significance values are: [1] = $P < 0.05$, [2] = $P < 0.01$.

Logging vs. Natural Disturbance

Does logging imitate nature? Natural disturbances (such as tree-falls) create forest heterogeneity and support species richness (Schemske and Brokaw 1981; Denslow 1987; Wunderle et al. 1987; Feinsinger et al. 1988). Because tropical forests are dynamic, many bird species have evolved to depend on these natural disturbances. In the rain forest of French Guiana, for example, over 40 percent of all species are associated with gaps, edges, upper canopy emergent trees, or natural openings (Thiollay 1990, 1994). Many of them are rare, local, or even absent from closed, mature, undisturbed forest. These species often increase—and in some instances become dominant—in secondary, fragmented, or disturbed forests. Because logging also creates gaps and habitat heterogeneity (mimicking natural disturbances to some extent), logging can favor these disturbance-adapted species.

Logging is additive to natural disturbances, however, and this is one reason why most logging operations result in the local loss of biodiversity (e.g., Johns 1991a; Thiollay 1992; Mason 1996). Highly selective logging operations may harvest far less than one tree per hectare, and the impact on some wildlife taxa can be negligible (see below under extraction rates). Most logging operations, however, are more intensive. While logging also creates treefall gaps, the overall density of logging gaps is higher and the average size larger than natural gaps (Riera and Alexandre 1988; Thiollay 1990; see chapter 2). The removal of these trees opens the canopy, allowing more light to reach the forest floor. The understory becomes hotter, drier, and eventually much denser—choked with plants responding to the increased light. The result is that most logging operations disrupt forest and bird community structures, with resulting decreases in bird diversity (Johns 1991a; Thiollay 1992; Mason 1996).

Logging in French Guiana illustrates these patterns. Among the 256 bird species recorded in all the sample plots used for a study in French Guiana, overall bird species richness and abundance were 27 to 34 percent lower in areas where an average of three trees/ha were logged, as compared to primary forest (Thiollay 1992). Only 15 percent of the species significantly increased in abundance following logging. Although a quarter of all species recorded in primary forest were associated with various types of natural gaps (Thiollay 1994), only 23 percent of these increased significantly after logging while many others decreased sharply (Thiollay 1992). This emphasizes the subtle requirements of gap species and the structural or functional differences between natural and artificial gaps.

Effects of Different Extraction Rates

Understory birds are among the most vulnerable to the effects of logging. The greater the extraction rate, the more the forest structure is changed, and the greater the impact of logging affects these birds (Johns 1991a; Mason 1996). While this response is not entirely surprising, it does provide a useful framework for evaluating and predicting the impacts of different forest management practices.

At one end of the disturbance spectrum are highly selective logging operations that may have little or no effect on understory species. Logging of mahogany (*Swietenia macrophylla*) tends to be very selective, as mature individuals occur at low densities. Loggers may harvest as few as one tree per 20 ha, leaving most of the forest essentially undisturbed (A. Whitman, personal communication). In Belize, where forests have a history of natural and anthropogenic disturbance, extraction of one mahogany tree per 4.5 ha had little immediate effect on bird communities (Whitman et. al. 1998). Highly selective logging of some Venezuelan forests (where extraction rates were less than one tree per hectare) also had negligible effects on understory birds (Mason 1995) (see photo 8-2). Selective logging may affect birds as time passes because many of the changes in forest structure that affect birds occur over time (discussed below; see chapter 2). Given enough time, however, forests can recover if the disturbances are not severe.

Logging intensity increases if operations harvest more abundant species or take a greater diversity of species. In parts of South America, these operations extract from two to 10 trees per hectare, roughly an order of magnitude less than is typical for Asian dipterocarp forests. Nonetheless, entire bird communities are strongly affected and most primary forest species decline (Johns 1991a; Thiollay 1992; Mason 1996). If extraction rates climb higher, the impacts of birds are likely to be greater unless efforts are made to reduce unnecessary collateral damage to the forest (see chapter 21).

Natural Succession Following Logging

The impacts of logging vary over time. Although the microclimate changes immediately following logging, many other effects develop

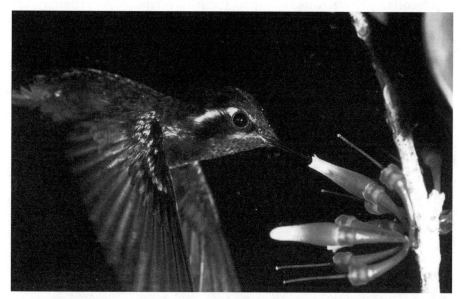

PHOTO 8-2 Logging has relatively little impact on hummingbirds. Some species become more common in clearings created by logging (e.g., purple-throated mountaingem, *Lampornis calolaema*), probably because increased light levels increase the abundance of understory flowers. (D. J. Mason)

long after the actual harvesting. During the initial years following logging, the structure of logged forests increasingly diverge from that of unlogged forests. Vegetation responds to the increased light levels in the understory. Although logged forests may eventually regain their original structure, succession initially creates structures that are very different as the canopy becomes lower and more open and the understory becomes choked with dense vegetation.

Four months after removing approximately three trees per ha in a Venezuelan forest, for example, the understory forest structure and its avifauna were similar to those of a primary forest, even though the canopy was much more open (Mason 1996). Vegetation responses to changes in light levels, temperature, humidity, compacted soil, and seed rain were not yet fully manifested. In sites logged five to six years earlier, the understory vegetation had become twice as dense as that of a primary forest, and the composition of the bird assemblages were significantly different (Mason 1996). A similar pattern was observed in Belize, where the immediate impacts on birds of logging cessation were minor.

A similar pattern was found in French Guiana, where understory bird communities were more impoverished 10 to 12 years after logging than one to two years following harvesting (Thiollay 1992). The early successional logged forest had a highly heterogeneous spatial structure, but when the gaps were closing, the older stages became increasingly dense and homogeneous. Regenerating stands had a vertical structure inverse of that of the primary forest (dense undergrowth and open canopy vs. open understory and closed canopy). For birds, changes in forest structure may have numerous effects on food resources, vulnerability to predators, patterns of foraging movements, and microclimatic conditions. Detailed information on species-specific responses, however, is rarely available.

Birds appear to respond to forest structure (or its correlates) (Mason 1996), but it may take many years for elements of the original forest structure to return. This has at least two implications for bird conservation. First, any efforts to reduce incidental damage to the forest will likely result in the conservation of more avian diversity. Second, the forest should be allowed sufficient time to recover between extractions. The longer the rotation between extractions for a given logging intensity, the greater the number of bird species that would likely be conserved.

Edge Effects and Fragmentation

In addition to changing its structure, logging fragments the forest into smaller blocks (see chapter 6). This has two important effects. First, it creates more forest/non-forest edge, which is ecologically different from *interior* forest far from these artificial boundaries. Second, the remaining forest blocks become more isolated.

Both of these changes affect birds, particularly those adapted to the forest interior. Birds of large, undisturbed forest blocks are far more sensitive to logging than are species associated with edges and large treefall gaps. In French Guiana, Thiollay (1992) defined interior species as those whose optimal abundance is reached in undisturbed forest understory less than 100 m from any edge. In this study area, 76 (60 percent) of these species declined following logging, while not a single interior species increased. In Venezuela, Mason defined interior species as those species found less than 1 km from the nearest anthropogenic edge (Mason 1995). He found that interior species, especially those that were naturally rare, were also sensitive to edge effects. The more rare the forest interior species was, the more its abundance dropped near a road edge. Even a relatively narrow (10 m wide) road can cause forest fragmentation and edge effects—at least for sensitive understory species.

Through the construction of extensive road networks, log yards, and skidder trails, logging creates edge habitat that separates formerly continuous forests with barriers that range from several to hundreds of meters or farther. The smallest gaps result from the construction of skidder trails, which are used to haul logs from the point they are felled to yards where they are sorted and loaded onto trucks. Because skidder trails are rarely planned (D. Mason, personal observation), operators wander through the forest in search of logs, creating a destructive maze of understory paths that significantly add to the over-all impact of log extraction.

The ecological impacts of roads are far greater than the relatively small area that they occupy. While canopy bird species are apparently little affected, many understory birds avoid road edges, resulting in roadside bird assemblages that are significantly different from those of the forest interior (Mason 1995). Birds may avoid roads because of their noise and human traffic, elevated risk of predation (from other fauna, and in the case of game species, from humans), or perhaps for physiological reasons—because forest edges can be hotter than the forest interior (Kapos 1989). Where forests along road edges are denser than the forest interior, specialized forest interior species may also forage less effectively or avoid them altogether.

Larger gaps can serve as total barriers to some species that physiologically should be capable of crossing such openings. Dramatic evidence of this comes from Barro Colorado Island, which was created when the Panama Canal filled with water. Over the past decades, several species were extirpated from apparently suitable habitat on the island, even though they remain common in larger blocks of similar habitat on the mainland 500 meters away (Willis 1974; Karr 1982). A similar pattern has occurred in experimentally isolated forest fragments near Manaus, Brazil. Sensitive bird species disappeared from small forest islands after newly created cattle pastures isolated them from large blocks of forest (Stouffer and Bierregaard 1995a, b). While the effects of fragmentation caused by logging have not been well studied, log yards coupled with roads and harvested areas can occasionally create gaps hundreds of meters wide. Efforts to reduce the impacts of logging should therefore consider steps to minimize forest fragmentation.

Effects of Silvicultural Treatments Other Than Logging

Where forest management is approached as a long-term enterprise rather than a simple one-time extraction, silvicultural treatments may be applied to increase the volume of commercially valuable wood. These postlogging treatments vary in their direct effects on the forest and its bird communities.

One of the least intrusive silvicultural treatments is cutting vines before and/or after logging (see chapter 2). Vines may be cut before logging to free intertwined tree crowns, or after logging to reduce competition with regenerating trees for light. By freeing intertwined trees and reducing logging damage, cutting vines may benefit birds. In Venezuela, cutting vines after logging did not significantly alter the forest structure or affect understory birds (Mason 1996). In Bolivia, the bird abundance was higher where vines were cut before logging, even though the number of bird species did not change (L. Zweede, unpublished data). Apparently, insectivorous birds prefer areas with dead vine tangles, which may harbor insects. Cutting vines may affect some canopy frugivores, however, as vines can be important fruit producers.

Other silvicultural treatments have much greater effects. In some logged-over forests in Venezuela, 3 x 1000 m *enrichment strips* are cleared and planted with nursery-grown saplings of commercial (although not necessarily native) tree species. Trees felled to create these strips are pushed into the adjacent forest, damaging other trees and widening the strips. As the parallel strips are placed 50 m apart and tend to widen as trees fall, very little of the original forest remains unaffected by this procedure. The understory changes to become nearly impenetrable. As the forest structure and microclimate are dramatically altered, so are the bird communities. Since changes in bird assemblages are proportional to the magnitude of forestry disturbances, enrichment strips reduce the populations of more primary forest birds than selective logging alone (Mason 1996).

Very little is known of the ecological effects of some other common silvicultural practices, such as liberation thinning, which is used to promote the growth rates of favored trees by poison-girdling their competitors (see chapter 2). Thinning reduces the structural and botanical diversity of a forest (Hendrison 1990), which is also likely to reduce bird diversity.

Direct Effects of Timber Management: Vulnerable Species

Forest Strata and the Vulnerability of Bird Species

The responses of birds to the direct impact of logging depend, to a large degree, on where they occur in the forest.

Canopy birds

Canopy birds descend into edges or gaps where they find conditions similar to those of the upper canopy. Compared to understory birds, they are less susceptible to the effects of forest fragmentation because they are more likely to cross nonforest gaps to forage opportunistically over large areas (Willis 1974; Lovejoy et al. 1986; Poulsen 1994). Macaws forage in the tops of canopy trees, for example, and regularly traverse deforested areas. They may be able to survive in fragmented forests, as long as food is available, adequate nesting sites remain, and they are protected from hunting.

In French Guiana, the overall richness of canopy species decreased slightly in logged forest—some species increased in abundance while others decreased (Thiollay 1992). Collectively, canopy birds accounted for 109 species, or 43 percent of all species recorded. The apparent increase in abundance of some species following logging may partly be because they became more visible (and hence more easily censused) from the forest floor. Others, such as medium and large frugivores, may have decreased in abundance because of hunting pressure. Neither assemblages of woodpeckers nor insectivores associated with vine tangles changed notably after logging.

Understory birds

Birds of lower forest strata are sensitive to timber management (see photo 8-3). Understory birds need relatively dark, humid undergrowth and many avoid canopy openings. In French Guiana, a total of 108 species were primarily associated with the understory of mature undisturbed forest (Thiollay 1992). Of these, 92 belonged to guilds or species groups (see box 8-1) that decreased significantly in logged forest. Twenty-four of these species were not recorded at all in forests one and ten years after logging—a nine percent net loss of total bird diversity. Selective logging also significantly affected understory birds in

PHOTO 8-3 Understory insectivores, such as this white-plumed antbird (*Pithys albifrons*), make up much of the avian diversity in lowland tropical forests. As a group, they are sensitive to forestry disturbances. (D. J. Mason)

Venezuela (Mason 1996). Many of these understory species are finely adapted to the continuously connected forest understory and will not cross nonforested areas (Karr 1982).

Terrestrial species are the understory groups most affected by logging. In French Guiana, they decreased in abundance by 91 to 98 percent, and at least 11 of 17 species disappeared (Thiollay 1992). In Venezuela, terrestrial species that both avoid edges and appear reluctant to cross them were also among the most vulnerable to fragmentation by roads (Mason 1995). The assemblage of terrestrial species declined significantly near edges, where eight of nine species were less abundant. Some terrestrial species are large and subject to hunting, which may explain their decrease in abundance.

Foraging Guilds and Sensitivity to Logging

Diet plays an important role in determining a species' response to timber management. As a group, nectarivores, frugivores, and carnivores are least affected by logging, whereas insectivores are most sensitive to forestry disturbances. These categories are useful simplifications, as all nectarivores and most frugivores also consume insects.

Nectarivores

The effect of selective logging on nectarivores is generally neutral to positive. In French Guiana, the impacts of logging had a minor impact on nectarivores, which decreased slightly (but insignificantly) both in number of species and abundance (Thiollay 1992). In Venezuela, nectarivores generally increased in abundance and were the feeding guild most favored by logging in the short-term (Mason 1996). In lowland Bolivian forests, nectarivores were more active in multiple tree gaps (of 400–1,000 m^2) than they were in either single tree gaps (< 400 m^2) or closed-canopy forests, even though variations in abundance were not significant (Fredericksen et. al. 1999a). The most likely explanation for the success of hummingbirds in clearings and logged forest is the abundance of flowers that result when light floods the forest understory. In Bolivia, the red- or violet-flowered plants of the genus (*Ruellia sp.*, Acanthaceae)—a favorite of hummingbirds—become more abundant following the creation of large gaps (Fredericksen et. al. 1999a). Research from Malaysia suggests that this result may be ephemeral, though. Wong (1986) found that the abundance of flowers visited by birds in second growth (23 to 25 years after logging) was considerably lower than in primary forest.

Frugivores

Frugivorous birds appear to be more resilient than insectivores to the effects of logging in the Neotropics (Johns 1985, 1988; Thiollay 1992; Mason 1996), a pattern also found in Asia (Lambert 1992). Frugivores tend to be generalists, taking a variety of fruit species. They are also adapted to foraging on a resource that is patchy and widely distributed, requiring foraging over large areas. This may explain their ability to exploit forest mosaics created by logging, provided that fruit is still available (Johns 1987). Many understory frugivores may also survive because second growth forest is often rich in small fruits, at least initially (Wong 1985). A reduction in large, fruit-producing tree species, however, may affect some specialist canopy frugivores.

Carnivores

The effect of logging on carnivores is mixed. Johns (1991a) reported that raptors were not seriously affected by logging. He concluded that timber management may actually benefit some species, as some prey are more visible and more easily captured in the open conditions created by roads and log yards. Thiollay (1992) also found that overall raptor richness tended to increase in secondary and fragmented forests. Secondary forests were colonized by species associated with forest gaps, edges, and large clearings, while primary forest species persisted in the large blocks of relatively undisturbed forest that remained (see box 8-2). In Thiollay's study, however, the overall abundance of raptors did change, decreasing by 17 to 27 percent. Primary forest species were the most affected, dropping in abundance as a group by 41 percent.

Box 8-2 *The effect of logging on raptors in French Guiana.*

French Guiana has a rich assemblage of 48 Falconiform species. Many species have a narrow habitat selection and are sensitive to human disturbance—serving as good indicators of habitat change (Thiollay 1985a, b). To assess the impacts of logging on raptors, four habitat types were censused:

— Type a) undisturbed primary forest

— Type b) large natural forest openings in primary forest (multiple tree fall gaps and large rocky outcrops with clumps of trees)

— Type c) selectively logged forest in which 1 to 3 trees/hectare were removed <15 years ago, with little or no disturbance afterward except hunting

— Type d) forest logged >20 years ago and subsequently fragmented by clearings, new pastures, or tree plantations and young regrowth

The number of 2x2 km quadrants (400 ha) surveyed in each habitat type (a–d noted above) were: a = 10, b = 4, c = 10, and d = 8 (see methods in Jullien and Thiollay 1996).

The 33 species of raptors recorded in the study areas were divided into five categories according to their distribution along the gradient, from undisturbed forests to logged and secondary forests (see box table 8-2). Of the 33 raptor species, all but the last eight in the table occurred in primary forest. Overall raptor richness increased in secondary and fragmented forests with the addition of some species associated with forest gaps, edges, or large clearings. Primary forest species persisted in habitat-type *d* (see above), in which old secondary forest still covered more than 66 percent of the area. The mean number of species recorded per 400-ha sample plot, however, was the same in primary forests (types a and b) and logged forests (types c and d).

The overall abundance of raptors did change, however—decreasing by 17 to 27 percent. More importantly, over half of the 25 primary forest species declined in logged and secondary forests, and the cumulative abundance of all the forest species dropped by 41 percent between natural and disturbed forests. While overall hawk diversity remained high in

BOX TABLE 8-2 *Habitat Distribution of Thirty-three Raptor Species in French Guiana*

Group	Species (% total)	Explanation
Absent from logged forest	2 (6%)	Rare species whose absence may be a sampling bias.
Present but less abundant in logged forest	12 (36%)	
Equally abundant in primary and logged forest	11 (33%)	Aerial hunters, or species associated with dense understories, gaps, or edges.
Only occur in old or open secondary forest	6 (18%)	Species of natural riparian, coastal, swamp or gallery forests and edges; rarely occur in newly logged forest.
Present only over large gaps in formerly logged forests	2 (6%)	Nonforest vultures that naturally occur in savannas and mangroves
Total	33	

logged forests, the abundance of hawks declined (Thiollay 1997; and in press). The species most vulnerable to logging fell into two groups:

⌐ Primary forest eagles, which may be more susceptible to hunting pressure than logging, *per se*

⌐ Species associated with the understory of tall, undisturbed, closed-canopy forests, such as large species of the genus *Accipiter*, specialized hawks of the genus *Leucopternis*, and the group-living red-throated caracara (*Daptrius americanus*)

Insectivores

Insectivores are the group most negatively affected by logging (Johns 1991a; Thiollay 1992; Mason 1996). This finding is important, as insectivores make up the majority of understory species in Neotropical forests and collectively decline dramatically following logging. The most affected groups include to mixed species flock members (primarily antbirds, Formicariidae), terrestrial insectivores (leaftossers [*Sclerurus* spp.], antpipits [*Corythopis* spp.], and antthrushes [*Formicarius* spp.]) and solitary sallying insectivores (Tyrannidae).

French Guianan insectivores illustrate this pattern. The species most vulnerable to logging were small solitary species (e.g., leaftossers, antthrushes) and army ant followers (*Pithys* spp. and *Gymnopithys* spp.)

that forage near to the ground. Their species richness decreased by 33 to 75 percent and their abundances by 72 to 93 percent (Thiollay 1992). The next most affected group included mixed flocks of insectivores, which included some of the most abundant species of the primary forest (e.g., foliage gleaners [*Automulus* spp.], woodcreepers [*Xyphorhynchus* spp.], and antshrikes [*Thamnomanes* spp.]), and the large solitary insectivores that also forage at mid- and upper-levels of the understory. In logged forests, the group of species that forage in mixed flocks decreased both in species richness (by 48 to 57 percent) and abundance (by 69 to 88 percent). Small bark gleaners (e.g., *Glyphorhynchus* spp.), slow moving foliage gleaners (e.g., antbirds [*Cercomacra* spp.), flycatchers (e.g., *Mionectes* spp., *Rhynchocyclus* spp.), and some ground foragers also significantly decreased, but to a lesser extent. Their overall species richness declined by 31 to 42 percent and their abundance by 37 to 63 percent. In Bolivia, Fredericksen and Fredericksen (1998) concluded that bark-gleaning birds (Pididae and Dendrocalaptidae) avoided tree species that are commonly harvested for their timber, suggesting that selective logging was therefore unlikely to significantly affect these bird species.

Ecological Characteristics of Vulnerable Bird Species

The sensitivity of birds to timber management varies along four gradients. These gradients reflect the ecological characteristics of bird species that appear to render them either resilient or sensitive to the effects of timber management.

- Vertical habitat use: Terrestrial and understory species tend to be the most sensitive to logging, while canopy species are usually more resilient
- Horizontal habitat use: Birds of the forest interior tend to be more sensitive to forestry disturbances than those that occur near anthropogenic forest edges
- Diet: Insectivores are the most sensitive, while nectarivores and frugivores tend to be more resilient, and
- Body size: Larger-bodied species tend to be more vulnerable to logging disturbance than smaller-bodied birds

The most seriously affected birds are primary forest interior taxa. Terrestrial insectivores and low- to mid-understory flycatchers are consistently the most vulnerable species. Members of mixed-species flocks, which are among the most abundant and widespread birds of the primary forest, are also some of the most affected. The frequent gaps or patches of dense regrowth may disrupt their traveling or foraging behavior through the understory, if not their ability to maintain flock bonds. Alternatively, understory insectivores may experience a drop in insect prey abundance as the understory becomes hotter and drier. Whatever the cause, the guilds that are the most vulnerable to logging are also among the most affected by forest fragmentation due to agricultural clearing in Amazonia (Stouffer and Bierregaard 1995b), Africa (see chapter 10), Asia (Johns 1986c, 1989a; Grieser Johns 1996; see chapter 9), and Papua New Guinea (Driscoll 1985).

Edge and upper-canopy species are almost always unaffected by logging, except some large frugivores which are probably dependent on a vulnerable set of fruit resources, lack large holes in big trees as nest sites, or are hunted to extinction. Nectarivores (hummingbirds) also persisted both in logged forest and forest fragments (Stouffer and Bierregaard 1995a).

Indirect Effects of Timber Management

The indirect effects of timber management are caused by people following logging roads into previously inaccessible forest. Two examples of how these actions impact forest bird communities include forest conversion to agriculture and hunting, with the latter being capable of rapidly eliminating sensitive species. Where little control exits over forest management, these indirect effects often have the greatest impacts on the persistence of the forest and its fauna (see chapters 4–7, 9–17).

Accelerated Colonization and Deforestation
of Previously Inaccessible Forests

In most of Latin America, the remaining forests are protected largely by their inaccessibility. With ambitious national plans to exploit natural resources, including timber, this is rapidly changing. Logging operations finance the construction of extensive road networks, but road-builders rarely control access to the forest. Spontaneous human colonization and deforestation are frequently the result (Sader and Joyce 1988).

The conversion of forests to agriculture has a far greater effect on tropical forest birds than does selective logging. In Brazil, Johns (1991a) compared birds in unlogged forests, selectively logged forests, regenerating scrub, and cultivated areas. Croplands supported the lowest overall bird diversity and the fewest number of forest species, while logged forests supported more species but less than that of unlogged forests. Lovejoy (1974) observed the same pattern in Peru and Panama. No primary forest species were found in scrub regenerating on cleared land, while selectively logged forests maintained many forest species. These findings support a more general rule: the more disturbed the habitat, the fewer species it holds in common with the original forest community. Where forests are converted to permanent agriculture, little or no forest bird diversity is maintained.

Increased Susceptibility of Forests to Fires

While undisturbed humid tropical forests rarely burn, the understory of logged forest may be much more fire prone. This transformation occurs as the canopy is opened and more sunlight reaches the lower forest strata, drying the slash left behind after logging (Kauffman et al. 1988; Uhl et al. 1988b; Uhl and Kauffman 1990). As the agricultural frontier advances, the risk increases that fires will enter areas of timber production, as fire is commonly used to clear forested land and to control weeds (Uhl and Buschbacher 1985; Kauffman and Uhl 1990). Fire can affect birds by damaging the forest structure, reducing the litter fauna foods, and destroying nests.

Hunting

Where logging is very selective and colonization is minimal, hunting is the greatest threat to some species. Logging brings increasing numbers of people, often with firearms, into previously inaccessible forests. Even where governments or logging operators formally prohibit hunting, they rarely control it in practice (see chapters 15–17 and 29).

Hunting targets certain classes of birds. In French Guiana, the largest species were more likely to disappear than were the smallest species from logged areas (Thiollay 1992). Because all of the larger species were hunted, hunting was the most likely cause of their decline. Throughout Latin America, the most affected species include the large, turkeylike birds of the Cracidae family. Many more species can be affected, however. In French Guiana, logging roads increase hunting pressure on all birds down to the size of pigeons or tucanets. Moderate hunting pressure can dramatically reduce populations of game birds, affect raptors, and even whole bird communities (Thiollay 1985b, 1986). In some areas, logging roads facilitate the capture of species valued in the pet trade (D. J. Mason, personal observation). Controlling hunting and the pet trade are, therefore, two important steps in minimizing the impacts of timber production on wildlife.

Potential Long-Term Effects of Logging

The studies on logging and birds in Latin America have focused on the short- to medium-term impacts of timber harvesting. Nevertheless, they suggest some potential long-term implications, both negative and positive, of this extraction.

Most species are able to persist in Latin American or Asian forests following logging, at least at low to moderate extraction intensities (Lambert 1992; Grieser Johns 1996; see chapter 9). Most primary forest species, however, are less abundant and many are locally extirpated (Thiollay 1992; Mason 1996).

Neither extirpated species nor species that persist in greatly reduced numbers will be able to fulfill their ecological roles. Predicting the long-term impacts of these losses is difficult, as forest ecosystems are likely to have a fair amount of redundancy (Mooney et. al. 1996; see chapter 3). Some species or groups of species play particularly important roles, though, and their loss may have great consequences for forest ecosystems (see photo 8-4). Species of special concern include those critical to the reproduction of plants, such as seed dispersers of canopy trees and understory plants (see chapter 3). The majority of tropical tree species have seeds that depend on animals for dispersal (Gentry 1982)—primarily by birds and bats. Seeds under parent trees suffer from high mortality (from pathogens and seed predators) and face stiff competition (see chapter 6). Dispersal enables plants to partially escape these constraints and to colonize the specific sites—such as tree fall gaps—needed for germination (Clark and Clark 1984; Murray 1988; Howe 1993; Schupp 1993). *Virola* fruits, for example, are designed to attract toucans and discourage other frugivores (Howe 1982). Toucans consume the fruit and carry the seeds, depositing them far enough away to increase (by 44 times) their chances of germinating

PHOTO 8-4 Seed dispersers, such as this Cuvier's Toucan (*Ramphastos cuvieri*), play key roles in dispersing the seeds of canopy trees. Seeds that fall directly under their parent trees often suffer from high mortality from pathogens and seed predators, and seedlings face stiff competition with each other. Dispersal by birds helps plants avoid some of these constraints while also colonizing particular sites, such as tree fall gaps, needed for germination. (D. J. Mason)

and becoming established—relative to seeds that fall under the parent plant (Howe et al. 1985).

Since an array of frugivorous birds usually consume the fruit of a given tree (Herrera 1984; Moermond and Denslow 1985; Wheelwright 1985; Fleming 1986), reductions in a single bird species may not dramatically affect the tree's survival. Reductions in the entire suite of seed dispersers may have a much more profound impact. In French Guiana, undispersed seeds accumulate below fruiting trees in hunted forests, whereas few such accumulations occur in forests that are not hunted (J. Thiollay, personal observation). Seeds that fall under their parent tree suffer high mortalities. This may at least partially explain why in French Guiana there have been marked reductions in some animal-dispersed trees (such as *Aniba rosaeodora*, Lauraceae) in heavily hunted forests of the more populated coastal zone—according to foresters (F. Gazel, personal communication).

Unfortunately, our knowledge of tropical forest systems is so rudi-mentary that we cannot reliably predict which processes are limiting—or which species are keystones—much less how to design management practices to conserve these species. In the absence of this information, the prudent response is to conserve as many species as is possible. In the context of forest management, this means reducing the structural impacts of logging (see chapter 21), controlling hunting (see chapters 15–17), and protecting the forest after harvesting.

There is little doubt that logged forests in Latin America, if they are allowed to recover and are not subject to intensive silvicultural treat-ments and hunting, conserve far more avian diversity than any com-peting agricultural system. Most bird species are able to persist following logging if the vegetation that remains resembles a forest. While logged forests cannot be considered alternatives to strictly pro-tected areas, they can play important complementary functions (Sayer et al. 1995). To do so requires a transition from careless log extraction to long-term forest management and stewardship.

Recommendations to Conserve Avian Diversity in Logged Forests

The greatest imperative is to protect forests following logging, allow-ing them time to regenerate. Too often, logged forests are eventually cleared because they are not protected after logging. Without this pro-tection, any additional steps to conserve biodiversity are futile. To pro-duce wood and conserve biodiversity, logged areas must be kept as part of a permanent forest estate.

If forest can be protected, then a number of actions will help to con-serve birds in areas managed for timber production.

Maintain Unlogged Patches of All Forest Types

The persistence of many bird species likely depends on patches of unlogged forests interspersed throughout logged landscapes (Thiollay 1992; Mason 1996). These patches should be sufficiently large to maintain populations of sensitive species. As the surrounding forest regrows and slowly regains its original structure, these blocks of unlogged forest could serve first as refuges, then as *sources* of understory bird species for recolonizing the surrounding *sink*. Until more information is available on the design of unlogged refugia, at least 10 percent of the commercial forest area should be protected in this manner (Fimbel et al. 1998; see chapter 23). To ensure the conservation of large game birds, hunting should not be allowed in these patches. Unfortunately, no studies have determined the optimal design and distribution of these unlogged patches, which makes this a research priority (see chapter 19).

To facilitate the movement of forest-interior species sensitive to logging, it may also be useful to connect these patches with corridors of unlogged forest. In some areas, the most practical way to do this may be to protect riparian areas (see photo 8-5).

Protect Riparian Areas

Leaving riparian forests intact can reduce the deleterious effects of selective logging on birds. In addition to maintaining productive stream ecosystems intact, the protection of unlogged swamp forests in

PHOTO 8-5 Protecting riparian corridors is an important step toward conserving avian biodiversity. This riparian forest in eastern Venezuela, if left undisturbed, helps conserve many of the bird and mammal species most sensitive to logging. (D. J. Mason)

Venezuelan riparian areas helped conserve 72 percent of the bird species sensitive to selective logging (Mason 1995).

Protected riparian forests may not only serve as habitat for sensitive birds species, they may also facilitate the movements of species that avoid logged areas (such as some understory species). A system with unlogged blocks (refugia), connected by a protected buffer of riparian forest, may form an effective combination of these two landscape elements (see Harris 1984 for a discussion of this approach in temperate forests). No studies have yet examined how wide these protected riparian corridors must be to conserve sensitive bird species. The patches Mason (1995) studied were usually 100 to 200+ m wide, suggesting that the width required to conserve birds may be far wider than the 10 to 30 meters sometimes advocated to protect streambanks (see chapter 21). The utility of riparian forests for bird conservation needs to be evaluated in other forests, however, before the generality of these findings can be determined.

Reduce Logging Damage

Careless selective logging causes far more damage to the residual forest than is necessary. Since changes in bird assemblages are proportional to the magnitude of forestry disturbances, as damage increases, avian diversity decreases. Fortunately, many opportunities exist to reduce the damage associated with forestry, including better planning of extractive operations, directional felling, more careful log extraction, and the choice of less intrusive silvicultural treatments (see chapters 21–25). Careful felling and skidding, for example, can reduce damage to understory trees by 50 percent while lowering the cost of extraction (Jonkers 1987; Wadsworth 1987; see chapter 28). Reducing this damage should contribute to the sustainability of forestry operations while simultaneously conserving more biological diversity. Policies and fiscal incentives need to be crafted to encourage forest producers to extract wood with greater care (see chapters 18, 19, 26–29).

Make Logging Rotations as Long as Possible

As postharvest forests regenerate, the original habitat structure slowly redevelops. It may be many decades, however, before the mature forest structure needed by the most specialized bird species is recovered. In general, the time required for forests to recover will depend upon the characteristics of the forest and the amount of disturbance associated with each extraction. To give the forest adequate time to recover between logging operations, unnecessary damage should be avoided and logging rotations should be as long as economically and politically possible. In some situations, this may be longer than the 20- to 40-year intervals sometimes recommended between cuts.

Control Hunting

Concessionaires should be required to ensure that their employees do not hunt or live trap for the animal trade. In addition, access to the forest should be limited once timber operations in an area have concluded. Small sections of roads accessing these areas can be bulldozed or blocked with logs to prevent motorized travel into these areas (see chapters 15–17).

Conduct Additional Research

A better understanding of the impacts of timber management on biodiversity and how to minimize them is clearly needed. It is useful to note, however, that biodiversity conservation is not a goal of most Latin American forestry programs. While efforts to elucidate the importance of biodiversity are needed, future research should also help forest managers maintain key ecological processes in managed landscapes. Forest managers are likely to be more interested in biodiversity conservation when they see its economic benefits. This has already occurred in Colombia, where foresters are promoting efforts to conserve the bats that disperse the seeds of important commercial tree species (see chapter 7). These efforts build upon what should be a common goal of both foresters and conservationists: the maintenance of the ecosystem functions that sustain healthy forests.

ACKNOWLEDGMENTS

The Wildlife Conservation Society, the Wild Wings Foundation, the Milwaukee Zoological Society, the World Nature Association, and the Vilas Foundation—with support from Industrias Tecnicas de Madera (INTECMACA), the Institute from Tropical Zoology (Central University of Venezuela), and the Venezuelan Audubon Society—funded Mason's research. Special thanks to José Ochoa, Clemencia Rodner, Gonzalo Morales, Tim Moermond, Eva Schulte, Jesus Mavarez, Laura Tate, Michelle Young, and Daniel Carillo.

Thiollay's research was part of a program on rain forest conservation in French Guiana, funded by the French ministry of Environment (SRAE department). The Office National des Forêts, in charge of all state forest management, provided information and some logistical support. Françoise Thiollay took part in field surveys and typed drafts of this study. Thiollay is grateful to them all.

Chapter 9

THE EFFECTS OF LOGGING ON BIRDS IN TROPICAL FORESTS OF INDO-AUSTRALIA

Mohamed Zakaria Bin Hussin and Charles M. Francis

The forests of tropical Asia and Australia support rich bird communities, with up to 200 or more bird species co-occurring in some areas of lowland rain forests (Wells 1988). Over the past century, many of these forests have been extensively cleared or exploited for timber. Only a relatively small proportion of the original forest in the region has been designated for protection in a pristine state (Dinerstein et al. 1995), and even less can be considered secure from disturbance. The long-term survival of most forest birds and other animals will depend largely upon their ability to persist in human-altered habitats.

Forest exploitation in the Indo-Australian region takes two forms: clear cutting and selective logging. and Where clear cutting is practiced, urban areas, agriculture, or tree plantations often replace the post-harvest site. Urban and agricultural lands are clearly of limited value for biodiversity conservation and result in the loss of virtually all birds (and other wildlife) that formerly occupied those areas. Even tree plantations, despite a superficial resemblance to forests, rarely support many bird species (see box 9-1)—although there are a few exceptions where the understory is well developed and diverse (Mitra and Sheldon 1993). In temperate forests, clear cutting is often used as a management tool, with forests allowed to regenerate after clearing. This type of forest management is rarely practiced in the Indo-Australian region, presumably because large clearcuts are unlikely to regenerate with commercially valuable trees—except for plantation or secondary forest species. Clear cutting, followed by deforestation, can also lead to fragmentation of the remaining forested habitat, which may lead to further loss of bird populations (see box 9-2).

Box 9-1 *Tree plantations and forest bird populations.*

In many parts of southeast Asia, much of the native forest has been cleared and replaced with tree plantation crops such as oil palm, rubber, or fast-growing softwoods. In large areas of Thailand and India, teak (*Tectonis grandis*) is also planted as a crop. Despite an external forest appearance, most of these plantations are of very little use for biodiversity preservation. Beehler et al. (1987) found many fewer bird species in mature teak plantations in India than nearby hill forests, despite a modest understory of shrubs and saplings. Most rubber plantations, and some comprised of other commercial species, are chemically sprayed to kill the understory vegetation. This practice eliminates almost all bird habitats and these plantations support virtually no forest species, and only a few edge species (Thiollay 1995; C. M. Francis, personal observations). Oil palm fruits are eaten by a wide variety of mammals—and a few birds such as black hornbills (*Anthracoceros malayanus*)—but palm plantations are generally not adequate for shelter or nesting sites. Among small insectivorous birds, oil palm plantations generally support only a few species that specialize in edge and secondary habitats, and virtually none of the species that normally inhabit the understory of primary forests (C. M. Francis, personal observations).

Mitra and Sheldon (1993) found that young plantations of fast-growing exotic softwoods, *Albizia falcataria* (in Sabah), supported a relatively rich community of forest birds. This richness was attributed to heavy infestations of insects, as well as a moderate understory growth. Even these plantations, however, only supported about 60 percent of primary forest bird species. Several guilds such as canopy frugivores and woodpeckers were poorly represented. The authors speculated that many species were actually dependent upon nearby primary forests for breeding sites. It was also thought likely that the plantation would lose many species as it matured and the undergrowth degenerated. Plantations of other tree species in the same area of eastern Sabah had many fewer bird species (Mitra and Sheldon 1993). Thiollay (1995) found that traditional agroforests in Sumatra were also relatively rich in bird species, although still substantially less diverse than natural forests. These areas were managed using silvicultural techniques that favored the preservation of a range of tree species. In some areas these *plantations* had been managed for 150 years and represented the only remaining lowland forests.

Overall, these studies indicate that some tree plantations—with extensive and moderately diverse tree species in the understory—may be useful for conservation of some forest bird species. The value of these habitats can be enhanced by planting mixed species of trees, leaving individual trees, or clumps of trees (especially fleshy, fruit-producing canopy species used by birds) minimizing clearance of the understory, and situating the plantation close to blocks of natural forest. Modern industrial plantations of cash crops, however, are of little value for conservation. Even the richest agroforests support smaller and less species-rich communities than forests that are allowed to regenerate naturally after selective logging.

In selective logging, the largest commercial trees are extracted. Up to 50 to 70 percent of the residual stand may be severely damaged or killed during this process (Johns 1988; Zakaria 1994). These postharvest forests tend to support rich bird communities (compared to areas of forest converted to other land uses), but the number of bird species is reduced in comparison to that of primary forests (Johns 1986c; Driscoll and Kikkawa 1989; Lambert 1992; Zakaria 1994). There is a need to identify exactly which species or groups of species are affected by logging, and whether logging practices can be modified to enhance populations of negatively-impacted species.

Box 9-2 *Deforestation, fragmentation, and the loss of forest bird populations.*

Clear cutting of forests, and replacement by agriculture or other land uses, reduces the total amount of habitat available for forest-dwelling species. Most resident bird populations are habitat limited, with deforestation leading to a proportionate reduction in bird populations. Forest clearing can also lead to the fragmentation and isolation of remaining forest patches (Webb et al. 1985). Small fragments may have insufficient area to support viable populations of some bird species, especially semi-nomadic species and large raptors (Thiollay 1989). Isolation also reduces the potential for recolonization, if species disappear due to stochastic events. Finally, small patches have an increased proportion of edge, which can lead to changes in micro-climate and/or invasion of non-forest species (both plants and animals) from adjacent habitats (Robinson et al. 1995).

Few studies have examined bird communities in southeast Asian forest isolates. Diamond et al. (1987) found that the bird community in an isolated 86 ha patch of forest in the Bogor botanical garden had lost most of the forest interior species. Those species that remained were largely those that also persisted in the surrounding countryside, which consisted largely of gardens and agricultural fields. Singapore, with very little remaining forest (the largest intact tract being 75 ha in Bukit Timah) has lost more than half of its resident forest species (Hails and Jarvis 1987). Ford and Davison (1995) found fewer bird species in isolated forest remnants of 550 to 830 ha in Selangor, peninsular Malaysia, than have been reported from similar sized contiguous areas of lowland forest elsewhere in Malaysia (Wells 1988). Although some species may have been overlooked in the isolated remnants because of reduced sampling efforts, the paucity of certain large conspicuous species (such as pheasants and hornbills) probably reflected species loss. Cooper and Francis (1998) set out artificial nests (using domestic quail eggs) on the ground around the Pasoh Forest Reserve in Malaysia, and found higher rates of egg predation near the forest edge than in the interior (suggesting one reason why forest fragments tend to support impoverished bird communities). These limited studies indicated that forest isolates, even as large as 500 to 1000 ha, are unlikely to support intact bird communities. This effect may be mitigated by ensuring that protected areas such as parks or reserves are integrated within a network of production forests to minimize fragmentation and maximize connectivity (see chapter 23). Thiollay (1989) suggested that areas of 1,000,000 ha or larger might be required to support viable populations of all bird species in South America over the long term. Because few protected areas in southeast Asia even approach this size, this clearly highlights the importance of maintaining the permanent forest estate, which supports a high proportion of wildlife species—as well as protected areas for wildlife conservation. A useful goal might be to encourage a mosaic of interconnected forest seral stages interspersed with a few protected areas.

In this chapter, we review existing data on the effects of forestry practices—primarily selective logging—on bird communities in the Indo-Australian region. We examine the effects of these activities on different guilds of birds, particularly those in the lowland rain forest communities of Malaysia. We conclude with research and management recommendations to reduce the impact of logging on forest birds.

Bird Community Responses to Logging

In the Indo-Australian region, tropical rain forests occur in western India, large areas of Burma, Thailand, Indochina, Malaysia, Indonesia, the Philippines, New Guinea, and northeastern Australia (Whitmore 1984). These forests are structurally diverse, with many distinct forest types—including lowland evergreen forests, montane forests (generally over 1200 m altitude), heath forests (stunted forests on sandy, acidic soils), peat swamp forests, and mangroves, as well as deciduous and semi-deciduous monsoon forests (Whitmore 1984). In general, the richest and most diverse bird and plant communities are found in the lowland rain forests, but each forest type supports a distinctive bird community with many species of birds restricted to particular forest types.

All forest types, in all countries in the region, are subject to some level of exploitation. Despite the diversity of habitat types and bird communities, the effects of forest exploitation on birds have been studied in relatively few areas or communities.

India and Indochina Region

In India, few studies have examined the impact of logging or other forestry practices on forest bird communities, perhaps because little or no undisturbed forest remains. Beehler et al. (1987) compared hill forests, a remnant ravine forest, and plantations of coffee and teak in the Eastern Ghats of India. The teak plantation was particularly poor in both bird species and individuals, due in part to the poorly developed understory in the relatively mature plantation. Daniels (1996) summarized research on the avifauna of the Western Ghats and noted that relatively few species have disappeared over the past century, despite considerable human interference. He suggested that most species in the region have become opportunistic in habitat selection due to the long history of disturbance in the region (ca. 1,000 years). Because little or no undisturbed forest remains, intolerant species may have already disappeared from the region. This implies that most remaining species would be tolerant of at least some level of disturbance associated with logging.

In Thailand, although much is known of the avifauna and its distribution (Lekagul and Round 1986), we have not found any published studies that explicitly examine the effects of selective logging on forest birds in that country. Virtually all the remaining natural forests are now in protected areas, and commercial logging of natural forests has now been stopped. Logging persists in many parts of Indochina, but this has been studied very little, particularly with respect to the impact on bird populations.

Malaysia and Indonesia

In contrast, several intensive studies have been carried out in peninsular Malaysia and Borneo on the effects of logging on forest bird communities. Wong (1985) used mist-nets and observations to compare bird communities at two sites in the Pasoh Forest Reserve—one that had never been logged, and another that had been selectively logged 25 years earlier in 1955 (logging intensity not reported). She found that the bird communities overlapped heavily, but that the logged forest site had lower species richness and fewer individuals. Johns (1986c, 1989a, 1992b) sampled bird populations through observations along transects in logged-over sites around Sungei Tekam in peninsular Malaysia. Sites logged between 1975 and 1981 were sampled in 1979 to 1981 and 1987 (see table 9-1). In his initial surveys, Johns found that most species known from primary forest were detected in one or more of the logged-over sites. The relative densities and numbers of species at any given site in the logged forests, however, remained quite different from those in primary forest sites. After a six-year interval, the community composition had not converged substantially upon that of the primary forest—although some understory species of babblers had increased slightly in abundance, reflecting establishment of a shaded understory (see table 9-1).

Several researchers have compared bird communities in logged and primary forest sites around the Ulu Segama Forest Reserve, including the Danum Valley, in Sabah, Malaysia (Lambert 1992; Zakaria 1994, 1996; Grieser Johns 1996; Nordin and Zakaria 1997; C. M. Francis, unpublished data). Most of these studies compared bird communities in primary forests (particularly near the Danum Valley Research Cen-

TABLE 9-1 *Changes in the Bird Community (Percentage of Individuals by Guild and Total Number of Species) in Response to Selective Logging Activities Based on Three Different Intensive Studies in Lowland Dipterocarp Forest in Malaysia*

Study Site	Sungei Tekam Forest Reserve, Pen. Malaysia[c]		Ulu Segama Forest Reserve, Sabah[d]		Ulu Segama Forest Reserve, Sabah[e]	
	Primary	Logged	Primary	Logged	Primary	Logged
Logging Intensity		18 trees/ha		79 m^3/ha		90 m^3/ha
Years of Recovery		6–7		3–4		9–10
Hunting Pressure	unknown	unknown	none	none	none	none
% of Individuals by Guild[a]						
Sallying Insectivore	16.2	2.0	7.2	8.0	7.4	3.5
Terrestrial Insectivore	3.6	0.0	10.3	5.6	12.9	9.2
Foliage-gleaning Insectivore	31.1	28.5	36.2	30.6	37.4	30.9
Bark-gleaning Insectivore	5.3	3.0	3.0	2.6	4.1	1.0
Nectarivore/ Insectivore	3.9	4.5	9.7	14.6	5.2	10.4
Arboreal Frugivore	10.6	14.9	4.2	2.5	7.6	9.9
Arboreal Frugivore/ Insectivore	15.1	39.2	24.4	30.5	15.3	27.1
Terrestrial Frugivore/ Insectivore	0.6	0.0	1.3	0.8	2.7	2.3
Carnivore/ Frugivore	11.5	5.8	2.4	2.3	5.3	4.3
Carnivore	2.1	2.0	1.3	2.6	2.1	1.3
Total # of Species[b]	225	181	222	188	207	199

[a]Based upon standardized observations along transects in all studies.

[b]Includes all species recorded from each study area, combining species from all plots and all sampling methods (including netting and observations), without regard for sampling effort.

[c]SOURCE Johns (1989a)

[d]SOURCE Zakaria (1994), Nordin and Zakaria (1997)

[e]SOURCE Lambert (1992)

ter), with a variety of logged sites. Most bird species known from the primary site have been recorded at one or more of the logged sites, although the total number of species known from primary forest remains slightly higher (see table 9-1). Zakaria (1994) reported similar numbers of species present in primary and logged forests, based on standardized observations along transects. The distribution of individuals among guilds, however, was substantially altered (see table 9-1).

Lambert (1992) observed fewer species of birds along transects in the logged forest, but he caught approximately the same number of species and individuals in secondary forest as primary forest using mist-nets, despite lower netting effort in the logged forest. Grieser Johns (1996) reported higher numbers of individuals and species in secondary forests, after standardizing for effort. Like Zakaria (1994), both of these studies found substantial shifts in the relative abundance of species in logged versus primary forests (see table 9-1). In general, edge and secondary-growth species increased (many of which are in the arboreal frugivore/insectivore, or the nectarivore/insectivore guilds), while forest-dependent species decreased in logged forests.

Grieser Johns (1996) statistically tested for differences in bird abundance, and concluded that only 15 species had declined at Sungei Tekam, while 46 species had increased as a result of logging. These numbers, however, must be treated cautiously for three reasons. Only species with a minimum sample size were included, so rarer species that might be particularly sensitive to logging were not represented. Several species detected more often in logged forests were canopy species that are more conspicuous along logging roads than in the understory of primary forests. Finally, the same transects were surveyed many times—with the same birds probably counted repeatedly—suggesting that the statistics may not be valid.

Bennett and Zainuddin (1995) compared large mammal and bird densities in forests before and after logging (which removed or destroyed 55 percent of the trees >10 cm dbh), and with varying levels of hunting pressure in Sarawak. Among birds, they found much lower densities of hornbills and pigeons in logged areas with moderate to heavy hunting pressure (see chapter 16). Logging roads in these areas were instrumental in providing access to hunters. Thiollay (1995) compared community structure in three types of managed agroforests and primary forests in Sumatra. These forests were planted for cash tree crops, but were maintained with a relatively diverse mixture of other species, thus more closely resembling regenerating secondary forest than large-scale commercial plantations such as rubber or oil palm. These forests were less diverse than primary forest, but still supported moderate bird diversity.

Papua New Guinea and Australia

In Papua New Guinea, Driscoll compared bird communities in logged and unlogged plots of swamp forest, terrace forest, and hill forest (Driscoll and Kikkawa 1989). The intensity of logging—which had taken place approximately five years before—was not indicated, but damage to the forest was suggested to be greater than 40 percent of the area. In lowland, river terrace forest, canopy species were more diverse at the primary site than at the logged site. Many of the species with reduced densities in logged forest were obligate frugivores. Among understory bird species in all three habitats (most of which were insectivores), some were relatively unaffected by logging, but many more had reduced numbers in the disturbed sites. Population turnover was also higher in the logged forest, with few individuals being recaptured on subsequent visits (in comparison with the pristine forest). Driscoll and Kikkawa (1989) also reported on studies in tropical Australia. In the lowlands, disturbed forests (largely regrowth after small clearcuts) had substantially fewer forest birds than undisturbed forests. On the tablelands, selective logging (removing up to 50 percent of canopy cover) had relatively little impact on bird communities. Crome and Moore (1989) found that bowerbirds (Ptilinorhynchidae) continued to use forests in tropical Queensland after logging, although some display courts were moved from disturbed areas into adjacent, unlogged forest.

Most studies have found some changes in the species composition of bird communities after logging, with retention of many of the original forest species and reductions or losses of others. At the same time, secondary or colonizing and edge species often increased. As a result, total diversity or numbers of species were often similar. To understand which species are affected by logging, we examined the response of various feeding guilds of birds to these forestry practices. We concentrated on bird communities in Malaysia, as these have been best studied—although even there, results vary among sites and most studies had limited replication.

Changes in Abundance of Individual Species by Feeding Guild

Insectivores

Most understory birds, as well as many that feed in the canopy, are insectivorous. They can be divided into guilds based on foraging strategies (Johns 1989a). The most species-rich groups are the sallying, foliage-gleaning, terrestrial, and bark-feeding insectivores.

Johns (1989a, 1992b) found lower abundance of many sallying insectivorous species (flycatchers and monarchs) after logging in peninsular Malaysia, as did Lambert (1992) in Sabah (see table 9-1). Zakaria (1994) found a large increase in one species—the black-naped monarch (*Hypothymis azurea*)—but found declines in most others.

Among foliage-gleaning babblers, some species have been reported to increase in response to logging, such as the tit-babblers (*Macronous* spp.) (see photo 9-1). Many other species decrease, although most still persist in logged forests. The quantitative results vary among study areas. The scaly-crowned babbler (*Malacopteron cinereum*), for example, decreased strongly in abundance in some logged areas (Lambert 1992), but less in others (Johns 1989a; Zakaria 1994). Other *Malacopteron* spp. remain fairly common in logged forest (see photo 9-2), but at lower densities in some areas (Lambert 1992; Nordin and Zakaria 1997) than others (Johns 1989a). These differences are not obviously related to differences in logging practices (see table 9-1), and may be due (at least in part) to sampling variation.

Several studies have reported declines in terrestrial insectivores after logging, especially wren-babblers (genera *Napothera*, *Kenopia*, *Ptilocichla*) and pittas (*Pitta* spp.; Johns 1989a; Lambert 1992). Many of these species are relatively rare or hard to locate, even in primary forest (Zakaria 1994), making it difficult to obtain conclusive results from observational surveys. One cause of these declines may be loss of a sheltered understory in large sections of the logged forests, which could affect food supply for these birds. Burghouts et al. (1992) and Chung (1993) found that resultant changes in microclimate were associated with substantial changes in invertebrate distribution and abundance. Behavioral avoidance of these areas could also be important—a few species such as wren-babblers appear to avoid crossing sunlit patches in forest gaps (M. Zakaria, personal observation).

PHOTO 9-1 Fluffy-backed tit-babbler (*Macronous ptilosus*) is one of the few babblers that appears to increase in abundance in logged forest, probably because it likes dense vine tangles and thick herbaceous growth associated with gaps. (C. Francis)

PHOTO 9-2 Grey-chested babbler (*Malacopteron albogulare*) is an understory babbler that declines in logged forest. (C. Francis)

Johns (1989a) found that the relative abundance of understory babblers had increased six years after logging—following the restoration of a closed canopy—although numbers were still much lower than in primary forests. Wong (1985) reported lower densities and diversity of understory insectivores in a site that had been logged 25 years earlier (logging intensity not reported) compared with that of nearby primary forest. Despite the difference in birds, she found few differences in arthropod abundance between the two sites (Wong 1986).

The bark-gleaning guild includes mainly woodpeckers, which decreased in relative abundance in all logged forest studies in Malaysia (see table 9-1). Presumably, these birds are affected by loss of large trees, although a short-term benefit might be expected due to an increase in standing dead trees shortly after logging. Little is known about the relative importance of different species of live or dead trees to these woodpecker species.

Nectarivore/Insectivores

Most species of nectarivore/insectivores appear to be more abundant after logging (Johns 1989a; Lambert 1992; Zakaria 1994; Grieser Johns 1996). Because many of these primarily feed in the canopy, this abundance could be partly an artifact of the increased conspicuousness of canopy species in logged forests, where the canopy is lower and more visible. Real increases, however, might be expected due to changes in the plant community. Enhanced growth of vines and flowering herbaceous plants along edges and openings could lead to direct increases in the food supply for nectarivorous birds (Lambert 1990). This enhanced growth could also lead to increased insect densities, which most nectarivores use to supplement their diet—especially during periods when flowers are scarce. Unfortunately, owing to the difficulties of observing and surveying canopy birds in tall, unlogged forest, changes in abundance are difficult to distinguish from changes in conspicuousness.

Arboreal Frugivores

Arboreal frugivores feed only on fruit and are largely forest-dependent. This group appears to be adversely affected by logging in some areas, though not in others (see table 9-1). Some of the variation among studies may be due to bias in transect surveys because most of these species forage in the canopy and may be more conspicuous in logged areas. Zakaria and Nordin (1998) used standardized observations of individual fruiting trees to avoid this bias, and found many fewer frugivores on fruiting trees in forests that had been selectively logged (intensity 79 m^3/ha) than in primary forest. Fruit-dependent species that also use the understory, such as the green broadbill (*Calyptomena viridis*) and Asian fairy bluebird (*Irena puella*), were also markedly reduced in logged forests in the Danum Valley (Zakaria 1994). With the exception of figs (*Ficus* spp.), these frugivores generally prefer to eat large-seeded fruits. Fruit trees important to these birds—many of which are large canopy trees—tend to be much reduced in logged forests (Lambert 1991; Zakaria 1994). These birds do not eat the fruits of pioneer plants, which are usually small-seeded and abundant in logged forests. Insufficient fruit resources, therefore, may be the main reason for the reduced numbers of these species.

Arboreal Frugivore/Insectivores

Many species of arboreal frugivore/insectivores increase in abundance after logging, particularly bulbuls (Pycnonotidae) (Johns 1989a; Lambert 1992; Zakaria 1994; Grieser Johns 1996). Most of these are generalist species that feed primarily at edges or gaps. Their increased abundance is probably associated with the increased availability of the small fruits of pioneer tree species. These birds are believed to play an important role in the dispersal of pioneer trees into logged areas (Zakaria 1994, 1996). The birds then benefit from a large food supply because many pioneer tree species grow rapidly and fruit when they are only a few years old (Zakaria 1994). Another factor influencing the abundance of these generalist frugivores in logged forests may be reduced competition from primary forest frugivores (Zakaria 1996). Some changes may also be due to changes in foraging heights because many of the bulbuls typically associated with edges are also found in the canopy of primary forests. The ability of these species to switch to an insectivorous diet enhances their ability to tolerate seasonal variations in fruit abundance in logged forests.

Terrestrial Frugivore/Insectivores

This guild includes many large terrestrial birds (such as pheasants, or phasianids) that generally decrease after logging (Johns 1989a;

Lambert 1992, Bennett and Zainuddin 1995; Grieser Johns 1996). In some areas, these declines are correlated with increased hunting pressure (e.g., Great Argus, *Argusianus argus*, in Sarawak; Bennett and Zainuddin 1995), although declines have also been noted around Danum Valley where hunting pressure is minimal. These species feed on the fallen fruits of canopy tree species that may have reduced abundance in logged forest. They also feed on terrestrial insects, which may be adversely affected by changes in the leaf litter layer (Burghouts et al. 1992).

Arboreal Frugivore/Carnivores

This guild refers mainly to hornbills, which decreased in relative abundance in some studies, but showed little change in others (see table 9-1). These species feed mainly on fruits of primary forest tree species, which may be less abundant in logged forests. Hornbills also require large mature trees with cavities for nesting, which are likely to be scarce in logged forests (Johns 1987). Because these species are long-lived, they may persist in logged forest for quite a long time even if they are unable to breed there. They also forage over large areas and, hence, may be observed flying over logged forest even if they spend little time feeding there (Zakaria 1994). As a result, the adverse effects of logging on hornbills may be underestimated by short-term studies.

Carnivores

Most studies have provided relatively little data on raptorial birds because these require specialized survey techniques. Grieser Johns (1996) suggested that two species of eagles were more abundant in logged than primary forest in Sabah, but his results may be biased by the increased conspicuousness of soaring raptors from logging roads. Zakaria (1994) noted that one of these species was observed more often in logged forest, but detected equally often in primary forest (data based on calls). Thiollay (1998) made extensive observations of diurnal raptors throughout southeast Asia. He suggested that Asian falconiformes could be divided into several categories, ranging from species that are relatively intolerant of logging to those that prefer disturbed habitats (see chapter 8). In each of five study areas (including India, Vietnam, Sumatra, Sri Lanka and the Andaman Islands), he reported higher densities and higher species richness per plot in sample areas of primary or little disturbed forests in comparison to degraded forests or tree plantations. Because he did not differentiate the latter two categories, it is difficult to be sure of the effects of selective logging alone.

The Conservation Value of Logged Forests to Birds

Most forest bird species appear capable of persisting in selectively logged forests. This suggests that forests managed for timber production serve as an important habitat for many forest birds, especially in areas where little primary forest remains. There are several reasons to be cautious when drawing this conclusion, however. First, some groups of species such as understory insectivores, woodpeckers, and (in the longer term) probably hornbills, are adversely affected by logging. All studies to date have also recorded the presence of species, but have not considered their demographic viability (see box 9-3). Populations of species that are still present a few years after logging are not necessarily self-sustaining. They may consist of individuals that were present before logging, or the populations may be dependent upon recruitment from remaining areas of undisturbed forests. Logged forests may well provide adequate resources for individuals to persist, yet insufficient resources to support successful breeding.

Bird habitats in logged forest are also vulnerable to degradation over time (see chapter 2). Logged forest is a mosaic of different habitats, including remnants of the original forests, regenerating patches of secondary trees, and heavily damaged areas such as skidder trails. While the stature of secondary trees may increase over time, the remnant patches of original forest trees tend to deteriorate through delayed mortality as a result of incidental damage during logging (allowing infection by disease agents) or increased vulnerability to being blown down by severe wind storms (Laurance and Bierregaard 1997). In many areas, the planned cutting cycle is 30 to 60 years, which is much too short for these forests to recover their original structure and composition. Regeneration may also be delayed in logged forest because of reduced dispersal of tree species due to fewer frugivorous birds (see box 9-4). Repeat logging in a previously harvested forest has the potential to create a cascading habitat degradation effect, eroding the habitat quality well below the conditions following the initial logging operation.

Box 9-3 *The demography of tropical forest birds.*

The presence of a species in an area does not necessarily imply that the population is viable or self-supporting. Reduced survival rates or reduced breeding success could be masked by immigration from other areas. Brawn and Robinson (1996) showed that populations of several bird species in fragmented woodlots in the mid-western United States were not self-sustaining (*sinks*). Instead, they were dependent upon recruitment from elsewhere (*sources*)—in this case, presumably populations breeding in large contiguous forest tracts farther north. Most studies of selectively logged forest in Malaysia have taken place within several kilometers of primary forest that could serve as a potential source of recruits. Species with high survival rates could persist for many years in a habitat that is not suitable for breeding. Conversely, Karr et al. (1990) suggested that species with relatively low survival rates were more likely to be extirpated in isolated patches of forest—as demonstrated by a study in Panama.

Very little is known about survival rates, recruitment rates, or movements (especially postjuvenile dispersal) of birds in tropical southeast Asia. The most intensive study of Malaysian bird demography so far published (Fogden 1972) suggests that annual survival rates of several species' breeding adults in a 20 ha plot in primary lowland forest (in Sarawak) may be 80 to 90 percent. This study was based on only a single year of data, however, and had very small sample sizes for most species. Karr et al. (1990) suggested that many of the early estimates of tropical survival rates may be biased, and that adult survival rates were lower than this. Preliminary analyses of three years of capture-recapture data from a 110 ha plot at Kuala Lompat in peninsular Malaysia suggests that territorial adults of many understory species may have survival rates around 65 to 70 percent (Francis 1994).

Very little is known about nesting success for most tropical southeast Asia forest birds, as few studies have found many nests. Recruitment rates do appear to be low, at least for many understory species. Age ratios in mist-net captures of most forest understory insectivores in Malaysia rarely include more than 20 percent juveniles at any time of year (D. R. Wells and C. M. Francis, unpublished data).

Further research is needed to compare the demography of birds in logged and pristine forests—to determine whether the populations in logged forests are viable, or dependent, at least in part, on recruitment from populations in nearby pristine forests.

Wildfire is another threat to bird communities in logged landscapes. The risk of fire is higher in secondary forests than in primary forests, owing to drier conditions—especially in years when rainfall is reduced. Huge fires demonstrated this in Borneo in 1983 (Beaman et al. 1985), and again in several areas of Indonesia in 1997 and 1998 (Simons 1998). Forests subject to catastrophic fire result in at least temporary loss of bird habitat, but often suffer long-term habitat loss because of conversion to other land uses. Many of the recent Indonesian fires appear to have been set deliberately in an effort to expand oil palm and other plantations in Indonesia (Simons 1998).

Box 9-4 *The effects of frugivorous birds on forest regeneration.*

Many canopy and emergent trees in Malaysia, especially dipterocarps, are wind-dispersed. Other species, including many pioneer species, understory species, and some canopy species (such as figs and nut-megs), are dispersed by birds or other animals (Chapman and Chapman 1995; Pannel 1989); (see chapter 3). Zakaria (1994, 1996) compared the seed rain from birds (and possibly bats) in forests that had been logged 3 to 4 years earlier (intensity 79 m³/ha) and unlogged forests near Danum Valley in Sabah. Both study sites were located adjacent to each other. Fifty seed traps, consisting of 1 m² of cloth suspended about 0.3 m above the ground, were placed at random in primary and logged forest study sites. The traps were emptied every two weeks over an 18-month period. All materials were dried and sieved to separate the seeds. Because it was difficult to identify species, seeds were sorted by size.

More small-seeded trees (seed width < 5.0 mm) were dispersed in logged forest (t = -5.3, p < 0.001), while more large-seeded trees were dispersed in primary forest (t = 2.6, p < 0.05). This was apparently related to lower densities of some of the primary forest frugivorous birds in the logged forest (see table 9-1). Most of the arboreal frugivores in primary forests are relatively large species, with a wide gape to swallow large fruits. They rarely consume small-seeded fruits, with the exception of the *Ficus* species. Seeds collected from the seed traps were not of *Ficus*, but from pioneer species. In contrast, most of the frugivores in secondary forests are smaller species, such as bulbuls (Pycnonotidae) that feed mainly on small fruits of pioneer tree species and insects.

One implication of this study is that the rate of natural regeneration in logged forest is likely to be lower than in primary forest because fewer seeds of primary forest trees are dispersed there. This means that the logged forest is less likely to regenerate fully (reach original species composition and abundance) before the next cycle of logging occurs (which, in some areas, is planned for as short a time frame as 30 years). Such repeated logging is likely to further degrade the forest such that many primary forest bird species that can presently use logged forest may be adversely affected (Zakaria 1996).

Increased hunting, as a result of increased forest access due to logging, can also lead to degradation of forest bird communities. The birds most heavily affected by hunting are the larger species—such as hornbills, pigeons, or pheasants (Bennett and Zainuddin 1995; see chapter 16)—which are also the species that are often at lowest densities and hence most vulnerable to forest loss. Increased trapping of birds for the pet trade could also potentially reduce populations of some bird species.

Finally, competition among bird species could also lead to long-term species loss following intensive logging and forest degradation. The high diversity of birds in tropical areas has been attributed to several factors, including greater niche packing, specialization (perhaps related to high floral diversity), and dependence on constant food resources (Terborgh 1980). If degraded forest allows reduced specialization, then species will need to shift their behavior to survive. Loss of species would be expected after a time lag, as species adjusting to altered niches compete with one another for limited resources.

Directions for Future Research

The most urgent need for further research is in-depth studies of the microhabitat requirements of each bird species, especially those that appear to be most adversely affected by logging—such as ground-foraging insectivores, flycatchers, and woodpeckers. Are these species being affected by changes in microclimate due to loss of canopy, by loss of certain understory plants, or due to damage to the leaf litter layer? Is restoration of a canopy of pioneer tree species sufficient to meet their needs, or are particular plant species or structures required to support them? For woodpeckers, Malaysia supports one of the most diverse faunas in the world, with up to 14 species in mature lowland forest. How do the requirements of each species differ? Do some of them need particular tree species for nesting sites that lead to specific management considerations? This information can help with the planning of logging practices, and may provide guidelines to help restore damaged habitats. Determining these requirements will require detailed behavioral studies of these species in conjunction with studies of the habitat.

Further research on the demography of forest birds in logged forests is also required, particularly to determine whether populations are self-supporting, or whether they are sinks dependent upon immigration from remaining tracts of pristine forest. For logistical reasons, this effort should initially be concentrated on a suite of tractable species, such as small understory species that can be caught in mist-nets. Large-scale capture-recapture studies are required to estimate survival and recruitment rates, both in pristine forest and in logged areas. They need to be carried out on a large scale (>100 ha) to achieve adequate sample sizes, and with moderate effort over at least five years (and preferably more) to obtain estimates with sufficient precision (see chapter 19).

Ideally, such capture studies should also be combined with studies of the nesting ecology of each species, but nests of most tropical birds are extremely difficult to find. As a surrogate, artificial nest experiments can be used to estimate relative predation rates in different areas (Cooper and Francis 1998). Comparisons of age ratios in primary and logged forests can also be used as an index of productivity.

Such long-term demographic studies are logistically difficult and expensive, but they provide information that cannot be obtained in any other way, and that is essential for understanding the true effects of forestry practices on tropical bird populations.

Another area for further research is the requirements of specialist canopy frugivores. Although many of the generalist frugivores (such as bulbuls) appear to do well in logged forests, at least some studies have noted declines in canopy-dwelling species (Zakaria 1996). These canopy frugivores are particularly important to forests, as they help to disperse many of the large-seeded primary forest trees (see box 9-4). Reduced densities of these birds could slow or hinder the regeneration of primary forest plant species (Chapman and Chapman 1995, 1996). There is a need to determine why these bird species have declined, and whether there are management practices—such as retention (or enrichment planting) of certain keystone tree species for feeding (or nesting)—that could better protect their habitat.

Management Recommendations

Logging practices need to be designed to maximize the value of logged forests for bird populations. As noted in the previous paragraph, the precise requirements for many bird species are not yet known but some general guidelines can be given that are likely to help most species. Incidental damage to nonharvested trees should be minimized, to ensure that at least small patches of the logged forest retain an intact canopy. Such patches are likely to be of particular importance for understory species, such as flycatchers and terrestrial insectivores and frugivores. Reduced Impact Logging (RIL) is one way to achieve this (see chapter 21).

Woodpeckers require mature trees, as well as standing snags, for both foraging and nesting sites. Hornbills also require mature trees for nesting. Standing snags are likely to increase in abundance immediately after logging, but the numbers of large trees are reduced. In the long-term, particularly after a second cut, care must be made to ensure that adequate mature or *overmature* trees are retained for these species (see chapter 23).

An effort should also be made to retain as many large fruiting trees as possible—particularly species such as figs—that fruit throughout the year and may be critically important for canopy frugivores during lean periods (Leighton and Leighton 1983: Lambert 1991). In severely damaged areas, enrichment plantings of fruiting trees may be useful. Protecting adequate numbers of frugivorous birds has the added advantage that these birds can aid with forest regeneration through increased seed dispersal.

Although logged forests appear to support substantial populations of many forest dwelling birds, some species are clearly adversely affected and there are uncertainties about the viability of other species' populations. As such, a sound conservation strategy must also include a network of undamaged forest reserves in the logged landscape (see chapter 23). At least a few of these uncut parcels should be large enough to ensure that viable populations of all bird species identified as *intolerant of logging* can persist until disturbed areas can recover to a state where they provide suitable habitat for these species. Reliable estimates of the required sizes are not available. Patches of at least 10 km^2 would be required to maintain populations of many of the smaller bird species, and much larger patches would be required for the larger species such as pheasants or hornbills. Thiollay (1989) suggested that reserves of 10,000 km^2 or larger would be required to protect viable populations of South American raptors and game birds, but for many species these could likely include a mosaic of logged and unlogged forests.

Finally, there is a need to protect the logged forests from further degeneration due to fires, or to inadequately-planned, repeat logging. Repeat logging should be postponed until the forest has regained at least a substantial portion of its original structure, and should again be designed to minimize damage to trees that are not extracted. It is also necessary to protect the forest from the hunting and trapping of birds for the pet trade. Reduced access (allowing logging roads to degrade after use, or deliberately closing them), combined with strict enforcement of wildlife regulations, will assist in this respect.

All of these recommendations are also appropriate and consistent with the need to manage the forests for a sustained yield. Even degraded logged forests support many more bird species than do tree plantations (see box 9-1). As such, it should be in the best interests of both foresters and conservationists to manage the forests in such a way as to protect the diversity of both the trees and the wildlife.

ACKNOWLEDGMENTS

The Economic Planning Unit of the Malaysian Prime Minister's Department, the Sabah Foundation, the Sabah Wildlife Department, the Forest Research Institute of Malaysia (FRIM) and the National Parks and Wildlife Department (PERHILITAN) gave permission to carry out our own research on birds in Malaysia.

The Danum Valley Rainforest Research and Training Programme, the Royal Society, the Wildlife Conservation Society, Bird Studies Canada, Universiti Malaya, Universiti Pertanian Malaysia, and Universiti Kebangsaan Malaysia all provided financial or logistic support. We thank Prof. Nordin Haji Hassan, Prof. Yong Hoi Sen, and Dr. David Wells for advice, ideas, and support.

Chapter 10

BIRD COMMUNITIES IN LOGGED AND UNLOGGED AFRICAN FORESTS

Lessons from Uganda and Beyond

Andrew Plumptre, Christine Dranzoa, and Isaiah Owiunji

Tropical forests are subjected to a variety of disturbances in Africa. These include the occasional tree fall, larger-scale wind damage, disturbance from fire (Hart et al. 1997), and human-initiated disturbances such as shifting cultivation and logging. Each of these types of disturbance has different effects on the forest and the bird communities within the forest (see chapter 22).

Many species of birds respond to small changes in habitat structure and composition, therefore they serve as good indicators of changes in the environment (Furness et al. 1993). Habitat heterogeneity within forests has been shown to increase bird species richness (Freemark and Merriam 1986). In addition, because there are many species of birds in forests, it is possible to analyze changes in guilds following changes in the habitat (guilds are assemblages of species utilizing a particular resource or group of resources in a functionally similar manner; Giller 1984). Analyses of guilds allow predictions to be made about the responses to the habitat change of other guild member species that have not been studied to date.

There have been few studies of the effects of logging on birds in Africa (see figure 10-1). Four studies were specifically designed to investigate the impacts of logging on communities of birds: two in Uganda (Kibale Forest—Dranzoa 1995; Budongo Forest—Owiunji 1996), one in Madagascar (Hawkins and Wilme 1996), and one in Kenya (Arabuko-Sokoke Forest—Fanshawe 1995). The two Uganda studies are described in detail below. The study in Madagascar involved walks along 500 m transects in unlogged and logged forest, near to and far from rivers (four study sites in the tropical, dry Kirindy Forest in western Madagascar). The logging intensity was not recorded for this forest, but in western

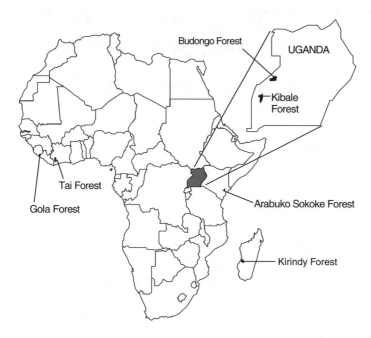

FIGURE 10-1 *A map of Africa showing the location of forests where comparisons of bird communities in logged and unlogged forest have been undertaken.*

Madagascar the intensity is generally very light—between 6 to 11 m³/ha (Cuvelier 1996). In Kenya, Fanshawe (1995) used a combination of mist netting and territory mapping of singing birds in eight 50 ha plots to study the bird communities in three forest types (*Afzelia quanzensis* forest, *Cynometra webberi* forest and *Brachystegia spiciformis* forest) within the Arabuko-Sokoke Forest. The effects of logging on each forest type were compared, but the intensity of logging was not recorded. Finally, several other studies looked at bird-logging interactions in Africa, but to a lesser extent than the studies noted above. There have been two surveys of birds in forests of West Africa—one in the Ivory Coast (Tai Forest—Gartshore et al. 1995) and one in Sierra Leone (Gola Forest—Allport et al. 1989). These surveys looked at bird species in different habitat types, including logged and unlogged forest. The intensity of logging is unknown. A few studies in east and southern Africa have investigated the effects of forestry plantations on bird communities (Allan et al. 1997) or species composition in forest fragments of different size (Newmark 1991), but these plantation studies will not be considered further as our focus is logging-bird interactions in natural forest. This chapter summarizes some of our work at two study sites in Uganda, looking at changes in bird guilds in response to logging, and then compares these findings with other logging-bird studies in Africa.

In Uganda, birds were studied in the Budongo Forest Reserve (Owiunji 1996; Owiunji and Plumptre 1998) and the Kibale National Park (Obua 1992a, 1992b; Dranzoa 1995). Hunting of birds by people is rare in Ugandan forests because only a few species (mainly guineafowl and francolins) are consumed in this region of Africa. The actual effects of logging disturbance on bird communities could, consequently, be ascertained without the confounding effects of hunting (see chapters 15–17).

Study Areas in Uganda

The two forests studied in Uganda form part of the eastern-most extension of the central Congo forest basin, which exists as a line of isolated forests between 3 to 700 km^2 wide along the western border of Uganda and the Democratic Republic of Congo (Hamilton 1982; Kingdon 1990).

Kibale and Budongo Forests

The Kibale Forest in Uganda (henceforth referred to simply as *Kibale*) is classified as a medium altitude, moist evergreen forest (Langdale-Brown et al. 1964) and covers an area of 432 km^2 with an additional 128 km^2 of grassland interspersed in and around the forest. The research reported here took place near the Makerere University Biological Field Station, which is situated at about 1,500 m elevation. Common trees included *Diospyros abyssinica, Parinari excelsa, Celtis durandii,* and *Markhamia platycalyx* (Skorupa 1988). *Parinari* forest dominated in the higher altitude areas, while *Cynometra alexandri* forest was more dominant at lower altitudes. The forest was selectively exploited for timber from 1950 to 1985 (Howard 1991), mostly in the *Parinari* dominated forest (see below).

The Budongo Forest Reserve (henceforth referred to as *Budongo*) has been the main source of high-quality timber for Uganda, and was well studied and managed on a sustainable yield basis from the mid-1920s to the early 1980s (Eggeling 1947; Philip 1965; Synnott 1985).

Budongo is classified as a medium altitude, semideciduous, moist tropical forest (Langdale-Brown et al. 1964) and lies between 1,000 to 1,200 m in altitude. Budongo contains 428 km^2 of forest, surrounded by 322 km^2 of grassland. It has areas of monodominant *Cynometra alexandri* forest, which is thought to form a climax forest type that replaces mixed-species forest (Eggeling 1947; Sheil 1996). Succession in Budongo is thought to progress from grassland, to colonizing forest (containing *Croton macrostachys*, *Maesopsis eminii*, and *Desplatsia dewevrei*), to mixed forest with mahoganies (*Celtis durandii* and *Celtis mildbraedii*), to *Cynometra* forest. Swamp forest occurs where the soil is seasonally waterlogged and contains *Raphia* spp. Mechanized logging started in the mid-1920s and was planned to allow sustainable harvesting at 60-year intervals (see next section). During the 1950s and 1960s, arboricide was applied to those trees with little economic value (particularly *Cynometra*) to open up the canopy and allow more light to germinating mahogany seedlings. Aerial photographs from the 1950s, 1960s, and 1990 have documented changes in forest composition as logging and arboricide treatments were applied (Plumptre 1996a). These changes show that the extent of the monodominant *Cynometra* forest was greatly reduced and mixed forest increased.

Logging Practices

Selective logging in the forests of Uganda has concentrated on a few species, removing those trees of commercial value that exceed 60 to 80 cm diameter at breast height (dbh). Forests were divided into 4 to 11 km^2 compartments for management purposes, with at least one compartment in each forest established as a nature reserve to preserve a representative area of the original forest. Volumes of removed timber varied between 20 to 40 m^3 per hectare for most compartments in the Budongo and Kibale forests (Skorupa 1988; Plumptre 1996a), although some felling reached 80 m^3 in Budongo (see table 10-1). This is approximately equivalent to harvesting three to seven trees per hectare. In Budongo, there was also arboricide treatment of trees.

TABLE 10-1 *The Management History of Five Bird-Logging Study Areas in the Budongo Forest Reserve and Two Study Areas in Kibale Forest Reserve, and the Methods Used to Study Birds in Each Area*

Compartment	Date Logged	Date Treated by Arboricide (m³/ha)	Volume of Timber Removed	Bird Methods
Kibale				
K30	unlogged	-	-	netting/transect counts
K15	1968–69	-	21.0	netting/transect counts
Budongo				
N15 Unlogged	unlogged	-	-	netting/point counts
K11–13[a]	unlogged	-	-	netting/point counts
N3	1947–52	1959–61	80.0	netting/point counts
W21	1963–64	1963–64	36.1	netting/point counts
B1[b]	1935	1957–58	19.9	netting/point counts
	1981–86	-	21.5	

[a]K11–13 is a 25 km² block of forest separated from the main forest block by grassland.
[b]B1 has been harvested twice.

Methods to Evaluate Bird-Logging Interactions in Uganda

Census Methods

In Kibale, two compartments were compared—one pitsawn lightly in the 1950s (K30) and one logged mechanically in 1968–69 (K15). The pitsawn compartment has been classified as an *unlogged* control site in previous studies of logging-primate interactions (Skorupa 1988), because the intensity of harvesting was similar to natural treefalls in the forest (Skorupa and Kasenene 1984). About 21 m³ (or seven stems) were recorded as being removed per hectare in K15 (see table 10-1). Measurements of the canopy, however, show that it was opened by over 60 percent (Dranzoa 1995), suggesting that measures of offtake were underestimated (C. Dranzoa, personal communication). K15 has many areas where trees have failed to regenerate and a dense matrix of understory shrubs and climbers have suppressed tree regeneration.

Mist netting (182,860 meter net hours) and transect counts (147 km of transect walked) were used to study the birds in Kibale (after Pomeroy 1992). Between October 1991 and June 1992, net lines were randomly established at least 200 m apart in both compartments K15 and K30, and understory birds were trapped each month. Nets were moved every three days to compensate for any net habituation. Transect counts—where all birds seen or heard are recorded while walking a transect at a steady speed of 1 km/hr (Pomeroy 1992)— were carried out to sample both the canopy and understory bird species. Transects measuring 3.5 km in length were walked, and all birds within a fixed width of 20 m were counted. From August 1992 to June 1993, comparisons within K15 (the logged compartment) were made between the bird community in isolated remnants of forest and the community in the surrounding matrix of climber tangles and shrubs— using both transect counts and mist netting. A study of breeding birds was made every month between November 1991 and August 1993, by searching a 20 m x 5,000 m transect for nests in each compartment.

In Budongo, five different compartments were compared—two unlogged and three logged (Plumptre and Reynolds 1994). Mist netting and point counts (Buckland et al. 1993) were used to census birds in each compartment (see photo 10-1). A total of 63,000-meter net hours and 2,200-point counts were made (Owiunji and Plumptre 1998). In each compartment, five, 2 km transects had been established in a stratified random manner (Plumptre and Reynolds 1994) for primate censuses. Points were marked at 100 m intervals along these transects. Four-point counts were made at 100 points in each compartment between June 1993 and October 1994, and October 1994 and January 1995. Mist-nets were placed at each point for one day between October 1993 and March 1994. Both mist netting and point counts sampled the compartments at different times of the year to include seasonal differences. Table 10-1 gives the management history of the compartments in both forests and summarizes the methods used to record birds.

Vegetation Structure and Composition

Measures of vegetation structure and composition were made in both forests. In Kibale, 10 x 20 m plots were established at the netting sites and all trees larger than 9.5 cm dbh were recorded. Tree height

PHOTO 10-1 A chocolate-backed kingfisher (*Halcyon badia*) caught in mist-nets in Budongo Forest. This species is found more commonly in unlogged forest, but survives in logged forest. (A. Plumptre)

and percent canopy cover were estimated (Dranzoa 1995). Ground cover was also estimated in five, 1 m^2 quadrants within the larger plots. In Budongo, permanent circular vegetation plots of 14 m radius were established at each point used for counts and mist netting. All trees larger than 10 cm dbh were recorded in the plot during 1992 and 1996. The following measures of vegetation structure were also made: canopy cover, vegetation density at ground level (Owiunji and Plumptre 1998), bole and crown height of trees in the plot, density of lianas with stems more than 1 cm in diameter, and each plot assigned a forest type (*Cynometra* forest; *Cynometra*-mixed forest, mixed forest, colonizing forest, and swamp forest) based on the composition data (Plumptre 1996a; Owunji and Plumptre 1998).

Statistical Analyses

Bird community diversity was measured using the *alpha* logarithmic series. Krebs (1989) has argued that alpha is a better measure of diversity than Simpson's index or the Shannon-Wiener index because it is not biased by sample size. Horn's overlap index (Krebs 1989) was used to measure the similarities between the bird communities in each compartment. Each species was assigned a guild based on our experience observing these birds in the field, and this was crosschecked with other ornothologists in east Africa (L. Bennun and C. Jackson at the Ornithology Department of the National Museum of Kenya, L. Fishpool at Birdlife International). Twelve guilds were determined (see table 10-2). Dranzoa (1995) lists all species in these guilds for Kibale, and Owiunji and Plumptre (1998) list all species in a similar guild classification using 10 categories (these were altered for the purpose of this chapter so that direct comparisons could be made between the two forests).

TABLE 10-2 *Bird Guilds Used to Distinguish Bird Communities in the Kibale and Budongo Forests of Uganda*

Frugivore	Large birds such as hornbills and turacos, but also smaller species such as Barbets
Frugivore/ Insectivore	Most greenbuls, tinkerbirds
Frugivore/ Gramnivore	Doves and pigeons, mousebirds
Sallying Insectivores	Flycatchers, wattle-eyes, Narina's trogon (*Apaloderma narina*), bee-eaters
Terrestrial Insectivores	Illadopses, alethes, robin chats, thrushes
Arboreal Foliage Gleaners	Cuckoos, apalises, honeyguides, and warblers
Understory Foliage Gleaners	Bristlebill (*Bleda syndactyla*), prinias, camaropteras, white-throated greenbul (*Phyllastrephus albigularis*)
Bark Gleaning Insectivores	Woodpeckers, Red-headed malimbe (*Malimbus rubricollis*)
Nectarivore/ Insectivore	Sunbirds
Gramnivore/ Insectivore	Francolins, Weavers
Predator	Birds of prey (Eagles, Hawks and Falcons)
Gramnivore	Mannikins, Green-backed Twinspot (*Mandingoa nitidula*), Bluebills

Densities of species were calculated using distance sampling techniques for the point counts in Budongo (Owiunji and Plumptre 1998). In Kibale, the densities were calculated using the number of birds seen/heard within a fixed width of 20 m of the transect, multiplied by the length of the transect, to obtain an area surveyed. The relationships between habitat structure and relative density were calculated from mist-net data in Kibale and point count data in Budongo. Mist-netting results were compared with results from point counts in Budongo because both were conducted at the same sites to see if similar trends were observed.

The different techniques used to determine bird densities in Budongo and Kibale make it difficult to directly compare results. Transect counts have some advantages, in that they cover more area of forest per unit effort so that rarer species are more likely to be seen. Their disadvantage, however, is that more secretive species are likely to be missed, whereas point counts are more likely to detect these species. The point counts in Budongo were also based on sightings only, unlike the transect counts in Kibale that also recorded bird calls. This chapter, therefore, focuses on the responses of guilds rather than individual species because it is difficult to compare between the two forests (except using mist-netting data).

Bird Responses to Logging

Bird Communities

Uganda

A total of 120 species of birds were recorded in the Kibale study and 104 species in the Budongo study, based on point counts, transect counts, and mist netting. These studies recorded approximately one third of the species known to these forests (Kibale: 325, Howard et al. 1996a; Budongo: 359, Howard et al. 1996b). Many birds that have been recorded in the total forest lists are odd migrants that include pelicans, storks, and herons; aerial fliers such as swallows and swifts; nocturnal species such as owls and nightjars (which were not counted in the point or transect counts), and grassland species such as vultures and *Cisticola*s. A number of rare forest species were also missed, although several of these were seen outside the surveys. In both forests, there was a slight trend towards higher alpha diversity in the logged forests (see table 10-3).

TABLE 10-3 *Bird Species Richness and Diversity in the Kibale and Budongo Forest Reserves of Uganda*

Compartment and Time Since Logging	Number of Species	Diversity (alpha)[a]
Kibale		
K30 (unlogged)	117	10.3
K15 (21 years)	120	14.8
Budongo		
N15 (unlogged)	64	15.5
K11-13 (unlogged)	65	15.8
N3 (43 years)	68	17.5
W21 (29 years)	64	15.9
B1 (8 years)	71	19.2

[a]Diversity was measured as alpha from the logarithmic series (Krebs 1989). Diversity data for Budongo are from point counts, and data for Kibale are from transect counts and mist-netting. Higher numbers represent higher diversity.

None of the bird species recorded by Dranzoa (1995) were IUCN Red Data Book species, although two—the white-naped Pigeon (*Columba albinucha*) and Prigogene's ground thrush (*Zoothera kibalensis*)—have been recorded in this forest in the past. In Budongo, Nahan's francolin (*Francolinus nahani*)—classed as *data deficient* by IUCN—was recorded in both logged and unlogged compartments (Plumptre 1996b).

The study in Kibale separated the logged area (K15) into patches of remaining forest and the surrounding matrix of shrubs and climbers. Patches of unlogged forest appear to support many of the species found in undisturbed forest, but climber tangles support a very different community than that in undisturbed forests (see table 10-4). Overlap in bird species composition was high (72–82 percent) between logged and unlogged compartments in Budongo (see table 10-4), indicating that many birds in Budongo do not respond directly to forest type or successional status of the forest.

Elsewhere in Africa

In the Arabuko-Sokoke Forest of Kenya, Fanshawe (1995) found that in *Afzelia* and *Cynometra* forests there were fewer species in logged areas when compared with primary forest areas of the same type, while

TABLE 10-4 *The Percentage Overlap in Species Composition Between the Bird Communities in Each Compartment, as Measured by Horn's Overlap Index (Krebs 1989)*

Kibale	Forest Patch (K15)		K30	
Climber tangle matrix (K15)[a]	66		17	
Forest patch (K15)[a]	-		85	
Budongo	**K11–13**	**N3**	**W21**	**B1**
N15 (unlogged)	75	78	78	82
K11–13 (unlogged)	-	75	72	72
N3 (1947–52)	-	-	80	77
W21 (1963–64)	-	-	-	78

[a]In Kibale the data from logged forest in K15 are separated into forest patches and matrix of climber tangles.

in *Brachystegia* forest (the most bird species-rich forest type) there was a trend towards a larger number of species in the logged areas. Of the 26 bird species that were seen regularly in all the forest types, 65 percent showed a change in abundance following habitat modification. The Sokoke pipit (*Anthus sokokensis*), a Red Data Book species, was very abundant in primary *Brachystegia* forest and also occurred in lower abundance in logged forests but, of 16 species identified as being forest-dependent in Madagascar, only the rufous vanga (*Schetba rufa*) was significantly less common in logged forest (Hawkins and Wilme 1996). The presence of rivers bordered by large trees seemed to affect species numbers more than logging. One Red Data Book species—the white-breasted mesite (*Mesitornis variegata*)—was found in both logged and unlogged forests, but it was not possible to test whether abundances differed due to the low number of sightings.

In the Tai Forest of the Ivory Coast, Gartshore et al. (1995) recorded 27 species unique to forests that were degraded by shifting cultivation. Twenty-nine species occurred only in logged forest, 21 species were common to both logged and unlogged forests, and 17 species were only found in unlogged forest. A total of 152 species were observed in degraded, logged, and unlogged forests.

In the Gola Forest of Sierra Leone, 159 bird species were recorded in logged forest, and 163 in unlogged forest. Twenty-seven bird species occurred only in logged forest, and 31 occurred only in unlogged forest. Seven out of the nine sighted IUCN Red Data Book species were also found in logged forest, although at lower numbers than in the unlogged forest (Allport et al. 1989).

These studies suggest that the majority of bird species can survive in selectively managed forests. A few primary forest-dependent species, however, appear to be at risk from logging activities. Many of the IUCN Red Data Book species were found to survive in logged forest, but often at a lower density.

Guilds

Uganda

Frugivores/Granivores More than 100 species of birds (comprising 12 guilds) were observed in both the Kibale and Budongo Forests. Species that eat fruit or seeds generally exhibited no change (or some increase) in density and/or biomass following logging (see table 10-5). Ganzhorn (1995) showed that fruit production was positively correlated with sun exposure to tree crowns—suggesting that selective logging *can* promote fruit production. Kalina (1988), however, argued that large-bodied frugivores such as the black-and-white casqued hornbill (*Bycanistes subcylindricus*) may require undisturbed forest if there is a loss of large fruiting trees. This was supported by Dranzoa's (1995) study, which showed a higher biomass of all large frugivores in unlogged forest (including two hornbill species, two turaco species, and two starling species). In Kibale, this may have been a function of the presence of large areas with no trees in the heavily logged compartment (K15)—hence the density of trees producing fruit was significantly reduced. This was not the case in Budongo, however, where the trees removed during logging were not fleshy fruiting species. In fact, more fleshy fruiting trees appeared following logging. In 25 ha plots (Latin square layout) established in 1953 to investigate the effects of arboricide treatment on the forest, for example, *Celtis durandii* (a tree that produces small fleshy fruits important in the diet of primates and birds) is presently much more abundant in the treated replicates than in the untreated replicates (Rukundo 1996; Plumptre et al. 1997). Granivores also increased in Kibale, which was perhaps a consequence of the greater degree of canopy opening that occurred in the logged K15 compartment—a change that allowed grasses to penetrate the forest.

Insectivores Understory foliage gleaners occur at higher densities and biomass in K15 in Kibale, possibly in response to the dense matrix of

TABLE 10-5 *The Responses of Different Bird Guilds to Logging Disturbance in the Kibale and Budongo Forests of Uganda*

	Kibale[a]		Budongo	
	Density	Biomass	Density	Biomass
Frugivore[b]	nc[c]	-	nc	+
Frugivore/Insectivore	+	+	nc	nc
Frugivore/Granivore	nc	+	nc	nc
Granivore	+	+	nc	nc
Granivore/Insectivore	+	+	nc	nc
Sallying Insectivores	-	-	-	-
Terrestrial Insectivores	nc	-	nc	nc
Arboreal Foliage Gleaners	nc	-	nc	nc
Understory Foliage Gleaners	+	+	-	nc
Bark-gleaning Insectivores	-	-	nc	nc
Nectarivore/Insectivore	nc	nc	nc	nc
Predator	+	+	nc	nc

[a]Results are based on point count data for Budongo and mist netting and transect counts for Kibale, and test between the mean number of birds in each guild per transect (point counts and mist-nets placed on the same transect were combined for these tests).

[b]Guilds are from Dranzoa (1995), and the Budongo data were altered to use the same guilds and biomass.

[c]+ = significantly more abundant following logging; nc = no significant change; - = significantly less abundant following logging ; $P < 0.05$.

vegetation at ground level that provides habitat for the insects that these birds prey upon. Sallying insectivores appear to be particularly vulnerable to logging, and probably are responding to changes in the structure of the forest following disturbance. For the Budongo data, Spearman rank correlations were made between guild abundance and measures of forest type and vegetation structure (see table 10-6). This table shows that in Budongo, sallying insectivore abundance was (significantly) negatively correlated with the density of vegetation at ground level, and positively correlated with canopy cover and percentage of *Cynometra* forest (Owiunji and Plumptre 1998).

The differences that occur between Kibale and Budongo forest are probably attributable to the much more open canopy in compartment K15 in Kibale, and the matrix of dense vegetation with few trees that exists in the large gaps.

TABLE 10-6 *The Mean Percentage Number of Bird Species, and Their Spearman Rank Correlation to Forest Habitat, for Nine Guilds in Logged and Unlogged Forest of the Budongo Reserve, Uganda*

Guild	Number of Species			Spearman Rank Correlations[a]	
	Unlogged Forest	Logged Forest	T-test[b]	Positively Correlated	Negatively Correlated
Frugivore[c]	3.5	10.6	*	VD, MT	CF, CC
Frugivore/Insectivore	22.1	22.6	ns		
Granivore/Insectivore	1.7	4.0	ns		
Leaf-gleaning Insectivore	39.5	34.4	*	CC	
Bark-gleaning Insectivore	1.0	2.2	*		CC
Terrestrial-feeding Insectivore	5.8	7.1	ns	LD, MF	CF
Sallying Insectivore	21.2	10.4	*	CC, CF	SF, VD, MT
Nectarivore	4.7	6.9	*		CC
Predator	0.0	0.2	ns		

[a]Spearman rank correlations between bird abundance and measures of forest type and forest structure are listed where they were significant (P<0.05).
LD = Liana density; VD = vegetation density at ground level; CC = % canopy cover; CF = % *Cynometra* forest; MF = % mixed forest; SF = % swamp forest; MT = density of mahogany trees

[b]*= significantly different at P < 0.05, ns = nonsignificant.

[c]Fewer guilds were defined for these correlations because of low sample sizes in some of those in table 10-5 (frugivore and frugivore-granivore were combined; arboreal and understory leaf gleaners were combined, granivores and granivore-insectivores were combined).

NOTE Percentages do not sum to 100 because mean values per transect were calculated.

Elsewhere in Africa

Using a relatively crude guild classification, Fanshawe (1995) found that forest specialist species, such as insectivorous gleaners and terrestrial feeders, had lower densities in logged forest in comparison to unlogged forest in the Arabuko-Sokoke Forest in Kenya. Frugivores and omnivores occurred at higher densities in logged areas of this forest. Sallying insectivores were more abundant in logged forest, which differs from the situation in the Ugandan forests. This result was probably due to high numbers of Drongo (*Dicrurus adsimilis*) observed in logged forest—a species also found in savanna woodlands. Allport et al. (1989) investigated the differences between guilds in the Gola Forest of Sierra Leone and found similar differences between primary forest and farm bush as those reported in Uganda and Kenya. Species that were only found in primary forest were predominantly sallying insectivores, ground-gleaning insectivores, and foliage gleaning insectivores.

Studies of other bird guilds support the findings that frugivore-insectivores and nectarivore-insectivores are more numerous in disturbed forests (Lambert 1990, 1992—Borneo; Levey 1988a—Costa Rica; Johns 1991a—Amazon; Thiollay 1991—French Guiana; Grieser Johns 1996—Borneo; see chapters 8 and 9), while specialist insectivores—particularly sallying species and ground-feeding species—are more abundant in undisturbed forests (see chapters 8 and 9). It appears that sallying and certain ground-feeding insectivores are at greatest risk from changes in forest structure following logging. In both Budongo and Kibale, the understory vegetation became more dense following logging as more light penetrated to the ground level (Dranzoa 1995; Plumptre 1996a). There was little difference in the densities of ground-feeding insectivores between logged and unlogged forest in Budongo (see table 10-6), but differences were significant in Kibale (see table 10-5)—in part because the density of ground vegetation is much higher in Kibale in the logged forest compartment (K15). Certain ground-feeding insectivore species, such as the black-eared ground thrush (*Zoothera camaronensis*), were only found in undisturbed forest in Budongo (see photo 10-2).

Dominance by a Few Species

The dominant bird species (in terms of biomass) for each compartment in Kibale and Budongo, based on point count data in Budongo and transect counts in Kibale, are listed in table 10-7 (see Dranzoa 1995; Owiunji and Plumptre 1998 for scientific names). Large frugivores—such as hornbills and turacos—dominate the avian biomass in most compartments, while other groups vary with logging history. This table shows, however, that after hornbills, very different species can dominate the biomass of a community even in compartments that are only 1 to 2 kilometers apart.

Mist netting studies from Kibale and Budongo suggest that a few species come to dominate the abundance of the understory bird community in disturbed areas. In Budongo, white-throated greenbuls (*Phyllastrephus albigularis*) (13.8 percent of captures), olive sunbirds (*Nectarinia olivacea*) (10.5 percent of captures), and Kibale yellow-

PHOTO 10-2 Ground feeding insectivores such as this Oberlaender's Ground-thrush (*Zoothera oberlaenderi*) tend to be particularly affected by logging, probably due to changes in vegetation structure at ground level. (A. Plumptre)

whiskered greenbuls (*Andropadus latirostris*) (27 percent of captures) were the most commonly captured species. Skewed abundance distributions were also found for understory birds in disturbed forest within the Ituri forest of the eastern Democratic Republic of Congo (Plumptre 1997), where olive sunbirds (27 percent of captures) were very numerous. The responses of particular species were not consistent across logged forests. Yellow-whiskered greenbuls, for example, occurred at higher densities in Kibale logged forest (also in Kakamega Forest, Kenya—L. Bennun personal communication), but at higher densities in Budongo unlogged forest (also in Trans-Mara Forest, Kenya—L. Bennun, personal communication). Care must, therefore, be taken when selecting indicator species. Their response to a disturbance event, such as logging, must be consistent across sites and the environmental variables influencing species-specific responses must be identified. Abundance distributions were not dominated so strongly by individual species in unlogged forests.

TABLE 10-7 *Ranked Biomass for the Most Abundant Species in the Logged and Unlogged Compartments of the Kibale and Budongo Forests, Uganda*

Species	Guild	Kibale Forest		Budongo Forest				
		K30 UL	K15 1968	N15 UL	KP UL	N3 1950	W21 1964	B1 1984
White-thighed hornbill *Ceratogymna albotibialis*	F[a]	_[b]	-	1	1	1	1	1
Black and white casqued hornbill *Ceratogymna subcylindricus*	F	1	2	3	4	2	2	2
Great blue turaco *Corythaeola cristata*	F	3	4	-	-	-	3	-
Black-billed turaco *Tauraco schuetti emini*	F	5	21	-	-	7	10	4
Purple-headed glossy starling *Lamprotornis purpureiceps*	F	7	19	16	10	9	34	8
Splendid glossy starling *Lamprotornis splendidus*	F	8	20	-	-	-	-	-
Grey parrot *Psittacus erithacus*	F	*	*	-	15	18	17	9
Joyful greenbul *Chlorocichla laetissima*	F	33	6	-	-	-	-	-
Little Greenbul *Andropadus virens*	F/I	26	25	6	21	4	5	6
Yellow-whiskered greenbul *Andropadus latirostris*	F/I	14	3	8	6	6	24	25
Red-tailed greenbul *Crinigar calurus*	F/I	-	-	7	13	8	14	26
Western black-headed oriole *Oriolus brachyrhynchus*	F/I	6	15	4	17	15	43	45
Common bulbul *Pycnonotus barbatus*	F/I	19	5	-	-	-	-	-
Tambourine dove *Turfur tympanistria*	F/G	13	9	14	14	26	6	7
Green pigeon *Treron australis*	F/G	9	17	-	-	-	-	-
Afep pigeon *Columba unicincta*	F/G	15	7	-	-	-	-	-

TABLE 10-7 *Ranked Biomass for the Most Abundant Species in the Logged and Unlogged Compartments of the Kibale and Budongo Forests, Uganda (continued)*

Crested guineafowl *Guttera pucherani*	G/I	2	1	-	7	3	-	3
Rufous thrush *Neocossyphus fraseri*	TI	27	45	2	5	11	13	17
White-tailed ant thrush *Neocossyphus poenis*	TI	12	32	38	35	-	-	-
Blue breasted kingfisher *Halcyon malimbica*	SI	*	*	9	9	29	8	20
Black-headed paradise fly-catcher *Terpsiphone rufiventer*	SI	*	*	15	8	30	40	40
Yellowbill *Ceuthmochares aereus*	AFGI	25	*	23	20	14	7	11
Red-chested cuckoo *Cuculus solitarius*	AFGI	16	*	11	11	12	15	37
White-throated greenbul *Phyllastrephus albigularis*	UFGI	*	*	13	3	17	12	14
Bristlebill *Bleda syndactyla*	UFGI	4	18	10	18	16	27	13
Green hylia *Hylia prasina*	UFGI	20	*	18	12	22	19	18
Black-faced rufous warbler *Bathmocercus fufus*	UFGI	23	8	-	-	-	-	-
Yellow-crested woodpecker *Dendropicos xantholophus*	BGI	*	*	22	19	13	16	16
White-headed wood hoo-poe *Phoeniculus bollei*	BGI	10	12	-	-	-	-	-
Olive sunbird *Nectarinia olivacea*	N/I	11	26	20	23	20	22	23
Nicator *Nicator chloris*	Pred	18	38	5	2	5	4	5
Luhder's bush shrike *Laniarius luehderi*	Pred	*	10	-	-	-	-	-

The guild for each species: F = frugivore; F/I = frugivore-insectivore; F/G = frugivore-granivore; TI = terrestrial insectivore; AFGI = arboreal foliage gleaning insectivore; UFGI = understory foliage gleaning insectivore; SI = sallying insectivore; Pred = predator; N/I = nectarivore-insectivore; BGI = bark gleaning insectivore.

* = ranked lower than 50th place; - = never recorded.

Breeding Sites

Many studies of the effects of logging on birds compare actual or relative densities in logged and unlogged areas, assuming that higher densities occur where the habitat is better (Lambert 1990; 1992—

Borneo; Levey 1988a—Costa Rica; Johns 1991a—Amazon; Thiollay 1991—French Guiana; Grieser Johns 1996—Borneo; Fanshawe 1995—Kenya; Allport et al. 1989—Sierra Leone). Relative density, however, might not be a good measure of their response to logging. What is important in maintaining viable bird populations is *breeding success* (see chapters 8 and 9), and studies should ideally look at breeding success in different species. Black- and white-casqued hornbills (*Ceratogymna subcylindricus*), for example, can be found feeding in disturbed forest in Kibale, but prefer to nest in undisturbed forest (Kalina 1988). It is theoretically possible to find lower densities of a species where breeding occurs than where breeding is not taking place—the result of territories around the breeding site being defended and other individuals being excluded. Many large raptors (see photo 10-3), such as the crowned eagle (*Stephanoaetus coronatus*), also require large trees to nest in but appear to feed over large areas of logged and unlogged forest in Budongo (A. Plumptre, personal observation). The assumption that bird densities reflect habitat quality needs to be tested for different species, with studies concentrating on breeding birds compared to density measures.

One measure of breeding can be obtained from mist-netted birds. In both Budongo and Kibale, the presence of brood patches was recorded during mist netting (see table 10-8). Brood patches are areas of bare skin that develop during the incubation of eggs where the birds lose feathers so that body warmth is transmitted directly to the eggs. Although some care must be taken when interpreting brood patch information—as some birds can develop brood patches but fail to hatch eggs successfully and other species do not develop patches—these data show that the majority of species captured that exhibited brood patches were breeding in both logged and unlogged forest compartments. Table 10-8 indicates that only half the species breed in the forests, however. This is because birds breed more commonly during the wetter months of the year in both Kibale and Budongo (Dranzoa 1995; Owiunji 1996) and if a species was trapped during the drier months, it was less likely to exhibit a brood patch. Those species that were rarely trapped, therefore, might not exhibit a brood patch because of the time of year, rather than because they did not breed. Those species that were only recorded breeding in unlogged forest were mostly uncommon species where captures of breeding individuals were few (see table 10-8), and again this may be a function of chance captures during the wetter seasons. The high number of species that were recorded as only breeding in logged forest in Kibale (16 spe-

PHOTO 10-3 *Raptors such as this juvenile buzzard require large areas of forest or woodland to survive and may require large trees for their nests in unlogged forest. (A. Plumptre)*

cies) were mainly due to birds from the forest edge and grasslands, such as Vieillots black weaver (*Ploceus nigerrimus*), entering the fragmented forest to breed.

In Africa, the only detailed study to investigate breeding of birds in logged and unlogged forests was conducted by Dranzoa (1995). She looked for nests within permanent plots in the logged and unlogged compartments of Kibale each month from November 1991 to August 1993, and also mapped territorial males during two wet seasons (when most birds breed in Kibale). From the territory mapping, she recorded 35 species, of which four species were not recorded as forming territories in logged forest and five species that did not form territories in unlogged forest. Of those species that formed territories in both logged and unlogged forest, eight had significantly higher densities of singing males in unlogged forest and five had higher densities in logged forest. Nesting success was relatively similar between the logged and unlogged sites—sixty-seven nests were followed in unlogged forest (21 percent successfully fledged), while 59 nests in logged forest were followed (23 percent successfully fledged). Fewer birds classified as forest specialist species bred in logged forest, despite a higher number of forest specialists being caught in mist-nets in logged forest. Crevice and hole nesters, in particular, failed to breed in logged forest (Dranzoa 1995).

Of the Red Data Book species (Collar et al. 1994), Nahan's francolin was recorded breeding in logged forest at two sites in Budongo, but no nests were found in unlogged forest (Plumptre 1996b). In other studies, Allport et al. (1989) made notes of breeding birds in different forest types in the Gola Forest in Sierra Leone. They noted that white-necked picathartes (*Picathartes gymncephalus*) and western wattled cuckoo shrikes (*Campephaga lobata*) (both Red Data Book species) were recorded as nesting in logged as well as unlogged forests.

Recommendations to Conserve Bird Communities in Logged Forest

Research Recommendations

Monitoring efforts should concentrate on the changes and recovery of the forest over long periods of time—before, during, and after the logging event (Johns 1992b; see chapters 18 and 19). The effects of repeated logging at sites also need to be assessed to evaluate the long-

TABLE 10-8 *Measures of Breeding from Brood Patch Data Taken from Mist-Net Captures in Each Compartment in Budongo and Kibale Forests*

	Kibale Forest		Budongo Forest				
	K30	K15	N15	KP	N3	W21	B1
	UL	1968	UL	UL	1950	1964	1984
Number of captures	1603	3809	328	478	498	427	354
Number of species	64	84	40	36	40	39	36
Number of captures with brood patch	160	381	133	102	190	97	90
Number of species with brood patch	31	40	24	22	31	27	24
Number of species in compartment breeding in unlogged forest only	7	-	3	5	-	-	-
Number of species in compartment breeding in logged forest only	-	16	-	-	6	4	3
Species breeding in unlogged forest only (n= no. captures)	Red-capped robin chat *Cossypha natalensis* (n=4)		Puvel's illadopsis *Trichastoma puveli* (n=10)				
	Lesser honeyguide *Indicator minor* (n=1)		African broadbill *Smithornis capensis* (n=1)				
	Red-bellied paradise flycatcher *Terpsiphone rufiventer* (n=1)		Crested guineafowl *Guttera pucherani* (n=1)				
	Black-crowned waxbill *Estrilda nonnula* (n=1)		Chestnut wattleye *Platysteria castanea* (n=1)				
	White-throated greenbul *Phyllastrephus albigularis* (n=9)		Forest flycatcher *Fraseria ocreata* (n=1)				
	Yellowbill *Ceuthmochares aereus* (n=1)		Grey-headed negrofinch *Nigrita canicapilla* (n=1)				
	Narina's trogon *Apaloderma narina* (n=1)		Grey headed sunbird *Nectarinia fraseri* (n=3)				
Species only breeding in logged forest (n=no. captures)	Afep pigeon *Columba unicincta* (n=1)		Blue-breasted kingfisher *Halcyon malimbica* (n=2)				

TABLE 10-8 *Measures of Breeding from Brood Patch Data Taken from Mist-Net Captures in Each Compartment in Budongo and Kibale Forests (continued)*

Kibale Forest	Budongo Forest
Dark-backed weaver *Ploceus bicolar* (n=2)	Brown illadopsis *Trichastoma fulvescens* (n=29)
Collared sunbird *Anthreptis collaris* (n=1)	Dusky long-tailed cuckoo *Cercococcyx mechowi* (n=1)
Green-headed sunbird *Nectarinia verticalis* (n=1)	Icturine greenbul *Phyllastrephus icterinus* (n=2)
Banded prinia *Prinia bairdii* (n=8)	Little grey greenbul *Andropadus gracilis* (n=1)
Luhder's bushshrike *Laniarius luehderi* (n=6)	Red-tailed anthrush *Neocossyphus rufus* (n=4)
African broadbill *Smithornis capensis* (n=3)	Red-tailed greenbul *Crinigar calurus* (n=2)
Yellow white-eye *Zosterops senegalensis* (n=4)	Slender-billed greenbul *Andropadus gracilirostris* (n=1)
Black-headed Oriole *Oriolus brachyrhynchus* (n=2)	White-tailed anthrush *Neocossyphus poenis* (n=2)
Vieillots black weaver *Ploceus nigerrimus* (n=1)	Yellow-billed barbet *Trachyphonus purpuratus* (n=1)
White-browed crombec *Sylvietta leucophrys* (n=2)	
White-chinned prinia *Prinia leucopogon* (n=3)	
Green-backed twinspot *Mandingoa nitidula* (n=1)	
Red-faced crimsonwing *Cryptospiza reichenovii* (n=2)	
Speckled tinkerbird *Pogoniolus scolopaceus* (n=1)	

term shifts in bird populations and their accompanying ecological processes (see chapter 3). Comparisons of logged and unlogged areas make the assumption that they were identical before logging took place, which is highly unlikely in spatially heterogeneous tropical rain forest. The results presented here, although based on relatively intensive field-work, can only be taken as preliminary until longitudinal studies show whether these findings are valid.

Future studies of the effects of logging on birds should aim to include the following:

- Several sites over time should be monitored beginning in advance of the logging event, and ideally continue for several decades after the event

- Replicated unlogged and logged sites should be monitored regularly (every 3–5 years) so that patterns in bird density can be detected through the noise of different censuses

- Sites that control for similar levels of harvesting intensity and forest damage

- Measurements beyond crude densities, including efforts to detect bird breeding success in logged and unlogged areas

- Measures of change in tree species composition and structure, so that effects of forest structure changes on bird densities can be assessed

- Species-specific studies on birds of national/international importance to assess their habitat requirements and how they respond to habitat change

General surveys of bird communities as reported here do not obtain enough data to assess the threats of logging to rare species. These points are discussed in greater detail in chapters 18 and 19.

Management Recommendations

Growth rates of trees in Budongo are of the order of 0.25 to 0.75 cm dbh (diameter at breast height) per year for timber species (Sheil 1996; A. Plumptre, unpublished data), and even with 45 years of recovery time, the structure of selectively logged forest is significantly different from mature undisturbed forest (Plumptre 1996a). The data presented here show that the bird communities in these areas are also different (Owiunji 1996). Consequently, if selective logging with rotations of 40 to 60 years is planned as a management option for tropical forests, large enough areas must be established to ensure the long-term survival of populations of those species that will never recolonize the logged areas. For small species such as flycatchers, these areas may only need to be 10 to 20 km^2. For larger species, however, such as the crowned eagle or rarer species like the Prigogine's ground thrush

(*Zoothera cameronensis kibalensis*), these reserves may need to be in the order of 100 to 500 km^2 to maintain viable populations of these species that occur at very low densities. If forests are not allowed to fully recover the structure found before logging because of the need for an economically-viable rotation period, then it is probable that bird communities will change further in subsequent harvests—leading to even fewer *disturbance sensitive species* (Johns 1992b). Logging companies and national governments, however, can take certain steps to mitigate the negative effects logging might have on bird communities (see chapters 21 and 24), including:

- Leaving old and dead standing trees in logging concessions, to provide dead wood and suitable sites for hole/crevice nesting birds (see chapter 23)
- Designating areas of the concession that will be left intact to provide refuge for those species of bird that are sensitive to logging (these can be along streams/rivers but should also include other forest types), and which include large emergent trees used for nesting by raptors (see chapter 23)
- Limiting the hunting of primates and small ungulates in the concession, as these are important for large birds of prey (see chapters 15–17)
- Reducing incidental damage when felling by cutting lianas before felling a tree and using directional felling so that the forest structure, particularly the understory structure, is less disturbed (chapters 21 and 24)
- Minimizing climber tangles that can smother regeneration by keeping gap sizes small (< 400 m^2) (Babweteera 1998) or cutting climbers later

Conclusions

The studies of the effects of logging on African bird communities are few, particularly in central Africa where the most intensive logging is taking place. In general, the studies to date show that where logging is fairly selective, many species of birds can survive and breed in logged forest. Changes in bird guilds varied between forests, and these changes are probably related to differences in forest structure and composition. Sallying insectivores and terrestrial insectivores seem to be particularly sensitive to forest change, probably because logging reduces the available space under the canopy for these species to hunt (dense understory vegetation makes hunting difficult). This condition is exacerbated where regeneration of the forest is hampered by climber tangles (as occurred in Kibale Forest). It is, therefore, important to try to reduce the prevalence of climber tangles over regenerating vegetation by keeping openings to a minimum.

Selective logging has the potential to protect many species of African forest birds in comparison with alternative uses of the forest, such as clearing for agriculture. There will always be species that suffer as a consequence of the logging activities, however. Selective logging cannot therefore be regarded as an alternative to the preservation of areas of tropical rain forest, but it can be used as a means to protect certain adaptable species which would otherwise become extinct if economic pressures were to cause the total removal of the forest.

ACKNOWLEDGMENTS

We would like to thank the Wildlife Conservation Society, USAID, ODA, and NUFU for funding the work reported in this chapter. Professor D. Pomeroy and Professor V. Reynolds have also supported these studies. We would also like to thank the Uganda Government, Forest Department, and National Parks for allowing us to carry out this research in Budongo and Kibale.

Chapter 11

THE EFFECTS OF LOGGING ON REPTILES AND AMPHIBIANS OF TROPICAL FORESTS

Laurie J. Vitt and Janalee P. Caldwell

Reptiles and amphibians comprise a significant portion of the vertebrate faunas of tropical forests throughout the world (Inger 1980a, b; Duellman 1990, 1993), where they have an impact on other organisms as predators and prey. The long-term effects of logging on tropical herpetofaunas (assemblages of reptiles and amphibians) are poorly understood, and, as a consequence, most of the discussion in this chapter is based on relatively short-term studies.

Logging in tropical forests occurs at several scales, each of which has the potential to differentially impact reptile and amphibian populations. Clearcut logging typically creates a patchwork landscape where microhabitats used by the interior forest fauna still exist in noncut patches. There are resulting edge effects, however, including the creation of nonforest microhabitat characteristics that have differential impacts depending on the size of the forest fragments (Marsh and Pearman 1997). Harvested sites tend to regenerate rapidly—provided the site conditions are not severely degraded during the logging process—giving rise to secondary forest habitat. In contrast, deforestation leads to land conversion (e.g., to pasture), with reduced potential for forest regeneration. Many animal populations, including reptiles and amphibians, are typically eliminated from these latter areas.

Selective logging, particularly in areas where the largest trees remain (in central Brazil, for example, where harvesting Brazil nut trees [*Bertholletia excelsa*] is prohibited), may have little negative impact on habitat structure from the perspective of amphibians and reptiles—particularly when incidental damage to other trees is minimized and trees are removed from the site prior to being processed (but see Macedo and Anderson 1993). Selective logging, however, often creates extensive, incidental damage to other trees. In Pará, Brazil, for example, felling individual mahogany trees resulted in a mean loss of 31 trees and gaps in the forest of 1,100 m^2 (Veríssimo et al. 1995). Harvesting of individual trees with processing on site, as occurs near villages and settlements, may also affect herpetofauna populations (Vitt et al. 1998, and discussion in next section), as prolonged changes in environmental conditions lead to competitive imbalances in native amphibian and reptile populations. We next discuss the potential effects of logging activities on reptile and amphibian assemblages (see table 11-1). These effects should be viewed with some caution because data are often lacking or inadequate to fully test these predictions.

Direct Effects of Tree Removal on Reptile and Amphibian Populations

Tree removal changes both the structural and microclimatic architecture of the habitat being cut. Direct effects of these activities on reptiles and amphibians include changes in the microclimate and microhabitats, such as reduction and/or elimination of arboreal perches or refuges that are important for feeding, defense, breeding, nesting, and other social behaviors. Forest converted to cacao plantation but abandoned five years previously in Costa Rica, for example, had higher abundance and biomass of reptiles and amphibians compared to converted plantations left undisturbed for 25 years. Species diversity, richness, and eveness, however, were higher in the latter sites (Heinen 1992). Furthermore, fragmentation of habitats may have a significant effect on the densities of reptiles and amphibians. Marsh and Pearman (1997) showed that two species of leptodactylid frogs, *Eleutherodactylus chloronatus* and *E. trepidotus,* were 3.5 and 2.1 times more abundant, respectively, in a large forest patch (200 ha) than in adjacent forest fragments (0.25–5.3 ha). One of these species moves primarily by crawling, rather than jumping, and may have limited dispersal ability. Other tropical arboreal species will not cross open patches (L. J. Vitt, personal observation), and populations may become isolated in areas of relatively intense spatial and temporal cutting activity.

TABLE 11-1 *Hypothesized Impact of Various Logging Practices on Abundance
and Diversity of Tropical Amphibians and Reptiles*

Extensive Logging	Patchy Logging	Selective Logging	Single Tree Harvest with On-Site Processing
Extensive areas containing intensive harvests such as clear cutting	*Group selection cuts at moderate spatial and temporal intensities*	*Low intensity single-tree selection cuts*	*Community forestry projects*

Habitat Alteration

Extensive Logging	Patchy Logging	Selective Logging	Single Tree Harvest with On-Site Processing
Drastic climatic change	Climatic change dependent on extent of patchy logging	Low to moderate climatic effect	Very point specific climatic effect
Loss of forest microhabitats	Loss of forest microhabitats within logged patches	Little overall change in types of microhabitats in forest	Little overall change in types of microhabitats in forest
Drastic reduction in microhabitat diversity at landscape level	Increase in microhabitat diversity at landscape level	Little change in microhabitat diversity at landscape level	Little change in microhabitat diversity at landscape level
Drastic reduction in microhabitat diversity within cleared patches at local level	Drastic reduction in microhabitat diversity within cleared patches	Increase in frequency of microhabitats associated with treefalls or edges	Increased frequency of completely open microhabitats
Drastic microclimatic change	Drastic microclimatic change in cleared patches and at forest edge	Possible microclimatic change at treefall site and along edges	Extreme microclimatic change at treefall site
Introduction of contact zones with open habitats	Introduction of corridors connecting forest with adjacent open habitats	Introduction of corridors connecting forest with adjacent open habitats	Introduction of corridors connecting forest with adjacent open habitats
Extensive introduction of open habitat species via contact zones	Limited introduction of open habitat species via contact zones	Limited introduction of open habitat species via contact zones	Limited introduction of open habitat species via contact zones

Species Alteration

Extensive Logging	Patchy Logging	Selective Logging	Single Tree Harvest with On-Site Processing
Loss of forest species, particularly microhabitat specialists	Loss of forest species in open patches, particularly microhabitat specialists	Little impact on forest species as long as canopy remains	Little direct impact on forest species

TABLE 11-1 *Hypothesized Impact of Various Logging Practices on Abundance and Diversity of Tropical Amphibians and Reptiles (continued)*

Extensive Logging	Patchy Logging	Selective Logging	Single Tree Harvest with On-Site Processing
Replacement of forest species by open habitat species	Replacement of forest species by open habitat species, but only in open patches	Possible introductions of open habitat species along corridors and edges	Possible introductions of open habitat species in tree cuts near edges
Reduction in amphibian and reptile diversity at landscape level	Increase in amphibian and reptile diversity at landscape level	Minimal increase in amphibian and reptile diversity at landscape level	Minimal increase in herpetofauna diversity at landscape level
Reduced abundance or local extinction of many species with increased abundance of some species	Increased abundance of some species, particularly in open patches and the open patch/forest interface	Increase in abundance of species using treefall or edge microhabitats	Increase in abundance of species using treefall or edge microhabitats
Long-term success of open habitat species due to habitat change	Medium-term success of open habitat species due to habitat change	Short-term or limited success of open habitat species due to ultimate gap closure	Success and impact of open habitat species dependent on distribution and frequency of tree cuts
Potential indirect effects on amphibian and reptile diversity of adjacent forest	Potential indirect effects on amphibian and reptile diversity of adjacent forest negative, but unexplored (see text)	Potential indirect effects on amphibian and reptile diversity may be positive, but unexplored (see text)	Potential indirect effects on herpetofauna diversity negative, but unexplored

SOURCE L. Vitt and J. Caldwell, personal observations.

Logging and Aquatic Herpetofauna Habitat

Streams, ponds, and temporary pools provide microhabitats for aquatic amphibians and reptiles. Most tropical amphibians require water for reproduction and some even use various plant parts that hold water (phytotelmata), on a temporary basis, for reproduction. These include leaf axils, Brazil nut fruit capsules (Caldwell 1993; Caldwell and Araújo 1998), vine holes (Caldwell 1997), and buttresses. Adults of aquatic amphibians and reptiles feed on aquatic organisms, and larvae of most amphibians depend on algae, detritus, or aquatic macroinvertebrates for food. For many amphibians and reptiles, the aquatic environment acts as a buffer to extremes of temperature and desiccation.

Logging may directly affect amphibians and reptiles through its impact on wetlands in the immediate vicinity of logging operations, as well as throughout watersheds impacted by logging. Species that rely on bromeliads, treeholes, and other phytotelmata as sites for depositing eggs or tadpoles would be heavily impacted or eliminated from areas of extensive logging. Changes in the nutrient composition and water chemistry of affected wetlands may impact larval amphibians, leading to malformations and death (Ouellet et al. 1997). During development and metamorphosis, amphibian larvae are sensitive to pH and other physical characteristics of aquatic environments. Many amphibian larvae also depend on algae or detritus for food, and as a consequence, nutrient reduction may reduce food availability or quality. This topic remains unexplored in tropical forests, even though considerable data exist on the effects of osmotic variation and temperature on amphibians (Feder and Burggren 1992).

In some cases, frog abundance may increase in disturbed areas, particularly when these areas supply needed resources (see box 11-1). The Amazonian treefrog (*Hyla geographica*), for example, typically requires small bodies of water in which they congregate in large breeding choruses. In some areas of Rondônia, Brazil, this species has colonized ponds formed along roads constructed through formerly forested areas (Caldwell 1989) (see photo 11-1). The large, black tadpoles of this species form huge aggregations (see photo 11-2a) that are apparently not susceptible to fish predation (Caldwell 1989). Because of the increased number of breeding sites, this species had become one of the most abundant in the area. In contrast, *H. geographica* was uncommon in nearby undisturbed forest (Caldwell 1989). In another area in Brazilian Amazonian forest, hylid frogs in the genus *Phyllomedusa* (see photo 11-2c) were as abundant in logged-over areas as they were in undisturbed forest (see box 11-2).

PHOTO 11-1 *Small pond created by the construction of logging road with improper drainage. (R. Fimbel)*

Box 11-1 *Low-intensity, selective timber harvest may enhance or decrease herpetofauna populations.*

The effects of selective logging with off-site processing may have little impact or, in some cases, even enhance populations of many reptiles and amphibians. During 1995, for example, we surveyed reptiles and amphibians in a forest near the Rio Curuá-Una, approximately 100 km south of Santarém in the Brazilian state of Pará. The forest was dominated by Brazil nut trees. Other select hardwood species such as *Hymenaea courbaril* and *Tabebuia* sp., were removed in 1987. The intensity of logging, however, was unknown. Our methods included recording the habitat and microhabitat of every reptile and amphibian encountered in logged forest, based on haphazard surveys over extended time periods (see Vitt et al. 1999). The lack of accessible, undisturbed nearby forest prevented direct collection of comparative data.

Compared to our observations of undisturbed forests at other Amazonian sites, most of the selectively logged forest at the Curuá-Una site appeared structurally similar to undisturbed forest—although the understory was denser in some areas. The nearly continuous canopy provided shade and partial sun microhabitats that are typical of undisturbed forest. The species richness of reptiles and amphibians appeared similar to that in undisturbed areas of the region, and the abundance of some species of frogs, lizards, and snakes appeared higher than at other Amazonian sites in which we have worked (Caldwell and Vitt 1996).

We hypothesized that a combination of factors led to the apparent increase in abundance of many species. Increased structural diversity of

PHOTO 11-2a-h *a) Tadpole school of the treefrog Hyla geographica, which becomes superabundant in areas where forest has been removed and artificial ponds are created by roadbed construction (J. P. Caldwell). b) The large-bodied teiid lizard Ameiva ameiva, which, because of its size and foraging mode, can be a major predator on invertebrates and bertebrates when it invades newly created gaps in forest (L. J. Vitt). c) The phyllomedusine treefrog Phyllomedusa tarsius, a forest species that successfully reproduces in pools formed in second-growth forest (L.J. Vitt). d) Tropidurus flaviceps, an arvoreal tropidurid lizard that feeds nearly exclusively on ants (L. J. Vitt). e) Dendrobates vanzolinii, a dendrobatid frog that lives in understory vegetation and vines, which carries out its entire life history above the ground (J. P. Caldwell). f) The gymnophthalmid lizard Alopoglossus artiventris, which inhabits leaf litter as a structural habitat and feeds on small invertebrates that live therin (L. J. Vitt). g) An undescribed dendrobatid frog in the genus Epipedobates, which carries out most of its life history in leaf litter, depositing tadpoles in small temporary pools on the shaded forest floor (J. P. Caldwell). h) Amplexing pair of Edalorhina perezi, a common frog dependent on the leaf litter microhabitat in western Amazonian forest. (J. P. Caldwell)*

the habitat was caused by small treefalls within the forest that were associated with tree felling and skidding. Small roads were constructed throughout the forest for access, and during the rainy season, extensive shallow pools formed in these roads—greatly increasing breeding sites for certain species of amphibians, particularly the smoky jungle frog (*Leptodactylus pentadactylus*), two leptodactylid frogs (*Physalaemus cuvieri* and *Physalaemus petersi*), the cane toad (*Bufo marinus*), two tree-frogs (*Osteocephalus taurinus*, and *Hyla multifasciata*), and the barred-leaf frog (*Phyllomedusa tomopterna*), and an undescribed species of *Hyla*. The latter four species are tree frogs and were abundant because of the presence of water (Pearman 1997; see box 11-2). The increases in species abundance using these pools may have actually enhanced the abundance of other species—many snakes, large lizards, and frogs prey on amphibians (Duellman and Lizana 1994).

Our observations that selective logging did not negatively affect reptile and amphibian species richness was corroborated by Pearman (1997), who studied twenty-three 1/2-ha sites in eastern Ecuadorian tropical forest, ranging from river floodplain to dissected hills reaching 35 m in elevation. Portions of this area were selectively cut five to ten years previously, whereas other areas were used for plantations—reliable data on the exact time passed since logging were not available for individual sites. Pearman found no difference in amphibian species richness between primary forest and selectively logged forest, but the composition of amphibian species changed with the distance of the measured site from mature forest edge. *Eleutherodactylus* species, which are dependent on forest, formed a greater proportion of total species richness at pristine sites further from pasture. As a result consideration must be given to individual groups of organisms and their particular biological characteristics (Pearman 1997), when using overall measures (such as species richness) in making management decisions. Selective logging (six to twenty trees/ha, or 4 to 15 m³/ha) also appeared not to influence the composition of the reptile community in a short-term (two-month) study in tropical forests of Madagascar (Bloxam et al. 1996).

Finally, single tree cutting associated with human settlements has received virtually no attention, even though it may extensively impact reptiles and amphibians. Different from selective logging, in which trees are removed from the forest and processed off-site, trees in community forests are felled and processed over an extended period at the site of the felling. This procedure results in the formation of a forest gap that is devoid of vegetation because sawdust from chain saws forms a thick mat on the ground that appears to inhibit plant growth in the clearing. This results in high exposure of the substratum to sun during the day. These treefalls receive more intense light and attain much higher temperatures than forest habitats or natural treefalls (see box figure 11-1). They are used only by large, highly mobile heliothermic lizards (Vitt et al. 1998) that move in and out of these extreme habitats and feed on both invertebrates and vertebrates (e.g., *Ameiva ameiva*, Vitt and Colli 1994; *Kentropyx* and *Mabuya*, Vitt et al. 1997; see box figure 11-2; see photo 11-2b). As the number of these barren patches increases, so does the accessibility of forest to these large predators. Presumably, once these patches have reached high enough densities, the entire forest becomes available to these lizards—leading to the potential for severe impacts on populations of invertebrates and other herpetofauna.

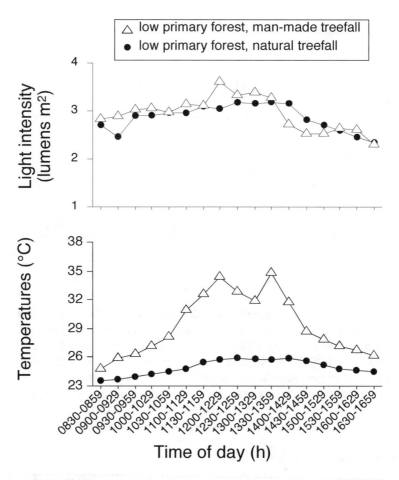

BOX FIGURE 11-1 *Comparison of light intensity and temperature in a natural and a man-made treefall near Porto Walter, Acre, Brazil. The natural treefall received more light during most of the day and achieved higher temperatures throughout the day. Both types of treefalls maintained higher temperatures than adjacent forest habitats (Vitt et al. 1998). Low forest was defined as forest in which rainfall remained pooled following downpours.*

The same phenomenon was generally found for hylid frogs in Amazonian Ecuador (Pearman 1997), where numerous species of hylid frogs were abundant in disturbed areas with small ponds, but were uncommon in pristine forest. The small hylid frog (*Scinax rubra*) is ubiquitous throughout the Amazon Basin. Although rare in undisturbed forest, it often becomes superabundant in converted areas, even moving into human residences (Rodríguez and Duellman 1994). The eggs and tad-

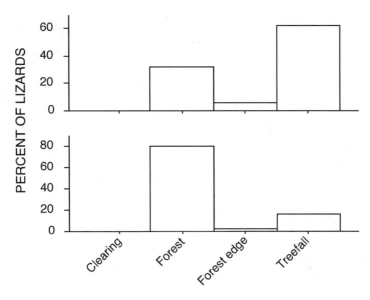

BOX FIGURE 11-2 *Habitat distribution of Amazonian lizards based on 906 observations taken near Porto Walter, Brazil (Vitt et al., unpublished data). Heliotherms (upper histogram) are those species that bask in sun to gain heat and typically are active at body temperatures well above environmental temperatures. Nonheliothermic species (lower histogram) rarely bask and are active at body temperatures at or near environmental temperatures. For presentation, we have grouped all habitats within the forest into one category—forest.*

poles of this species can tolerate the extremely high temperatures (above 40°C) found in tiny shallow pools exposed to open tropical sun (J. Caldwell, personal observation).

In contrast to these examples, most species of frogs are specialized for a forest existence (see photos 11-2, d-h) and will disappear following intense disturbance of the forest (Marsh and Pearman 1997). Pearman (1997) found that many species of *Eleutherodactylus*—which are leaf-litter frogs that deposit terrestrial eggs and are not dependent on water for breeding—were found only in pristine forest.

> **Box 11-2** *Logging and clear-cut activities may negatively and positively influence tropical frog populations. (Claude Gascon)*
>
> In the Manaus region of Brazil, the intensity of human disturbance and the level of forest recovery greatly influence the resident frog communities (Tocher 1997). Disturbed sites that had better forest recovery (i.e.,

more complex vegetation structure and higher plant species richness) showed higher frog species richness, and over 30 percent more community overlap with intact primary forest, than did sites with younger, less-developed secondary forest. This suggests that reduced-impact logging techniques (well-planned extraction of trees, cutting of vines to minimize damage, etc.; see chapters 21, 24, and 28), that better preserve the original forest structure and composition (compared to conventional harvesting techniques), should also help to maintain the integrity of interior-forest frog communities.

For species that depend on aquatic habitats for reproduction, the presence of appropriate aquatic breeding habitat is a major determinant of species distribution on a local scale. In second-growth forests, ponds often exist. Although these habitats are structurally very different from primary forest aquatic habitats, they harbor many aquatic-breeding frogs. In fact, surveys performed as part of the Biological Dynamics of Forest Fragments Project (north of Manaus) (C. Gascon, unpublished data) have shown over the last several years that there is a large overlap (35–50 percent) in species composition between disturbance-created ponds and aquatic habitat in primary forest. Although many forest-associated species are not found breeding outside primary forest ponds (and therefore appear to be affected by deforestation or intense logging), several frogs characterized as *primary forest species* before logging in the area can be found breeding in secondary forest. A more detailed investigation of the reproduction of a common tree frog in the area, *Phyllomedusa tarsius*, showed similar population sizes of males around breeding sites in disturbed-area ponds and primary forest ponds. Males were larger in primary forest, however, in comparison to males breeding in open-area sites. Hatching success did not differ between the two habitats. While there is no doubt that logging can negatively affect many forest-associated species, some species that depend on aquatic habitats for reproduction may survive in disturbed areas due to the presence of additional aquatic habitats such as pools in old skid trails.

Logging and Leaf Litter Herpetofauna Habitat

The diversity and abundance of litter amphibians and reptiles has been investigated in several different tropical forest habitats. Twenty-six of 55 (47.3 percent) frog species and thirteen of thirty (43.3 percent) lizard species in Brazilian Amazonian forest at Porto Walter, Acre, live in leaf litter (see photos 11-2e-h) and could be severely affected by changes in the leaf litter microhabitat. The number of litter species (excluding snakes and turtles) in Panama and Costa Rica varied from 17 to 35, whereas the number at a lowland site in Borneo was 38, and in the Philippine Islands was four to 19 species (Scott 1976). The latter study also demonstrated that drier sites had fewer species than wetter sites. Removal of leaf litter could effectively eliminate small-bodied vertebrates because of loss or reduction in resources (food) and loss of microhabitats (within the leaf litter) that provide refuges from desiccation and predators. Likewise, any disturbance that might affect nutrient cycling and, ultimately, the invertebrate fauna and/or the microclimate within leaf litter can be expected to have considerable negative effects on the vertebrates residing therein.

Logging drastically affects the ground-level environment of forests. The mat of undisturbed forest leaf litter may be replaced by various species of grasses or a thick tangle of pioneering vegetation. The removal or change in quality of leaf litter directly affects herpetofauna. Leaf litter in tropical forests serves as a structural, thermal, and hydric microhabitat for many small-bodied species of lizards and frogs (Scott 1976; Lieberman 1986; Allmon 1991; Vitt and Caldwell 1994). Many leaf litter reptiles and amphibians also feed on tiny arthropods, including mites and springtails (Toft 1980, 1981; Vitt and Caldwell 1994; Caldwell 1996b), which also depend on the physical properties of leaf litter microhabitats.

Few forest reptiles and amphibians survive after the opening of the forest and disturbance of the leaf litter. Virtually all leaf litter species depend on a diversity of leaf litter arthropods for food and on the structural diversity of the leaf litter for predator escape, nesting, and maintenance of relatively constant microclimate. In Puerto Rico, lower montane rain forest was selectively cut or converted to plantations before being abandoned in the 1930s. Even after 50 years of secondary succession, decomposition of leaf litter still occurred faster in this mid-successional forest compared to a mature forest in Puerto Rico (Zou et al. 1995; see chapter 13). The greatest diversity of leaf-litter arthropods, overall, was in the more mature forest with the least amount of logging. The lowest was in the clearcut areas. The intermediate areas and plantations were indistinguishable from each other. Abundance of certain springtails (Collembola), an insect group that forms a significant part of the diet of many leaf-litter species, was at least two to three orders of magnitude greater in the plantation (G. Camilo, personal communication; see chapter 13). The effect of this variation in leaf-litter arthropods—the primary food of leaf-litter reptiles and amphibians—is unstudied in the tropics (see Toft 1981). In a study comparing salamander densities in a temperate, deciduous hardwood forest in Virginia, Mitchell et al. (1996) found that salamanders from mature forest consumed significantly more springtails (a soft-bodied, high-quality prey item) than salamanders from recent clearcuts and shelterwood cuts. Whether this difference in diet quality translates into faster growth or reproduction is unknown.

Indirect Effects of Logging on Reptiles and Amphibians: Cascading Effects of Energy Flow Through Food Webs

Reptiles and amphibians are important components of tropical food webs (Reagan and Waide 1996). Most frogs and lizards feed on a myriad of invertebrates—mostly insects and spiders—and each species feeds on a different composition of animals (Vitt and Caldwell 1994; Caldwell 1996b; Reagan 1996; Stewart and Woolbright 1996; Thomas and Kessler 1996; Vitt and Zani 1996a). Those species that are specialists, feeding on a limited number of types of food items, may be at greater risk from logging disturbance than species with a generalist diet.

One of the difficulties in understanding tropical food webs is the extreme complexity involved (Schoener 1989). No analyses exist for an entire mainland tropical community, but a comprehensive analysis of the food web of a 40 ha site in the Luquillo Experimental Forest on the island of Puerto Rico has been published (Reagan and Waide 1996; see chapter 13). The amphibian fauna of this island is small by mainland standards—eleven species of leptodactylid frogs in the genus *Eleutherodactylus* and one species of *Leptodactylus*. Among the 11 *Eleutherodactylus*, only four are described as common, and only one of these (*E. coqui*) is a widespread and abundant generalist. Nevertheless, amphibians in this system are described as providing a major pathway of energy flow through the system (Stewart and Woolbright 1996). *E. coqui*, like many frogs, is a sit-and-wait predator that feeds largely on arthropods (Woolbright and Stewart 1987). Thirty-nine major taxonomic groups of prey were found in an analysis of 173 frogs, and these frogs or their eggs served as prey for 38 types of predators, ranging from ants, millipedes, and spiders to snakes, lizards, birds, and mammals (Stewart and Woolbright 1996). Similar results were found in analyses of lizards in the genus *Anolis* (Reagan 1996) and other reptiles (Thomas and Kessler 1996). Considering the large numbers of prey eaten by one species of Puerto Rican frog, *E. coqui*, and the large number of predators that consume these frogs, the high species richness of amphibians and arthropods found in mainland tropical forest systems reflects the magnitude and importance of herpetofauna-related interactions in these systems. We caution readers, however, against using these results to make predictions about other systems. The vertebrate species composition at the Puerto Rico site is different from many sites, in that there are no large herbivores or large predators. One frog species (*E. coqui*) and one lizard genus (*Anolis*) also reach unusually high densities at the site. Consequently, applying the results of research on Puerto Rican amphibians and reptiles to other sites may not be possible due to gross differences in the overall vertebrate faunas and the heavy weighting of amphibian and reptile data by two abundant species. It is clear that amphibians and reptiles are key elements of the food webs in tropical forests. These small- to medium-size organisms may provide significant biomass in tropical forests (Stewart and Woolbright 1996), and they form a link between tiny prey species (e.g., ants, mites, termites; Caldwell and Vitt 1999) and the larger organisms (e.g., snakes, birds) that consume them. Logging practices that change the density of these species can significantly affect the entire food web of the impacted area.

Perturbations of even relatively simple food webs have been shown to produce surprising responses (Schoener and Spiller 1987). In a study on the effects of the removal of lizards from enclosed areas, spider density increased threefold. When present, however, lizards reduced the abundance of large aerial arthropods used by spiders as food. This reduction in spider food by lizards translated into smaller spider size and reduced spider fecundity. This, of course, affected fitness (Spiller and Schoener 1990a, b; 1994). Subtle changes in forest structure (and, subsequently, reptile or amphibian densities) could, therefore, have cascading effects on many other kinds of forest inhabitants and the ecological processes they perform.

Challenges to the Study of Logging Effects on Tropical Reptile and Amphibian Populations

Determining the long-term effects of logging on tropical herpetofaunas requires at least the following:

- Pre- /postlogging assessments of the herpetofauna for comparisons (see chapters 18 and 19)
- Data on basic ecology of resident species
- Some indication of the resource base necessary to sustain the herpetofauna
- An understanding of physiological requirements of resident species
- At least a basic understanding of species interactions (e.g., food web structure)

All of these are problematic in that logging operations have taken place and, in most areas, will continue to take place, before these basic data can be collected. Determination of the prelogging herpetofauna is confounded by insufficient taxonomic work necessary to determine exactly what species are present in a given area. Finally, a lack of sufficient funding for the kinds of studies that are needed in ecology, physiology, and food web interactions further hampers our ability to make reliable predictions about the effects of logging on amphibians and reptiles (Domenici 1996).

Taxonomy

Tropical forest herpetofaunas are among the richest in the world and among the most poorly known taxonomically and ecologically (Crump 1971; Laurent 1973; Inger and Colwell 1977; Duellman 1978, 1979, 1990, 1993; Duellman and Mendelson 1995; Avila-Pires 1995; Caldwell 1996a; Vitt and Zani 1996a). Most faunal surveys conducted in little known tropical regions produce numerous reptile and amphibian species new to science (e.g., Inger et al. 1987; Duellman and Lynch 1988; Caldwell and Myers 1990; Duellman and Salas 1991; Duellman and Hoogmoed 1992; Duellman and Schulte 1992; Duellman and Mendelson 1995; Caldwell 1996a). Even though we know that amphibian and reptile species richness is high in tropical forests (Dixon and Soini 1975, 1977; Inger and Colwell 1977; Duellman 1978; Inger 1980a, b; Rodríguez and Cadle 1990; Zimmerman and Rodrigues 1990; Vitt 1996), the number of sites sampled sufficiently to accurately determine the composition of the herpetofauna is small (Heyer 1988). Most studies are based on relatively short sampling periods (but see Zimmerman and Simberloff 1996 for a long-term example), or are restricted to just reptiles (e.g., Dixon and Soini 1975, 1977) or just amphibians (Rodríguez and Duellman 1994; Duellman and Lynch 1988, 1997). Consequently, generalizations about patterns of tropical species richness of amphibians and reptiles, and their response to disturbances such as logging, must be put forth with caution. The database is simply poor. Amphibian and reptile species richness and abundance varies considerably among tropical forest habitats (e.g., Scott 1976; Inger and Colwell 1977; Brown and Alcala 1986) and through time within forested habitats (e.g., Inger and Voris 1993; Duellman 1995)—in part because forests themselves are mosaics of habitat types (Culotta 1995; Tuomisto et al. 1995). Differences in methodology and observations also have the potential to influence estimates of species diversity and abundances. Finally, even the timing and duration of sampling may significantly affect actual species counts (see Duellman 1995; Zimmerman and Simberloff 1996).

Food Resources

Reptiles and amphibians are ectothermic vertebrates that, for the most part, feed on invertebrates—primarily arthropods. Body temperatures, activity times, and prey types and sizes vary among species. Many species feed on a diversity of arthropods and small vertebrates (e.g., the large teiid lizard [*Ameiva ameiva*] Vitt and Colli 1994), and others have highly specialized diets (e.g., poison frogs in the genus *Dendrobates*, Caldwell 1996b; some tropidurid lizards, Vitt and Zani 1996b). Changes in arthropod density, due to alterations in forest composition and structure, can seriously alter reptile and amphibian populations (Robinson et al. 1992; Zou et al. 1995; see chapter 12).

Temperature Requirements

Because their body temperatures are affected by immediate surroundings (Cowles and Bogert 1944), and because nearly every aspect of the behavior, physiology, and ecology of reptiles and amphibians is influenced by temperature (Pianka 1986; Duellman and Trueb 1986; Hutchison and Dupré 1992), changes in the thermal environment of reptile and amphibian populations will have drastic effects on community organization, population densities, and microdistribution.

Reproductive Cycles

Reptile and amphibian life histories are such that modification of natural habitats may drastically reduce, or in some instances increase, reproductive output. Most reptiles that deposit eggs, for example, require egg deposition sites that provide appropriate moisture and temperature for development. Some entire groups (e.g., *Eleutherodactylus* and dendrobatids) deposit terrestrial eggs and are not dependent on bodies of water for reproduction to the degree that other groups (treefrogs, bufonids) are. In some species, sex may be determined by temperature during development (Bull 1980). Tropical amphibians display the largest diversity of reproductive modes known for vertebrates, with the possible exception of teleost fishes (Duellman and Trueb 1986; Hutchison and Dupré 1992). Many species of amphibians in tropical forests carry out their entire life cycles in leaf litter, including deposition of terrestrial eggs that undergo direct development. Other species inhabit the lower canopy only, using small vines and other water-holding plants (such as bromeliads and leaf axils) for egg deposition and tadpole development. Species that are restricted to specific areas in forests that are subject to timber-harvesting operations may be at risk as a result of structural and compositional changes in their habitats.

Behavior

Herpetofaunal studies are inherently more difficult to carry out than studies on many other animal groups because reptiles and amphibians are frequently secretive, cryptic, fossorial, and nocturnal, and most cannot easily be trapped. Some species may also occur at low densities and other species are highly seasonal. Even during a season when they are active, some species may be active only at times of specific climatic conditions. The study of microhylid species richness and abundance, for example, is very difficult because most species are fossorial (subterranean), and they come above ground to breed only after very heavy rainfall (Gascon 1991; Wild 1995; Caldwell 1996a). In 230 hours of survey work on amphibians relative to habitat type, Pearman (1997) found 51 percent of a total of 75 species of amphibians known to occur in the Amazonian basin of eastern Ecuador. Goldingay et al. (1996) found that plot-based surveys of amphibians and reptiles in New South Wales, Australia, yielded two of 15 frog species known from the region and 15 of 28 known reptile species. In the case of reptiles, snakes always present difficulties in surveys because they are typically found in low numbers and at unpredictable times (Goldingay et al. 1996). These examples illustrate that long time periods are necessary for adequate surveys in tropical forests, and that the method of survey can greatly influence the outcome of the results. Further knowledge of the biology of individual groups must be considered when conducting surveys.

Future Research and Management Recommendations

A Need for Long-Term Studies

The few existing observations and studies available indicate that selective logging, followed by a recovery period, can have variable effects on herpetofaunal populations. One of the major influences appears to be whether road construction necessary to reach selected trees creates new areas of standing water, which in turn provides new breeding areas for amphibians. Densities of certain amphibians may increase under these circumstances, but no quantitative data have been obtained.

Carefully designed long-term studies are needed to confirm the effects of different logging regimes on reptile and amphibian species richness and abundance (see chapters 18 and 19). Local (Vitt 1996), seasonal, and yearly (Duellman 1995) variation in species composition and abundance may change by orders of magnitude, and short-term studies will not provide adequate information for addressing questions about the effects of logging in tropical forests on herpetofaunas. Long-term studies are required in order to separate natural variation in reptile and amphibian populations from variation induced by logging or other anthropogenic activities (e.g., Pechmann et al. 1991). Determining cryptic indirect effects of habitat change on herpetofaunas also requires integrated study by a diverse group of highly trained scientists—including systematists and ecologists.

Although such experiments are heuristically attractive, they would require large teams of experienced researchers to spend extended periods in the field to produce conclusive results (see Zimmerman and Simberloff 1996). Such experiments could not be designed without considerable background data on the natural history, activity, and ease of capture for resident species—most of which does not exist.

Techniques to Minimize the Negative Effects
of Logging on Herpetofauna

In areas that are slated for logging, negative influences on the herpetofauna can be lessened by a variety of practices that could be implemented within existing logging technology. We make the following recommendations with regard to conserving as much of the herpetofauna and its habitat as possible.

Protect Interior Forest Conditions

Many amphibians and reptiles require moist, relatively shady conditions. By maintaining as much of the intact canopy as possible, lower levels of the forest will not be subject to the drying influences of intense sunlight. In tropical forests, a significant component of the herpetofauna is composed of small, leaf-litter-inhabiting species. These species rely on a dense mat of moist litter for shelter and food (tiny to small arthropods). Preserving the forest canopy will also aid in preserving the leaf litter habitat.

Minimize Incidental Damage

Silvicultural techniques that minimize the amount of incidental damage to forests should be used to the extent possible (see chapter 21). While most tropical logging is selective because a relatively small proportion of the trees have commercial value (Grieser Johns 1997), the incidental damage to other forest components can be great. Minimizing this damage in terms of reducing the size of gaps and the number of other trees killed could have a significant positive effect on the herpetofauna.

Avoid Soil Compaction

Whenever possible, soil compaction should be reduced by logging during dry periods. Many species of tropical reptiles and amphibians are fossorial or burrowing or spend significant periods of time below ground.

Maintain Interior-Forest Connectivity

Maintaining corridors of habitat throughout forested areas (see chapter 23), especially along watercourses, can have positive effects on the herpetofauna. Corridors of habitat connect patches of undisturbed forest, thus facilitating dispersal of animals that will not enter open areas. Studies to determine how wide corridors must be to allow the exchange of reptile and amphibian species among forest fragments are needed.

Minimize Edge Effects

Many problems are associated with edge effects. Unfavorable microhabitat conditions (high temperature, low humidity) can extend a large distance from the forest edge into the undisturbed forest (Laurance and Bierregaard 1997), resulting in a large area of habitat unsuitable for reptiles and amphibians. One possible method for reducing these effects would be to establish a tree plantation at the edge of the natural forest (Grieser Johns 1997)—using this method could ensure that favorable conditions for organisms would exist in the natural forest.

Provide Adequate Recovery Periods

Logging schedules should be adjusted to allow adequate recovery times of cutover areas or areas where selective logging has occurred. Length of time for regeneration is variable, depending on many factors, with projections ranging from 35 to over 100 years for full structural diversity to occur (Grieser Johns 1997). Each area is unique, and a balance between economic interests and conservation of biodiversity must be struck.

These recommendations, if carried out in areas slated for logging, should help considerably in conserving populations of reptiles and amphibians in tropical forests. These recommendations will not be successful without an overall plan for sustainable management of tropical forests, such as the guidelines for conservation of biological diversity proposed by Blockhus et al. (1992). The cornerstones of this plan include creation of a protected system linked by natural forest corridors, and the establishment of biodiversity monitoring programs to closely follow changes in diversity, with the implementation of remedial action before changes are irreversible. Long-term goals should balance economic benefits with environmental costs to ensure that all components of tropical biodiversity, including reptiles and amphibians, are maintained.

ACKNOWLEDGMENTS

Four peer reviewers and the editors added considerably to the manuscript's organization and content. The National Science Foundation (DEB-9200779 and DEB-9505518) supported the data collection and preparation of this paper. Permits to conduct research and to collect reptiles and amphibians in Brazil were issued by the Conselho Nacional de Desenvolvimento Científico e Tecnológico (CNPq, Portaria MCT no. 170, de 28/09/94) and the Instituto Brasileiro do Meio Ambiente e dos Recursos Naturais Renováveis (IBAMA. no. 073/94-DIFAS), respectively, under a research agreement between the Oklahoma Museum of Natural History and the Museu Paraense E. Goeldi (MPEG). We particularly thank Dr. T. C. S. Avila-Pires of the MPEG for facilitating our work in Brazil.

Chapter 12

THE IMPACTS OF SELECTIVE LOGGING ON TROPICAL FOREST INVERTEBRATES

Jaboury Ghazoul and Jane Hill

Invertebrates dominate the animal component of tropical forest habitats in diversity, abundance and biomass. The estimated global number of invertebrate species is five to 15 million (May 1988; Stork 1988; Gaston 1991) compared with less than 45,000 vertebrate species. Described invertebrate species alone outnumber vertebrates by a factor of 30. Invertebrates also comprise over 90 percent of the animal biomass of tropical rain forests—a third of which is composed of ants and termites (Wilson 1987).

Invertebrates fill a wide variety of niches, occupying all parts of the forest from the canopy to the soil and streams. Many tropical forest invertebrates are spatially and temporally highly segregated with some species having very restricted distributions, feeding preferences, and activity periods (Willmer and Corbet 1981; Hanski 1987). Their small body sizes and highly fluctuating populations makes it difficult to predict the degree and direction of invertebrate responses to changing biophysical factors, and population responses to disturbance may be highly sensitive to population size or stage of the life cycle at the time of the disturbance (Pimm 1991).

Little is known about invertebrates in tropical forests, despite their dominance of the animal community. Most are undescribed scientifically, and for many named species little is known about their trophic associations, habitat requirements, or interactions with other community members. Nevertheless, it is widely recognized that invertebrates are the primary mediators of a large array of forest ecosystem processes and have central roles in decomposition, nutrient cycling, pollination, and seed predation—all of which contribute to forest regeneration and resilience to disturbance (Janzen 1987; LaSalle and Gauld 1993; Bond

1994). The taxonomic complexity and high diversity of invertebrates makes them challenging subjects to study, and there is an urgent need to understand how disturbance-induced changes in invertebrate assemblages affect ecological functioning (Didham et al. 1996).

This chapter reviews our knowledge of the impacts of selective logging on tropical rain forest invertebrate communities. We summarize studies that investigate the impacts of logging on specific taxa and then consider impacts on invertebrate-mediated processes and trophic structure. Invertebrate responses at the landscape scale are explored. The use of insects as indicators of biodiversity and disturbance in logged forest is evaluated. Finally, we outline options for improved forest management that recognizes the roles of invertebrates, and helps to conserve their populations.

Consequences of Selective Logging for Invertebrate Microhabitats and Resources

Impacts of Logging on Forest Microclimate and Insect Microhabitat

In undisturbed rain forest, closed canopy shades the forest interior and produces a relatively dark, cool, and humid microclimate (Fetcher et al. 1985). Following canopy fragmentation, increased irradiance on the forest floor elevates daytime temperatures, while reducing humidity (conditions are reversed at night due to high convection) (Bastable et al. 1993). Tolerance of invertebrates to widely variable local climates is limited, as most invertebrates are incapable of physiological temperature regulation. Changing climatic regime in selectively logged areas, therefore, may cause a decline in the populations of invertebrate taxa occurring in microhabitats that normally have restricted climatic variation—such as the understory, leaf litter, and soil (e.g., Alvarez and Willig 1993).

Selective logging also affects the condition and variety of invertebrate microhabitats, through structural changes such as soil compaction, decreased density of trees and associated epiphytes, increased woody debris and siltation of watercourses (Ewel and Conde 1978; Uhl and Vieira 1989), and alterations of the local climatic regime (Howden and Nealis 1975; Chiariello 1984). High habitat heterogeneity at the landscape scale is typical of selectively logged areas throughout the tropics (see next section), but structural complexity at local scales may be reduced. Where logging simplifies structural complexity, arthropod diversity may decrease correspondingly (Holloway et al. 1992; Gardner et al. 1995; Rith Najarian 1998). To date, no detailed attempt to quantify the abundance and variety of invertebrate micro-habitats in disturbed and undisturbed areas has been made.

Short generation times and high potential fecundity provide inverte-brate populations with high resilience to disturbance. Changes in population densities and species diversity of invertebrates due to microhabitat loss and alteration are part of long-term forest regeneration dynamics. Butterflies, for example, are sensitive to environmental gradients (Kremen 1992) and in dry conditions many species, particularly satyrids beetles, undergo marked but temporary reduction in abundance and extent of local distribution (Spitzer et al. 1997; Braby 1995; Hill 1999). The extent to which invertebrate populations recover from the impacts of selective logging depends on how closely logging mimics natural disturbance and how regeneration proceeds (see chapter 22).

Impacts of Logging on Invertebrate Resources

Due to short generation times and high reproductive rates, the potential for invertebrate populations to recover following logging is rapid, provided sufficient resources are available. Changes in the abundance of wood-boring beetles and invertebrates belonging to the wood decomposer guild, for example, are correlated with the amount of woody debris following logging in temperate forests (Heliovaara and Vaisanen 1984; Martikainen et al. 1999, 2000). Similar responses might be expected in the tropics with, for example, increases in the wood boring scolytid (bark beetles) and cerambycid (longhorn beetles) beetles and increased abundance of wood-feeding termites. Given innumerable invertebrates exploiting a wide range of forest resources, any single disturbance is likely to decrease the resources used by some, while increasing resources required by others. To date, invertebrate responses to increased resources resulting from selective logging in tropical regions has not been tracked, but studies from temperate regions have successfully correlated invertebrate abundance with the availability of resources over periods of several years following a logging event (Huhta 1976). In these populations, invertebrate abundance gradually returned to the original predisturbance state as the resource returned to pre-harvest levels. In tropical forests, similar responses might be expected, but a return to the original state might be faster given higher turnover rates. The ways in which disturbance can increase or decrease resource availability, at least in the short term, are by:

- Influencing the rate at which resources are supplied
- Affecting the quality of the resource
- Affecting competitor or predator populations
- Causing previously unavailable resources to appear

Increased competition for resources may arise directly by reduction in the amounts of resources produced, or indirectly by the invasion of competitively superior species into disturbed forest areas. Particularly susceptible to disturbance-induced resource competition are invertebrates that depend on patchy and ephemeral resources, such as coprophagous dung beetles and nectar feeding insects (MacArthur and Pianka 1966; Hanski 1987). Changing environmental conditions such as increased water stress arising from logging-induced microclimatic changes (Bastable et al. 1993), for example, may result in reduced flower and nectar production (Willmer and Corbet 1981), while elimination of large vertebrates may reduce the availability of dung to coprophages. Competitively superior species that invade disturbed forests include generalist honeybees (*Apis mellifera*) that can deplete nectar and pollen resources to the detriment of other flower visitors, and often do not facilitate effective pollination (Buchmann and Nabhan 1996, Roubik 1996; but see Butz Huryn 1997). The foraging behavior and abundance of primary forest flower visitors may be altered as a result of resource depletion by honeybees (Butz Huryn 1997), leading to decreased fruit set of forest plants (Paton 1993). Other examples include ants (Hoffmann et al. 1999) and wasps (Barr et al. 1996).

Increased abundance of pioneer species and the changing composition of the forest plant community will lead to changes in the quantity and quality of resources, including food and nesting site availability. Habitat destruction, fire, and the spread of invasive grasses due to logging of forest habitats are the primary causes for the decline in nesting site availability and the corresponding decline in solitary bee abundance in Costa Rica (Vinson et al. 1993). Changing conditions can also affect the quality of plant resources for invertebrate herbivores. Increased light levels in the understory might allow previously shaded plants to produce greater quantities of chemical defenses, thereby reducing the quality of the resources for insect herbivores (Newbery and de Foresta 1985). The chemical composition and nutrient content of pioneer plant species tissues may be very different from late successional species (Cates and Orians 1975), increasing or decreasing their palatability to different invertebrate species.

Following selective logging, the invasion of disturbed areas by early successional plant species can provide a new resource base that may lead to changes in the abundance and behavior patterns of forest invertebrates (Ghazoul, in review). It may also impact the spread of invertebrates associated with open areas into forest habitats (Spitzer et al. 1993, 1997). The cascading impacts of changing invertebrate community structure—resulting from changes in the resource base—have yet to be studied in tropical forest environments.

Invertebrate Population Dynamics and Diversity in Selectively Logged Forest

In contrast to vertebrate research (see chapters 4–11 and 14), few studies have considered the direct effects of logging on forest invertebrate communities. The increasing appreciation of the immense richness and diversity of invertebrates in tropical forests, coupled with the extent of human forest resource exploitation, has led some authors to predict global invertebrate extinction rates many times higher than for plant and animal groups (Wilson 1987). Small species ranges, narrow niche breadths, and little adaptive ability of forest organisms to environmental changes, have been attributed as causes of the purported high extinction rates among tropical forest invertebrates (Ehrlich and Ehrlich 1981; Samways 1993a; New 1999). Consequently, calls have been made to place a new emphasis on the conservation of invertebrates (Wilson 1987; Samways 1993b), especially in light of the important functional roles invertebrates play in forest ecological processes (see below). Despite this call for action, the impact of logging activities on invertebrate population dynamics remains virtually unknown. This is due in part to the complex multispecies interactions involved, but also because most studies are conducted over ecologically brief time periods that represent only a small part of the postdisturbance community dynamic. Extrapolation from existing studies is difficult, as the relationship between the disturbance magnitude (or time since the perturbation) and the magnitude of the response is rarely known and may not necessarily be linear. Nevertheless, certain predictions about population recovery following disturbance events

can be based on features of the life cycle and feeding guild of invertebrate groups, and these might be tested using results from the increasing number of studies on the diversity and abundance of invertebrates
in selectively logged forests. Logging impact studies on the diversity
and richness of several invertebrate taxa—notably butterflies and termites—provide the basis for a preliminary analysis (see below).

The impacts of selective logging operate on invertebrate community
composition at different time scales:

- Logging changes the local climate and soil structure which can cause
  rapid mortality among invertebrates that are particularly sensitive to
  environmental conditions
- Changes caused by disturbance in the availability of resources (e.g., food
  and nesting sites) can alter species diversity more slowly through their
  effects on growth, reproduction, and competitive interactions
- On an evolutionary time scale the proportion of species in the community that are preadapted to survive the disturbance or that adapt to the
  changes it causes will increase as natural selection eliminates sensitive
  species and favors resistant species

Biological communities in forests that have been subject to a long
history of anthropogenic disturbance may be more resistant to disturbances as a result of the evolutionary filtering and prior elimination of
sensitive species (Balmford 1996; see chapter 22). Although many
tropical forests have been subject to a history of natural and human
disturbance (Bush et al. 1992; Foster 1999), the scale in recent
decades, in contrast to many temperate broadleaved forests, far
exceeds previous disturbance intensities (see chapter 1). Owing to the
long time periods over which evolutionary changes occur, and the relatively brief period of commercial logging, it is difficult to accurately
assess logging impacts on invertebrate extinction and future community structure. We begin to answer this question, however, by studying
community changes in the short and medium term, and by determining patterns of responses from different invertebrate taxa.

In conducting studies on invertebrate community dynamics, care has
to be taken in the selection of taxa and the taxonomic level of the analysis. The complexity of community dynamics has frequently been
approximated by considering species and genera collectively according
to some higher taxonomic grouping. While this makes for easier analysis, it undoubtedly obscures many wider patterns. Observations on
ground beetles (Lenski 1982) and butterflies (Hill et al. 1995; Ghazoul

in review), and to a lesser extent, spiders (Coyle 1981), reveal contrasting responses of genera within single families. They also illustrate how it may be misleading to consider responses according to a taxonomic grouping rather than by guild or functional association. It is important, therefore, to consider ecological as well as taxonomic associations in determining invertebrate responses to perturbation. The rest of this section describes short- and medium-term responses of a few invertebrate groups to selective logging—in a taxon-by-taxon review—and illustrates the often-contrasting responses of species occurring within the same higher taxon.

Butterflies

Tropical butterfly assemblages are particularly diverse, with many endemic species, most of which are dependent to some extent, on closed-canopy forest (Collins and Morris 1985; Sutton and Collins 1991). Compared with other invertebrate groups, the high visibility of butterflies, together with their relatively well known taxonomy, has resulted in this group receiving a reasonable amount of attention throughout the tropics. Data on the effects of selective logging and other forms of intermediate forest disturbance on butterfly assemblages are accumulating (Bowman et al. 1990; Raguso and Llorente-Bousquets 1990; Kremen 1992; Spitzer et al. 1993, 1997; Hill et al. 1995; Hill 1999; Hamer et al. 1997; Lawton et al. 1998). These studies have concentrated on the effects of selective logging on species richness and relative abundance, and have used a variety of methods—including transect counts (modified from techniques used in temperate regions, Pollard 1977) and collecting with hand nets and baited traps.

Selective logging generally results in more dense forest with smaller trees, with greater vegetation cover at ground level, and decreased canopy cover (e.g., Hill et al. 1995). Changes in butterfly assemblages have been related to these changes in vegetation structure (Hamer et al. 1997), with both increased (Spitzer et al. 1993, 1997; Hamer et al. 1997; Lawton et al. 1998) and decreased (Bowman et al. 1990; Hill eet al. 1995) butterfly diversity reported (see table 12-1). There are many possible explanations for these contrasting results (e.g., differences in severity of disturbance, or differences in forest recovery rates), but currently there are insufficient studies to resolve these confounding effects. Butterfly flight behavior also often changes as logging alters forest structure, and it is difficult to disentangle sampling effects from the actual effects of vegetation changes on butterfly community composition.

TABLE 12-1 *Responses of Forest Butterfly Communities to Disturbance*

Location	Logging Intensity or Habitats Monitored	Response to Disturbance or Canopy Openness	Correlation Between Species Range and Abundance Along a Disturbance Gradient	Reference
North Vietnam	Forest, transitional scrub, and ruderal habitat	Increase in species richness and species diversity	+	Spitzer et al. 1993
North Vietnam	Closed canopy forest and forest gaps	Increase in species richness and species diversity	+	Spitzer et al. 1997
Buru, Indonesia	Unlogged (31% cover) and logged (19% cover) forest	Decrease in species richness and species diversity	+	Hill et al. 1995
Sumba, Indonesia	Canopy cover range from 54% to 21%	Increase in species richness and species diversity	+	Hamer et al. 1997
Mbalmayo, Cameroon	Six sites along a disturbance gradient from "near primary" to "farm fallow"	Decrease in species richness	no data	Lawton et al. 1998

It has been assumed that the presence of adult butterflies in an area reflects local availability of butterfly resources (e.g., larval host plants) (deVries 1988; Beccaloni 1997). The availability of these resources, however, and the effects of habitat disturbance on them, have not been investigated (see Ghazoul, in review, for an exception involving butterflies and nectar resources; see photo 12-1). A more consistent response to disturbance is the loss of species with restricted distributions (see photo 12-2), with high biogeographic and/or taxonomic distinctiveness (Spitzer et al. 1993, 1997; Hill et al. 1995; Hamer et al. 1997)—probably because these species have narrow habitat requirements (Thomas 1991b). In terms of global biodiversity conservation, the loss of restricted range or endemic species is more important than the loss

PHOTO 12-1 *Catopsilia pomona* is common in the canopy of undisturbed dry forests in southeast Asia, but in open disturbed forests it occurs more frequently in the understory, responding to the increased profusion of herbaceous flowering plants and nectar resources as a result of increased light levels in the understory. (R. Davies)

PHOTO 12-2 *Delias descombesi*, a relatively rare butterfly that decreases in abundance with increasing disturbance (Ghazoul 1999). (R. Davies)

of diversity *per se* (Vane-Wright et al. 1991). Further work is needed to see if butterfly assemblages come to resemble those in unlogged forests after longer periods of habitat recovery, or whether repeated selective logging will result in the loss of these species with high conservation value.

Moths

Seven years after intensive (unquantified) logging of moist, lowland evergreen forest in Sabah, Borneo, Holloway et al. (1992) recorded significant losses of diversity and changes in the taxonomic composition of moths on local and geographic scales. Moths are specialist herbivores, and their assemblages are likely to change in concert with changes in floristic composition following disturbance events. On the larger spatial scale, Holloway et al. (1992) argued that different moth assemblages (whose species have restricted geographic distributions) are associated with each of the components of the complex mosaic of habitats in primary forest (which includes successional phases, riparian corridors, ridgetop forest, and high canopy forest) (see also Willott 1999). Logging events that are followed by the widespread invasion of pioneer plant species, and subsequent reductions in habitat heterogeneity, are expected to negatively impact vulnerable habitat-specific species, even though overall diversity may increase (due to an invasion of opportunistic species. Small taxonomic groups consisting of a high proportion of endemics and localized species (such as the Bombycoidea family), for example, are replaced by geographically widespread species associated with disturbed and open habitats (e.g., Sphingidae; see photo 12-3) (Holloway et al. 1992).

Beetles

Little research has been conducted on the impacts of logging on beetles (see box 12-1). They are the most diverse group of insects in tropical forests, and their ecological and taxonomic complexity may contribute to their apparent neglect. The studies reported here show that different groups (e.g., dung beetles [scarabids] and ladybird beetles [coccinellids]) respond to disturbance in divergent ways, which re-emphasizes the difficulties in how best to group organisms for ecological interpretation (see box 12-1).

PHOTO 12-3 Hummingbird hawkmoths appear to be associated with disturbed and open habitats. This *Macroglossum* sp. is pictured visiting the flowers of *Sindora siamensis* in a dry forest in Thailand that has been subject to repeated logging and burning. (J. Ghazoul)

Box 12-1 *Responses of the invertebrate community fifteen to twenty years after selective logging in the Kibale Forest of Western Uganda. (Matti Nummelin)*

Tropical invertebrate responses to logging during the initial years following the disturbance event are better documented than responses over longer time scales, which are largely unknown. Long-term assessments of ecosystem recovery and invertebrate community resilience are needed for conservation planning and sustainable use management.

This case study is drawn from a series of studies conducted in the Kibale Forest of western Uganda (Nummelin 1989, 1992, 1996; Nummelin and Hanski 1989; Nummelin and Borowiec 1991; Nummelin and Fürsch 1992), and summarizes aspects of invertebrate community responses fifteen to twenty years postlogging. The Kibale forest is a moist, evergreen, midelevation (1,500 m) forest with an average yearly rainfall of 1,500 mm, most of which falls in a distinct wet season (see chapters 4 and 10). About 35 percent of the northern and central forest blocks were selectively harvested in 1969, and large gaps in logged areas remain over twenty years later. The intensity of logging ranged from 14 m^3ha^{-1} to 21 m^3ha^{-1} of timber extracted. Invertebrates at both logged and unlogged sites were sampled from the forest floor vegetation by sweep netting and pitfall trapping.

Community Structure

Over 95 percent of the arthropod fauna caught by sweep netting the forest floor vegetation consisted of eight invertebrate orders (Araneae,
Hymenoptera, Hemiptera, Homoptera, Coleoptera, Orthoptera, Lepidoptera, and Diptera). The proportions of these invertebrate taxa were
similar in logged and unlogged forests, and remained similar throughout
the wet and dry seasons. Canonical correspondence analysis, however,
showed that logging significantly affected the relative proportions of
invertebrate taxa within the taxonomic groupings of a forest community.

At the species level, tortoise beetles, grasshoppers and ladybirds were
more species-rich in logged sites than in virgin forest (see box table
12-1). Only dung beetles declined in richness in disturbed areas. Grasshoppers, for example, were represented by thirteen species in the disturbed sites, but only eight of these occurred in undisturbed sites. All
grasshopper species in the undisturbed sites were flightless (e.g.,
Auloserpusia synpicta and *Gemeneta* spp.), indicating specialization to a
constant environment with little need for dispersal. Invertebrate communities in open and densely vegetated logged sites very likely include,
in addition to forest specialists, widely dispersing species which favor
open and edge areas.

A difference in response to logging among species within a taxonomic
grouping was also defined along trophic guilds. Coccinellid beetles
(ladybirds) were more abundant in logged forest than unlogged sites,
and although predaceous species (e.g., scymnid ladybirds) dominated all
the samples, herbivores (e.g., *Epilachnini* spp.) were much more abundant in logged sites.

Seasonality

High light levels at the forest floor, in combination with the seasonal rainfall pattern in the Kibale Forest, results in a lush forest floor vegetation in
logged sites during the late rainy and early dry seasons (January to
March). As the dry season progresses, the vegetation dries in the high
insolation environment of logged sites, leaving few resources for herbivorous arthropods. In unlogged forests, the density of forest floor vegetation remains relatively constant. Invertebrate abundance tracks resource
availability. Arthropod numbers in logged forests peaked during the late
rainy and early dry season (Nummelin 1989, 1996) and exceeded late dry
season samples by four to five times. Peaks in arthropod abundance in
undisturbed forests were also evident, but these did not coincide with
rainfall patterns. Different patterns of invertebrate seasonal abundance
across logged and unlogged sites may largely be due to the presence in
logged forest of invertebrates normally associated with open and highly-
seasonal grassland environments. Many of these species respond to
ephemeral resources by rapidly increasing adult population densities and
passing resource-limited periods as dormant pupae.

Large between-year variations in the coccinellid community in the
Kibale Forest, in both logged and unlogged sites, are also evident. During 1983, two small predaceous species—*Scymnus kibonotensis* and *S.
nummelini*—were very common and dominated the coccinellid community. These species disappeared from samples in 1984 and were not

found two years later. Small numbers of these species were found in samples eight and ten years later from the same sites.

Conclusions

These studies illustrate two key points about invertebrate response to logging. First, analyses conducted only on higher taxonomic categories can be misleading and might not reflect the often contrasting and highly variable responses at lower taxonomic levels. Fifteen years after logging, the relative proportions of invertebrate orders remained similar to undisturbed sites, but differences among proportions of species within these higher taxa were observed.

Secondly, logging affects the temporal variation of populations. More pronounced annual environmental extremes in disturbed site climate and vegetation density results in far larger variance amplitudes in invertebrate abundance than that of undisturbed sites. This clearly has important implications for the timing of invertebrate monitoring programs in disturbed sites and for pest management in secondary forests.

Fifteen years after selective logging, the forest arthropod community in the Kibale Forest shows no gross changes at higher taxonomic levels, but does show changes in the proportion and seasonal abundance of component species. Some invertebrates associated with primary forests have become much less abundant—or have even disappeared entirely—from disturbed areas. Other invertebrate groups (notably insect herbivores), increased in abundance in the logged sites, and this may have long-term impacts on sapling growth and forest regeneration. Further study is needed to determine whether changes in the forest invertebrate community have affected the functioning of ecological processes.

BOX TABLE 12-1 *Numbers of Insect Species Recorded from Sweep Net Samples in Logged and Unlogged Forest at Kibale Forest, Uganda*

Insect Group	Trophic Guild	Unlogged	Logged	Shared Species
Ladybird beetles	Predator	28	44	20
Tortoise beetles	Herbivore	7	13	6
Dung beetles	Detritivore	19	17	15
Grasshoppers	Herbivore	8	13	6

Howden and Nealis (1975) reported an almost total absence of forest dung beetles in man-made clearings (size unspecified) located within undisturbed rain forest in the Colombian Amazon, and attributed this result to microclimatic effects on the quality of the dung resource. Selective logging, however, had no apparent affect on the composition or richness of dung beetle assemblages in Sabah six years after logging (Holloway et al. 1992, extraction intensity not specified), or in Uganda

16 years after logging at $21m^3ha^{-1}$ and $14\ m^3ha^{-1}$ extraction intensities (Nummelin and Hanski 1989). Provided there is a sufficient resource supply (i.e., dung from vertebrates) selective logging may have little impact on the abundance and species richness of dung beetles.

Nummelin and Fürsch (1992) report higher species richness of ladybird beetles in logged sites ($21m^3ha^{-1}$ and $14\ m^3ha^{-1}$ timber extraction in 1968 to 1969) than unlogged forest sites at Kibale in Uganda (38 and 44 species compared to 28 and 12 in unlogged forest), although large differences in the numbers of individuals caught (1,146 in logged compared to 595 in unlogged forest) cast doubt on the validity of this conclusion. Clear differences in species composition in the two habitat types were expressed among trophic guilds, with herbivores almost totally restricted to disturbed sites—apparently due to the increased herbaceous vegetation (see box 12-1).

Earthworms

Earthworms may comprise the highest biomass among tropical soil macrofauna (Odum and Pigeon 1970; Fragoso and Lavelle 1972), and they have important roles in regulating soil processes (Marinissen and De Ruiter 1993). Despite their apparent importance, however, there are no studies that directly assess the impacts of logging on earthworms. Lavelle and Pashanasi (1989) reported a decline in earthworm density and biomass from primary forest to 15-year-old secondary forest. Changes in earthworm community structure during secondary succession have been recently studied in tropical pastures in Puerto Rico (Zou and Gonzalez 1997; see chapter 13) where earthworm density decreased, but diversity increased as succession proceeded from abandoned pasture to regenerated secondary forest. This was associated with an increase in the sources of organic inputs to the soil from a belowground dominated pattern to a more balanced below- and aboveground litter input. Similar studies in the Western Ghats of India showed an increase in species diversity and richness with increasing forest cover, but unchanged biomass and density (Blanchart and Julka 1997). These contrasting results illustrate the need for further studies and the likely importance of local environmental conditions affecting earthworm communities.

Termites

Termites, which are important mediators of decomposition in tropical forests and occur as soil feeders, wood feeders, or litter foragers, generally are affected by habitat change (Burghouts et al. 1992; Holt and Coventry 1988; Lepage 1984). Soil termite communities in natural forests in the Western Ghats of southern India, for example, were greatly affected when the habitat was disturbed by even a small amount of logging (described as *slightly logged in the past*). Maximum termite species richness declined from 10 species in undisturbed natural forests to four species in disturbed forests, and termite density and biomass showed similar trends (Basu et al. 1996). In Cameroon, where termites reach their highest diversity and richness, they appear to be relatively resistant to a moderate amount of disturbance (site logged 30 years prior to study; tree density 133 ha^{-1} compared to 533 ha^{-1} in near primary site) (Eggleton et al. 1995, 1996). Indeed, selective logging at old secondary forest sites in Cameroon resulted in an increase in termite species, probably due to the increased supply of dead wood (Eggleton et al. 1995). More recent data from undisturbed sites in Sumatra and Cameroon, however, show that species richness of soil feeding termites is sensitive to changes in canopy cover and forest structure (P. D. Eggleton, personal communication). Termites in Cameroon do not appear to be limited by tree species number or type, and there is little change to the original termite assemblage so long as they are provided with a more or less closed canopy and adequate resources. More intense logging practices that lead to forest clearance do, however, result in the gradual depletion of soil feeding termites to almost nothing over a period of months and years (Eggleton et al. 1996). Recovery of termite populations is enhanced, however, when dead wood is left on the ground following perturbation (Davies et al. 1999). The correlation between overall abundance of wood-feeding termites and the availability of large logs indicates that high clearance intensities may also cause declines in the wood-feeding termite biomass.

Ants

Ants comprise one third or more of the total insect biomass and abundance in tropical forests (Fittkau and Klinge 1973, Levings and Windsor 1982), and are important components of the forest ecosystem through their physical and chemical alteration of the environment

(Folgarait 1998). Despite this, only two published studies in Ghana (Belshaw and Bolton 1993) and Madagascar (Olson and Andriami-adana, in press; quoted in Belshaw and Bolton 1993) have compared leaf-litter ant communities in primary and regenerating secondary forest (cleared eight to 50 years earlier; no further information on logging intensities given). Both found no significant differences in diversity or composition. Likewise, ant communities from forest understory and natural treefall gaps in primary moist lowland forest in Panama were not different in abundance, richness, composition, or diversity (Feener and Schupp 1998). There are no data available on canopy ant fauna, which are taxonomically very different from leaf litter ants (Belshaw and Bolton 1993). Sociality undoubtedly contributes to the ecological success of ants among terrestrial invertebrates, but Wilson (1990) found no evidence that sociality may contribute to reduced vulnerability to disturbance, and indeed suggested that requirements for colony establishment may make ants less resistant to perturbations. Extrapolation from very few field studies is clearly inappropriate and more research is required before reaching conclusions about disturbance impacts on this very important invertebrate group. Nevertheless, early indications suggest that ants may be more resistant to anthropogenic disturbances than other arthropod groups.

The Impact of Selective Logging on Insect-Mediated Ecological Processes in Tropical Forests

Invertebrates mediate several important ecological processes, including decomposition, pollination, and plant competition, and there has recently been a shift from the study of the impact of logging on populations to its impact on ecological processes (Didham et al. 1996; Ghazoul et al. 1998; Guariguata and Pinard 1998). Attention has been focussed on the ways in which organisms interact and how disturbance affects these interactions and the processes they regulate. Loss of biological diversity has been correlated with a decrease in ecosystem functional efficiency in experimental microhabitats where the diversity and abundance of species are carefully controlled and measures of ecosystem function are monitored (Naeem et al. 1994; Symstad et al. 1998). These experiments suggest that events that affect community diversity can potentially disrupt ecological processes such as nutrient cycling and tree regeneration. How well experimental systems translate to real life habitats still remains uncertain. Compensation by previously insignificant species, for example, may result in no change to the ecological functioning of the community, but the extent of compensation in biologically complex habitats is largely unknown (Pimm 1991). This compensation is most clearly illustrated in plants that have a few specialist and efficient pollinators (i.e., the predominantly moth-pollinated dipterocarp tree [*Dipterocarpus obtusifolius*] in southest Asia [Ghazoul 1997]), but are also visited by a wide variety of generalist flower visitors—many of which effect pollination. Recent studies of natural, managed, and disturbed ecosystems have shown that individual species population dynamics are more sensitive to stress than ecosystem processes (Vitousek 1990). This implies that changes in species composition might not have a large impact on ecosystem functional performance, and that even the limited number of *keystone* species that drive the ecological processes controlling ecosystem functions may change over time (Lawton and Brown 1993).

Insect Pollination Systems and Plant Reproductive Success

The reproductive dependence of plants on pollination is mediated by the probability of failure of particular pollinator-plant mutualisms, the degree of dependence on the mutualisms, and the importance of seeds in the demography of the plant. Many plants are self-pollinated or have a range of diverse pollinators, ensuring that perturbations in pollinator species populations have little impact on seed set. Reproduction by vegetative propagation can also help to compensate for reduced seed set where pollinator mutualisms are disrupted. These compensatory mechanisms may be less apparent in tropical forests where there are high levels of self-incompatibility or dioecy (Bawa et al. 1985), and more specialized plant-pollinator mutualisms (Regal 1982).

Plants in lowland tropical forests tend to have more specialized pollinator and disperser relationships than temperate regions (e.g., Regal 1982), and pollination by a single species remains infrequent (notable exceptions include *Yucca*, orchids, and figs). Most species of trees are pollinated by a few pollinator species that belong to the same taxonomic group (Bawa 1990), but within each pollinator group, one or two species contribute a disproportionate number of pollinations. The dependence of some tropical plants on relatively few pollinators, and the comparative lack of compensatory mechanisms, makes them particularly susceptible to disturbance-induced breakdowns in pollinator mutualisms (e.g., Ghazoul et al. 1998, see below). No study, however, has unequivocally demonstrated plant reproductive failure by the extinction or population decline of an invertebrate pollinator species through human disturbance.

Disturbance by logging need not impact directly on pollinator species populations to affect pollinator mutualisms. By causing changes in forest structure and composition, logging can cause changes in the foraging behavior of pollinators, thus affecting plant reproductive output. *Shorea siamensis*, a selectively logged dipterocarp tree in Thailand, had reduced reproductive output as a result of changes in the foraging behavior of its small bee pollinators (see photo 12-4) (Ghazoul et al. 1998). Bees remained longer within single trees in areas where logging had reduced *S. siamensis* density from 86 to 22 trees per hectare and created large distances (exceeding 20 m) between adjacent flowering trees. Consequently, little cross-pollination was effected and fruit set was considerably lower than at nearby unlogged sites (Ghazoul et al. 1998). Reduced fruit set has long-term implications for the ability of *S. siamensis* populations to recover in disturbed areas, and the genetic structure of future populations is likely to be affected as reduced outcrossing rates among trees in disturbed regions results in relatively inbred seed. Extraction of *S. siamensis* also led to increased abundance of understory herbaceous plants, which attracted canopy-foraging butterflies to the understory. The pollination success of the butterfly-pollinated tree, *Dipterocarpus obtusifolius*—which occurs in the same geographic area as *S. siamensis*—was negatively affected as a result (Ghazoul, in review). This illustrates the indirect effects of selective logging on the reproductive ecology of a nontarget species.

Seed Predation and Seedling Establishment

Insect seed predators destroy a large portion of the seed crop in tropical forests, and by doing so prevent numerical dominance by one or a few species of highly productive plants, while contributing to the local distribution of plant populations. Herbivores of flowers and fruits of a rain forest herb, for example, caused more damage in closed forest than natural forest gaps and limited seedling recruitment under mature forest (Calvo-Irabien and Islas-Luna 1999). In tropical southeast Asia, weevils can destroy up to 90 percent of the seed crop of the dominant dipterocarp trees (see photo 12-5) (Toy 1988). In Central America, another small weevil destroyed 80 percent of the seed crop of *Ateleia herbert-smithii*—a fast-growing legume that would otherwise quickly dominate secondary forest (Janzen 1987). The loss of insect seed preda-

PHOTO 12-4 *Trigona fimbriata* and other small bee pollinators of the tree *Shorea siamensis* spend more time visiting flowers of isolated trees than clumped trees, causing reduced fruit set of isolated trees due to high selfing rates. (J. Ghazoul)

PHOTO 12-5 *Alcidodes sp. nova*, a weevil seed predator from western Thailand, recently emerged from the fruit of a dipterocarp tree (*Dipterocarpus obtusifolius*). (R. Davies)

tors is most likely to occur through the destruction of the sites in which the adult beetles pass their time in the periods between seed crops. For many beetles, these sites are largely unknown and it is impossible to predict the impact of disturbances without this information.

Nutrient Cycling and Decomposition

Invertebrate soil fauna, together with fungi and bacterial microorganisms, are important regulating agents of soil processes and perform a variety of functions that contribute to the decomposition of organic matter. Soil microorganisms, the primary decomposers, are dependent on the soil fauna to convert organic matter to a form that they can use (Didden 1990). Organic matter is initially broken down by macrofauna such as earthworms, termites, ants, millipedes, and other groups. Their direct effect on organic matter decomposition occurs through the burial of plant detritus, improving its availability to microbes. They also contribute to the physical rearrangement of soil particles and, as a result, patterns of water and nutrient infiltration. Mesofauna, which includes mites (*Acari*) and springtails (*Collembola*), have a largely indirect effect on decomposition by changing soil pore sizes, which affects water transport and oxygen availability (Didden 1990)—although they also feed on the primary decomposers (fungi and bacteria) and plant detritus. The microfauna are composed mainly of protozoa and nematodes, which feed on microbial tissue and excrete mineral nutrients. The abundance and structure of the decomposer community can vary greatly with climate, soil conditions, and vegetation type (Collins 1980; Lavelle and Pashanasi 1989; Leaky and Proctor 1987). Land management that causes dramatic changes in the vegetation will affect each of these factors.

Primary forest in the Peruvian Amazon had higher decomposer taxonomic richness and biomass than 15-year-old secondary forest, but population densities remained the same (Lavelle and Pashanasi 1989). Following removal of underbrush and small- and medium-sized trees from a forest plot in Nigeria, total insects and oribatid mites declined severely in soils over a one-year period (Lasebikan 1975). Several insect groups (e.g., *Coleoptera* [beetles] and *Psocoptera* [booklice]) were unaffected, while others fluctuated erratically at cleared and control sites. Different responses were observed from families of the same group and among the springtails. Entomobryidae decreased, but Sminthuridae remained unchanged. In Zimbabwe, the abundance and biomass of spiders, millipedes, and beetles in deciduous miombo savanna woodland increased with disturbance (resulting in an open canopy), but Orthop-tera (grasshoppers and crickets) and Dermaptera (earwigs) decreased (Dangerfield 1990). Overall, invertebrate macrofaunal abundance was unchanged, although biomass was higher in disturbed areas. In Argentine semi-arid forest, where human activity has resulted in the conversion of mature forest to selectively logged mixed forest and logged and grazed shrubland (logging intensity unreported), the taxonomic composition of the ground fauna differed at each habitat. There was, however, no difference in overall diversity or abundance (Gardner et al. 1995). These few studies indicate that selective logging most often changes the taxonomic composition rather than the abundance or biomass of the decomposer community. A decline in the abundance of mesofauna and macrofauna has been shown to severely impact overall decomposition rates (Reddy et al. 1995; Tian et al. 1992, 1995; Tian 1998). The significance of taxonomic changes to the functional efficiency of the decomposer biota, which was not measured in any of these studies, remains uncertain. Evidence from controlled environments indicates that changes in the composition of the arthropod community may significantly alter ecosystem performance (Naeem et al. 1994), although this conclusion is contradicted by other studies (Vitousek 1990; Lawton and Brown 1993; Burghouts et al. 1992, see below). Studies that monitor changes in the decomposer fauna, together with decomposition rates, are clearly needed.

Burghouts et al. (1992) monitored both decomposition rates and decomposer community structure in logged (ca. 20 trees exceeding 40 cm dbh removed per hectare, 10 years prior to the study) and unlogged primary dipterocarp forest plots in Sabah, Malaysia. The ground-level invertebrate community in logged plots differed from unlogged plots only in that there was reduced abundance of mites and pseudoscorpions. No difference in decomposition rates between sites was recorded. At both these sites, decomposer invertebrate abundance was correlated with the mass of leaf litter and woody debris, but a further correlation with litter humidity in the logged forest suggests that the decomposer community is affected by changes in the microclimate caused by logging. Leaf quality and flushing patterns of many emergent trees differ from that of pioneers, and are likely to affect forest floor decomposition processes in different ways (Golley 1983; see chapter 13).

Insects and Their Role in the Food Web

Insects and other invertebrates occupy the critical middle links in many ecosystem food webs. The consumption of plant material by insects, particularly during massive defoliation events following insect outbreaks, can stimulate plant growth by mobilizing nitrogen (Lerdau 1996) and prevent dominance by saplings from competitively superior species. Insects are also important sources of food for a variety of other invertebrates and vertebrates. Ants, for example, comprise a high proportion of invertebrate biomass in most tropical forests and are a rich food source for other animals. In Neotropical forests, several bird species are closely associated with army ants and feed on the insects that the ants disturb. The disappearance of army ants from small forest fragments results in the rapid loss of this small and specialized bird community (Willis 1974). Many arthropod groups, including spiders, beetles, mites, flies, woodlice, springtails, bugs, and moths and butterflies, take advantage of the huge food resource that ant nests represent (Hölldobler and Wilson 1990). Other abundant invertebrate taxa are also important food sources to vertebrates and invertebrates alike. Vertebrates are often quite specific in the invertebrates they prey on (e.g., Janzen 1987; Canaday 1997), and although they might compensate for a decline in their main food resource by finding other types of invertebrate food, this would probably result in a loss of foraging efficiency (Janzen 1987).

It is an oversimplification to consider invertebrates as only consumers of leaf material and prey for vertebrates. Food chain lengths among invertebrates can vary from three in carrion systems to as much as seven in gall webs, and nearly half of insect families have members that are top predators in certain habitats (Schoenly et al. 1991). As food chain lengths increase, resilience to invasion or perturbation decreases (Pimm 1991). Food webs from many different ecological systems and geographic regions appear to share a common structure—the ratios of trophic levels are constant regardless of the size of the web (Pimm et al. 1991). It follows that the loss of invertebrate species at the bottom of the food web may have cascading effects on predators at the top, reducing the size and diversity of the resource base (Schoenly et al. 1991).

Issues of Scale

At the landscape scale, selective logging results in a mosaic of disturbance intensities ranging from areas of severe extraction along logging roads and around timber collection points, to areas of minimal disturbance and patches of undisturbed forest (Whitmore 1990). Invertebrates, as poorly mobile, habitat-specific organisms, are likely to be sensitive to habitat heterogeneity on this scale—butterflies have also been studied in this respect (Blau 1980; Kremen 1992; Spitzer et al. 1997). Significant differences in butterfly diversity have been observed over distances of only a few hundred meters, in response to habitat heterogeneity caused by illegal, selective logging (Spitzer et al. 1997) (intensity of logging is unreported). The importance of sampling widely to take account of this heterogeneity has been recognized (Sparrow et al. 1994), and although species-accumulation curves are often used to monitor sampling effort (e.g., Eggleton et al. 1995), the size of a *sufficient* sampling area has not been explicitly stated and is likely to vary both among and within taxa in relation to species' mobility. Although habitat heterogeneity is usually viewed as a problem within the context of sampling, it is an important aspect of the forest for some butterfly species—particularly those which use early-successional host plants and which develop immediately after the formation of canopy gaps (Blau 1980).

Degradation and fragmentation of forest landscapes have restricted previously wide-ranging and continuous distributions of many invertebrates to a series of smaller and spatially disjunct populations (Nieminen 1996; Davies and Margules 1998). The vulnerability of each isolated population to extinction is a function of the size of the population and the degree of connectivity with neighboring populations. Previously widespread and common species, therefore, may persist as small isolated populations linked by occasional dispersal (metapopulations). Metapopulation models have been crucial in understanding the persistence of populations in temperate regions (Hanski 1991; Hanski et al. 1996), and have shown that long-term population persistence can be heavily dependent on immigration from surrounding source populations (Pimm 1991). Conservationists now appreciate that management at regional scales, involving the conservation of several populations, is more appropriate and successful than the conservation of single isolated populations on local scales (Samways 1990; Samways 1993a; Thomas et al. 1996; Roderick 1996; see chapter 23 for a discussion of protected areas within the production forest landscape). The spatial distribution and integral management of suitable forest habitats within the landscape matrix is, consequently, crucial to the conservation of many tropical invertebrate species. To date the application of metapopulation principals to tropical forest management systems has not been attempted, partly because basic information on invertebrate habitat requirements and dispersal rates is not yet available.

Unlike vertebrates and flowering plants, invertebrate conservation cannot be undertaken by concentrating on single flagship species. A suitable range of habitats needs to be preserved, in which a wide variety of ecologically important invertebrates might be expected to persist. It is necessary to consider the spatial distribution of these suitable habitats in the landscape to ensure that persistence is assured in the long-term by establishing healthy populations linked by dispersal corridors.

The impact of selective logging on invertebrates must, therefore, be considered at both local and regional scales. On a local scale, the preservation of the mediators of ecological processes should be a primary aim of sustainable forestry. On a regional scale, the conservation of viable populations of rare and vulnerable species in a sustainable habitat or landscape matrix is the objective of biodiversity conservation. Sound forest management is based upon the understanding of these principles and their interdependencies.

Insects as Indicators of Biodiversity and Forest Disturbance in Logged Forest

Insects have attracted much attention as indicators of biodiversity because they are diverse, interact with a wide variety of plants, are sensitive to microclimate and other environmental conditions, and are easy to collect in large samples (Kremen et al. 1993; McGeoch 1998). Given the role of insects as herbivores, seed predators, pollinators, and decomposers, it is easy to understand that data on insect abundance and diversity relates to the functioning of ecological processes that contribute to ecosystem resilience. How to interpret changing community structure in terms of ecosystem function and stability remains, however, an unsolved problem at the center of current ecological research.

Other requirements of ideal indicators include a stable and tractable taxonomy, a well-known natural history, and that they are readily observable. It is doubtful whether the majority of invertebrates satisfy all or even some of these criteria, but the most serious criticism is that there is little evidence that diversity patterns among invertebrate taxa are observed in other plant, animal, or invertebrate taxa (Lawton et al. 1998). Recording insect diversity as a surrogate for all biodiversity is subject to the largely untested assumption that there are correlations between these groups. Throughout this review we have seen that taxa respond to disturbance in different and often contrasting ways (for examples see Lawton et al. 1998). In a recent study comparing the indicator qualities of birds, butterflies, beetles, ants, termites, and nematodes, only about 10 percent of the variation in species richness of one group was predicted by changes in the species richness of other groups (Lawton et al. 1998) and some groups were negatively correlated. Prendergast et al. (1993) found no evidence for common distribution patterns among several plant, bird, and insect taxa in 10 x 10 km grid squares in the United Kingdom. Attempts to assess the impacts of tropical forest modification using changes in the richness of one or a few taxa, therefore, may be highly misleading. How well insects represent local habitat heterogeneity, richness of the plant biota, or the degree of anthropogenic disturbance, remains to be tested (Kremen 1992).

Given the high spatial and temporal fluctuations of tropical insect population dynamics (Wolda 1992), the reliability of insects as indicators is also diminished due to difficulty in distinguishing perturbation-induced changes in population numbers from natural fluctuations. This will remain problematic as long as we continue to conduct short-term research over one to three seasons. The spatial scale or sample size area will also influence the effectiveness of one species to represent another (Currie 1991; Curnutt et al. 1994), and there are problems of non-independence of adjacent samples at all spatial scales.

High richness of indicator taxa is purported to be an important condition of its utility as an indicator of biodiversity, yet high richness also entails high effort in the processing of the samples, many of which cannot be ascribed confidently to separate species without expert taxonomic assistance. The proportion of invertebrates that cannot be assigned to described species is inversely related to body size (Lawton et al. 1998), indicating that the smallest and most abundant invertebrates are the least reliable as indicators.

Finally there is little point in using indicators to measure environmental changes that can be measured directly (e.g., litter decomposition). The purpose and value of biological indicators is to observe the biotic responses to environmental stress and to provide early warnings of natural responses to environmental impacts. Species assemblages may be able to do this if they include insects that cover a wide range of habitat specificities and mutualistic interactions, and provided there is sufficient information about their taxonomy and natural demography. There remains, however, a further problem of how to interpret data on the composition of insect assemblages. Thresholds beyond which forest managers can be sure that sustainability is no longer achieved have yet to be determined. We are still a long way from establishing what these thresholds are, in part because the integration of ecological, economic, and social concepts of sustainability has yet to be achieved.

Management of Selective Logging for Ecological Sustainability

Due to the huge diversity of invertebrate species, behaviors, and microhabitats, specific management recommendations for the conservation of invertebrate biodiversity can only target a small proportion of the invertebrate community. It is unrealistic to expect that a list of such recommendations will ever be adopted in a production forest. Invertebrates are clearly important components of the forest system, however, and the ecological processes they mediate must be maintained, if timber production is to be sustainable. The short generation times and high fecundity of most invertebrates makes them relatively resilient to disturbances, provided refuges exist from where they can re-colonize disturbed areas (see next section; see chapter 23). Recolonization of disturbed areas is more rapid if structural damage to forests caused by logging is minimized. Reduced impact logging techniques that include directional felling, removal of lianas and creepers, and careful location of skid trails, minimize the extent of canopy gaps, retain much of the original forest vertical structure, and avoid extensive soil exposure and compaction (see chapters 21 and 24). These techniques generally minimize changes to forest microhabitats and microclimates.

Conservation of maximal diversity under selective logging systems can also be achieved by retaining a mosaic of habitats, which include tracts of primary and old growth secondary forest (see chapter 23). This principal can be extended to smaller scales such that stand structure consists of many different age classes, tree sizes, and regeneration phases. Habitat diversity can be further enhanced by leaving dead wood and woody debris on the site that has the multiple functions of providing shelter (see chapter 23), favorable microclimate, and resources to many invertebrates. Preparation of the site for regeneration by burning should be avoided as it reduces habitat heterogeneity and destroys many resources and microhabitats used by invertebrates. Areas that are particularly sensitive to damage or that are important as refugia for invertebrates and vertebrates alike should be set aside. In the seasonal tropics, there are few leaf or insect resources during the dry season, and much of the animal biota—including invertebrates— retreat to the evergreen riparian forests (Foster 1980; see chapter 8).

Indicators of seed production and seedling establishment provide information on the preservation of pollination guilds and soil conditions necessary for regeneration, while simple measures of leaf decomposition rates can indicate the state of decomposer communities. These monitoring schemes should be used to help guide forest management practices towards achieving ecological sustainability of environmental resources (see chapters 18 and 19 for a more complete discussion of monitoring practices).

Conclusion and Future Research Directions

Two issues need to be addressed when considering the impacts of logging on the biotic community: 1) the continued existence of native species, and 2) ecosystem function. Local extinction may not have a detrimental effect on ecosystem function if other ecologically equivalent species persist to replace those that are lost (Walker 1992). Conversely, some ecological processes may be disrupted by disturbances that affect only the behavior of process-mediating invertebrates without causing a decline in abundance or richness (see examples above). Where dramatic changes in the invertebrate community assemblage occur, associated changes in ecosystem processes should be expected. Our poor understanding of invertebrate responses to logging makes it difficult to propose sound management strategies. The task is further complicated by the short-term nature of most studies, the confounding influence of other disturbances associated with logging, and, particularly, the complex nature of responses of various invertebrate taxa (see box 12-1). Species richness and diversity data alone are inadequate for the determination of the importance of logging impacts on invertebrate biodiversity. High diversity assemblages in logged areas also cannot be viewed as beneficial or desirable if composed of generalist species with wide geographic ranges that interact weakly with other trophic and functional groups. More attention to taxonomic composition and geographic distributions is called for if invertebrate biodiversity is to be adequately conserved for the purposes of ecosystem conservation and resource sustainability. Future research attention should also be focused on the long-term impacts of logging, and particularly the dynamics of species loss and reestablishment. This requires deeper understanding of the variety of invertebrate-plant mutualisms, their sensitivity to disruption, and the thresholds of disturbance beyond which mutualisms break down. The existence, extent, and effectiveness of ecological compensatory mechanisms (by which ecological processes are maintained through the substitution of component species) needs further investigation. Ecologists should, therefore, be aiming to answer questions not only on which species are most likely to disappear from disturbed stands, but also why, and what species-level features facilitate rapid recovery.

Finally, simple correlation of stand history with species composition is a first step in developing useful strategies for forest management. A finer understanding of how species live in the forest environment, and demographic responses of species to natural and human disturbances, are needed to clarify the impacts on ecosystem functions and processes (Didham et al. 1996). This is a difficult task, as complex biological systems show considerable temporal variability and frequently behave entirely differently in isolated cases (Pimm 1991). It is not surprising, therefore, to discover that contradictory results may be obtained from similar studies, and many more long-term data sets that incorporate information on taxa with different ecologies, resource requirements, and trophic links are needed before we can confidently relate human disturbance-induced causes of invertebrate dynamics in forest systems.

ACKNOWLEDGMENTS

Thanks to Dr. Paul Eggleton and several anonymous reviewers for commenting on an earlier draft of this manuscript.

Chapter 13

SOIL FAUNA IN MANAGED FORESTS

Lessons from the Luquillo Experimental Forest, Puerto Rico

Gerardo R. Camilo and Xiaoming Zou

Soil organisms play crucial roles in nutrient cycling, food webs and soil properties of terrestrial ecosystems. These processes are the result of intricate ecological interactions among soil fauna, bacteria, fungi, plants, and the stratum of soil in which they are found (Moore et al. 1988, 1993). Modeling studies indicate that the resilience and persistence of ecosystems is closely tied to the recycling of materials through these soil food webs (Moore et al. 1993). The diversity of soil organisms is of great importance not only to ecosystems, but also to the health of people dependent on the land. Structural (Zou et al. 1995) and biogeochemical (Ananthakrishnan 1996; Silver et al. 1996a) processes that maintain the fertility of soils are directly related to the biodiversity found within them (Silver et al. 1996a, b). While many soil organisms are considered to be redundant (i.e., substitutable) in soil processes (Andrén et al. 1995), their actual functional roles in maintaining soil fertility and properties remain poorly understood (Swift and Anderson 1993).

The responses of soil fauna to anthropogenic disturbances (frequency, intensity, and scale) are poorly understood, and assumed to be a function of their adaptability to natural disturbance regimes (Silver et al. 1996; see chapter 22). Logging practices usually stretch the resiliency of the system beyond the range encountered in natural disturbance events. In the wake of harvesting operations, organisms in the litter and topmost layer of the soil are highly susceptible to erosion and changes in microclimate, while species that live deeper in the soil compartment are vulnerable to alterations in the physical parameters of the soil (Vos and Kooistra 1994; Stirzaker et al. 1996). The paucity of empirical evidence about the effects of logging on soil fauna severely limits our capacity to predict the impacts associated with silvicultural interventions and options to mitigate them.

In this chapter, we explore how soil fauna may be affected by logging practices, and the consequences to the animal community and soil ecosystem processes. We specifically describe how logging may influence soil food webs, litter decomposition, and nutrient cycling. To illustrate these conditions, we report on soil fauna studies conducted at the Luquillo Experimental Forest in Puerto Rico. We also compare these studies with those in the literature. We conclude with basic management recommendations to mitigate logging effects on soil fauna, and suggest areas of future research on logging-soil fauna interactions.

Logging–Soil Organism Interactions: A General Overview

Natural disturbance events are integral parts of ecosystem processes, disrupting organism populations, communities, and ecosystem structure, through changes in resource availability and the physical environment (White and Pickett 1985). A main difference between natural and human disturbances of similar scales in tropical forests (see table 13-1) is that human disturbances such as logging remove biomass from the system. Elimination of resources (i.e., decomposing wood; see photo 13-1) from the basal trophic levels (bacteria and fungi), may have cascading effects that delay habitat restoration after a logging event (Zimmerman et al. 1995b). Logging activities can also directly influence the abundance and diversity of soil fauna through alteration of soil physical properties (e.g., porosity, specific gravity), microclimate conditions, the quantity or quality of plant litter inputs, and the abundance of predators and other soil organisms (see photo 13-2; Gonzalez et al. 1996; Gonzalez and Zou 1999a, b; see chapter 12). Soil fauna serve as important prey items for nonsoil animals (Pfiffer 1996), and many birds, mammals, reptiles, and amphibians regularly consume invertebrates that live, at least part of their life, in the soil (Stewart and Woolbright 1996; Thomas and Gaa 1996). As logging disturbance intensifies from selective logging to clearcutting, the impacts of logging on fauna populations and their interactions intensifies (Schowalter et al. 1981).

Finally, logging imposes indirect effects on soil fauna by changing the scale, frequency, and intensity of natural disturbance such as land

PHOTO 13-1 *Large, coarse, woody debris on the forest floor serves as a food resource for decomposers, as well as a seedbed for many plants, and cover for terrestrial vertebrates and invertebrates. (R. Fimbel)*

slides, fires, and treefall gaps. Logging in steep slopes with heavy machinery, for example, may trigger landslides during storms. Plant debris generated from logging may enhance fire intensity during the dry season, and logging activity may make residual trees more susceptible to blow-down from hurricanes or windstorms, thus creating additional treefall gaps (Myster et al. 1997).

Long-Term Soil Fauna Studies in the Luquillo Forest

The Long-Term Ecological Research Site

The Luquillo Experimental Forest is one of the long-term ecological research (LTER) sites funded by the National Science Foundation and managed by the U.S. Forest Service. A 16 ha plot was established at the site in 1990 to monitor the long-term dynamics of forest structure and processes (see figure 13-1). The forest is located in the northeastern part of the island, and contains four major types of forest ecosystems: tabonuco forest (<600m and site of the research plot), palo colorado forest (600–900 m), cloud forest (>900 m), and sierra palm forest where soils have poor drainage (Brown et al. 1983). For the past decade, a major research focus in the Luquillo Forest has been the quantification of natural and anthropogenic disturbances in this subtropical montane forest (Waide and Lugo 1992; Zimmerman et al. 1996).

Natural and human-caused disturbances are relatively common within the Luquillo Experimental Forest. Natural disturbance events include treefalls, landslides, and hurricanes (Walker et al. 1991a), while past human activities in the forest include agriculture (Zou et al. 1995), cattle grazing (Zimmermann et al. 1995a), managed tree plantations (Lugo 1992), and selective logging (García-Montiel and Scatena 1994). Most logging practices stopped several decades ago. Data on ecosystem recovery, and the long-term effects of human activity on forest structure and function, are producing an emerging picture of the ecological imprint of human activity on the forest (García-Montiel and Scatena 1994; Zimmermann et al. 1995a; Zou et al. 1995). After more than 50

PHOTO 13-2 *Soil fauna are important for their roles in decomposing litter, altering soil physical properties, and serving as both predators and prey in soil food webs. (C. Pearl)*

TABLE 13-1 *Comparison of Logging Activities and Natural Disturbance Events with Similar Scales of Disturbance in the Luquillo Experimental Forest, Puerto Rico*

Scale	Logging	Natural
Small (< 500m^2)	Selective logging — low intensity, low frequency	Treefall — low intensity, high frequency
Intermediate (< 2 hectares)	Small-scale clear-cutting — high intensity, intermediate frequency	Small Landslide — high intensity, frequency Wind Throw — intermediate intensity, intermediate frequency
Large (2 – 5 hectares)	Strip logging (20 - 50 m width) — high intensity, intermediate frequency	Tropical Storm — intermediate intensity, intermediate frequency Large Landslide — high intensity, low frequency
Very Large (> 5 hectares)	Clear-cutting — high intensity, low to intermediate frequency	Hurricane — high intensity, low frequency

years, and the passage of three major hurricanes, the signature of past human activity still permeates the secondary forest of the Luquillo Forest (see figure 13-1).

Soil Food Webs and Logging

The El Verde Food Web in the Luquillo Forest

The El Verde food web from the tabonuco forest in the Luquillo Forest is the best-studied food web from a tropical rain forest (Reagan and Waide 1996). There are over 2,600 animal species in the community, and of these, more than 2,500 are invertebrates (see table 13-2) (Garrison and Willig 1996; Pfiffer 1996). These figures are very likely an underestimation (G. Camilo, unpublished data; R. L. Edwards, personal communication regarding spiders). Of the 2,500$^+$ invertebrates, nearly 70 percent of these species spend some or all of their life cycle in

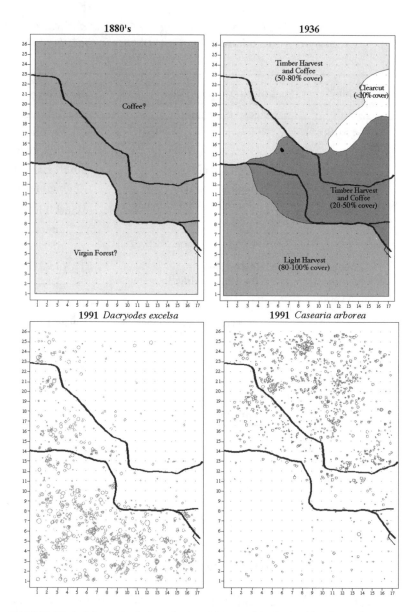

FIGURE 13-1 *Maps depicting the land history use in the Hurricane Recovery Plot, Luquillo Experimental Forest, Puerto Rico. The plot is 16 ha in area, and the maps are oriented north. By the 1880s, the north part of the plot had been cleared for coffee planting. By 1936 the U.S. Forest had decided to buy the land in order to annex it to the Luquillo Experimental Forest. By that time the coffee had been eliminated from most of the area, and timber harvest had become the principal activity. The 1991 distribution of two common tree species in the plot points to disjunct distributions as a result of past human land use, rather than historical disturbance regime or soil attributes.*

the soil or leaf litter (Pfeiffer 1996; Garrison and Willig 1996; G. Camilo, unpublished data). Their specific functional attributes are not well documented (for complete description, see Reagan and Waide 1996). These species constitute the greatest proportion of the diet of the most abundant vertebrate predators above ground (Reagan et al. 1996).

TABLE 13-2 Functional Groups, Taxonomic Identities, and Mean Annual Species Densities for the Soil Food Web from the Tabonuco Forest in the Luquillo Experimental Forest, Puerto Rico (modified from Pfeiffer 1996 and Reagan et al. 1996)

Functional Group	Taxa	Individuals/m^2
Seprotrophic and microbitrophic	Acari	15,960[a]
		1,390[b]
	Collembola	1,245
	Diplopoda	300
	Isopoda	549
	Dictyoptera and Grillidae	115
	Isoptera	374
	Diptera	618
	Formicidae	26
Seed predators and root feeders	Hemiptera	28
	Homoptera	493
	Coleoptera	75–125
	Formicidae	358
Predators	Aranae	356
	Pseudoscorpiones	262–340
	Opiliones	146
	Schizomida	3.1
	Amblipygi	1.8
	Chilopoda	8
	Hemiptera	83
	Coleoptera	89
	Neuroptera	1.7
	Formicidae	650

[a]density in litter; [b]density in soil

Functional Diversity of Soil Food Webs

Food webs can be considered *road maps* that illustrate several concepts of community organization and function (Winemiller and Polis 1996), such as predator-prey interactions (DeAngelis 1992; Polis and Strong 1996), and can provide ways to quantify the amount of complexity in the system. Traditionally, food web research has been the realm of population and community ecology. More recently, however, this concept has been applied to the functional ecology of species within the community and the role they play in ecosystem persistence (Coleman 1996). The functional diversity of species has also become a general framework to understand how species richness relates to ecosystem function (Schultze and Mooney 1994; Lawton 1994).

The role of soil invertebrates has historically been considered as aiding in decomposition processes, but this is an oversimplification (Pfiffer 1996). Soil macrofauna are essential in the mechanical breakdown of wood and leaf litter (Ananthakrishnan 1996), and also in the inoculation of necromass (i.e., dead biomass) with microorganisms that will decompose it biochemically (Lodge 1996). Soil invertebrates and microbiota also cause changes in nutrient pools that facilitate the establishment of vegetation and further successional stages (Lodge et al. 1991, 1994).

Resource quality has been shown to have a profound effect on the diversity and trophic structure of litter communities. Arthropod diversity in India, for example, was lower in monospecific tree stands than in mixed species stands (Ananthakrishnan 1996). The age of the stand, and thus its successional stage, was also positively correlated with soil faunal diversity. The conclusion is that increasing tree stand diversity increases litter chemical diversity, which, in turn, allows an increased functional diversity of soil invertebrates (Ananthakrishnan 1996.) This tropical regime favors groups of organisms that have high population turnover rates and, thus, respond quickly to brusque resource inputs.

Logging and Soil Food Webs

Ecosystem stability is tied to the recycling of materials through the decomposer food web (Moore et al. 1993). In the tropics, the response of soil processes to forestry practices is largely unknown. Few studies have dealt with the effects of logging on soil food webs, and most are from temperate systems (Moore et al. 1988; Moore and de Ruiter 1991; Coleman 1996). From the limited theoretical and empirical work that has accumulated, it appears that high disturbance regimes (such as those characterized by logging) will: a) simplify food webs, b) shorten food chains, c) decrease connectance, and d) eventually eliminate higher trophic levels from the soil ecosystem (Pimm 1982, 1991; Moore and de Ruiter 1991; Moore et al. 1993). Such a simplification in soil trophic webs can profoundly alter the numerous interactions between soil organisms comprising complex tropical food webs, which, in turn, influences cross-predation, omnivory (Moore et al. 1988; Strong and Polis 1996), and positive feedback mechanisms (Bengston et al. 1996). Logged-over secondary forests in Puerto Rico, for example, exhibit decreased population densities of predatory arthropods in the leaf litter layer (Pfiffer 1996), which is predicted to decreased food web functionality (Moore et al. 1993; Polis and Strong 1996). The end result is a disruption in the flow of energy (Coleman 1996) and nutrients (i.e., nitrogen, see Moore et al. 1988) throughout the impacted portion of the forest. The magnitude of this disruption, and the recovery from it, will depend on the intensity of the logging in relation to the natural disturbance regime (see table 13-1).

Earthworms, Litter Decomposition, and Logging

One of the most obvious ecosystem processes associated with soil fauna is the breakdown of litter into simpler organic constituents, which then recycle through the system. Within the decomposer-detrital food webs, up to 60 percent of the biomass in the above-ground food web of most terrestrial ecosystems comes from the detrital circuit (Polis and Strong 1996). Earthworms are a major component of this circuit.

Earthworms and Their Role in the Decomposition Process

Earthworms dominate fauna biomass in moist and wet tropical forests. Odum (1970) reported that up to one third of fauna biomass is comprised of earthworms in a tropical wet forest in Puerto Rico. There are three ecological groups of earthworms (Bouchè 1977):

- Epigaeic earthworms (species that feed on and live in plant litter on the soil surface)
- Endogeic earthworms (species that feed on soil organic matter and live in soils)

— Anecic earthworms (those that live in soils but feed on plant litter on soil surface)

Epigaeic and anecic earthworms consume plant material directly from the forest floor. Anecic earthworms often accumulate plant leaves around the entrance of their burrows. Both endogeic and anecic worms exude feces that bury plant organic matter in the soil. Microbial activity in earthworm feces is often higher than in the surrounding soil (Lee 1985).

One of the most important ecological roles that earthworms may play is the regulation of plant litter decomposition (Darwin 1881; Lee 1985; Lavelle and Martin 1992; Edwards and Bohlen 1996). Earthworms increase plant litter decomposition rates by 20 to 50 percent in agroecosystems (e.g., Parmelee et al. 1990). While similarly controlled experiments have not been performed in tropical forests, available information comparing logged forestlands with tree plantations and secondary forests shows that plant litter decomposition is linearly correlated with earthworm abundance (see figure 13-2). Earthworms influence the decomposition process in several ways, including:

a) comminuting plant litter, b) redistributing and burying plant litter (Lavelle and Martin 1992), and c) regulating microbial activity (Doube and Schmidt 1997). While affecting the decomposition process, earthworms also play an important role in regulating soil carbon and nutrient levels, and thus nutrient-use efficiency by plants (Blair et al. 1995). Soil carbon concentration, for example, correlated positively with earthworm density under similar climate and soil condition in Hawaiian tree plantations (see figure 13-3; Zou and Gonzalez 1997). On the other hand, the efficiency of plant phosphorus use—defined as the ratio of net above-ground primary productivity to annual litterfall phosphorus content—correlated negatively and exponentially with earthworm density, suggesting earthworms increase soil phosphate availability. The mechanism for increased phosphorous availability may result from the accelerated decomposition of soil organic matter (Zou and Gonzalez 1997).

Logging and Earthworms

Few studies have measured the direct effects of logging on earthworms. Earthworms that live and feed on the soil surface (epigaeic) generally appear to be most sensitive to logging disturbance because plant litter on the soil surface is most prone to logging disturbance (see table

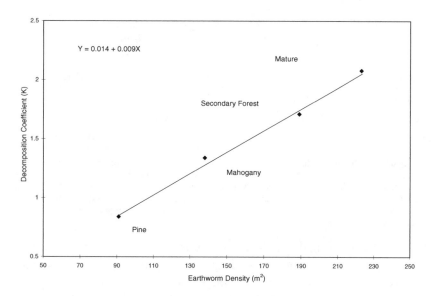

FIGURE 13-2 Relationship between litter decomposition coefficient (*k*), the ratio of annual litterfall to litter biomass on the forest floor, and earthworm density in plantations of *Pinus caribaea* and *Swietenia macrophylla*, a thirty-year-old secondary forest, and a mature tabonuco forest in Puerto Rico.

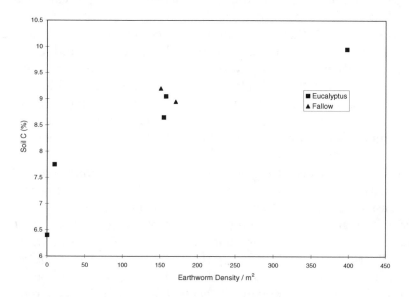

FIGURE 13-3 Soil carbon concentration in relation to earthworm density in secondary fallow forest and adjacent *Eucalyptus* plantations in Hawaii.

TABLE 13-3 *Relative Impact of Logging on the Abundance of Epigeic,*
Anecic, and Endogeic Earthworm Groups

Logging[1]	Epigeic	Anecic	Endogeic
Selective	Medium	Low	Minimum
Small-scale	High	Low-medium	Low
Large-scale	High	Medium-high	Medium-high

[1] There are four habitual categories of impact: the "minimum" category refers to less than 30% change in group density and no change in species composition; the "low" category refers to more than 30% change in group density and no change in species composition; the "medium" category refers to less than 50% change in group species composition; and the "high" category refers to greater than 50% change in group species composition. Logging categories are based on area values as in table 13-1.

13-3) (Zou and Gonzalez 1997). Earthworms that live and feed in the soil (endogeic) are most resistant to logging disturbance, as they spend the majority of their lives deep within the soil compartment.

A study in a tropical Puerto Rico rain forest (Devoe 1989) showed that seven years after small-scale clearcutting, the density and fresh weight of endogeic species *Pontoscolex corethrurus* increased, while those of the anecic species *Amynthas rodericensis* decreased more than 30 percent (compared with adjacent secondary forest). There were no changes in species composition. In a 50-year-old, large-scale clearcut area (Zou et al. 1995), however, the density of endogeic and anecic earthworms were 120 and four individuals per m^2 respectively, and did not differ from those in the undisturbed forest (Zou and Gonzalez 1997; Gonzalez et al. 1999; Gonzalez and Zou 1999b). These studies suggest that both small-scale clear cutting (see table 13-1), and low frequency, large-scale clearcutting (>50 years interval; see table 13-1), have a limited long-term impact on endogeic and anecic earthworm communities. While we lack available data, the impact of clear cutting on epigaeic earthworms may be greater.

The indirect effects of logging on earthworm abundance and community structure are often more pronounced because logging is commonly associated with other land management practices. Converting forest to pastures after logging, for example, triggered an eightfold increase in the density of endogeic *Pontoscolex corethrurus* in northeastern Puerto Rico, but eliminated the anecic species *Amynthas rodericensis* (Zou and Gonzalez 1997). Where plantations of Caribbean pine (*Pinus caribaea*) and mahogany (*Swietenia macrophylla*) have replaced natural forest, earthworm richness and density are half those in adjacent secondary forests that regenerated on previously logged land

(Gonzalez et al. 1996). Logging also changes tree species composition in tropical forests, and different tree species may trigger changes in earthworm abundance and species composition (Zou 1993).

Earthworms represent a single, albeit important, functional group found in and on soils (Bengston et al. 1996). They are also good indicators of the status of the ecosystem (Edwards and Bohlen 1996). Monitoring a specific group for the purpose of evaluating the integrity of the forest ecosystem will depend on the forest type and resources available. We provide some general suggestions on how to use earthworm and soil fauna diversity to monitor ecosystem function (see box 13-1).

Box 13-1 *Soil fauna as indicators, and a rapid assessment technique of soil fauna diversity: morphospecies and the jackknife estimate.*

Conservation efforts depend on estimates of diversity, based on monitoring of indicator groups. Soil organisms can provide such indicators (Majer and Beeston 1996). The principal problem with assessing faunal diversity in soils is the staggering species richness of some taxa such as mites (over 200 families, and estimates of more than 1,600,000 species), insects (estimates range up to 30 million species), and nematodes (as diverse as insects by some accounts). This is exacerbated by the lack of taxonomic expertise for these groups. So the question is, how best can one invest limited resources to monitor soil diversity? There is no single strategy that will work for all systems, but some general procedures are outlined below.

1. Identifying indicators. Extensive work by J.D. Majer and co-workers in Australia has shown that ants are an excellent indicator group (Majer and Beeston 1996). Ants are highly diverse, the general taxonomy is well known (Hölldobler and Wilson 1990), and regional keys are available for the generic level (Bolton 1994). Other groups that have shown relevance as indicators are earthworms (Doube and Schmidt 1997), and other arthropods (van Straalen 1997), although the taxonomy is not as equally known as with ants. Which group or groups are best suited for a forest type will depend on the specific taxa and conditions prevailing at the site. Greater levels of community diversity are generally correlated with increased food web complexity (Reagan et al. 1996) and ecosystem health (Pankhurst 1997).

2. Once an indicator group is selected, trying to identify individuals at the species level can prove to be a daunting task. When taxonomic keys are available, people conducting the identifications may not have the specialized knowledge required to find and elucidate key morphological characters. Nonetheless, experiments have shown that individuals with general knowledge of the morphology and taxonomy of invertebrates can correctly separate species based on their morphology, although they can not correctly identify them (Oliver and Beattie 1996b). This is known as morphospecies identification.

The establishment of voucher specimens and continued referral of new material to established collections aids this procedure.

3. Biodiversity assessments of soil fauna are complicated by the heterogeneous spatial and temporal distribution of species. The bulk of the diversity in the tropics also consists of uncommon and rare species. Trying to use parametric statistical estimates to determine the species richness is therefore not recommended. A more recent approach is the use of non-parametric estimates of species richness. A simple estimate that has been applied with some success is the jackknife estimate, developed by Heltsche and Forrester (1983) and Edwards (1992). The equation is:

$$J(ESR) = s + [(n - 1)/n] * k$$

where J(ESR) is the jackknife estimate of species richness, s is the total number of species (or morphospecies) collected in all samples, n is the total number of samples collected, and k is the number of unique species (i.e., those found only in one sample). The variance of the estimate is given by the equation:

$$\text{var } J (ESR) = (n - 1)/n * [E (j^2 fj) - \frac{k^2}{n}]$$

where var J(ESR) is the variance of the jackknife estimate, j is the number of samples with f unique species per quadrant (i.e., spatial sample), and f is the number of unique species in a quadrant. Confidence limits can then be calculated as:

$$J(ESR) \pm t \, {}^{\alpha}\!\!\sqrt{\text{var } (J(ESR))}$$

where t is the student's t value for (n − 1) degrees of freedom, and α is the error rate of the estimate.

Conclusions and Recommendations

Soil fauna is a large and heterogeneous collection of organisms that spend at least part of their life cycle in the soil. The species richness and diversity of soil invertebrates are of crucial importance to the maintenance of numerous ecosystem functions, yet we understand very little about the composition and interactions between members of the soil fauna community. Some functional groups, such as earthworms, influence the structure and chemical properties of soils which, in turn, is directly tied to the recovery and conservation of managed tropical forests. Within the arthropod group, increasing diversity leads to increased complexity in the decomposer food web, with subsequent increases in decomposition rates (Torres 1994). The energy and biomass that enters the soil food web via decomposers is also a significant contribution to the above-ground food web. The management of forest resources must consider that silvicultural interventions are likely to disrupt these processes—at least in the short term.

Logging impacts soil fauna through the removal of resources (i.e., decomposing wood) from the system, along with disrupting food webs in the wake of soil compaction, erosion, changes in litter composition, and alterations in microclimate conditions (Zou et al. 1995; Zimmerman 1995a). As the intensity of logging increases, a concomitant reduction of energy and biomass can slow the recovery and decrease the persistence of a forest system. The exact mechanisms by which these processes operate are poorly understood. Further research is needed to identify the effects of logging disturbance on soil fauna diversity, as well as the roles that these complex food webs play in tropical forest succession and recovery. Understanding how food webs work below the ground is a major research challenge for ecologists, but it has important repercussions on the practical management of tropical forests.

Finally, logging management strategies that mimic the frequency and intensity of natural disturbance regimes, combined with reduced-impact logging practices (see chapters 14, 21, and 24), will likely minimize the impacts of these operations on soil fauna. Increasing divergence from the natural regime can produce long-term changes in tree community composition, which subsequently brings a change in litter quality. This, in turn, has the potential of affecting both taxonomic and functional diversity of the soil fauna, and degrading the productivity potential of the forest.

ACKNOWLEDGMENTS

We wish to thank the editors for their valuable comments and recommendations, as well as their persistent efforts to bring this volume to fruition. Gerardo acknowledges the valuable discussions in the ecology and conservation groups at Saint Louis University. Part of this research was funded under grant BSR-8811902 from the National Science Foundation to the Institute of Tropical Ecosystem Studies, University of Puerto Rico and the International Institute of Tropical Forestry, USDA Forest Service—as part of the Long-Term Ecological Research Program in the Luquillo Experimental Forest. The University of Puerto Rico and Saint Louis University provided additional support.

Chapter 14

THE EFFECTS OF LOGGING ON TROPICAL RIVER ECOSYSTEMS

Catherine M. Pringle and Jonathan P. Benstead

While there is a growing body of literature on logging and its effects on the terrestrial components of tropical ecosystems (e.g., Uhl et al. 1991; Barros and Uhl 1995; Frumhoff 1995; other chapters in this volume), very little is known about logging and effects on tropical aquatic systems. Studies that relate logging practices in tropical regions to changes in surface hydrology and fluvial geomorphology are less common than in temperate regions, as are ecological studies on the effects of logging on tropical aquatic biota. This focus contrasts sharply with that in temperate areas such as North America, where the conservation of aquatic biodiversity in timber production forests is a central issue in conservation biology (e.g., Salo and Cundy 1986; Adams et al. 1988; Carlson et al. 1990; Beschta 1991; Naiman 1992; Adams et al. 1994). The effects of logging practices on stream biota are particularly well documented for temperate zone fish (Gibbons and Salo 1973; Wydowski 1978; Garman and Moring 1993).

The dramatic losses in aquatic biodiversity documented in many tropical regions (Goulding et al. 1996; Lowe-McConnell 1987) can be attributed to several factors—logging is only one. Recent studies in the Gombak River basin of Southeast Asia show that 41 percent of native fish species were lost from 1969 to 1990. This was attributed to a combination of logging, highway construction, and land clearing for agriculture (Mohd 1994), which resulted in loss of riverine habitat. Long-term shifts in the abundance and species composition of aquatic communities can also occur as a result of timber harvesting—as in the increased percentage of cyprinid fish observed in the Kerlig River in Peninsular Malaysia. This second-order *stream* (small river) runs through a forest that was logged 40 years ago but has since been left to regenerate (Mohd 1994).

Logging in tropical areas is often localized along major river systems because of the abundance of timber in floodplain forests, low relative costs of wood extraction and transport (i.e., transport in barges is only one third the cost per cubic meter of truck transport; see photo 14-1), and good access to markets (Barros and Uhl 1995). Over the last several centuries, for example, most of the logging in the Brazilian Amazon has occurred along the lower Amazon River and estuary. Barros and Uhl (1995) recorded 1,295 functioning wood industries along the lower Amazon in their 1990–91 survey. Small-, medium-, and large-sized companies all depend on the river and its tributaries to transport cut timber and/or processed wood. Once streamside forests have been cleared, logging operations typically move inland. Where streamside riparian vegetation has been removed, riverine systems are further vulnerable to the effects of logging within the watershed due to the loss of the buffering capacity of the riparian forest.

In this chapter, we concentrate on the immediate and local effects of logging on tropical riverine ecosystems, since effects on lakes are virtually unexplored. We also examine the potential effects of logging on estuarine ecosystems. While deforestation is known to have direct effects on marine ecosystems, such as increased erosion and consequent silting of coral reefs (Huber 1994; Hutchings et al. 1994), we do not focus on this topic. Our objectives are to:

- Review the existing literature that focuses on effects of deforestation/ logging on tropical river ecosystems
- Consider how studies of logging effects on temperate riverine systems potentially relate to tropical systems
- Make recommendations for the conservation of aquatic ecosystems in tropical forests managed for timber

Whenever possible, we present logging examples. The paucity of studies on the effects of logging on tropical aquatic species and ecosystems, however, has forced us to include more general literature on the effects of deforestation on tropical rivers. While we appreciate that the effects of logging on rivers will depend on its intensity, in the absence of better quality data, we hope that deforestation examples will throw light on the potential responses of tropical river ecosystems to logging activities.

PHOTO 14-1 *Raft of logs being towed by a tug to timber ships, Sabah, East Malaysia. (G. Davies)*

The Direct Effects of Logging on Riverine Ecosystems

Increased Erosion and Sedimentation

Logging and related activities, such as road building and slash burning, can affect many of the factors that control the delivery of water and sediment into river systems (Dunne 1979, Bosch and Hewlett 1982; Bruijnzeel and Critchley 1994). Any removal of vegetation decreases the rates of evapotranspiration and the interception of precipitation, thereby increasing the rate of runoff (see photo 14-2). Soils may become compacted, particularly along logging roads and in the vicinity of log landings and skid pans (Nussbaum et al. 1995; see photo 14-3). This compaction reduces local roots and soil fauna communities that promote porosity (Webster et al. 1992; see chapter 13), dramatically reducing soil infiltration (Malmer and Grip 1990). Reduction of the canopy and the leaf litter layer can increase erosive rain splash and cause disaggregation of the surface soil material and blockage of macropores (Ross et al. 1990). These changes further reduce infiltration capacity and increase overland flow, potentially increasing erosion of surface material and its subsequent transport into river systems (El-Swaify and Dangler 1982; Ross et al. 1990). Douglas et al (1990) stressed the quantitative importance of storms in determining sediment inputs into streams during and after logging. Finally, riverbank erosion may increase as a result of increased storm flows (see next section) and loss of stabilizing riparian vegetation.

Selective logging of a catchment within the Jengka Experimental Basin, Malaysia, increased sediment yield from 100 to 277 kg/ha the first year following logging, and to 397 kg/ha by the second year. Sediment yield had recovered to its prelogging level by the fourth year

PHOTO 14-2 *Improperly designed logging roads cause excessive vegetation removal, leading to decreases in precipitation interception, and increases in soil runoff to stream channels. (R. Fimbel)*

PHOTO 14-3 *Roads and log landings cause severe soil compaction that lead to poor water infiltration and concentrated precipitation runoff. (R. Fimbel)*

(Kasran and Nik 1994). The logging intensity was 25 percent of timber stocking removed (60 and 45 cm dbh limit of dipterocarps and non-dipterocarps, respectively). Several skid trails and logging roads crossed the stream that was sampled in the study.

Two studies from Peninsular Malaysia illustrate the effects of different logging practices on sediment yield, concentrations of suspended solids, and turbidity levels in streams draining dipterocarp forest (Kasran 1988; Yusop and Suki 1994). Two small catchments in the Berembun Forest Reserve were selectively logged (25 percent of stocking removed). One catchment was logged using conventional selective logging methods while the other was harvested using modified *close supervision* (or Reduced Impact Logging [RIL]) practices. This latter approach included carefully designed roads, with cross drains on steep slopes to minimize the amount of runoff entering the stream channel. A third catchment of similar size was left unlogged as a control. Immediately following logging, suspended sediment yield increased by 97 percent and 70 percent in the conventional and RIL catchments, respectively (Kasran 1988). In the first year after logging, values for suspended solids in the conventionally logged catchment were 12 times those of the control catchment. High levels of suspended solids and turbidity persisted until the fifth year after logging. In contrast, the catchment in which reduced impact logging showed only a doubling of suspended solids one year after logging. These recovered to pre-disturbance levels within two years of logging (Yusop and Suki 1994). These studies demonstrate that sedimentation effects on streams can be significantly reduced by the application of RIL practices (see chapters 21 and 24).

Quantifying the effects of logging on erosion and sedimentation are difficult (Bruijnzeel and Critchley 1994). These difficulties include controlling for the inherent differences between catchments, especially with regard to natural sediment load (which depends on topography and soil erodibility), and interannual differences in precipitation. Bruijnzeel and Critchley (1994) concluded that "no single study has, so far, achieved an accurate picture of the true extent of erosion and sediment delivery for logged forests." Rates of recovery are particularly poorly understood, due to rates of regrowth by vegetation (which will depend on the degree of soil disturbance), patterns of precipitation, and the role of localized, temporary sediment storage. Further studies on the effects of logging on erosion, and subsequent sedimentation in tropical forest rivers, must clearly be made a high priority for research. Nonetheless, general recommendations are available for limiting the erosion and stream sedimentation associated with logging. These include the minimization of soil disturbance, the judicious locating of logging roads and stream crossings, and the maintenance of buffer strips. Measures for mitigating the adverse effects of logging on tropical aquatic systems are discussed more fully at the end of this chapter, and elsewhere in this volume (particularly chapters 21, 23, and 24). We are not aware of any studies that deal specifically with the effects of increased sedimentation on tropical river organisms. Sedimentation and associated turbidity, however, are known to have deleterious effects on aquatic insect and fish populations in temperate streams (as reviewed by Waters 1995)—sedimentation effects similar to those observed in temperate areas are expected in tropical rivers. The negative effects of sediment deposition on fish include:

- A reduction of interstitial spaces between cobbles in riffle areas of streams, smothering eggs and eliminating habitat essential for fish fry
- A reduction of water depth in pools, including the complete loss of pools, which decreases the carrying capacity for fish (e.g., Salo and Cundy 1986; Chapman 1988; Meehan 1991)

In a comparison of the fish communities in streams draining primary tropical forest (and forest logged three to 18 years previously),

Martin-Smith (1998) found stream size to be more important than logging history in determining community structure. They also concluded that logging effects had been fairly limited. Logging regimes at this site (Danum Valley Field Centre, Sabah, Malaysia) involved the removal of trees 60 cm dbh (8–15 trees/ha or 50–170 m^3/ha). The trees were removed by a skidder, and 30 to 50 percent of the trees 1 to 60 cm dbh were killed during extraction.

The vitality and health of stream invertebrate populations are tied closely to the particle size of streambed sediments (Cummins et al. 1966; Waters 1995). The principal effect of excess sediment in the streambed is an increased filling of interstitial spaces among cobbles and other large particles by fine sediments. This effect reduces habitable area for major insect groups such as Ephemeroptera, Plecoptera and Trichoptera. Severe habitat modification results in shifts to invertebrate communities dominated by small burrowing forms such as larval Chironomidae (Diptera) and oligochaetes. The burrowing habit of these smaller animals may reduce their availability as food for fish (Waters 1995).

Alterations in River Discharge

Logging can also affect the hydrology and discharge regimes of rivers. Decreases in evapotranspiration (from reduced canopy) increases mean discharge and rates of surface runoff, leading to more rapid and extreme discharge response to storms (with lower base flow between storm events; see Ross et al. 1990; Bruijnzeel and Critchley 1994), and increases total annual water yield. The effects of different logging practices on the discharge regimes of tropical rivers have been less widely studied than temperate areas, and much information is observational or anecdotal. Logging in tropical regions, however, is expected to increase stormflow volume, peak flows, and stormflow duration (Hamilton and King 1983). The magnitude of these changes will depend on many factors, including logging intensity, logging methods, and the extent of logging roads (Bruijnzeel and Critchley 1994). In Borneo, water yield by regenerating, logged-over forest stabilized after five to 10 years (Bruijnzeel 1990). Recovery to original conditions depends to a large extent on the speed of regeneration. Bruijnzeel (1990) states that "carefully executed light selective logging will have little (if any) effect on streamflow, whilst the effect increases with the amount of timber removed."

Morphological changes in river channels often occur as a result of increased sediment yields and changes in river discharge regimes.

Meijerink et al. (1988) used aerial photos, sequential satellite images, hydrologic data, and field surveys to study the effects of deforestation and associated land degradation in the Komering River basin of Indonesia (no control basin was used). The study found that deforestation (for agriculture) in steep areas of the upper catchment resulted in erosion of large amounts of sediment into the upper river system. This caused massive increases in bedload and deposition of sands, subsequently giving rise to increased flooding of the lower basin as well as chronic waterlogging of soils. Farmers were forced to abandon cultivation in large areas of the basin. Very little data exists regarding whether the rapid expansion of forestry practices in tropical watersheds has caused channel changes that affect the integrity of riparian and river communities (Douglas et al. 1993).

Riverine hydrology is integrally tied to the biology of the system, therefore alteration of discharge regimes can have serious negative consequences on stream biota. While natural extended peaks in discharge are often important in tropical systems (since they lead to inundation of river floodplains and significant land-water interactions), unnatural sporadic peaks in discharge caused by widespread intensive logging could potentially result in local depletions in algal and insect biomass (which constitute the food source for higher trophic levels). As discussed below, alterations of the hydrological regime can also affect community composition and biodiversity by altering the breeding and feeding migrations of fish and invertebrates (Goulding et al. 1996).

Increased Light and Water Temperatures

Terrestrial vegetation has profound effects on the structure and function of river communities, particularly in forested headwater sections (Hynes 1975; Vannote et al. 1980). Primary production in forest streams is limited by the shading effect of riparian trees, and leaf litter consequently forms the base of the stream food web. Removal of riparian vegetation, therefore, leads to fundamental shifts in the resource base of stream food webs. Increased insolation resulting from the loss or reduction of canopy cover leads to higher standing stocks of algae (Feminella et al. 1989; Ulrich et al. 1993). This shift is reflected in the structure of riverine macroinvertebrate communities (e.g., the relative importance of insect trophic groups; Wallace and Gurtz 1986), and in components of stream ecosystem function—such as the ratio of photosynthesis to respiration. Species diversity may also be affected. Preliminary results from research in Madagascar suggest that species richness of stream macroinvertebrate communities in catchments cleared for agriculture may be less than half that of communities in streams draining primary forest (J. P. Benstead, unpublished data).

Unfortunately, data on the cumulative effects of logging on freshwater communities are rare even in temperate ecosystems. Wallace and Gurtz (1986) studied the responses of grazing invertebrates to the removal of riparian vegetation in a temperate forest catchment. Production (amount of biomass produced per unit area in a given amount of time) of the generalist mayfly (*Baetis* spp. [Ephemeroptera]) was up to 28 times higher in the logged catchment than in a similar unlogged control site—presumably as a result of simultaneous increases in primary productivity (chiefly diatoms). Concurrently, species that specialize in feeding on decaying leaf litter often decline in streams draining logged catchments. This research points to a possible general response of stream communities to catchment disturbance—removal of vegetation may favor generalist *weedy* species that have short generation times, high fecundity, and less specialized feeding habits. In some regions, these generalist species may be non-indigenous introductions (exotics), including fish (see box 14-1).

Few studies have examined the role of vegetation in the mediation of water temperatures in temperate, forested stream ecosystems, and none are known from the tropics. In the temperate zone, removal of riparian shading leads to an increase in average water temperature and to larger fluctuations in the daily temperature regime (Swift and Messer 1971; Swift 1983). Clearly, the impact of logging on water temperature will depend largely on the proportion of riparian trees harvested, discharge, and pool volume. Typically, elevations in mean stream temperature are short-lived (often less than five years), since temperatures return to pre-disturbance levels once the canopy of secondary vegetation has closed over the stream channel (Swift 1983).

Box 14-1 *The indirect effects of logging on river systems: potential facilitation of invasion by exotic species.*

While commercial logging is not an important component of land-use change in Madagascar, examples from the island serve to illustrate an insidious and often overlooked effect of degraded riparian zones on stream ecosystems: the facilitation of invasion by exotic species of fish. Many of Madagascar's river systems have been severely degraded by sedimentation due to erosion that has been induced by clearance of natural vegetation. The fish that inhabit these rivers are considered to be the most endangered vertebrate taxa on the island (Reinthal and

Stiassny 1991). Several factors have combined to create this high degree of threat, including the removal of catchment vegetation, overfishing, and replacement of native fish by introduced exotic species. The interaction of these factors is complex, but what has occurred is essentially a sequential process initiated by degradation of riparian habitat (Benstead et al. 2000).

Many of Madagascar's native fish are adapted to forest stream habitats (Reinthal and Stiassny 1991), where inputs of terrestrial insects from the forest canopy form a significant component of the diets of several endemic species (e.g., *Bedotia* and *Rheocles* spp.). Removal of riparian forests drastically reduces these inputs, while increasing algal product-ivity through increased insolation and nutrient levels. These changes in conditions tend to favor many of the physiologically plastic, generalist species of fish that have been introduced into the island's rivers during the last one hundred years (e.g., *Tilapia* and *Oreochromis* spp.). After the passage of a tropical cyclone, for example, riparian forests bordering the streams of Perinet-Andasibe Reserve were significantly thinned. A marked increase was subsequently reported in the abundance of an exotic poecilid—the green swordtail (*Xiphophorus helleri*)—which feeds on attached algae. A simultaneous decrease in numbers of the endemic bedotiid, *Rheocles alaotrensis*—which feeds on canopy insects—was also reported (Benstead et al. 2000). Such shifts in the relative abundance of native and exotic species are likely to be widespread in the tropics, and logging activities that degrade the structure and composition of riparian forest can facilitate this replacement process. The long-term consequences of such shifts in species composition to the health and productivity of aquatic systems and adjacent forestlands is unknown.

Changes in Stream Solute Chemistry

Very few investigations have focused on the effects of logging on the solute composition of tropical rivers and streams. In the few studies that do exist, logistical difficulties have often hampered the collection of good quality data (Douglas et al. 1992). Insights can be gained from temperate research, however. Logging affects nutrient inputs to streams in several ways (Webster et al. 1992; Bruijnzeel 1990). First, it temporarily reduces the uptake of nutrients by vegetation and their subsequent storage in biomass (following this initial response, however, regrowth may result in lowered nutrient export). Second, it can increase the rate of water movement through soils. Lastly, removal of vegetation can promote soil conditions, which accelerate mineralization of organic matter (often increased by the presence of slash). Released nutrients are subsequently lost from the catchment. Typically, the net effect of logging is increased levels of Ca, K, Mg, Na, and NO_3-N in stream water. Nutrients that are most affected are those that are mobile in soil solution and that cycle biologically. The nitrogen cycle is particularly sensitive, with peaks in streamwater concentration usually observed within five years of disturbance (Webster et al. 1992). Solute concentrations in streams typically decrease with the onset of forest regeneration and may be lower than reference levels during subsequent stages of succession (Webster et al. 1992).

Departures from this temperate model can be expected in logged areas of tropical forest. Growth of secondary vegetation, for example, may be relatively rapid compared to temperate forest ecosystems—resulting in a faster return to predisturbance solute concentrations. Responses to logging, however, are likely to be broadly similar. Douglas et al. (1992) cited two studies that report a 300 percent increase in nitrogen export after selective logging in an area of peninsular Malaysia. More studies are needed in tropical areas to identify how tropical river systems respond to logging events, how these responses differ from those of temperate systems, and the subsequent effects of timber harvesting on aquatic fauna.

Effects of Logging on Critical Riverine Habitats

Floodplain Forests

In many large rivers in the tropics, the breeding and feeding cycles of economically important fish species are often closely tied to the seasonal inundation of floodplain habitats (some which extend up to 30 km inland from the main watercourse). The lakes that form in the floodplain serve as settling areas for alluvial materials, resulting in greater water transparency, and enhanced algal and zooplankton production. These environments are extremely important for young fish. This explains why migratory fish species move down the clearwater and blackwater tributaries of the Amazon to spawn near floodplain habitats (Goulding et al. 1996). Intensive feeding upon allochthonous (i.e., terrestrial) materials (Goulding 1980; Goulding et al. 1996) often accompanies lateral migrations of adult fish into flooded forest areas. Inundated forest habitats are also important for commercially important fish species in Africa and southeast Asia (Lowe-McConnell 1987). In the Zaire River (Democratic Republic of Congo; formerly Zaire), riparian forest is considered to be of vital importance to the continued productivity of artisanal fisheries (Lowe-McConnell 1987; also see examples from Borneo, Sumatra and Cambodia in the same volume). Many adult fish have special adaptations for feeding on fruits and seeds that fall into the water of flooded forests (Welcomme 1985). The Neo-tropical tambaqui (*Colossoma macropomum*), for example, has huge crushing molars (Marshall 1995) and is well adapted to feeding in the flooded forest (Araujo-Lima and Goulding 1997). Welcomme and Hagborg (1977) deduced that in rivers with extensive floodplains, the most significant factor to influence the growth of fish in any single year was the area of land flooded, since this was an index of food availabil-ity during the prime growing season. During this period, fish can achieve 75 percent of their annual growth.

In addition to species that rely on fruits and seeds, many fish are adapted for feeding on fine detritus. Detritivory is especially prevalent in Neotropical fish communities and many of these species (e.g., *Prochilodus* spp.) are commercially important (Bowen 1983). Removal or degradation of riparian vegetation could lead to reductions in the resource base (organic matter) of these fish communities.

The floodplains of the Amazon River are also critical habitat for many arboreal and terrestrial animals because they supply fruits and seeds at times when they are scarce in the uplands (Goulding et al. 1996). Over half of the 200 mammal species in the Amazonian low-lands, for example, inhabit floodplain areas at some point during the year. The world's smallest monkey, the pygmy marmoset (*Cebuella pygmaea*), feeds largely on the sap of floodplain trees. Uacari monkeys (*Cacajao* spp.) are also restricted to the Amazon floodplains (Goulding et al. 1996).

Logging of critical floodplain habitat is considered to be one of the major forces behind the decline of Amazonian fisheries (Goulding et al. 1996). Although tropical floodplain habitats may constitute a rela-tively small percent of a given watershed, their accessibility to logging operations has led to destruction of floodplain forests—often with consequently negative effects on fish and other aquatic biota.

Logging threatens the world's largest wetland—the Brazilian Panta-nal. As the world's largest wetland, the Pantanal is one of the largest breeding grounds for waterfowl and one of the last refuges for many of Brazil's threatened or endangered species, including jaguars (*Pan-thera onca*), giant anteaters (*Myrmecophaga tridactyla*) and swamp deer (*Blastocerus dichotomus*). The President of Brazil declared the Pantanal Matogrossense a priority area for research and conservation in 1985 as part of the opening ceremonies of a worldwide campaign directed at the conservation of wetlands (Alho et al. 1988). Nonethe-less, this wetland is currently threatened by logging operations and general deforestation. An official survey of the state of Mato Grosso do Sul indicated that the area authorized for logging has increased yearly in the southern Pantanal (Alho et al. 1988).

The floodplain of the Pantanal contains patches or strips of forest that are integral to the ecosystem. Gallery forests line many rivers and there are dense, semideciduous forests in the more elevated areas. Logging has increased dramatically in recent years, with illegal sawmills in operation in Mato Grosso and Mato Grosso do Sul (Cunha 1986). Floodplain and upland forests are biologically extremely important and logging is increasing erosion since most elevated areas have sandy soil that is easily washed away. The Rio Taquari and Rio Coxim are clogged with eroded material that has altered stream discharge (Ministerio do Interior 1974; Ministerio das Minas e Energia 1982). Apparently the Rio Taquari was once navigable, but it now overflows its banks as the confluence of the Rio Taquari with the Rio Paraguai is nearly blocked with sand (Alho et al. 1988).

The potential for the logging of estuarine floodplains to disrupt the integrity of tropical rivers can be illustrated by the many taxa of tropical fish and crustaceans that migrate longitudinally along the river—spending a portion of their life cycle in the estuary. Most taxa of freshwater shrimps (Decapoda), for example, migrate to the estuary as larvae and spend up to several months of their life in the estuarine habitats associated with floodplain forests before they migrate back upstream to adult habitat as postlarvae (Hunte 1978; Hobbs and Harte 1982; March et al 1998). Migratory catfish of the Amazon use tidal estuaries as nursery habitat, while adult and pre-adult schools migrate upstream from the estuaries where commercial interests exploit them from the river mouth upstream to Peru and Bolivia (Goulding et al. 1996; Barthem and Goulding 1997). Disruption of floodplain and mangrove forest by logging has the potential to affect the ability of these important nursery habitats to support many commercially important species.

Finally, massive logging of riparian and floodplain forests also results in dwindling supplies of coarse woody debris, which plays an important role in maintaining the stability of the stream channel and in creating habitat heterogeneity for fish and other aquatic biota along the entire stream continuum (from small tributaries, to large rivers, estuaries, and oceans). While the ecological importance of this wood (and the debris dams it creates) has been documented for temperate streams (Bilby and Likens 1980; Adams et al. 1988; Gregory 1992; Maser and Sedell 1994), few studies have examined its role in tropical rivers. Goulding et al. (1996) note that there is a decrease in the wood material that is being carried by the Amazon. Now that much of the middle Amazonian floodplain and its tributaries is already deforested, (most likely permanently), logging operations are targeting the tidal forests of the estuary. This will undoubtedly have severe ecological implications for the ecology of the Amazon and its estuary (Goulding et al. 1996).

Mangroves

Approximately 240,000 km^2 of mangrove vegetation covers the river deltas, lagoons, and estuarine ecosystems of the tropics. Mangrove forests play critical roles in these areas, including the maintenance of water quality and shoreline stability (through the control of nutrient and sediment delivery and distribution), and the provision of food and refugia for a multitude of organisms of different life stages and trophic levels (Twilley et al. 1995). These include the proboscis monkey (*Nasalis larvatus*), a species endemic to the mangrove forests of Borneo, and the tiger (*Panthera tigris*), for which the mangroves of the Indian and Bangladeshi Sunderbans form a critically important habitat (FAO 1982).

Direct exploitation of mangrove forests takes many forms, including cutting for fuelwood, manufacture of charcoal for domestic and light industrial use, timber production, and cutting roundwood for chips and pulp. Pulp production is an increasingly important market (Ong 1995). Large areas of mangrove in southeast Asia are cut for this purpose (e.g., 40 percent of mangrove in Sabah, Malaysia). Wood chips are exported primarily to Japan for production of dissolving pulp and rayon (Saenger et al. 1983).

Few quantitative studies exist on the effects of logging on the aquatic communities of mangroves. Logging of mangrove habitat, however, will inevitably affect the ecological *services* provided by forest and riverine ecosystems. The most important of these are coastal stability and protection from erosion, and the provision of nursery areas for juvenile fish and invertebrates. Effects of logging on mangrove soils are better understood. Clear cutting mangrove forests (see photo 14-4) often leads to significant and rapid accumulation of sulfides, which acidify soils. This lowering of soil quality limits seedling growth and subsequent regeneration of forest (Hamilton and Snedaker 1984). Poor regeneration, combined with the use of cleared mangroves for alternative land uses (e.g., mining, aquaculture, development), means that clear cutting is often irreversible (Saenger et al. 1983).

PHOTO 14-4 *Recent mangrove clearcut as part of a forest management operation. (R. Fimbel)*

In many areas of the tropics, and especially in southeast Asia, management of mangrove forests consists of selective logging practices with the retention of buffer strips to prevent or reduce erosion. The Matang Mangrove Forest Reserve in Malaysia is an example of efforts towards sustained yield management of mangrove forests (FAO 1985; Ong 1995). This 40,000-ha reserve operates on a 30-year cutting cycle. Felled areas are no more than a few hectares in extent and standards (7 trees/ha) are retained to provide seed sources for regeneration. Narrow belts of trees (3 m width) are preserved alongside the river channel to prevent erosion. Slash is left to decompose and seedlings are planted after two years if natural regeneration is poor. Trees are thinned after 15 and 20 years. Similar systems are operated in Thailand and Indonesia (Ong 1995). Unfortunately, little is known of the potential effects of RIL practices on the aquatic communities of mangrove habitats. Such attempts at sustainable use are more likely to result in decreased negative effects on mangrove habitat and the ecological processes they support, as compared to more conventional practices characterizing poorly regulated forest exploitation. More research is needed that investigates the effects of logging practices on mangrove communities, since information that could guide management is extremely scarce.

Conclusions and Recommendations for Conservation of River Ecosystems in Managed Tropical Forests

The physical and chemical effects of logging on streams and rivers depend on many factors. These include the percentage of the catchment logged, the amount of canopy removed, the amount of biomass removed, the removal methods, timing with regard to precipitation, soil type and topography, extent and siting of roads, speed of regeneration, and presence of buffer strips, among others (Hamilton and King 1983). Tropical climates can be expected to result in departures from conditions typical of temperate areas. High total annual precipitation in tropical areas has high erosive intensity and can cause relatively high levels of erosion in logged forests (El-Swaify and Dangler 1982). Erosion problems are also exacerbated by the current trend of logging on steep slopes in tropical areas. Subsequent high sediment loads and deposition may result in increased local flooding and water logging of floodplain soils with concomitant effects on aquatic biota. These effects may range from light limitation of algal primary producers (due to high turbidity) to reductions in numbers and diversity of stream fish and benthic invertebrates intolerant of high sedimentation.

Little baseline information is available on the effects of logging and related development activities (roads, settlement, agriculture, etc.) on tropical aquatic systems, as well as the complex socioeconomic influences that underlie these effects—making resource management and policy decisions difficult. There is clearly a critical need to obtain information on tropical aquatic systems (Pringle and Scatena 1999a, b) in order to make the necessary management and policy decisions in tropical forests managed for timber. The current extent of aquatic degradation in tropical areas, however, demands that management and policy decisions be made based on existing information. Despite the paucity of detailed research on the effects of logging on tropical river ecosystems, recommendations for management and policy can be made. These recommendations are based on knowledge that has already been acquired or extrapolated from experience gained in both temperate and tropical logged forests. Perhaps the most significant finding is that the method of extraction is the single most important determinant of the effects of logging practices on aquatic ecosystems, and that "at least half of the damage caused by selective logging can be avoided by careful execution of well-planned harvesting operations" (Bruijnzeel and Critchley 1994).

Priority Management and Policy Needs

Application of Reduced-Impact Logging Practices

Much damage to tropical river ecosystems could be prevented by responsible logging practices (Hamilton and King 1983; Bruijnzeel and Critchley 1994; see chapters 21 and 24), including:

- The proper planning of extraction that avoids wet periods and areas that are particularly sensitive to disturbance (e.g., low-lying areas subject to water logging)
- The careful planning of roads and log landings, including establishing roads along ridge crests, providing adequate drainage to prevent erosion and runoff reaching stream channels, and limiting maximum road gradients
- The minimization of stream crossings and the construction of bridges where technically and economically feasible
- The use of vehicles with reduced tire pressure in order to reduce soil compaction and disturbance
- The uphill extraction of logs (see figure 14-1)
- The reduction of soil disturbance through skyline or winch rope systems where possible (see chapter 2)
- The mulching and reseeding of logging roads and skid trails as soon as possible after logging is completed
- The maintenance of riparian buffer strips (10–40 m wide) along perennial stream channels

Maintenance of Riparian Buffer Strips

We strongly recommend the maintenance of riparian buffer strips in managed tropical forests. Buffer strips serve several important functions. They filter sediment and nutrients from runoff, prevent extreme fluctuations in water temperature, provide inputs of organic matter into stream ecosystems, and maintain a source of large woody debris for stabilizing channels (Hamilton and King, 1983). The cost of excluding riparian areas from logging will depend on whether commercial tree species are found in large quantities in these areas. Although the protection of intermittent stream channels may be prohibitively expen-

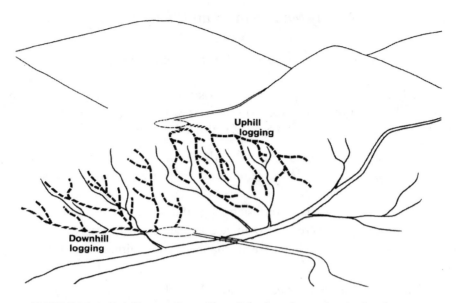

FIGURE 14-1 *Uphill extraction of logs is far less damaging to riparian areas than downhill extraction, which concentrates runoff toward lower slopes and stream channels. (Hamilton and King 1983)*

sive (Scatena 1990), buffer strips along perennial streams should be considered a minimum standard in managed forests. The amount of streamside land and vegetation needed to realize the full benefits of a riparian buffer zone vary according to goal and location. The maintenance of ambient water temperatures and sediment control in the temperate zone, for example, may require riparian buffer strips ranging from 10 to 45+ meters, depending on slope, soil, forest type, and cutting intensity (Dewberry 1992). High levels of erosive rainfall in many tropical areas will probably necessitate wider riparian buffer zones than those recommended for many temperate areas, however no studies have been conducted. We recommend a minimum width of 10 meters on level terrain, and wider buffer strips (30–40m) for habitat maintenance in flat areas and stream protection on steep sites.

Codes of Practice for Logging

We recommend the adoption and policing of detailed regional/ national codes of practice for logging (see Dykstra and Heinrich 1996 for examples from both temperate and tropical areas). These codes should set out environmentally acceptable practices for each component of logging (e.g., planning, road construction, cutting, extraction) and include guidelines for workforce training (see chapters 18, 19, 21, and 24). These codes of practice should include the application of integrated watershed management (i.e., management efforts that observe watershed boundaries and consider the entire river basin) when planning and implementing logging operations. Watershed management has not been widely applied in many tropical countries and national water quality monitoring programs are often nonexistent (Ongley 1993). Multiscale management strategies that consider the longitudinal and lateral dimensions of aquatic habitats and the role of riparian and floodplain vegetation are crucial (Dudgeon 1994).

Establishment of Foodplain Parks and Reserves

Very few floodplain reserves exist in tropical areas. There currently are no floodplain parks or reserves, for example, in the lower 2,500 km of the Amazon River. The first reserve (as one moves upstream) is the Mamiraua Ecological Reserve (11,000 km^2), which was established in 1990. At least one third of the reserve is flooded forest (Alexander 1994). Establishment of floodplain reserves where logging is prohibited or significantly limited would be a first step in the conservation of these unique ecosystems. Representative pristine forest habitats must be conserved in concert with sustainable management of other forests for timber. These pristine forests would provide a baseline for evaluating and refining ecologically sound forest management practices.

Restoration of Degraded Floodplain Ecosystems

Allowing natural succession to take place in degraded floodplain and riparian zones, and/or replanting floodplain timber species, could decrease further loss of floodplain habitat. We have been unable to find any information regarding efforts of this kind. In order to be effective, these areas would have to be given some degree of protection from further development (degradation) activities, such as agriculture and cattle grazing (see photo 14-5). This will require legislation and effective policing of protection laws.

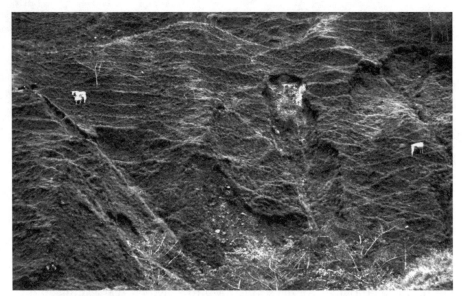

PHOTO 14-5 *Logged forest is often converted to agriculture or cattle grazing, leading to increases in soil erosion and changes in the microclimate of the stream channel. Such forests are unlikely to recover without restoration efforts and protection policies. (C. M. Pringle).*

Priority Research Areas

Basic Research on the Effects of Logging on Tropical River Ecosystems

Assessment of the effects of different types of logging practices and other related activities (road building, burning, provision of buffer strips) on aquatic biota and ecosystem-level processes—such as primary production and decomposition—is needed. We suggest that sites be established for intensive basic and applied research on the effects of forest management practices on tropical aquatic ecosystems (see chapter 19). Sites could be established in southeast Asia (which has a long history of managed forestry and selective logging), and also in representative managed forests in Central Africa and the Neotropics. Results could be extrapolated to other systems and used to formulate general management guidelines for logging in tropical forests.

Determination of Optimum Sizes of Riparian and Floodplain Zones

We need to determine the optimum (or minimum critical) sizes of riparian and floodplain zones for different tropical regions/life zones that will (as far as possible) maintain the biotic integrity of stream ecosystems. By biotic integrity, we mean the ability to support and maintain a "balanced, integrated, adaptive community of organisms having a species composition, diversity, and functional organization comparable to that of a natural habitat of the region" (Karr and Dudley 1981). While attempts at defining the *effective* size of riparian zones have been made in some temperate zone areas (e.g., Welsch 1991; Dewberry 1992), virtually no information exists for tropical areas (see Scatena 1990; see chapters 21 and 23). Well-designed studies are needed that explore the effects of different width riparian zones on channel stability, water temperature, rates of water, nutrient, and sediment delivery to stream channels—and their subsequent effects on tropical aquatic biota. These studies need to be conducted under a variety of logging regimes and in different tropical regions.

Identification of Priority Areas for Conservation

Identification and protection of tropical watersheds that exhibit high biodiversity/endemism of aquatic biota (e.g., Papua New Guinea, the Pantanal in Brazil) or important ecosystem services are needed (see chapter 20). The most critical regions for conducting surveys of potential floodplain reserves are the middle and lower Amazon because of the deforestation that has already taken place there (Goulding et al. 1996). Such surveys could be used to identify areas that are particularly sensitive to logging activities (with a view to their subsequent protection), and areas in which responsible logging could take place with minimal impact. As such, we favor an approach in which ecologically sound logging of forests is balanced by the conservation of representative pristine forest habitats (see chapters 20–24).

ACKNOWLEDGMENTS

We gratefully acknowledge the National Science Foundation, grant DEB-95-28434 to C. M. Pringle and F. J. Triska for support in writing this paper. A University-Wide Assistantship awarded by the University of Georgia helped support J. P. Benstead. We thank the editors and anonymous peer reviewers for the comments and suggestions that substantially improved the manuscript.

Part III

HUNTING: A MAJOR INDIRECT IMPACT OF LOGGING ON GAME SPECIES

Hunting in tropical forests, for subsistence and commerce, often has a far greater impact on certain kinds of wildlife populations than the direct habitat perturbations associated with silvicultural interventions (see parts I and II). Road networks created during timber harvesting operations substantially increase access to game, while facilitating their transport to markets. These markets are local (created by the influx of concession personnel and their families into rural areas) and external (where truckers and commercial hunters team up to supply urban centers).

In this section of the book, we examine case studies of hunting associated with logging from three continents and their impacts on wildlife. The scale of logging activities ranges from the community level to large, industrial operations. In all three case studies, hunting is either prohibited or strictly regulated by law, yet hunting violations are blatant and pervasive. The following three chapters describe the hunting activities and factors contributing to them, their impact on select wildlife species, and recommended measures to minimize unsustainable wildlife harvesting in timber concessions.

Hunting in Timber-Harvesting Concessions

In chapter 15, Damian Rumiz, Daniel Guinart S., Luciano Solar R., and José C. Herrera F. compare hunting techniques, offtake rates, and potential impacts on game populations from two timber management areas in Bolivia: one a commercial operation on government lands, and the other an indigenous community forest operation. In the commercial timber concession, a few valuable and widely dispersed tree species are extracted eight months of the year, during which time personnel from the logging company live and hunt in the forest. In the indigenous forestlands, about 15 tree species are extracted year-round using low-impact techniques—along with nontimber forest products including game. Both sites support similar vertebrate fauna, and all hunting is done for personal consumption. Hunting techniques differed between the two sites, with company personnel using vehicles and a variety of guns, and the indigenous peoples using .22 rifles or shotguns, dogs, and traps. Hunting was more selective in the commercial concession, where collared peccary and deer were favored species. The native Indians recorded a much more diverse catch, and higher hunting pressure on many species. Recommendations are presented in this chapter to strengthen the enforcement of laws designed to protect wildlife, and to assist local populations in managing their wildlife resources in a more sustainable manner.

In chapter 16, Elizabeth Bennett and Melvin Gumal describe the interrelationships of commercial logging, hunting, and wildlife in Sarawak—with wildlife products a $3.75 million-per-year enterprise. Logging creates hunting access to new areas, with hunters falling into three main categories:

- Local people: 29 percent of their rural meals contain wild meat
- Logging company employees: hunt mainly for subsistence (one logging camp of 167 workers and their families consumed 29,000 kg of wild meat in a year)
- Outsiders: use four-wheel-drive vehicles on logging roads and hunt for sport and the wild meat trade

The main species hunted are ungulates, but once these species become rare, smaller animals are taken. The end result of over-hunting is a cascading decline in wildlife populations, where some species are threatened with extinction within large areas of the permanent forest estate. Conservation measures range from leaving unlogged (roadless) blocks within the forest estate, to providing employees of timber firms with incentives to reduce hunting practices within the logging concession.

In chapter 17, David Wilkie, John Sidle, George Boundzanga, Philippe Auzel, and Stephan Blake wrap up part III by describing commercial logging and market hunting practices in Northern Congo, and the forest *defaunation* that follows in the wake of these activities. Logging is highly selective in these forests, removing less than one tree/ha, and leaving over 90 percent of the canopy intact. Nonetheless, wildlife species suffer under these logging conditions—not from loss of habitat, but from hunting. Tree prospecting transects crisscross the forest in a regular lattice, providing hunters with an efficient network of easily traversed trails, from which they can kill wildlife with shotguns and spotlights. Logging company roads and vehicles dramatically enhance travel and transport opportunities for hunters, permitting intense wild meat exploitation in the farthest reaches of the forest, and greater access to buyers in urban markets. Logging companies either tacitly support bushmeat hunting by allowing hunters to transport themselves and their game on logging trucks, or explicitly support them by creating a demand for cheap meat (for consumption by concession workers). The end result is that logging concessions are creating empty forests, filled with trees, but devoid of medium- to large-bodied animals. Supply- and demand-side conservation options, ranging from conservation bonds to alternative protein sources, are presented in this chapter to reduce the adverse impacts of market hunting in timber concessions.

Wildlife Conservation in Timber-Harvesting Concessions Subject to Hunting Pressures

Hunting provides an essential source of protein, and a major source of income, for the inhabitants of logging concessions and their surrounding communities. Facilitated by the concession infrastructure and transportation system, hunting at both the subsistence and market levels creates a serious threat to the wildlife in many areas. In most instances, the problem is not a function of weak conservation legislation, but of ineffective enforcement, few alternative sources of protein, and a general lack of education related to managing wildlife resources in the face of high hunting pressure.

A number of options exist to promote the conservation of wildlife in and around timber harvesting areas. All the authors in this section have proposed closing abandoned roads and monitoring vehicles along logging road arteries as inexpensive and highly effective tools for discouraging market hunting. The threat of vehicle detainment and the resultant high cost in lost productivity for the company, coupled with the risk of fines and the loss of meat to the drivers (via confiscation or the meat perishing), deters illegal hunters from entering the logging concession. These contributors also deemed the provision of alternative sources of protein to logging crews as important to wildlife conservation—because a domesticated source of protein reduces demand for wild meat. Programs to develop alternative protein sources—and demands for it—include:

- Promoting livestock husbandry (see chapters 15 and 17)
- Requiring concession owners to provide fresh sources of domestic meat (see chapter 16)
- Instituting a high tax on bush meat to add incentives to use domesticated sources of meat (see chapter 17)

Several other options were also considered to be available in the effort to conserve wildlife in areas exploited for timber. Conservation bonds, as part of the government's contract with a logger, could supply the funds required to support conservation officers in the field and minimize their corruptibility (see chapter 17). Modifying the work schedules of workers in the concession to make salaries dependent on output, decentralizing and dispersing crews to reduce socially facilitated hunting, and banning all firearms by concession personnel, are three measures likely to reduce the time and energy employees expend on hunting (see chapter 16). Finally, encouraging silvicultural practices that increase profits to the concessionaire, such as directional felling and road planning (these practices are discussed in part V), reduce the need for logging firms to exploit bush meat resources in an effort to decrease logging costs (see chapter 15). Incentives for promoting these activities are reviewed in part VI.

Hunting may become more sustainable in areas adjacent to logging concessions, and on lands managed by local communities for timber production, provided communities are organized, committed to conserving the resources they manage and willing to enforce community-based restrictions on resource exploitation (see chapter 15). In some instances, the wildlife harvest can be enhanced through habitat manipulation procedures (e.g., artificial nesting sites within agricultural areas). Other conservation programs, such as the promotion of domestic meat sources within rural communities, will also help to alleviate pressures on wildlife arising from human population growth and encroachment from outsiders (see chapter 15).

Lastly, hunting associated with logging can effectively preclude wildlife migrations from reserve areas (see chapter 17). To combat this situation, efforts in addition to those mentioned above should include expanding the system of totally protected areas in managed forest landscapes—including new reserves and blocks within concessions (Bennett and Gumal in chapter 16 recommend 200+ ha blocks of unlogged forest scattered throughout timber concessions), in addition to areas already legally protected as riverine reserves and/or very steep terrain. A network of protected areas is addressed in part V. These areas can serve as sources from which wildlife can recolonize logged and hunted areas, and must have their boundaries clearly marked so that infringements on their protection are easy to detect. Programs to monitor these sites (see part IV), and educate the population of their conservation importance (see chapter 16), should be promoted and supported by donor organizations and NGOs (see chapter 17). These institutions have the means to fund these conservation activities, the capacity to bring media attention to bear on them, and the expertise to train the personnel to carry out wildlife protection measures. Reducing hunting pressures in forests managed for timber production is a multi-level, multidisciplinary approach, with a great potential for conserving wildlife.

Chapter 15

LOGGING AND HUNTING IN COMMUNITY FORESTS AND CORPORATE CONCESSIONS

Two Contrasting Case Studies in Bolivia

Damián I. Rumiz, Daniel Guinart S., Luciano Solar R., and José C. Herrera F.

The hunting of wildlife in forests is a common practice associated with timber extraction in Bolivia (López 1993; Ribera 1996a; Rumiz personal observation). Hunting occurs in community forests exploited by indigenous peoples and in forests awarded as concessions to logging companies. Hunting reduces the density of forest wildlife (Fragoso 1991a; Glanz 1991; Redford 1992) and can be the most detrimental impact that selective logging activities have on biodiversity in Bolivia (Rumiz and Taber 1994).

A series of policy changes have been taking place in Bolivia during this decade, with the aim to promote a better use of natural resources and protect the environment (see chapter 29). As part of this policy, Bolivia negotiated for significant international donor aid to develop projects of sustainable development and conservation. The seven-year, USAID-funded Sustainable Forestry Management Project (BOLFOR) seeks to conserve forest, soil, and water resources, and to protect the biological diversity of Bolivia's forests by strengthening the local public and private sector capacity to develop programs for sustainable forest use. To provide guidelines for reducing the impacts of forestry activities on biodiversity, a series of studies on forest ecology, wildlife abundance, and human use of forest resources are being conducted under this project. These studies address several controversial issues, including the widespread practice of illegal hunting associated with timber activities in forest concessions, and the potential unsustainability of subsistence hunting by indigenous people.

In this chapter, we describe the impacts of hunting both within a commercial logging concession and in an indigenous community forest subject to timber management. The results of wildlife surveys and hunting assessments at both sites are used to evaluate wildlife abundance, hunting techniques, the species harvested, and the potential long-term effects of the harvest. Finally, we propose measures to reduce the negative impacts of hunting on animal populations.

Community Forest Management (Lomerío Site)

The Lomerío Community Forest covers an area of approximately 300,000 hectares south of Concepción, Santa Cruz Department (see figure 15-1), and is inhabited by some 25 indigenous communities of Chiquitano ethnic origin (about 5,306 people; Fischermann 1996). The communities are spread throughout moderately hilly terrain, which is mostly covered by semi-deciduous forest, wooded savanna, and less extensive plant communities associated with streams and rock outcrops (Navarro 1995). The climate is strongly seasonal, with a mean annual rainfall of 1100 mm occurring mainly from October to April (Killeen et al. 1990). This seasonality produces a dry period (August-September) in which fires can spread from the savannas into the forest.

Community Logging

The forest supports several valuable timber species (e.g., *Machaerium scleroxylon*, *Cedrela fissilis*, and *Amburana cearensis*), some of which have been selectively extracted by logging companies in past decades. In addition to high-value species, other species of construction quality and export quality timber occur. Since 1983, the communities have been grouped through their indigenous organization, the Central Intercomunal Campesina del Oriente de Lomerío (CICOL), and with external assistance have initiated programs to promote the sustainable use of natural resources. The sustainable forestry management project of CICOL extracts timber of some 12 species from two 400 ha units belonging to different communities each year, and processes the logs in their own mill. This logging operation has relatively low impact, with roads opened by local crews using chainsaws, machetes, and shovels, and logs extracted with a modified farm tractor. In 1994, the extraction rate in a 400-ha plot was 1.34 trees/ha (or 2.26 m^3/ha), representing 12 timber species—with a road density of 12.5 m/ha for primary roads and 45 m/ha for skidding trails (Peña 1996).

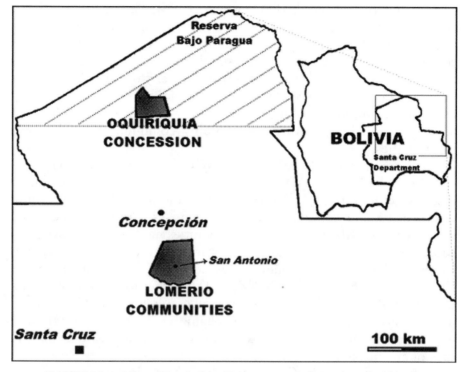

FIGURE 15-1 *Map of Santa Cruz Department, Bolivia, showing the two forestry operations studied.*

CICOL pays salaries to community people involved in forest management in the field and at the mill. Profits from the operation are shared between CICOL and the communities to improve local living conditions and cover operating expenses. The forestry operation of CICOL was certified in 1995 as *reasonably managed* by Smartwood (Rainforest Alliance Timber Certification Program, Richmond, VT, USA)—the first certification of this kind in the country. The certification approval indicates that the Lomerío operation complies with internationally set standards of environmental, social, and economic aspects of forest management, which favor its long-term sustainability (see chapter 26).

Community Subsistence

Families in each community have small agricultural plots where they grow rice, corn, yucca, plantain, peanuts, and other crops. They use forest resources as building materials, food, fibers, and medicine to satisfy most of their needs (Fischermann 1996). They also raise chickens, ducks, pigs, and a few cattle, but many people depend on hunting and fishing as their primary meat supply (see photo 15-1). Most of the communities have a basic school, a small church, and are connected by dirt roads to the town of Concepción and the city of Santa Cruz. A bus line joins Lomerío with Santa Cruz, while radio and TV broadcasting from San Antonio—the largest community and base of the Roman Catholic Church in Lomerío—maintains communications between communities.

The Catholic Church, APCOB (an NGO for aid to indigenous communities), SNV (the Dutch international aid agency) and, more recently, BOLFOR are all involved in programs related to natural resource use and community development. As a result of these and other programs, the Chiquitano communities have improved access to education, external markets, and information which, in turn, has increased their need for money to buy new goods. To obtain cash, the Chiquitanos may work for CICOL and the other local programs (at a daily rate of U.S. $5), or find temporary employment in Concepción, on neighboring ranches, and with logging companies. The minority who own cattle or smaller livestock can sell their domestic animals in case of an emergency.

In the last decade, the Chiquitanos have been changing their way of life from a relatively isolated subsistence to an economy more integrated with external markets. This, in addition to the population growth, has increased the pressure on natural resources and the environment. In the opinion of local hunters, game is becoming more and more scarce and they worry that it may disappear.

Concession Forest Exploitation (Oquiriquia Site)

The Oquiriquia timber concession encompasses 312,000 hectares and is located in the Bajo Paraguá Forest Reserve in northern Santa Cruz Department, on the San Martín River (see figure 15-1). The terrain is less hilly than Lomerío, which is approximately 200 km to the south. The nearest climatic data comes from Concepción, which receives approximately 1,300 mm of rainfall (Guamán and Montaño 1988). The dominant forest is semievergreen, with a higher proportion of deciduous species on the high ground and rocky outcrops (Navarro 1992). Long stretches of forest along the river and streams are flooded

PHOTO 15-1 *Chiquitano family butchers a collared peccary for subsistence, Todos Santos, Lomerio. (D. Rumiz)*

in the wet season, limiting forestry activities to no more than eight months a year (May-December).

The logging operation described below is one of the last of its kind in Santa Cruz, and perhaps Bolivia, and illustrates the traditional mahogany logging practices of the 1980s and early 1990s. Changes in the forestry sector due to the depletion of commercial stocks of mahogany and the shift from *per volume* to *per hectare* tax of the new forestry law (see chapter 29), forced this company to dissolve and reduce its concession area in 1997. In 1998, the more accessible parts of the concession were exploited by a former partner of the company.

Concession Logging

Like similar concessions in the area, operations in Oquiriquia flourished during the early 1990s when large patches of mara or mahogany (*Swietenia macrophylla*) were still common. The other two sought-after timber species were *C. fissilis* and *A. cearensis*. This highly selective extraction of dispersed, valuable trees in remote sites called for large investments in heavy machinery (such as bulldozers for road building, skidder tractors for retrieving logs, loaders, and logging trucks). This equipment, plus installed machinery at the mill, was owned by the timber concessionaire and operated by its personnel. In the forest, however, tree finding and felling was performed by independent contractors who worked in assigned areas with their own hired crew of wood prospectors, chainsaw operators, apprentices, and cooks (Solar 1996). Forest workers usually had only a basic education and came from poor families but received higher wages than ranch or farm workers (between U.S. $75 and U.S. $130 per month, according to their experience). Contractors were paid by timber volume sent to the mill, and were charged for the supplies provided by the company.

Timber operations started in the last unlogged portion of the concession in 1994. Wood prospectors scouted on foot from the road-ends looking for concentrations of valuable timber in their assigned forest areas. Carrying minimal equipment and supplies, they established temporary camps from which to find, mark, and fell trees. Construction of primary and secondary roads came after, with an influx of vehicles and personnel to extract the logs. Extraction rate for mahogany (and very few cedar trees) in a sample of 3,200 ha area was only 0.08 tree/ha (0.293 m^3/ha). Mahogany trees were not uniformly distributed, however, and their densities varied between 0.17 and 1.06 trees/ha in one 750 ha mahogany patch near our hunting study site

(Santillán 1996). Road density was 84.5 m/ha in the timber patches and 12.5 m/ha in the general area. Logging as described above ended at this site in 1996, when all the identified mahogany clumps were felled. Other sites in Oquiriquia (and other concessions), which had been already logged once, were subject to second- and third-round searches for trees of these three species that may have been missed in previous harvests. These reentries took advantage of preexisting roads and continued high grading the forest, taking a few other species of lesser value until there was no marketable timber left.

Concession Hunting

The Oquiriquia study site had been empty of roads and human settlements prior to logging, and therefore suffered no hunting pressure in the recent past (10–20 years). Wildlife was abundant, not wary of people, and relatively easy to kill. During the timber exploration phase, meat from game became a vital resource for wood prospectors because of the difficulty of obtaining food supplies from outside the concession. Later, road building crews, truck drivers, and other company personnel increased the hunting pressure on wildlife near roads and camps. In accessible sites of this and other concessions, the effect of continuous hunting has reduced wildlife populations until, according to workers, "it was not worth the effort to hunt."

The forest workers traditionally viewed hunting as a legitimate way to obtain food, although law forbids it. This activity was allowed and even encouraged by some company administrators because hunting reduces operation costs, provides recreation, and may fill requests for wild meat, skins, or pets from friends and relatives. Evidence of game used in logging camps of different concessions shows that hunting is widespread in all the Bajo Paraguá Forest Reserve (Rumiz and Taber 1994; Frumhoff 1995).

Wildlife Studies and the Monitoring of Hunting

The Lomerío Community Forest

Since 1994, a series of studies had been conducted in Lomerío with the objective of assessing the status of wildlife in a community site subjected to forest management. Research camps with a system of trails, trapping plots, and track plots were established at two forest sites 10 km away from the nearest community. Diversity and the abundance of mammals were assessed during bimonthly field trips in 1995 to 1996, recording all animal encounters along 2,888 km of walking trails in the forest and by noting animal signs in track plots (Guinart 1997). Local hunters were temporarily employed to help with fieldwork. They demonstrated great knowledge and concern for wildlife issues.

To approach the issue of wildlife use in the local communities, several slide talks and two four-day courses on the monitoring of hunting and basics of wildlife management were offered to local hunters from eight communities in Lomerío. A booklet was designed and field-tested, with the participation of the hunters, to record every hunted animal. After attending the courses, 20 hunters from five communities began monitoring their game intake in an effort to start managing their wildlife resource. The hunter recorded species, sex, weight, reproductive condition, hunting technique, trip duration, and killing site in the booklet for each animal (see photo 15-2). Records in booklets were reviewed, discussed with the hunter if unclear, and transcribed by a researcher every one or two months. This chapter reports on data collected from July 1995 to June 1996, involving 20 trained hunters who recorded their own catch plus part of the catch of 55 other hunters. These 75 hunters represented about 80 percent of the 95 families of these five villages, although the exact proportion of the sample to the total game harvested cannot be evaluated because not all the animals from each hunter were recorded. For the comparisons made in this chapter, however, the year-round species composition of the harvest and the relative contribution of the most important game species are representative of the general situation in Lomerío. To further complete the picture, hunters were interviewed about species taken during exploration and forest inventory activities far away from their communities, and two remote camps were searched for animal remains.

The Oquiriquia Concession

Beginning in 1995, a semipermanent research camp was established on the San Martín River in Oquiriquia, following the opening of roads in that area. Studies on forest ecology, wildlife diversity, and logging efficiency were conducted in the production forest surrounding that river. Mammal diversity and abundance were assessed through

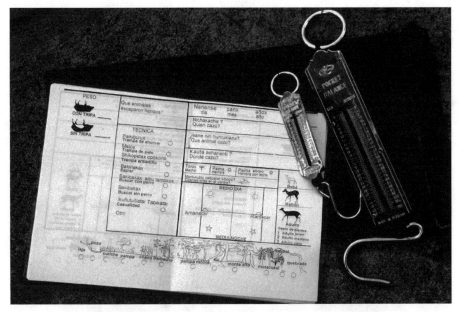

PHOTO 15-2 *Balances and booklet used by trained Chiquitano hunters to record their game. (D. Rumiz)*

transect censuses and track plot records in a 600 ha forest block adjacent to the research camp.

Four camps of wood prospectors and one of the road building crew located between 1 to 20 km of our camp were monitored daily, or visited weekly, to record wildlife use during a total of 75 days in June to August 1995 (Solar 1996). Monitoring at each camp was performed during three to five consecutive days. The time spent hunting during wood prospecting activities such as exploration, trail opening, tree marking, and tree felling was also recorded for each of the four camps. All wildlife brought to camp was identified by species, weighed, and recorded with additional information on hunting technique and killing site. Weekly visits to the other camps, questioning hunters and cooks about the week's game harvest, and examining remains of recent kills (see photo 15-3), gave an estimate of all hunting throughout the study period. Opportunistic observations about wildlife use at the company's mill and in other logging areas of Bajo Paraguá were also made by BOLFOR researchers during these and other studies.

PHOTO 15-3 *Land tortoise and giant armadillo shells found at a logging camp in Bajo Praraguá, northern Santa Cruz. (D. Rumiz)*

Hunting in Community Forests and Corporate Concessions

Hunting Techniques of Indigenous Hunters in Lomerío

In each family of the sampled communities, the father and sometimes an older son usually hunted regularly for domestic consumption. The most commonly used weapons—.22 caliber, one-shot rifles—were owned by half of the hunters and borrowed by the other half. Game was consumed entirely by the hunter and his close family, or shared with other hunters, the owner of the gun or dogs if borrowed, and extended family—depending on size of catch. No commerce in wild meat appeared to exist in Lomerío, although exchange for other goods and reciprocity seemed to be common among neighbors.

Animals were harvested during active searches through forest and savannas, while waiting (stalking) at water holes or fruit trees, in casual encounters while engaged in other activities, and by trapping. About 85 percent of searching trips were done with dogs, and in half of the casual encounters, dogs were also involved. Most of the prey was taken by active searches, although waiting was very important during the dry season and trapping was widespread during the wet season. Most traps were traditionally crafted, consisting of a heavy falling log (see photo 15-4)—although some hunters used steel leg-traps. Trapping took place mostly in or around farm plots when crops were ripening, with the additional purpose of eliminating potentially harmful animals such as agoutis, armadillos, and foxes.

While Chiquitanos use a range of animal products, such as skins for musical instruments or fat and hooves for medicinal purposes, this did not constitute the primary incentive to hunt. Wild animals as pets were rare in their communities.

Hunting Techniques of Logging Crews in Oquiriquia

Wood prospectors usually carried guns for the potential encounter with edible or threatening (such as jaguars) wildlife. The most commonly used guns were single-shot, 16-gauge shotguns and .22 caliber rifles; .22 caliber and .38 caliber revolvers, semi-automatic .22 caliber rifles, and

PHOTO 15-4 *Traditional Chiquitano trap using a heavy falling log to immobilize prey. (R. Fimbel)*

even a military Mauser 7.65 rifle were used by company personnel. There were usually three or four guns per camp of seven to 17 people.

Most animals (65.7 percent of total numbers) were harvested during the day while prospectors were engaged in the exploration and marking of trees. Nocturnal searches on foot with flashlights yielded 20.4 percent of the catch, and stalking from tree platforms at fruiting trees, saltlicks, or waterholes produced the remaining 13.9 percent of the harvest. Diurnal hunts yielded most of the cracids (guans and curassows), while nocturnal hunts and stalking produced most of the large ungulates (tapir and deer). Opportunistic shootings from vehicles (both day and night) were conducted primarily by drivers and supervisor-level staff living at the road-building camp. All of the reported harvest in the road camp was taken on or near roads, and by using vehicles to reach the killing sites. No trapping was recorded, and workers reported that the practice of setting up guns in trails with a cord attached to the trigger was not used for safety reasons. Target shooting of animals not eaten in camps—such as monkeys and foxes—occurred when workers were bored.

Apart from the wildlife consumption in forest camps, it was common to find wild meat at the rustic dining areas, which served three meals and charged a daily fee to each employee. Employees who hunted in their free time could cover the fee by providing game to the kitchen operator. Skins from cats, monkeys, and otters, as well as river turtle eggs, pet monkeys, and parrots were observed around mill facilities during this study. Although forbidden by law, these products could be transported by logging trucks and sold or traded outside the concession.

Composition of the Wildlife Harvest in Lomerío

The Chiquitanos recorded 32 hunted species during the 12-month study—24 mammals, six birds, and two reptiles—although it was known that at least two mammal, two bird, and one reptile species were also hunted during the study period (see table 15-1). Armadillos were the most abundant prey group taken (35 percent of capture frequency and 12 percent of biomass), with the nine-banded armadillo (*Dasypus novemcinctus*) the most numerous mammal species hunted. The most important prey group by biomass (accounting for 66 percent of the estimated 3,529 kg harvested) was the ungulates, with gray brocket deer (*Mazama gouazoubira,* 40 percent) and collared peccaries (*Tayassu tajacu,* 13 percent) dominating this category. Tapirs (*Tapirus terrestris*) were rare among the catch, with only two killed during the study period. Coatis (*Nasua nasua*), agoutis (*Dasyprocta variegata*), and pacas (*Agouti paca*) were caught between three and a dozen times each. A number of small mammals and birds—such as squirrels, rabbits, doves, and toucans—were also hunted for food, although they represented a small return in meat for effort. Cracids were scarce in the sample. Rare carnivores that apparently were not very tasteful, like otters (*Lutra longicaudis*) and raccoons (*Procyon cancrivorus*), were also consumed. Land tortoises (*Geochelone spp.*) and tegu lizards (*Tupinambis teguixin*) were hunted and regarded as a good meat source.

TABLE 15-1 *List of Selected Vertebrate Species Present in Lomerío and Oquiriquia, Individuals (ind) and Biomass (kg) Harvested at Each Site, and Conservation Status in the Red Data Book of Vertebrates of Bolivia and CITES*

Species	Common Name	LOMERÍO		OQUIRI-QUIA		Conservation Status	
		ind	kg	ind	kg	R.D.B	CITES
MAMMALS							
Agouti paca	Paca	12	84	5	35	CT	
Alouatta caraya	Black howler	1	7	x		V	II
Alouatta seniculus	Red howler	1	7	x		I	II
Aotus azarae	Night monkey	1	1	x			II
Ateles paniscus	Spider monkey	1	8	4	32	V	II
Callithrix argentata	Silvery marmoset	x		y		V	II
Cebus apella	Brown capuchin	4	12	y			
Cerdocyon thous	Crab-eating fox	12	60	y			
Coendou prehensilis	Tree porcupine	6	24	x			
Dasyprocta variegata	Agouti	32	128	1	4		
Dasypus novemcinctus	Nine-banded armadillo	96	384	6	24		
Eira barbara	Tayra	2	4	x			
Euphractus sexcinctus	Yellow armadillo	12	48	?			
Felis concolor	Puma	1	26	y		I	II
Felis pardalis	Ocelot	x		y		V	I

Species	Common Name	LOMERIO ind	LOMERIO kg	OQUIRI-QUIA ind	OQUIRI-QUIA kg	Cons. Status R.D.B	Cons. Status CITES
Felis wiedi	Margay	x		y		V	I
Felis yagouaroundi	Jaguarundi	1	4	x		I	II
Hydrochaeris hydrochaeris	Capybara	3	120	x			
Lutra longicaudis	River otter	y		x		V	I
Mazama americana	Red brocket deer	3	90	12	360	CT	
Mazama gouazoubira	Gray brocket deer	71	1420	2	40	K(CT)	
Myrmecophaga tridactyla	Giant anteater	x		x		V	II
Nasua nasua	Coati	36	144	1	4		
Panthera onca	Jaguar	x		y		V	I
Potos flavus	Kinkajou	x				y	
Priodontes maximus	Giant armadillo	y		2	80	V	I
Procyon cancrivorus	Crab-eating raccoon	1	7	x			
Pteronura brasiliensis	Giant otter	no		y		E	I
Sciurus spadiceus	Red squirrel	1	0.4	x			
Sylvilagus brasiliensis	Brazilian rabbit	1	0.8				
Tamandua tetradactyla	Collared anteater	2	10	x			
Tapirus terrestris	Tapir	2	320	10	1600	V(CT)	II
Tayassu pecari	White-lipped peccary	2	60			V(CT)	II
Tayassu tajacu	Collared peccary	23	460	1	20	V(CT)	II

BIRDS

Species	Common Name	LOMERIO ind	LOMERIO kg	OQUIRI-QUIA ind	OQUIRI-QUIA kg	Cons. Status R.D.B	Cons. Status CITES
Ara ararauna	Blue and yellow macaw	x		y		CT	II
Brotogeris sp.	Parakeets	x		y			

TABLE 15-1 *List of Selected Vertebrate Species Present in Lomerío and Oquiriquia, Individuals (ind) and Biomass (kg) Harvested at Each Site, and Conservation Status in the Red Data Book of Vertebrates of Bolivia and CITES (continued)*

Species	Common Name	LOMERIO		OQUIRI-QUIA		Cons. Status	
		ind	kg	ind	kg	R.D.B	CITES
Cairina moschata	Muscovy duck	?		1	3	V	
Crax fasciolata	Bare-faced curassow	1	2.9	3	8.7		
Crypturellus sp.	Tinamou	y		y			
Leptoptila verreauxi	White-tipped dove	1	0.1	x			
Mitu tuberosa	Razor-billed curassow	?		52	182	CT	I
Ortalis guttata	Chalchalaca	1	0.6	x			
Penelope sp.	Guans	1	1.5	13	19.5		
Pipile pipile	Piping guan	?		14	18.2		
Ramphastos sp.	Toucan	2	0.8	x			
Rhea americana	Rhea	1	20	no		V	II
Tinamus sp.	Tinamou	y		x			

REPTILES

Species	Common Name	LOMERIO		OQUIRI-QUIA		Cons. Status	
		ind	kg	ind	kg	R.D.B	CITES
Caiman yacare	Spectacled caiman	y		3	45	CT	II
Geochelone sp.	Land tortoises	11	44	4	16	V	II
Melanosuchus niger	Black caiman	no		y		E	I
Tupinambis teguixin	Tegu lizard	13	39	x		K	II

Sampling period		365 days		75 days			
Harvested individuals in the sample		358 ind.		134 ind.			
Harvested biomass in the sample		3,538 kg		2,491 kg			
Total spp. hunted during sampling		32 spp.		17 spp.			
Total spp. known as hunted		37 spp.		30 spp.			
Total game spp. confirmed per site		45 spp.		48 spp.			

ind: number of individuals in the sample, kg: harvested biomass per species in the sample, x: present but not recorded as hunted in the area; ?: possible but unconfirmed presence; no: absent; y: confirmed use outside sampling period; R.D.B.: Status in the Red Data Book of Vertebrates of Bolivia; E: Endangered; V: Vulnerable; I: indeterminate; K: Insufficiently known; CT: Commercially threatened CITES; Appendix I: Endangered; Appendix II: Vulnerable

Hunting around temporary logging camps of Lomerío, as reported in interviews, did not seem to be selective, and almost any available mammal, bird, or reptile was eaten. Rare species not found near villages, such as giant armadillos (*Priodontes maximus*), tapirs, and white-lipped peccaries (*Tayassu pecari*) were hunted in two distant camps.

Thirteen species from the 32 hunted during the sample (or 16 out of 37 in the extended list of prey) were listed in the Red Data Book of Vertebrates of Bolivia (column RDB in table 15-1). Of these, giant armadillos, tapirs, white-lipped peccaries, monkeys, river otters, cats, rheas (*Rhea americana*), and cracids (*Crax fasciolata, Penelope superciliaris*) are species of concern because of their conservation status and their low numbers in the sample (which suggest low populations). Other listed species which were hunted in larger numbers, such as collared peccaries, brocket deer, tortoises, and tegus also need attention because of their high hunting pressure.

Composition of the Wildlife Harvest in Oquiriquia

From the five monitored camps in Oquiriquia, a total of 134 individual animals of 17 vertebrate species (weighing approximately 2,491 kg) were recorded as hunted (see table 15-1). Of these, cracids (*Mitu tuberosa* and others) were the most frequent prey item (82 individuals or 61 percent of the total), but made up only nine percent of the biomass. Ungulates accounted for 81 percent of the biomass, the most important being *Tapirus terrestris* and *Mazama americana*. Large rodents (*Agouti* and *Dasyprocta*) and armadillos (mostly *Dasypus* but also *Priodontes*) were of lesser importance, as were caimans (*Caiman yacare*) and tortoises. According to hunters, spider monkeys (*Ateles*) and coatis (*Nasua*) were not preferred food, but were shot if nothing else was available.

Eleven of the 17 species hunted during sampling, or 19 out of the 30 known to have been killed in the concession, were listed in the Red Data Book of Vertebrates of Bolivia (see table 15-1). Among these, giant armadillos, tapirs, cracids, and monkeys are species of concern because of their high conservation priority and because they were hunted in large numbers.

Wildlife Diversity, Abundance, and Use in the Two Study Areas

Of 51 game species (see table 15-1), 44 (86 percent) were confirmed at both sites and five more are probably shared, but could not be confirmed for one site. This suggests that up to 96 percent of the hunted species were shared between the sites. Giant otters (*Pteronura brasiliensis*) are currently absent from the study area in Lomerío, although they may still survive in some remote lagoons. Greater rheas (*Rhea americana*) and yellow armadillos (*Euphractus sexinctus*) did not occur in Oquiriquia—probably because of the lack of open habitats—while white-lipped peccaries are known 50 km north of the site, but not seen at Oquiriquia. Two cracids (*Mitu tuberosa* and *Pipile pipile*) were quite common in Oquiriquia, but may have disappeared from or never existed in our site in Lomerío.

The relative abundance of 26 game species was calculated based on encounter rates during diurnal census walks totaling 2,888 km in Lomerío and 458 km in Oquiriquia (see table 15-2). Eighteen species (69 percent) were encountered more frequently in Oquiriquia than in Lomerío (Wilcoxon Signed Rank Test, p = 0.007). This list included tapirs, red brocket deer, most monkeys, porcupines, agoutis, squirrels, guans, curassows and tortoises. Some species were not found during censuses. The curassow (*M. tuberosa*), for example, may have been absent from Lomerío due to unsuitable habitat, while two guans and spider monkeys were reported as present but rare by locals. In contrast, two armadillos, collared peccaries, gray brocket deer, and coatis appeared more abundant in Lomerío.

TABLE 15-2 *Relative Abundance of 26 Selected Game Species[a] Based on Individuals Encountered per 100 km Walked in Lomerío (2,888 km) and Oquiriquia (458 km) During 1995–96 Studies*

Species	Common Name	Relative Abundance ind/100km	
		Lomerío	Oquiriquia
MAMMALS			
Alouatta caraya / Alouatta seniculus	Black/Red Howler monkeys	1.21	0.22
Ateles paniscus	Spider monkey	0	37.27
Callithrix argentata	Silvery marmoset	0.38	2.61
Cebus apella	Brown capuchin	11.39	12.2

TABLE 15-2 *Relative Abundance of 26 Selected Game Species[a] Based on Individuals Encountered per 100 km Walked in Lomerío (2,888 km) and Oquiriquia (458 km) During 1995–96 Studies (continued)*

Species	Common Name	Relative abundance ind/100km	
		Lomerío	Oquiriquia
Cerdocyon thous	Crab-eating fox	0.35	0.44
Coendou prehensilis	Tree porcupine	0.04	2.83
Dasyprocta variegata	Agouti	0.52	10.24
Dasypus novemcinctus	Nine-banded armadillo	0.28	0.22
Eira barbara	Tayra	0.73	0.87
Euphractus sexcinctus	Yellow armadillo	0.1	0
Felis spp.	Small cats	0.07	0.22
Mazama americana	Gray brocket deer	0.42	6.97
Mazama gouazoubira	Red brocket deer	2.18	0
Nasua nasua	Coati	4.4	0.87
Sciurus spp.	Squirrels	1.11	10.02
Sylvilagus brasiliensis	Brazilian rabbit	0.55	0.22
Tamandua tetradactyla	Collared anteater	0.04	0.44
Tapirus terrestris	Tapir	0.48	3.92
Tayassu tajacu	Collared peccary	0.52	0

BIRDS

Species	Common Name	Lomerío	Oquiriquia
Crax fasciolata	Bared-faced curassow	0.47	1.74
Mitu tuberosa	Razor-billed curassow	0	11.55
Ortalis guttata	Chalchalaca	0	1.09
Penelope sp.	Guans	0	17.22
Pipile pipile	Piping guan	0	7.19

REPTILES

Species	Common Name	Lomerío	Oquiriquia
Geochelone spp.	Land tortoise	0.13	0.65
Tupinambis teguixin	Tegu lizard	0.08	0

[a]Abundance data are limited to species recorded in both sites (or represented by an equivalent species, e.g., howler monkeys) and to available census results (e.g., for some mammals and most birds transect observation was not reliable)

In spite of the similarity of the game species lists, hunting patterns differed between the sites. Of the game species present at each site, 37 (82 percent) were known to be hunted by the Chiquitanos and 30 (62 percent) by logging company personnel. Wood prospectors were more selective than the Chiquitanos in their game intake—this is evident from the diversity indices of the harvest. A Shannon H index of 0.93 for harvested wildlife in Oquiriquia was lower ($p < 0.02$, Zar 1984) than the 1.05 computed for comparable sampling periods in Lomerío.

Although sampling methods were not the same in the two sites, we are confident that the lower diversity in the concession harvest is real and not a sampling artifact. Harvest diversity in the Lomerío area is probably also underestimated because community hunters did not record small prey such as birds or rodents.

Hunting Differences Between Lomerío and Oquiriquia

As mentioned above, the type of guns and other aids used to hunt in the two study sites were different (see table 15-3), due in part to:

- The greater economic capacity of company employees who had access to vehicles and a variety of guns
- The time availability and hunting traditions of the Chiquitanos (longer hunts that often used dogs, and handcrafted traps near villages)

The ambush technique used to hunt ungulates at fruiting trees, salt licks, or water holes was common among hunters from both sites

Diversity of prey taken in both sites conforms with general differences reported for Indians and colonist hunters in the Neotropics (Redford and Robinson 1987). Wood prospectors in Oquiriquia focused on few animal species, which represented high returns in meat, and on other abundant and easy to find species with high quality meat. The Chiquitanos had a more varied list of prey, and took species of smaller size such as tegus, doves, toucans, squirrels, and rabbits. Elder Chiquitano hunters mentioned that species such as anteaters, tayras, and foxes were not hunted before, but that overall wildlife scarcity was leading people to forget traditional restrictions.

Wildlife products in Lomerío were used exclusively at the family level, with food or medicine shared between relatives and neighbors. Wild meat consumption in one Chiquitano community throughout a year—estimated from a subset of the data presented here (Guinart 1997)—was 48 g of meat (or 13.8 g of protein) per person per day.

TABLE 15-3 *Summary of Differences in Hunting Between the Chiquitanos in Lomerío and Logging Crews in Oquiriquia Logging Concession*

Site	Tools/ Aids	Techniques	Use of Products	Main Prey Species	Bushmeat Use Rate
LOMERÍO	one-shot rifle/ shotgun, dogs, traps	day search, ambush, trapping, casual /in crops	family food, meat sharing, local exchange, medicinal, skins,	armadillo, gray brocket, agouti, coati, peccari, (+27 spp.) H'= 1.05	38 g/day/ person
OQUIRIQUIA	various guns, vehicles	day/night search, ambush, driving	camp/mill food, out-side sale, pets, skins	cracids, red brocket, tapir, (+11 spp.) H'= 0.93	524.6 to 450.7 g/ day/person

This consumption level is far below the FAO's recommended daily intake for protein (37–62 g), although beans, peanuts, and occasional fish and livestock products supplement the Chiquitano diet. Local people repeatedly pointed out the scarcity of wildlife resources.

Game harvested at Oquiriquia logging camps was not consumed all *in situ*, but shipped out to other camps, the mill, or town. Assuming all meat was consumed in the camp, two closely monitored camps (from which no much meat could have left unseen) yielded means of 524.6 and 450.7 g/person/day over approximately 38 days (Rumiz and Solar 1996). This daily amount of meat seemed more than adequate, and people did not complain about scarcity. In other camps for which game harvest was estimated by regular interviews and counts of animal remains, the estimated daily meat intake had a mean of 1,909 grams per person. It is unlikely that all of this meat was consumed on site, suggesting that some of the meat was shipped out.

Effects of Hunting by Indigenous Hunters Compared with Commercial Logging Crews

Wildlife Abundance and Hunting Patterns in Lomerío

Game abundance influences harvest patterns, but that abundance is also influenced by hunting pressure (Redford and Robinson 1987; Robinson and Redford 1991). Lomerío seems to have suffered increasing rates of hunting in the last decade as a result of human population growth and encroachment from outside areas. The concern expressed by local people about diminishing game is probably related to over-hunting and habitat modification—the latter through increased burning to improve pasture. An apparent conclusion of this study is that hunting for domestic consumption, where human population is about 2 people/km^2, has resulted in a decline of large mammals and a shift to consumption of a more diverse prey base dominated by generalist species having high reproductive rates.

Tapirs were rare prey in Lomerío, as were cracids, monkeys, and red brocket deer. Species such as tapirs and red brocket deer have probably diminished due to overhunting, although spider monkeys (*Ateles paniscus*) and razor-billed curassows (*Mitu tuberosa*) may never have been abundant because of the lack of tall, humid forest in the region. Other species, such as gray brocket deer, collared peccary, armadillos, coati, and agouti, were hunted in high numbers and their relative abundances seemed to remain high. It is not known if current harvest rates for these species have been stable over the years, and if the area from which they were taken is large enough to allow for recovery of depleted stocks. Preliminary sustainability analyses for two communities (Guinart 1997) suggest that coatis and gray brocket deer are being over-hunted, while nine-banded armadillos and agoutis are not. Using production estimates (based on observed female productivity) and estimated or assumed population densities for these four species (local densities for the two first species and lowest values from the literature for the other two), hunting in these communities seemed to be taking one to three times the production of brocket deer and one to five times that of coatis. Agoutis were estimated to be hunted at seven percent and 40 percent of their annual production, and armadillos at about 30 percent—values that might be within sustainable levels (Robinson and Redford 1991b). Long-term monitoring is needed to confirm the trend of these populations since wildlife demographics and hunting pressure may change year to year due to rainfall pattern, fruit availability, and fires.

Since the start of communal forest management activities, crews of Chiquitanos have worked in the forest conducting timber inventories, marking trees, opening trails and roads, and felling and extracting timber. When the sites are within a couple of hours walking distance from the villages, workers return daily to their homes to eat and sleep. When they are further away, however, they have to cook and set up temporary camps in the forest. Each group of workers takes local food into the for-

est (rice, yucca, corn, peanuts) and carries at least one gun to hunt wildlife. Meat from domestic sources is scarce, expensive, and difficult to
transport or preserve in the forest. Hunting does not seem to be selective,
and almost any available prey is eaten. Because camps are relatively
small and used for only a short time, hunting conducted from these
camps may not substantially increase the hunting pressure over that of
hunting that occurs near villages. The opportunity to hunt rare species
and bring meat home from game-rich sites, however, does still exist.

The decrease in game abundance, together with changes influenced
by rural development plans, have induced a decline in the relative
importance of wild meat as a source of protein in the Chiquitano's
diet. The larger communities—more connected to a market economy—rely more on the regular butchering of cattle and pigs, while
smaller or more isolated communities partially substitute wild meat
with chicken and plant protein. The Chiquitano families' need to hunt
should decrease if more sources of protein were available. This should
also allow adoption of wildlife management practices to protect some
species and allow increased harvesting of others.

Wildlife Abundance and Hunting Patterns in Oquiriquia

The high harvest of wildlife in areas recently opened to logging—
like Oquiriquia—depended mostly on the relatively high ungulate
abundances. Tapirs and red brocket deer were the main prey for
immediate consumption or for dried meat in logging camps. Logging
crews consumed over ten times the daily amount of meat as compared
to the Chiquitano people.

Harvest rates per km^2 were also high and indicated that these
extraction levels were not sustainable in the long term. For tapirs
(0.247 individuals/km^2), it was 24 times the estimated sustainable
yield (using Bodmer's [1994] equation), and was almost eight times
the sustainable estimate made with the same source (W. R. Townsend,
unpublished data) for red brocket deer (0.394 individuals/km^2). The
hunting of guans and curassows was also at a high rate in Oquiriquia.
Considering the slow growth and reproduction of most of these large
birds (Silva and Strahl 1991), current levels of harvest are probably
not sustainable. In addition to the harvest of relatively abundant species, rare and vulnerable species such as giant armadillos were killed
for food—and monkeys were killed for fun.

Hunters in Oquiriquia reported that the most important game species
had vanished from around camps that were used repeatedly and saw

mill areas. Wood prospectors did not like to be assigned to work in these accessible areas because the scarcity of game affected food quality and cost. We believe that the abundance of tapir, deer, and curassows at the Oquiriquia site was due to the previous absence of hunting this site, and that overhunting explains the scarcity or absence of tapirs in heavily-hunted places—similar to the conditions reported by Fragoso (1991a) for Belize. If large mammal and bird populations are to be conserved in forest concessions, new management practices and hunting restrictions must be implemented.

Proposed Measures to Reduce the Impact of Hunting on Wildlife

Managing Subsistence Hunting with Indigenous Communities

In Bolivia, indigenous peoples are allowed by law to use natural resources for subsistence—including the right to hunt wild game. The Chiquitanos exploit this right, but it may be threatening the future of several wildlife species. The people of Lomerío are concerned about local wildlife depletion and seem willing to make some changes in the use of their wildlife resources. To this end they are amenable to developing a participatory wildlife management plan, in which the communities would agree that wildlife is a communal resource, and that measures or restrictions governing its use will be enforced through their traditional channels. If hunting restrictions are established (such as a hunting ban for tapir, giant armadillo, and rhea or the setting aside of refuge areas), then alternative sources of meat must be developed (such as improved fish production, use of micro-livestock, and management of armadillos and agoutis). Another promising option is the manipulation of the habitat to increase resources for wildlife—such as planting or helping natural regeneration of trees important for frugivores in abandoned farm plots—and protecting or increasing the number of permanent water holes. Armadillo and agouti populations could probably improve their growing potential and support more intense harvests.

Territory claims by indigenous groups to the government of Bolivia recently reached 17 million hectares, significantly increasing the land that potentially will control native communities. The ways that indigenous groups manage their resources (wildlife included), therefore, will affect much of the biodiversity of Bolivian forests. There is clearly a need to disseminate the idea of sustainable use of wildlife, and to develop participatory plans for wildlife management in indigenous territories.

Corporate Forest Management with Reduced Environmental Impact

The logging concessionaires have generally maximized the exploitation of any forest resource on their timber concessions, including wildlife for food, pets, or skins although expressly forbidden by law. Wildlife harvest in concessions, which has never been controlled by government officials, has been an illegal way to decrease logging costs. The recently passed forestry law contains regulations that address this situation (see chapter 29). New enforcement mechanisms should control some of these illegal activities, while other regulations and guidelines should promote a more ecologically sustainable forest management. One change brought about by this law is that the number of people in the forest, and the area of forest in Bolivia to be logged each year, is being reduced and concentrated in area—a move away from the traditional mahogany logging practices. Detailed maps of annual harvest areas—with the location of commercial trees—streams, and fragile habitats favor the reduction of logging impacts through fewer roads and better protection of critical habitats for wildlife. Mandatory identification of endangered or threatened species, and areas set aside for reserves, improve the chances for biodiversity conservation.

While effective government control of illegal hunting will still be difficult, voluntary management certification initiatives are another mechanism to improve the wildlife situation (see chapter 26). Several logging companies have passed evaluations of certification programs and showed significant improvements compared to the traditional practices. These improvements include clear, no hunting policies and adequate supplying of domestic meat for field crews. The pressure from international *green* timber markets, as well as political support from the government and technical assistance from environmental programs, may all contribute to a change in traditional logging and hunting practices. If new wildlife legislation allows for harvesting selected animal species, and adequate studies provide the basis for their sustainable use, specific wildlife management plans could be applied in logging concessions.

ACKNOWLEDGMENTS

We want to thank Wendy Townsend for helping to start the hunting monitoring program in Lomerío and for her advice during the studies in both sites. We are indebted to the Chiquitano communities and CICOL, but especially to concerned hunters such as Francisco Cuasase, Jesús Cuasase, Santiago Rodríguez, Eusebio Ribera, Francisco Aguilar, José Chuvé, Pedro Rodríguez, Ignacio Cuasace, José Soriocó, Juan Parapaino, Miguel Bailaba, Juan de Dios Surubí, Benito Chuvé, Pedro Surubí and Esteban Bailaba. At the Oquiriquia concession, Mr. Ciro Quiroga was of invaluable assistance. Robert Fimbel, John Robinson, and Alejandro Grajal reviewed the manuscript and provided useful suggestions to improve it. USAID, PL-480, and the Government of Bolivia financed the study.

Chapter 16

THE INTERRELATIONSHIPS OF COMMERCIAL LOGGING, HUNTING, AND WILDLIFE IN SARAWAK

Recommendations for Forest Management

Elizabeth L. Bennett and Melvin T. Gumal

Since the 1950s, southeast Asia has been the world's main supplier of tropical hardwoods. The source has changed over time as declining resources follow booms in each country. In the 1950s, most logs came from the Philippines, followed by Malaysia in the 1960s. In the early 1970s, Indonesia was a major source of round logs, but Malaysia again took over as the world's main supplier (IUCN 1991) when the Indonesian export of round logs was stopped. By 1986, 89 percent of Japan's tropical timber imports were from Malaysia (Nectoux and Kuroda 1989). Within Malaysia, the early exports were from Peninsular Malaysia and Sabah. These exports soon declined due to depleted supply, and Sarawak's exports rose. By 1986, therefore, the total world production of tropical logs was about 25.2 million m^3 (IUCN 1991), of which 11.4 million m^3 was from Sarawak (Caldecott 1996). Exports from Sarawak increased further to peak at more than 19 million m^3 in 1989 (Bruenig 1996), but have been declining since due to increasingly strict government restrictions—including quotas on exports.

Sarawak lies along the northwest of Borneo, and covers 24,450 km^2. It is close to the equator (from 0° 50' N to 7° 50' N) and rain falls in every month, with a pronounced peak during the northeast monsoon season (from November to February). Annual rainfall in coastal areas is about 3,000 mm, but reaches 6,000 mm in parts of the interior (anonymous 1995). The three main types of natural vegetation are: mangrove forest (1.3 percent of the state's land area), peat swamp forest (10.0 percent of land area) and hill mixed-dipterocarp forest (58.7 percent of land area) (Sarawak Forest Department 1994). Other less widespread forest types are tropical heath forest, montane forest, beach forest and forest over limestone (Whitmore 1984; Sarawak Forest Department

1994). In the 1950s, timber for export was initially supplied from peat swamp forests, but the majority, by far, has come from hill mixed-dipterocarp forest since the early 1970s.

At present, 8.3 percent of Sarawak's land is in current or proposed Totally Protected Areas (TPAs). These areas are comprised of wildlife sanctuaries and national parks, and exist strictly for conservation—all commercial extraction of resources is banned. In addition, 36.2 percent of the state's land is gazetted as Permanent Forest Estate, and there are proposals to increase this to 48 percent (Sarawak Forest Department 1994). The Permanent Forest Estate is land allocated to remain under forest in perpetuity, but commercial extraction of timber is allowed under license. A wide-ranging series of rules (that are detailed and well-planned) govern this, although there are enforcement problems (Kavanagh et al. 1989; ITTO 1990a).

The value of exports from forestry is some U.S. $2.4 billion per year, which constitutes 15 percent of the gross domestic product (GDP). Royalties from logs total U.S. $360 to U.S. $400 million per year, and provide about 50 percent of the state's revenue (Bugo 1995; Sarawak Forest Department Information 1996). In 1992, 10.7 percent of Sarawak's labor force worked directly in the timber industry, and another 21.3 percent in jobs related to it (Sarawak Forest Department 1994). Seventy-five percent of the direct jobs are in timber extraction, which draws most of its labor force from the rural areas. As a result, logging is the single largest source of jobs and income in rural areas throughout much of Sarawak (Bugo 1995).

Logging affects the forest and its wildlife in a variety of ways. Some are the direct results of timber extraction such as loss of timber trees and damage to the forest during felling. Others are indirect effects such as the presence of large numbers of timber workers in the forest, and the opening up of the interior by logging roads—which allow hunters and cultivators into previously inaccessible areas.

In this chapter, we examine the effects of logging on wildlife in Sarawak. We present a summary of the direct effects of logging, and then look in more detail at the indirect effects. We concentrate on those because, for most large mammals and birds in Sarawak, the indirect effects are more damaging than the initial logging itself. We conclude with a section on the implications for future forest management.

This chapter only examines selective logging in hill mixed-dipterocarp forest. It does not consider reentry logging, clear-felling, or logging in swamp forests where timber practices, extraction rates and access are very different.

The Direct Effects of Selective Logging on Wildlife

In Sarawak, selective logging results in a loss of about 54 percent of trees greater than 30 cm girth-at-breast-height, and damage to another 13 percent. Most of this is due to incidental damage such as making roads and skid tracks (Dahaban 1996). This is similar to damage by selective logging in Peninsular Malaysia (Burgess 1971; Johns 1992a).

In the absence of hunting, the effect of forest damage on wildlife varies greatly with species. Studies in Sarawak and Sabah show that some species disappear or become extremely rare following logging, such as the Asian giant tortoise (*Geochelone emys*), Malaysian rail babbler (*Eupetes macrocerus*), some partridges and pittas (Johns 1989b; Lambert 1992; Dahaban 1996). Others survive but their numbers often decline greatly, such as gibbons (Hylobatidae), colobine monkeys (Colobinae), giant squirrel (*Ratufa affinis*), barking deer (*Muntiacus* spp.), hornbills, flycatchers, woodpeckers and trogons (Johns 1992a; Lambert 1992; Bennett and Dahaban 1995; Dahaban 1996). Others increase in number, such as sambar deer (*Cervus unicolor*) (Stuebing 1995), medium and small squirrels (Dahaban 1996), and some bulbuls and spiderhunters (Lambert 1992). Some new species also enter areas after logging, such as magpie robins (*Copyschus saularis*) and lesser tree shrews (*Tupaia minor*) (Bennett and Dahaban 1995). Many species do, therefore, survive in logged forests although the general trend is for edge or colonizer species to replace those that depend on primary forest and cannot survive elsewhere (Bennett and Dahaban 1995). For Sarawak as a whole, therefore, logging will result in decreasing diversity, even though diversity values are similar before and after felling in each individual logged area.

Many species probably also survive logging because of unlogged patches within the concessions (Johns 1989b; Lambert 1992). These areas act as refuges, especially for smaller species that recolonize the logged forest as it regenerates. Under normal conditions, 20 to 25 percent of forest remains unlogged in many concessions because of steep terrain or poor forest (L. Chai, Former Director of Forests, Sarawak Forest Department, personal communication). As technology improves, especially the advent of helicopter logging, the number and size of unlogged patches will decrease. The effects of technological changes on wildlife survival are unknown but likely to be significant. There are moves to classify all forests on steep terrain as *protection forests*, thereby making it illegal to log them (J. Dawos Mamit, Controller,

National Resources and Environment Board, Sarawak Ministry of Resource Planning, personal communication).

The Indirect Effects of Logging on Wildlife

Opening up a forest via logging has two main indirect effects: allowing access to cultivators and increased hunting.

Shifting Cultivation

Each year, about 1,000 km^2 of land are cleared for shifting cultivation in Sarawak, of which about one third comprises primary forest (Sarawak Forest Department 1994). Increased access by logging roads undoubtedly facilitates this, although the extent to which it does so is not quantified. The diversity and density of plant species present in fallowed areas depend on age since cultivation and distance from primary forest. As long as hunting levels are low, older fallow areas can be important habitats for certain species of animals (Bennett and Dahaban 1995). Recent and intensive cultivation eradicates all large animals, but regenerated land more than thirty years old and close to primary forest has many large mammals and birds—including orangutans (*Pongo pygmaeus*), Bornean gibbons (*Hylobates muelleri*), sun bears (*Helarctos malayanus*) and large hornbills (Bucerotidae). In general, shifting cultivation inside the Permanent Forest Estate is relatively limited since it is illegal and strongly discouraged by the authorities.

Hunting

A major indirect effect of logging is opening the forest to hunters. The hunters fall into three main categories: local people, logging company employees, and outsiders.

Hunting by Local People

About 30 different racial groups populate Sarawak, many of whom still live throughout the interior in longhouses—a whole village under one roof. Traditionally, they plant rice through shifting cultivation, and obtain their protein by hunting and fishing. Most interior people still do so (Caldecott 1988; Bennett et al. 1999.). In the past, hunters used blowpipes, dogs, and spears. Today, most hunting is done using shotguns—88 percent of hunted animals die by gunfire (Bennett et al. 1995; 1999). There are about 60,000 legally registered shotguns in Sarawak (Caldecott 1988), or one shotgun per 15 adults. Prior to 1997, about 3.5 million cartridges were imported every year (Sarawak Statistics Department information), although recent strict regulations have greatly reduced this number.

Wildlife is one of the main sources of protein for rural people in Sarawak. All interior communities studied eat wild meat in at least 20 percent of their meals (Chin 1985; Strickland 1986). Throughout the state, an average of 29 percent of the evening meals eaten by rural farmers contain wild meat, and the proportion rises to 67 percent in remote parts of the interior (Bennett et al. 1999). On average, each rural family eats about 350 kg of wild meat per year. They continue to depend on wild meat even if the surrounding forest has been logged. In one logging area in Sabah, 10 percent of the evening meals of local people contained wild meat (Bennett et al. 1985). In logging areas in Sabah and Sarawak, local hunters tended to concentrate their hunting in unlogged patches within the logged forest (Gumal and Bennett 1995; Bennett and Sompud, unpublished data).

The most important dietary species by far is ungulates: 72 percent of the dressed weight of animals hunted in Sarawak was bearded pig (*Sus barbatus*; see photo 16-1), and an additional 10.9 percent was deer (Bennett et al. 1999). In the absence of hunting, these species survive well in logging areas, and deer even increase in numbers, due to increasing browse in regenerating forest (Stuebing 1995). The prevalence of ungulates in the diet, however, masks the fact that local people hunt many animals that, because of their small size, only comprise a minor proportion of the diet. Ungulates, therefore, only comprise 29.7 percent of the animals hunted in Sarawak, and the remainder comprises a very wide range of smaller mammals, birds, reptiles, and amphibians (Bennett et al. 1999). This means that hunting greatly reduces the numbers of many species that only form a minor part of peoples' diets.

Hunting by Logging Company Employees

Workers in most logging camps throughout Sarawak are rural Sarawakians, often recruited from the local area. Most are traditional hunters by background, therefore, and many bring their shotguns to

PHOTO 16-1 Bearded pig (*Sus barbatus*) is the most important game species, by dress weight, in Sarawak, Malaysia. (M. F. Kinnaird)

the logging camp. They hunt mainly for subsistence, and this is often their main source of protein in remote camps.

Loggers in one camp in Nanga Gaat served as a case study of hunting. For a summary of the methods, see box 16-1. The main transit camp was comprised of 167 workers, many of whom had families with them. All were Ibans (the single largest indigenous group in Sarawak), and most were recruited from the local area. Fifty percent of the workers had shotguns. They hunted on foot at the logging face during the day between work shifts, and along the logging roads at night—sometimes using company vehicles. All hunting was done using shotguns—no animals were killed by other means.

Forty-nine percent of evening meals in the camp contained wild meat, and it was the main source of protein for the workers and their families. The camp consumed about 29,000 kg of wild meat per year (see table 16-1). Of that, the majority was bearded pig, which made up 71.4 percent of all animals hunted. Hunters, however, shot any large animals seen, irrespective of whether or not they were legally protected.

Box 16-1 *Summary of methods used in the Nanga
Gaat logging camp case study.*

The transit camp for the logging workers and their families was visited at
three different times of the year—sampling hunting patterns in different
seasons. Each visit lasted nine nights (eight full days), and was spent living in the camp so people could become comfortable with the survey
staff. We also wanted to be sure that we were accurately observing the
number of people and vehicles involved in hunting and all animals coming into the camp. On each visit, the following data were collected:

1. *General hunting interviews.* General hunting interviews were conducted to obtain an overall picture of hunting in the camp, its importance to the hunters, and broad impact on the wildlife. All hunters
   were interviewed and results recorded on a standard form. Questions
   included: a) the frequency of hunting; b) the methods used; c) the
   types of animals normally hunted; d) the prey selection process used
   by hunters (were size and sex a factor?); e) hunting success rates (e.g.,
   out of every four hunting trips, how many were successful; out of
   every ten cartridges fired, how often would an animal be killed); f) the
   reason for hunting (for subsistence, trade, or killing crop pests); and
   g) the sale of wild meat (if the hunter ever sold the meat or other animal parts). Thirty-six hunters were interviewed (identifying hunters
   also determined the number of men in the community who hunted).

2. *Interviews about individual hunts.* Interviews about individual hunts
   were used to obtain a detailed picture of the hunting offtake and
   effort. All recorded hunting trips lasted for less than 24 hours. Every
   time a hunter returned from a hunting trip, he was interviewed and
   results recorded on a standard form—this included both successful
   and unsuccessful hunts. Data collected included: the hunt duration,
   number of hunters participating, number of guns or other weapons
   used, number and species of animals hunted, number of shots fired
   (thereby allowing us to subsequently calculate the number of cartridges used per animal killed or wounded), reason for going out (e.g.,
   logging, hunting), use of the hunted animal, and location of the hunt.
   Records were obtained from 72 hunting trips.

3. *Measurements of hunted animals.* Measurements of hunted animals
   help to determine the importance of different wildlife species to the
   camp, and the possible effects of the offtake on wildlife populations
   (e.g., if the hunt was strongly sex- or age-biased). Any hunted animals,
   or parts of them, brought back to the camp were measured. As many
   as possible of the following measurements were made: total weight,
   head length, total length, tail length, ear length and hind foot length.
   Sex and approximate age class were also noted. A total of 31 animals
   were measured.

4. *Record of diets.* Diet records were used to assess the importance of
   wild meat in the diet, and to crosscheck with the number and species
   of animals being hunted in the camp. Each family in the camp was visited on a daily basis to ensure that all animals brought back to the
   camp were detected. Each family was also asked what they were eating every evening, and the results recorded on a standardized form.
   The composition of 341 evening meals was recorded.

For further details about the methods, see Bennett et al. 2000.

Other killed animals included sambar deer (see photo 16-2), barking deer (*Muntiacus* spp.), mouse deer (*Tragulus* spp.), langurs (*Presbytis* spp.), Bornean gibbon (*Hylobates muelleri*), marbled cat (*Felis marmorata*), colugo (*Cynocephalus variegatus*), rhinoceros hornbill (*Buceros rhinoceros*), Bulwer's pheasant (*Lophura bulweri*) and great argus pheasant (*Argusianus argus*).

The hunting was wasteful, with animals being killed on sight, even if the worker did not need the meat. During a bearded pig migration, for example, workers would kill a pig or sambar deer, take the prime cuts of meat, leave the rest to rot in the forest, and kill another animal the next day.

Most animals were hunted for food, but occasionally animals were captured alive and kept as pets. The most common instance of this occurred with gibbons: if the hunter shot a female carrying an infant, and the infant was still alive, it was kept as a pet, and usually sent to the hunter's longhouse.

This camp is probably typical of many throughout Sarawak, in that workers hunted a wide variety of species. There are few or no taboos on what can be taken, and generally any large animals seen are

TABLE 16-1 *Number of Wild Animals of Different Species Killed Per Year, and Total Weight of Dressed Meat, in One Logging Transit Camp in Nanga Gaat, Sarawak*

Species	Average Dressed Weight (kg)	Total # Killed/Year	Total Weight of Wild Meat/Year (kg)
Bearded pig, *Sus barbatus*	30	879	26,370
Sambar deer, *Cervus unicolor*	60	27	1,620
Barking deer, *Muntiacus* spp.	8.25	68	561
Mouse deer, *Tragulus* spp.	2.2	41	90.2
Primates	3.3	135	445.5
Total		1,150	29,086.7

SOURCE E. L. Bennett and A. J. Nyaoi, unpublished data.

NOTE The camp comprised 167 workers and their families. Many other species were also hunted, but they were small animals that did not contribute significant amounts of meat to the diet.

PHOTO 16-2 Sambar deer (*Cervus unicolor*) are an important source of wild meat for logging company employees. (E. Bennett)

hunted. This is typical of longhouse hunters throughout Sarawak (Bennett et al. 1999). Hunting patterns by logging company employees and rural hunters are very similar. The main differences are that:

- Logging workers are concentrated in larger numbers in one patch of forest than would otherwise be the case
- The roads and vehicles give employees greater access to more areas
- Employees hunt at the logging face where animals are being disturbed by logging, and so might be easier to hunt

Thus, the impact of the hunting is likely to be greater in logging areas than usually occurs in unlogged forests.

The amount of hunting by company employees varies between logging camps, and depends largely on company policy. The following practices tend to reduce hunting levels:

- *Employing workers on contract, with payment dependent on work output.* This is designed to make logging more efficient, but it has the indirect benefit of reducing hunting—staff work harder, so have less time and energy to hunt (B. R. Ibbotson, Former Managing Director, Batang

Balleh Forest Enterprises, Sarawak, personal communication; R. Lee, Regional Manager, Samling Corporation, Sarawak, personal communication). Almost all logging camps in Sarawak now employ workers on contract.

- *Decentralizing and dispersing the cutting crews.* The primary aim is increasing work efficiency, but it also reduces hunting: when workers are in large camps, social facilitation means that they tend to hunt more (see photo 16-3; B. R. Ibbotson, personal communication)

- *Selling fresh meat and vegetables in the transit camps* several times a week at the company store, thereby reducing the need for wild meat. This is done by at least one large company, (Gumal and Bennett 1995)

PHOTO 16-3 *Decentralizing and dispersing cutting crews helps to reduce socially facilitated hunts by employees in logging concessions. (J. G. Robinson)*

Hunting by Outsiders

The access provided by logging roads opens forests up to hunters from outside the immediate area who use four-wheel-drive vehicles and hunt for sport and the wild meat trade. Because of the climate and distances involved, meat goes bad rapidly, so hunters and traders frequently carry refrigerated boxes on the back of their vehicles. Traders either hunt the meat themselves, or buy it from local people. The trade is facilitated by canteens and shops along main access routes, where sizeable freezers are used to store the wild meat before it can be sold locally or transported to larger towns. In one of Sarawak's major river systems (the Baram), there are at least six such centers of distribution (Danys Lang, senior government administrator for the region, personal communication).

The main species hunted for the trade are ungulates: bearded pigs, sambar deer, and barking deer make up the majority of the kill, and are the species most often sought by markets and restaurants throughout Sarawak. This has wider repercussions on populations of other species, however. Once the large ungulates have become rare, local people hunt smaller animals for food. Since they provide less meat, more are killed. The effect of overhunting large species, therefore, has cascading effects throughout the wildlife community until eventually even large squirrels become rare (Caldecott 1988; Bennett et al. 1995; Gumal and Bennett 1995).

Wild meat is widely sold in towns, villages, restaurants, and logging camps throughout Sarawak. Local markets and Chinese restaurants anywhere in the state almost invariably sell wild, bearded pig meat, and often one or more species of wild deer meat. In the mid-1980s, the value of the wild meat traded along the Rejang River was about U.S. $1.6 million, and involved about 16,000 bearded pigs, 1,500 sambar deer, and 250 barking deer per year (Caldecott 1988). In 1996, there were an estimated 250 outlets in Sarawak selling wild meat, and the estimated value of the wild meat trade was about U.S. $3.75 million per year (Bennett and Gumal, unpublished data). Much of this is meat coming out of the forest along logging roads directly to towns or river access points.

There are no legal or other controls on commercial hunting, provided that the hunted species are not protected by law. The market for the meat is great and probably expanding. The drain that hunting for trade puts on populations of wild animals, therefore, is considerable. In parts of Ulu Belaga and Ulu Balui, for example, Penan hunters were successful on about 90 percent of their hunts (Brosius 1986) before logging roads entered the area in about 1990. Wildlife is now so scarce that Penan subsistence hunters rarely manage to find wild meat (Gumal and Bennett 1995). Of seven Penan headmen and three other hunters interviewed, 100 percent said that hunting was difficult and wild meat extremely scarce since logging came to their area. Hunters said that they only find meat on one out of ten hunts, where previously they caught something every time. One hunter had not caught anything for three months. There seemed to be a genuine lack of animals—not merely increasing shyness—because hunting dogs could not find them using scent, and animal tracks were rarely seen (Gumal and Bennett 1995). All interviewees in the area, including logging company managers, reported frequent hunting by outsiders from major towns using 4-wheel drive vehicles and shotguns. Hunters stayed several days—until they had enough wild meat. Logging itself does not generally drive commercially sold animals such as pigs and deer to extinction (Johns 1992a; Stuebing 1995; Dahaban 1996). The fact that those species are so severely depleted in the area is at least partly due to excessive commercial and sport hunting.

Conservation Implications

Even if all proposed new Totally Protected Areas are gazetted, they will only cover 8.3 percent of Sarawak's land area. Most such areas are small—the average current size is only 241.8 km^2. If all new areas were gazetted, then the average size would increase to 444.5 km^2 (this figure excludes marine parks, and also nature reserves which are, by definition, less than 10 km^2 and aimed mainly at protecting sites of special botanical or geological importance) (Forest Department information 1996). By contrast, the Permanent Forest Estate currently covers 36.2 percent of the state's land area, and this might increase to 48 percent. The permanent forest estate is made up of forest reserves and protected forests, both of which can be licensed for logging and are designated to remain under forest in perpetuity. Individual reserves in the Permanent Forest Estate are usually many times larger than individual Totally Protected Areas, and different Permanent Forest Estate reserves are frequently contiguous. For many wide ranging and rare animals, therefore, their only long-term hope for survival in Sarawak is in the Permanent Forest Estate.

Not only do animals need the Permanent Forest Estate but the forests of the Permanent Forest Estate need the animals. The relationship between forest diversity and the presence of large mammals has been well documented (for review, see Redford 1992). Ninety-one percent of animals hunted in Sarawak are those whose diets contain a high proportion of fruits and seeds (data in Bennett et al. 1999), and all large frugivores and browsers are hunted. If they are no longer in the forest, the effect on tree diversity could be enormous. Loss of fruit dispersers such as hornbills, pigeons, primates, flying foxes and civets is also likely to hamper forest regeneration after logging (see chapter 3). All studies of the effects of logging on wildlife in Malaysia have deliberately been done in areas where hunting levels have been relatively low—the forests have retained significant populations of large animals. This is not typical of most of Sarawak, however, and while the full effect of the loss of large fruit dispersers on forest regeneration is not known, it is probably significant. Unfortunately, due to the long time scales involved, it will be many more years before this is quantified.

The effects of over-hunting in logging areas also have repercussions for local communities. In some areas, it means that they have little wild meat left. Many communities make the transition to a cash economy as areas are opened up, so they have alternatives and can buy food. The logging industry also provides a major source of jobs and income to rural people, and the cash means that they can buy many commodities, including food. Others, such as the Penan and some other rural communities, still depend strongly on wild meat. Penan culture inhibits them from eating animals that they have reared themselves, and older people also still will not eat domestic meat believing it to be unclean (Jayl Langub, Secretary, Sarawak Council for Customs and Traditions, personal communication; Matu anak Tugang, Penan, and son of headman, Long Jaik, Belaga, Sarawak, personal communication). Thus, loss of wild meat is more of a problem for them.

Recommendations for Forest Management

Two forest management improvements needed to improve wildlife conservation in production forests include: 1) enlargement of primary forest-protected areas, and 2) control of hunting in logged zones.

Protecting Primary Forest

To protect species which do not survive well in logged forests, and which need primary forest for their breeding and survival, two main steps are needed:

1. *Expansion* of the Totally Protected Area System as formally proposed by the Forest Department (Ngui 1991; Wildlife Conservation Society

and Sarawak Forest Department 1996), to include new areas and also increase the size of many existing reserves.

2. *Retention* of unlogged blocks within the Permanent Forest Estate and within each concession. In Peninsular Malaysia, blocks of unlogged forest within larger logged areas succeeded in retaining many species of plants and animals if they were at least 200 ha in size, and if they were situated in the middle of the logging area rather than the edge (Laidlaw 1994). In Sarawak, therefore, unlogged blocks should be a minimum of 200 ha each, and together should comprise at least 10 percent of the area in each concession. This should be in addition to areas already legally protected as riverine reserves or on very steep terrain. The unlogged blocks should be sited through the whole of each concession, and their boundaries clearly marked so infringements are easily detected.

Many species can survive logging, but their populations are negatively affected by hunting. Unless such animals are protected in the Permanent Forest Estate, many species will become extinct—either locally, or throughout Sarawak. The Permanent Forest Estate will, in the long term, also comprise the only large areas of forest outside Totally Protected Areas. It will, therefore, be the only source of wild meat for rural people who depend on it. Ensuring that such hunting is sustainable is critical to long-term protein supplies for remote rural communities.

Controlling Hunting in Logging Areas

To ensure the survival of large animals in the Permanent Forest Estate, a series of measures is needed to reduce hunting, both by the loggers themselves and by outside hunters using the logging roads for access. These include:

1. *Control of access.* This is probably the single most effective way to control hunting in logging areas. It would prevent hunting in most of the concession by people in vehicles, or using vehicles for access—be they loggers, local people or outsiders coming in to hunt for sport or trade. Closing logging roads as soon as a cutting area is completed is simple, feasible, and effective. Bulldozing the road to form a ditch can easily accomplish this. This acts as a silt trap, thereby reducing erosion of silt into rivers. Monitoring of access is also needed along main roads, to prevent their use as a major wild meat trade route out of the Permanent Forest Estate. This is also relatively simple, requiring that a manned gate

be placed across the main access point(s) for all logging roads which link to the main outside road systems. These check points would minimize the current, huge loss of wild animals from the Permanent Forest Estate through hunting for sport and market.

2. *Banning possession of firearms.* Prohibiting guns among logging company employees, except for those who need them for security (e.g., staff dispensing wages).

3. *Enforcing regulations concerning protected tree species.* Some of the main food trees of bearded pigs are already protected by law, but not always in practice (e.g., *Shorea macrophylla, S. splendida, S. hemsleyana, S. seminis, S. palembanica, S. stenoptera*). If felling and other damage to those trees is minimized, it will help to ensure a continuous food supply for bearded pigs, and thus ultimately for rural people.

4. *Providing alternative sources of animal protein for logging company employees.* The exact means of doing this depends on the company, and the degree of access. The best solution in most areas would be for the company to establish a mobile shop that sells to workers living in the forest. Since there is a constant flow of vehicles up to the logging face, this should be feasible in almost all areas—it is already practiced in some.

5. *Establishing a statewide moratorium on the sale of wildlife and wildlife products, including wild meat.* This has been proposed by the Sarawak Government as the single greatest step toward protecting wildlife throughout interior Sarawak, including in the Permanent Forest Estate. This involves two main steps: 1) banning all sales, to be enforced at the points of sale (markets, restaurants etc.); and 2) banning possession of commercial quantities of wildlife (including wild meat), except under strict permits.

6. *Greatly increased conservation education.* Long-term conservation of wildlife depends on public awareness and support (Wells and Brandon 1992; Jacobson 1995). For this to be achieved, conservation education programs are needed for all levels of society in Sarawak, including decision makers, hunters, loggers, and consumers of wild meat. Such a program exists in Sarawak through the formal education system, and for rural communities in and near Totally Protected Areas. It needs to be expanded greatly to other sectors.

All of these recommendations are detailed as part of a Master Plan for Wildlife in Sarawak (Wildlife Conservation Society and Sarawak

Forest Department 1996). Initiated by state government, this plan includes a comprehensive strategy for conserving Sarawak's wildlife. Many of the recommendations aim to protect wildlife throughout the Permanent Forest Estate, by controlling hunting and stopping the commercial wildlife trade. The aim of such measures is to ensure that the wildlife in Sarawak's forests survive, that forest diversity and regenerative ability are maintained, and that rural people will still have a supply of wild meat into the next generation.

Update

Since 1997 when this paper was written, the Sarawak state government has been taking extremely rapid strides to implement the Master Plan for Wildlife in Sarawak (E. L. Bennett, personal observation) and to put into effect many of the recommendations summarized in this chapter. In particular, a new law banning all trade in wildlife and wildlife products—including all wild meat—is now in effect and being enforced throughout Sarawak. The same law makes it easier to gazette new protected areas, and several are now being created. The number of shotgun cartridges being issued is now strictly controlled throughout the state. There are new regulations banning all shotguns in the possession of, and hunting by, logging company workers. This is already in effect in one trial concession. It is too early to assess the effects of all this on wildlife populations, but it is likely to be dramatic. In taking such steps, Sarawak is showing that scientific theory can indeed be put into practice to conserve wildlife in tropical forests.

ACKNOWLEDGMENTS

Our understanding of logging issues in Sarawak has been expanded immeasurably by the work of Dr Zainuddin Dahaban, and the help, discussions, and ideas from B. Ross Ibbotson, David Manjah, and Patrick Braganza—to whom we are extremely grateful. Many thanks go to all of the other staff of Batang Balleh Forest Enterprises for their unfailingly cheerful help. None of the work could have been done without the constant support of Datuk Leo Chai, former Sarawak Director of Forests, to whom we are very grateful. Mr Cheong Ek Choon, current Director of Forests, has also been extremely helpful. We would like to thank Dr. John G. Robinson for discussions that helped to formulate many of the ideas of this paper, and Drs. Robert Fimbel and Alejandro Grajal for their comments on a previous draft. Elizabeth L. Bennett's work is funded by the Wildlife Conservation Society, and Melvin T. Gumal's work is funded by the Sarawak Forest Department.

Chapter 17

DEFAUNATION, NOT DEFORESTATION

Commercial Logging and Market Hunting in Northern Congo

David S. Wilkie, J. G. Sidle, G. C. Boundzanga, P. Auzel, and S. Blake

Concerns about the unsustainable use of tropical forests have focused primarily on deforestation by commercial loggers, charcoal makers, and by ranchers and farmers (Myers 1980; Barnes 1990; Rudel and Horowitz 1993). Tree felling is responsible for the decline in forest cover as well as forest-dependent plants and animals everywhere that tropical forests exist (Friends of the Earth 1991). Focusing efforts solely on keeping the trees standing, however, may not be sufficient to ensure that healthy, productive, tropical forest ecosystems survive into the future. Defaunation by subsistence and market hunters, though leaving the trees intact, may have as strong an impact on the long-term viability of tropical forest ecosystems as does deforestation.

Birds and mammals are the primary predators and dispersers of the seeds of many tropical forest trees, shrubs, and vines (Gautier-Hion et al. 1982; Emmons et al. 1983; Gautier-Hion 1984; White 1994c; Feer 1995; see chapters 3-10). Wild animals are also nutritionally and culturally important components of the diet of both forest-dwelling and recently urbanized human populations throughout the tropics. This is particularly true in central Africa, where wild meat often constitutes the primary source of animal protein in the diet (Mbalele 1978; Hladik et al. 1990; Chardonnet et al. 1995).

Previous studies have shown that both market and subsistence hunting can result in the unsustainable exploitation of forest game (Caldecott 1987; Geist 1988; Ludwig et al. 1993; Alvard 1994; Bodner et al. 1994; Joanen et al. 1994; Fitzgibbon et al. 1995; Noss 1995; see chapters 15 and 16). Even when human population densities are low and technologies are traditional, hunters can locally extirpate large, slow-reproducing species (Redford and Robinson 1991; Fitzgibbon et al. 1995). Introduction of new technologies that dramatically increase hunting efficiency (Saffirio and Scaglion 1982; Yost and Kelley 1983; Fragoso 1991a; Noss 1995) has the potential for rapid overexploitation of wild game, particularly amongst cultures that historically have had little selection pressure to evolve a conservation ethic (Redford 1990; Alvard 1994). Widespread declines in forest animal populations not only increase the risk of local and global species extinction, but more perniciously are likely to have adverse affects on seed dispersion and seed survivorship—thus potentially altering forest plant composition and ecosystem function over the mid- to long term (see chapters 3 and 6).

In this paper, we used data collected within three commercial logging concessions to show how the direct (deforestation) impacts of logging on wildlife are slight in comparison to the indirect (facilitation of market hunting) impacts. While mechanized, selective logging may leave most of the trees intact, construction of logging roads, and hunters' access to transportation on logging vehicles so greatly facilitates market hunting that logging may indirectly result in the collapse of large mammal populations in logged areas of the Congo Basin.

Study Area

Congo covers some 342,000 km^2—about the size of the state of New Mexico (USA)—and is inhabited by about 2.5 million people. Over 60 percent of Congolese live in urban areas (45 percent in the capital Brazzaville, and the coastal port Pointe Noire, and 12 percent along the railway that links these two principal cities). The remaining 40 percent are scattered in towns and small villages, primarily in the central and southern portions of the country. Northern Congo is sparsely populated (< 3 persons/km^2) with most settlements located along rivers (CIA World Factbook, http://www.odci.gov/cia/publications/factbook/cf.html; official Congo Web site, http://www.amb-congo.fr/congo/_index.htm).

About 60 percent of Congo is covered by tropical, moist forests. By the end of the 1960s, timber had become the country's principal export (between 1947 and 1980, 14.5 million cubic meters of timber were exploited—approximately 1.5–2 million trees) and much of the southern forests had been logged. Offshore oil now provides over 75 percent of the estimated $1.1 billion in annual export revenues. Though timber sales now generate less than 5 percent of export earnings (U.S. Department of Commerce 1988), logging remains a major focus of Congo's economic development plans. European and Mediterranean wood products companies have traditionally dominated the logging industry in Congo. Faced with dwindling domestic resources, however, southeast Asian companies are now beginning to invest in Congo's logging industry. Logging in Congo has been promoted not only as a way of generating funds for the national treasury, but also as a sustained means of economic development of rain forest communities—the inhabitants of which are isolated, barely assimilated into national culture, and the last to be provided with social services or included in market economies (Wilkie 1996). Establishment of timber concessions in northern Congo (see figure 17-1) has thus been heralded as an effective way of providing much needed income, education, and health care to local communities.

Northern Congo has been divided into 15 forestry management units (UFAs), covering approximately one sixth of the country. In 1990, logging was underway by seven companies. Since that time, and despite loans to some concessions from the African Development Bank and the Caisse Central de la Cooperation Francaise, all but three have gone bankrupt or have simply abandoned their logging concessions. Transportation costs make even the most valuable logs uncompetitive on the world market for all but vertically integrated timber companies. In January 1994, the UFA (Unité Forestière d'Aménagement—forestry management area) Nouabalé (see figure 17-1), as a result of the efforts of the Ministère des Eaux et Fôrets (MEF) and the Wildlife Conservation Society (WCS), was gazetted as a new national park. MEF and WCS are also attempting to have the bankrupt Société Nouvelle des Bois de la Sangha (SNBS) Kabo concession that lies due south of the Nouabalé-Ndoki National Park declared as a reserve (Wilkie et al. 1994).

Logging Operations

Société Forestière Algero-Congolaise (SFAC) was allocated an 855,000 ha concession in the Sangha region in 1983 (see figure 17-1). Selective logging operations of SFAC are targeted at only the most profitable species (which are often highly spatially dispersed) and typical of logging practices employed throughout tropical moist forest regions (Centre Technique Forestier Tropical 1989). In 1990, SFAC hired six to 10 timber inventory teams, each composed of eight men. Inventory teams, comprised of local BaNgombe foragers and BaKouele farmers

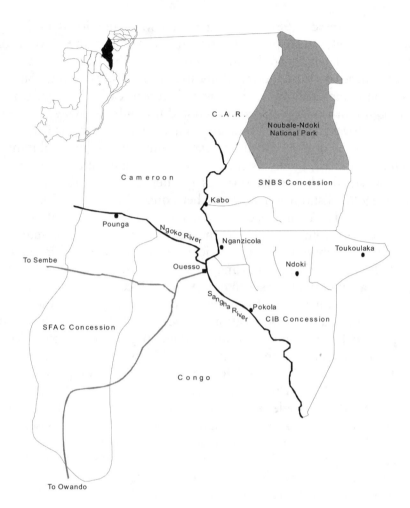

FIGURE 17-1 *Location of the SFAC, CIB, and SNBS logging concessions in northern Congo.*

knowledgeable in the identification of trees, were responsible for mapping all exploitable trees (> 80 cm dbh) one year in advance of road construction, felling, and timber extraction within a sector of the concession. To inventory all trees of the few commercial species, inventory teams cleared a series of primary transects (north-south oriented, 1 km apart, 3 m wide) and secondary transects (east-west oriented, 250 m apart, 1 m wide), forming a lattice that divided the future logging area into 0.25 km² blocks. Rivers, marshes, and topography were also roughly mapped during the tree inventory to help guide the location of log stock yards, and to design a minimum distance road network to transport logs (see figure 17-2). As road construction and transportation costs comprise the most significant fraction of the costs of logging, systematic inventories are essential if selective logging is to be profitable in northern Congo.

Teams of three to five men felled exploitable trees. Prior to felling, trees within the fall path were cleared, as were attached lianas. After trees were felled, they were cut into sections (logs) and dragged by bulldozers along extraction tracks to secondary roads. Although as many as 10 to 15 tree species could legally be exploited within the concession, the cost of transporting cut timber to Pointe Noire and the world price for tropical hardwoods made it economically feasible to extract only the most valuable species: Sipo (*Entandrophragma utile*), Sapelli (*E. cylindricum*), Afrormosia (*Pericopsis elata*), and Wengue (*Millettia laurenti*).

Assessing the Effects of Logging on Wildlife

Canopy Removal

Damage to forest canopy increases with: a) the number of trees felled, b) the area damaged at the felling site, and c) the area cleared for secondary and primary roads and log stockyards. Wilkie, Sidle and Boundzanga spent three weeks in the SFAC concession from December 1989 to January 1990, and collected data on these three variables. The total number of trees removed was determined by an examination of the detailed records SFAC keeps of the number of commercial trees felled in each 2 km^2 logging block. The average area of canopy lost when one commercial tree was felled was determined by measuring the area of canopy removed at the site of 61 trees felled in 1989. Canopy removal at each site was determined by mapping the perimeter of the area exposed to the sky. Bulldozer trails generally affected stems $\leq$ 10 cm dbh, and canopy removal assessments above these trails were not made. The average width of forest cleared for both secondary and primary roads was measured. Total length of constructed primary and secondary roads was determined using concession inventory maps and a vehicle odometer. The total surface area of canopy removal due to secondary and primary roads was calculated by multiplying the average width cleared by the total length of roads constructed. The cleared area where logs are stacked prior to transport was also measured at several areas in several logging blocks. The total area of canopy

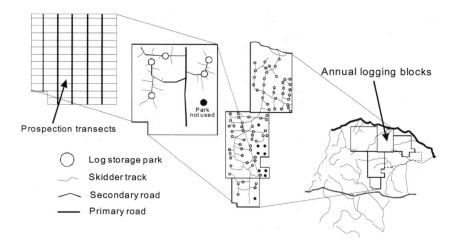

FIGURE 17-2 *Road and prospection transects within logging blocks provide hunters with easy access to wild meat in the SFAC concession.*

affected by logging (excluding the housing area located at the SFAC base camp) was calculated as the sum of the above factors assessed within each 2 km² logging sector.

Direct and Indirect Impacts

The impact of logging on forest fauna was assessed directly by conducting a series of line transects within the SFAC concession, and indirectly by estimating the average rate of return from hunting within the SNBS and CIB concessions. Direct assessment of wildlife focused on primates and elephants. In general, primates are the most easily seen or heard diurnal fauna, and often decline in relation to the intensity of logging (Skorupa 1986; but see chapter 4) and hunting (Lahm 1994a, b; Fa et al. 1995). Elephant droppings can be observed readily in the forest (Fay 1991). Though duikers are the most commonly hunted animals in the forest, their size, coloration, and behavior patterns make them difficult to census reliably over the short term (Payne 1992). Time constraints, therefore, precluded duiker population density surveys within the SFAC concession. Direct assessment methods were adapted from those used by Hart and Kiyengo (1989) during their brief fauna assessment of the Maiko National Park in the Democratic Republic of Congo (formerly Zaire), and are comparable to those used during other surveys in forested areas of central Africa (Tutin and

Fernandez 1984; Fay and Agnagna 1989, Fay et al. 1990; Stromayer and Ekobo 1991a, b, and c).

Arboreal primate sightings, nests of terrestrial primates, and elephant droppings were recorded along primary and secondary transects in the yet-to-be-logged 1990 logging sector, skid trails in the 1989 logging sector, active secondary roads in the 1989 logging sector, and abandoned secondary roads in the 1986, 1987, and 1988 logging sectors. A census was also conducted along a trail that connected the Sembe-Ouesso road and the active sections of the concession. Surveys were conducted during the morning and early afternoon. Survey routes traversed a variety of forest habitats, including swamp forest, mixed forest, and monodominant Limbali (*Gilbertiodendron dewevrei*) forest.

For six weeks in 1994, Blake (1994, 1995) used participatory observation to estimate the rate of return (ERR) of local hunters during 40 hunts conducted in the northern section of the SNBS Kabo concession (see figure 17-1). Assuming that all hunters used comparable technologies and modes of travel, and were of equal competence, differences in the estimated rate of return to hunting (measured in kg of game captured per man hour from different areas of the forest) should have reflected differences in animal abundance. Auzel (1996a, b) used the same methods to estimate the rates of return of 223 shotgun hunts conducted in 1995 and 1996 in sections of forest that were zero to 11 years post logging by the CIB concession (see figure 17-1). Blake's estimated rate of return included travel time, whereas Auzel reported estimated rates of return for hunting time only, and for hunting plus travel time.

Like all other methods for estimating animal densities in tropical dense forests, estimated rate of returns are likely to be influenced by possible changes in animal behavior and detectability in logged and unlogged areas. Although variability in hunter expertise may contribute to estimation errors, those errors are likely to be of the same magnitude as those introduced by variances in line-transect observers' expertise. While estimated rates of return are not error-free, the sources and magnitude of errors and the cost of data collection (particularly when hunter diaries are used) are likely to be less than direct observation methods.

Modeling Hunter Access to the Forest

We assume that one major factor facilitating forest resource exploitation is access to the forest. Prior to the establishment of logging concessions, anyone interested in exploiting forest resources such as game, honey, or timber had to walk into the forest from the nearest access point (e.g., in the SFAC concession the Ngoko River, or the Ouesso-Liouesso, Ouesso-Sembe roads), and hand-carry the harvested goods out of the forest. To assess the degree to which establishment of a logging concession increases access to forest, we modeled the impact of road construction and access to logging concession vehicles on the travel and transport costs of hunters within the SFAC concession, using digital geographic information systems (GIS) tools. By digitizing SFAC forest survey and exploitation base maps, which included the primary and secondary road network, the time required for hunters to travel from Pounga on the Ngoko River to all parts of the forest was determined. Travel costs used to generate the travel time pre- and postroad construction were set in minutes/km as follows: primary roads—one minute (i.e., 60 km/hr), secondary roads—two minutes, forest with survey transects—20 minutes, and undisturbed forest—40 minutes. A travel cost surface is generated within a GIS by reclassifying the pixel values (0.16 ha in this study) within a land cover layer (where the value of each pixel represents a land cover class such as forest, river, or primary road) to the time that it takes to travel across each respective pixel. The GIS then uses the travel cost surface to calculate the minimum travel time from any source pixel to all other pixels within the map. We assumed that hunters traveled by concession vehicle along the roads, and by foot within the forest.

Importance of Wild Meat to the Local Economy

Auzel and Congolese field assistants collected data between June and August of 1995 and 1996 in three villages: Ndoki (within the CIB concession), Nganzicolo (near the Sangha River), and Toukoulaka (near the Likouala swamp) (see figure 17-1). The 1995 field season was used primarily for qualitative data collection, to train local field assistants, and to habituate human subjects to the presence of the researchers. Data were collected to determine the quantity of wild meat eaten, the proportion sold at regional markets by rural households living within and outside logging concessions, and the value of wild meat sales to rural households.

A complete household census was undertaken in 1996 in all settlements other than Ndoki, where 73 percent of all households were surveyed. The importance of wild game to rural populations was determined using irregular interviews of the female head of household (in a sample of five households per village during the period of time that a researcher was resident). Women were asked to recall the composition of the primary meal of the day that they had prepared—or were planning to prepare—for their household. The frequency that animal protein and fresh or smoked wild meat were components of households' meals was then collated from the interview data. Community-level consumption of wild meat was estimated using methods developed in Cameroon (Koppert and Hladik 1990; Dounias et al. 1996). The species, gender, capture method (shotgun, snare, net, crossbow, and bow and arrow), source, destination (domestic consumption or market), and extent of processing (none or smoked) were recorded for all wild meat entering a village during each day that a researcher was resident. An attempt was made to validate destination data by monitoring all wild meat leaving the village, although this approach fails to measure wild meat that was sold and consumed within the settlement.

The biomass of captured wild meat was estimated using literature values for the mean weight for each species, and an average carcass butchering weight loss of 40 percent. Consumption estimations obtained with these methods were comparable to those obtained by nutritionists who systematically weighed all food consumed by households (Koppert 1996). The price of each wild meat species that was exported from each settlement was determined by interviewing wild meat traders.

The Effects of Logging on Wildlife

Direct Impacts of Logging on Forest Structure and Wildlife

Selective logging within the SFAC concession removed, on average, one tree every 6.6 hectares. Sapelli (*E. cylindricum*) accounted for over 75 percent of all trees felled. A comparison of inventory and logging records showed that over 90 percent of all Sapelli trees that exceeded minimum statutory size limits (> 80 cm dbh) were removed by logging within the SFAC concession.

Highly selective logging of tree species that occur at low densities within the forests of northern Congo resulted in less than seven percent of the canopy being cleared. The extent of disturbance at felling sites was likely to be less than that of natural treefalls (Hart et al.

1989) because any attached lianas were cleared prior to tree cutting, thus reducing pull-downs of neighboring trees or tree limbs.

Reduction in canopy cover was considerably below the 30 percent threshold considered by Skorupa (1986) to adversely affect primate species in a Ugandan forest, and the 38 percent considered by Thiollay (1992) to cause a 25 to 30 percent loss of bird species richness in a Guyanan forest. Canopy removal as a result of one cycle of highly selective logging, as practiced in the SFAC concession, may not, therefore, have major adverse impacts on forest primates and birds. This contention is further bolstered by the fact that commercially exploited tree species in northern Congo are primarily wind-dispersed emergents and therefore are not major food sources for frugivorous primates and birds. In fact, vegetation regrowth that occurs after a logged area is abandoned may increase food availability for folivorous species such as elephants (Fay 1991), gorillas, and duikers (Nummelin 1990; McCoy 1995; Carroll 1996). Almost complete removal of old canopy-height Sapelli trees may, however, adversely affect species-dependent folivores and their predators, and may reduce the availability of some species of caterpillars that are seasonally collected by humans as a food source.

Survey results from SFAC indicated a low occurrence of primates (0.15 groups/km) relative to other tropical moist forests of central Africa (see table 17-1). Stromayer and Ekobo (1991b) recorded over twice as many primate groups over a comparable total transect length in the unlogged Boumba Bek (0.32 groups/km) and Mongokele (0.38 groups/km) areas, and almost five times as many primates (0.77 groups/km) in the Lake Lobéké area (Stromayer and Ekabo 1991a) of southeastern Cameroon. Similarly, Fay et al. (1990) recorded much higher densities of primates (0.29 groups/km) in the unlogged Ndoki-Nouabalé region of northern Congo. Gorillas were present and, based on nest counts, interviews with local hunters, and prospecting teams, they were probably in quite high numbers throughout the SFAC concession—particularly in swampy areas that are not logged and were less accessible to hunters (Blake et al. 1995). Gorilla sign was most common at disturbed sites such as alongside abandoned secondary roads. Few elephant droppings were observed and chimpanzee sign was rarely encountered. While the hundreds of kilometers of transects, trails, and roads created within the logging concession facilitated primate censusing, they also allowed easy and systematic exploitation of primates (Pearce and Ammann 1995).

TABLE 17-1 *West Africa Forest Primate Groups Encountered per km of Transect Surveyed*

Location	Forest State	Primate Groups/km	Reference
SFAC concession, northern Congo	logged	0.15	Wilkie et al. 1992
Boumba Bek, southeast Cameroon	unlogged	0.32	Stromayer and Ekobo, 1991c
Mongokele, southeast Cameroon	unlogged	0.38	Stromayer and Ekobo, 1991c
Lake Lobéké, southeast Cameroon	unlogged	0.77	Stromayer and Ekobo, 1991a
Nouabalé-Ndoki, northern Congo	unlogged	0.29	Fay et al., 1990

Indirect Impact of Logging Operations on Wildlife

Hunting in all three study sites was done primarily with 12-gauge single-barrel or double-barrel shotguns and wire-cable snares. Snares were set most often within a one day walk of settlements, anywhere near logging roads, transects, and tracks, and less frequently in the more distant reaches of the forest not yet inventoried for trees. Shotguns were used both for daytime hunting of arboreal primates, duikers, and pangolins, and at night using a battery-powered lamp to *spotlight* any game encountered. Animals most commonly observed in settlements were blue (*Cephalophus monticola*) and red duikers (*C. callipygus, C. nigrifrons,* and *C. dorsalis*), brush-tailed porcupine (*Atherurus spp.*), and the four primates, *Cercocebus albigena, Cercopithecus cephus, C. nictitans,* and *C. pogonias* (see photo 17-1). A complete list of species hunted in the SFAC concession is shown in table 17-2.

The range of species hunted within concessions was similar to that reported by Mitani (1990) and Bennett-Hennessey (1995). Elephant and gorilla hunting was common in the SFAC concession, and for some hunters it was their primary source of income. Local hunters developed a technique for hunting elephant with shotguns rather than 0.458 caliber rifles, dramatically increasing the number of hunters able to kill elephant. The technique used spent shotgun cartridges that were recharged with a double shot of powder and a solid round of lead or machined steel.

Although Congolese law (No. 48/83 of April 21, 1984) prohibits spotlighting, and the use of nontraditional material (wire cable) in the

PHOTO 17-1 *Typical daily catch of primates and duikers by wild meat hunters in the SFAC concession. (D. Wilkie)*

construction of snares, both techniques were openly employed and were acknowledged as the most common, preferred, and effective techniques for obtaining game. The number of spent shotgun shells observed along forest trails (0.24 cartridges/km) demonstrated the intensity of hunting with firearms within the SFAC concession. Cartridge counts actually underestimate hunting pressure because not all cartridges on the trail are located by the observer, and hunters who follow trails, transects, and roads often leave them to discharge their weapons.

An examination of the differences in rates of return to hunting in the SNBS Kabo concession quantitatively shows the impact of hunting on forest fauna, and emphasizes the role logging companies play in facilitating and participating in hunting (see table 17-3). Daytime shotgun hunts conducted on foot from settlements located outside of the SNBS concession resulted in an estimated rate of return (ERR) that is higher than for hunts conducted on foot from within the concession (1.9 kg/man-hr versus 1.3 kg/man-hr). This suggests that animals are more abundant outside of the concession. The ERR for hunters using concession roads and motorized vehicles to travel deep into the forest was also higher than for hunts conducted on foot from settlements within the concession (daytime hunts—1.9 kg/man-hr versus 1.3 kg/man-hr; nighttime spotlighting—3.7 kg/man-hr versus 2.0 kg/man-hr).

TABLE 17-2 *Animals Captured by Hunters in the SFAC Concession*

Latin Name	Common Name
Cercocebus galeritis	Crested mangabey
Cercocebus albigena	Grey-cheeked mangabey
Cercopithecus nictitans	Greater white-nosed monkey
Cercopithecus cephus	Moustached monkey
Cercopithecus pogonias	Crowned guenon
Colobus guereza	Black and white colobus
Manis tricuspis	Tree pangolin
Atherurus spp.	Brush-tailed porcupine
Cricetomys emini	Giant rat
Genetta servalina	Small spotted genet
Crossarchus spp.	Dark mongoose
Felis aurata	Golden cat
Panthera pardus	Leopard
Potamochoerus porcus	Bush pig
Boocercus euryceros	Bongo
Tragelaphus spekei	Sitatunga
Cephalophus monticola	Blue duiker
Cephalophus sylvicultor	Yellow-backed duiker
Cephalophus dorsalis	Bay duiker
Cephalophus callipygus	Peter's duiker
Cephalophus nigrifrons	Black-fronted duiker
Hyemoschus aquaticus	Water chevrotain
Python sebae	Rock python
Varanus spp.	Monitor lizard

These data demonstrate that animal densities increased with increased distance from settlements, and that hunting at night with a spotlight was more efficient and intensive than simple daytime shotgun hunting.

If we assume that logging results in overexploitation of forest fauna during the period that timber is being surveyed and extracted, but that post-exploitation populations will recover, then expected rates of return should show a U-shaped distribution in relation to time since logging. Auzel's data from the CIB concession did not, however, show a relationship between rates of return and the number of years since logging. One site that had been logged 12 years previously and was isolated from villages and roads showed an estimated rate of return of 8.49; almost twice that of the next most productive site. Although this single case suggests recovery from logging-facilitated hunting, other sites that had been logged 10 to 13 years prior to the study did not show higher than average return rates (range 1.49–4.81 kg/man-hr; this study).

A higher estimated rate of return outside of concession areas and with increasing distance from concession settlements indicates that hunting within concessions reduced game densities, and they may be unsustainable during the time that the area is being logged. Shotgun hunting appears to result in reduced rates of return in areas of forest that were hunted regularly. This should not be surprising as rates of return to shotgun hunting are seven to 25 times higher than for hunts using traditional weapons (Wilkie and Curran 1991) such as bows (0.12 kg/man-hr) and nets (0.18 kg/man-hr). Hunters themselves commented, "It will be good when SFAC opens up a new logging area, because then there will be more game; the forest around Pounga is empty." The impact of intensive shotgun hunting in the CIB concession is shown in figure 17-3. ERR for hunters declined by more than 25 percent over a three-week period, when hunters were forced to return to exploited areas as they waited for a logging bridge to be built and new areas of the forest to be opened up for hunting.

Access to the Forest

Between 1986 and 1987 in the SFAC concession, over 60 km of primary roads and 80 km of secondary roads were constructed, and over 3,000 km of primary and secondary transects were cut within the forest. Road construction and access to transportation on logging vehicles cut the average distance that hunters had to travel from an access point (i.e., a river or a logging road) to any section of forest within the 1986 to 1987 logging sectors (see figure 17-2) from 9.2 km (max 34.6 km) to 360 meters (max 3.1 km). Average round-trip travel time for hunters declined from 12 hours (max 46 hours) to under two hours (max eight hours). What was once a three- or four-day trip for a hunter on foot to reach and hunt the farthest tracks of the forest (within the 1986 to 1987 logging sectors), had become a one-day event once logging roads and logging vehicles were available.

TABLE 17-3 *Impact of Transportation and Hunting Method on Estimated Rate of Return (ERR)*

Hunting Method	Location	Transportation	ERR (kg/man-hr)
Daytime shotgun	Outside concession	Walking	1.9
Daytime shotgun	Inside concession	Walking	1.3
Daytime shotgun	Inside concession	Logging vehicle	1.9
Nighttime spotlight	Inside concession	Walking	2.0
Nighttime spotlight	Inside concession	Logging vehicle	3.7

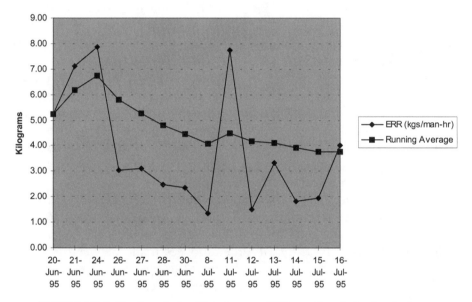

FIGURE 17-3 *Declines in hunting success (ERR or estimated rate of return from hunting) in a forest block over time.*

Value of Wild Meat to Local Economies

Wild meat consumption in three villages within the CIB logging concession ranged from 0.16 to 0.29 kg/person/day. These data are within the range reported by Koppert and Hladik (1990) for forest hunter-gatherers (0.29 kg/person/day), by Anstay (1991) for rural Liberians (0.28 kg/person/day), and for foragers (0.16 kg/person/day) and farmers (0.12 kg/person/day) in the Ituri forest of Zaire (Wilkie et al. 1998).

Residents of the most logging-integrated settlement, the Ndoki camp, ate the most wild meat (0.29 kg/person/day), and captured in absolute terms the most wild meat in comparison to two villages that did not have logging employees (see table 17-4). Wild meat captured per capita in the Ndoki camp, however, was not unlike Nganzicolo—with access to logging roads and vehicles (0.35 vs. 0.35 kg/person/day respectively)—and was over twice that of Toukoulaka (0.17 kg/person/day), which is isolated from roads and markets. Most interesting is the contrast between the isolated settlement of Toukoulaka and the settlement of Nganzicolo, which has access to roads and markets. Toukoulaka eats about as much wild meat per capita as Nganzicolo (0.16 vs. 0.17 kg/person/day respectively), but trades only five percent of the animals it captures, and hunts only half as intensively. It appears that the Nganzicolo households focus on wild meat hunting and marketing is based on a combination of access to markets and absence of logging-based salaries.

In summary, logging company employees (i.e., salaried individuals) eat more wild meat than their village counterparts both in absolute and relative terms, but sell less than those households that have equal access to markets. Households not employed by the logging company, but that have access to market, eat no more wild meat than the residents of isolated villages, but they hunt more intensively and market a much higher proportion of all game captured.

Both the logging camp village and the village on the Sangha had access to markets for wild meat, and on average sold between 36 to 52 percent of all wild meat captured, generating income of U.S. $0.7 to 0.8/household/day. In 1989, SFAC employees made between U.S. $4–12/day. For nonlogger families, wild meat sales generate income that is only six to 20 percent of the average logger's wage.

TABLE 17-4 Bush Meat Captured and Eaten by CIB Concession Villages

Settlement	Population	Survey days	Loggers	Access to market	Bush meat captured (kg/person/day)	Bush meat eaten (kg/person/day)
Ndoki	562	58	Yes	Yes	0.45	0.29
Nganzicolo	114	43	No	Yes	0.35	0.17
Toukoulaka	197	38	No	No	0.17	0.16

Logging Roads, Logging Trucks, and the Wild Meat Trade

Market hunting, whether using traditional (bows, crossbows, nets, and dead-fall traps) or modern methods (firearms and wire snares), lowers the densities of exploited fauna (Freese 1996). Even domestic wild meat consumption can potentially result in overexploitation of forest fauna (Alvard 1994; Wilkie et al. 1998). In the absence of logging, the scale and impact of market hunting is limited by hunters' access to the forest and the relatively high costs of transporting wild meat to the market (Lahm 1994a). Logging operations dramatically overcome these constraints through the creation of tree inventory transects and primary and secondary roads that provide hunters access to the furthest reaches of the forest, and by transporting market hunters' wild meat on logging vehicles. Data from these studies suggest that if logging company-facilitated market hunting is not curbed, the future of K-selected animal populations (slow-reproduction) within concessions in central and west Africa is bleak—particularly when logging occurs as an advancing wave, leaving no undisturbed areas of forest in its wake.

The presence of concessions in the area certainly augmented the market for meat. Employees and their families, the majority of which are local residents (other than technical staff), were not only consumers of game, but many also became game traders. People used their concession salaries to buy shotguns (that were legally available) and/or ammunition, which were then given to BaNgombe (pygmy) hunters in return for game. This game was subsequently sold for a substantial profit. Though wild meat was sold locally, Ouesso and ultimately Brazzaville and Pointe Noire were the major sources of demand (Stromayer and Ekobo 1991c; Assitou and Sidle 1995). Several hunters said that they preferred trading their meat at Ouesso as it commanded a far higher price than at SFAC or other nearby concessions.

In addition to providing a market for meat, the concessions provide hunters with easy access to large areas of the forest, and a means of transporting game to market. The system of transects in a logging area provided hundreds of kilometers of easy access. Abandoned and active roads function similarly, unless the former is properly barricaded (limiting vehicles, but not walking access). Logging vehicles are routinely used to transport hunters, weapons, and wild game. SFAC boats are used to transport fresh or smoked game to Cameroon, and downstream to Ouesso, where hunters received at least double what they would sell the meat locally (Bennett Hennessey 1995). Qualitative observations from typical days in the Congo concessions (see box 17-1) illustrate how logging companies often facilitated poaching by employees and non-employees. Succinctly, systematic commercial exploitation of the forest for timber results in systematic commercial exploitation of the forest's fauna, leading to a dramatic reduction in game animal densities (Assitou and Sidle 1995; Eves 1995).

Box 17-1 *Qualitative observations of timber concession personnel facilitating wildlife hunting.*

All three concessions facilitate hunting by transporting hunters and game on concession vehicles. For example, SNBS actively promotes hunting by organizing a weekly hunt on Sundays to provide meat for concession workers. Over 20 hunters with shotguns participate in the SNBS Sunday hunt that takes hunters as far as 80 km into the forest from the concession base camp.

In the SFAC concession, vehicles leave the Pounga compound each day, taking workers out to the active logging area. On the way, each vehicle typically picks up one or more BaNgombe pygmies and shotguns. The shotguns' owner, a SFAC employee, gives three shotgun shells to the hunter with instructions to kill three animals—most often primates or duikers. The gun owner receives two animals and the hunter the other. At the end of the day, the hunter and game are transported in a SFAC vehicle back to Pounga or another nearby village, where the game is smoked or otherwise prepared for shipment out of the concession to Cameroon (or downstream to Ouesso). Rarely did we see a vehicle returning to Pounga at the end of the day that was not carrying a load of game meat (see box photo 17-2).

CIB moves logs from Pokola to the Sangha River north of Ouesso by truck, then by ferry to Sucambo in Cameroon, and then along the road to Douala. CIB trucks, traveling between Pokola and the Sangha loading dock, almost always carry passengers and wild meat. In late 1998, CIB entered into a formal agreement with the Wildlife Conservation Society to stop the export of wild meat from their Kabo concession in northern Congo. CIB appears to be committed to the agreement as it has already fired two truck drivers for transporting wild meat.

On a trip across the river to Cameroon, the departure of a SFAC log canoe was delayed until a SFAC employee with two shot monkeys arrived. On our departure trip to Ouesso, five freshly killed blue duikers were transported in the pirogue. Sacks of fresh game meat accompanied our flight from Ouesso to Brazzaville. SFAC vehicles make the rounds of primary and secondary roads during the early morning and evening to pick up game shot by local hunters. SFAC roads and vehicles are also used to facilitate hunting at night. The above qualitative observations highlight the logging company's facilitation of market hunting.

BOX PHOTO 17-2 *SFAC logging vehicle being used to transport wild meat from isolated sections of the concession to Pounga on the Ngoki River. (D. Wilke)*

Recommendations to Limit Logging-Facilitated Hunting

Hunting provides an essential source of protein and a major source of income for the inhabitants of logging concessions (6–40 percent of household revenue). Facilitated by the concession infrastructure and transportation, market hunting has become a serious threat to the region's wildlife. Since the data for this paper were collected, logging has become progressively less economically viable in northern Congo, and SFAC and several other companies have gone bankrupt and their concessions have been closed. Logging is continuing in other regions of Congo and in other central African countries. Our recommendations, though based on our understanding of three logging operations in northern Congo, appear generally applicable to most logging concessions throughout central and western Africa, where logging practices are comparable.

Given the importance of hunting to local household economies, and the fact that the market for wild meat is primarily in urban centers, strategies to reduce incentives for (and the impact of) market hunting will have to address economic, demand, and law enforcement issues (Wilkie 2001).

Supply-Side Mitigation Options

The importance of wild meat to the diet and income of forest dwelling families, the huge areas of forest involved, and the shortage of well-paid (i.e., less corruptible) forestry officers are likely to preclude the use of command and control measures to limit market hunting outside of logging concessions. Some options do exist, however, for controlling hunting within the networks of logging concessions.

Using Conservation Bonds

Command and control measures may work, within the confines of logging concessions because logging companies could be required to pay for sufficient numbers of incorruptible law enforcers, and to provide them with the transportation and equipment necessary for monitoring hunting. Wildlife law enforcers should not be paid directly by the logging concessions. Instead, the companies should be required to post a bond, paid to the appropriate government ministry, for an amount indexed to the area of forest to be exploited that year. These monies would be earmarked for natural resource conservation within logging concessions, and could only be used to support forestry and wildlife law enforcers, and plant and animal surveyors who are stationed in logging concessions. Repayment of the bond to the logging concession could be indexed to the ratio of pre- and postlogging game survey figures, with the highest rebates occurring at parity. If the bond was set high enough, logging companies might comply with recommendations that wildlife and firearms laws of the country be respected by personnel of logging companies, and that vehicles, roads, facilities, and company time should not be used to support poaching.

Using a logging company, bond-financed fund—earmarked for natural resource conservation within logging concessions—would allow the development of wildlife management plans and a regulated harvest of forest protein. A conservation bond would also help strengthen Congo's capacity and institutions to enforce wildlife protection, as Verschuren (1989) urges. This approach will, of course, only work if a) the logging companies do not attempt to bribe forestry and wildlife officers, and

b) the forestry ministry establishes and enforces wildlife conservation bond legislation and uses the earmarked fund appropriately.

Logging companies may be persuaded to comply with conservation measures associated with paying a conservation bond, if consumer demand for green-labeled timber was sufficient to significantly alter their profit margins (see chapter 26). A recent study by CIFOR (D. Kaimowitz, personal communication), however, demonstrated that although several logging companies in Central and South America improved their market positions by selling certified wood, there was no evidence that timber purchasers were willing to pay substantially higher prices for certified timber.

Curbing the Transportation of Wild Meat

An alternative or additional approach to regulating hunting is to control the shipment of wild meat from the concessions to the point of sale, which is illegal according to Congolese law (Loi No. 48/83; Decret No. 85/879). This directly impacts the profitability of market hunting, which is largely determined by access to—and the cost of—transportation. When CIB started transporting logs to Douala (Cameroon) from the Sangha River port at Sucambo (near Ouesso), wild meat from Cameroon soon comprised over 13 percent of the game sold in Ouesso markets (Bennett Hennessey 1995). A dispute between the trucking company and the concession, however, halted traffic from Congo through Cameroon, resulting in the temporary collapse of the wild meat market and the closure of hunting camps that border the roads during August 1995 (Pearce and Ammann 1995). Wild meat marketing is, therefore, a risky business. If the truck does not arrive to ship meat to the market, the hunters' produce may rot and become worthless. The key to reducing market hunting is, therefore, curbing transportation of wild meat on logging vehicles, and could be accomplished using roadblocks for wild meat (an incorruptible constabulary is not an absolute necessity as described below), associated with conservation bonds or green-labeling. Wildlife Conservation Society and the CIB concession have recently signed an agreement with the government of Congo to halt exportation of wild meat from the CIB Kabo concession in northern Congo. This agreement signifies the first test of the approaches suggested above.

Demand-Side Mitigation Options

Rather than looking solely toward supply-side command and control measures to reduce the supply of wild meat, mitigation efforts might focus more on altering household demand for wild meat. Assuming that demand for wild meat is elastic, consumption of wild meat will decline if the price of wild meat rises relative to substitutes. Preliminary results from a comparative study of the income and price elasticity of wild meat demand by forest farmer-hunters in Bolivia and Honduras show that a) demand follows an inverted U-shaped curve with increasing household income, and b) that a one percent reduction in the price of beef is associated with an eight percent reduction in the consumption of wild meat (R. Godoy, personal communication).

Manipulating the Price of Wild Meat

Though wild meat could be confiscated at roadblocks or in local markets, taxing the transportation of wild meat (at roadblocks and/or in markets) is likely to be a more effective approach to increasing the scarcity and thus the price of wild meat. Taxes should be targeted at the wild meat traders, and set at a value per kilo that exceeds the price differential between wild meat and chicken sold in urban markets. Wild meat does not need to be confiscated at roadblocks because, as the costs of taxes are transferred to consumers, the price of wild meat will rise above that of substitutes driving down demand for wild meat. Not confiscating wild meat avoids the need to dispose of the game in a way that: a) does not encourage corruption (assuming that guards are not going to purloin the taxes), and b) prevents the sale of confiscated game at reduced prices, thus fueling demand. Corrupt guards could sell confiscated meat at lower prices than that of wild meat traders because with zero costs any price constitutes pure profit. As traders' costs increase with taxation (even if guards steal the tax monies or impose lesser "fines"), their profits will fall if wild meat consumption is elastic. As demand and profits fall, the price that traders are willing to pay hunters will decline and the income-generating incentive for hunting will also decline. Setting the wild meat tax at over double the price differential per kilo attempts to mitigate cultural preferences for wild meat and consumer willingness to pay a price premium for wild meat. Market prices of wild meat and domesticated alternatives should be monitored regularly so that the level of taxation can be maintained high enough to curb consumer demand for wild meat.

Reducing wild meat sales through taxation may have an adverse impact on local economies dependent on market hunting if, as we hope, consumers switch to buying the meat of domesticated animals. The costs of transportation will also make it unlikely that isolated rural communities can market domesticated animals at prices competitive with animals raised closer to centers of demand—assuming that livestock food prices are comparable. To compensate local communities for the loss of revenue-generating opportunities associated with a tax on wild meat transportation, the portion of taxes remaining after interdiction costs are paid for could be returned to the local community in the form of social services. Admittedly this assumes legal authority and political will to return taxes to local communities, and is not equitable as individual hunters pay the full cost of the tax yet only receive a share in the percentage returned to the community.

Providing Substitutes: Small-Livestock Production

Cane rat (*Thyronomys swinderianus*) and giant rat (*Cricetomys emini*) production is possible using domestic food scraps and agricultural waste (Asibey 1974; Tewe and Ajaji 1982; Jori and Noel 1996). Promoting small domestic animal production, such as rabbit-raising, has proven effective in Cameroon in areas were wild meat is already scarce (HPI 1996a, b). Several pilot projects are underway in Gabon to raise cane rat, brush-tailed porcupine, and bush pig/domestic pig hybrids to reduce the demand for wild meat in cities (Steel 1994). Small-game raising activities are also part of a UNDP/GEF project in Gabon that focuses on the commercial use of forest flora and fauna (Steel 1994). Raising small, domesticated animals such as rabbits is attractive in that methods of husbandry and veterinary care are well-known. Domestic animals are also generally more productive than other wild counterparts. Feer (1993) argues that pigs, zebu cattle, cane rats, and duikers exhibit a decreasing scale of meat productivity, with pigs being most productive and duikers being the least. Consequently, it makes more sense to promote pig or cane rat production—both of which are relatively well understood—than to attempt to raise duikers for meat. Consumer tradition and tastes, however, may prefer wild meat over the meat of domestic animals. Steel (1994) found in Libreville, Gabon that the average price for the most popular wild meat species was $3.7/kg—over 1.6 times the price of the most popular cut of beef. In contrast, Gally and Jeanmart (1996) found that the price of wild meat per kilo was 0.10 to 0.25 times the price of available substitutes (i.e., goats, chickens, and caterpillars) in three markets in Cameroon, Congo, and the Central African Republic.

Rabbit, porcupine, or cane rat rearing as an alternative to wild meat hunting is unlikely to be successful, however, if the labor and capital costs of production exceed the costs of wild meat hunting (i.e., when game becomes too scarce to be worth searching for) and marketing. Of course, if domestic production of meat only becomes economically viable after wild game have become so scarce as to be unprofitable to hunt, the strategy is clearly ineffective as a conservation measure. This is why it is important to institute a wild meat tax to reduce the price differential between domestic meat and wild meat.

Small-domestic animal raising should be promoted in peri-urban areas in an attempt to lower demand for wild meat (Jori and Noel 1996). To encourage game raising—even if the wild meat hunting/ domestic production cost ratio favors hunting—could be promoted as a child health and development measure rather than as a game management initiative. Promotion of small-game raising in peri-urban areas will, of course, disrupt the flow of economic benefits from urban consumers to poor rural producers of wild meat. Depending on the proportion of rural households involved in wild meat marketing, and the contribution of wild meat sales to household income, demand-based approaches to controlling wild meat hunting may warrant short-term interventions to offset their adverse impacts on the living standards of the rural poor.

Role of Donors and Conservation NGOs

Congo remains bankrupt with massive unemployment, and ranks in the top ten of the poorest nations. Regardless of whatever solutions to wild meat overexploitation are selected, it does not appear that Congo can afford them without external aid, and requests for assistance are inevitable. Even areas devoted to wildlife conservation, such as the Nouabalé-Ndoki and Odzala national parks, are completely subsidized by external funds. They would collapse in the absence of foreign aid. The Léfini Reserve, Congo's oldest protected area close to Brazzaville, is essentially unprotected and poaching is rampant. The reserve, within a few hours drive of the headquarters of Congo's Direction de la Conservation de la Faune in Brazzaville, waits for foreign assistance. The upshot is that whether one is discussing problems and potentials for Congo wildlife conservation—inside or outside protected areas—logging concessions, or elsewhere—foreign involvement appears to be a perpetual undertaking.

International donors and conservation NGOs should focus their efforts on:

- Using the media to place pressure on European companies to ensure that their logging subsidiaries comply with national forestry and wildlife laws
- Conducting surveys of market hunting in logging concessions throughout the region
- Establishing whether consumer demand for green-labeling of timber harvested from concessions that conserve wildlife would raise profit margins sufficiently to offset the costs that concessions would incur to stop wild meat hunting
- Training national wildlife conservation officers and biologists to conduct field censuses of plants and animals, and to monitor domestic and market hunting within logging concessions

ACKNOWLEDGMENTS

Thanks to the editors for their comments, and to the directors of SFAC, CIB, and SNBS for allowing us to work within their concessions. The U.S. Department of Agriculture's Office of International Cooperation and Development, the Wildlife Conservation Society, and the Global Environment Facility supported this study.

Part IV

RESEARCH TO INTEGRATE NATURAL FOREST MANAGEMENT AND WILDLIFE CONSERVATION

Our understanding of the direct and indirect effects of forest management practices on wildlife (parts I through III) is very incomplete because of the limited number of studies in this field, and problems associated with their experimental designs. In this section of the book, contributors address how applied research efforts should be focused to achieve ecological and economic sustainability of our natural resources. Discussions range from improvements in experimental designs to a demonstration forest model where reduced impact logging practices, protected area networks, and silvicultural systems based upon natural disturbance models are evaluated and refined. The recommendations suggest avenues to conserve wildlife in future forestry practices. Specific approaches to reduce the environmental impacts associated with timber harvesting are addressed in parts V and VI.

Forest Research Programs to Conserve Wildlife in Managed Tropical Forests

In chapter 18, Andrew Grieser Johns reviews the experimental designs used in research studies examining the impacts of logging systems on wildlife and their habitat. His findings suggest that the effects of many important forest management systems have not yet been studied, and where they have, there is a great inequality in their scope and geographical distribution. Most studies also describe differences in species composition between areas of forest with different logging histories. Such comparisons may identify broad changes at the level of species assemblages or guilds, but may have very low utility at the level of individual species since they cannot control for spatial heterogeneity within tropical forests. This is important because differences due to spatial heterogeneity are frequently greater than changes caused by the logging process. Finally, few studies demonstrate causal relationships between species abundances and habitat changes resulting from logging. The author concludes chapter 18 with a call for long-term monitoring under conditions where spatial heterogeneity is accounted for in the design.

In chapter 19, Robert Fimbel, Elizabeth Bennett, and Claire Kremen close part IV by discussing a two-level approach to assessing the direct and indirect impacts of timber harvesting practices on wildlife. At the major forest type level, a design is presented for a model demonstration forest where reduced impact logging practices, protected area networks, and silvicultural systems based upon natural disturbance models, are evaluated for their wildlife conservation values. Other aspects of sustainable forest management, such as financial costs/benefits and growth-and-yield, are also appraised at these sites. A complementary research program, using rapid, low-intensity studies of forestry impacts on wildlife (the second level of the design), is proposed for the numerous harvest sites occurring throughout a major forest type. The goal of this latter approach is to provide immediate feedback to resource managers on the impacts associated with their prescriptions. The chapter closes with a discussion of the obstacles and opportunities influencing the adoption of these forestry research programs.

The Road to Sustainable Forest Management and Wildlife Conservation Through Research

A resounding message from the contributors to this section, and others throughout the volume, is the need to advance efforts to achieve sustainable forest management. Sustainable forest management, by definition, conserves wildlife species and their habitats. To promote these practices in all major forest types where the conservation of native fauna and flora is considered of equal importance to the production of timber products, resource managers need to better understand the environmental impacts and economic options associated with forest management practices. In the absence of this knowledge, old paradigms of forest domestication and conversion will continue to dominate silvicultural prescriptions, with subsequent losses in tropical forest biodiversity.

The demonstration forest and concession level studies proposed by Fimbel and his coauthors (see chapter 19) incorporate the replicated, long-term, pre- and postharvesting design proposed by Grieser Johns (see chapter 18) and others in this volume, in an effort to clearly identify the response of wildlife to harvesting practices. This framework also compares silvicultural prescriptions to natural disturbance events (see chapter 22), evaluates the contributions of reduced impact logging (see chapters 21 and 24) and protected area networks (see chapter 23) to wildlife conservation, and assesses the economics associated with these management practices (see chapter 28). Finally, the potential of the demonstration forest as a training center and a catalyst for input by local stakeholders (see chapter 19) offers a unique opportunity to unite all participants in the sustainable forest management/wildlife conservation process. The challenge facing conservationists is to find a way to implement this comprehensive research, support training of forest staff, and enhance institutional capability to incorporate the research data into forest management policy.

Chapter 18

NATURAL FOREST MANAGEMENT AND BIODIVERSITY CONSERVATION

Field Study Design and Integration at the Operational Level

Andrew Grieser Johns

Until the beginning of the 1980s, it was a maxim of practical conservation that areas of disturbed tropical forest were not considered useful in maintaining biodiversity (Johns 1985). This attitude has changed very rapidly. Following the study by Johns (1983b), which illustrated the potential for retention of biodiversity in managed rain forests, there has been a movement toward the development of ecological guidelines for the management of tropical forests that incorporate biodiversity conservation measures (e.g., Poore and Sayer 1987; Blockhus et al. 1992; Dykstra and Heinrich 1994; Sist et al. 1998a, b). These guidelines tend to reiterate the general principles of low-damage, polycyclic management systems (particularly the Queensland selective management system—see Part V of this volume for a review of this system and other reduced-impact logging practices). While providing a framework for the development of improved forest planning and implementation—which can be elaborated on a site-by-site basis by suitably trained foresters—current reduced-impact logging guidelines do not provide specific prescriptions for biodiversity retention.

The introduction of specific actions to promote biodiversity retention within best-practice tropical forest management strategies has occurred rarely to date, except in Australian forests. This condition is a function of several factors, of which the most important are the lack of:

- Knowledge related to the ecology of tropical organisms and how they respond to logging
- Political will for change
- Suitably trained foresters to oversee the changes
- Control of the forest resource by foresters

This situation is changing, albeit slowly, with a growing trend toward privatization of forestry departments (e.g., ODA/FRR 1996), political reform, and retraining to satisfy new and *greener* tropical timber markets.

Applied ecological research on the effects of logging on tree regeneration and growth has provided information of direct interest and benefit to forestry (e.g., Brown and Press 1992) and is being *taken up*, or has the potential to be taken up, in field operations (Pinard et al. 1995). So far, biodiversity studies have, to a much lesser extent, been conducted and presented in a form that allows their incorporation into field planning or harvesting methodologies (Grieser Johns 1997).

The inability of biodiversity conservation research to take its place in shaping tropical forest management strategies urgently needs to be addressed. This should be approached in two ways. First, the design of research projects needs to be reexamined to provide more data of direct relevance to forest planning and management, in a form usable by forest managers. Second, research needs to be directed to areas where it is likely to be incorporated into forest policy.

The Effects of Logging on Tropical Forest Biodiversity— Past and Present Field Studies

The geographic distribution of studies of the interaction between tropical forestry and biodiversity has been rather uneven. Most early studies were located in Australia and southeast Asia (see figure 18-1), but with a recent expansion into East Africa and the Neotropics. This parallels, to a large extent, the focus of intensive logging activity and the extent to which forestry authorities are willing to embrace conservation-related research programs (Bruenig 1996).

Past Studies of Logging and Biodiversity

Intensive logging of tropical forests dates back to at least the development of power saws and dedicated hauling and transport machinery in the 1940s and 1950s (Grieser Johns 1997). The earliest studies addressing the issue of logging impacts on tropical plant and animal species tended to consider the effects of clear felling of forest (e.g., McClure and Othman 1965). This remains an important focus, partic-

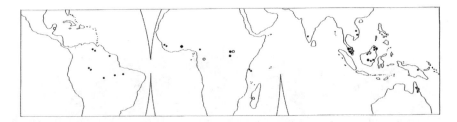

FIGURE 18-1 *Locations of research studies on the effects of timber logging on tropical and subtropical forest biodiversity. Studies are according to a bibliography prepared by the author and available from the author on request.*

ularly in Australia, and is well researched in that country (Saunders et al. 1987). Clear felling as a forestry option is not widespread in the tropics (see photo 18-1), although there are some notable exceptions (e.g., Gogol in Papua New Guinea: Lamb 1990). In most tropical forests, selective felling systems are applied.

The impacts of selective felling on specific animal groups began to be investigated in the 1970s (e.g., Pattemore and Kikkawa 1974; Kemp and Kemp 1975; Wilson and Wilson 1975; Asibey 1978). The early studies, and indeed most since, have been short-term surveys using comparative sampling of unlogged forests and forests logged at some previous stage. Many studies were opportunistic, taking advantage of logging activity in the region of established primary forest research sites to produce comparative data. Research of this type usually examined differences in the abundance of animals in different forest types. While the accuracy of sampling methodologies and the statistical treatment of results has improved markedly over the last 20 years, such studies remain prone to severe sampling biases (see part V).

A development of the above approach has been to conduct detailed ecological research comparing the ecology of selected species within unlogged and logged sites. Research of this type has been undertaken in a number of sites, notably the Kibale Forest and Budongo Forest of Uganda (Skorupa 1986; Plumptre et al. 1994; Struhsaker 1997) and Ulu Segama Forest Reserve of Sabah (Marshall and Swaine 1992). Studies of this type require a detailed knowledge of the felling operations (numbers and types of trees cut, subsequent damage levels, etc.) as a foundation on which to build a discussion of the implications of forestry on animal species. Results are difficult to interpret if forestry data are inaccurate or absent (this is a particular problem in the Kibale Forest studies: Grieser Johns 1997). Studies of this type, however, can

PHOTO 18-1 *Large-scale clearcut in Malaysian forest. (R. Z. Donovan)*

produce useful information where they are able to correlate changes in animal abundances with changes in the vegetation resulting from defined forestry practices (e.g., Plumptre and Reynolds 1994).

Present Studies of Logging and Biodiversity

The optimal design for logging-biodiversity research studies is to survey the fauna and flora of a forest prior to the onset of logging, and then to monitor the long-term responses of the plant and animal communities subsequent to logging (see chapter 19). Very few such studies (here referred to as *long-term monitoring studies*) have been established, however, because of the time and financial commitment involved. Johns (1983b) established the first such study at Tekam in Peninsular Malaysia. This remains one of the most detailed works of its kind. During the 1980s and 1990s, however, a number of additional before/after logging studies were undertaken—some incorporating the monitoring of biodiversity recovery (e.g., Crome and Moore 1989; Crome 1991; White 1992; Plumptre et al. 1994; ADB 1994; ODA 1995; Dahaban 1996). It will always be argued that studies of this type take too long and are overtaken by events (improved or changed management systems, etc.). In fact, this was already being argued in the late 1970s when Johns (1983b) began the work at Tekam. The fact remains that studies of this type produce results extremely useful to planners and forest managers.

Research programs have recently been proposed or established at a number of sites, with the specific aim of providing biodiversity information for incorporation into improved or reduced-impact logging practices (e.g., ADB 1994; ODA 1995; Lee et al. 1998). These programs aim for interlinkage between management and research on habitat requirements for plants and animals (of the type successfully developed in the sub-tropical forests of Australia: e.g., Lindenmayer et al. 1993; Kavanagh and Bamkin 1995). Unfortunately, the research conducted by most of these donor-funded programs is short-term, intersite comparative work, rather than the long-term monitoring required to produce definitive results.

Research and Its Application to Forest Management

There are a number of problems affecting the establishment of improved tropical forest management (see box 18-1). In particular, better training of forest staff is required at all levels and institutional capability requires general improvement (see photo 18-2). Forestry institutions need an increased capacity to research, develop, and manage improved harvesting systems (Palmer and Synnott 1992). The incentives for improved forest management also need to be shared equitably—a crucial criterion in this respect is that political and economic leaders set examples of high ethical and moral standards (Bruenig 1996).

Box 18-1 *Principal problems preventing the establishment of efficient tropical forest management (after Grieser Johns 1997).*

- *Insufficient reinvestment in the resource.* Concession agreements and taxation/fee structures generally do not offer incentives for investment in the long-term productivity of the forest (Barbier 1993).

- *Inadequate infrastructures.* There has long been a shortage of trained tropical foresters and little commitment on the part of governments to invest in forestry. The forest sector has typically been regarded as a source of central revenue—their activities have been underfunded while their profits have been absorbed by national treasuries (Palmer and Synnott 1992).

- *Inadequate information base.* Much research is outdated, and there is little knowledge of the breadth of current forest management techniques. Field practice manuals are generally unavailable or of poor quality. There is little understanding of the requirements for reduced-impact logging and biodiversity conservation.

- *Forest invasion.* During the 1980s, 300 to 400 million people moved into tropical forestlands. In African countries, for example, more than 50 percent of logged forests are subsequently deforested by agricultural invasion (Barbier et al. 1994). Settlers may further damage forest regeneration by removing pole-sized trees for construction purposes (which has been a main cause of the failure of forest management in coastal Colombia), or setting fires to create pasture or agricultural land.

- *Inefficient species use.* Most tropical logging operations select only a few tree species, which can lead to a great deal of waste. There is a great deal of resistance to change in the marketplace since importing countries will not consider lesser-known species while stocks of preferred species continue to be available. Increasing the price of preferred timber species may not cause a shift toward the use of lesser-known tropical species, but may facilitate their substitution by temperate timbers or solid synthetics (Vincent et al. 1991; Barbier et al. 1994).

- *Timber pricing systems.* Current tax structures are poorly constructed; they do not favor the implementation of improved forest management. Rent captures are also very low, and can be manipulated by a variety of illegal practices. Governments have generally allowed excessive profit margins for concessionaires.

- *Inequitability of profit sharing.* The motivation for the adoption of sustainable management policies will only arise if incentives are strong and all people concerned share the benefits (e.g., Sargent et al. 1994). This includes government, civil servants, the private and public sector, their laborers, and local populations. Government and industry leaders frequently do not set high ethical and moral standards and a climate for responsible financial management and profit sharing does not exist (Bruenig 1996).

— *Political instability.* Forestry requires long management cycles and is particularly susceptible to disruption through macroeconomic and political instability. In the former case, hyperinflation of the kind common in developing countries can quickly influence the cost-effectiveness of management through higher discount rates and reduced incentives for long-term investment. In the latter case, politicians too often predominate in forest policy such that the foresters themselves lose control of the management of the resource. Tropical forestry has in many areas also been targeted by environmentalist movements that negatively affect government or donor funds for the research and development of sustainable management systems (e.g., Moulds 1988).

Biodiversity research has enormous practical value and should not be conducted in isolation from decision-makers within the forest sector. There is a growing tie-in between donor-funded forest sector development and the researching of improved management strategies, including the incorporation of conservation criteria (Van Bueren and Duivenvoorden 1996). A platform exists from which research tailored to the needs of foresters can be developed and integrated into management practice.

This next section examines the extent to which different types of research can provide appropriate information for foresters and how such information may be usefully incorporated into forest management planning and implementation.

Field Study Design and Information Output

Variation in the densities of plants and animals can be very high over quite short distances (e.g., Braithwaite et al. 1984). Studies designed to assess this phenomenon objectively have shown that patchiness of animals within logging concessions can relate more to historical factors, elevation, and soil fertility, than to logging intensities or damage levels (Johns 1989a; Kavanagh and Bamkin 1995; see discussion in Grieser Johns 1997). Variation due to original spatial heterogeneity has been shown to be greater than variation due to the effects of timber logging—at least up to damage levels representing losses of 50 percent of trees (see next paragraph).

A number of research studies designed to interpret the effects of logging operations have encountered this problem. In the dry forests of western Madagascar, variation in forest structure eight to nine years following logging (with less than 10 percent canopy damage) was smaller

PHOTO 18-2 *To improve tropical forest management, better training of forest staff is required at all levels (forest operators practicing directional felling in Indonesia). (R. A. Fimbel)*

than variation due to natural causes where large study plots were considered (Ganzhorn et al. 1990). In the Budongo Forest of Uganda, variation in bird populations between unlogged and nine to 48 years postlogged (with more than 50 percent canopy mortality) forest sites could not be correlated with logging history or any linked variables (I. Owiunji and A. Plumptre, personal communication). In Budongo, both bird and primate populations varied spatially to such an extent that studies of the impact of logging would give entirely different results depending on which unlogged and logged compartments were actually compared (Plumptre et al. 1994). Similarly, at the Tekam Forest Reserve in Peninsular Malaysia, primate densities have been shown to vary more along a 20 km transect through uniform habitat than at the same site pre- and up to 18 years postlogging (with 51 percent canopy mortality) (Johns 1989a; Grieser Johns and Grieser Johns 1995).

Controlling the effects of spatial heterogeneity is difficult in intersite comparative studies. Increasing the number of replications or the size of the sampling units may narrow the variance, but these still cannot address the overall uncertainty caused by a lack of controls. It has to be assumed that biological communities of the different sites were not significantly different prior to logging, and that any significant differences observed are due in some way to the logging process. This assumption can clearly be called into question, especially when sample sizes are small.

It may be argued that there is also considerable temporal variation within biological communities at a single site, which could make single site studies equally unreliable. Variation in the relative abundance of birds in unlogged forest at Tekam, however, showed a maximum variation of 30 percent between months (measured by indices of similarity), whereas variation between equally aged logged sites during the same season was 80 percent (Johns 1989b; Grieser Johns 1997). This again indicates the predominance of variation due to spatial heterogeneity.

Levels of Analysis and Their Contributions to Management

The level of uncertainty resulting from spatial heterogeneity is expressed to a greater or lesser extent according to the detail with which results are analyzed. Spatial heterogeneity may not affect underlying ecological trends, but may seriously bias data on the abundance of individual species (Grieser Johns 1997). This needs to be borne in mind when developing guidelines for management activities—detailed guidelines should not be presented based on dubious data. To analyze the possible uses of different types of information collected from inter-site comparisons, it is appropriate to consider three levels of analysis: community level, guild level, and individual species.

The Community Level

The avifauna at the Tekam Forest Reserve of Peninsular Malaysia can be used to compare *real* and *predicted* changes in various statistical parameters resulting from logging (Grieser Johns 1997). Real changes are those resulting from long-term monitoring of the same site since a logging event, and predicted changes are those resulting from a comparison of unlogged forest and other sites with different logging histories (1–18 years post logging; 51 percent canopy mortality) located along a 20 kilometer transect. The latter imitates the more typical inter-site comparison.

At this level of analysis there is general agreement between the results of inter-site comparisons and long-term monitoring projects. Species richness (but not species diversity) tends to decrease in logged forests, and logged areas remain distinctly dissimilar to each other (in terms of comparative species abundances) for at least eighteen years after logging. These patterns indicate changes, but not the underlying causes of changes, how they might efficiently be countered, or if the changes that occur are of real significance in conservation terms.

The Guild Level

Studies from various sites have reached a general consensus on certain guilds of animals that tend to be adversely or favorably affected by logging (see part II of this volume). Among primates, for example, large-bodied frugivores have long been known to respond adversely to habitat disturbance, often further exacerbated by hunting pressure (Johns and Skorupa 1987; see chapters 4, 15–17). Bird studies typically show drops in the abundance of understory and terrestrial insectivores compared to increases in the abundance of generalist species associated with edge habitats (see chapters 8–10). Generalities of this type are apparent, both from long-term monitoring and inter-site comparisons. Results for the avifauna at Tekam Forest Reserve show differences of this type within the long-term study at site C13C (a 13-year postharvest site) and also where unlogged forest is compared with other sites (see table 18-1). Differences in the latter are much greater due to additional effects of spatial heterogeneity.

TABLE 18-1 A Comparison of Feeding Guild Membership (% total individuals) Within a Bird Species Assemblage—Tekam Forest Reserve, Peninsular Malaysia

Feeding Guild	Years Post-logging Area C13C			Area C5A	Area C2
	0 (UL)	7	13	8	12
Frugivores					
⁓ terrestrial	0	0	0	0	0.3
⁓ arboreal	10.6	14.9	11.9	31.7	22.0
Faunivore-frugivores	11.5	5.8	6.6	4.3	4.5
Insectivore-frugivores					
⁓ terrestrial	0.6	0	0.2	0	0
⁓ arboreal	15.1	39.2	31.0	30.6	23.9
Insectivore-nectarivores	3.9	4.5	8.9	3.9	2.5
Insectivores					
⁓ terrestrial	3.6	0	2.4	0.1	0.5
⁓ bark-gleaning	5.3	3.0	3.8	1.1	3.0
⁓ foliage-gleaning	31.1	28.5	28.4	14.6	33.1
⁓ sallying	16.2	2.0	5.2	10.6	6.5
Carnivores					
⁓ raptors	2.0	2.0	1.4	2.9	3.5
⁓ piscivores	0.1	0	0	0	0.3
Comparison with C13C					
⁓ unlogged (X^2 value)	-	43.2	84.6	242.7	131.6
⁓ p	-	<0.001	<0.001	<0.001	<0.001

Sample size is 800 individuals in all cases, except C13C (13 years postlogging) where sample size is 500 individuals.

UL = unlogged

Despite their rather broad base, guild comparisons do not always agree, even between closely-linked studies. At the Ulu Segama Forest Reserve in Sabah, some inter-site comparisons have shown that bark-gleaning insectivorous birds are less abundant in forest logged eight years earlier (57 percent canopy mortality) (Lambert 1992), whereas other comparisons have not (Grieser Johns 1996). In this case, the birds were reacting primarily to the local abundance of snags (standing deadwood), which provide both foraging substrates and breeding sites. The local abundance of snags is a largely unpredictable parameter in recently logged forests. Relationships of this sort can be quite subtle and may not be detected by simply comparing sites of different logging histories.

The Individual Species Level

The problems inherent in interpreting inter-site data on individual species for management purposes can be illustrated using an example from the Ulu Segama Forest Reserve, Sabah. In 1987 and 1988, Johns (1989b) surveyed and compared the bird communities in a primary forest and a nearby forest site that was logged six to seven years prior (57 percent canopy mortality). In 1989 and 1990, the study was duplicated by Lambert (1992), using the same primary forest site, the same logged compartment (albeit a section 1 km from that surveyed by Johns), and the same sampling techniques. Johns identified 15 species as decreasing significantly in logged forest and 46 as increasing. Lambert considered that 37 species may be decreasing and 32 increasing. The studies agreed on only six adversely affected species and 19 that benefited from logging.

At the level of individual species, the effects of logging operations are very site-specific. This is due to both the high degree of spatial heterogeneity in species abundances and the essentially random nature of logging damage (Johns 1988). Intersite comparisons cannot be sufficiently accurate in their documentation of single species responses to provide data interpretable in terms of management interventions, unless correlations can be drawn between specific population and habitat variables (which requires extensive replications and has proven elusive—no examples exist). Since more variables can be controlled in long-term monitoring studies, these will be best able to identify critical parameters within the ecosystem. While precise interventions may be site-specific, an understanding of the effects of logging on the ecosystem (which can only really be determined from long-term monitoring studies) is the basis for concession-wide management.

Applying Research Data to Forest Management

It may be tempting to develop blanket prescriptions based on the results of observations made in logged forest, but these may be neither justified nor viable. As an example, it is often stated by conservationists that all figs (*Ficus* spp.) should be marked for retention during logging operations and avoided during felling—with all the cost implications involved in surveying, marking, realigning skidroads, etc. This is based on the observation that many animals eat figs. In only two cases, however, has the abundance of figs through a range of unlogged and logged forest sites actually been correlated with the abundance of an animal species: a) redtail monkeys (*Cercopithecus ascanius*) in the Kibale Forest (Skorupa 1986), and b) lesser mousedeer (*Tragulus napu*) in the Ulu Segama Forest Reserve (Heydon 1994). The abundance of many fig-eating primates and birds has definitely been shown to not be affected by the loss of 50 percent or more of fig trees one to 20 years after logging (less than 60 percent canopy mortality) (e.g., primates and hornbills in Peninsular Malaysia and Sabah: Grieser Johns 1997). A more rigorous approach would be to determine from long-term monitoring whether reductions in animal density actually occur as a result of typical levels of fig tree loss, and only apply expensive protection measures in felling blocks where the local density of figs prior to logging is lower.

Similar arguments could be applied to interventions that retain snags, deadwood, and other forest resources. This type of research has been developed in determining the precise requirements of cavity-dwelling marsupials in sub-tropical forests in Australia (Lindenmayer et al. 1993), or of deadwood specialists in temperate forests (Sandstrom 1992), but is extremely rare. The main point is that research is needed to determine the critical point at which interventions should be applied. The research required by forest managers is highly specific—i.e., precisely what densities of which trees or other resources need to be retained in which logging blocks. If this is known, and if biodiversity conservation is to be incorporated into the forest management strategy, then forest managers can ensure that the necessary field and budgetary planning is made. Ideally, this information should be considered on a regional basis so the costs of intervention can be balanced by the importance of conserving the species. An expensive intervention might be justifiable at a particular site if the species concerned is rare and local, but not if it is widespread.

Research and Management Directions

Van Bueren and Duivenvoorden (1996) recently summarized the priorities for biodiversity research in supporting improved forest management. The following discussion considers specific recommendations for the implementation of research on the effects of logging on forest biodiversity, and management activities to reduce these impacts.

Geographical Priorities

It is important that baseline information gathering be focused in poorly known areas undergoing rapid expansion in logging activities. Central Indochina is a good example of a region where there are very few data on logging-biodiversity interactions (Kemp et al. 1995). Most studies of logging-biodiversity interactions have also been continental rather than insular, and the potentially different responses of the simpler island ecosystems require examination. Southern Melanesia and the Solomon Islands are particularly important in this respect, both in terms of their endemic biota and because they are currently a focus of logging activity (see chapter 24).

Long-Term Monitoring Programs

It is of paramount importance to establish more long-term monitoring programs to begin to generate the type of data needed by forest managers. Very few monitoring programs exist, although there are several baseline studies in areas that have since been logged that could be developed into long-term monitoring programs (see box 18-2).

Long-term monitoring programs (or pre-/postlogging biological inventories) have been recommended for forest managers as a means of monitoring the success of management measures (e.g., Blockhus et al. 1992; chapter 19). They do not, however, appear to have made their way into actual field manuals or forestry practices. Prospects for including biodiversity-oriented management procedures within forestry are currently limited since manuals need to outline specific activities while conservation-oriented guidelines are very general. This can only be addressed by designing manuals responsive to data inputs from long-term monitoring as they become available. The establishment of long-term monitoring programs and the production or updating of field manuals, therefore, should be a priority in many regions (see chapter 19).

Important sites for monitoring the effects of logging on biodiversity are:

- *Tekam Forest Reserve, Peninsular Malaysia.* Before/after study conducted in 1979 to 81, with subsequent periodic monitoring (51 percent tree loss during logging)

- *Pasoh Forest Reserve, Peninsular Malaysia.* Some comparative work, particularly on plant species, with a before/after study and subsequent monitoring initiated in 1996

- *Ulu Segama Forest Reserve, Sabah.* Much comparative work and scope for re-surveying recently logged areas for which baseline data were collected prior to logging (32–58 percent tree loss during logging)

- *Nanga Gaat, Sarawak.* Before/after study conducted in 1992 to 94 with scope for subsequent periodic monitoring, although results are liable to be confused by hunting (54 percent tree loss during logging)

- *Mt. Windsor, Queensland.* Before/after study conducted in 1984–87 with subsequent periodic monitoring (16–23 percent tree loss during logging)

- *Budongo Forest Reserve, Uganda.* Much comparative work and a before/after study with subsequent monitoring initiated in 1992 (more than 50 percent tree loss during logging)

- *Lopé, Gabon.* Before/after study initiated in 1990 with scope for subsequent monitoring (11 percent tree loss during logging)

- *Forêt de Kirindy, Madagascar.* Before/after study with subsequent monitoring initiated in 1990. [Actually a dry deciduous forest rather than a rain forest, but offering relevant data.] (10 percent tree loss during logging)

- *Chiquibul Forest Reserve, Belize.* Before/after study of conventional and reduced impact logging conducted in 1995 to 96, with subsequent monitoring planned (around 37 percent tree loss during logging)

The monitoring system discussed earlier for the Tekam Forest Reserve of Peninsular Malaysia, is an example of the type of system that can provide useful results for managers (Grieser Johns 1997; see chapter 19 for a detailed design of a model long-term monitoring program). Sufficient information is available from Tekam, and perhaps from other sites (see box 18-2), such that more directed monitoring could be applied to those taxa that respond adversely to logging. At

Tekam, data are available showing specific responses of species to specific environmental changes. Key indicators that could form a basis for management through the concession area and beyond (such as levels of logging damage that do not result in the loss of species) are identifiable. Based on the results of pre- and postfelling inventories, any required mitigatory measures may be devised and costed. If economically feasible, an argument exists for the application of such measures. It is then up to the forest authorities to amend concession agreements accordingly. This should be the ultimate aim of a monitoring program.

Reduced-Impact Logging

In establishing new monitoring programs in tropical rain forest regions, it should be kept in mind that the most useful new studies would record the effects of reduced-impact logging operations in locations where the political will and level of training available to forestry staff will allow any interventions that may be identified to be put into practice (chapters 19 and 20). An example of this type of work is the integrated approach adopted in logging experiments in the Chiquibul Forest Reserve of Belize, where reduced-impact logging plots are also monitored for changes in various animal groups (Manomet Observatory 1996a). Similar integrated work is proposed for reduced-impact logging experiments at Kribi in Cameroon (M. Parren, personal communication).

While the relationship between damage levels caused by logging and changes in biodiversity is by no means linear (Grieser Johns 1997), the introduction of reduced-impact logging is fundamental to both the improvement of forest growth and yield (Palmer and Synnott 1992; see chapter 28) and the improved retention of biodiversity. The basic principles of reduced-impact logging are well known and widely explored (see photo 18-3a and 18-3b; Bruenig 1996). Their immediate implementation would provide an effective base for improving conservation of forest biodiversity, although this should not in itself be regarded as sufficient to conserve all species, as is sometimes the case (e.g., U.K. government 1994). In terms of research input, opportunities for the fine-tuning of management's response to data collected on the habitat requirements of vulnerable plant and animal species may well exist—given that the implementation of such systems suggests a willingness on the part of forest authorities to embrace changes in logging techniques.

PHOTO 18-3a High- vs. low-impact harvesting: a) high-impact, unplanned "pretzel" skid trails on step forest terrain in Malyasia. (R. Z. Donovan)

PHOTO 18-3b *High- vs. low-impact harvesting: b) low-impact skid road in Malaysia (narrow width and minimal soil disturbance on sloping terrain). (R.Z. Donovan)*

ACKNOWLEDGMENTS

I am grateful to FRR, Ltd. for the use of its facilities while writing this paper. The manuscript benefited from the comments of the editors and four anonymous reviewers. Bettina Grieser Johns kindly reviewed and corrected the manuscript prior to final submission.

Chapter 19

PROGRAMS TO ASSESS THE IMPACTS OF TIMBER HARVESTING ON TROPICAL FOREST WILDLIFE AND THEIR HABITAT

Robert A. Fimbel, Elizabeth L. Bennett, and Claire Kremen

Efforts to assess the impacts of tropical logging on wildlife have provided many insights into the direct and indirect effects of harvesting activities on select forest fauna and their habitat (Grieser Johns 1997; Struhsaker 1997; see chapters 2–17 and 24). Most of these studies, however, have shortcomings associated with their designs (see chapter 18) that limit their capacity to predict and mitigate the impacts of logging on wildlife. Few include longitudinal (pre- and post-treatment) measurements at the same site, making it difficult to separate the responses to forest management activities from the normal spatial variation found in heterogeneous tropical forests. The absence of replication in many studies, short investigation periods, and limited geographic scales, compound the problem. Many examinations also focus on only one or a few species or taxonomic groups, leaving one to speculate how other organisms in the community are responding to forest management. Much research also continues to be performed by academic scientists who often fail to translate their findings to management practices. This condition manifests itself from their short-term involvement in projects, limited interactions with foresters, and failure to consider management needs and constraints in the design and recommendations of their studies. Finally, the value of including protected areas within concessions (see chapter 23), while mentioned as important refuges for some species, has received very limited quantitative evaluation. While rich in methods and instructive at a coarse level, most impact assessment studies are generally weak in their capacity to predict, track, and subsequently mitigate negative ecological impacts associated with forest management practices (Bawa and Seidler 1998; see chapter 18). The result is that foresters and logging firms have limited information with which they can evaluate and refine silvicultural interventions to ensure that impacts to wildlife and their habitat are minimized.

Recently, studies focusing on logging-biodiversity issues have progressed towards more complex pre- and post-investigations. They track the response of one or more taxonomic groups of animals and/or plants in an effort to overcome many of the shortcomings of past assessment research designs. In some cases, multiple silvicultural techniques are examined. These include comparisons of conventional harvests versus reduced-impact logging systems (see chapters 21 and 24), intermediate practices (thinnings, weedings; see chapters 2 and 21), and treatments mimicking natural disturbance events (see chapter 22). These research programs are most developed for temperate forest systems (e.g., Ristau et al. 1995, Manomet Observatory 1996b, National Audubon Society 1996, WCS 1998b). Current initiatives in Australia (Crome 1991), Belize (Manomet 1996a), Brazil (TFF 1995), Ecuador (WCS 1998a), Gabon (White 1992), Guyana (Tropenbos 1994), Indonesia (Sist el al. 1998b), Madagascar (Ganzhorn and Sorg 1996; Kremen et al. 1998a, b; Merenlender et al. 1998), Malaysia (Pinard et al. 1995; Dahaban 1996; Grieser Johns 1996), Venezuela (Mason 1996), and other sites (see chapter 18), however, are beginning to address these topics in the complex forest systems of the tropics (see table 19-1).

In an ideal world, detailed ecological, social, and economic evaluations of logging would be instituted in all concessions, with adjustments made to silvicultural prescriptions as practices detrimental to fauna and flora populations were identified. With few exceptions, however, management practices in tropical production forests make little effort to integrate biodiversity into their planning. This is due to financial, methodological, political, and staffing constraints (see chapters 18 and 28). To overcome these obstacles, impact assessments must be developed to provide resource managers with: a) a reasonable level of information about the impacts of their practices on biodiversity; and b) economically viable options for mitigating these treatments.

In this chapter, we outline a national program to assess the impacts of forest management practices on wildlife and their habitat within production forests. This approach proposes the use of a two-tiered system: one detailed *demonstration forest* in each major forest type of the country (e.g., White 1983, Letouzey 1985), complemented by *coarse-grained* assessment programs within the timber concessions comprising these forest types (see table 19-2).

TABLE 19-1 Biodiversity Assessment Studies Using Pre-/Postlogging (Logitudinal) Multitaxonomic Approaches to Evaluating Forest Management–Biodiversity Interactions, Including Components of the Demonstration Forest Model

Site Name	Chiquibul, Hill Bank	Several Sites	Cotacachi Cayapas Buffer	Lopé	Bulungan Model Forest	Masoala	Kalabakan and Ulu Segama	Tekam Forest Reserve	Demonstration Forest
Location	Belize	Brazil	Ecuador	Gabon	Indonesia	Madagascar	Malaysia	Malaysia	Anywhere
Forest Type	Subtrop. lower, montane moist	Lowland rain	Lowland rain	Lowland rain	Dipterocarp hill	Humid evergreen	Dipterocarp hill	Dipterocarp hill	Production forests
Managing Agency	Ministry N.R., Manomet, Prog. for Belize	TFF, IBAMA	USAID, CARE, WCS, Eco-Ciencia Jatun Sacha	WCS, ECO-FAC	CIFOR, WCS, ITTO, MOF, Inhu-tani II	WCS, CARE, ANGAP, DEF, TPF	ICSB, NEES, CIFOR	FRIM	Govt., NGOs, universities
Initiation of Studies	1994	1992	1998	1989	1998	1991	1992	1979	
Studies continue?	intermittently	yes	yes	yes	yes	yes	yes	intermittently	
Silvicultural Harvesting Systems	mechanized	mechanized	mechanized/ unmechanized	mechanized/ mechanized	mechanized	unmechanized	mechanized	mechanized	
Selection	x	x	x	x	x	x	x	x	possible
Seedtree									possible
Shelterwood	planned								possible
Clearcut									possible

TABLE 19-1 Biodiversity Assessment Studies Using Pre-/Postlogging (Logitudinal) Multitaxonomic Approaches to Evaluating Forest Management–Biodiversity Interactions, Including Components of the Demonstration Forest Model (continued)

Site Name	Chiquibul, Hill Bank	Several Sites	Cotacachi Cayapas Buffer	Lopé	Bulungan Model Forest	Masoala	Kalabakan and Ulu Segama	Tekam Forest Reserve	Demonstration Forest
Silvicultural Intermediate Practices									
Thinning		planned	planned		planned		climber cutting		possible
Weeding		climber cutting	planned			climber cutting	climber cutting		possible
Enrichment	x	planned	planned		x	x	x	on damaged sites	possible
Reduced Impact Logging (RIL)	x	x	x		x	x	x		x
Taxonomic Groups Investigated									
Animals									
Lg. terrestrial animals	x		x	x	x	x	exploratory		x
Primates	x		x	x	x	x	exploratory		x
Rodents			x	x	x		exploratory		x
Birds	x		x		x	x	exploratory		x
Reptiles					planned				x

TABLE 19-1 *Biodiversity Assessment Studies Using Pre-/Postlogging (Logitudinal) Multitaxonomic Approaches to Evaluating Forest Management–Biodiversity Interactions, Including Components of the Demonstration Forest Model (continued)*

Site Name	Chiquibul, Hill Bank	Several Sites	Cotacachi Cayapas Buffer	Lopé	Bulungan Model Forest	Masoala	Kalabakan and Ulu Segama	Tekam Forest Reserve	Demonstration Forest
Amphibians			x						x
Invertebrates	butterflies		earthworms		planned insects				x
Other		aquatic invert.							x
Vegetation									
Trees	x	x	x	x	x	x	x	x	x
Shrubs			x	x	x	x	x		x
Herbaceous			x	x	x		x		x
Woody regeneration	x	x	x	x	x	x	x		x
Other			ephiphytes		plt. funct. attrib.	phenology	lianas		phenology
Habitat Structure									
Gaps	x			x	x	x			x
Canopy Cover	x		x	x	x	remote sensing	roads and yards	x	x
Related Studies									
NTFP	x		x		x	x	limited		x

TABLE 19-1 Biodiversity Assessment Studies Using Pre-/Postlogging (Logitudinal) Multitaxonomic Approaches to Evaluating Forest Management–Biodiversity Interactions, Including Components of the Demonstration Forest Model (continued)

Site Name	Chiquibul, Hill Bank	Several Sites	Cotacachi Cayapas Buffer	Lopé	Bulungan Model Forest	Masoala	Kalabakan and Ulu Segama	Tekam Forest Reserve	Demonstration Forest
Rare/threatened spp			x		x	x			x
Keystone/Indicator spp			x		x	x			x
Hunting			x		x	x			x
Economics		x	x		x	x	limited	x	x
Other		training center	training		training	socio-economics	C-offset	hydrology studies	training
References	Manomet 1996a	TFF 1995	WCS 1998a	White 1992	Sist et al. 1998	Kremen et al. 1998	Pinard et al. 1995	Greiser Johns 1997	This chapter

Acronyms

ANGAP	Association National des Aires Protégées	MOF	Ministry of Forestry
CIFOR	Center for International Forestry Research	NEES	New England Electric Systems
DEF	Direction des Eaux et Forêts	NTFP	Non-timber Forest Products
ECOFAC	Conservaton et Utlisation Rationnelle des ÉCOsystemes Forestiers d' Afrique Centrale	TFF	Tropical Forestry Foundation
		TPF	The Peregrine Fund
FRIM	Forest Research Institute Malaysia	WCS	Wildlife Conservation Society
IBAMA	Instituto Brazileriro do Meio Ambiente e dos Recurso Naturais Renováveis		

TABLE 19-2 *Two-Tiered Approach to Wildlife Assessment in Production Forests, with Brief Descriptions of the Goals, Size, Duration, Methods, and Outputs Associated with These Programs*

	Major Forest-Type Level: The Demonstration Forest	The Concession Level: Two Wildlife Conservation Programs	
		Wildlife Reconnaissance for Protection Area Planning	Wildlife Inventory and Monitoring Program for Impact Assessment and Management Evaluation
Goal:	To assess (through inventories and monitoring) the impacts of forestry practices on biodiversity, and to develop management practices that effectively reduce ecological impacts while maintaining the economic viability of the silvicultural treatments	To identify potential areas for protection in the concession/management unit	To track the responses of target species and habitat variables to site-specific silvicultural treatments
Size:	5,000 to 10,000 hectare mix of production forest and reserve areas. One site per major forest type within production forests	Entire concession (above a minimum size specified by the host country)	Entire concession (above a minimum size specified by the host country)
Duration:	Long-term permanent plots	One-time evaluation of area during management planning phase. Actual assessments may occur prior to or in conjunction with timber inventories	Pre-/postsilvicultural treatment areas are inventoried and monitored during life of concession

	Major Forest-Type Level: The Demonstration Forest	The Concession Level: Two Wildlife Conservation Programs	
		Wildlife Reconnaissance for Protection Area Planning	Wildlife Inventory and Monitoring Program for Impact Assessment and Management Evaluation
Methods:	Pre-/post-quantitative evaluations of select plant and animal taxonomic groups to conventional and RIL forest management techniques, and control areas, using replicated design.	Remote sensing and on-the-ground rapid appraisal techniques that utilize 'indicators' (species, habitat types, etc. developed at the demonstration forest) to identify areas that are of special conservation value and thus potential reserve status.	All diameter classes of important tree (timber and non-timber forest product species), wildlife (game and rare/threatened species), and indicator species (known keystone and sensitive species), are inventoried in advance of cutting activities. Sampling designs are simple and done in conjunction with timber inventories. Sites are revisited at periodic intervals to track species responses to disturbances.

TABLE 19-2 *Two-Tiered Approach to Wildlife Assessment in Production Forests, with Brief Descriptions of the Goals, Size, Duration, Methods, and Outputs Associated with These Programs (continued)*

	Major Forest-Type Level: The Demonstration Forest	The Concession Level: Two Wildlife Conservation Programs	
		Wildlife Reconnaissance for Protection Area Planning	Wildlife Inventory and Monitoring Program for Impact Assessment and Management Evaluation
Outputs:	Documentation of fauna and flora structure/composition in representative logged and unlogged forest; evaluation of different taxonomic groups as "indicators" of disturbance; development of relatively simple yet effective techniques to evaluate the impacts of forest management practices on biodiversity; refine silvicultural treatments so that they maximize biodiversity conservation while still remaining economically attractive; and, train nationals to implement, analyze, and interpret forest conservation research	Identification of potential reserve sites within the concession that will serve as refugia during harvesting activities and sites from which recolonization of post-disturbance area can occur. They also serve as control areas to evaluate the impacts of the management plan prescriptions on economically important and/or rare species	Quantification of impacts of management prescriptions on select important biological species allowing planners, foresters, and certification groups to modify existing silvicultural practices where impacts from these treatments are found to negatively affect important commercial, ecological, or rare plant and animal species. Information is complementary to the demonstration forest data, helping to broaden the level of understanding about forest dynamics in that forest type

A Two-Tiered Approach to Assessing Logging Impacts on Wildlife and Their Habitat in Production Forests

To assess the impacts of different forest management practices on wildlife and their habitat, we suggest establishing one demonstration forest per major forest type, across a national forest system. Detailed studies of forestry-wildlife interactions at the demonstration forest should be complemented with less-intensive impact assessments designed to describe and track select taxa within concessions above a minimum threshold size. The minimum concession size would be determined by each country (e.g., areas 1,000–5,000 hectares or larger might be an appropriate starting point), and depend on the average size of concessions, diversity of forest communities, potential to contain threatened and endangered species, and the extent of the national forest system and adjacent protected areas.

The Demonstration Forest

Each demonstration forest would represent an in-depth biological and economic examination of one representative area of a major forest type (see table 19-2). It would be designed to reflect the *typical* biotic, edaphic, topographic, climatic, and socio-economic factors that characterize management activities in that forest. Within the demonstration forest, researchers and foresters would attempt to identify *what* impacts conventional and reduced-impact logging (RIL) systems have on biodiversity, *why* the animal and plant populations respond in the ways that they do to these treatments, and *how* to modify future forestry prescriptions to better conserve the biological richness within the forest. Box 19-1 outlines the research objectives of a model demonstration forest. The complexity of these studies, coupled with long-term commitments of time, trained personnel, and funds, negate the likelihood of their application to more than a very few sites within a country. Priorities would be given to forest types with: a) few/no totally protected areas, b) a high percentage of their area slated for silvicultural interventions, and/or c) known or suspected areas supporting high levels of biodiversity and/or endemism (see chapter 20). In time, the goal is for all major production forests to support a demonstration forest.

Box 19-1 *Objectives of wildlife-forestry studies within a model demonstration forest.*

The goal of the demonstration forest is to provide information about the impacts of natural forest management practices on biological communities to resource managers, ecologists, conservation biologists, certification officials, and policy makers. To achieve this goal, longitudinal (pre-/ post-treatment), multitaxa studies are conducted in an effort to:

- Quantify the *structure* and *composition* (flora and fauna) of managed and unlogged (control) areas of a country's major forest types

- Establish *pre-treatment baseline inventory* data on mammals, birds, herpetofauna, invertebrate communities (e.g., butterflies, earthworms, ground beetles, and spiders), and the vegetation supporting these organisms, and to monitor their long-term (minimum of one rotation period) *post-treatment responses* to conventional and reduced impact management systems (compared to natural background variation in the control)

- Evaluate the relative merits of different taxonomic groups as primary and secondary "indicators" of environmental impacts associated with silvicultural activities

- Clarify the contribution of a *protected area network* (corridors and reserves) to maintain viable populations of sensitive species, and their capacity to recolonize harvested forests from these reserves during stand recovery

- Develop relatively *simple yet effective techniques* to evaluate the impacts of logging and other silvicultural treatments on forest fauna and flora, thereby assisting foresters and "eco"-certification organizations in their capacity to evaluate environmental impacts associated with forest management activities

- Provide forest departments with *silvicultural options* that maximize the conservation of biodiversity across the landscape, while maintaining commercially important timber species, rare/threatened/endangered species, and other animal or plant species of special interest in the forest composition

- Serve as a *demonstration area* where forest stakeholders have an opportunity to evaluate first-hand efforts at achieving sustainable forest management activities and their potential to conserve forest resources

A Model Demonstration Forest at the Major Forest-Type Level

Within a demonstration forest, a suite of studies would be designed to evaluate the responses of animal and plant communities to silvicultural prescriptions. Treatments would be applied to a patchwork landscape containing areas of active management interspersed with small reserves (see figure 19-1)—a concept modified from the *old-growth islands* design introduced by Harris (1984). Control sites would also be established in undisturbed areas to monitor background variation in study parameters due to climate and natural disturbances processes.

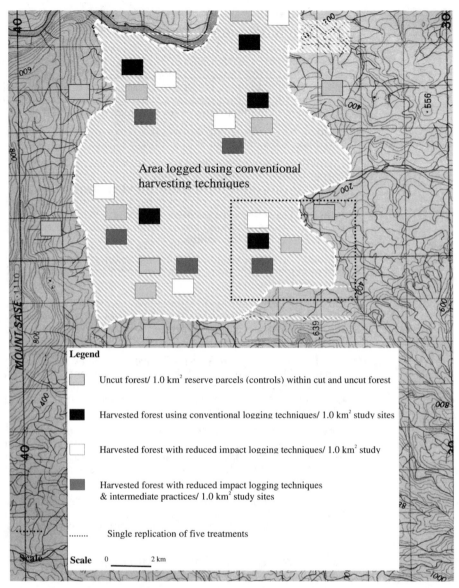

FIGURE 19-1 *Theoretical Demonstration Forest showing conventional and reduced-impact management areas, and control sites in adjacent reserve areas. In addition to studies evaluating the impacts of silvicultural prescriptions on wildlife and their habitat, the design also allows for an assessment of small "within" concession area reserves, and an evaluation of protected areas along stream corridors. Map modified from Royal Australian Survey Corps. 1978. Karimui Papau New Guinea 1:100,000 topographic survey.*

There is no single experimental design that will fit all research and management criteria. A reasonable design, however, might include 1.0 km^2 (100 hectare) treatment parcels as the core of the impact assessment program (see figure 19-1). Using this design, five treatments might initially be evaluated:

— Control (area of unharvested forest near to harvest sites)

— Reserve within the harvest landscape

— Areas harvested using conventional logging techniques

— Areas harvested using reduced-impact logging techniques (see chapters 21 and 24)

— Areas harvested using reduced-impact logging techniques, with follow-up silvicultural treatments (e.g., enrichment plantings and/or thinnings; see chapter 2)

All monitoring activities would be nested within the above parcels, at least 100 m from the edge of the treatment boundary (an area of approximately 64 hectares, at the center of each parcel, is used to evaluate that particular treatment while minimizing edge effect). Treatments would be replicated across the demonstration forest landscape in advance of silvicultural activities (see figure 19-1), with sites as uniform as possible in edaphic features and biotic composition.

Ideally, parcels and their interior monitoring measurements would be established at least two to three years in advance of silvicultural treatments, to allow adequate time to monitor plant and animal populations under *normal* (undisturbed) environmental conditions. The monitoring process would continue during the period of silvicultural treatment, and for two to three years directly following these activities. After this point in time, monitoring might proceed at intervals of approximately once every five years, for a period of at least one cutting cycle (20–40 years). This would allow adequate time to assess the impacts of the initial silvicultural treatment, and the response of the habitat and taxonomic groups under evaluation during the post-treatment recovery period.

Within the parcels, a number of habitat characteristics and fauna taxa would be inventoried and monitored to assess their response and recovery to natural and forestry-related disturbance. These characteristics would include:

— Composition of the habitat (abundance and size distributions of trees, lianas, shrubs, epiphyte loads, and regeneration)

- Structural attributes of the habitat (coarse woody debris, canopy cover and closure, vertical strata, depth of soil litter and decomposition gradient, soil characteristics, and topographic features)
- Processes influencing the habitat (phenology, seed dispersal and predation, seed germination, herbivory and litter decomposition)
- Select fauna taxa, including mammals, birds, herpetofauna, and components of terrestrial and aquatic invertebrate community (e.g., earthworms, butterflies, carabid beetles, damselflies, etc.)

Methodologies to inventory and monitor select fauna taxa are species- and season-dependent, with the taxonomic groups in table 19-3 recommended for evaluation because they are relatively well described, have demonstrated responses to silvicultural treatments in previous studies (Ganzhorn and Sorg 1996; Grieser Johns 1997; Struhsaker 1997; see chapters 4–14), and trained personnel are often available to identify specimens. An example of an inventory and monitoring program to assess the impacts of sustainable forest management activities on forest biodiversity, using the variables noted above, is discussed in WCS (1998a).

Wild Meat Evaluations

In most tropical countries, laws exist that regulate the taking of wild meat from public lands. In reality, they are seldom enforced, and hunting may have a greater impact on game species than habitat disturbance caused by silvicultural practices (see chapters 15–17). Distinguishing between the direct impacts resulting from forestry and the indirect impacts due to hunting may be difficult for many bird and mammal species. To remove this confounding effect, the demonstration forest must be either intensively protected from hunting (all impacts will then be associated with management actions), or wildlife harvests must be monitored. The former scenario is preferred, as it will allow for a clear evaluation of silvicultural impacts. Where the latter scenario prevails, discussions within Robinson and Bennett (1999) describe methods for monitoring hunting activities (hunter follows, village surveys, etc.).

Wildlife Assessment Programs in Timber Concessions That Complement the Demonstration Forest

The strengths of the demonstration forest approach are also its weaknesses—detailed studies create financial and logistical constraints that restrict their widespread application. This limits the information available to managers, certifiers, planners, and policy makers striving to manage forests sustainably at both the landscape and concession levels. We, therefore, recommend that demonstration forests be complemented by biodiversity conservation programs at the concession level (see table 19-2). These *within-concession* programs would be

TABLE 19-3 *Fauna Taxa to Be Assessed Within the Demonstration Forest, and Suggested References for Protocols to Conduct These Inventory and Monitoring Activities*

Taxonomic Group	Inventory and Monitoring	Select References[a]
Large mammals	linear transects for sign (tracks, dung, trails, nests), calls, and sightings	Rabinowitz (1993), Wilson et al. (1996), White and Edwards (in press), Fimbel et al. (1999)
Small mammals and marsupials	trapping, mark and recapture	Rabinowitz (1993), Ristau et al. (1994), Wilson et al. (1996), Fimbel et al. (1999)
Birds	aural point or transect counts, mist-netting	Keskimies and Vaisanen (1986), Baillie (1991), Roberts (1991), Ralph et al. (1993), Fimbel et al. (1999)
Herpetofauna	pitfall traps, visual encounter surveys, litter searches	Heyer et al. (1994), Ristau et al. (1994)
Select terrestrial invertebrate groups	sweep nets, litter and soil searches, baited and pitfall traps	Thomas (1983), Klein (1989), Kremen (1992), Kremen et al. (1993), Oliver and Beattie (1996), Rykken et al. (1997), Fimbel et al. (1999)
Select aquatic invertebrate groups	kick nets, bottom samplers	Kerans and Karr (1994), Fore et al. (1996)

[a]The reader is encouraged to review chapters 4 to 14 of this volume for a more thorough review of methods used to assess the impaces of silvicultural prescriptions on the different taxonomic groups listed above.

designed to gather information on practices that minimize the impacts of silvicultural activities on target and nontarget species of flora and fauna—*prior to and during* timber management operations.

Biodiversity Reconnaissance for Protection Area Planning in a Concession

Prior to, or shortly following, the awarding of a concession, forest departments should initiate a planning process to assess areas to harvest and reserve within that concession (see table 19-2). Using a combination of remote sensing (aerial photographs), topographic maps, and historical inventories, potential sites that might be placed in reserves include: steep slopes (more than 30 percent for skidder operations, more than 60 percent for cable systems), riparian areas (typically 10–50 m wide buffer strips for all permanent streams and rivers), areas of low commercial stocking, and areas supporting biologically or socially important sites (Sist et al. 1998b; see chapters 14, 21, 23, and 24). From a biological perspective, important areas for protection are those supporting populations of rare and endangered species; high concentrations of endemic species and/or exceptional species richness; unusual vegetation, landforms, geology, or other physical features not well represented in nearby protected areas, and areas serving as corridors between reserves (Sayer and Wegge 1992; see chapter 23). Corridors should include some areas of commercial stocking, in an effort to represent a portion of all forest-types with a natural condition. In total, reserves might account for ten to 30 percent or more of the total concession area (incentives for the establishment of these areas are discussed in chapters 23, 27, and 29).

Where remote information is limited, on-the-ground rapid appraisals that focus on known habitat-types for rare/threatened/endangered species, keystone species, important pollinators, and seed dispersers, should be undertaken. Government wildlife biologists, scientists affiliated with a demonstration forest, and conservation NGOs could contribute to both the planning and ground reconnaissance portions of this land-use mapping process.

Preharvest Inventories of Annual Harvest Sites

In most tropical countries, loggers are required to conduct stock surveys of all commercial timber trees over the minimum cutting diameter, thereby quantifying the merchantable (or usable) volume and helping to determine the location of access routes (R.A. Fimbel, personal observation). These inventories create a framework within which baseline data on select wildlife species (threatened, endangered, game species, and species of ecological importance for forest health and productivity), their habitat, and select ecological processes (see chapter 3) could be established. Simple, rapid appraisal techniques, using transects and/or plots (the same transects that are created during the stocking surveys), would be used to collect information on animal sightings, their sign (tracks, scat, nests, trails, etc.), and habitat (general structure and composition). Measuring ecological processes will be a bit more difficult. Regeneration and decomposition are two processes that indirectly measure the health of numerous organisms and their habitat (e.g., pollinators and dispersers involved in regeneration; earthworms, soil fauna, and microbes involved in decomposition; see photo 19-1). These two processes are relatively easy and inexpensive to measure (phenology studies and regeneration sampling, litter bags to access decomposition rates), and are of direct importance to resource managers as they impact the quality and productivity of the forest. This information serves two purposes: a) to help refine the selection of harvest and reserve sites (water courses, wetlands, rock outcrops, areas of unique wildlife habitat, sites of religious and cultural importance, etc.) begun in the concession-wide evaluation (see above), and b) to help establish baseline biological information against which responses and recoveries of select wildlife parameters to logging can be evaluated. The latter step is critical to refining silvicultural prescriptions so that they minimize their impact to biodiversity. Finally, methods to conduct these procedures are discussed in Sobrevila and Bath (1992); Tangley (1992); Rabinowitz (1993); Beattie and Oliver (1994); Crump and Scott (1994); Colwell and Coddington (1995); Oliver and Beattie (1996a); WCS (1998a); Fimbel et al (1999); and White and Edwards (in press).

In some forests, wildlife species may be difficult to survey along transects, especially if they are subject to high hunting pressure. At these locations, resource managers could conduct basic sociological surveys of local game consumption to complement the biological surveys noted above (Marks 1996). Efforts at refining these assessment techniques continue (Stork et al. 1997; Boyle et al. 1998; Feinsinger in press).

PHOTO 19-1   *Wildlife impact assessment methods and biodiversity conservation principles should be a part of every forestry student's education (students at fauna field techniques training course in Sabah). (WCS)*

Postharvest Monitoring

It is important that post-harvested areas be monitored to assess the response of select commercial, ecological, and threatened species, their habitat, and associated ecological process. The information generated in these programs will help planners, foresters, concession holders, and wood certifiers assess the impacts of silvicultural prescriptions on the general health of the forest concession. This, in turn, would identify where modifications to existing forestry practices are needed to enhance the conservation value of post-treatment areas.

There are several options available for developing a post-harvest monitoring program. The easiest way to accomplish this would be to permanently mark the transects/plots used in the pretreatment survey, and then periodically remeasure them using the same techniques. A second option would be to simply remeasure selected sites in the same areas of the initial surveys, reapplying the methods used during the pretreatment period. While different microsites can contribute to different observations (thus confounding treatment locations with normal background environmental variation), numerous replications across the concession can help to minimize this *noise* in the design.

The frequency at which one would remeasure an area varies with the variable under evaluation and the postcut use of the area. Sampling in the short-term (one to two years postharvest), followed by periodic monitoring at punctual intervals (every three to five years), would be sufficient to track the response of most target species and processes to the recovery process. For animals, it is important to re-sample during the same season to remove this potential confounding factor from the data interpretation (the dry season is preferable for mammal and bird surveys as these animals tend to be inactive—and therefore difficult to detect—when it rains). Species of special interest (e.g., those listed as threatened or endangered) might require more frequent monitoring, and during several seasons throughout the year, depending on their biological habits. As animal and plant populations (and ecological processes) appear to stabilize, the frequency of surveys might be further reduced to once every five to ten years. Changes in land use activities and/or hunting pressure, however, would necessitate a resumption of more intensive monitoring activities.

Intensity and Interpretation of Concession-Level Wildlife Assessments

The effort put into concession-level wildlife assessment programs will vary with the forest type, the location of the concession within it, the proximity of management activities to totally protected areas with similar biotic and edaphic (soil and topography) characteristics, and adequate incentives to motivate the implementation of these activities. Concessions having a high conservation value (high endemism, serve as a corridor between two large protected areas, or contain rare/ threatened species) would be more closely inventoried and monitored than concessions supporting forest types that are well-represented in the demonstration forest or totally protected areas. The selected impact assessment design would need to reflect direction from the demonstration forest staff, with input from foresters, the concession-aire, and local stakeholders (see photo 19-2). This last step is integral to refining sustainable forest management techniques at the local level.

Incentives will be necessary to promote wildlife assessment programs in concessions. Where concessions occur in areas having a low biodiversity conservation priority (see chapter 20), reconnaissance and monitoring efforts would be minimal and supported by the concessionaire as part of their contract with the forest owner (along with appropriate penalties if the activities are not properly undertaken). Interactions between the concession and demonstration forest staff

PHOTO 19-2 *Local stakeholders can greatly influence the indirect effects of timber harvesting operations on wildlife. Their involvement in management activities is a critical step in building a long-term program of sustainable forest resource use. (D. Gillison)*

would help to maintain transparency in this process. Where concessions are expected to serve a higher biodiversity conservation role, these areas might be targeted for certification (see chapter 26), and the incentives they can impart on concession holders. Finally, in areas of high conservation value, direct donor support (carbon-offset programs supporting natural forest projects; see chapter 27), and/or tax incentives (see chapter 29) could help to offset costs incurred from more intensive monitoring designs and the loss of volume associated with the creation of reserves *within* the concessions. These options are discussed further in chapters 26 to 30.

Data analysis and interpretation would be done in conjunction with personnel at the demonstration forest to assure that the maximum number of management options are explored, and to maintain strong communication links between the two tiers of the impact assessment program. The demonstration forest could also act as an independent source for verifying the quality of data collected and as a final repository for the long-term storage of biodiversity information within the forest type. This information would complement the findings of the demonstration forest evaluations, broadening the level of understanding about forest dynamics in the wake of silvicultural treatments.

Translation of Wildlife Assessment Program Findings Into Forest Management Prescriptions

"The best-laid schemes o' mice an' men" (from Robert Burns' poem *To a Mouse*, Burns 1966) are worthless unless they can be translated into action. At present, ecologists and foresters possess the technology to institute reasonably sophisticated animal and plant assessment programs in tropical forests. These sampling procedures are capable of determining the impacts of silvicultural treatments on biological communities, and generating the information necessary to refine forestry practices toward a more sustainably managed condition. Obstacles to these processes, however, exist in most tropical forestry programs.

Foresters must receive training and support so that they can be integral partners in the biodiversity assessment process. This will help to ensure that research findings are translated into management actions. Training and involvement should begin during university training, with biometric and silviculture classes modified to incorporate biodiversity assessment procedures and conservation biology principles (see photo 19-3). Students should also be encouraged to work seasonal positions or undertake short-term projects at a demonstration forest site. Following their formal training, all foresters should spend time collecting, analyzing, and interpreting data from a demonstration forest or a concession biodiversity survey. Finally, once a forester begins working in a concession, that individual should be encouraged to visit the nearest demonstration forest several times each year for assistance with their biological data analysis/interpretation and get further training in new sampling and application techniques, to re-instill in them a sense that their participation is intricate to the success of the sustainable forest management in their country.

Also key to translating wildlife and habitat assessment findings into forest management is the involvement of forest supervisors, influential government officials, and concession owners in all aspects of the biological assessment and conservation effort. Research performed in a vacuum remains in a vacuum. What motivates people to action is a feeling that they have a vested interest in the process. Researchers and foresters, therefore, must *market* efforts to improve forest management practices. An important step in this direction is to bring forest stakeholders to demonstration forests, where they have an opportunity to witness and evaluate efforts at balancing economic and ecological conditions through good forestry.

PHOTO 19-3 *Measuring animal sign and soil decomposition processes in India. (U. Karanth)*

Lastly, a partnership of government, industry, and researchers should be capable of implementing the necessary biodiversity assessment programs and their translation to management. Such a trio, however, lacks a system of checks and balances on management, and therefore may fall short of accomplishing its objectives of conserving biodiversity across the national forest system. An independent group is needed to periodically inject new perspectives, and if necessary, watch over this process (see chapter 26). Certification organizations might offer the best opportunity to provide this independent input. Their current focus is on promoting SFM, of which biodiversity assessment is one tool by which to measure compliance with their criteria. As a part of this effort, these organizations are encouraging the certification of domestic NGOs and foresters to serve as their representatives. With limited additional effort, their charge could be expanded to include involvement at the assessment stages as well. An independent host country NGO linked to an international network of certification and conservation-oriented groups, with a mission to advance sustainable forestry and biological conservation, would probably be in the best position to serve as a catalyst to motivate the implementation of biodiversity assessment programs at all levels (from promoting policy in government agencies to certifying local technicians working in concessions). Other incentives for promoting sustainable forest management and wildlife conservation in production forests are outlined in chapters 21 and 26 to 30.

Summary

A two-tiered approach to inventorying and monitoring forest fauna and their habitat seeks to overcome design limitations in many historical and current assessment programs. This comprises one demonstration forest site per major production forest type (level I), in combination with less intensive biological surveys at each concession within the same forest (level II). This strategy draws strength from several quarters. Scientifically, it presents an opportunity to evaluate numerous sites within a production forest, using replicated designs that incorporate a diverse array of plant and animal groups, and which allow comparisons between different areas of the forest receiving silvicultural treatments and those protected from forestry interventions. On the management level, it integrates foresters, researchers, concessionaires, local communities, government officials, and NGOs into all facets of the biodiversity assessment and conservation program. Finally, the program begins to address a growing global demand for sustainable forestry practices, offering opportunities to evaluate the impacts of current management prescriptions while identifying methods to bring economic and ecological considerations into greater balance.

The success of a national forest biodiversity assessment program relies on several conditions:

- Biodiversity is essential to the health of natural forest systems and the overall value of the resource to humans
- Governments are interested in monitoring the impacts of logging on biodiversity, with the goal of improving forest management practices, provided adequate funds are available to train personnel and oversee the assessment process
- Logging firms will comply with best logging practices and conserving biodiversity, provided costs associated with these activities do not make logging unprofitable (see chapter 28)

In the absence of these conditions, it will be extremely difficult to conduct sustainable forestry and to conserve biological diversity within production forests (see chapter 20). Fortunately, there is a growing trend towards recognition that the diverse commercial and nonmarket values associated with natural forests (Wilcox 1995, Myers 1996, Bawa and

Seidler 1998) are compatible with forestry practices (Dickinson et al. 1996), beginning the process of sustainable forest management and resource conservation for future generations.

ACKNOWLEDGMENTS

We wish to thank participants in the 1996 Bolivia Logging-Wildlife Workshop for the thoughts and comments that lead to the development of this chapter (especially Eliana Flores, Alejandro Grajal, Daniel Guinart, Hugo Hurtado, Lilian Painter, Andrew Plumptre, Steven Rosholt, Andrew Taber, and Andrew Whitman). We also would like to acknowledge the anonymous reviewers and contributors to this volume who took the time to share their insights and suggestions on early versions of this chapter. This work was possible with the support of the Wildlife Conservation Society.

Part V

FOREST MANAGEMENT PROGRAMS TO CONSERVE WILDLIFE IN PRODUCTION FOREST LANDSCAPES

Wildlife habitat in tropical forests is being degraded at an unprecedented rate, due in large part to timber exploitation activities. Direct
and indirect environmental impacts associated with these forest management policies (detailed in parts I through III) can be greatly reduced
by applying existing forestry technologies. In many areas, economic and
political incentives currently exist for promoting the adoption of sustainable forest management (SFM) and wildlife conservation measures
within tropical forest landscapes managed for timber production (see
part VI). In this section of the book, contributors discuss resource management techniques available to accomplish these conservation goals
and where they should be focused within the tropical forest landscape.
Options range from technical silvicultural interventions, to *set aside*
areas of natural forest as reserves and corridors for wildlife. Examples
of programs that integrate these conservation alternatives at the landscape and local level are furnished. The presented recommendations
should be viewed as possible models, rather than absolute guidelines,
and will require adaptation and refinement to local conditions.

Options for SFM and Wildlife Conservation in Managed Tropical Forests

Peter Frumhoff and Elizabeth Losos open Part V with chapter 20
and point out that no *one-size-fits-all* solution exists to the problem of
conserving wildlife in tropical production forests. They emphasize that
site-specific ecological and socioeconomic conditions must be taken into
account when designing effective conservation strategies. Their model
for setting priorities for conserving wildlife and its habitat moves
beyond current debate on the importance of sustainable forest management, and identifies criteria for areas where natural forest management
and wildlife conservation are likely to succeed. Conservation options
for production forests where sustainable forest management is not likely
to be achievable in the short term are also discussed.

Douglas Mason and Jack Putz review a suite of silvicultural options
in chapter 21 that are effective in reducing the negative environmental
impacts of forestry practices on wildlife communities. The chapter
describes some of the approaches that forest managers, educators,
conservationists, and governments can take to facilitate the transition
from unplanned to well-managed timber harvesting, including:

⁻ Planning to minimize forestry-wildlife conflicts at the landscape level (e.g., logging vs. reserve areas within the management matrix, rotations that allow adequate stand recovery, etc.)

⁻ A step-by-step guide to reduce the environmental impacts of management practices at the timber concession level

Recommendations offered by the authors apply to a wide range of silvicultural treatments—from planning, to felling and extraction. Finally, the conservation benefits and constraints of these actions for wildlife are outlined.

Colin Chapman and Robert Fimbel present a conceptual framework in chapter 22 that links the evolutionary and ecological processes of natural disturbance to forest recovery following timber harvesting in tropical forests. Three factors influence the recovery of the forest after logging: the evolutionary adaptability of the species, the local ecological conditions, and the level of impact associated with the logging operation. The authors propose that the ability of forest plant and animal species to rebound following logging is the result of natural selection that has operated in response to the types of disturbances these species have experienced throughout their evolutionary history. The potential of this evolutionary framework to help resource managers select silvicultural interventions that benefit harvestable species while maintaining key ecosystem processes is discussed along with future research needs in this area.

In chapter 23, Bruce Marcot, Ted Gullison, and James Barborak complement the technological conservation recommendations presented in chapters 21 and 24 with a discussion of low-intensity forestry conservation measures involving small reserves and wildlife corridors within the production forest matrix. The importance of these *within* and *between* forest concession refugia for biodiversity is examined, along with current efforts at protecting these habitats. Research priorities for refining reserve designs within managed landscapes are presented.

William Laurance presents a model forestry program in chapter 24, using many of the principles described in chapters 21 and 23. The Queensland Selective Logging System—the only reduced-impact logging method ever used on a long-term commercial basis in the tropics—is examined along with its relevance for wildlife conservation. Research on terrestrial vertebrates suggests that most species can persist in selectively logged forests, although many exhibit short- or medium-term declines in population density. Certain species of opossums, birds, frogs, reptiles, insectivorous bats, and cavity-nesting vertebrates appear to be especially sensitive to logging. Little is known of the long-term effects of repeated logging on these species, and recommendations are provided for improving upon the silvicultural practices in this system.

In chapter 25, Nick Salafsky, Max Henderson, and Mark Leighton present the opposite end of the timber-harvesting spectrum from conditions outlined in chapter 24. They compare and contrast the wildlife conservation potential of small-scale, community-based logging to large-scale, industrial operations. Issues range from resource ownership to direct and indirect environmental impacts associated with these activities. The authors argue that using chainsaws as a tool for conservation makes sense in the context of large-scale external threats to the forests, provided that communities are well organized and have a strong sense of stewardship towards their natural resources. Communities also need to better monitor and evaluate their management actions.

Practices That Promote Sustainable Forest Management and Wildlife Conservation in Commercial and Community Managed Tropical Forests

The chapters in this section consider a number of policy and management options that resource managers can apply *today* to promote wildlife conservation within managed tropical forest landscapes. Frumhoff and Losos begin in chapter 20 by pointing out that well-designed and implemented natural forest management programs are one of several possible conservation strategies available for application to tropical timber production forests. They also stress that efforts to promote these well-managed programs everywhere are likely to be counter-productive and costly. They recommend that improvements in forest management practices be limited to those locations where the forest does not warrant total protection for biodiversity conservation, and where sufficient economic incentives and/or enforcement capacity exist to promote long-term timber production.

In chapter 21, Mason and Putz provide an exhaustive review of cost-effective silvicultural measures for protecting wildlife populations and their habitat in areas where support exists for implementing well-managed timber harvesting practices. Actions that foresters and logging operators should consider include:

⁓ Planning road systems to minimize their width and length

⁓ Using directional felling to maximize residual stand structure

⁓ Deferring re-entry harvests until the treated area has recovered

⁓ Controlling hunter access to the management area

⁓ Ensuring on-site forester supervision of all silvicultural treatments

Details on how to initiate these practices are also discussed.

The reduced-impact logging practices discussed in chapter 21 must fit within an overall silvicultural prescription designed to mimic the natural disturbance events that gave rise to the current forest (plants and animals in an area are adapted to such disturbances and the recovery periods that follow). In chapter 22, Chapman and Fimbel discuss *cues* that foresters can use in assessing whether monocyclic or polycyclic harvest-regeneration systems are the most appropriate for an area, and the approximate intensity with which these treatments should be applied. The need to consider stand structure, composition, soils, topography, and precipitation patterns are discussed. The long-term productivity and value (economic and ecological) of a production forest relies on cutting systems that regenerate a diversity of plant species and habitat types.

Protected areas are critical to wildlife conservation within areas of the forest receiving silvicultural prescriptions, according to Bruce Marcot, Ted Gullison, and James Barborak in chapter 23, and a network of these reserves should be a part of all management plans for several reasons. First, some species of plants and animals are unable to persist in disturbed environments—at least until there is substantial recovery of the forest structure. Second, some habitats are too fragile to recover from forest management practices, or too biologically valuable to risk, and should thus be exempt from treatments. Finally, protected areas maintain reserves of locally adapted species, thus ensuring a genetically fit source to help perpetuate the forest's productivity and value. It is proposed that at least 10 percent of treatment areas be maintained in a natural state (this often occurs by default due to access difficulties), with additional areas between these *default* reserves set aside to serve as travel corridors for wildlife (especially along water courses) and to provide a broader range of available undisturbed habitat. The more intensive the silvicultural intervention is, the more important these connective corridors will be.

Model forestry-conservation programs that incorporate the best available, low environmental-impact harvesting procedures (see chapter 21) can significantly reduce the negative impacts of logging on wildlife populations and their habitat—regardless of their scale (for large scale, see Laurance in chapter 24; for small-scale, see Salafsky and coauthors in chapter 25). Maintenance of the forest's structural integrity, and control of hunting within harvest areas, appear to be the two over-riding factors leading to these conservation improvements. Even under the best forest management conditions, however, it is evident that some wildlife species have difficulty surviving in areas subject to timber harvesting (see chapter 24). While technical solutions are needed to refine silvicultural treatments that minimize impacts on the forest structure and composition, the urgency to address socioeconomic threats to the forest (part VI) is of equal importance in conserving wildlife in these production forest landscapes. This less visible half of the conservation equation includes commitments to maintaining the forest cover, training personnel to oversee the management and protection of these resources, and forging partnerships with forest stakeholders to ensure their continued support for forest conservation efforts (see parts IV and VI).

Chapter 20

WHERE SHOULD NATURAL FOREST MANAGEMENT BE PROMOTED TO CONSERVE WILDLIFE?

Peter C. Frumhoff and Elizabeth C. Losos

Current patterns of tropical timber production pose a considerable challenge for all concerned with the conservation of tropical forests and their biological diversity. Timber concessionaires widely engage in destructive logging practices, rapidly harvesting trees of marketable species without attempting to protect the forest from structural damage, fire, overhunting, or subsequent conversion to agriculture or pasture (Johnson and Cabarle 1993; Frumhoff 1995; Bawa and Seidler 1998). The area of tropical forest affected by such practices is vast—more than four million hectares of tropical broadleaf forests were logged annually through the 1980s (FAO 1993), and Grieser Johns (1997) estimates that 31 percent of tropical rain forests are currently gazetted for commercial timber production. By contrast, less than 10 percent of these forests are slated for stricter protection (Grieser Johns 1997). The area of tropical forest affected by commercial logging is growing rapidly, particularly in Latin America where expanding domestic and international markets for tropical hardwood products are fueling rapid growth (FAO 1993; Uhl et al. 1997; Bowles et al. 1998a).

How can donors, forest policymakers, conservation scientists, and foresters help bring about a more ecologically sound use of tropical production forests? One approach highlighted in this book is to train and motivate loggers to switch from current *liquidation* logging practices to what is variably termed *natural* or *sustainable* forest management for the long-term production of timber. Under this approach, loggers are encouraged to adopt a combination of silvicultural techniques, harvesting guidelines, and reduced-impact logging measures that are intended to both sustain timber production and reduce the impacts of timber harvests on forest structure and biodiversity as far as is practicable (see chapters 18,

This chapter is adapted from P. C. Frumhoff and E .C. Losos (1998), *Setting Priorities for Conserving Biological Diversity in Tropical Timber Production Forests* (Union of Concerned Scientists, Cambridge Mass. USA and Center for Tropical Forest Science, Washington, DC USA).

19, and 21–24). The overarching conservation goal of natural forest management is to reduce deforestation by making timber production an economically viable alternative to clearing the forest for crops or livestock (Johnson and Cabarle 1993; Dickinson et al. 1996).

A highly polarized debate is currently taking place between proponents and critics of efforts to promote natural forest management as a strategy for conserving tropical forests and their biological diversity (e.g., Dickinson et al. 1996; Rice et al. 1997; Bawa and Seidler 1998; Bowles et al. 1998b). This chapter seeks to address and move past this debate by:

- Emphasizing that natural forest management is one of several conservation strategies that might be applied to tropical forests currently gazetted for commercial timber production

- Assessing the conditions under which it may be the most viable strategy to pursue

We stress that production forests vary in ways that render efforts to apply any single strategy—including natural forest management— to all forests counterproductive. In some cases, motivating a shift from current logging practices to more sustainable natural forest management will be the best means to conserve a forest and its biological diversity. But in others, doing so will neither be feasible nor desirable from a conservation perspective. The vast scale of commercial logging operations, and the limited availability of resources to alter current practices, challenge us to identify the most appropriate conservation strategies for production forests and to carefully prioritize investments across them.

We propose a simple conceptual model for identifying the conservation strategies best suited for different tropical production forest landscapes. We begin by describing how production forests vary along three key dimensions:

- The value of the forest for conserving biodiversity

- The availability of sufficient financial incentives and enforcement capacity to alter current logging practices

- The current pressures for converting the forest to agricultural land

Production forests also vary in other important dimensions, such as their *stocking value* or available volume of harvestable timber. Stocking value is a measure of a forest's production potential and, as described below, may significantly affect the ease with which different conservation strategies can be applied. We suggest in this model, however, that it should not be a primary determinant of the best conservation strategy.

The results of how production forests fall along these three dimensions are then used to select one of three site-specific conservation strategies:

1. Natural forest management (as described above for sustained timber production)

2. A modified version of natural forest management, in which loggers are encouraged or required to adopt measures that reduce the ecological impacts of logging, but are not aimed at protecting the growth and regenerating commercial timber species (e.g., enrichment plantings and thinnings)

3. Regazetting production forests for nontimber multiple use or for stricter protection

Finally, we apply this model to three different production forest landscapes and evaluate its strengths and limitations.

Factors Influencing Conservation Strategies for Tropical Production Forests

We propose that selection of an appropriate conservation strategy for a tropical production forest should begin with an assessment of the forest's characteristics along the three dimensions listed above.

The Value of a Production Forest for Biodiversity Conservation

All tropical forests have valuable biological diversity. The magnitude of their potential contribution to biodiversity conservation, however, varies widely. Surveys of butterflies, birds, vascular plants, and other taxa reveal that some forests contain a far more distinctive biota than others. They vary, for example, in:

- Species richness (the number of species per unit area)
- Endemism (the extent to which species exist only within a locally circumscribed area)
- Beta diversity (the rate of change in species composition across the forest landscape)

Comparing forests using these measures requires a common biogeographic scale. Dinerstein et al. (1995) used an eco-regional scale to compare the biological distinctiveness of tropical broadleaf moist forests (and other major types of natural habitat) in Latin America. Comparing and combining survey data from several different taxa, they concluded that of the 55 broadleaf moist forest eco-regions evaluated for biological distinctiveness, sixteen were *globally outstanding*, thirteen were *regionally outstanding*, eighteen were *bioregionally outstanding*, and eight were *locally important*. The Napo moist forests, for example, which run along the western arc of the Amazon, form an eco-region with globally outstanding biological distinctiveness because of their extremely high levels of species richness and endemism (e.g., Prance 1977).

Decisions about where to apply particular conservation strategies depend not only on the biological distinctiveness of a forest or other ecosystem, but also on the extent to which timber harvests (and other human activities) threaten its biota. In forests under pressure for conversion to agricultural land, for example, the establishment of active timber concessions may accelerate deforestation by providing access to colonists along logging roads (see section on agricultural land use). Even where pressures for human colonization are low, however, logged forests often undergo a marked loss of biological distinctiveness. The spread of commercial-scale harvests through a previously unlogged region tends to produce a more homogenized and biologically impoverished forest landscape (Frumhoff 1995; Bawa and Seidler 1998). As detailed in several chapters in this book, the specific impacts vary widely across taxa and with the intensity and scale of logging, the history of previous forest disturbance, the extent of logging-associated hunting, and other factors (see chapters 2–17 and 24).

Tropical forests with highly distinctive biological communities that are currently gazetted for commercial timber production merit priority attention for conservation investments and the strongest possible conservation measures. In practice, however, obtaining information about which forests have distinctive biotas is often difficult. Eco-regional analyses are an essential first step. Timber concessions, however, rarely cover an entire eco-region, and the biological distinctiveness of forests within an eco-region can differ considerably as a result of natural variation among sites and variation in the intensity, scale, and frequency of past human disturbance. In addition, such basic information as the locations of current and planned concessions is often difficult for researchers to obtain in many tropical countries.

Donor agencies are well positioned to play an important role in helping to fill these critical information gaps. They can, for example, encourage forestry departments in tropical countries to make information on the locations of all concessions publicly available—as a condition of forestry sector assistance, if necessary. They can also support standardized, objective assessments of the biological distinctiveness and conservation status of all tropical forest eco-regions, building on work already completed (e.g., Dinerstein et al. 1995; Green et al. 1996; Olson and Dinerstein 1998). Within eco-regions identified as having globally and regionally outstanding biota, the biological distinctiveness of tropical production forests should be evaluated at the concession scale. Donors can most effectively target support for assessments that combine remotely sensed evaluations of forest cover and type with on-the-ground *rapid assessments* of endemism, beta diversity, and other key factors.

Within other ecoregions, priority sites for assessments should include those where the area of production forest is expanding rapidly, such as some terra firme forest in the Brazilian Amazon (Uhl et al. 1997; Laurence 1998), as well as concessions adjacent to protected areas. In the latter case, a primary concern is to assess whether resident populations of threatened or endangered fauna regularly move into the production forest, and whether they, or key resources upon which they depend, merit further protection (Frumhoff 1995).

The Availability of Sufficient Financial Incentives or Enforcement Capacity to Alter Current Logging Practices

Logging companies in the tropics widely engage in unsustainably rapid harvesting of high-value timber because it is highly profitable to do so—often far more so than the delayed harvest and investment in commercial regeneration that is necessary for long-term timber production under natural forest management systems (see chapter 28). The return from investments in future timber production is largely a function of the growth rate of commercial species, the cost of silvicultural treatments to promote regeneration, the expected market prices of timber, and the discount rate. This return is generally lower than that gained by rapidly harvesting all marketable trees and investing the profits elsewhere (Kishor and Constantino 1993; Vincent 1995; Reid and Rice 1997; Bawa and Seidler 1998). At the same time, the administrative capacity and political will of government forestry departments to restrict harvests by enforcing diameter cutting limits and other regulations are often extremely limited (Hardner and Rice in press; see chapters 18 and 24).

Where these conditions prevail, it will be difficult to motivate timber concessionaires to invest in future harvests (Reid and Rice 1997; Hardner and Rice in press). Where financial incentives and limited enforcement capacity favor rapid logging with little or no investment in commercial regeneration, such logging is likely to continue despite the best efforts of internationally funded programs to provide technical assistance and training in natural forest management. Limited conservation dollars would, therefore, be better invested in other strategies or on other forests where conditions are more favorable to long-term timber production.

Conditions *are* favorable for natural forest management in some production forests. In Costa Rica, for example, the economic environment is relatively conducive to long-term forest management (Howard and Valerio 1996; cf. Kishor and Constantino 1993). Across the tropics, some individual landowners and communities that manage small-scale production forests may also be more willing to invest in long-term forest management for social or cultural reasons and accept more modest profits than industrial logging firms. Several efforts to improve the financial attractiveness of natural forest management are also under way. These include expanding the market for certified timber from well-managed forests (Putz and Viana 1996; see chapter 26), and carbon-offset payments for the incremental costs of adopting reduced-impact harvesting practices (Putz and Pinard 1993; Boscolo et al. 1997; Brown et al. 1997; Pinard and Putz 1997; Frumhoff et al. 1998; see chapter 27). These financial mechanisms show modest promise (see chapter 28), but their potential remains highly uncertain (FAO 1997; see chapters 27 and 28). We believe that it would be a mistake, therefore, to assume that they would be able to motivate natural forest management on a broad geographic scale in the short term.

The Degree of Pressure to Convert the Forest to Agricultural Land

The conservation value of natural forest management lies in the prospect for sustained timber production to provide an economically viable alternative to converting the forest to agricultural land. The degree of pressure on forest areas for conversion, however, varies widely. In much of Suriname and the Congo Basin, for instance, the pressure for conversion is extremely low, while forests in the Petèn region of Guatemala and in Kalimantan Indonesia are being rapidly converted to agriculture. Predicting the conservation benefits of instituting natural forest management in a given production forest, therefore, requires assessment not only of the likelihood that concessionaires will actually harvest timber on a long-term basis, but also of the current and expected pressure for land conversion.

Consider the following three scenarios:

First Scenario The pressure to convert the forest is high, resulting perhaps from high population density or in-migration rates, weak or inappropriate land tenure and agricultural policies, and the relatively high fertility of the forest soil. High interest rates and weak enforcement of forestry policies also provide strong incentives for the concessionaire to log rapidly with minimal investment in long-term production. Under these conditions, concessionaires have little incentive or ability to protect the forest from conversion. The presence of logging roads, particularly in Latin America, often accelerates colonization and the conversion of forests to agriculture (e.g., Uhl et al. 1997), making road improvements intended to increase the profitability of timber harvests potentially counterproductive.

Second Scenario Pressure for agricultural land conversion is also relatively high. Here, however, a combination of low interest rates, tax incentives, timber certification, and/or well-enforced forestry regulations significantly increase the likelihood that the concessonaire will be willing to engage in long-term forest management. In this case, the concessionaire has a clear interest in protecting the forest from land conversion. If natural forest management techniques were adopted, deforestation rates would presumably decrease relative to poorly managed or unmanaged forests.

Final Scenario Pressure for agricultural land conversion is low, perhaps because population density in the region and in-migration rates are low, access roads are limited, or soil fertility is poor. In short, this is an area that is experiencing a low *background* rate of deforestation. As noted above, much of Suriname and the Congo Basin fits this description, including areas with substantial timber concessions. In such areas, the successful adoption of natural forest management practices could thwart a future expansion of the colonization frontier. The attraction of doing so, however, must be balanced against the risk that unsuccessful efforts may promote more rapid forest and biodiversity loss by providing greater forest access to hunters and colonists along newly constructed or improved logging roads. As described below, we propose that alternative conservation strategies be fully explored before promoting natural forest management in forest regions with low pressure for land conversion.

Identifying Conservation Strategies for Tropical Production Forests

The characteristics of a production forest with respect to its value for biodiversity conservation, the availability of incentives and enforcement capacity to alter logging practices, and the pressure for agricultural land conversion, can be usefully mapped onto a three-dimensional graph that describes the range of possible variation in these characteristics (see figures 20-1 through 20-3). We propose that forests with different combinations of these characteristics merit different conservation strategies.

Where to Promote Natural Forest Management for Sustained Timber Production

Natural forest management for timber is an appropriate conservation strategy for some, but not all, tropical production forests. To indicate why this is so, and to determine the cases in which it should be promoted, we begin by reviewing two principal components of this practice: a) *reduced-impact logging techniques* aimed principally at reducing the ecological impacts of timber harvests, and b) *silvicultural treatments* and *harvesting guidelines* designed primarily to increase post-harvest growth, regeneration, and the density of commercial timber species for future harvests. Each has consequences for forest structure and biological diversity.

Reduced-impact logging (RIL) techniques include planning roads and skid trails to minimize loss of forest cover and soil erosion, and selecting the direction in which harvested trees fall to minimize damage to unlogged trees of both commercial and noncommercial species (Grieser Johns 1997; see chapters 21 and 24). They may also include protecting riparian forest and resources such as nest sites and trees that produce fruits consumed by vertebrate frugivores in seasons when edible fruits are generally scarce (see chapter 23).

The silvicultural techniques and harvesting guidelines that comprise the other major component of natural forest management vary considerably across managed forests. They often include liberation-thinning (poison girdling and/or removing trees of noncommercial species), harvesting of trees of lesser-known species, protection of selected, mature *seed trees* of commercial species, and, much less commonly, enrichment planting (planting seedlings of commercial species; see chapters 2 and 21). Limiting the current harvest of high-value species to trees of large size classes is also critical to the long-term success of natural forest management systems.

Because these silvicultural practices favor specific size classes of selected timber species, they will also reduce the structural diversity of a forest and alter its species composition—often considerably—relative to both unlogged forests and logged forests in which only reduced-impact logging techniques are applied. The effects of a silvicultural treatment on noncommercial forest species depend on both the intensity and the geographic scale at which it is applied (e.g., Mason 1996; Bawa and Seidler 1998). Under programs intended to promote the successful adoption of natural forest management, this intensity and scale will be primarily determined by the need to sustain an economically attractive rate of timber production.

The decision to promote natural forest management, encompassing both types of techniques described above, needs to take several considerations into account:

- The direct ecological costs of adopting these practices on a commercial scale
- The prospective site-specific conservation benefits of natural forest management as an alternative to forest degradation and possible conversion
- The prospects that resource managers and loggers will have sufficient incentives to continue this management strategy over the long term
- The relative costs, benefits, and risks of alternative conservation strategies

We propose that natural forest management is best suited as a conservation strategy for those production forest sites with the following characteristics:

- Relatively low-to-moderate biodiversity conservation value (e.g., sites that have *locally important* or *bio-regionally outstanding* biological distinctiveness, *sensu* Dinerstein et al. 1995)
- Moderate-to-high pressure for conversion to agricultural land (Dickinson et al. 1996)
- Sufficient incentives and enforcement capacity in place to motivate long-term management for timber (Rice et al. 1997; see figure 20-1)

Natural forest management may also be a viable conservation strategy for some production forests of high biological distinctiveness, provided that sufficient financial incentives and enforcement capacity are clearly present (see figure 20-1). Under these conditions, forestry practices should be instituted only after first carefully assessing whether

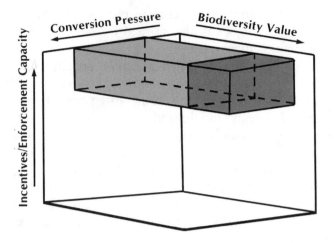

FIGURE 20-1 *A three-dimensional representation of the proposed model for selecting appropriate conservation strategies for tropical production forests. The site-specific characteristics of a forest with respect to its biological distinctiveness, the availability of incentives and enforcement capacity to alter current logging practices, and the pressure for conversion to agricultural land determine the location of the forest on the figure. The light-shaded box represents production forest areas with low to moderate value for conserving biological diversity, moderate to high pressure for conversion to agricultural land, and sufficient economic incentives and/or enforcement capacity to ensure long-term timber production. For these forests, we recommend natural forest management for timber. The dark-shaded box represents areas that have high biodiversity value. Here, natural forest management should be instituted only after first assessing whether stricter conservation measures are feasible, and be employed with harvest intensity and methods set by conservation rather than optimal profitability criteria.*

stricter conservation measures are feasible (see next section). They should also be employed with a harvest volume and recovery period determined by conservation rather than optimal profitability and productivity criteria.

In many production forests, it may make sense to decouple the two major components of natural forest management and seek to apply reduced-impact harvesting techniques (but not silvicultural management) for long-term timber production. In the following paragraph we propose conditions under which doing so may be a viable conservation strategy.

Where to Promote Reduced-Impact Logging but Not Commercial Regeneration Practices

In many countries, rapid liquidation harvesting occurs until concessionaires find it no longer financially attractive to continue. The result is often considerable residual damage to these forests. In the worst cases, roads left in place by loggers become conduits for colonists, increasing resource exploitation and deforestation. Clearly, this status quo is not desirable. Objective assessments of the ecological and socioeconomic characteristics of tropical production forests, however, are likely to find that, given competing land uses and competing demands on conservation and forestry resources, many concessions currently do not merit priority attention for intensive forest conservation and management measures. Such concessions might include those in areas that have less critical biological distinctiveness, where financial incentives or enforcement capacity to motivate sustained timber production are limited, and where pressure for conversion to agriculture is relatively low (see figure 20-2).

For production forests with these characteristics, investments in silvicultural treatments to promote commercial regeneration and sustained timber production are likely to simplify forest structure and reduce biodiversity with a low probability of gaining conservation benefits from avoided deforestation. In such cases, a better conservation strategy may be to promote only those low-cost forest management measures that reduce the ecological impacts of current harvesting practices (see chapters 21 and 24). Particular emphasis might be placed on techniques that produce cost savings, such as extensive stock surveys or pre-harvest mapping of roads, or on those that add value by enabling the harvested timber to become a target for *green* market certification or other fiscal incentives (Rice et al. 1997; see chapters 26 and 28). Eliminating sustained timber production as a criterion for certifying these production forests would facilitate this.

Limiting harvest rates is also important and could be accomplished by restricting the number and size of mills and, consequently, the processing capacity to a targeted level below the maximum capacity of the local concessions (R. E. Gullison, personal communication). Marginal lands for timber production within these concessions might also be identified and protected at relatively low cost, by taxing timber producers on the total concession area under active harvest rather than on the volume of timber produced (Hardner and Rice in press; see chapter 29). Donors and forest policy makers should also work with con-

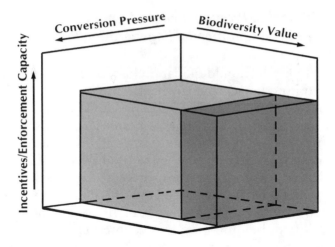

FIGURE 20-2 *The light-shaded box represents production forest areas in which neither sustained timber production nor protection are currently viable options. These are forests that have relatively low to moderate biodiversity value, conversion pressure and capacity to alter current logging practices. For these forests, we recommend that several low-cost steps be taken to reduce the ecological impacts of current timber harvests, but that no measures be taken to promote growth and regeneration of commercial timber stocks for future harvests (e.g., thinnings and enrichment plantings). The dark-shaded box represents production forest areas of high biodiversity value where this approach might be taken if stricter protection is not a viable near-term option.*

servation biologists to identify low-cost means of reducing hunting pressure in these concessions, as well as in concessions where natural forest management is instituted (see chapters 15–17).

The above measures to reduce the impacts of timber harvests on forest structure and biodiversity might be usefully extended to some production forests with high biological distinctiveness, where stricter protection is not a viable near-term option (see figure 20-2). Near-term protection will often be difficult in areas where high biological distinctiveness and high volumes of commercially attractive timber overlap (see next section). Under these conditions, prospects for conservation could be enhanced by allowing selective high-grading of high-value timber, thereby reducing the value of the resource to the logger and using the proceeds to help finance subsequent protection (R. Rice, personal communication).

Where to Regazette Production Forests for Protection or Nontimber Multiple Use

Tropical timber production forests that are identified, through an objective process, as having the highest value for biodiversity conservation merit the fullest possible protection (see figure 20-3). That is, they merit protection not only from deforestation and fragmentation, but also from the inherent direct and indirect impacts of even well planned logging and silvicultural treatments on forest structure and species composition. Wherever possible, these forests should be swiftly identified and regazetted for full protection or nontimber, multiple-use management. Such management could include nontimber forest product extraction, ecotourism, biodiversity prospecting, and carbon sequestration (see chapter 27). Whole concessions might be regazetted in this manner, as appropriate to the size of the concession and consistent with conservation needs.

Alternatively, priority habitat for conservation might be re-gazetted for protection within a larger concession (Hardner and Rice in press; see chapter 23). In some countries, such *conservation set-asides* are already being developed. In Bolivia, for example, timber producers now receive tax benefits for leaving up to 30 percent of a concession unlogged (Gaceta Oficial de Bolivia, 1996; see chapter 29). Yet these forests may not contain the highest value for biodiversity conservation because the concessionaire generally selects the areas to protect. The size and locations of conservation set-asides should instead be based on independent, site-specific assessments as described above.

Areas of production forests with high conservation value, but marginal timber value should be relatively straightforward to regazette for protection or nontimber multiple use. Such areas may include steep slopes, ridges, and swamp forest that have biologically distinct communities, but are difficult to harvest. They may also include areas of production forest that have already been graded *high* for valued timber species—such as mahogany (*Swietenia macrophylla*)—and are, therefore, of low current commercial value (R. Rice, personal communication).

By contrast, regazetting will be far more challenging in areas of production forest that have critical value for biodiversity conservation, economically attractive stocks of commercial timber, and limited government capacity for enforcement of harvest restrictions. One possibility is for donors to buy out such concessions to finance a transition from production to lower-impact use or protection. Carbon-offset investments from industrialized countries might also finance some of

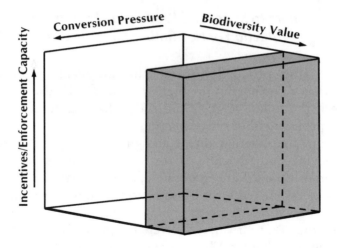

FIGURE 20-3 *The shaded box represents production forest areas with high value for biodiversity conservation as determined by an assessment of their biological distinctiveness and condition. For these forests, we recommend regazetting for strict protection or lower-impact multiple use.*

these transitions, such as when three U.S. energy companies recently financed the conversion of a 634,000-hectare timber concession into an extension of Bolivia's Noel Kempff Mercado National Park (USIJI 1997). Their investment of U.S.$7 million helped buy out the concessionaire, establish a park endowment, and support ecotourism and other nontimber economic development initiatives with local communities.

Applying the Model to Specific Production Forest Landscapes

To illustrate how the previous strategies might be applied to identify site-specific conservation systems, we consider timber production forests within three landscapes that differ substantially along one or more of the dimensions described above. The specific recommendations for these forest landscapes should be viewed as preliminary, as the model requires further development and some of the socioeconomic and ecological data for these sites are lacking.

Production Forests Within the Northern Zone of Costa Rica

Virtually all timber production in Costa Rica takes place on privately owned lands that are under strong pressure for conversion to agricultural land. This is a result of both the historically high rate of deforestation in Costa Rica (2.9 percent per year during the 1980s: FAO 1993), and the fact that most of the country's remaining forests now lie within a national system of parks and protected areas. One principal timber production region is the northern zone, located along the Nicaraguan border. Timber in this zone is harvested in seasonally moist primary and secondary forest, both in fragments up to several hundred hectares that are embedded within agricultural land or pasture, and in areas that are relatively extensive and contiguous with protected forest lands.

What is a suitable conservation strategy for these production forests? Dinerstein et al. (1995) classified the biological distinctiveness of this Costa Rican seasonally moist forest eco-region as *locally important* (relatively low distinctiveness according to their classifications). Given limited donor resources, therefore, this area may not be a high priority for international investments in stricter protection. More detailed assessments, however, may reveal local conservation priorities. Some forest patches, for example, contain mature *Dipteryx panamensis* trees—a commercially valued timber species that also provides essential nest and feeding sites for the locally endangered green macaw (*Ara ambigua*; Rivera et al. 1996; see photo 20-1). Strong pressure for agricultural land conversion makes natural forest management a potential means of retaining forest cover in this region. Some recent analyses also indicate that long-term timber production may be an economically viable alternative to agriculture in this and other regions of Costa Rica (Howard and Valerio 1996; cf. Kishor and Constantino 1993). This is, in part, a result of Costa Rica's restriction on log imports, which has led to high market prices for domestic timber (Reid and Rice 1997, as well as efforts on the part of the Costa Rican government to attract carbon offset investments for maintaining and regenerating forests on private lands.

This combination of relatively low conservation value, high pressure for conversion to agriculture, economic feasibility, and substantial government support, appears to make natural forest management a viable conservation strategy for private forests in Costa Rica's northern zone. This is the strategy that Costa Rica is now pursuing, with

PHOTO 20-1 Miskito Indian woman with a locally endangered green macaw (*Ara ambigua*). (J. Barborak)

considerable international donor financing for technical support, training, research, and implementation.

Production Forests Within the Coastal Zone of Cameroon

Like Costa Rica, Cameroon experienced a rapid rate of deforestation in the 1970s and 1980s, an annual loss of 0.6 percent (FAO 1993; Dixon et al. 1996). The ecological and socioeconomic characteristics of production forests within the Congolese coastal forest eco-region of western Cameroon differ considerably from those in the northern zone of Costa Rica. This ecoregion is species-rich and a center of endemism for several taxa. The World Conservation Monitoring Center has identified it as having "highest global priority for conservation action," and the World Wildlife Fund recognizes it as one of the 200 most biologically distinctive and threatened eco-regions of the world (Green et al. 1996; Olson and Dinerstein 1998). Much of the remaining forest is also rich in timber species, including *Entandrophragma* spp., *Khaya* spp., and *Dyospyros* spp. (I. Itoua, personal communication). Traditionally, commercial timber exploitation has been mostly on a small scale and dominated by local landowners (Dixon et al 1996). More recently, though, large-scale exploitation—often through foreign enterprises—has become increasingly common.

A substantial area of Cameroon's remaining forest is designated for timber production, with production forests actually overlapping national parks along the central and southern coast. At present, Cameroon has little institutional capacity for enforcing forestry laws, and rapid liquidation harvesting is the norm (R. Fimbel, personal communication). Pressure on coastal forests for conversion to agricultural land ranges from moderate to high in some areas (I. Itoua, personal communication).

For the past several years, the government of Cameroon has been embarking on an ambitious effort to institute natural forest management practices throughout the country's production forests. The high biodiversity value of Cameroon's coastal forests, however, indicates that sustained harvesting through natural forest management should not be promoted as a conservation strategy. We suggest instead that the government be strongly encouraged to support concession-level assessments of biological distinctiveness, and to reclassify the biologically critical areas into formally designated protected areas. If funds for such actions are not directly available, innovative tools such as carbon-offset funding could be used to help set aside areas of high conservation value (Frumhoff et al. 1998). If the potential timber revenue is too great to allow strict protection, then conservation efforts should focus on requiring RIL techniques, minimizing harvest rates, and maximizing recovery periods. Donor and government support, for assessing the concessions' biodiversity value and for strengthening enforcement capacity, appear to be major prerequisites for effective, long-term management.

Production Forests of the Department of Beni, Bolivia

Because market access is poor in the remote Beni region (along the northwestern edge of the Bolivian Amazon), timber production consists almost exclusively of selective harvesting of high-value mahogany (*S. macrophylla*). Low-density stands of mahogany are currently harvested in seven industrial timber concessions within the 449,000 ha Chimanes production forest in the Department of Beni (Gullison et al. 1996; see chapter 29). As in Cameroon, concessionaires are responding to strong economic incentives to liquidate the harvestable stands of valuable timber. Because interest rates are high and mahogany grows slowly, they have little incentive to invest in commercial regeneration.

Once the marketable mahogany trees have been removed from a forest, the forest's value to logging companies drops appreciably. Despite a long history of international technical assistance, political will to enforce forestry laws in this region is also virtually non-existent, and regulations governing diameter cutting limits and protection of seed trees for regeneration are commonly ignored (Gullison et al. 1996).

What is the best overall conservation strategy for the Chimanes production forests? Dinerstein et al. (1995) characterize the biological distinctiveness of the swamp and gallery forest eco-region as relatively modest (locally important). Because the location is remote, current pressure for conversion to agricultural land is low (R. E. Gullison, personal communication). As a result, neither natural forest management nor stricter protection are likely to be viable near-term conservation strategies for this forest. The substantial investment in infrastructure needed for a long-term forestry commitment might in and of itself have a negative impact on biodiversity, by actually facilitating human colonization and the exploitation of a wider variety of timber species. The Chimanes appears to be an area where donors and forest policy-makers should encourage concessionaires to adopt low-cost measures to reduce the impact of production forestry, but should not invest in long-term, natural forest management practices that are likely to continue to be resisted or ignored by concession holders.

As the northern Costa Rica example shows, the concurrence of several factors—noncritical biodiversity value, high colonization pressure, and viable economic incentives and enforcement—can create conditions that are ripe for the successful adoption of natural forest management practices (see chapters 21–25). A systematic assessment (that is beyond the scope of this paper) would be necessary to identify the production forest landscapes under which these conditions currently apply. It is our perception, however, that they align more rarely in tropical timber-producing countries than is commonly assumed by advocates for the adoption of natural forest management on a broad geographic scale (e.g., Grieser Johns 1997). Where these conditions do not align, our model suggests that donors and forest policy makers should give higher priority to considering investments in alternative conservation strategies.

Conclusions

The approach that we sketch out here explicitly recognizes that there is no "one-size-fits-all" solution to the problem of conserving biodiversity in tropical production forests. By acknowledging that natural forest management has an important but circumscribed role, we seek to move beyond the debate now taking place between proponents and critics of natural forest management (e.g., Dickinson et al. 1996; Rice et al. 1997; Bawa and Seidler 1998; Bowles et al. 1998b). We attempt, instead, to give natural forest management appropriate weight as one of several conservation strategies that can be applied to tropical forests currently gazetted for commercial timber production. The proposed model emphasizes that site-specific ecological and socioeconomic conditions must be taken into account in selecting and designing effective conservation strategies, and identifies what some of the key conditions are likely to be.

Our model further suggests that the elements of a natural forest management strategy designed to reduce the ecological impacts of timber harvesting should, under some conditions, be decoupled from those designed to promote commercial regeneration. Under these conditions (low-to-moderate biological distinctiveness, pressure for agricultural land conversion, and limited incentives or enforcement capacity for long-term timber production), we suggest that efforts to promote commercial regeneration and sustained timber production are not warranted as part of a conservation strategy. It may still be appropriate, however, to:

- Apply reduced-impact logging techniques to minimize residual harvest damage
- Promote market certification for timber harvested using ecologically sound methods
- Build long-term enforcement capacity, whether for future protection, non-timber multiple-use, protection, or timber production

The information necessary to institute these conservation strategies, such as concession-level data on biological distinctiveness, is widely lacking. Some may argue that in its absence it is more prudent to try to broadly institute natural forest management measures than to let destructive patterns of exploitation logging continue unabated. To date, however, destructive patterns of exploitation logging have continued largely unabated *in spite of* substantial investments in fostering sustained timber management. We encourage donors, therefore, to support targeted investments to map and assess the biodiversity values of tropical production forests. This will make it possible to more systematically identify those production forests where natural forest management is the most appropriate conservation strategy, and where other strategies are more appropriate.

Finally, we recognize that the approaches discussed in this chapter focus on conservation goals, and do not fully take into account other goals for tropical production forests such as timber yield, or social goals such as when local communities have ownership rights or claims to production forest areas (see chapter 25). As conservation biologists, we assume (and hope that policy makers and donors agree) that timber production goals should be secondary to conservation goals for concessions that contain globally and regionally outstanding biological diversity. Because conservation efforts that ignore the social and economic needs of local communities are unlikely to succeed, we recognize that site-specific efforts to regazette production forests for lower-impact multiple-use or protection will need to incorporate the lessons of community-based conservation programs (e.g. Western and Wright 1994; see chapter 25).

ACKNOWLEDGMENTS

We thank Jose Joaquin Campos, Rob Fimbel, Ted Gullison, Jack Putz, Dick Rice, and Argelis Roman for helpful comments on manuscript drafts. Thanks also to Jake Brunner, Illanga Itoua, Emma Underwood, and David Olson for providing information on Cameroon forest concessions and biodiversity. Stan Byers and Julia Petipas were helpful with research assistance; and Anita Spiess with editorial assistance. Peter Frumhoff's work on this chapter was supported in part by Wallace Global Fund and Summit Foundation grants to the Union of Concerned Scientists' Global Resources Program.

Chapter 21

REDUCING THE IMPACTS OF TROPICAL FORESTRY ON WILDLIFE

Douglas J. Mason and Francis E. Putz

Some loss of biodiversity is likely when humans extract resources from natural areas. Forestry is not exempt from this rule. While managed tropical forests may not conserve every species, they do have an important role to play in an overall strategy to conserve biodiversity. Production forests cannot replace national parks, but they can complement strictly protected areas that are often too small or too isolated to conserve all species. Because many developing countries are unwilling, or economically unable, to strictly protect all of their remaining forests, some of the largest remaining forested landscapes in the tropics will be managed for timber production. If properly planned and executed, tropical forestry may be the most appropriate land use, conserving more biodiversity than the most common alternatives—intensive agriculture, cattle pastures, biomass plantations, or oil palm estates.

The problem is that, far too often, *uncontrolled logging* (extraction) occurs in place of *forestry* (long-term forest stewardship). Forests are mined for timber instead of managed for a renewable, diverse collection of resources. While forest mining can be financially lucrative, it is economically costly. Careless timber extraction realizes only a tiny fraction of the value of the forest while degrading its natural resources, diminishing ecosystem services (such as flood protection and water production), impoverishing its biological diversity (see chapters 4–14), and producing little lasting benefit for local residents or for society as a whole. Furthermore, timber harvesting finances the construction of roads that penetrate remote areas, leading to colonization, excessive hunting (see chapters 15–17), and eventually deforestation. As a result, commercial timber extraction has catalyzed far more forest destruction than forest conservation. Even when governments enforce forestry regulations, they often place too much emphasis on repairing forest damage once it occurs, rather than minimizing it in the first place (see photo 21-1).

PHOTO 21-1 *Even where extraction rates are low, such as this Venezuelan forest where an average of three trees per hectare are harvested, significant damage occurs to the forest and its wildlife because of careless logging. Low-impact logging practices can significantly reduce this damage. (D. J. Mason)*

This chapter focuses on the technical aspects of how to implement environmentally sound forestry to reduce its impact on wildlife. It describes steps that forest managers, conservationists, and governments can take to facilitate the transition from forest mining to forest stewardship. Despite our limited understanding of tropical forest ecology, we know enough to make significant improvements today, where there is the will to do so (see chapter 20). To paraphrase Aldo Leopold (1949), we need not relinquish the saw, but we do need "gentler and more objective criteria for its successful use."

Key Principles for Conserving Biodiversity in Logged Forests

Tropical forestry is a broad category under which much falls. The size and type of management units range from modest private holdings to huge concessions on public land. Tropical forests themselves vary from humid dipterocarp forests to dry African woodlands. Who is managing the forest is equally diverse—ranging from individuals and communities to governments and corporations. Given this diversity, it is beyond the scope of this chapter to develop a list of detailed prescriptions (e.g., the density of seed trees to reserve, the appropriate level of public participation, etc.) for all forests. We provide some general principles, instead, on which more detailed, locally adapted guidelines can be based. We begin with the broadest principles, which are refined in the remainder of this chapter.

Protect the Forest from Preventable Degradation or Outright Elimination

The forest cannot produce wood, conserve biodiversity, or maintain environmental services if it is severely degraded or eliminated. Reduced-impact logging (RIL) practices can assist this process during logging. Working together, forest managers, government officials, local residents, and other stakeholders must ensure that forests are harvested properly and protected following harvesting. Public participation in forest management can facilitate these partnerships, particularly on public lands.

Control the Harvesting of Targeted Species

Targeted species, whether they are timber trees, non-timber forest products, or wildlife species, should be harvested at levels that maintain viable local populations and also allow them to perform their ecological role in the forest (Redford 1992, Peters 1994). Given our ignorance of tropical ecology, this implies that we take a conservative, precautionary approach and that we adjust harvesting rates as needed—based on regeneration data derived from population monitoring. The less that is known about a particular species, the greater is the care needed to avoid overexploitation.

Mimic Natural Disturbances to Foster Regeneration

Silvicultural approaches should build upon natural succession. The more a forest management system causes disturbances ecologically similar to *natural* disturbances (in terms of their frequency, duration, and severity), the more likely that the impacts will be environmentally benign (see chapter 22). Forestry operations that mimic natural disturbances reduce the risk of regeneration failures. The principle is to follow nature—perhaps guiding regeneration in useful directions—but always using natural processes as the foundation for forest management (Wadsworth 1997). It is important to recall, however, that human disturbances are usually additional to the natural disturbances affecting a forest, and intensive logging operations can likely cause disturbances that are far more severe and frequent than those to which some forests are already adapted. A substantial percentage of forests considered *primary* actually regenerated after the abandonment of lands manipulated by humans within historical times.

Maintain Key Ecosystem Processes

Key processes include water flows, nutrient cycles (that depends on a diversity of forest life forms, structures, and trophic levels), and natural regeneration processes (including pollination and seed dispersal) (see chapters 3-14). Proper forest management has always sought to maintain the processes on which forest health depends. Tangible steps include:

1. Properly zoning the forest to avoid damage to sensitive areas (such as steep slopes)

2. Carefully siting roads to avoid watercourses and erodible soils

3. Employing harvesting techniques and equipment adapted to local environmental conditions

4. Maintaining populations of native plants and animals—particularly those that play important ecological roles (e.g., pollinators and seed dispersers)

Match Forestry Operations to Their Social Context

Attempts at forestry have often failed when forestry operations were not appropriate for the socioeconomic context in which they operated. Examples span the spectrum, from community-based operations that lacked the ability to meet the quantity and quality requirements of international markets, to large companies that failed to properly take into account the needs of local residents. The solutions vary with the context. Substantive participation by key stakeholders, for example, is very important, but the kind of participation will vary from private to public land, local to international stakeholders, and community to corporate management.

Monitor the Impacts of Forestry, Share Information, and Manage Adaptively

Tropical forest management is more complex biologically, and often socially, than its temperate counterpart. Nevertheless, management decisions must be made today, often based on little information. This situation makes it particularly important to regularly monitor forestry operations and their impacts on the forest. Even more importantly, the information generated by monitoring must be used to make more informed decisions in the future (see chapter 19). Management must be adapted as new biological, economic, and social information becomes available. Since many of the challenges facing forestry lie outside the forest sector (colonization, for example), forest managers will not be able to address all the problems. This problem is one additional reason why monitoring information must be shared between the private sector, government, and civil society—a challenge in many countries.

The following discussion is divided somewhat arbitrarily into three sections:

- Proper harvest planning
- Road networks, felling, and extraction
- Long-term stewardship

The majority of this chapter focuses on choosing the most appropriate technology and techniques to employ in a particular forest management unit (e.g., a concession, a privately held tract of land, etc.). The last section briefly highlights the national enabling conditions for improved forest management. Appendix 21-1 summarizes the advantages/benefits and the disadvantages/constraints of a number of these practices, both in terms of timber production and biodiversity conservation. It allows for general comparisons between management improvements or options that forest managers may consider.

Proper Harvest Planning

Proper planning is the key to reducing the impacts of forest harvesting (see photo 21-2). While planning imposes additional up front costs on forest management, these costs can be recovered. Planning leads to more efficient road networks; machine time is reduced through more efficient harvesting, skidding, and yarding, and fewer valuable trees

PHOTO 21-2 *A reduced-impact skid trail in Malaysia four months after 30 logs were skidded using a bulldozer. (F. E. Putz)*

are damaged (Hendrison 1990; Pinard et al. 1995; Johns et al. 1996; Uhl et al. 1997; Baretto et. al. 1998). There are four steps to the planning process.

Choosing an Appropriate Harvest-Regeneration System

The harvest-regeneration system best suited to a site depends on:

- Characteristics of the forest (such as regeneration requirements of the harvested tree species)
- The goals of forest managers (e.g., how intensive an approach to management is required)
- Timber markets
- Local social conditions (e.g., how great is the threat of spontaneous colonization?)
- Values and needs of society (e.g., how much value is placed on wood versus other forest products?)

We focus on timber harvesting because logging is generally the only silvicultural intervention applied during natural forest management in the tropics (see chapter 2).

Although there is a growing body of literature on the role of large-scale and intensive natural and anthropogenic disturbances in the regeneration of timber tree species, the selection of the appropriate silvicultural approach is difficult (Snook 1996; Whigham et al.1998). The challenge is to mimic disturbances (e.g., hurricanes, meandering rivers, fire, treefall gaps, or slash and burn agriculture) in the gentlest means and on the smallest possible scale. It is important to recognize that there may be several means by which the historical disturbance regime can be mimicked to secure regeneration. Mechanical scarification of the soil can sometimes substitute for fires where using controlled burns is not an option.

Broadly, there are several harvest-regeneration systems that often intergrade with one another (also see chapter 2).

Selective Logging (Single Tree and Group Selection)

Selective logging is by far the most common harvesting system in the tropics. When properly applied it is a *polycyclic* forest management system in which few trees are harvested during each cutting cycle. Logging intensities vary by continent and are driven primarily by the density of commercially valuable trees:

- Low-intensity harvests (less than three trees per hectare, and in some places less than one tree per hectare) occur in much of Africa and parts of Latin America
- Medium-intensity harvests (four to five trees per hectare) occur in much of Latin America
- High-intensity harvests (greater than seven trees per hectare) occurs in southeast Asia

Harvests are repeated at relatively short intervals, generally from 15 to 40 years. Where single tree or group selection is appropriately applied, the harvested trees are replaced in the canopy by seedlings, saplings, or pole-size trees that grow up in the gaps created by logging. If extraction rates are low and/or harvesting is done with care, the remaining forest can retain much of its original structure (Smith et al. 1997).

Where densities of commercial species are low, single tree selection causes the least disturbance to the forest. Wildlife species characteristic of closed forests are likely to benefit from maintenance of prelogging forest structure after carefully controlled single tree selection (see chapters 4-14). For this reason, it has been suggested that lightly logged forests should be purchased and converted into wildlife reserves (Rice et al. 1997; Bowles et al. 1998). This may be a cost-effective means to purchase forested land for conservation. The road network created by selective logging operations unfortunately increases the risk of illegal hunting and colonization—a very common problem in frontier forests (see chapters 15–17). If logged-over forests are both unprotected and accessible, they and their wildlife are likely to be at risk. Some management presence is almost always needed for conservation to occur.

Selective logging faces several potential problems. While single tree and group selection causes only minor damage relative to clearcuts, the damage per volume of wood harvested is high, given the extensive roads and skidder trails required to provide access. Some valuable tropical timber species (such as mahogany in Latin America or Okoumé in Africa) also do not regenerate under the small gaps created by low-intensity selective logging (Lamb 1966; Snook 1996; Whigham et. al. 1998). Timber *highgrading* or *creaming* often masquerades as single tree selection, especially where it does not result in conditions appropriate for continued production of the harvested species (Wadsworth 1997). Where the density of commercial trees is high (e.g., Southeast Asia), even *selective* logging can substantially open the forest, favoring disturbance-adapted and light-demanding species over those that depend on intact forest. In dipterocarp forests of Sabah, Malaysia, for example, felling clusters of trees that exceed the minimum diameter (60 cm) may promote the growth of moderately light-demanding trees, but more often results in the development of liana tangles (F. E. Putz, personal observation). Finally, while undisturbed humid tropical forests rarely burn, heavily logged forests are much more fire prone (Kauffman et al. 1988; Uhl et al. 1988; Uhl and Kauffman 1990). This transformation occurs as the canopy is opened, allowing more sunlight to reach the lower forest strata and drying the slash left behind after logging.

When practical, the least damaging silvicultural treatments that will promote the regeneration of the species being harvested should be applied. We do not expect any targeted tree species will require more than a 0.5 to 1.0 ha openings to regenerate. In some situations, light-demanding species may be most easily regenerated in plantations.

Clearcuts

Clearcuts are *monocyclic* systems in which all marketable and non-marketable trees above a minimum diameter (usually 10 cm dbh) are harvested in an area during a single operation. Future commercial trees will be grown from seeds and seedlings rather than advanced regeneration that remains following selective logging. Clearcuts allow large amounts of wood to be harvested from a small area and promote regeneration of light-demanding species.

Large clearcuts can have both ecological and economic problems, including:

- Clearcuts can be expensive to create if most of the felled trees are not commercially valuable
- Seed dispersal and seed rain from adjacent trees may be inadequate to regenerate the centers of large clearcuts
- Birds and mammals, which disperse the seeds of many tree species, are less likely to cross larger gaps in the forest
- Weed infestations can occur and inhibit (at least temporarily) tree regeneration
- The centers of clearcuts can have interrupted nutrient cycles, and plant growth may suffer from temperature and moisture stress, all of which can inhibit regeneration of many canopy tree species

Jordan et al. (1987) found that in artificially-cleared forest openings that are less than 50 m on a side, there was no detectable increase in nutrient leaching. Larger openings did experience nutrient losses, however. Clearcuts have the potential to be benign, when they mimic aspects of natural disturbances (see below and chapter 22). At present, clearcuts are rarely used in tropical natural forest management except when the site is being prepared for conversion (either to agriculture, forest plantations, etc.).

Shelterwoods

Shelterwood systems represent a compromise between the extremes of large clearcuts and isolated, single tree selection. Canopy trees are removed in stages to promote the regeneration of the species being harvested. In typical shelterwoods, initial commercial thinning of the canopy (i.e., the regeneration cut) is followed several years later by harvesting of the remaining overstory trees. By the time of the final cut, the next cohort of trees should be well established. The Tropical Shelterwood System, as applied in Trinidad, is a good example of this silvicultural approach (Lamprecht 1989). Shelterwoods are controversial because the strips are likely to expand as trees along the edge fall due to windthrow.

The strip shelterwood system creates long, narrow (20–50 m) clearcuts. These artificial gaps are designed to promote the natural regeneration of fast-growing, shade intolerant trees (Hartshorn 1990b). A notable experiment with strip shelterwoods was carried out in Amazonian Peru in the 1980s (Hartshorn 1990a). The strip shelterwood system is most profitable when the markets will accept a wide range of species and tree sizes.

Other Silvicultural Treatments

Although rarely applied to date in the tropics, silvicultural treatments can also be used to increase the stocking and improve the growth of timber trees. Treatments to improve stocking include enrichment plantings in cleared strips and site preparation treatments such as mechanical scarification of the soil to promote seedling germination and establishment. Controlled burns and mechanical treatment of slash and residual vegetation can substantially increase the establishment of light-demanding commercial timber trees—many of which have small seeds. Other treatments, such as thinning, are designed to improve the growth rates of commercial trees once they are established (see chapter 2).

These silvicultural treatments merit special attention because they can have even greater impacts on wildlife habitat (e.g., Mason 1996) than logging itself. In the future, these silvicultural treatments may also become more common as primary tropical forests are exhausted and secondary forests must provide a larger proportion of commercial wood. Where high-intensity forest management is necessary, caution is required. All else being equal, the more trees harvested per hectare, the greater the change will be in the structure and microclimate of the forest, and the greater the impact will be on forest biodiversity (Ewel and Conde 1980). Practices that reduce the deleterious impacts of harvesting become increasing important as extraction rates rise (Sist et al. 1998b). Wildlife biologists and other stakeholders need to help provide answers to where, how, and at what intensity these treatments should be applied (see chapter 2).

Establishing Ecologically Viable Cutting Rules

Forest management systems should favor natural regeneration. Most forestry laws, however, direct concessionaires to harvest trees above a minimum diameter (often 40, 50, or 60 cm dbh or diameter at breast height) where minimum diameter limits may be below the size at which trees become reproductive. Very large trees produce the greatest volume of seeds, and therefore serve as the most prodigious seed trees for natural regeneration. In addition, large trees often occur in clusters, and minimum diameter limit systems seldom specify intertree felling distances (see photo 21-3). As a result, postharvesting gaps can be much larger than is silviculturally desirable for natural regeneration.

Minimum tree diameter for harvesting (cutting limits) and other rules governing tree selection (tree marking rules), therefore, should be set to ensure that residual stems are capable of reproducing and maintaining the ecological functions within the forest. This is a challenge because of our poor knowledge of the biology of even the most common commercial species. Nonetheless, cutting (tree marking) rules should strive to:

Reflect Local Forest Ecology

Tree marking rules should be species- and site-specific and reflect the reproductive biology of the harvested species, which are usually poorly known (see chapter 22). Harvesting and extraction activities should seek to create the conditions necessary for commercial species to regenerate (e.g., very selective cutting of shade-tolerant species, heavier selective logging, or monocyclic techniques for shade-intolerant species). Local site-specific information is critical, as tree species can display different regeneration requirements over different parts of their ranges (Hartshorn 1990b).

Often, little information is available about the ecological requirements of many tree species (Martini et al. 1994; Pinard et al. in press). This lack of information makes it essential for forest managers to monitor regeneration in managed forests (looking for early warning signs of reproductive failure) and modify their management practices if regeneration is insufficient.

Encourage Natural Regeneration

Natural regeneration is far more economical and likely to succeed than intensive silvicultural treatments that involve planting (Dawkins and Philip 1998). It also offers the potential of producing more diverse

PHOTO 21-3 *Large trees often occur in clusters, which makes simple cutting rules based on diameter limits problematic when large gaps are not silviculturally desirable. (F. E. Putz)*

forests in which trees are locally adapted to environmental conditions. To promote natural regeneration:

- Protect advanced regeneration. During logging and other silvicultural operations, efforts should be made to ensure established seedlings, saplings, poles, and small trees of commercial species (i.e., advanced regeneration) survive unscathed. Clearly marking advanced regeneration in damage-prone areas (e.g., along skid trails) can reduce avoidable losses

- Maintain adequate densities of seed trees. If logging reduces the density of reproductive trees to the point that pollen does not regularly travel from one individual to the next, or seeds are unable to reach the majority of potential establishment sites, reproductive failures may occur. Our limited knowledge of reproductive tree ecology makes it impossible to specify the exact number and distribution of seed trees required to reforest a logged site. A conservative approach, therefore, might be to leave 25 to 50 percent of the preharvest, seed-bearing stems of a target species distributed in a relatively uniform manner across the harvested site

- Provide conditions favorable for seedling establishment and growth. Natural regeneration will not be successful if the trees that remain following logging produce viable seeds but logging practices create conditions in which their seeds fail to germinate, seedlings fail to establish, or saplings fail to grow. Based on the available knowledge of a target species' ecology, seedbeds should be created to maximize the germination and establishment success of the target species. Small-seeded, wind-dispersed, shade-intolerant species, for example, will likely establish in highest numbers where the soil litter has been disturbed to expose mineral soil, and where openings average 0.25 to 1 ha in size to ensure abundant sunlight

- Preserve genetically superior seed trees. The harvesting of all well-formed and genetically superior trees may lead to local genetic erosion of that species. The commonly held belief that big trees with hollow trunks are genetically defective is false. On the contrary, these large trees are survivors and may be left (with little opportunity cost to loggers and great benefits to wildlife) as seed trees

- Conserve *key* wildlife resources. The first priority is to maintain critical resources required by species that play key ecological roles, and/or those species that are threatened or endangered. Pollinators and seed dispersers are examples of species that play key roles (see chapters 3–14). In many forests, the majority of tropical tree species have seeds that depend on animals for dispersal (Gentry 1982). By consuming fruit and depositing seeds away from the parent tree, these animals greatly increase the chance that seeds will survive and become established as seedlings (see chapters 3 and 6). Threatened or endangered species are a high conservation priority and can be identified by local surveys or listings in the IUCN Red Data Book. Key resources that may require protection may include nest and denning trees, figs, important feeding areas, water sources, and gallery forests

Develop Proper Management Plan to Zone the Forest

Forests should have a long-term management plan (greater than 20 years) and annual operation plans. The long-term plan focuses on the forest or concession and identifies what harvesting will be done, why, where, and when (Dykstra and Heinrich 1996). Since forests should contribute to meeting environmental and social objectives, the management plan should not be limited to harvesting. Managers should outline how they will maintain environmental services, protect sensitive areas, and accommodate the needs of local communities, etc. (Klein and Heuveldop 1993). Where people live in or around the forest, involving them in the forest management process can be very important. Socioeconomic assessments are a prerequisite to understanding the needs and interests of stakeholders, particularly those in and around the management unit (or concession). Not all stakeholders extract resources from the forest or even live within the boundaries of the concession, however. Farmers located lower in a forested watershed, for example, may depend on clean water supplies produced by a forest (see photo 21-4).

Management units should be clearly zoned to identify production and conservation areas (see chapters 20 and 23). The location of these areas must then be clearly communicated to the loggers, who should be held accountable for following the plan. Protected zones may include the following areas:

Riparian Areas

Riparian areas should be protected to reduce erosion and sediment load in streams, maintain water temperature and microclimate, and serve as habitat for sensitive species (Phillips 1989). To reduce erosion, it is particularly important to protect the small headwater streams (although most regulations protect only larger streams). All streams should have a buffer. Relatively narrow buffers (from 10–30+ m) where no logging is permitted have been advocated to protect larger streams from sedimentation (APFC 1997; Sist et al. 1998b). Larger buffers, however, may be required if the riparian forest is to serve as

PHOTO 21-4 *Protecting riparian areas is an important step to conserving forest biodiversity and maintaining ecosystem functions, including water cycles. This riparian forest in eastern Venezuela, if left undisturbed, helps conserve many of the bird and mammal species most sensitive to logging. (D. J. Mason)*

habitat for sensitive species. In Venezuela, for example, unlogged riparian buffers larger than 100 m harbored many of the bird and mammal species most sensitive to logging (Mason 1995; Ochoa 1997). While more information is needed to accurately design riparian buffers, wider is clearly better—especially in areas of steep topography.

Other Sensitive Areas

Logging should generally not occur in swamps (but see Webb 1997), on very steep slopes (greater than 30–70 percent, depending on equipment) (Sist et al. 1998b), or in ecologically unique or key wildlife areas. If exceptions are made to log-sensitive areas, special extraction techniques (such as helicopter logging or cable logging of steep slopes ranging from 30–60 percent) may be required to minimize environmental damage.

Set-Asides, Genetic Reserves, and Protected Areas

Concessionaires should set aside areas that are representative of each forest type (Sayer et al. 1995). We recommend that at least 10 percent of each concession be protected (see chapter 29). This area should be in addition to that protected as streamside buffers. Protected areas should ideally be identified with the assistance of professional foresters/biologists and demarcated in the field (see chapter 23 for a more detailed discussion of protected area strategies).

Because set-asides take some forests out of production, one option to pay for these protected areas is to waive any area-based taxes that would normally apply (even though areas occur inside the concession). This incentive for conservation is a component of Bolivia's forestry law. Concessionaires can set aside up to 30 percent of the concession area, exempting those areas from the $1/year/ha area tax (Pavez and Bojanic 1998; see chapter 29). Under this system, concessionaires have an obvious financial incentive to set aside forests that have limited commercial value. Protected areas would ideally be representative of all forest types, including those that are commercially valuable.

Develop Annual Operation Plans

Annual operation plans are tactical, detailing how the broader prescriptions of the management plan will be implemented in a given part of the forest in a given year. They answer the questions of how forest harvesting will be done (in great detail), by whom, and when (Dykstra and Heinrich 1996; Sist et al. 1998b). These plans should include accurate topographic maps on a scale of 1:2,000 to 1:5,000 that identify the boundaries of harvest areas, water courses, swamps, rock outcrops, sites of religious and cultural importance, and communities and settlements (Dykstra and Heinrich 1996; Sist et al. 1998b). The maps should delineate streamside buffers and other areas where cutting is prohibited. The maps should be used to plan log extraction, to avoid stream buffers to the extent possible, and to minimize the total length of roads, skid trails, or cableways. Tactical plans must also give information on vegetation types and stand volume. In tropical forests that are selectively logged, harvesting maps should indicate the location of commercial trees, roads, and skid trails and the direction that each tree will be felled to minimize damage (see figure 21-1).

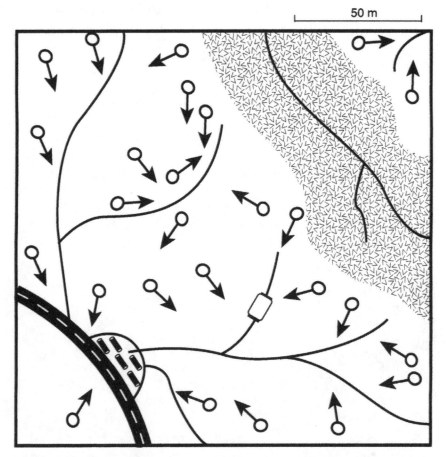

FIGURE 21-1 *An illustration of some of the elements captured in an annual harvesting map. Circles represent commercial trees to be harvested and arrows indicate the direction that they should be felled. The direction of felling is determined in the field to avoid damage to future crop trees (not shown) and to facilitate yarding. Thin lines represent skid trails that lead to major haul roads. The stippled area represents a riparian buffer surrounding a stream.*

Road Networks, Felling, and Extraction

Reduce the Direct Impacts of Roads

Roads are one of the greatest environmental impacts associated with forestry practices because they reduce forest cover, damage trees, compact soils, cause soil erosion, fragment the forest, and provide access to the forest interior (often leading to hunting, mining, and clearing for agriculture) (see photo 21-5). To minimize these environmental impacts, roads should be designed and constructed as described below.

Minimize the Total Length of New Roads

Shorter roads can cause less disturbance and hence less soil erosion. Extensive road networks fragment the forest and are difficult to control, increasing the risk of illegal extraction, hunting, and colonization. Minimizing the length of roads through proper planning (e.g., mapping the location of trees to harvest and topographic features in advance of the road system design) reduces these threats and lowers road construction and maintenance costs.

PHOTO 21-5 *Roads can cause some of the greatest impacts of forestry. They reduce forest cover, damage trees, fragment the forest, compact soils, and cause soil erosion. They also provide access to the forest interior, which often leads to hunting, mining, and clearing for agriculture. Properly planning and constructing roads, coupled with adequately controlling the access they create, are among the most important steps toward reducing the impacts of forestry. (D. J. Mason)*

Minimize the Area Disturbed by Roads

Road margins are often excessively wide in an effort to allow the sun to reach the road surface. As a result, an area of forest cleared is far wider than the road surface itself. The result is a loss of forest, increased erosion, weed infestations, and the creation of partial barriers to the dispersal of gap-sensitive wildlife species (Mason 1995). With proper construction—especially proper contouring and the application of sufficient gravel—these wide road clearings are not necessary. Dykstra and Heinrich (1996) suggest, for example, that the maximum clearing width should be 7.5 m for major haul roads and 5 m for minor roads—far less than the 18.4 m average they cite for forests in Papua, New Guinea.

Properly Site and Build Roads

Roads cause considerable erosion, resulting in up to 90 percent of the sediment loss in logged catchments (Bruijnzeel 1990). This is especially true where roads cross sensitive areas, such as steep slopes and streams. If stream crossings lack proper drainage and the upstream forest is flooded, the road bed stays wet (thereby reducing trafficability). When the blockage eventually gives way, substantial sediment loading of the streams results.

Planners should avoid placing roads on steep slopes, keeping road grades at 10 percent (6°) and constructing properly-spaced cross drains to channel water away from the road to reduce maintenance costs and soil erosion (Keller et al. 1995; Dykstra and Heinrich 1996; Sist et al. 1998b). When steep roads are necessary, the amount of cutting and filling should be minimized and the excavated material should be hauled to disposal areas that are not near streams or steep slopes (Dykstra and Heinrich 1996). Roads should also avoid wet areas, keeping the number of stream crossings to a minimum.

Appropriate road siting can reduce road construction and maintenance costs, increase the number of days per year and the number of years the road will be in service, and protect the environment. Unfortunately, road standards are often absent or not enforced because road builders are paid by the length of road they complete rather than their compliance with good road-building standards. Owners of a road—whether the government, concessionaire, or private landowner—should directly supervise road construction to ensure that the road is properly built, respecting all environmental guidelines to minimize damage. Well-trained forest engineers have a great deal to contribute to sustainable forest management.

Keep Vehicles Off Roads During Wet Weather

Substantially more damage by vehicular traffic is done when roads are wet. Restricting wet-weather use reduces erosion, while also reducing maintenance costs.

Control Forest Access During and After Logging

The greatest impacts of roads are often not the roads themselves, but the uncontrolled access they provide to previously inaccessible areas. *Controlling road access is often the single most important step that forest managers can take to conserve biodiversity.* Given that the least protected forests are usually those on public lands, concessionaires should be responsible for controlling access to their concession for the life of their contract in the following ways:

- **Plan road systems to facilitate control of access.** Road systems in concessions should be designed to facilitate control of access, minimizing the number of access points. A single entrance with a permanent camp nearby facilitates control by making it difficult for unauthorized personnel to enter the area
- **Close roads when not in use.** Roads should be gated and, when no longer in use, closed (blocked or partially destroyed). Installation of substantial cross drains can reduce both erosion and vehicular access. If the stand will be treated silviculturally at a future date, the road can be gated and opened only when treatments are underway. Because gates can be easily circumvented, removing cross drains more effectively prohibits access. Another possibility is to use portable bridges that are removed following timber extraction

When roads entering the concession are public, or when villages are located in or near the concession, the forest management process needs to become more participatory if access is to be controlled. Where logging roads cross village lands, entrances should be jointly monitored to limit activities that are agreed to be illegal by all parties.

Adopt Proper Felling and Extraction Procedures

Most of the damage caused by forest operations occurs during two steps: when trees are felled and when the logs are skidded from felling sites to roadside landings (F. Putz, personal observation). RIL practices seek to reduce the environmental damage associated with timber harvesting (Pinard et al. 1995; Bertault and Sist 1997; Sist et al. 1998b). In Malaysia, for example, RIL practices reduced damaged to unharvested trees less than 60 cm from 41 percent to 15 percent (Pinard and Putz 1996). In Indonesia, implementation of RIL reduced the damage from 48 percent to 25 percent, saving the equivalent of 100 stems greater than 10 cm dbh per hectare (Bertault and Sist 1997). In the eastern Amazon, unplanned felling damaged nearly twice as many trees per hectare as planned felling (see chapter 28), while the implementation of RIL spared more than 80 trees greater than 10 cm dbh per hectare (Johns et al. 1996). By reducing damage to the forest structure, these practices are also likely to reduce the impact of harvesting on biodiversity (Sayer et al. 1995). Although the principal components of RIL have been recognized for many decades (e.g., Bryant 1914), they are still rarely applied in tropical forests.

Directional Felling

As the name implies, directional felling involves careful planning and then felling of trees to facilitate skidding and avoid damage to the remaining forest vegetation. Proper felling reduces damage to soils and vegetation, protects the next tree crop, provides a greater volume of recoverable timber, and reduces the number of accidents (e.g., Stokes and McNeel 1990; Bertault and Sist 1997). To facilitate winching to the skid trail, trees should be felled either toward or away from skid trails or cableways at an oblique angle, preferably at 30 degrees (see Dykstra and Heinrich 1996; Tanner 1996; or Klasson and Cedergen 1996 for additional guidance on directional felling). Trees should be felled into existing gaps where possible. Trees should not be felled down slopes or into streamside buffers.

Well-Planned and Executed Extraction

Timber extraction damages the forest by disturbing and compacting the soil, impounding streams and loading them with sediments (see chapter 14), and damaging residual trees and other vegetation (Dykstra and Heinrich 1996). The impacts can impede regeneration and lead to declines in wildlife populations that depend on less disturbed forests (see chapters 4–14). To increase yarding efficiency and reduce damage to the soil and advanced regeneration, skid trails should be planned in advance of harvesting (using maps of the crop trees) and should be marked on the ground (Marn et al. 1981b; Hendrison 1990; Sist et al. 1998b). Skid trails should be as straight as possible and should avoid stream buffers (trees should be felled away from streams). Logs should be winched to the skidding machine, because driving the skidder to every log is uneconomical and unnecessarily damaging to soils and vegetation (e.g., Hendrison 1990) (see photo 21-6). The blade of skidding machines should be used rarely or not at all to minimize soil disturbance. Ground skidding should be avoided during very wet periods and on steep slopes.

When the slopes exceed 30 percent (17°) or in swamp forests, skyline cable extraction systems should be used (see Aulerich 1995 and Dykstra and Heinrich 1996 for additional details). These systems use a series of suspended cables to lift logs off the ground and then haul them from the felling site to landings, where they are loaded onto trucks. Skyline cable systems allow extraction on steep slopes where ground skidding (including *high lead* cable systems in which the log is skidded on the ground) cause excessive environmental damage.

PHOTO 21-6 *By simply pulling out the winch cable rather than driving skidders up to every log, damage to soil and advanced regeneration can be reduced substantially. (R. Fimbel)*

Where extraction distances are short (less than 200 m), logs are small, and slopes are gentle (less than 30 percent downhill and 15 percent uphill), log extraction by draft animals can significantly reduce soil disturbance, soil compaction, and damage to vegetation compared with mechanized ground-skidding equipment (Cordero 1995; Jayasekera 1995).

The least damaging of the modern extraction systems use balloons or helicopters to lift logs from the felling site to log yards (Arentz 1992). Extraction roads are still needed, but skid trails are not. At present, balloon logging is probably only economical with clear cutting operations and is rarely employed due to high costs and unfamiliar technology. Helicopters are also costly, but are used more widely to remove high-value trees. Helicopter yarding results in far less damage than any other timber extraction method available. One environmental problem associated with helicopter yarding is that logs can be harvested from steep slopes and swamps that might otherwise be spared from logging (see chapter 16).

A substantial portion of timber harvesting in the tropics is done by operators who saw logs into manageable pieces where they fall in the forest. The lumber is then carried out on the backs of these operators or their draft animals. These *pit sawing* operations result in relatively little damage to the forest structure, but they are difficult to monitor or control. Hunting by pit sawyers can also affect the forest.

Long-Term Stewardship

Encourage Adaptive Management

Given the complexities of tropical forests, our lack of silvicultural information, and rapidly changing social and economic conditions, forest plans must be flexible. Management must be responsive to new information and changing social, political, and environmental conditions. Adaptive management requires administrative agility, in which managers are willing and able to take risks and make course corrections in management plans (Walters 1986). In areas where a government authority controls management, local forest managers should be able to alter their management plans when justified and adopt new practices that contribute to sound forest management.

Improve Forest Monitoring

Adaptive management depends on a free flow of information to decision-makers, whether they are forest managers, government officials, or the general public. Two general types of information are needed:

Forest Health Following Logging and Other Silvicultural Treatments

Data must be collected repeatedly after logging and other silvicultural treatments to measure forest damage, tree regeneration, and growth. Permanent plots are necessary to monitor tree establishment and growth to better understand a forest's ecology, to design appropriate harvesting regimes, and to prescribe silvicultural treatments. Postlogging surveys should determine whether seed trees, future crop trees, riparian buffers, and other key elements of the landscape were protected during harvesting (see chapters 18 and 19).

A key question is who should manage and monitor permanent plots. Some countries have rules requiring forest concessionaires to establish and maintain permanent plots—in others, forest departments bear this responsibility. Unfortunately, the data generated by these plots are seldom analyzed and, when they are, the information is often so inaccurate as to be unusable (D. Mason, personal observation). Since concessionaires and the workers they hire are seldom researchers, they often need assistance from trained forest ecologists and foresters to develop estimates on forest damage, tree regeneration, and growth (see chapter 19).

Compliance with Forestry Agreements

Forest management involves a number of rights and responsibilities, particularly when conducted on public land. Better information, coupled with a willingness and ability to enforce rules and regulations, is needed to ensure that forestry management agreements are respected. Information on which areas are harvested and the width and siting of roads, skid trails, and log landings, for example, should be checked against the standards established in the harvest plan. This information should be used immediately to inform management decisions, and these decisions must be communicated to logging crews, with incentives for good performance and penalties for bad. On public lands, management plans and information on the concessionaire's performance should be available to all stakeholders—not just government officials. Bolivia's new forestry law, for example, states that any citizen accompanied by a certified forester is allowed access to forest concessions to check compliance with the law (see chapter 29; J. Nittler, personal communication). Every forest management document is also public information. For these checks to work, more nongovernmental organizations must become interested and able to evaluate management plans and compliance with forestry regulations (see chapters 19 and 26).

Control Hunting

Hunting is a major threat to large forest vertebrates and must be controlled (see chapters 15–17). Hunting is a particular concern because the most likely targets of hunting may play key roles in forest dynamics. Large frugivorous birds, monkeys, and even fish distribute the seeds of many canopy trees (see chapters 3–14). The genus *Virola*, for example, is harvested commercially in parts of its range, such as Brazil. *Virola* fruits are designed to attract toucans and to discourage other frugivores (Howe 1982). Toucans consume the fruit and carry these seeds and deposit them far enough away to increase their chances of germinating and becoming established by 44 times, relative to seeds that fall under the parent plant (Howe et al. 1985). In many tropical forests, the majority of tropical tree species have seeds that depend on animals for dispersal (Gentry 1982) (see chapter 3), primarily by birds and bats (see chapters 6–7). In the moist forests of Guyana, Suriname, and French Guiana, most timber species are dispersed by mammals (51 percent), and birds (21 percent), with a few being wind dispersed (20 percent) (Hammond et al. 1996). Animal dispersal also reduces the likelihood of inbreeding depression, maintains tree population structures, and facilitates the colonization of new sites, such as treefall gaps, needed for germination (Clark and Clark 1984; Murray 1988; Howe 1993; Schupp 1993).

Maintaining these ecological processes that structure and maintain forests is critical for long-term forest management. In French Guiana, for example, there is some evidence that hunting has reduced seed dispersal. Undispersed seeds accumulate below fruiting trees in hunted forests, whereas few such accumulations occur in forests that are not hunted (J. M. Thiollay, personal communication). Because seeds that have fallen under their parent trees suffer high mortalities, this may partially explain reductions of some animal-dispersed trees (such as

Aniba Rosaeodora, Lauracae) in heavily hunted forests along the populated coastal zone.

Local extirpation of large animals can have other cascading results. Large predators, such as cats, can control populations on mid-sized animals. Populations of these smaller animals can explode when *apex* predators are removed. As these animals become artificially common, they consume elevated levels of their favorite foods, ranging from the eggs of songbirds to favored tree seeds (Terborgh 1992a). Forest may change over time as a result.

The most appropriate mechanism for controlling hunting depends on the context:

Hold concessionaires accountable for controlling hunting, fishing, and the pet trade. Concessionaires should provide their employees with adequate food of domestic origin, ensure that their employees (or contractors) do not hunt, and control access to the forest so that market or sport hunters cannot enter. The last two requirements should be part of the concession agreement and it should be enforced.

Help communities and private landowners manage their wildlife. Where traditional residents practice subsistence hunting, a more appropriate goal may be to help communities manage their hunting so that it does not exceed sustainable levels. Reducing hunting pressure will obviously require cooperation between local residents and forest managers (see chapter 15). Efforts to assist local residents should:

- Document current use of wildlife (and where hunted species are becoming scarcer, use this as motivation for improved management)
- Help rural people to manage their hunting, through zoning of hunting areas, catch limits, gear limits, and so forth
- Discourage hunting of species that do not tolerate hunting pressure (such as primates)
- Develop alternative protein sources (ideally avoiding systems like extensive cattle ranching that require forest clearing) (see chapter 16)

Support Practical Research to Answer Management Questions

Practical research is needed to answer crucial management questions and fill key information gaps at both stand and landscape scales. Some of the most urgently needed research should address the ecologies of harvested species, their regeneration or reproduction requirements, and how to design improved incentives and disincentive systems to encourage improved forest conservation. Forest managers and conservationists also need answers to very basic questions on how to conserve biodiversity in production forests. What rules can we use in designing a system of unlogged protected areas within forest concessions? Should we recommend several large reserves or many small reserves? Do the reserves need to be connected by corridors? What should the spatial pattern of the reserves be? Which is better for wildlife conservation: harvesting a little over a large area or harvesting a lot in a little area—if both systems produce the same volume of wood (Crow and Gustafson 1997)? Should areas designated for logging each year (annual coupes) be clustered or isolated from one another? For a more thorough discussion of priority research topics for wildlife conservation, see chapters 18 and 19, and research recommendations at the end of most chapters.

Implementing Better Tropical Forestry

Forest managers are aware of many changes that they could implement to reduce the impacts of timber harvesting to promote forest stewardship. The question, therefore, is what needs to change for better forestry to be more widely applied?

The challenge of improving forestry is often more political than technical (Bawa and Seidler 1998). A number of excellent reviews discuss the pre-requisites to the adoption of better forestry practices (e.g., Johnson and Cabarle 1993; see chapter 18).

Policy Changes, Financial Support, and Improvements in Training

National Policy and Legal Systems Needed to Support Forest Stewardship

As a general rule, forest management is most likely to conserve wildlife when it creates an economic incentive for maintaining forest cover following logging operations. When large-scale development activities provide uncontrolled access to forest frontiers, deforestation is often accelerated. By encouraging rapid development on forest frontiers and flooding markets with cheap timber, government policies have discouraged long-term forest management (which cannot compete with a glut of cheap wood from areas being deforested). Policies must be changed to promote environmentally sound forest management. Policies that allow forest managers to plan over the long-term, for example (e.g., longer-term concessions or secure land title), are a necessary (but not sufficient) pre-requisite for forest conservation.

Forestry Institutions Need to Be Functional and Properly Financed

Most government forestry institutions are relatively poor. More revenues from forestry must go toward better training, monitoring, and enforcement of forest rules and regulations (Palmer and Synnott 1992; see chapter 18). Forest management agencies should be allowed to retain a greater portion of the receipts of taxes they collect (typically these are transferred directly to the government's central funds) for re-investment in forest management. Where forest management authority is devolving from central government to local stakeholders, forest service may also require decentralization.

Forestry Information Needs to Facilitate Public Participation

Transparency and participation in forest management decisions is key in engendering public support for forestry and in reducing waste and corruption. Information on public forestlands should be genuinely accessible to the public. This information should include who has development rights to an area, details of management plans, evaluations of concessionaire compliance with the terms of management plans, and so on. In very few countries is this type of information easily available, making it nearly impossible for the general public to know whether forest resources are being well managed (certified forestry operations are an exception—see chapter 26). This can foster corruption and poor management. Bolivia may become an exception insofar as it has taken great strides to make such data available to all interested parties (Lizarraga and Helbingen 1998; see chapter 29). Private environmental watchdog organizations in Bolivia are even authorized to check compliance with forest management plans themselves. Forestry that does not benefit people living in or near forests is often unethical (Pelusa 1992) and less likely to be successful, in part because of uncontrolled encroachment (Moad and Whitmore 1994).

Appropriate Education and Training Opportunities Needed in Forest Management

Three complimentary types of educational programs may be required:

- Vocational training for loggers, road builders, foresters, and non-governmental organization (NGO) staff interested in monitoring concessions
- University training on subjects that contribute to forest management (forestry courses should include modules on forest ecology, environmental services of forests, and conservation of biological diversity/wildlife management)
- Public environmental education/outreach to help the public better understand the importance of forests and their proper management

Universities, non-governmental organizations, governments, forest managers, international development organizations and other groups all have prospective roles to play in education/training programs.

Incentives for Better Forestry

The key to implementing better tropical forestry is an incentive system that includes rewards for good forest management and enforceable penalties for poor management. Proper tax systems, coupled with functional government forestry agencies, can provide both.

There are a number of potential incentives for good forest management. Conservation-oriented silvicultural prescriptions that are most likely to be adopted are those that have an immediate payback: road and skid trail planning, directional felling, careful extraction, and controlled access (Barreto et al. 1998; A. Moad, personal communication). The costs associated with planned logging, for example, can be more than offset by reduced machine time, reduced labor per volume of wood harvested, and by less waste (Hendrison 1990; Johns et al. 1996; Barreto et al. 1998; see chapter 28). Other postharvest silvicultural treatments, such as those designed to increase the density and quality of commercial species (e.g., thinnings), are less financially profitable in the short-term because their benefits are realized far into the future. These silvicultural treatments are also less likely to benefit wildlife. The good news for biodiversity conservation, therefore, is that techniques aimed at reducing harvest damage are among the most profitable for forest managers and the most beneficial for wildlife conservation.

How can better forest management be more efficiently and effectively promoted so as to avoid the degradation/deforestation alternative? Under some conditions, penalties for poor forestry practices, like fining individual loggers or boycotting timber from an entire country, might motivate improvements. Fines have proven difficult to collect, however, and boycotts may serve to drive down forest product values and thereby stimulate increased harvesting rates (Viana et al. 1996). Forest regulations certainly have a major role to play in improving management, but they are more likely to prove effective if coupled with positive incentives and increased environmental awareness of the public. Of the available incentives for better forestry, third-party forest product certification (Viana et al. 1996; see chapter 26) and forest-based carbon offsets (Faeth et al. 1994; Trexler and Haugen 1995; Moura-Costa 1996; USIJI 1996a; Brown et al. 1997; Pinard and Putz 1997; see chapter 27) seem to be some of the most promising.

Focusing attention on the need for proper incentives is crucial. The technology to dramatically improve forestry is at hand, but unfortunately it is too infrequently applied. This does not diminish the real need for significantly more research on tropical forests, the needs of people that live in and around them, and the economic and political contexts in which they are managed. We have much to learn, as natural forests in the tropics are under-studied. But even today, we know enough to make very substantial improvements, where there is the will to do so.

ACKNOWLEDGMENTS

Initial ideas for this chapter were developed by a working group composed of the authors and Jan Falk, Robert Wallace, Rene Boot, Jose Carlos Cornesos, Betty Flores, Ivo Kraljevc, Alicia Grimes, Johan Zweede, and David Hammond. Rob Fimbel, Todd Fredericksen, Damian Rumiz, and two anonymous reviewers provided useful comments on the manuscript.

APPENDIX 21-1 *The Benefits and Costs of Select Forestry Practices in Terms of Timber Production and Wildlife Conservation*

Forestry Practice		Advantages/Benefits		Disadvantages/Constraints	
		Timber Production	Conservation	Timber Production	Conservation
Proper Planning: Choosing an appropriate harvest-regeneration system	Monocyclic systems (e.g., clearcuts, shelterwood cuts)	Large volume of wood harvested per unit area; can create the large disturbances needed for some shade-intolerant tree species	Loggers may enter the forest less frequently; may conserve species that require large disturbances	Where the density of commercial tree species is low, clear cutting is uneconomical	Creates more forest disturbance than selective logging, changing the forest structure upon which some mature forest wildlife species depend
	Polycyclic system (e.g., selective logging)	Where the density of commercial tree species is low, selective logging may be more economical. Low-intensity selective logging creates small disturbances that can favor shade-tolerant tree species	Where density of commercial trees is low, creates less forest disturbance than monocyclic systems; this maintains more original forest structure and probably more wildlife diversity	Potential loss of commercial tree species that require larger disturbances than those created by selective logging (i.e., high grading)	Particularly extensive road networks required can facilitate hunting and forest conversion. Selective logging can cause considerable damage, particularly where extraction rates are high
	Silviculture that seeks to maintain all tree species, including those that do not regenerate under current management practices (e.g., selectively logged mahogany and pine)	Valuable species regenerated for future harvests, maintaining value of forest. Potential for secondary species to enter market in the future	Conservation of tree diversity and associated species; logging operations may mimic natural disturbances	Silviculture required may be difficult and costly	Silviculture may dramatically change forest structure (compared to selective logging), causing more damage and affecting sensitive wildlife species; keeping access open for silviculturalists increases the risk that hunters and agriculturalists will enter the management area

APPENDIX 21-1 The Benefits and Costs of Select Forestry Practices in Terms of Timber Production and Wildlife Conservation (continued)

Forestry Practice	Advantages/Benefits		Disadvantages/Constraints	
	Timber Production	Conservation	Timber Production	Conservation
Developing Management Plans — Improved management unit zoning to identify areas that should not be logged	May reduce tax burden on concessionaire. There may be reduced "obligations" to cut difficult areas	Increase area of unlogged forest, maintaining habitat for sensitive species that could later recolonize logged areas; critical areas protected and environmental services (e.g., water flows) maintained	Costs of monitoring and enforcing the zoning (ideally done by an independent body) and the potential loss of commercial trees that would not be logged	Lack of information—are a few large reserves better than many small ones? (Note: large reserves are easier to manage). What size? How many? What arrangement?
Improved management plans (management plans and annual operating plans)	More effective extraction, better coordination of activities, less damage, overall harvesting cost reduced, systematic planning of transportation system reduces road costs	Reduced forest damage means reduced impacts on many sensitive wildlife species	Increased cost of planning and the need to monitor any recommendations in the plan. Lack of trained personnel to conduct planning and supervise implementation	Increased monitoring costs; lack of information (e.g., should logging be dispersed or concentrated?)
Protecting trees of special ecological value (keystone species, seed trees, snags, etc.)	May contribute to natural regeneration	Maintaining key tree resources will help conserve biological diversity	Opportunity cost of not harvesting a species; additional costs associated with planning, directional felling	Lack of information needed to identify which tree resources are key
Silvicultural Practices to Increase Stocking — Site preparation using fire	Inexpensive where fires natural; quick flush of nutrients; controls insect pests	Can mimic natural disturbance; can be less damaging than herbicides or mechanical site preparation	Risk that fire will spread; postfire weed problems; smoke	Risk that fire will spread (risk usually small in intact humid forest, but may be high in logged forest due to increased fuel); loss of slow-moving animals; smoke can kill arboreal animals

APPENDIX 21-1 *The Benefits and Costs of Select Forestry Practices in Terms of Timber Production and Wildlife Conservation (continued)*

Forestry Practice		Advantages/Benefits		Disadvantages/Constraints	
		Timber Production	Conservation	Timber Production	Conservation
Silvicultural Practices to Increase Stocking	Site preparation using herbicides	Quicker and cheaper than mechanical treatments when labor or machine costs high	Less soil disturbance than mechanical treatments	Costly compared to mechanical treatments where labor is cheap; potential health problems; requires well trained personnel and equipment; may not prepare seed bed	Risk of contamination
	Mechanical site preparation	Favors seedling establishment of species that require exposed mineral soil (such as mahogany and pine)	No herbicides are used	Labor and machine time are required and expensive	More disturbance to the soil and potential loss of soil productivity through the removal of topsoil; erosion; potential loss of non-timber forest products
	Enrichment planting (planting seedlings of commercial species in logged areas, often after additional clearing)	Long-term investment to increase stocking	Can rehabilitate very degraded forests, providing an incentive to maintain them as forest	High maintenance costs first 5 to 10 years; often ineffective, usually because of a lack of tending	Can cause homogenization of forest (in terms of both species composition and genetic diversity) and high levels of collateral damage

Forestry Practice		Advantages/Benefits		Disadvantages/Constraints	
		Timber Production	Conservation	Timber Production	Conservation
Silvicultural Treatments to Improve the Growth of Timber Trees	Vine cutting	Reduced collateral damage during felling and increased postharvesting timber production; increased worker safety during felling	Decreased tree damage during felling; vines regenerate vigorously, reducing the risk they will be eliminated from forest	Labor costs; logistics of coordinating vine cutting in advance of the harvest	Some vines produce nectar and fruit that is important to some species; in dry forests, vines produce leaves during more of year, providing folivores food; vines provide access to trees for nonflying canopy animals
Silvicultural Treatments to Improve the Growth of Timber Trees	Liberation thinning by girdling (killing noncommercial trees that overtop future crop trees)	Competitive release and increased growth may increase profitability, providing an incentive to maintain forest	Less damaging, when properly practiced, than thinning treatments, as only areas near crop trees are treated and no trees are felled; can help create "old growth" conditions (e.g., snags), enhancing forest structure	Cost; removes trees that may become valuable as markets change; not always effective silviculturally, in part because long-suppressed trees may not accelerate growth; requires trained personnel	If not carefully controlled, can domesticate forest, eliminating species that are ecologically valuable; potential for pollution when herbicides are used improperly
Roads and Access	Proper planning of roads to minimize overall length and to avoid sensitive areas, such as streams and steep slopes	Lower aggregate road length results in fewer unnecessary roads and lower construction, maintenance, and transportation costs; increased number of days per year and increased number of years that roads are in service	Reduced sedimentation, forest fragmentation, and, in some cases, reduced colonization and/or hunting	Lack of trained personnel to conduct planning and supervise implementation	None apparent.

Forestry Practice	Advantages/Benefits		Disadvantages/Constraints	
	Timber Production	Conservation	Timber Production	Conservation
Improved road standards (restrictions on road siting and width, culverts for proper drainage, erosion control, etc.)	Lower road maintenance costs; less productive forest lost to roadways	Improved drainage (fewer flooded forests), reduced siltation (better aquatic conservation), and lower overall habitat disturbance	Initial investment is high, as roads with good drainage cost more to build	Forestry activities may have to intensify to justify the added costs
Roads and access Mechanisms to control hunting	May increase natural regeneration because some hunted species are seed dispersers	Hunting is a major threat to large vertebrates; controlling hunting would greatly contribute to the conservation of these species	Would remove recreational opportunities and the meat "subsidy" provided to workers; added cost of importing domestic supplies; poor communities may depend on subsistence hunting[1]	None
Mechanisms to control access	Risk of forest conversion and illegal logging greatly reduced	Risk of forest conversion and hunting can be greatly reduced	Concessionaires may be reluctant or unable to control access; limiting the number of entrances to large concessions may be inconvenient and costly	None

Forestry Practice		Advantages/Benefits		Disadvantages/Constraints	
		Timber Production	Conservation	Timber Production	Conservation
Felling and Extraction	Directional felling	Increased skidding efficiency as trees felled are oriented towards preplanned skid trails; reduced damage to soils, streams, and vegetation; increased utilizable timber volume by reduced breakage; reduced damage to advanced regeneration; reduced logging injuries	Reduced damage to forest structure and water courses; less soil erosion; reduced damage to ecologically important tree species	Requires trained workers (often felling is done by unskilled workers with little training or supervision). Skilled workers may demand better salaries and benefit	None apparent
	Pit sawing and portable sawmills	Low cost and low tech; can be used in a wide variety of terrain; high potential for local employment	Less damage from extraction; unused portion of tree left in forest, exporting fewer nutrients	Low productivity; conversion efficiency into saw timber may be low	Difficult to monitor and control; sawyers often live off land and hunt
Felling and Extraction	Use of rubber-tired skidders	Faster and less expensive to purchase and to operate than bulldozers; require less maintenance	Cause less damage to soils and trees than bulldozers when used properly	Cannot be used where the topography exceeds 30 percent; less versatile than bulldozers as the blade is small	Can cause more compaction than tracked vehicles
	Winching logs to skidders or bulldozers	Reduces machine time, which is expensive	Reduces forest damage as length of skid trails is reduced; need to maneuver skidder around log is eliminated	Properly trained personnel required; machines with winches are more expensive; dangerous if cable snaps	Can lead to logging of areas inaccessible to skidders or bulldozers (e.g., wet or steep sites)

Forestry Practice		Advantages/Benefits		Disadvantages/Constraints	
		Timber Production	Conservation	Timber Production	Conservation
	Animal traction	Low cost; generally low damage to forest; provides income to local people	Generally little damage to forest	Large logs cannot be removed whole; limited skidding distances; extraction is slow, and animals must be rotated	Draft animals may introduce ungulate diseases into forest, infecting wildlife. Also, draft animals may require grazing land, which may imply further deforestation
	Semimanual extraction systems (converted light trucks with manual winches)	Low cost; appropriate for small-scale use	Low impact, as roads are built by hand and volumes harvested are small	Low and erratic productivity (equipment breakdowns common)	Difficult to monitor
	Skyline cable extraction	Ability to log steep slopes and swamps	Requires a road network only 1/2 to 1/3 as long as if ground skidding used	Usually more costly than ground-skidding except on steep slopes; the need for highly trained technicians	Logging steep slopes may cause soil erosion; logging swamps disturbs critical habitat
Felling and Extraction	Aerial (helicopter or balloon) extraction	Greater access over a larger area; can reduce road density and eliminates the need for skid trails; little collateral damage and almost no soil disturbance; very selective logging possible	No skid trails required, and road network reduced (even more than with skylines), leading to less road damage and lower risk of colonization than with traditional logging (to be economical, roads must still be built, usually within 2 to 5 km of felled trees)	High initial and recurrent costs of operating the machinery; the need for highly trained people, most of whom may be expatriates; need for high-value crop trees; logging is usually a high grading	Makes it possible to harvest steep slopes and wet areas that would otherwise be protected; difficult to monitor; may not reduce hunting pressure, as roads are still built

[1]Generally, it is not recommended to attempt to domesticate wild (hunted) species. Traditional domestic species are much more productive than wild species, as they are selectively bred to reproduce and gain weight. For efficient, cheap replacement of meat supply, domestic animals are often better. Also, if the animals are used in any way for trade, then having domestically reared individuals of wild species makes enforcing laws controlling the trade of hunted individuals of that species nearly impossible. There is also the potential for disease to spread from domesticated to wild populations.

Chapter 22

AN EVOLUTIONARY PERSPECTIVE ON NATURAL DISTURBANCE AND LOGGING

Implications for Forest Management and Habitat Restoration

Colin A. Chapman and Robert A. Fimbel

Natural disturbances alter the structure and composition of tropical forests. The intensity and frequency of such events vary from daily limb and treefalls (Denslow 1980, 1987; Brown and Whitmore, 1992; Whitmore and Brown, 1996) to periodic large-scale disturbances, such as landslides, hurricanes, and fires that can damage hundreds of square kilometers (Garwood et al. 1979, Whitmore 1991). Recovery times from these perturbations may range from less than a year (e.g., limbfalls), to a number of centuries (e.g., after landslides that remove topsoil). Where geographical variations occur in the nature or extent of large forest disturbances, forest communities appear to have evolved to perpetuate themselves in the wake of such disturbances (Pickett and McDonnell 1989; Oliver and Larson 1990; Whitmore 1991, Perry and Amaranthus 1997). Silvicultural practices that mimic the local frequency and intensity of natural disturbance patterns have the potential to conserve large proportions of the evolutionarily adapted biodiversity found in dynamic tropical forest landscapes.

The aim of this chapter is to consider a conceptual framework linking the evolutionary and ecological processes of natural disturbance to management prescriptions for tropical forests (Whitmore 1991). The discussion begins by reviewing the concept that forest recovery following a disturbance event is a function of the evolutionary adaptations by the species in that area, local site conditions, and the intensity of the disturbance. Linkages between these factors and their influence on the type of forest community that might develop in an area are explored. The chapter closes with a discussion of how foresters can incorporate evolutionary and ecological information about a site into the development of silvicultural treatments—thereby somewhat *mimicking* the historical natural disturbance conditions that gave rise to the forest community in that area. Specific research recommendations on how to refine this process are noted.

An Overview of the Hierarchy of Factors Influencing Habitat Recovery

There are three hierarchical levels influencing forest recovery following a timber harvest: evolution, local environmental conditions, and modifications to the natural environment induced by the logging operation (see figure 22-1). Plant and animal communities experience a variety of natural disturbances that operate on different scales from isolated treefall gaps to hurricanes and landscape-level fires. The ability of rainforest plants and animals to recover from a disturbance event (natural or human-induced) is partially the result of natural selection that has operated in response to the disturbances a forest community has experienced over its evolutionary history (Skorupa and Kasenene 1984; Stocker 1985; Hartshorn 1989, 1990; Uhl et al. 1990; Whitmore 1991). Ecological conditions occurring at the local level temper this natural selection process, thereby influencing the speed of the recovery after a disturbance event. All other factors being equal, for example, a site located on poor soils will recover at a slower rate than a neighboring site on rich soils (Grubb 1995; Medina 1995). Finally, specific features of the logging operation will affect the recovery process (see chapter 2), with areas experiencing extensive logging road and skid trail networks recovering more slowly than areas with fewer disturbances of this kind (Sessions and Heinrich 1993a; Frumhoff 1995).

The importance of a forest's evolutionary history to its post-logging recovery should be emphasized. The rate of forest regeneration and the speed of recovery of wildlife populations vary greatly between sites, and this variation cannot be predicted based on the knowledge managers would typically have. The different evolutionary pressures operating in different forests probably account for some of the variability observed. An example of the recovery of primate populations following logging clearly illustrates the level of variability in how wildlife responds to logging. Blue monkeys (*Cercopithecus mitis*) have been classified as extreme generalists based on studies conducted in the Kibale National Park of Uganda (Johns and Skorupa 1987; Butynski 1990), and therefore should adapt relatively well to changes in their habitat following a

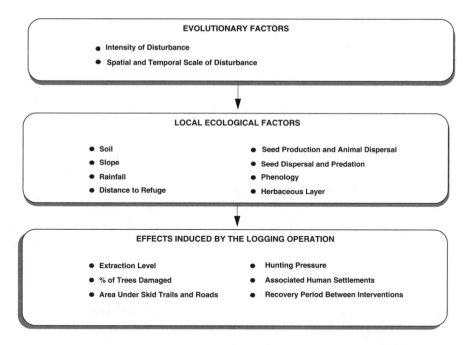

FIGURE 22-1 *Three hierarchical levels, and some of the factors within each level, influencing forest recovery following a timber harvest.*

disturbance such as logging. At the same site supporting this description, however, logging had a severe negative influence on this species (Skorupa 1988). Fifteen years after logging, logged areas had 20 to 30 percent fewer blue monkeys than unlogged areas (Skorupa 1988). Their populations continue to decline 28 years after the logging (C. Chapman, unpublished data). In contrast, blue monkeys in the Budongo Forest Reserve of Uganda are 3.7 times more abundant in logged areas than in unlogged areas (Plumptre and Reynolds 1994; see chapter 4).

Natural Disturbance Regimes and Their Influence on Forest Composition

Are there large-scale natural disturbance patterns that occur across continental or regional scales? Are there geographical variations in the frequency and nature of such disturbances? Do these conditions give rise to specific forest- and stand-types? The most complete evidence of geographical variation in large-scale disturbance events, and their subsequent influence on forest structure and composition, comes from the temperate zone (Oliver and Larson 1990; Barnes et al. 1998). As one example, the estimated recurrence interval between natural fires in North America (pre-European settlement) varied from two to 1,000 years (see table 4-1 in Oliver and Larson 1990). High fire frequencies usually give rise to forest-types dominated by fast-growing, early seed-producing, shade-intolerant species such as certain pines (e.g., *Pinus banksiana, P. rigida, P. taeda*), while low intervals of fire recurrence tend to support slower-growing, later-seed producing, shade-tolerant species such as sugar maple (*Acer saccharum*) and red spruce (*Picea rubens*). There are numerous exceptions to this simplistic cause-and-effect relationship, however, as a multitude of disturbance (wind, floods, erosion, siltation, landslides, etc.), topographic, edaphic (soils), and biotic (pests, competitors, mutualistic organisms, etc.) factors influence what species are capable of establishing and developing on a given site. Foresters are becoming increasingly aware of the multitude of variables shaping forest development, and continually seek ways to incorporate this knowledge into their silvicultural prescriptions (Kohn and Franklin 1997; Smith et al. 1997).

A similar suite of environmental variables determines the composition and structure of major tropical forest-types, and stands within them (Letouzey 1968,1985). Information on the natural disturbance regimes and the basic ecology of most tropical forest plants and animals is very limited, and often quite general. Broad regional assessments of natural disturbance events exist for the tropics (see box 22-1), however the high heterogeneity characterizing these large land areas permits only the most general insight into how tropical forests are likely to respond to timber harvesting practices. The ecology of most tropical species, including those exploited for timber, is also still largely unknown. There are exceptions to this rule. For a few select tropical tree species, information exists on their ecology, the forest communities in which they grow, and the forest management recommendations necessary to maintain them in a healthy and productive state (e.g., Greenhart [*Chlorocardium rodiei*] dominated tropical rain forest in Guyana—Steege et al. 1996; Ek 1997; Zagt 1997). Unfortunately, these examples are limited in number, and the skills necessary to implement them often lag far behind the research leading to these recommendations (see chapter 18).

Box 22-1 *Natural disturbance regimes at the continental and regional levels in the tropics, and their influence on forest composition.*

The most complete evidence of geographical variation in tropical forests comes from studies of seismic activity. Approximately 18 percent of tropical rain forests lie in zones of high seismic activity. This includes 38 percent of the Indo-Malayan forests, 14 percent of the South-Central American forests, and less than one percent of the African rain forests (Garwood et al. 1979). In American and Indo-Malayan forests, the effects of earthquakes can be dramatic. In 1976, for example, two earthquakes struck off the coast of Panama triggering extensive landslides that affected at least 450 km^2 of forested land and denuded approximately 54 km^2 (Garwood et al. 1979). In 1935, two similar earthquakes hit Papua New Guinea and denuded 130 km^2 of forested land (Garwood et al. 1979; Johns 1986d, 1992b).

Hurricane and cyclone activities also vary geographically. Hurricanes cause frequent natural disturbances affecting Atlantic coastal areas from northern South America, through the Caribbean and Gulf Coast region, to eastern North America (Boose et al. 1994). Between 1871 and 1964, an average of 4.6 hurricanes affected the Caribbean each year, with hurricanes hitting select areas once every 15 to 22 years (Tanner et al. 1991; Walker et al. 1991b). Between 1920 and 1972, 362 cyclones hit Madagascar (Ganzhorn 1995). The effect of hurricanes and cyclones on forest systems can be dramatic. When Hurricane Joan touched land in Nicaragua, for example, 80 percent of the trees were felled (Boucher 1990). African forests do not appear to be similarly influenced by cyclones or hurricanes.

Outside hurricane and cyclone belts, large windthrows can be common. Large windthrows have been reported from South America (Dyer 1988), southeast Asia (Whitmore 1984), and Africa (Thomas 1991b). Windthrows ranging in size from 30 to 3,370 ha have been documented, as widespread occurrences in mature Amazonian forest (Nelson et al. 1994).

Rivers often play large roles in changing forest dynamics in tropical forest areas with flat topography (Foster et al. 1986). In the Amazon basin, for example, 26 percent of the forest has characteristics of recent erosional and depositional activity caused by changes in the paths of water courses (Salo et al. 1986; Salo and Kalliola 1990).

There has been a great deal of debate over the historical frequency and extent of large-scale fires in moist tropical forests. Leighton and Wirawan (1986) reported on fires in southeast Asia that damaged 3.7 million ha, and suggested that massive die-offs from drought and fire occur in East Kalimantan once every several hundred years. Sanford et al. (1985) discovered that charcoal was common in the soils of mature Amazonian rain forest, and that these forests have repeatedly experienced fires over the last 6000 years. Based on charcoal deposits, Hart et al. (1997) describe that fires were relatively common but small (i.e., less than 1 ha) in the Northeast Democratic Republic of Congo starting 4,000 years ago. Hart et al. (1997) attributes the greater frequency of fires during the last two millennia to increased human activity. While conclusive evidence is lacking, and the role of human activity in initiating fires remains unclear, it appears that large-scale natural fires in moist tropical forests are more common in South America (Bush and Colinvaux 1994) and southeast Asia (Whitmore 1984) than in Africa (Tutin et al. 1996).

Continental and broad regional assessments permit only the most general insights into how tropical forests are likely to respond to timber-harvesting practices, owing to the very heterogeneous conditions characterizing these large land masses. Following the framework outlined earlier, investigations are best focused on a smaller spatial scale (e.g., forest type and stand levels) if efforts to reduce habitat disturbance and expedite forest recovery following logging are to succeed. One must always keep in mind, however, that the processes operating at a higher level may constrain what happens at a lower level.

The Hierarchical Concept and Silvicultural Refinement

There are a variety of studies that detail how the extraction level (Ward and Kanowski 1983; Wilkie et al. 1992) and road and skid trail design (Sessions and Heinrich 1993a; Frumhoff 1995) influence forest structure, composition, and rates of recovery in forests after logging (see table 22-1).

The interactive nature and extent of variability in these factors make it difficult to construct predictive recovery models. This is not to suggest that nothing can be done to expedite the postharvest recovery of logged forests. Techniques exist to minimize disturbances to forest structure, composition, and ecological processes during logging operations, thereby facilitating the recovery process (Pancel 1993; Pinard et al. 1995; Pinard and Putz 1996; part 5 of this volume). Many of these recommendations are also well known to foresters and harvesters. Further research in this area is a lower immediate priority compared with identifying the means of implementing what is already known (Palmer 1975; Struhsaker 1997).

Where information is limited about the ecology of a forest, including its historic exposure to disturbance events, what *cues* can foresters take from the existing forest to help in designing silvicultural prescriptions that are likely to expedite the recovery of the treatment area after logging? To begin, the regeneration and growth cycles of plants (and the animals that use this habitat) appear to be compatible with the frequency and intensity of disturbance events (Grime 1977; Pickett and White 1985; Oliver and Larson 1990). Knowing this, foresters must access any available information on the shade-tolerance and growth rates of important commercial timber species. Diameter distributions from forest inventories should help to confirm suspected shade-tolerance levels for species in questions (i.e., reverse *j* distributions indicate

TABLE 22-1 *Select Factors Influencing Forest Habitat Recovery After Logging in Tropical Forests*

Factor	Reference
Soil fertility	Ewel 1980
Soil nutrients and pH	Jordon 1990; Grubb 1995; Medina 1995
Water stress associated with logging	Bazzaz 1990
Changes in mcorrhizae associated with timber extraction	Herrera et al. 1990
Nature and extent of skidding, soil compaction, and erosion following timber yarding	Uhl et al. 1982; Douglas et al. 1992; Cannon et al. 1994; chapters 2, 13, 14
Extent and rate of herb and shrub growth	Burgess 1975; Fitzgerald and Selden 1975; Brokaw 1983; Kasenene 1987; Yap et al. 1995
Extent of liana growth	Putz 1992
Proximity to undisturbed refugia and other seed sources	Viana 1990; Bierregaard et al. 1992; Marcot et al. (see chapter 3); Jansen and Zuidema (see chapter 23)
Availability of seeds (seedbank and seed rain) and seed dispersers	Fox 1976; Viana 1990; Whitmore 1991; chapters 3 -14
Ability of the trees to coppice	Whitmore 1991
Level of extraction and incidental damage	Uhl and Viera 1989; Wilkie et al. 1992; White 1994b; chapter 21
Death of damaged trees	Putz 1993; Putz and Chan 1986
Probability of fire inhibiting regeneration	Uhl et al. 1981; 1988c; Uhl and Buschbacher 1985; Janzen 1988; Woods 1989; Uhl and Kauffman 1990; Nepstad et al. 1991

shade tolerance, while distributions skewed towards a few large stems suggest shade intolerance). Growth rates usually have to be extrapolated from plantation trials or a few natural forest-monitoring sites (see box 22-2). These two variables, combined with the distribution of commercial stems across the forest (rare, clumped, and mono-dominant), should give a reasonable estimate of: a) when these trees established, and b) the intensity of the disturbance event that lead to their establishment. Large (100–200+ cm diameter) stems of the relatively shade-intolerant Ayous (*Triplochiton scleroxylon*), Sapelli (*Entandrophragma cylindricum*), Frake (*Terminala superba*), Kapok (*Ceiba pentandra*), and others (see box 22-2) dominate areas of the timber-rich, semi-deciduous forests of southeast Cameroon. The size and abundance of these species suggests that they became established under a canopy that was much more open than the near-closed conditions characterizing these forests today. Foresters should, therefore, consider refining silvicultural prescriptions to create the necessary conditions that these species have evolved to require for their establishment and growth. These might include small clearcuts, shelterwood cuts, or group selection cuts (Smith et al. 1997). This approach should not only help to re-establish the harvested stems, thereby maintaining the value of the forest by helping to prevent its conversion to other land uses (see chapter 20), but it should also create environmental conditions that other plants and wildlife in this forest type are pre-adapted to.

Box 22-2 *Treading lightly may not always be the best silvicultural prescription.*

Several semi-deciduous forest types dominate the forest of southeast Cameroon (Letouzey 1968, 1985). Selective timber harvesting is common in these forests, focusing on a few high-value mahoganies (*Entandrophragma* spp., *Guarea* spp., and *Pericopsis elata*; see box photo 22-1), and several lower-value species that occur in very high densities (*Triplochiton scleroxylon*, see box photo 22-2, and to a lesser extent *Terminalia superba*). Within these species, merchantable stems are selected based on a tree's minimum diameter (60–100 cm), bole quality, and accessibility.

In 1995, 55 ha of upland forest were inventoried within a 400 km^2 area of unlogged forest (CTFT 1985; Fimbel et al. 1996, Fimbel and Runge, submitted). Extensive areas of forest appear to have established within the last 200 to 800 years, given the ecology of the forest dominants: *Albizia adianthifolia, Alstonia boonei, Ceiba pentandra, E. cylindricum, P. elata, T. superba, T. scleroxylon,* and others. These shade-intolerant, wind-dispersed species (Hall and Swaine 1981, Vivien and Faure 1985, Keay 1989, Hawthorne 1993), which currently lack representatives in their small-medium diameter classes (see box figure 22-1), must have been established at a time when shade conditions (i.e., forest basal area) were significantly lower than they are today. Regional diameter growth rates for the three most common canopy species—*T. scleroxylon, T. superba,* and *E. cylindricum* (which, combined, accounted for 24 percent of the recorded basal area)—suggest that the largest individuals of these species established 140 to 880 years BP (growth rates range from 1.3-13.8 mm/yr, n=1142; see Maitre and Hermeline 1985; API Dimako 1995). Radiocarbon-dated samples of *E. cylindricum* trees in northern Congo support these growth estimations. Thirty trees, ranging in diameter from 88 to 197 cm (in the survey, the largest individuals of this species approached 220 cm diameters), exhibited a mean age of 272 years (ages varied from 80 to 510 years, with low correlation between size and age of individual stems) (Fay 1997).

Current timber-harvesting prescriptions in this region call for selective cutting to remove all merchantable stems above the minimum allowable diameter limit. To date, impacts to the forest structure have been relatively limited, owing to forest planning (stocking surveys with subse-

BOX PHOTO 22-1 One meter in diameter Iroko (*Pericopsis elata*) log awaiting log buyer in southeast Cameroon. (R. Fimbel)

BOX PHOTO 22-2 Large 2m dbh Ayous (*Triplochiton scleroxylon*) felled across main haul road in southeast Cameroon. (R. Fimbel)

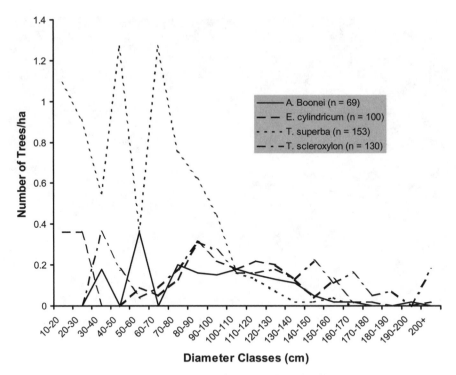

BOX FIGURE 22-1 *Stems/ha, by diameter class, for four moderate- to shade-intolerant canopy tree species in the semideciduous forest of southeast Cameroon.*

quent road minimization), gradual slopes, low-to-moderate stocking of merchantable stems (mean of 3 to 4/ha), and limited pressure to convert the forest to other land uses (Fimbel et al. 1996). This approach appears to represent good forest management and the conservation of the forest resource, as most of the forest habitat remains intact. These conditions, however, may be relatively superficial. The majority of seed trees are being removed in the first and second cuts of this forest while creating small openings that are unsatisfactory seedbeds for the majority of commercial species harvested (Fimbel, personal observation). This approach amounts to little more than *mining* the forest resource. Continuing this silvicultural course is likely to lead to the long-term degradation of the timber resource (and habitat in general) as important commercial species fail to establish and recruit into the merchantable-size classes, thereby increasing the risk of forest conversion to more profitable land-use activities (e.g., oil palm plantations).

Foresters and conservationists would do well to look closely at what the pre-logged forest composition and structure indicate about how these forests became established. The dominance of shade-intolerant species suggests that even-aged silvicultural prescriptions, while initially higher-impact than selection harvest-regeneration systems, may in the longterm be the most suitable management strategy for maintaining the diversity and productivity of these semideciduous forests. Time scales appropriate for the regeneration and recovery of these long-lived species must also be considered.

The above example is quite *coarse* in its approach, and foresters need to consider any other available information (soil properties, rainfall patterns, phenology, etc.) to help them in refining the size, shape, distribution, and frequency of their silvicultural interventions. Lacking extensive evolutionary or ecological information about a forest type, however, this approach does present an opportunity for resource managers to consider a wider range of silvicultural options than the selection cutting system that is applied with near uniformity across tropical production landscapes today.

On a cautionary note, foresters must keep in mind that while logging operations can be designed to mimic natural disturbance events, they cannot duplicate these conditions. Difficulties associated with seedling establishment in compacted soils on logging landings and skid trails, for example, are unique to logging with ground-based yarding (Pinard et al. 1996). Here the soil is compacted in a fashion that the flora (and fauna) is not equipped to combat (Whitmore 1991). Many of the trees damaged during the harvest also subsequently die, expanding the size and connectivity of gaps to sizes that are atypical in natural settings (Putz and Brokaw 1989; Cannon et al. 1994; see chapter 2). Putz and Chan (1986) documented that in an old-growth mangrove forest, all trees that suffered even slight mechanical damage die within a decade of being damaged. Logged areas in Uganda had up to 47 percent greater mortality than unlogged areas 30 years after logging (Chapman and Chapman 1997). Much work remains, therefore, in documenting the effects of timber-harvesting operations on forest structure and composition (see chapter 19), and refining silvicultural interventions to mitigate these environmental impacts (part V this volume), before prescriptions will exist that approach conditions characterizing natural disturbance events.

Research Recommendations to Promote Silvicultural Treatments That Mimic Natural Disturbances

There are a multitude of research studies needed to help minimize the environmental impacts associated with timber harvesting, while maintaining the economic bottom line of these operations (see other contributors to this volume). Where efforts are focused at refining silvicultural prescriptions so that they better mimic natural disturbance events and cycles, important areas for further research include the assessment of:

- Current and historic disturbance events that gave rise to present stand conditions; this includes the identification of disturbance type, intensity, frequency, and spatial impact across an area of forest

- How one disturbance event can predispose the forest to other disturbances. Logging or blow-downs, for example, can lead to increases in fuelloads that may predispose a forest to a catastrophic fire

- How the frequency of silvicultural interventions lead to a deviation from the natural forest recovery process; the arrangement of silvicultural treatments in the landscape over time is especially important for the maintenance of interior forest conditions

- Strategies used by species to survive disturbance events; this basic ecological information is critical to the perpetuation of commercial, threatened and endangered, and other species of interest in managed tropical landscapes

- Seasonal effects on the recovery process, including phenological conditions prior to and immediately after the logging event

- Techniques to assess how quickly the forest habitat and ecological processes recover after logging

ACKNOWLEDGMENTS

The work described here from Kibale National Park was supported by the Wildlife Conservation Society, Lindbergh Foundation, National Geographic Grants, an NSF Grant, and USAID PSTC Funding. We thank the Government of Uganda, the Forest Department, and Makerere University for permission to work in Kibale Forest. The work in Cameroon was supported by the Wildlife Conservation Society. We thank the Ministry of the Environment and Forests for their permission to work in the Lobeke region, and the numerous Baka and Bangando field technicians that helped in the procurement of the field data. We thank A. Grajal, J. Robinson, and A. Vedder for asking us to participate in this project. D. Livingstone provided helpful insights into understanding patterns of large-scale habitat disturbance in Africa. L. J. Chapman, A. Johns, A. Kaplan, L. Kaufman, D. Onderdonk, C. Peres, J. Putz, L. Sirot, T. T. Struhsaker, and A. Treves provided helpful comments on this work.

Chapter 23

PROTECTING HABITAT ELEMENTS AND NATURAL AREAS IN THE MANAGED FOREST MATRIX

Bruce G. Marcot, R. E. Gullison, and James R. Barborak

How can plant and animal species, ecological communities and processes, and ecosystems be maintained in areas also being used by people for their natural resources commodities? For most developing societies, extensive preservation of natural resources or reverting to indigenous resource use habits and smaller human population sizes (e.g., Alcorn 1993, Gadgil and Berkes 1993) are simply not options. Rather, the answer may lie in managing forests as ecosystems and for long-term conservation of wildlife habitat, as well as for human use.

Much scientific literature has highlighted the value of natural areas for conserving the biodiversity of a region (DellaSala et al. 1996; Falkner and Stohlgren 1997) and the need for linking natural areas with habitat corridors or connections (Harrison 1992). Biodiversity is defined as the variety of life and its processes—here our focus is on wildlife and their habitats. Natural areas can serve as refuges for hunted, threatened, or imperiled species of plants and animals, as demonstrated by Neotropical migrant birds in Mexico (Gram and Faaborg 1997) and large vertebrates in Paraguay (Hill et al. 1997).

In the tropics, protecting at least small patches of native forest and linking them can greatly help provide for populations of game and nongame wildlife (Glanz 1991; Grieser Johns 1996). Protecting biodiversity in tropical forests, however, entails more than simply reserving specific forest sites. Bawa and Seidler (1998), for example, concluded that retaining natural tropical forest diversity also entails conserving some wildlife habitat within secondary forests, restoring degraded lands, managing plantation forestry, providing for nontimber uses for some forests, changing accounting procedures to reflect the true value of natural forests, and supporting forestry agencies charged with protecting forest reserves. Obviously, protecting biodiversity in managed tropical forest landscapes is a challenge that spans ecological as well as economic, social, and political arenas (Panayotou and Ashton 1992).

Managing for non-timber forest products in the tropics also can be integrated with conserving habitat for wildlife and managing natural forests, such as has been demonstrated in Rio San Juan, Nicaragua (Salick et al. 1995). In other examples (Petén, Guatemala, and West Kalimantan, Indonesia), non-timber extractive reserves were found by Salafsky et al. (1993; see chapter 25) to play a key role in forest protection, but their conservation effectiveness was dependent on local ecological, socio-economic, and political conditions.

In some cases, natural areas that provide for rare or endemic species or high biodiversity coincide with sites of high religious or cultural significance. These can include sacred forest groves and sites with spiritual meaning. In India, China, Myanmar, and Thailand, for example, sacred groves and cemeteries with banyans (*Ficus* spp.) and other trees sometimes provide the last local bastion for indigenous wildlife (Wachtel 1993). In West Africa, sacred groves can help protect habitats for small mammals (Decher 1997). In the Garo Hills of Meghalaya, India, the state government has designated entire national parks (e.g., Balpakram National Park) and wildlife sanctuaries (e.g., Siju Wildlife Sanctuary) based not only on their biodiversity value, but also on their significance to the animistic religious beliefs of the local Garo Hills tribes (B. Marcot, personal observation). Sacred groves and spiritual sites can survive without official designation, but their formal recognition in a conservation network can help ensure their future protection.

Protecting wildlife habitat is one facet of maintaining the ecological integrity of managed tropical forests (see chapters 3–14). By planning protected areas as a network, their value to wildlife can be strengthened. Protected areas of various types and sizes, within managed landscapes, serve as important refugia and corridors for wildlife.

The purpose of this chapter is to recommend where and how wildlife (vertebrates, invertebrates, and their habitat) can be protected within tropical production forest landscapes. We begin by defining terms and providing an overall framework for understanding wildlife habitat. We then briefly review what is known about the elements of that framework, including habitat corridors, habitat connections, and natural areas, and their ecological, economic, or social values. We next outline a set of habitat elements and landscape conditions that managers can map and delineate in the field—using existing information— for wildlife and ecosystem conservation. We end with a discussion of research needs, and we offer some conclusions on a concerted approach to wildlife habitat conservation that spans ownership, species, and ecosystems.

Protected Areas and Their Objectives

The conservation biology and wildlife literature extensively discusses the utility of delineating protected areas (Harris 1984; Hunter 1990; Meffe and Carroll 1994; Beier and Noss 1998). Authors have focused on a wide variety of factors when recommending protected areas, including maintaining the genetics or dynamics of specific plant or animal populations, and protecting rare species, unique species assemblages, rare plant communities, rare successional stages, ecosystem dynamics, global climate change, and economics (see box 23-1). Other objectives of protected areas include the protection of a representative sample of all vegetation types (representativeness) and capturing the full range of habitat conditions within a region (complementarity).

Box 23-1 *Reasons for delineating protected areas and protected area networks.*

The following is a list of reasons for protecting natural areas, and providing for natural elements in the managed forest matrix. In many cases, such protection can provide multiple benefits simultaneously. The manager can use this checklist to determine which objectives might pertain to their needs, and then determine the specific habitat elements to include in a management plan (see text).

1. Protection of scarce terrestrial vegetation conditions or communities

— Protection of old-growth, ancient, or primary forests (Barnes 1989)

— Protection of degraded ecosystems (Schmidt 1996)

— Providing for rare or sensitive plants (Lesica 1992)

— Protection of floral diversity (Rebelo and Siegfried 1992)

2. Protecting vertebrate and invertebrate wildlife

— Providing for rare or viable populations of individual species (Ryti 1992)

— Providing for multiple species (assemblages, guilds, communities, richness) (Kershaw et al. 1995)

— Providing for multiple species (island biogeography theory) (Higgs 1981)

— Providing for invertebrates (Stokland 1997)

3. Protecting aquatic and riparian systems (Rabe and Savage 1979)

4. Providing for research and management (chapters 18–22, 24, 25)

— Protection of baseline conditions for measuring change and effects of management (Arcese and Sinclair 1997)

— Providing for other research use, including experimental forestry (Stoms et al. 1998)

5. Providing for specific ecological conditions or processes

— Soil conservation (Burke et al. 1995)

— Providing for natural disturbance events and regimes (e.g., fire, hydrology; Baker 1989)

— Accounting for climate change effects (Halpin 1997)

— Representing ecological conditions ("representativeness") (Bedward et al. 1992)

— Staving off invasion by exotic species (Soule 1994)

— Accounting for natural spatial and temporal variation in resources and in the distribution and abundance of species that use them (Brown et al. 1995)

6. Providing for, or consideration of, human uses, interests, and needs

— Areas for economic income and incentives (Pressey et al. 1993)

— Areas to offset adverse effects of urbanization on habitat loss (Bolger et al. 1997)

— Areas for other uses by nonindigenous peoples (esp. recreation; Gotmark and Nilsson 1992)

— Areas for traditional (indigenous) resource use:

—Ethnobotanical or ethnobiomedical uses (also benefits nonindigenous peoples) (Peres 1994)

—Forestry (Pinedo-Vasquez et al. 1990)

—Providing for other cultural or economic values (Gotmark and Nilsson 1992)

—Accounting for poaching (Lewis et al. 1990)

—Sites of religious or ritual significance (e.g., spirit groves; Decher 1997)

— Providing for a long-term source of plants and animals that may contribute to the productivity of adjacent managed forests used mostly for commodity production (no examples found in the literature)

SOURCE This list was derived from an extensive review of the literature conducted by the senior author. An example reference is given for each category; a fuller list of literature is available from the senior author upon request.

Protected natural areas in timber management landscapes can serve as refugia for native plants and animals and as means of conserving biodiversity (Hunter 1990; Sayer et al. 1995). Within such production forests, protected natural areas can be used to conserve sensitive sites such as wetlands, grottos, and steep slopes, and act as source areas from which organisms can recolonize postlogged sites. These refugia can range in size from small zones of uncut forest along watercourses and hillsides, to larger blocks reserved from commercial timber harvests (e.g., parks and nontimber extraction areas). Collectively, the set of conserved habitat elements and the larger refugia may be called a *protected area network* (PAN) (see table 23-1).

TABLE 23-1 *A Protected Area Network or PAN Can Consist of Four Major Elements, Listed Here with Corresponding IUCN (International Union for the Conservation of Nature and Natural Resources) Categories of Protected Areas*

PAN Element	Examples	Corresponding IUCN Category (see appendix for detailed description)
(1) Individual habitat elements conserved within a portion of the cutting block (concession, management unit, or whatever term is locally used) that is actually open to harvesting	Large overstory trees bearing many bromeliad epiphytes; standing dead or hollow trees; select fruit-bearing shrubs; large down wood; concentrated feeding areas such as salt licks; wildlife watering holes	These elements are usually so small and common as to not require specific zoning; rather they are located on detailed silvicultural management maps and plotted through GPS systems where available; they occur within IUCN categories IV and VI
(2) Small natural area zones (that is, small parcels and corridors reserved from cutting within a concession, cutting block or management unit)	Grottos with rare plants; groves of old-growth trees; riparian buffers along streams or around springs; wetlands; nest sites of rare birds	Specific management zones within extractive categories, primarily VI (Managed Resource Protected Area), but also category IV (Conservation through active management)
(3) Reserves of large size between cutting blocks (but still within the production forest landscape)	Riparian gallery forests along major river courses; travel corridors for migrating wildlife; pristine, baseline research natural areas; unique natural phenomena	Where they have stricter statutory protection than the surrounding production forest matrix, they can be IUCN categories I, II or III; otherwise, they are zones or management units within categories IV and VI, and primarily the latter
(4) Parks or similarly protected areas such as wildlife sanctuaries outside of the production forest landscape	Strict nature reserves, wilderness areas, national parks, ecological reserves, and natural monuments	Categories Ia, Ib, II, and III

The four PAN elements, from smallest to largest, are:

1 = Local habitat elements within the managed forest matrix
2 = Zones of protected forest conditions within the managed forest matrix
3 = Habitat connections and corridors between and among managed forests and larger protected areas
4 = Large protected areas such as parks

A schematic of how these four elements might occur on the landscape is shown in figures 23-1 and 23-2.

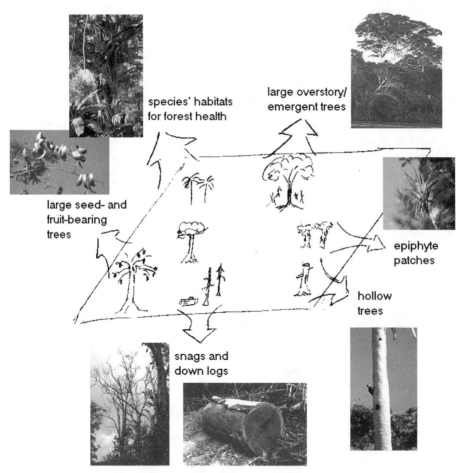

FIGURE 23-1 *Examples of protected area network (PAN) element 1 components—individual habitat elements that can be protected during forestry operations. These elements do not necessarily require specific zoning of protection areas within the managed forest matrix. (B. G. Marcot)*

PAN elements 1 and 2 are within the purview of planners, field foresters, and logging operators. These habitat elements and natural reserves within timber concessions (or cutting blocks—that is, the managed forest matrix) may include sites or stands that are of limited interest for timber harvesting and forest production, such as steep sites, wet areas, areas of low commercial tree stocking and low productivity, and individual trees of low or no commercial value. Natural areas within cutting blocks also can include some areas of special interest to a particular region or country, such as areas with unique habitats, fragile soils, high cultural values, and rare plants. Conserving specific habitat elements in the managed forest matrix (PAN element 1), and providing small natural areas as zones within timber concessions (PAN element 2), should help conserve wildlife and their ecological processes by:

- Preserving sites important for protected or rare plants and animals that do not occur elsewhere, or that have been severely impacted elsewhere
- Safeguarding specific locations of resources—such as key fruit trees or open water—important to maintaining populations of some wildlife species throughout a broader landscape or region
- Contributing to conserving the full variety of ecological communities and native biodiversity of a broader region (Zuidema et al. 1997)

The larger PAN elements 3 and 4 are vital parts of the overall protected area network, although their specific discussion is beyond the scope of this chapter. Delineating PAN element 3 areas—the larger habitat linkages and reserves occurring between the cutting blocks—can follow many of the guidelines presented below for PAN element 2. PAN element 3 areas are typically larger than the natural area reserves within the cutting blocks. This element protects habitats important for wider-ranging wildlife, maintains genetic diversity of plants and animals, and provides natural ecological processes such as fruit production. To the extent possible, PAN element 3 should be interconnected to within- and between-cutting block reserves and corridors of older secondary forests or uncut primary forests in upland or riparian situations. The larger, between-cutting unit reserves of PAN element 3 should also be connected to large parks and preserves (PAN element 4) outside the production forest area.

The four PAN elements can be expressed in the complementary roles of federal, state, private forest lands, and associated parks, preserves, wildernesses, and conservation forests established for preserving biodi-

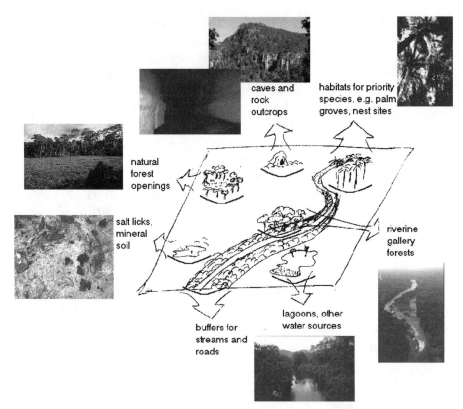

FIGURE 23-2 *Examples of protected area network (PAN) element 2 components—natural areas. These elements can be mapped and zoned as protection areas within the managed forest matrix prior to harvest operations. Specific sizes, orientations, and patterns of these components will vary widely among tropical forests. (B. G. Marcot)*

versity in the United States (Loomis 1993). They also can be related to traditional IUCN land allocation categories (see table 23-1). All elements must be designed in concert for them to function correctly, that is, to help provide for productive forests and to conserve or restore native biodiversity and ecosystem processes.

What Information Is Needed for Delineating Protected Area Networks?

Information useful for delineating protected areas and identifying specific habitat elements within a timber harvest area or concession includes the following:

Distribution of Terrestrial Vegetation Conditions or Communities

The distribution of terrestrial vegetation communities and conditions can be described as:

- Maps of vegetation types
- Maps of vegetation structural stages, age classes, and/or disturbance classes
- Information on the effects of forest management activities on vegetation types, structural stages, age classes, and/or disturbance classes

Such information might be obtained from a variety of sources including aerial photos, stocking surveys, and forest inventories. This information describes the location and extent of scarce terrestrial vegetation types or communities that may be adversely changed by forest management conditions. Where published scientific studies are lacking, much of this information may have to come from local experts such as in-field biologists and ecologists, zoologists, and habitat managers.

Wildlife Species

The distribution and abundance of wildlife species can be described by:

- Lists of species of amphibians, reptiles, birds, and mammals (and selected plants and invertebrates of conservation interest)
- Information on the general abundance and distribution of each *focal* species
- Information on area requirements of the focal species
- Information on specific resource requirements of focal species (Ryti 1992)

Focal species are those of special conservation interest which can include species of known viability concern, species with large body size or large home ranges (that is, large habitat area requirements—

e.g., the Group 2 or 3 species in figure 23-3 and table 23-2), species reliant on rare conditions or resources, and species hunted heavily and extirpated elsewhere. Much of this information on wildlife species may have to come from local experts or be extrapolated from other geographic areas with similar vegetation and environmental conditions (see chapters 4-14).

Plants and Animals Important to Forest Productivity

Identifying and describing the importance of plant and animal species to the productivity of managed forests requires:

- Information on ecology of soil productivity and tree production
- Lists of species and ecological conditions providing such services, particularly those more closely associated with unmanaged forest conditions

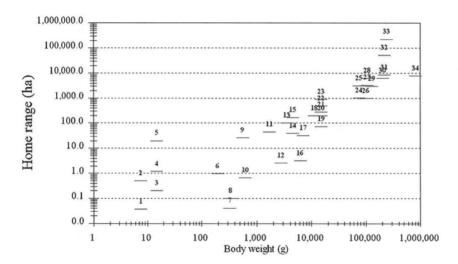

FIGURE 23-3 *What sizes of natural areas are needed to provide for various species? This may be answered by the size of home ranges of wildlife. Here, home range size of selected tropical mammals is plotted as a function of their body weight. Note the logarithmic axes. Species are coded as shown in table 23-2. Three major groups of species can be identified based on home range area (note that such body size and home range groups were also identified for northern temperate wildlife species by Holling (1992): species with home ranges up to 20 ha, up to 1,000 ha, and over 1,000 ha. Each group can be provided by a different PAN element or combinations thereof, as discussed in the text. Note that carnivorous species (e.g., tiger, species 31-33) often have larger home ranges and area needs, than do herbivorous species (e.g., gorilla, species 30, and gaur, species 34) of the same body weight class. (Nowak 1991)*

TABLE 23-2 *Key to Species Shown in Figure 23-3, with Species Listed in Order of Increasing Mean Body Size and Home Range Area (see figure 23-3 for source)*

Entry No.[1]	Common Name	Scientific Name	Location[2]
GROUP 1 (body weight < 20 g, home range up to 20 ha):			
1 f	Pygmy mice	*Baiomys taylori*	P
2 m	Pygmy mice	*Baiomys taylori*	P
3	Northern pygmy gerbil	*Gerbillus gerbillus*	P
4	Tarsier	*Tarsius bancanus*	P
5	African swamp rat	*Otomys irroratus*	P
GROUP 2 (body weight 200 to 20,000 g, home range up to 1,000 ha):			
6 f and m	Tree shrew	*Tupaia glis*	P
7 f and m	Wood rat	*Neotoma lepida*	N
8	Spiny rat	*Proechimys trinitatis*	N
9 g	Tamarin	*Saguinus nigricollis*	N
10	Titi monkey	*Callicebus moloch*	N
11 f	African palm civet	*Nandinia binotata*	P
12 g	Brush-tailed porcupine	*Atherurus africanus*	P
13 m	African palm civet	*Nandinia binotata*	P
14 g	Coati	*Nasua nasua*	N
15 m	Coati	*Nasua nasua*	N
16 p	Duiker	*Cephalophus monticola*	P
17	Howler monkey	*Alouatta spp*	N
18 f	Ocelot	*Felis pardalis*	N
19	Macaque	*Macaca fascicularis*	P
20	Macaque	*Macaca nemestrina*	P
21	Macaque	*Macaca radiata*	P
22	Macaque	*Macaca silenus*	P
23	Macaque	*Macaca mulatia*	P
GROUP 3 (body weight > 50,000 g, home range over 1,000 ha):			
24 f	Jaguar	*Panthera onca*	N
25 f	Jaguar	*Panthera onca*	N
26	Sloth bear	*Ursus ursinus*	P

TABLE 23-2 *Key to Species Shown in Figure 23-3, with Species Listed in Order of Increasing Mean Body Size and Home Range Area (see figure 23-3 for source) (continued)*

Entry No.[1]	Common Name	Scientific Name	Location[2]
27 m	Jaguar	*Panthera onca*	N
28 m	Jaguar	*Panthera onca*	N
29 f	Tiger	*Panthera tigris*	P
30 g	Gorilla	*Gorilla gorilla*	P
31 m	Tiger	*Panthera tigris*	P
32 m	Tiger	*Panthera tigris*	P
33 m	Tiger	*Panthera tigris altaica*	P
34 g	Gaur	*Bos gaurus*	P

[1]m = males; f = females; p = pair; g = group; blank = not specified

[2]P = occurs in the Paleotropics (Old World); N = occurs in the Neotropics (New World)

These species are those that play key ecological roles, such as plant pollination, seed dispersal, soil turnover, creation of tree cavities and burrows, decay and physical breakdown of wood, and nutrient cycling (see chapters 3–14). This kind of information is hard to quantify, but ultimately may prove vital in maintaining long-term productivity of managed tropical forests.

Habitat Connectivity

The extent to which the habitat for focal plants and animals is connected across the landscape can be assessed by use of vegetation and habitat maps, and regional maps showing locations of parks and other reserves (PAN elements 3 and 4), managed forests, and general information on their conditions. The vegetation and habitat maps can be related to general habitat requirements of the focal species, and the regional maps can show where such conditions occur that could provide important connections across the landscape.

Guidelines for Identifying Protected Zones and Habitat Elements Within Cutting Areas

Key Habitat Elements to Be Protected in the Managed Forest Matrix (PAN Element 1)

This section lists key habitat elements that can be protected during timber harvest and other activities in the managed forest matrix outside protected zones.

Snags and down logs. Snags (standing dead or partially dead trees) and down logs (logs on the forest floor) provide critical habitats for many species of plants, invertebrates, lizards, snakes, cavity-using birds, and other kinds of wildlife (Grieser Johns 1997). In turn, many of these organisms aid in the breakdown of wood and its eventual return to productive soil. Such organisms, including ants and termites, also serve as a major prey source and base for food chains in the forest (Sazima 1989; see chapters 12–13). Primary cavity-excavating birds such as woodpeckers, barbets, and toucans often create hollows in snags, which are used in turn by a variety of secondary cavity-using species such as swallows and small owls (see chapters 8–10). In general, larger diameter snags and down logs are used more extensively by wildlife than smaller ones. At a minimum, snags 25 to 30 cm diameter and 3 m tall, and down logs at least 25 cm diameter and 3 m long, provide habitat for many species. Large species of woodpeckers, owls, hornbills, flying squirrels, and other species may require snags or hollow trees at least 40 to 50 cm diameter, and large ground-dwelling mammals may require logs over 35 to 40 cm diameter for den sites. Wherever possible, snags and down logs should be left undamaged—whether standing or on the forest floor—to provide these services.

Trees prone to hollowing. Trees prone to hollowing provide cavities for a wide variety of wildlife (Grieser Johns 1997). Such trees include figs and palms, which also provide important fruit food sources. Some trees with physical defects, such as large dead branches, lightning strikes, and butt rot, might be prone to hollowing and can serve as indicators of potential value to wildlife.

Epiphyte patches. Patches of epiphytes, such as terrestrial or epiphytic bromeliads, occur in many forest types and are of particular value to wildlife (Remsen 1985). These patches provide important food sources (fruits and stalks) for birds, tapirs (*Tapirus* spp.), deer, monkeys, peccaries or wild boar (*Sus scrofa*), and other species. The patches also provide shelter for wildlife species that disperse seeds. The water-filled interiors of individual epiphytes often provide habitat for many frogs and invertebrates, some of which occur or breed nowhere else in the forest. Epiphytes can be protected for wildlife as individual habitat elements within managed forest situations, such as in the coffee plantations of Mexico (Moguel and Toledo 1999).

Large seed- and fruit-bearing trees. Large seed- and fruit-bearing trees (see photo 23-1) provide important food sources for toucans, hornbills, birds of the family Cracidae, many monkeys including spider monkeys and langurs, and many other arboreal and terrestrial fruit-eating wildlife species (Johns 1988). Figs (*Ficus* spp.) are keystone resources that appear to be high in calcium and provide fruit throughout the year. Although timber managers may not desire seed predation, the ecological function of seed dispersal by wildlife is critical to maintaining forest diversity and distribution of some desirable trees (Howe 1977, 1980).

Protect large overstory/emergent trees. Where possible, protection of large overstory or emergent trees can greatly add to the overall vertical structure of forest vegetation (see photo 23-2). These trees provide feeding, resting, cover, and nesting habitats for a very wide variety of birds, monkeys, invertebrates, plants, sloths, and many other species (Pearson 1971). Large overstory trees, even a few per hectare, can help serve as wildlife habitat connectors linking protected areas of primary forests within a managed, secondary forest landscape (see photo 23-3; B. Marcot, personal observation).

Sites and Habitat Elements for Species Critical to Forest Health and Productivity

The following are key habitats and sites for groups of species that play critical roles in maintaining the health and productivity of forest ecosystems. For many of these species groups, specific publications and the local knowledge of experts can be sought to help identify key types of habitats and specific sites for protection.

Habitat for essential plant pollinators. In the Neotropics, nectar-feeding or long-tongued bats (family Phyllostomidae, subfamily Glossophaginae) pollinate nectar-producing plants bearing large, pale flowers with musky odors that open at night. Key habitats for these bat species include caves, rock outcrops, tree hollows, and patches of dense pri-

PHOTO 23-1 Example of a PAN Element 1 component: Large seed- and fruit-bearing trees. This mature tree (*Aspidosperma* sp.) in a timber management area of lowland Amazonian Bolivia has deep fluting of its bark and is prone to hollowing. It provides not just fruits and overstory canopy cover for arboreal wildlife such as monkeys and toucans, but its bark crevices and hollows can provide nesting or roosting habitat for bats, birds, reptiles, and other wildlife.
(B. G. Marcot)

mary forest and deciduous forest. Bat-pollinated New World plants include balsa trees, ceiba (silk-cotton) trees, and jicaro (calabash) (Emmons and Feer 1990).

PHOTO 23-2 Example of a PAN Element 1 component: Large overstory/
emergent trees. In some forests, large overstory or emergent trees can be
retained during timber harvest operations to provide high foliage, branches,
and food resources used by some wildlife. This example shows a mature
Dalbergia paniculata retained in a managed teak forest of central India.
Dalbergia also is prone to fluting of the bole and thus can also provide bark
crevices for roosting and nesting wildlife. (B. G. Marcot)

PHOTO 23-3 Example of a PAN Element 1 component: Large overstory/ emergent trees across the managed forest matrix. Retaining at least selected large overstory or emergent trees across a landscape of managed forest can help maintain suitability of the overall managed forest matrix for some species of canopy-using birds and other wildlife. In this photo, many overstory trees have been retained in this managed forest just outside Reserva de Amigos de la Naturaleza de Mindo on the Pacific slope of the Andes Mountains in Ecuador. This type of management helps to retain key overstory lek (breeding) sites for the rare Andean cock-of-the-rock (*Rupicola peruviana*). However, wildlife species requiring extensively interconnected tree canopies, such as many monkeys, sloths, tree squirrels, and others, will not be able to persist in such circumstances; for them, it may be better to provide the retained trees in locally dense clumps at least 0.1 to 0.5 ha in size (see home range sizes in figure 23-3). (B. G. Marcot)

Hummingbirds, orioles, sunbirds, and a multitude of invertebrates serve as important pollinators for many plant species not served by bats. Collectively, this bird and invertebrate group occupies a wide array of nesting or breeding substrates, including dense low shrubs, riparian thickets, understory tree canopies, tree bark, and other sites. Individual nest or breeding sites—as known or as discovered opportunistically or by surveys—can be protected during forest management operations.

Habitat for essential seed dispersers. In the Neotropics, short-tailed fruit bats of the genus *Carollia* (family Phyllostomidae, subfamily Carolliinae) feed on fruits of shrubs and small trees (see chapter 7), especially Piper species. They are one of the most important seed dispersers for many plants, carrying fruits into cleared areas, and stimulating forest regeneration after logging. In some instances, these bats feed on mangos and banana crops and may become pests (Hill and Smith 1984), but they serve critical seed dispersal functions in forest environments (Emmons and Feer 1990). They do well in disturbed habitats but need tree hollows, caves, overhanging banks, tunnels, culverts, or abandoned buildings for roosting, as well as understory forest vegetation for cover and feeding.

Although many of the forest plants pollinated by these bats are not of commercial value, the array of other wildlife associated with such plants can provide pollination or dispersal services to commercial tree species. It is this array of ecological services that constitutes an ecosystem, and which foresters may wish to conserve to ensure the overall ecological health and integrity of their managed forests. To this end, foresters may wish to protect known or potential roost and feeding sites for short-tailed fruit bats in order to maintain their ecological functions within a forest ecosystem. For these species, this would entail maintaining such elements within mature or disturbed forests, gardens and plantations, deciduous forests, and gallery forests.

All species of Neotropical fruit bats of the subfamily Stenodermatinae (family Phyllostomidae) feed on fruit, supplemented by flower nectar in the dry season when fruit is scarce (Emmons and Feer 1990). These bats are the main dispersers of seeds for many plants. They carry seeds for early secondary or early successional growth into forest gaps caused by natural disturbances or cutting, thus restoring new forests and maintaining plant species richness of the forest. They can play roles of dispersing seeds of plants into newly burned or cut sites, thereby helping to stabilize soils or add quickly to soil organic matter—helping maintain soil productivity for future forestry use. These bats, like many other Neotropical and Old World species, require hollow trees for their roosting sites. Foresters can identify and protect key roost and feeding sites of these species. Roosts are used many times and are identified by piles of seeds, feces, and wads of fibers below large leaves that may be several meters above the ground (Emmons and Feer 1990).

In both Old and New World tropics, many species of monkeys disperse seeds of plant species—especially canopy trees and lianas (see chapters 3 and 4). Sites known to be heavily used by monkey troops can be identified by field surveys and protected to provide for monkeys, maintain the continuity of the tree canopy, maintain the diversity of fruit and vegetation food resources, and control hunting.

In many forests, birds (including toucans, aracaries, trogons, cotingas in the New World, and hornbills and fruit pigeons in the Old World serve as key seed dispersers for many canopy plant species, including trees of commercial value (see chapters 8–10). Individual nests, often in cavities and hollows of trees of limited to no commercial value, can be located and protected during forest management operations.

Habitat for fungal spore dispersers. Some wildlife species play critical roles in dispersing spores of underground fungi, including mychorrizal fungi, which can be critical to the growth, productivity, and health of many commercial forest trees (see chapter 6). Examples in the New World are the spiny rats (*Proechimys* spp.) and rice rats (*Oryzomys* spp.) (Janos et al. 1995), which also are important prey for small carnivores including large snakes and birds of prey. These rats also depend on fruits, seeds, and insects and likely also disperse seeds of palms and other trees of commercial interest. To provide for such species, den sites and ground cover could be maintained as well as appropriate fruit- and seed-bearing plants. Spiny rats and rice rats shelter under dense fallen brush, in hollow logs, or in holes in the ground (Emmons and Feer 1990)—such substrates could be provided (brush piles) or protected (hollow logs, ground burrows) during ground-disturbing forestry operations.

In another New World example, brocket deer (*Mazama* spp.) feed on fungi on decaying logs and may serve to disperse underground fungi, including mychorrizal fungi. Specific bedding, birthing, and key foraging sites for brocket deer could be protected. In general, for bedding and birthing, deer use forest stands with hiding cover. Hiding cover can be any foliage up to two meters high that largely obstructs the visibility of a standing deer within a few dozen meters distance. Deer foraging areas include dense brush less than two meters high that show obvious sign of pruning. Where possible in areas of known deer occurrence, such understory foliage for hiding and foraging could be provided for at least a portion of a forest stand. This would often not entail any special management activities or directives other than ensuring that the entire understory cover is not removed during operations.

Habitat for predators of seeds, seedlings, and animals. In both the New and Old World tropics, pigs (wild boar) and peccaries have an important influence on the spatial distribution of plants, including some of the commercial palm species (e.g., heart-of-palm). Pigs and peccaries can decline from hunting and from habitat destruction (see chapters 5, 15, and 16). In Neotropical forests, pacas (*Agouti paca*) and agoutis (*Dasyprocta* spp.) eat large seeds of trees and aid their dispersal by caching (burying and storing) seeds (see chapter 6). In Paleotropical forests, such seed eating is done by wild boar and other species. Large carnivores—including puma (*Felis concolor*) and jaguar (*Panthera onca*) in the New World, and tiger (*P. tigris*) and leopard (*P. pardus*) in the Old World—control the herbivory impacts of their prey species. Their absence can signal potentially undesirable changes in plant communities and plant diversity.

Conservation of large carnivores is relegated more to PAN elements 3 and 4—large parks and preserves and their connections. Some countries have established management programs for specific large carnivores, such as India's system of tiger reserves (Panwar 1978). Most large carnivores do not enjoy the luxury of such species-specific protection. As a result, their prey and the food of their prey could be provided in managed forest landscapes. Many species of large carnivores are not choosy about specific plant species or vegetation conditions, but rather depend on some kind of general cover such as rocks and dense brush, and on the availability of prey—including herbivorous mammals such as boar and deer. Foresters could provide the habitats and foods on which such prey depend, including low (less than 2 m tall) shrub foliage, seed- and nut-bearing shrubs and trees, tuber-producing forest floor plants, and other substrates and foods as discussed in sections above.

Habitat for nutrient cyclers and insectivore prey. Termites, ants, and earthworms play major roles in transferring organic material, altering vegetation structure and composition, and cycling nutrients throughout forest ecosystems (see chapters 12 and 13). They include leaf-cutter ants in the Neotropics, which appear resilient to the effects of selective timber harvesting (Johns 1988). Termites serve to break down dead forest wood and thereby recycle organic matter back into the soil. Smaller armadillos, birds, pangolins, and many other species eat termites and termite nests are often used by birds—such as trogons—for nest holes. Protection of soils from undue erosion, compaction, burning, and scarification would help termite populations. Termite mounds and ground ant nests should be protected during ground-disturbing operations to the extent possible.

Protect individual trees or forest groves supporting colonies of birds. Some tropical birds nest or roost in colonies. Examples include some parrots and crows. Protection of individual trees or small forest groves with colonial nest or roost sites can greatly help to maintain these species, some of which can provide important benefits to foresters by dispersing seeds of desirable plants.

Universal Considerations for Natural Areas (PAN element 2)

This section lists key habitat elements that could be protected from forest and timber management operations within zones located in timber concessions (also variously termed *cutting blocks* or *management units* in some areas).

Preserve Habitat for Priority Wildlife Species

Determine habitats for priority species. One of the major steps in providing wildlife habitat within timber concessions is to identify priority wildlife species and their habitats, and then to protect such key habitats. Priority species can include threatened, endangered, and IUCN Red Data Book species, as well as locally and regionally endemic species (species found nowhere else). Depending on local management objectives, these species also could include those of great economic or social importance, species of traditional value, or sacred organisms.

One example of a priority species is the endangered Sokoke scops owl (*Otus ireneae*), which is endemic to the *Cynometra* woodland of the Arabuko-Sokoke forest of East Kenya, Africa (Virani 1994). The Sokoke scops owl is threatened by loss of its forest habitat and by specimen collectors (Everett 1977). Its total population was reported by Clark et al. (1978) to be 1,300 to 1,500 pairs, and estimated by Virani (1994) to be only 1,000 pairs within the Arabuko-Sokoke forest (Virani 1994). Between 1956 and 1966, its forest habitat was halved to 350 km^2, and by 1991 only one forest reserve of 40 km^2 had been established for its protection (Hume and Boyer 1991). Few studies have been done on the species, but its habitat requirements likely include cavities in hollow or dead trees and lack of disturbance in its foraging territory, which is about 12 to 14 ha (Virani 1994). Field surveys could help identify locations of cavity trees and breeding pairs, and such areas of *Cynometra* woodland could be protected at least in local, small forest reserves of 12 to 14 ha.

Habitats of other priority species can include nest or den sites, important resting sites, and important feeding sites, and are too diverse to annotate here. We encourage foresters to consult with local biological experts to determine presence of priority species and to identify their habitat requirements. Presence of threatened, endangered, Red Data Book, or endemic species also can be determined from existing or new biological surveys (see chapter 19), and by consulting wildlife occurrence records for the area.

Size of habitats. The size of individual forest protected zones can vary according to the site-specific objective or conditions. These zones may range from a tiny patch of a few hectares or less for protecting endangered or rare plants, to forming a part of a broader forest area that protects wide-ranging, disturbance-sensitive species such as spider monkeys (*Ateles paniscus*) (McFarland-Symington 1988) in the Neotropics. It is important to realize that no single size will meet all conservation needs and all species' requirements. Protecting undisturbed forest blocks of 100 to 200 ha—or 10+ percent of the total forest area—from timber harvest, has been recommended (Blockhus et al. 1992; see chapters 16 and 21).

The size of habitats for PAN element 2 protection zones would typically focus on key breeding or roosting sites for smaller-ranging animals, such as the Group 1 species listed in table 23-2, and depicted in figure 23-3. Local management objectives and conditions would dictate the specific sizes and distribution of PAN element 2 protection zones. An overall objective, for example, may be to place ten percent of a timber concession (plus areas uneconomical to cut) into protection zone status. Arrangements should follow stream courses, dense interlinked tree canopies, and other natural features important to the species of conservation interest. Protection zones within a timber concession should be as contiguous as possible to provide habitat connections for organisms. As far as is feasible, protection zones also should generally be:

- Larger than smaller, so as to encompass more resources and cover
- Rounder than more linear (except for gallery type forests in riparian buffers), so as to reduce adverse edge effects

~ Closer together rather than more spread apart, so as to increase opportunities for wildlife to move among areas

A buffer sufficient to shield protected areas from forestry activities that could unduly disrupt vegetation conditions or cause the site to be abandoned by wildlife also needs to be considered (Grieser Johns 1997). The width of the buffer should be defined on a site-specific basis—depending on the type and intensity of the forest management activities, local topography, age of the edge, and other factors—but more than 30 m of width on each side would begin to protect interior habitat from the adverse effects associated with forest edge environments. Depending on the kinds, some limited activities could be allowed within buffer areas, such as collection of non-timber forest products or reduced cutting intensities. The objective, however, should primarily be to buffer the adverse effects of more intensive activities outside the area.

One study of anthropogenic edges in a Neotropical montane forest in southwestern Columbia (Restrepo and Gómez 1998) suggests that edges of various ages influence understory birds differently—by species and ecological functional group such as feeding guilds—and that some effects can extend up to 200 m from the edge. Studies at La Selva, Costa Rica report that avian predation effects occur at distances greater than 30 meters from the edge (Gibbs 1991). The Smithsonian Minimum Critical Area studies in the Brazilian Amazon also demonstrated impacts of wind throw, sunlight, and other conditions farther than 30 meters from the forest edge (Bierregaard et al 1992). A wonderful synthesis by Laurance et al. (1997) of Amazonian edge effects studies suggests that some effects (leaf-litter invertebrate species composition and invasion by disturbance-adapted beetles and butterflies) penetrate up to 250 meters into a forest. Most biological and physical effects, however, occur within 100 meters—including bird-density effects up to 50 meters. On this basis, forest buffers for vertebrate wildlife habitat could be targeted to be at least 50 to 100 meters wide.

Provide Buffers for Streams and Roads

Another means of habitat protection for species is to protect streamside riparian vegetation and provide a no-cut buffer along all perennial, intermittent, and ephemeral streams (see chapter 21). Riparian buffers help lessen undue erosion and sedimentation of the stream channel from up-land clearing and adjacent road building (see chapter 14). Buffers also help protect vegetation cover close to water sources, which provides essential habitat for primates (Wallace et al. 1996) and other wildlife species (e.g., Mason 1995; Machtans et al. 1996; Ochoa 1997).

Riparian buffers should be large enough to protect the integrity of the stream system, and to provide sufficient shading, vegetation canopy structure for canopy-using organisms, and natural sources of dead wood for standing, down, and in-stream coarse woody debris. Buffers would vary in width according to local topography and stream channel morphology. At the least, several tree heights are typically needed to buffer stream temperatures, but wider buffers are needed to provide longer-term habitat structure such as sustainable sources of large down wood (Reeves et al. 1991). Finally, as part of riparian protection, only temporary bridges should be built if necessary and when feasible. If roads must cross streams, crossing should be at right angles to the stream to the extent possible, and culverts or other structures should be provided along roads to minimize erosion damage to the stream channel (Sist et al. 1998).

Protect Scarce and Declining Habitats

Scarce and declining habitats include any plant communities or wildlife habitats that are unique, rare, or greatly declining locally or regionally. Representative examples should be conserved of each type of community or habitat, including a sufficient buffer to help maintain the integrity of the protected area. Local biologists can be consulted to identify such species, communities, and the locations of key sites.

Protect Unique Wildlife Habitats

The following special wildlife habitats typically cover a small land area, but are especially important to a wide variety of wildlife on a per-hectare basis. Such habitats can be identified in the field or from aerial photographs, and delineated as small, protected areas. They should be delineated during, and excluded from, stocking surveys both on maps and on the ground.

Riverine gallery forests. Forests alongside creeks, streams, and rivers often hold exceptionally high levels of biodiversity (Naiman 1993) (see photo 23-4). Riverine forests and wet valley bottoms are particularly important during the dry season in seasonally dry forest types (B. Mar-

PHOTO 23-4 *Example of a PAN Element 2 component: Riverine gallery forests. Dense riparian or streamside forests, such as this riparian gallery forest in Caribbean lowland Costa Rica, provide habitat for a strikingly wide array of wildlife species, and can help to link other habitats by acting as a habitat corridor throughout a managed forest matrix. Plants in such habitats provide an array of fruits and other food and cover resources for wildlife. (B. Marcot)*

cot, personal observation). In dry forest types, during very dry years with many fires, riverine gallery forests may be the only green foliage and open water available as habitat and cover for much of the wildlife.

Riverine forests also provide important sources of plants with fleshy fruits, including palms and other plant families (e.g., in the Neotropics the families Moraceae, Sapotaceae, Annonaceae) (B. Marcot, personal observation). Fleshy fruits are important resources for a wide variety of wildlife species.

Salt licks and mineral soil. Areas of bare mineral soil, particularly where soil salts have evaporated to the surface, usually occur in wet areas near riverine forests. They are frequently used by peccaries, tapirs, deer, parrots (if on cliffs), butterflies, bees, and a variety of other wildlife (D. Rumiz, personal communication; B. Marcot, personal observation).

Caves and rock outcrops. Caves are important roosting and hibernation habitats for many bats (see chapter 7). They also provide denning sites for carnivores (including cats, dogs, and foxes), owls, oilbirds (*Steatornis caripensis*), swifts, rodents, porcupines, bears, and a variety of other medium-to-large mammals, as well as habitat for unique invertebrate communities.

Rock outcrops may support unique and scarce plant communities, often containing plants with fleshy fruits eaten by many kinds of wildlife. After short rainfalls, rock outcrops often hold temporary ponds that become important watering holes for nightjars and other birds, and for amphibian reproduction. Rock outcrops also provide crevices for bats and specialized rodents and lizards, and seem to attract an especially high concentration of carnivores, rabbits, and other species. Heat captured by these areas during the day provides warmth in the evening that attracts invertebrates, reptiles, and nocturnal birds during the cool dry season (D. Rumiz, personal communication). Rare cliff-dwelling organisms can serve as sensitive indicators of environmental change (Everett and Robson 1991).

Natural forest openings. Some tropical forests have savanna openings (usually adjacent to semideciduous forests such as those found in southeast Cameroon; Letouzey 1985). Savannas may look *poorer* than forest environments, but because they occur as inclusions in generally forested regions, they often provide unique environments for plants and animals (Fimbel et al. 1996). Some plant species are found only in savannas, such as the totai palm (*Acrocomia aculeata*) in lowland Bolivia. Natural gaps in forest cover often provide ground vegetation cover and seed-bearing herbs and shrubs that are used by a variety of wildlife. In lowland Bolivia, large savannas in northern Santa Cruz harbor the endangered marsh deer (*Blastocerus dichotomus*), pampas deer (*Ozotoceros bezoarticus*), and maned wolf (*Chrysocyon brachyurus*)—and birds such as rheas (*Rhea americana*) (see photo 23-5). Insectivorous bats feed over natural forest openings. Because savannas may burn during the dry season, providing some adjacent forest

PHOTO 23-5 *Example of a PAN Element 2 component: Natural forest openings. Natural savannas (such as shown in the right half of this photo of lowland Bolivia) provide unique habitats for many wildlife species not found in more closed-canopy, managed forests (such as in the left). Savannas can be protected for deer, birds, bats, and other species. (B. Marcot)*

cover—especially riverine forests—for wildlife is an important conservation measure.

Natural forest openings are not at risk of being logged, but may be degraded if used as landings, campsites, or converted to forest cover.

Palm groves. In some tropical forests, palm groves can provide special habitats for wildlife (see photo 23-6). In the Upper Amazon, Mauritia palms (*Mauritia flexuosa*) have large trunks and grow large fruits that are eaten by many kinds of wildlife, including tapirs and peccaries (D. Rumiz, personal communication). As the trunks decay, they provide cavity sites for macaws, toucans, and other species. In some cases, palm groves are logged—such as the heart-of-palm in the upper Amazon Basin and in south India. In other cases, palm groves may be felled during forestry operations aimed at accessing other commercial trees.

Care can be taken, when designing forestry operations, to retain palm groves. Roads and skid trails should avoid palm groves. Palm groves can be protected as PAN element 2 zones, and individual palms suspected or known to be important for wildlife use can be protected as PAN element 1 habitat elements.

PHOTO 23-6 *Example of a PAN Element 2 component: Unique wildlife habitats. One type of unique wildlife habitat is palm groves, such as the heart-of-palm grove shown here in Taruma Timber Concession in the Amazon Basin in Bolivia near the border with Brazil. Such palm groves provide fruits for both canopy- and ground-dwelling wildlife which can help disperse their seeds, thus helping ensure a sustainable crop, and can be delineated and retained during timber harvest layout and operations. The palm trunks also are used by many birds for nesting. (B. Marcot)*

Lagoons and other water sources. Lagoons and other water sources such as springs provide critical sources of water during the dry season for a wide variety of wildlife, especially in seasonally dry forests. Gallery forests often contain pools well into the dry season. These are important wildlife habitats needing protection.

Low hills within seasonally flooded forests. In the wet season within seasonally flooded forests, low hills provide critical refuges for many wildlife species. On the north coast of Honduras, for example, the most critical habitats to protect for wildlife are low hills, which serve as wildlife refugia during occasional major floods (J. Barborak, personal observation). During monsoons in the terai (foothill wet grassland) habitats along the base of the Himalayas in northern India and southern Nepal and Buthan, great Indian one-horned rhinoceroses (*Rhinoceros unicornis*) move to the security of locally higher ground covered by sal (*Shorea robusta*) forests. Maintaining forest cover within such high ground conditions can help protect wildlife.

Other Considerations and Design Criteria for Delineating Protection Zones

Habitat corridors and connections. PAN element 2 protection zones also can be delineated as wildlife corridors or connections between other important wildlife sites. In many cases, streamside riparian buffers can help serve as habitat corridors and connectors (Grieser Johns 1997). It is important to protect mosaics of contiguous, different habitats (Wong 1985; but see Beier and Noss 1998).

Natural disturbances. Delineating protected areas should account for major disturbances—even if intermittent or periodic—such as drought or inundation. In such circumstances, the distribution and abundance of habitats, resources, and wildlife populations may be vastly different— possibly confined to more important refugia—than in normal years. Protected areas should be designed based on those less frequent conditions when wildlife needs are more extreme.

Logging effects and general suitability of the logged forest matrix. Effects of forest management activities, particularly logging, on wildlife can vary greatly according to harvest rates, reentry schedules, prior stand condition, and additional silvicultural activities such as enrichment planting or thinning. These will affect the suitability of the logged matrix for wildlife, and this in turn will influence the need for and design of PAN element 2. The less suitable the logged matrix is for wildlife, the more comprehensive the PAN element 2 needs to be. Suitability of the matrix for wildlife can be determined by comparing the list of wildlife habitats presented in the sections in this chapter on PAN 1 and 2 elements.

Meeting other objectives. Protected areas are established not just to protect biological diversity, but for other objectives as well (see box 23-1 and appendix 23-1). In impoverished nations with weak institutional governance mechanisms and great land scarcity, efforts to conserve biological diversity often are most successful when other societal concerns are considered while designating protected zones. Important wildlife habitat often occurs on sites for which societal support for protection might be forthcoming anyway. Examples include potable water catchments for downstream communities; sites with historic, spiritual, or traditional recreational value; areas with steep slopes and fragile soils; and wetlands where, because of past natural disasters (floods, mudslides, etc.), local inhabitants already respect site limitations. Finally, providing PAN element 2 (and element 1, above) is likely to benefit timber production by providing mature seed trees and the ecological services of wildlife that contribute to regeneration of timber and other forest products. Reserving elements or patches of timber species can be an economically effective way to improve natural regeneration in the logged matrix.

Consideration of the Human Element

This section pertains to identifying areas of wildlife habitat that can also coincide with sites of cultural, religious, or other non-timber interests.

Protect Sites of Cultural or Religious Interest

In some cases, wildlife habitats can be conserved concomitant with conserving sites of cultural or religious significance for people. These habitats can include sacred groves, spirit groves, or sites of traditional and ritualistic use. In some cases, habitat protection can coincide with controlled or limited traditional hunting or gathering of native animals and plants.

Protect Sites of Other Use to Humans

Some sites are of value to people for tourism, non-timber forest products, seed trees, and other uses. In particular, superior quality seed trees (or groves of them) for important timber species often provide seed sources for use in enrichment plantings and reforestation projects. At La Selva, Costa Rica, for example, where considerable research is ongoing regarding native timber trees suitable for reforestation projects, a major limitation is that some of the most promising tree species have seeds that rapidly lose viability or are produced very erratically among individual trees or stands (J. Barborak, personal observation). Many of those seeds also are important to fruit-eating and seed-dispersing wildlife.

Many of these sites or elements provide habitat or resources for wildlife. If protected as PAN elements 1 or 2, they also help to bridge planned protected zones within and around the production forest matrix (PAN elements 3 and 4).

Socioeconomic Considerations and Incentives for Managing a Protected Area Network

Why should forest managers exempt even a small portion of their harvestable timber concession lands or commercial trees for protecting natural areas and habitat elements? In some areas, local tax laws might provide an incentive. Bolivia, for example, has a law where up to 30 percent of the concession can be reserved from cutting (for the reasons noted above), and the contractor does not have to pay taxes on those areas (there is a flat tax per hectare in Bolivia, rather than stumpage pricing) (see chapter 29). In other countries, protecting natural areas and habitat elements can contribute to local economies through eco-development or by encouraging an ecotourism economy with community forests (see chapter 25). An example is the La Selva ecotourism lodge along Rio Napo in lowland Ecuador. As in the example of managed protected areas on some Pacific Islands, there may be several social and economic goals met at the same time (Gilman 1997). In these cases, a PAN approach—particularly the PAN elements 1 and 2—can complement timber harvest.

Privately owned nature reserves in sub-Saharan Africa and Latin America can help stave off regional depletion of native biodiversity

(Langholz 1996). Such reserves could be established as PAN element 2 zones, or as PAN element 3 or 4 conservation areas or parks. They could provide profitable ventures for their owners and a basis for involving the private sector in conservation. At the site level, private nature reserves are functioning quite well profitably and ecologically—such as in parts of New Zealand—and might fit well into several of the PAN elements described in this paper for tropical forest regions. If integrated with private timber management sites and objectives, they can provide useful examples of successfully meeting several, very different objectives of forest use in an area. Private forest management need not be antithetical to private nature reserve conservation.

The different facets of PAN elements 1 and 2 vary significantly in their opportunity cost, making some far cheaper to implement than others. The main cost to forest managers of implementing PAN elements 1 and 2 is the opportunity cost of excluding merchantable timber from harvest, although many of the habitat elements that comprise these two PAN elements are not necessarily of commercial value (e.g., dead standing trees and steep slopes). The relative cost would vary by site, forest type, and management objectives. All else equal, areas with the highest access cost (i.e., least accessible) can be chosen over easily accessible areas for protection to minimize reductions in profit. This approach could actually save operators money if they were not required to exploit all areas of their concession, including those of low value, as is required in many countries (R. Fimbel, personal communication). Some types of PAN elements, such as rocky outcrops and natural grasslands, have no opportunity costs associated with protecting them, although they may incur some limited costs if forest protection buffers are needed. Within the logged matrix, habitat elements such as dead trees and live rotten or hollow trees will have no timber value and thus no opportunity costs associated with their protection.

The largest opportunity costs will occur if important wildlife tree species are also valuable timber species. A classic example is the depletion of sandalwood (*Santalum* spp.) from tropical Pacific islands that occurred during the early 1800s. Sandalwood was one of the native tree species that comprised the natural mixed mesic and lowland forests and wildlife habitats of these islands (Sohmer and Gustafson 1987). It also had high commercial value for timber, joss sticks, aromatic tapers, incense, medicines, and perfumes. Decimation of sandalwood and other native trees of commercial value on Hawaii, the Society Islands, New Caledonia, the New Hebrides (Vanuatu), and Fiji coincided with degradation of the islands' native lowland wildlife communities (Mitchell 1990).

If important wildlife tree species are not commercial, then they can be favored over other less valuable noncommercial species—with little or no opportunity cost. Where commercial tree species are involved, it may still be to the advantage of timber production to protect at least some individual or patches of mature trees to serve as seed trees and help ensure long-term, locally adapted genetic resources.

A final incentive for the PAN approach is that it may be possible and desirable to include PANs in eligibility criteria for receiving funding from carbon offsets. This could ease any opportunity costs of excluding lands from the timber estate, help provide funds for active management, and help protect genetic and wildlife resources for overall protection of timber production, forest resource sustainability, and biodiversity (see chapter 27).

PAN Research Priorities and a Way to Fill the Knowledge Gaps

PAN elements provide a baseline from which to monitor natural changes and forestry induced within the production forest landscape (Arcese and Sinclair 1997; see chapter 19). This can help tell us the degree to which refuges protect native biodiversity, such as populations of threatened species, and the differences between natural and anthropogenic disturbances and their impacts on wildlife. PAN elements 1 and 2 within active timber concessions can be monitored to determine the degree to which logged areas can protect key environments and wildlife habitats, and especially how individual elements and forest structures can be created silviculturally.

Overall, monitoring PAN elements can help managers and ecologists determine how each element best contributes to the overall goals of conserving the ecological integrity of tropical forests on a variety of spatial scales. Local ecological communities, broader eco-regions, and even entire biomes can be monitored as to the degree to which their indigenous plants and animals and associated ecological processes are represented and maintained over time. Research can also be conducted on the population distribution, density, and dynamics of species related to key ecological functions of forests—such as listed in the guidelines above. Such research would help identify reserve dimensions and habitat characteristics to be maintained within managed forests.

Priority areas for future research, to fill in major information gaps related to a PAN approach, include the following:

- Research on the optimal size, shape, and spatial distribution (connectivity) of PAN element 2 zones, for specific wildlife conservation objectives
- Improvements in the rapid assessment techniques for inventory and survey of biodiversity elements in a timber concession area (Cannon 1997; Oliver and Beattie 1993, 1996a)
- Research on effective corridor and buffer zone widths
- The capacity of all four PAN elements—particularly 1 and 2—to serve as refugia and dispersal centers for sensitive organisms. Some research on tropical forest fragments has provided some understanding (e.g., Laurance and Bierregaard 1997), but more needs to be done
- Need for better understanding of how PANs benefit timber production objectives. This would include research on how PANs—particularly PAN elements 1 and 2 within the logged forest matrix—provide sources of seeds of commercial tree species, as well as habitat for wildlife critical to pollination, seed dispersal, soil health, and many other aspects of ecosystems that contribute to sustainable production of timber and non-timber forest products

In the end, reserving natural areas of various scales (PAN elements 2, 3, and 4) and providing habitat elements (PAN element 1) in the managed forest landscape by mimicking natural disturbance events (see chapter 22) and using indigenous long-term forest management practices (Gadgil and Berkes 1993) are complementary approaches to protecting forest productivity and biodiversity for future generations.

ACKNOWLEDGMENTS

Our thanks to Robert Fimbel, Alejandro Grajal, John Robinson, and an anonymous reviewer for providing helpful reviews of the manuscript. Our appreciation also goes to Todd Fredrickson for providing the references on reserve size for carnivores. Very special thanks to Damián Rumiz, Todd Fredrickson, and Dick Rice for discussions on conservation approaches in tropical forests and identifying specific Neotropical habitat elements and species for conservation focus. Ted Gullison devised the term and acronym PAN.

Appendix 23-1 *The IUCN International Protected Area Classification System (excerpted from IUCN 1994)*

Protected areas are defined as "areas of land and/or sea especially dedicated to the protection and maintenance of biological diversity, and of natural and associated cultural resources, and managed through legal or other effective means."

Protected areas are designated for numerous reasons, but usually for a mix of the following management purposes:

— Scientific research

— Wilderness protection

— Preservation of species and genetic diversity

— Maintenance of environmental services

— Protection of specific natural and cultural features

— Tourism and recreation

— Education

— Sustainable use of resources from natural ecosystems

— Maintenance of cultural and traditional attributes

According to the specific mix and priority of the management objectives for a given protected area, the following distinct categories have been identified:

1. Category I	Strict Protection	
— Category Ia	Strict Nature Reserve	
— Category Ib	Wilderness Area	
2. Category II	Ecosystem Conservation and Recreation (i.e., National Park)	
3. Category III	Conservation of Natural Features (i.e., Natural Monument)	
4. Category IV	Conservation Through Active Management (i.e., Habitat/Species Management Area)	
5. Category V	Landscape/Seascape Conservation and Recreation (i.e., Protected Landscape/Seascape)	
6. Category VI	Sustainable Use of Natural Ecosystems (i.e., Managed Resource Protection Area)	

Though the primary purposes of management determine the category that will be assigned, management plans will often contain management zones for a variety of purposes that take account of local conditions. In order to establish the appropriate category, however, at least three quarters of the area must be managed for the primary purpose, and the management of the remaining area must not be in conflict with that primary purpose.

Protected areas of different categories are often contiguous—sometimes one category *nests* within another. This is entirely consistent with the application of the system, provided such areas are identified separately for accounting and reporting purposes. Although there are obvious benefits in having the entire area within the responsibility of one management authority, this may not always be appropriate. In such cases, close cooperation between authorities will be essential.

Chapter 24

LOGGING AND WILDLIFE RESEARCH IN AUSTRALASIA

Implications for Tropical Forest Management

William F. Laurance

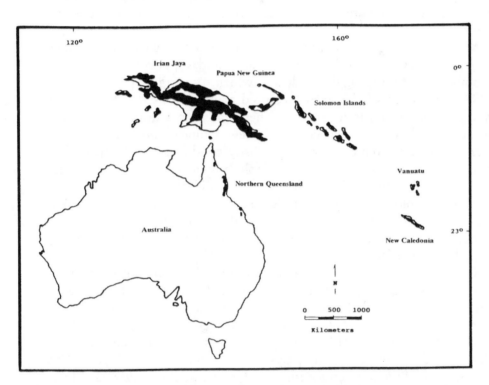

FIGURE 24-1 *Distribution of closed tropical forests (shaded areas) in Australasia (after Collins et al. 1991). Smaller islands in the region are not shown.*

Because of its long history of biogeographic isolation, Australasia supports a unique biota characterized by exceptionally high levels of endemism. Wet tropical forests occur extensively in the region, especially in New Guinea, with smaller forest areas in the Solomon Islands, New Caledonia, Vanuatu, and northern Queensland.

In socioeconomic terms, the Australasian region contains two distinctive components. Australia is a developed nation with a low population density, a high literacy rate, and a tradition of multiple-use land management. Tropical rain forests are confined to north Queensland and occupy only a small fraction (0.3 percent) of the continent's land area. This is in stark contrast to Papua New Guinea, Irian Jaya (Indonesian New Guinea), the Solomon Islands, and New Caledonia, which contain much higher proportions of forest and are beset by rapid population growth, low literacy, and pressing needs for economic development.

Mechanized logging in the region has also followed two distinct trends. For several decades, north Queensland supported a small but regionally important logging industry, which was largely phased out in 1988 when much of the forest acquired protection as the 900,000 ha Wet Tropics World Heritage Area. There has been a dramatic increase in logging elsewhere—especially in New Guinea and the Solomon Islands—with mechanized harvests mostly conducted by Asian timber firms. In Papua New Guinea, for example, annual log exports nearly tripled from 1991 to 1994—from 1.4 to 3.5 million m^3 (Arentz 1994). The rapid exploitation of forests has been the subject of considerable controversy, characterized by frequent allegations that foreign timber companies were involved in graft and environmental mis-management (Marshall 1991; Shenon 1994).

This chapter reviews available research on the effects of logging on Australasian wildlife, and identifies some technical and socio-economic measures needed to limit logging damage in developing regions. It focuses to a considerable extent on north Queensland, because this is where both the bulk of logging research has been conducted and the Queensland Selective Logging System (QSLS) was developed. The QSLS is one of the few reduced-impact logging methods ever used on a long-term commercial basis in the tropics, and its tenets and their implications for wildlife management will be reviewed in some detail.

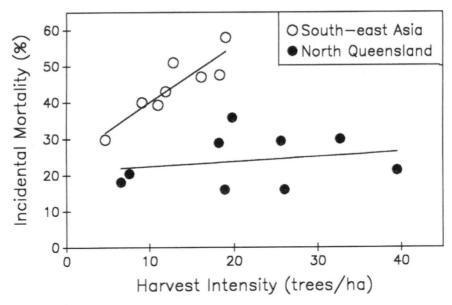

FIGURE 24-2 *Relationship between harvest intensity and incidental damage to trees in Asia and north Queensland (after Crome et al. 1992). Incidental mortality is defined as the percentage of all trees of > 10 cm diameter-at-breast-height (dbh) present before logging that were accidentally killed during the harvest operation.*

The Queensland Selective Logging System

In brief, the QSLS seeks to minimize forest damage by controlling many aspects of logging operations (see box 24-1). As implemented in north Queensland, it was not a concession system, but used yield calculations (Vanclay 1989) to determine an annual cut, which was allocated among milling companies. Timber sales were confined to small areas (e.g., 20–200 ha) with the company paying a royalty calculated for each log.

> **Box 24-1** *Some guidelines from the Queensland Selective Logging System designed to reduce the environmental impact of logging operations (after Anonymous 1991).*
>
> 1. Logging Plan
> - A preoperational harvest plan is used to plan the logging operation and minimize its environmental and visual impacts

— Prior to logging, the Forestry Officer responsible for the timber sale and a timber company representative jointly inspect the sale area on foot to discuss the harvest plan and any special environmental features or restrictions

2. Buffer Strips and Special Management Areas

— Unlogged buffer strips of 10 to 30 m width are required along both banks of all permanent and temporary watercourses; wider (20-30 m) strips are required for steeper (> 15°) slopes

— Harvest machinery may not enter any part of a buffer strip except at designated crossing points

— "Special Management Areas" (sites with particular uses or environmental values) may be designated by the Forestry Officer; harvesting may be prohibited in these areas or permitted under specified conditions

3. Logging Roads

— The locations of all logging roads are determined by the Forest Officer in consultation with the timber company representative; permitted road locations are marked on a map of the harvest area and in the field

— The road network is designed by the Forestry Officer to minimize the total length of roads, earth-works and the number of stream crossings, and for ease of drainage; roads are usually confined to ridgetops and adjoining moderate slopes

— The widths of major and minor roads should not exceed 7.5 m and 5 m, respectively

— To reduce soil erosion, road grades should normally not exceed eight degrees (steeper grades may be used for very short distances if adequate drainage can be provided)

— Cross-drains or culverts are required to minimize erosion on logging roads, with the frequency of drains specified according to slope and soil erodibility

4. Felling and Log Removal

— Directional felling is used to fell trees toward logging roads and reduce incidental damage to nearby trees

— Trees may not be felled into watercourses or their banks

— Wherever possible, winches are used to drag logs to roads, rather than creating unnecessary tracks which increase incidental forest damage

— On soils with low erodibility, harvest machinery is confined to areas with slopes of 30° or less; sites with moderate or highly erodible soils have maximum slopes of 25° and 20°, respectively

— Log ramps (for loading logs onto trucks) should not exceed 500 m² in moderate terrain (maximum of 1 ramp per 5 ha) or 750 m² in steep terrain (maximum of 1 ramp per 10 ha)

— Harvest operations are prohibited during the height of the wet season, and logging is stopped in wet conditions if surface runoff occurs or the soil becomes saturated

Well-trained staff from the Queensland Forest Service supervised all logging operations. Trees to be harvested were marked prior to logging, using size-limits that varied among classes of species (Anonymous 1983). A set of environmental guidelines introduced in 1982 established procedures for planning logging operations, constructing roads, felling and hauling logs (see photo 24-1), controlling erosion, and other measures (detailed in Anonymous 1983, 1991). A number of these guidelines are listed in box 24-1.

Effects of Logging on Wildlife Habitats

The impact of a logging regime on wildlife populations will be influenced by both the intensity of habitat disturbance during logging, and the frequency of repeat logging. Logging can cause a myriad of changes in forest structure, microclimate, floristics, and food

PHOTO 24-1 *Under the Queensland Selective Logging System, heavy winches on bulldozers are used to drag trees from where they are felled to the nearest logging road, limiting the extent of roads and forest damage in harvest areas. (W. F. Laurance)*

resources—any one or a combination of these factors may influence wildlife populations.

In Australasia, most research on the effects of logging on tropical forest structure and floristics has been conducted in north Queensland. The influence of logging on hydrology and soils is detailed elsewhere (Gilmour 1971, 1977; Cassells et al. 1985; Gillman et al. 1985).

Forest Structure

Many wildlife species are sensitive to changes in forest structure, and logging methods that minimize forest disturbance will tend to have lower and shorter-term effects on affected species. A rigorous assessment of logging effects on forest architecture was conducted by Crome et al. (1992), who compared up-land rain forest on the Windsor Tableland in north Queensland before and after selective logging. Their experimental study site was logged under QSLS guidelines at a low-to-moderate harvest intensity (mean log volume = 37 m^3/ha, averaging 6.6 trees/ha). They assessed changes in forest canopy cover, three-dimensional canopy profiles, and tree size distributions, and enumerated and identified all harvested trees. They also mapped all logging roads and minor tracks, pin-pointed areas prone to erosion, and recorded incidental losses of trees during the logging operation.

Crome et al.'s (1992) study yielded some important conclusions. Logging caused complex changes in forest structure at all vertical levels of the forest. Light intensities on the forest floor were markedly increased, even in areas that had no loss of overhead canopy cover. As is typical of selective logging, damage was highly patchy and concentrated on ridge tops and upper slopes where most trees were harvested. Loss of trees was greatest in the largest (> 60 cm dbh) and smallest (10–30 cm dbh) size classes, with the large trees removed for harvesting and smaller trees destroyed incidentally by bulldozers and skidders. Within 18 months of logging, conspicuous soil erosion had occurred on major roads and in a small clearing used for loading logs, leading to siltation of a nearby stream.

Despite these changes, however, the net level of forest damage was limited (Crome et al. 1992). It was estimated that only 22 percent of the canopy cover was removed, and that logging roads and minor tracks covered a mere five percent of the sale area. There were no significant changes in the density of trees in five different size classes, and no tree species was lost as a result of the harvest. Drains on roads and a prohibition on the use of heavy machinery in wet conditions reduced soil erosion.

As discussed by Crome et al. (1992), the overall level of forest damage in Queensland compares very favorably with logging operations studied in southeast Asia. One way to compare forest damage is to assess the number of trees killed incidentally during harvest operations (see figure 24-2). Logging operations studied in north Queensland consistently had lower levels of incidental tree death than those in southeast Asia. Surprisingly, tree mortality was not significantly correlated with harvest intensity in north Queensland (this may be partly due to methodological differences among studies; Crome et al. 1992), while in southeast Asia there was a strong, positive relationship between harvest intensity and incidental mortality (P = 0.003, linear regression analysis; Crome et al. 1992; see figure 24-2).

Despite the clear trend for lower collateral damage in Queensland forests, it should be noted that logged trees in Asia (mainly dipterocarps) are generally larger than those in Queensland. This could contribute to forest damage if, for example, larger bulldozers and wider roads were needed for log removal.

Floristic Composition

Floristic changes in logged forests may affect species such as herbivores, frugivores, and nectarivores. One notable study is Nicholson et al.'s (1988) assessment of changes in stand composition in logged and unlogged rain forest in north Queensland. The authors summarized data from 25 small (usually 0.1–0.4 ha), permanent monitoring plots in which the richness, mortality, and growth of trees (> 10 cm dbh) were assessed for periods of up to 32 years. They concluded that selective logging caused only a limited, short-term reduction in tree diversity, which was soon offset by the in-growth of new species on the plot. In fact, the temporary loss of tree species was more apparent than real, they suggested, because many species were still present after logging, but were too small to be censused.

Indeed, Nicholson et al. (1988) asserted that logging—especially logging plus silvicultural treatments involving removal of noncommercial trees and saplings—could enhance tree diversity in rain forests. This idea was based on the observation that tree richness on some plots in virgin forest declined over time, while such effects were not evident in the silvicultural plots and some logged plots. They attributed this pattern to Connell's (1978) *intermediate disturbance hypothesis*, which states that species diversity should peak in sites with a moderate frequency or intensity of disturbance because both old-growth and pioneer species can co-exist under such conditions while only one or the other can flourish under low or high levels of disturbance.

The notion of logging or silviculture-enhanced tree diversity was challenged by Saxon (1990), who reevaluated Nicholson et al.'s (1988) data and asserted that their conclusions were unjustified because they assessed only postlogging changes in tree diversity. He argued that when pre- and postlogging data were compared, tree species richness was consistently lower on logged and silviculturally-treated plots than before disturbance.

This debate notwithstanding, it does not appear that selective logging has caused either a large or long-term reduction in local tree diversity in Queensland, especially when silvicultural treatments were not applied. This is not to imply, however, that the proportions of different plant taxa remain constant. Logging leads to an increase in pioneer and early successional plants at the expense of some mature phase species (Shugart et al. 1980; Johns 1988), and alters the abundance of fruits, flowers, and foliage needed by wildlife (Johns 1988). In addition, the effects of repeated logging on plant diversity and composition will be more important than a single logging cycle (Horne and Hickey 1991), and these effects have not yet been evaluated in any detail.

Few data are available for areas outside Queensland. R. J. Johns (1992c) discusses some general effects of selective logging on plant communities in the Gogol Valley of Papua New Guinea, and suggests that a single cycle of moderate-intensity selective logging (harvests of 20–45 m^3/ha, equivalent to 3–8 trees/ha) does not reduce plant diversity and probably favors regeneration of commercially valuable timber species. He emphasizes, however, the general paucity of information on effects of logging on plant and animal communities in New Guinea.

Tree Cavities

Denning or nesting cavities in trees are a crucial resource for species such as possums, lorikeets, and cockatoos. Circumstantial evidence suggests cavities could be a key limiting resource in some Australasian tropical forests because many tropical species require cavities (e.g., Laurance 1990; Gibbs et al. 1993) and because the region lacks woodpeckers (family Picidae), which are prime excavators of tree cavities elsewhere.

No published study has assessed the effect of logging on cavity availability in tropical Australasia. A theoretical growth model developed by Shugart et al. (1980), however, suggests that repeated selective logging of Australian subtropical rain forests at 30-year intervals would cause a reduction in timber volume and a shift toward earlier successional species. Such changes in stand characteristics could lead to a decline in the mature and overmature trees that often contain natural cavities. Some species might be strongly disadvantaged by repeated logging, especially if tropical species partition cavity types as strongly as do some temperate species (e.g., Lindenmayer et al. 1990).

Effects of Logging on Wildlife Populations

Research on faunal responses to logging in the tropics of Australasia has usually focused on birds or mammals (see table 24-1). There have been no studies of logging effects on invertebrate populations. Work has been largely concentrated in north Queensland, with limited research in Papua New Guinea. It is noteworthy that virtually nothing is known about the effects of logging on biotas of oceanic and smaller continental islands, which often support many locally endemic species.

Amphibians and Reptiles

No published studies of the effects of logging on herpetofauna are available. As part of an undergraduate honors thesis, however, Preen (1981) assessed terrestrial vertebrate populations in a recently logged (2 years earlier; harvest = 10–15 trees/ha) and adjoining unlogged site on the Atherton Tableland, in north Queensland. He found that several frog and reptile species declined sharply in logged forest, and concluded that these were species requiring cool, moist microclimatic conditions typical of undisturbed rain forest. In contrast, certain reptiles that depend on basking to maintain a high metabolic rate, increased in logged forest. Frogs and reptiles were particularly sensitive to changes in microclimate that had both direct (physiological) and indirect (altering invertebrate communities that comprise their food source) effects on their populations.

TABLE 24-1 *Principal Studies (Including Theses and Reviews) of the Effects of Selective Logging on Wildlife, Forest Structure, and Floristics in Australasian Tropical Torests*

Author	Topic	Location
Wildlife		
Preen 1981	Vertebrates	North Queensland
Driscoll 1985	Birds	Papua New Guinea
Driscoll 1986	Birds	Papua New Guinea
Crome and Richards 1988	Bats	North Queensland
Crome and Moore 1989	Bowerbirds	North Queensland
Crome et al. 1996	Birds, mammals	North Queensland
Laurance and Laurance 1996	Arboreal mammals	North Queensland
Forest Structure and Floristics		
Nicholson et al. 1988	Tree diversity	North Queensland
Lamb 1990	Regeneration	Papua New Guinea
Saxon 1990	Tree diversity	North Queensland
Crome et al. 1992	Forest structure	North Queensland
Johns 1992c	Plant diversity	Papua New Guinea
Literature Reviews		
Winter 1979	Arboreal mammals	North Queensland
Winter 1984	Arboreal mammals	North Queensland
Laurance 1986	Wildlife	North Queensland
Crome 1991	Wildlife	North Queensland
Horne and Hickey 1991	General	Eastern Australia

Birds

As part of the Queensland logging experiment described above, Crome and Moore (1989) examined the responses of three bowerbird species to moderate-intensity timber harvesting conducted under QSLS guidelines (logged area = 18.4 ha; harvest = 6.6 trees/ha). Males of all three bowerbird species they studied build conspicuous bowers or courts for attracting females. One of these, the tooth-billed catbird (*Ailuroedus dentirostris*), constructs its courts on ridge tops where logging damage was heaviest.

Beginning two years before logging, Crome and Moore (1989) mapped the display sites of all bowerbirds in the logging and adjoining control areas, then monitored the bowers and courts over the next four years. Only a single bower of the golden bowerbird (*Prionodura newtoniana*) occurred in the logging area. This was destroyed during logging, but another bower was immediately built only 8 m from the original— apparently by the same bird. Tooth-billed catbirds seemed more adversely affected. Five of 12 courts disappeared after logging and were never rebuilt on the logging site, although remaining courts were located as close as 10 m from logging tracks. Additional courts were built on the adjacent unlogged site in the following breeding season and may have been made by the displaced birds. A single bower of the satin bowerbird (*Ptilonorhynchus violaceus*) disappeared soon after logging and was not replaced.

Under normal conditions, bowerbirds exhibit very strong fidelity to their display sites (Crome and Moore 1989; Grant and Laurance 1991), and these results suggest that moderate-intensity harvests can reduce their habitat quality for at least a few years after logging. Habitat quality for these birds would probably have been more strongly reduced if damage had not been limited by the QSLS guidelines.

In the Madang region of northern Papua New Guinea, Driscoll (1985, 1986) assessed the effects of clear cutting for woodchips on forest bird assemblages, using both mistnetting and visual bird counts. Although clear cutting usually results in near or complete forest removal, some of Driscoll's sites were only partially felled in a manner quite similar to heavy selective logging. His study sites, which included both well-drained sites and swamp forests, had been logged up to five years before the commencement of his study.

Driscoll (1985, 1986) documented several differences between bird assemblages in primary and logged forest. After logging, bird activity became concentrated around remnant trees, leading to both vertical and horizontal compression of the foraging ranges of many species. Regenerating forests had about 40 percent lower bird densities than unlogged forest, including a large component of migratory or transitory species that rarely occurred in unlogged forest. Although many forest species were detected in logged forest, at least 22 species (14 obligate herbivores, five mixed feeders, and three insectivores) had lower densities there than in unlogged forest. Understory birds, in particular, seemed most affected by logging.

Bats

Crome and Richards (1988) assessed the effects of logging on microchiropteran bats, again as part of their large-scale experiment on the Windsor Tableland. They used ultrasonic detectors to compare bat species that use canopy gaps in logged forest with bats in intact forest. Bat assemblages in gaps (dominated by fast-flying species with long, narrow wings) differed markedly from those in sites having a continuous canopy. These bats are common in woodlands and open forests but colonize large treefall gaps in disturbed rain forest. Bat assemblages in intact forest, however, were dominated by species with broad, short wings specialized for flying and hovering in a cluttered environment.

These findings suggest closed-canopy specialists are particularly vulnerable to rain forest logging. Treefall gaps up to six years of age were still dominated by open-forest bats (Crome and Richards 1988)—a significant finding indicating that bat communities recover quite slowly from even low-to-moderate logging damage (harvest = 6.6 trees/ha). In areas with heavy structural damage, bat communities might require one to two decades or longer to completely recover.

Nonflying Mammals

In north Queensland, the effects of logging have been assessed for both scansorial and arboreal mammals. As part of his honors thesis, Preen (1981) used live trapping to compare small mammal populations in recently logged (2 years earlier; harvest = 10–15 trees/ha) and unlogged rain forest. He observed lower abundances of several rare, rain forest-dependent marsupials in logged forest, and a marked increase in the fawn-footed melomys (*Melomys cervinipes*)—a semiarboreal rodent known to favor disturbed forest (Laurance 1994). The apparent decline of rare marsupials in logged forest, however, may be temporary because all of the affected species have been recorded in sites logged 10 to 30 years previously (Laurance 1994).

Several authors have focused on arboreal mammals. Preen (1981) used spotlighting to compare arboreal mammal abundances in recently logged and unlogged rain forest. He concluded that logging caused a decline in the Herbert River ringtail possum (*Pseudocheirus herbertensis*), which requires a separate den site for each adult animal. The coppery brushtail possum (*Trichosurus vulpecula*), which has flexible denning requirements, increased in logged forest.

Winter (1979, 1984) assessed the conservation status of seven possum species in north Queensland and concluded that all could persist in logged forest. Even the lemuroid ringtail possum (*Hemibelideus lemuroides*), which is strongly rain forest-dependent, was detected (often at relatively high densities) in areas that had been selectively logged two to three times on 15- to 20- year cycles (see photo 24-2).

Laurance and Laurance (1996), however, found that lemuroid ringtails declined by almost 70 percent in a forest logged at moderate intensity (harvest = 50–55 m^3 / ha, equivalent to 8-10 trees/ha) three years previously, relative to pre-logging abundances at the same site. Three other possum species and a tree-kangaroo, however, did not differ significantly between pre- and post-logging censuses. Arboreal mammal populations monitored at nearby reference sites over the same period exhibited no significant changes, confirming that logging was responsible for the declines in lemuroid numbers. Based on population assessments at other logged sites, Laurance and Laurance (1996) estimated

PHOTO 24-2 Some arboreal mammals, such as this lemuroid ringtail possum (*Hemibelideus lemuroides*) in north Queensland, decline sharply in recently logged forest. (W. F. Laurance)

that lemuroid populations would require at least six to nine years to recover to pre-logging densities in forests logged at moderate intensity.

Summary of Research on Australasian Logging and Wildlife

In Australasia, industrial logging operations generally fall into two distinct categories: moderate-impact logging typified by the QSLS in Queensland; and minimally controlled, often heavy-impact logging in New Guinea and the Solomon Islands. A third strategy—the so-called *walkabout sawmills*—involves the use of portable milling equipment by local communities and results in far less forest disturbance than industrial operations (see chapter 25). Non-governmental and foreign-aid organizations are promoting walkabout sawmills in Papua New Guinea, the Solomons, and Vanuatu as a sustainable and socially-beneficial alternative to industrial logging (Senn et al. 1993), but they currently comprise only a minor component of the region's logging industry.

Because the QSLS was generally effective in limiting logging damage, the relatively large body of logging research from Queensland may not be broadly applicable throughout the Australasian region. Studies in Queensland indicate that, over the short term, some wildlife species appear very sensitive to selective logging, even at moderate intensities. Some forest-dependent frogs and reptiles respond negatively to microclimatic changes in recently logged forest (Preen 1981). Understory insectivorous birds (Crome et al. 1996) and bats (Crome and Richards 1988) are often sensitive to logging-induced changes in forest structure. Certain species of small mammals (Burnett 1992) and possums (Laurance and Laurance 1996) appear to exhibit a strong psychological avoidance of open areas, and hence decline in logged forest. Finally, bowerbirds and other species, which require microhabitats prone to logging damage (such as ridge-tops), could also be strongly affected (Crome and Moore 1989).

Despite the negative short-term effects of logging, available evidence suggests that populations of most wildlife species in Queensland can recover to levels approaching the prelogging condition in one to two decades or less. Recovery times may very often be longer in other parts of Australasia, however, because logged sites in Queensland are typically small (< 200 ha), surrounded by mosaics of unlogged and previously logged forest, and protected from excessive harvest damage and other forms of exploitation.

In Australasia, most research to date has considered only a single logging cycle, and the longer-term effects of repeated logging are much more poorly understood (Grieser Johns 1997). The impacts of polycyclic logging are influenced by the interval between logging cycles and the level of forest damage incurred during each harvest. In areas such as Papua New Guinea, a major concern is that forests are being over harvested in an unsustainable manner (Arentz 1994). As timber supplies dwindle, there will be increasing pressure to re-log forests rapidly, or to convert them to plantations or agriculture. The repeated, heavy logging of forests at sub-optimal intervals will likely have serious effects on some wildlife species. It must be emphasized, however, that even heavy, uncontrolled logging is vastly superior to deforestation from a wildlife conservation perspective.

Of course, logging can have important, synergistic effects with other forms of land exploitation. In New Guinea, for example, hunting pressure is often severe, and logging roads commonly lead to a marked influx of hunters into forests (W. F. Laurance, personal observation). Swidden agriculturalists use logging roads to gain access to new areas, and can convert large tracts of logged forest to mosaics of clearings and degraded second growth. Logged forests are also prone to desiccation and wildfires, especially during drought years (Laurance 1998). The effects of logging, therefore, cannot be considered independently from secondary forms of forest exploitation, which can be far more damaging to forests and wildlife than logging itself (cf. Bowles et al. 1998; see chapters 15–17). Management practices such as placing barriers across disused logging roads could be used to prevent vehicle access after the harvest, but hunters and others on foot are unlikely to be deterred.

Challenges to Reduced-Impact Logging

There seems to be little question that reduced-impact logging systems such as the QSLS and other methods (e.g., Hendrison 1990; Dykstra and Heinrich 1992; Putz and Pinard 1993; Pinard 1994; Putz 1994; Sist et al. 1998a; see chapter 21) can make a valuable contribution to wildlife management in the tropics. There are, however, potentially serious impediments that must be overcome before such methods can be successfully implemented in developing nations.

First, reduced-impact harvests may be less profitable than unregulated logging, in part because fewer logs are harvested, but mainly because the costs of managing logging operations are relatively high. In Queensland, the QSLS was viable because the state government used tax revenues to underwrite the Queensland Forest Service (QFS), which in turn subsidized the costs of major logging roads for timber companies. Revenues generated from timber royalties were relatively low in Queensland and failed to offset the real operating costs of the QFS. It should be noted, however, that the QFS was responsible for many aspects of forest management (e.g., road maintenance, tourism, fire control), not just timber harvests.

Second, reduced-impact logging requires an adequate knowledge base and infrastructure. Probably the single most important feature of reduced-impact logging methods is the supervision of timber harvests by dedicated and professional forestry officers (Crome et al. 1992). Forestry officers require effective training programs as well as vehicles, offices, support staff, and other infrastructure—even with adequate funding it can take years to develop these resources.

Third, there must be a political and social commitment to responsible forest management. Such commitment is crucial to ensure that forestry departments receive adequate funding and that a professional atmosphere prevails in which excessive environmental damage, graft, and other excesses are not tolerated. Unfortunately, toleration of these kinds of excesses are occurring with alarming frequency in parts of Australasia, and seem especially prevalent in those areas experiencing the most rapid expansion of logging.

ACKNOWLEDGMENTS

I thank Rod Keenan for providing information on the Queensland Selective Logging System and Bill Magnussen, Susan Laurance, the book editors, and two anonymous referees for commenting on drafts of the manuscript.

COMMUNITY-BASED TIMBER PRODUCTION

A Viable Strategy for Promoting Wildlife Conservation?

Nick Salafsky, Max Henderson, and Mark Leighton

The conservation and development community is increasingly interested in linking conservation with sustainable use of biological resources. Integrated conservation and development projects seek to meet these twin goals of promoting conservation and enhancing community economic development by assisting communities in their development of sustainable uses for the wealth contained in their natural resources. From an income generation perspective, it makes sense for local communities in forested areas to develop their own timber resources. But does this make sense from a conservation perspective? Should conservation organizations invest in helping communities to develop their timber resources? At face value, these projects offer something of a paradox—conservation groups helping local communities to obtain chainsaws and sawmills to *save the rain forest*.

The idea of community-based timber harvesting as a conservation tool is largely untested in practice. This field is so new that most existing efforts by conservation groups to establish community-based timber enterprises are now just getting under way. In this chapter, we examine community-based timber harvesting projects from around the world in an exploration of the issues and challenges confronting these projects. We specifically consider the questions:

- What is community-based timber production and how does it differ from industrial logging?
- What are the issues involved in using community-based timber production as a tool for wildlife conservation?
- Can the development of community-based timber production enterprises be a viable strategy for wildlife conservation?

What Is Community-Based Timber Production?

People do all timber harvesting and management, and all people live in communities. So how do community-based timber production systems differ from industrial ones?

Determining a precise definition of community-based systems is difficult owing to the different ecological, social, economic, and institutional settings in which these systems are based (Salafsky et al. 1998). In Papua New Guinea and other countries of the South Pacific, for example, local people are constitutionally guaranteed ownership of surface land rights. In Indonesia, however, most forested lands belong to the government. Thus, a community-based timber harvesting system cannot simply be defined by local ownership of forests. There are also often gray areas between strictly community-based and strictly industrial timber production systems.

Despite these difficulties, there are a few generalizations that can be drawn regarding community-based timber management systems compared to industrial-managed ones:

- Resource rights are either owned by or assigned to local community members
- The people harvesting the timber live locally and depend on the forest for other goods and services
- Harvesting is on a smaller scale with limited capital investment
- Enterprises seek to add value to raw materials on or close to the harvesting site
- Capital is re-invested locally and there is a greater long-term incentive for sustainability

These points are discussed in greater detail in table 25-1.

Community-Based Timber Production as a Tool for Wildlife Conservation

Developing a community-based timber production enterprise as a strategy for wildlife conservation requires that the potential direct and indirect impacts on wildlife populations and their habitats be considered.

TABLE 25-1 *Conditions Defining Community-Based Timber Management Systems Compared to Industrial Logging Operations*

Condition	Explanation
1. Resource rights are either owned by or assigned to local community members.	In industrial timber harvesting systems, timber lands or the rights to harvest timber are held by companies whose owners often live far away. In community-based systems, these rights are owed or held by local community members.
2. People harvesting the timber live locally and depend on the forest for other goods and services.	In industrial timber harvesting systems, the people involved in logging efforts can live either near the forest or far from it. In community-based systems, timber is harvested by people who live in or near the forest. These people generally depend on the forest for other products such as food and construction materials, and for its cultural and spiritual values.
3. Harvesting is on a smaller scale with limited capital investment.	Industrial timber harvesting enterprises employ a wide range of tools, from heavy machinery and mechanized skidders to hand tools and draft animals. As a rule, however, industrial timber production tends to be fairly large scale and capital intensive, and thus less reliant on human labor. In community-based systems, timber harvesting is disposed to be on a smaller scale. There tends to be a higher reliance on human labor and thus a lower degree of reliance on machinery and other capital intensive techniques.
4. Enterprises seek to add value to raw materials on or close to the harvesting site.	Industrial timber harvesting enterprises generally seek to harvest raw materials which are then transferred to large centralized processing mills that are located in cities or even abroad. Community-based systems, by contrast, often seek to add value to the raw materials locally by producing finished (furniture, toys, tools) or semi-finished (window frames, flooring, moldings) products, or processed lumber (wood that has been planed or sanded, dried, and sometimes chemically treated). This added value provides greater income to the local residents who own the enterprise.
5. Capital is reinvested locally and there is a greater long-term incentive for sustainability.	Industrial timber harvesting enterprises often move the capital for, and profits from, logging to other localities or sectors of the economy. These systems thus limit incentives for long-term, sustainable harvests. Community-based systems, however, generally will invest their profits locally, and thus provide more of an incentive for maintaining forest stocks that provide employment and income for local residents over the long term.

SOURCE Salafsky et al. (1998), based on a survey of the four systems described in the boxes in this chapter and other similar projects: Quintana Roo, Mexico (Bray et al., 1993); Eastern India (Poffenberger 1994); Michochan, Mexico (Sanchez Pego 1995); Vanuatu (Wyatt 1996); Lae, Papua New Guinea (Louman 1996; Van Helden 1996); Lak, New Ireland, Papua New Guinea—(Orsak 1996; McCallum and Sekhran 1996); Western Islands, Solomon Islands (Schep 1996); and New Britain, Papua New Guinea (Salafsky 1997).

Direct Impacts of Community-Based Timber Production on Wildlife Populations and Their Habitats

The direct effects of community-based timber production on wildlife populations are generally not very different from industrial timber harvesting operation efforts. Direct impacts on wildlife populations from these operations include cutting down trees in which animals are nesting, scaring animals off with human presence, disrupting travel, causing social conflicts, removing important food trees, and generally lowering the carrying capacity of the habitat for many primary forest species (see chapters 4-14). Community-based timber production, however, may have less impact on wildlife populations compared to industrial ventures because, by definition, community timber enterprises are staffed by local people. Local people tend to be more knowledgeable about wildlife and their needs, and may value the wildlife as a food or aesthetic/spiritual resource (Stevens 1997). They may be willing to take the steps necessary to reduce the impacts of logging on wildlife populations. In one community in the Papua New Guinea case study (see box 25-1), for example, we observed the community members taking care not to cut down a large tree in which a hornbill pair was nesting until after the chicks were fledged. The group also left a buffer ring of a few trees around the nesting tree so as not to overly disturb the birds. This accommodation of wildlife could also be enhanced by community-run enterprises operating at a smaller and less capital-intensive scale compared to industrial operations. Community-based operations, therefore, can afford the opportunity costs inherent in tailoring harvesting patterns to meet wildlife needs.

Box 25-1 *Working in the face of massive threats: Walkabout sawmills in East New Britain, Papua New Guinea. (M. Henderson and N. Salafsky)*

The Gazelle Peninsula of East New Britain contains large tracts of lowland and upland tropical forests that are among the most endangered parts of Papua New Guinea (Government of PNG, Department of Environment and Conservation 1993). These forests contain a wide diversity of plants, birds, bats, and insects (Balun et al. 1996).

The people of the Gazelle peninsula live in small villages that contain members of one or more different clans. Each family typically controls small pieces of land near the village that are used on a rotating basis for shifting garden plots. Larger tracts of forest located farther away from the village are typically owned by all members of a certain clan, and are used primarily as hunting grounds. Villages that contain about 100 households control forest areas ranging between 5,000 and 20,000 ha. These traditional land rights are recognized by the Papua New Guinea constitution, which assigns land and surface resource rights to the traditional land owners.

Along with land ownership, comes the ability for clans to sell their timber rights to commercial interests. The forests of East New Britain are currently under massive threats from large, foreign-owned companies that are purchasing timber rights from local people at a fraction of their value (typically paying $5 for a cubic meter of timber that is later sold for over $400), and then stripping the land of all salable timber—often in violation of existing forestry regulations (Henderson 1997).

Timber Harvesting System

To combat the threat posed by commercial logging, the Pacific Heritage Foundation has been working with residents of seven local communities since 1992 to help them develop small-scale timber enterprises that will enable the communities to develop their timber resource on a more sustainable basis (BCN 1997a and b). The Pacific Heritage Foundation and community members survey forest areas jointly. Each community enterprise has one or more chainsaws that is used to fell designated trees and remove their branches. The boles are then cut into 3- to 4-meter-long sections that are moved into position using hand winches. They are then rough sawn into planks using a small portable (*walkabout*) sawmill. These planks are carried by hand to a central transport site from which they are either sold to local markets or shipped by truck or barge to a central sawmill run by the project.

A typical walkabout sawmill can process about 0.5 to 1.2 cubic meters of timber per day, which is equivalent to roughly one or two trees per week. In 1997, groups were selling wood for US $90 to 135 per cubic meter in local markets, with the higher prices being paid for higher quality woods. Benefits thus come to communities in the form of wages (around US $3/day/person), profits (a couple of enterprises are putting thousands of dollars in clan bank accounts), and timber for house construction.

Impact on Wildlife Populations and Their Habitats

In this case study, using merely the potential rate of harvest of the walkabout sawmills, it is clear that the community sawmills have far less of an impact on the forest and on wildlife than the alternative of communities selling their timber rights to foreign firms capable of harvesting dozens of trees per week. At this point, however, it is not clear whether the community timber harvesting practices will be sustainable over the long-term. Most of the groups are currently using their mills to clearcut areas near the village that would be felled anyway to produce subsistence gardens. Only a few groups are now starting to move into the primary forest, as they are currently limited by transport problems.

The project is developing long-term biological monitoring techniques to assess sustainability. In one site, for example, staff worked with an outside researcher to measure tree species diversity in three types of forest—primary forest, forest that had been harvested using walkabout mills, and forest that had been harvested by a foreign-owned industrial firm. The study found that tree biodiversity decreased in the forest logged with the walkabout sawmill when compared to the primary forest, but was much higher than the level of tree biodiversity in the forest logged by the foreign firm (Lindemalm 1997). Assessments of wildlife responses to these practices are still in the planning phase. Pacific Heritage Foundation will have to determine over the long-term whether the projects will be ecologically sustainable

The local perspective, however, can also just as easily have a deleterious effect on wildlife. In another community in the Papua New Guinea case study (see box 25-1), the community deliberately went out of their way to cut down an emergent walnut tree in which large numbers of parrots traditionally roosted, citing the need to keep the parrots from eating the cacao beans in their nearby smallholder plantations as their incentive.

Harvesting Technologies

The direct impacts of community-based timber harvesting on forest habitat depend to a large extent on the type of technology being employed to harvest timber. Since community-based systems by definition generally use simpler and more labor intensive technologies than industrial ones, they may have less impact on the forest. Since communities by definition have control over the forest rights and are presumably interested in multiple harvests, they may also be more inclined to practice sustainable forestry techniques such as directional felling, cutting lianas, maintaining seed trees, following through on replanting efforts, and cutting trees on long rotation periods sufficient to guarantee future harvests (see chapters 21 and 24). To do so, however, local people have to have the knowledge and incentives to act in a sustainable fashion.

In the Papua New Guinea case study (see box 25-1), for example, the groups are using small, portable sawmills (*walkabout sawmills*; see photo 25-1) that in theory enable the groups to move around in the forest taking a tree here and another there, thus mimicking natural gaps and reducing the impact on the habitat (see chapter 22). This is also the plan of the groups in the Madagascar case study (see box 25-2), except that they are using pit saws. Sawing the timber *in situ* (using either walkabout sawmills or pit saws) also has the potential to minimize the impacts associated with log removal and nutrient loss from the soil.. In practice, however, difficulties in transporting timber to ponding sites at road or beach heads makes it hard for community-based timber opera-

PHOTO 25-1 *Sequence of photos showing the operation of a small portable sawmill (walkabout sawmill) in Papua New Guinea: a) carrying the mill into the forest, b) setting up the mill, c) cutting boards, d) removing sawn boards, and e) the finished product. (N. Salafsky)*

tions to move around the forest. Most of the groups in the Papua New Guinea project, for example, are now considering building roads to improve truck access to the sites and will probably have to intensify their harvesting efforts to cover the increased costs. In the Indonesian case study (see box 25-3), the enterprise is planning to make use of a labor-intensive traditional wooden rail system for transporting the logs from the felling site to the river (see photo 25-2). The group plans to construct the rails out of noncommercial, non-fleshy-fruit tree species, thus enhancing the quality of the habitat both for future harvests and key animals such as the orangutan (*Pongo pygmaeus*). The latter is achieved by planting and releasing important food resources for orangutans. This practice appears less destructive than building a road into

PHOTO 25-2 *Railed skid trail (a) and sled used for skidding logs (b) in Indonesian community forest operation. (N. Salafsky)*

these areas. It is conceivable, however, that these community systems might actually promote habitat degradation by exploiting pockets of forest on swampy or steep terrain which would have been passed over by industrial logging operations as not being profitable to harvest.

Box 25-2 *Buffering a new national park: Community-based forest management on the Masoala Peninsula, Madagascar.*
(R. P. Guillery and C. Kremen)

MADAGASCAI

The Masoala Peninsula is the last remaining large area of lowland tropical forest in Madagascar. It contains lowland rain forest as well as isolated fragments of littoral forest and coastal evergreen forest—two of the most threatened habitats in the region. The forests of the peninsula are home to countless species endemic to Madagascar, including many that are known only from the immediate region.

The indigenous residents of the Masoala Peninsula primarily make their living by practicing shifting cultivation. The forests of Masoala are constitutionally designated as belonging to the state and permits are required for land clearing or to purchase land. Only three people of the approximately 45,000 who live on the peninsula, however, legally own land. The vast majority of residents employ a traditional land tenure system that assigns rights to a person who clears and cultivates a plot of land. This tenure system is not recognized by the state. There is serious pressure being placed on the forest resource by the local community for agricultural land. Pure protection of the highly threatened areas outside the park's boundaries is unlikely to succeed given this pressure.

Timber Harvesting System

To relieve this pressure, the Masoala Integrated Conservation and Development Project adopted a dual strategy. First, the project worked with the government to set aside 2,100 km² of forest as a national park. Second, the project is promoting sustainable uses of the remaining forest that surrounds the park. Community-based timber harvesting constitutes a major element of the strategy for this effort (at the time this case study was written) and began in 1997 with one community in an area of about 1,000 ha. The Malagasy government is writing new laws that will allow communities to legally manage and control forests around villages for timber and nonforest timber products on the condition that the forests are not cleared for agriculture. Timber harvesting will focus on selective felling of ten commercially valuable tree species, which will be harvested from small plots (5 to 20 ha) on a 60-year rotation cycle. About two to six trees (or 10–30 m³) of roundwood will be removed per ha. We expect that the use of benign techniques, including cutting of lianas prior to logging, directional felling, pit sawing at the felling site, absence of mechanized equipment, stream and slope buffering, and the exclusive use of hand labor, will greatly minimize ecological impacts. These same factors also increase the number of community members who are directly involved and benefit from forest management. Rough cut timber will then be sold to domestic and international markets.

Impact on Wildlife Populations and Their Habitats

In this case study, the primary issue driving conservation is the ability of the community forestlands to serve as a buffer for the newly created national park. Forest management within the highly threatened forests should discourage shifting cultivation and help prevent farmers from penetrating into the park. Sustained use of the forests outside the park will also protect habitat for species tolerant to small-scale, unmechanized selective logging.

Forest management plans have been designed to minimize impacts on the forest habitat and wildlife and maximize economic benefits to the local population. Logging practices are the most important factor in determining harvest impact on wildlife (local subsistence hunting appears to have a very low impact on wildlife, and few market hunters operate in the area). To our knowledge, harvest of the ten timber species will not disrupt ecological interactions of the largest lemur species (*Eulemur fulvus albifrons, Varecia variegata rubra* and *Daubentonia madagascariensis*) found within the forest, since these trees are not keystone resources for the lemurs. The disturbance from selectively logging these ten species should fall within the range of natural disturbances (treefalls to cyclones) encountered in the evergreen humid forest zone of Madagascar. Selective logging will mimic treefall gaps created by tree falls. The density of gaps, however, will increase from 7.5 to 30 gaps per hectare over background levels.

To evaluate the effects of logging on habitats and wildlife adequately, monitoring of timber species, impacts on biodiversity, and social and economic factors are explicit activities within the community-based forest management plans. The results of the monitoring program will be used to further develop these links by refining management plans and educating communities. This will not only allow the forests around the park to be sustained for timber, NTFP and wildlife, but it will help determine if community-based forestry can be financially, culturally, and ecologically sustainable.

Box 25-3 *Setting policy precedents for community control of resources: community management of buffer zone forests in West Kalimantan, Indonesia. (M. Leighton, R. Cherry, and N. Salafsky)*

Gunung Palung National Park in West Kalimantan contains 90,000 hectares of forest representing a wide range of habitats including mangrove, peat swamp, freshwater swamp, bench, hill, and cloud forests (MacKinnon and Warsito 1982). These habitats contain a wide range of endangered species, including orangutans (*Pongo pygmaeus*), proboscis monkeys (*Nasalis larvatus*), gibbons, flying foxes, six species of hornbills, and dozens of other bird species.

The villages surrounding the park are inhabited by a mixture of Dayak and Melayu peoples living along with a growing population of transmigrants from Java and Bali. Village residents are primarily small-scale farmers, growing rice and other crops in small plots. Many villagers also own small forest garden plots which produce durian *(Durio spp.)* and other products for commercial sale. Forested areas in and around the Park are owned by the government and are zoned for different uses. Many of the parcels immediately bordering the park have been classified as production forest and have been logged over the past few decades by concession holders who use mechanized equipment in the upland sites, and who hire community members to do hand logging in the swamp habitats.

This industrial logging has generally been conducted in a nonsustainable fashion. Sites are not regulated and indeed there is often illegal harvesting of timber and nontimber forest products from within the national park itself. In the upland areas where mechanized logging takes place, local people get little or no benefits. In swamp sites, the local villagers are in a *debt-peonage* system in which they are given food and other supplies (at high interest rates) prior to setting out into the forest, and then end up making little or no money when they sell the timber they have harvested. In either case, most of the profits are flowing to the middlemen and especially to the concession holder.

Timber Harvesting System

To combat the threat posed by industrial logging, Harvard University's Laboratory of Tropical Forest Ecology has been working with the Indonesian Ministry of Forestry to create the first community-managed timber concession in Indonesia. The community project, operational since 1997, is slated to take over the management of an 8,000-hectare peat swamp forest site on the northwest border of the Gunung Palung National Park. Political and economic problems in the country have delayed this process, however, recent government and donor (United States Agency for International Development) support suggests that the project remains viable and the community will assume management rights in the foreseeable future. The community will harvest timber from 50-hectare strips (4 km x 125 meters), cutting 80 percent of the trees greater than 40 cm dbh in each strip. Felled trees will be cut into 4.2 meter sections, which will then be hauled to the river using traditional wooden rail techniques. Logs will then be rafted down to the village where they will be processed using a band sawmill. Sawn wood will be shipped to domestic markets in Java or to international buyers. At projected extraction rates, communities will be able to harvest the timber on a sixty to 100 year rotation cycle. Direct processing and marketing will generate both employment and revenue for local community members.

Impact on Wildlife Populations and Their Habitats

The major factor affecting the project to date is the ability of the project staff to work out a precedent-setting agreement with the Indonesian government that will enable communities to gain access to the forests. Developing and implementing this agreement has been a long process, involving several years of discussion and negotiation, and has not yet been finalized. If an agreement can be reached, then it seems likely that the project will have a substantial positive impact on wildlife since both the selection of the site and the forest management system are designed to directly and dramatically contribute to the conservation of wildlife in the park.

The site is comprised of peat swamp forest, which ten years of research at the park by project staff has shown to be a primary habitat used by many species—including the orangutan population at Gunung Palung National Park, which is the largest in all of Kalimantan (BCN 1997). The site also provides an important buffer to the national park and should help to minimize the impact of illegal logging occurring in the park since it restricts access to the park and employs the villagers who have been doing much of the illegal logging. The controlled timber harvesting will also hopefully protect the peat forest from forest fires.

Cutting timber from thin strips should disturb one to two percent of the forest in the 8,000-ha managed area in any year. Residual trees will also be left in the strips to facilitate crossings by arboreal animals. The rail system will minimize impact on the soils. The strip-cutting system should allow for natural regeneration of the forest species, which will also be augmented by re-planting of select wildlings including *Tetramerista glabra*—a valuable timber species and a keystone resource for orangutans and other large vertebrates.

The project will draw on existing project research at the site to monitor long-term effects on key wildlife populations. In particular, the project will be able to compare seasonal population densities of orangutans and other key indicator species in the buffer zone with baseline levels in the park. If these efforts can be shown to work, then the project may also contribute to conservation by providing a precedent for changing policies within Indonesia and allow other communities to manage the forests in their areas for the benefit of both themselves and wildlife.

Conservation Organizations

Finally, the direct impacts of community-based timber production on both wildlife populations and their habitats may be reduced, compared to industrial timber harvesting, because community-based enterprises are often initiated or supported by national or international conservation organizations. These organizations can propose or even insist that local people minimize impacts on wildlife as a condition for their support (Salafsky 1997; see chapter 26). In the Indonesian case study (see box 25-3), for example, the project will leave corridors of standing trees for arboreal animals to use to cross the strip cuts, and will replant species that are keystone resources for orangutans and other endangered species in the harvest areas. In the Madagascar case study (see box 25-2), the project plans to cut only trees that are not keystone resources for the endangered lemurs in the area. It is not clear, however, whether the community members themselves would initiate or even support these efforts if conservation NGOs were not working directly with the communities.

Indirect Impacts of Community-Based Timber Production on Wildlife Populations and Their Habitats

Hunting

A major indirect impact on wildlife populations from timber production is hunting (see chapters 15–17). Because community-based timber enterprises are (by definition) staffed by local people, they could potentially differ from industrial ventures in that local people tend to have the knowledge and incentives to not overhunt—thereby preserving game populations for long-term use. Community-based timber production enterprises can also reduce hunting impacts by keeping local people employed in the timber industry and providing money to purchase alternate sources of food. Finally, community-based timber production systems do not need to bring in outside workers, and thus do not increase the human population density in the area and the associated pressures placed on game animals.

The degree to which community-based timber production reduces hunting impacts, however, depends on local conditions and customs. In the Papua New Guinea case study (see box 25-1), for example, most of the community groups are expert hunters who traditionally rely on wild meat in their diet. The community-based timber enterprises are more likely to reduce hunting pressure by providing employment opportunities that reduce the time available for hunting and giving the communities a cash income with which they can purchase tinned fish and other substitutes for bush meat. In the Indonesian case study (see box 25-3), the project is working primarily with the Islamic Melayu populations of the region who have a taboo on eating most wild meat other than deer. There are high densities of primates and other large game species living in the forests near the village and even in people's forest gardens. It is, thus, not likely that the project will have much of an impact on animal populations, unless a disproportional number of the logging workers are of Dayak origin. This group has a long cultural tradition of hunting, and could dramatically increase hunting pressure around the field camps.

Habitat Conservation

Perhaps the greatest inherent difference between community-based and industrial timber production efforts is their potential indirect impact on conserving habitat in the long term. Community-based enterprises both persuade and empower communities to take the steps required to reduce internal and external threats to the forest resource that their enterprise depends on (BCN 1997a, b). In effect, community-based timber production provides *incentives* for communities to provide *stewardship* over the forest—similar to the community stewardship initiatives that conserve forest resources in other areas of the tropics (e.g., rubber tappers in Brazil [Fearnside 1989]; forest gardens in Indonesia [Salafsky 1994]; forest products in India [Poffenberger 1994]; and wildlife in Amazonia [Bodmer 1994]). State- and privately-managed production forests do not always share a similar fate, with many areas of the *permanent forest estate* succumbing to short-term, high-return land conversion schemes following logging (e.g., oil palm plantations, cattle ranching, mining, and small-scale agricultural cropping). This stewardship ethic can be a strong deterrent to short-term forest exploitation and conversion (see boxes 25-1 and 25-2), thereby helping to conserve valuable habitat for forest wildlife.

Finally, low cutting intensities, combined with long-term stewardship and adaptive management strategies (Holling 1978; Lee 1993; Margoluis and Salafsky 1998), make community-based timber production enterprises excellent buffer zones between areas of high-intensity management and parks. Efforts such as those outlined in the Indonesia case study (see box 25-3) and elsewhere (SUBIR Project in Ecuador; R. A. Fimbel, personal communication), seek to buffer wildlife populations from abrupt habitat fragmentation and structural/compositional degradation. Furthermore, community timber management can lower the risk of wildfire within community lands and adjacent holdings (fuel loads are lower following light, selective logging compared to the more aggressive cuts characterizing commercial ventures). Finally, community enterprises can even reduce hunting and habitat degradation impacts on wildlife within parks, as the community now employs those villagers who traditionally performed illegal logging and hunting (MacKinnon et al. 1986).

Community-Based Timber Production and Wildlife Conservation

Technical, Financial, Political, and Social Factors

At the most basic level, before we worry about the impacts of community-based, timber-harvesting enterprises on wildlife, we first have to determine if they can even work as enterprises. These enterprises are beset by a vast array of challenges (Salafsky et al. 1998):

- Legal problems in gaining access to forest resources
- Training problems in developing the skills of workers employed by the enterprise
- Technical problems regarding equipment purchase and how to keep it working
- Transport problems in getting the timber out of the forest
- Marketing problems in finding buyers who will pay fair prices for the timber and require only limited quantities
- Social problems in keeping community members involved in the enterprise
- Political problems in dealing with economies that provide perverse incentives for activities of large companies that the projects are competing with
- Financial problems in keeping the enterprises solvent

Finally, even if all these problems can be solved, community-based, timber-harvesting enterprises can be destroyed by success if there are disagreements as to what to do with the profits.

Solving these challenges requires work that is not directly related to wildlife conservation. But if the enterprises are not financially, culturally, and institutionally sustainable, then there is no chance for them to contribute to conservation. At this point, the evidence is, at best, mixed as to whether these enterprises can solve these problems—especially without support from outside organizations. In the Papua New Guinea case study (see box 25-1), for example, a few of the groups seem to operate self-sustainable enterprises, but others are stymied by the many problems that they face. The Madagascar project (see box 25-2) is also experiencing many technical and logistical problems that have led to the cessation of timber-harvesting activities. In the Indonesian case study (see box 25-3) the project has been working for over four years to arrange an agreement with the government that will grant the community title to the forestlands. In the Peruvian case study (see box 25-4), the project ultimately failed because they could not solve these problems.

Box 25-4 *Institutional and economic success as a prerequisite for conservation: the Yánesha Forestry Cooperative in Central Peru. (G. Hartshorn and N. Salafsky)*

The Palcazú Valley in the Central Selva region is in the heartland of the Peruvian Amazon. The lower valley contains a mixture of agricultural and forestlands. In the early 1980s, the valley was inhabited by approximately 2,200 *colonos* of European or Andean origin, who practiced commercial agriculture (especially cattle ranching), and 2,800 Yánesha, an indigenous group of people who make their living by practicing subsistence agriculture, hunting and fishing, and working for the *colonos* (Stocks and Hartshorn 1993). Under Peruvian law, nonagricultural lands (i.e., forests) belong to the government, with timber extraction contracts obtained through the Ministry of Agriculture. At that time, the forests of the valley were under severe threat from a massive agricultural development plan to encourage immigration of 150,000 colonists to the valley as part of government efforts to develop the Amazonian frontier. With the 1980 return to a democratically-elected government, the United States Agency for International Development (USAIDS) agreed to assist with development of the valley—primarily via the building of a spur road into the valley as part of the Marginal Highway at the base of the Andes.

An environmental assessment of the valley revealed that, due to very high rainfall (~7,000 mm/yr) and extremely poor soils, very little of the valley would be suitable for traditional agriculture, but that the lower valley could be used sustainably for natural forest management. A companion social soundness analysis indicated that Yánesha land tenure was still insecure and at grave risk from agricultural colonists. As a condition precedent for the USAID loan to the Peruvian government, all native communities' land claims in the valley had to be officially recognized and legally titled. In 1984–85, the USAID technical advisors completely redesigned the development project to focus on production forestry by the native communities in the southern lower valley.

The Yánesha Forestry Cooperative (COFYAL) was created and officially recognized in 1986 to help provide employment to members of indigenous communities, to manage the natural forests in a sustainable fashion, and to protect the cultural integrity of the Yánesha people. The creation of the Yanachaga-Chemillén National Park, the San Matías and San José protection forests, a communal forest reserve, and the recognition of all native communities' land claims effectively excluded most of the valley's forests from agricultural development by colonists. Deteriorating political conditions in Peru eventually halted road construction part way through the valley (as far as Iscozacín) and forced the USAID project to close in 1988. Due to guerrilla uprisings, COFYAL shut down in 1990 and closed in 1993.

Timber-Harvesting System

Forestlands managed by COYFAL were classified for production forestry, protection forests, or swidden agriculture. The innovative strip-cut forest management system (Hartshorn 1989) was adopted by COFYAL. Trees were harvested in 25 to 35 m wide strips that were positioned parallel to rivers and topography wherever possible. The plan required total felling of all trees and harvesting of all trunks and large branches, leaving only smaller branches and organic debris on the ground. The project successfully relied on natural regeneration from seed trees on either side of the strip, augmented by some silvicultural thinning treatments in the regenerating forest. It was projected that production forests would be managed on a 40-year rotation (Hartshorn 1995). Timber was skidded out of the strip using oxen to a primary collection point along a woods road and from there, by wagon or truck to the COFYAL facilities. Timber was processed at the cooperative's plant, which had a portable sawmill, a re-saw, a carpentry shop, and a pressure treatment facility. Small-dimension boles were preserved for fence posts and utility poles, while larger logs produced sawn timber that was marketed primarily to local and regional buyers.

Impact on Wildlife Populations and Their Habitats

There is evidence that the strip-cut technique employed by the COFYAL project promoted outstanding regeneration of native tree species (Hartshorn and Pariona 1993). Even with occasional thinning, very few tree species were lost from the regenerating strips. There are few data available, however, on the impact of the system on wildlife species. Most of the natural forests in the lower valley were under serious hunting pressure by the Yánesha as well as the colonists, therefore wild meat species were very scarce.

Although natural forest management may have inhibited the movements of some non-volant arboreal mammals (see chapter 2), understory birds (see chapter 8) and select small animals (see chapters 6 and 11), this is a far better alternative than conversion to cattle pastures. Habitat for some species was eliminated in the cut strips (e.g., large cavity trees) and the projected 40-year rotation may have led to their permanent removal from the forest structure in managed areas of the forest. On the positive side, very few tree species were lost (diversity probably increased overall), and the strip harvests were distributed through time and space to minimize impacts to the forest. It is important to note that only about 50 percent of any production forest was scheduled for strip-cutting over the entire rotation—i.e., not only did the forest management system maintain the natural forest matrix, but it designated appreciable areas of forest that would remain for other uses or purposes such as wildlife habitat. Finally, hunting probably had the greatest impact on select mammal and bird species. Increases in forest access and demand for wild meat by immigrating *colonos* eventually led to overexploitation of preferred species in the southern lower valley (G. Hartshorn, personal communication).

The wildlife conservation potential of the land tenure and low-impact logging initiatives were undermined when USAID support for the COFYAL project halted in 1988. Benavides and Pariona (1995) list a number of reasons for the collapse of COFYAL, including that the cooperative was:

- Dependent on outside support
- Too complex for the local people to manage
- Hindered in being able to transport timber to market
- Unable to successfully find markets for its timber

⁓ Subject to the political and economic instabilities of the region

The bottom line from this project, therefore, is that even if the project might have been an ecological and social success (Hartshorn 1995), the fact that it proved to be economically and politically unfeasible resulted in it being incapable of contributing to wildlife conservation over the long term.

Despite the challenges that these projects are facing, there are signs—in both these projects and from a few (but growing) number of other cases from around that world—that these enterprises can succeed, especially if they receive suitable nurturing and support from conservation organizations and other interested parties (BCN 1997a; Salafsky 1997). Before making a final judgment on this strategy for conservation, we need to continue to work with these projects and monitor them to see what it takes to make them technically, financially, socially, and institutionally sustainable.

Ecological Sustainability

Assuming that viable community-based timber production enterprises can be established, the next question is, can they deliver on the elusive grail of *ecological sustainability*? Sustainability certainly seems to have escaped industrial loggers. In the early 1990s, it was estimated that less than one tenth of a percent of tropical logging was done on a sustained yield basis (Worldwatch Institute cited in Bray 1991). Hartshorn (1995) has also stated that "almost by definition, industrial harvesting of tropical timber is not sustainable." Should we, therefore, expect community-based, timber-harvesting enterprises to do any better?

There are certainly tremendous challenges to overcome. In addition to those noted above, people may not know what steps are required to achieve sustainability and compatibility with wildlife. Even if they have the knowledge, despite the best plans and intentions, economics and human nature often conspire to make it tempting to take short cuts that decrease sustainability. In the words of Schep (1996) describing a community-based timber production project in the Solomon Islands: "Although the program was about 'sustainable' forest management, there was a natural preference for 'convenient' forest management—for instance milling the nearest big tree and then the next nearest one."

Finally, even if people are willing to take the extra steps to try to set up sustainable systems, there are still problems with monitoring the projects in a cost-effective fashion—especially within the relatively short time frames over which these projects are funded and implemented (Margoluis and Salafsky 1998). There are several approaches that can be taken to monitoring ecological sustainability and compatibility with wildlife needs. One is to conduct extensive monitoring of habitats and indicator species at the site to determine whether wildlife is being affected and modify practices accordingly (see chapters 18 and 19). Even with the support of outside donor funding, this is a difficult and expensive undertaking. A second approach is to find an outside party to inspect and *certify* the timber operations (see chapter 26). This is also expensive, and it is likely that few community-based enterprises can afford the costs, especially given that it is very difficult for these small-scale operations to produce the quantity and quality of timber necessary to command a *green* price premium that in theory comes with certification. Many groups, therefore, may have to resort to a third approach that involves following a few basic rules of thumb for timber harvesting: trying to maintain a long rotation interval, monitoring the health of a few key plant and animal species (extrapolating from methods discussed in chapter 19), and hoping for the best.

The Role of Community-Based Timber Production Enterprises in Tropical Forest Wildlife Conservation

The preliminary evidence from the case studies considered in this chapter, combined with our understanding of similar projects around the world, suggests that in places community-based timber production systems can be the foundation for adaptive management regimes that conserve wildlife. These enterprises appear to have their highest conservation value where:

- Forests are faced with massive *outside* threats (e.g., commercial logging followed by land conversion)
- Communities are well-organized and have at least some recognized land rights to the forest in question
- Opportunities exist to develop sustainable timber harvesting practices that yield incomes in balance with community needs and expectations

Continued applied research on how to improve the economic and ecological properties associated with community-based timber production practices, and their integration with other conservation strategies, are of the utmost importance. Although these enterprises are financially, culturally, and institutionally difficult to develop, and may not be 100 percent ecologically sustainable, they can greatly contribute to conservation by helping to reduce large and immediate threats to forests in many areas. By reducing these threats, community-based timber production enterprises maintain forest habitat and thus conserve tropical wildlife populations.

ACKNOWLEDGMENTS

We would like to thank our colleagues who worked with us on the Papua New Guinea and Indonesia case studies. We would also like to thank Phil Guillery and Claire Kremen for providing information for the Madagascar case study and Gary Hartshorn for comments on the Peru case study. We would also like to thank Amy Shannon for providing various references. Additional thanks go to Hank Cauley, Robert Fimbel, Alejandro Grajal, and John Robinson for reviewing and commenting on various drafts of this manuscript.

Our work in this chapter was in part supported by the Biodiversity Conservation Network (BCN), which is administered by the Biodiversity Support Program (BSP)—a consortium of World Wildlife Fund (WWF), The Nature Conservancy (TNC), and the World Resources Institute (WRI), with funding by the United States Agency for International Development (USAID). The opinions expressed herein are those of the authors and do not necessarily reflect the views of USAID or BSP and its consortium partners.

Part VI

INCENTIVES FOR INTEGRATING NATURAL FOREST MANAGEMENT AND WILDLIFE CONSERVATION

Technical (part V) and research (part IV) solutions to the direct and indirect environmental impacts associated with tropical timber-harvests (parts I to III) can only succeed where inducements exist for their adoption. This section reviews several financial, social, and political incentives for advancing wildlife conservation measures into forest management programs. The presentations, while grounded in economics, also touch on public relations and policy issues.

Incentive Mechanisms for Wildlife Conservation Within Managed Tropical Forest Landscapes

In chapter 26, Richard Donovan opens this part of the book with an historical summary of the wood certification program, and its utility in assisting forest management organizations in the integration of wildlife concerns into commercial forestry operations. Current certification guidelines by the SmartWood program and other certification groups contain criteria directly related to wildlife protection within managed forests. These efforts have been very effective in encouraging forest stakeholders to communicate with one another; moderately successful at protecting threatened and endangered species (particularly mammals and birds), and weak in incorporating landscape-level biological concerns and assuring the protection of less charismatic animal taxonomic groups (e.g., reptiles, amphibians, and invertebrates). The author discusses efforts underway to strengthen the wildlife conservation components of the certification process and reduce the costs associated with the certification process.

In chapter 27, Elizabeth Losos discusses the potential of carbon-offset projects to promote the conservation of tropical forest wildlife and their habitat. The addition of a *carbon component* to otherwise non-profitable forest conservation projects (whether protection, natural forest management, or reforestation-oriented) can make them operational ventures. This is especially true where detailed ecological research is required to assess the impacts of forest management practices on wildlife. Carbon-offset projects in forest environments, however, also have their own unique set of problems—including the difficulty of monitoring carbon sequestration through forest growth and the uncertainty of long-term land tenure, among others. Carbon-offset programs, therefore, are not a conservation panacea. They should be viewed as one additional tool for the advancement of wild-life conservation in forested environments.

In chapter 28, Neil Byron reviews the economics of sustainable forest management (principally reduced-impact logging measures), and evaluates its potential to contribute to the conservation of wildlife in tropical landscapes managed for timber production. Traditional efforts at conserving the forest integrity through mandated reductions in timber harvests and increases in management investments have met with strong opposition from timber producers. Recent initiatives employing reduced-impact logging techniques, however, have the potential to create win-win scenarios in the areas of conservation and economics in forests managed for timber. Technological improvements in wood harvesting and processing are providing concession holders with the financial incentives to incorporate conservation measures into their operations. Strong regulatory mechanisms to complement these technical improvements—via the market or penalties—are deemed essential to ensure long-term conservation of wildlife and their habitat in managed tropical forests.

In chapter 29, Damián Rumiz and Fernando Aguilar close part VI with a discussion of the legal framework that directs the conservation of wildlife within the production forest landscape of Bolivia. Forest management in Bolivia reflects conditions characterizing many developing tropical nations where a large proportion of the land mass is gazetted as permanent forest estate. Logging, colonization, de-forestation, hunting, and the trade of wildlife products have long been regulated by Bolivian law, but with little or no enforcement of these statutes. Factors that contribute to the breakdown of conservation measures stem from a shortage of financial and human resources, ill-planned colonization programs, the corrupt awarding of land titles and concessions, and political instability, which creates an atmosphere of high job insecurity. An atmosphere of limited enforcement bred the growth of illicit operations, which increased revenue losses and accelerated the process of forest degradation. To arrest the increasing conflicts developing between timber firms, peasant settlers, indigenous groups, and other forest users, the government of Bolivia placed a five-year moratorium on new logging permits and strictly enforced the regulations governing existing harvesting activities in 1990. Where gaps in the regulatory process were identified, new legislation was enacted to preserve the integrity of the forest estate. The potential of these new government initiatives to advance sustainable forest management in Bolivia's production forests, and to serve as a model for other countries, is discussed.

The Feasibility of Sustainable Forest Management and Wildlife Conservation in Managed Tropical Forests

A critical step in advancing sustainable forest management and its potential to conserve wildlife and their habitat, is to create an atmosphere where forest stakeholders can come together to share ideas and concerns about the management of the resource. Timber certification programs (see chapter 26) and carbon-offset projects (see chapter 27) are two important vehicles for opening lines of communication between concerned parties. They also provide a framework within which specific management guidelines (especially certification) reflecting stakeholder interests can be developed and advanced to conserve a forest's resources. At present, however, these programs have limited application due to the relatively scarce funds available to carbon-offset projects and the costs associated with certification. The growing concern over global warming, and efforts to reduce certification costs through the training of local resource managers, suggest that these processes may play a larger conservation role in the near future.

Despite years of biophysical research, remarkably little is known about the full ecological impacts associated with timber exploitation, and even less about the long-term impacts of commercial harvesting of non-timber forest products or sustained hunting of select wildlife. This makes dollar-value assessments of forestry impacts extremely difficult. Nonetheless, examples are emerging that show logging operations having low ecological impacts, can yield increased operator profits (see chapter 28). These initiatives tend to increase planning, training, and supervisory costs—however, the costs associated with these activities can be re-captured through increases in efficiency and greater utilization of the resources. These efforts need to be widely publicized and evaluated elsewhere throughout the tropics.

Reduced-impact logging measures, and regulation of wildlife hunting and trade, are of little conservation value to tropical forests if they are not implemented and enforced. Like Bolivia in the late 1980s, the natural resources programs in many tropical countries are not very effective. Bolivia is attempting to break this cycle by enacting harvesting moratoriums and new legislation that clearly defines acceptable forest management practices and how they will be enforced. This process proposes a win-win situation for the logging sector and the public (see chapter 29). Tax incentives to concession holders provide economic incentives for loggers to implement sound silvicultural practices, thereby protecting the forest resource and its amenities for future generations. Enforcement is strengthened through *social control*, where concerned citizens (and organizations) participate in verifying that logging operations comply with management plans. This latter step empowers forest stakeholders to accept responsibility for the conservation of their forest resources. Finally, a percentage of tax revenues generated from the leasing of timber concessions will help local communities participate in the enforcement of these forestry practices. This reform process is still in its infancy, and only time will tell if this approach will serve as a model for other nations confronted with non-sustainable forest exploitation practices.

Chapter 26

TROPICAL FOREST MANAGEMENT CERTIFICATION AND WILDLIFE CONSERVATION

Richard Z. Donovan

Certification of forest management enterprises was started on a worldwide basis in 1989. In various regions of the world, this type of forest management certification is sometimes referred to as *eco-certification*, *green certification*, or in Latin American countries as *green stamp* or *sello verde* programs. Initially, certification resulted from a frustration with the limitations of tropical timber bans or boycotts, and a perception that a positive incentive was needed to promote good forestry operations. Certification provides an independent third party seal of approval for forest management operations that are strongly committed to sustainable forest management practices at the field level. These operations include large-scale commercial forest products companies, indigenous groups, consulting foresters, or community forestry cooperatives. Once certified, forest products are sold with the independent seal of approval, allowing consumers to choose an environmentally safe product.

Written certification standards, which often vary by region, are used as a basis for assessing each candidate operation. These standards represent what forest management certifiers believe is a path towards sustainable forest management—"a set of objectives and outcomes consistent with maintaining or improving the forest's ecological integrity and contributing to people's well-being both now and in the future" (Woodley et al. 1998). These certification standards incorporate criteria, indicators, and thresholds aimed at improving biological conservation, forest management planning, silviculture, and relationships with local communities (FSC 1996; Ervin and Pierce 1997). Because the certification program places a strong emphasis on biological conservation, certification has recently gained attention as a useful tool for improving wildlife conservation in commercial forestry operations. Commercial forest managers and forest management certifiers have expressed a need for wildlife managers and biodiversity researchers to help identify practical methods of maintaining, restoring, or improving biological diversity in commercially-managed natural forests (Cabarle et al. 1995; International Tropical Timber Organization 1993; Viana et al. 1996). This chapter describes the certification process, the role it has played in biodiversity conservation, and the opportunities and limitations of certification in the conservation of wildlife and their habitat in well-managed tropical forests.

Certification History and Processes

Historical Perspective

The first worldwide forest management certification initiative was started by the nonprofit Rainforest Alliance in 1989 with the development of the SmartWood Certification Program (Elliot and Donovan 1996). The Forest Stewardship Council (FSC) was established in 1993 as the international accrediting body for certification programs, granting certification based on the FSC's Principles and Criteria of sustainable forest management (see table 26-1). FSC-approved certification programs, in addition to SmartWood, include the Forest Conservation Program of Scientific Certification Systems, Inc. (SCS), the Qualifor Programme of SGS Forestry, Ltd., and the Woodmark Programme of the Soil Association (Elliot and Donovan 1996). Currently there are similar or complementary forest management certification programs being developed by organizations such as the International Standardization Organization (ISO) (Elliot and Viana 1996), however these programs are still evolving and not yet operational. As of late 1998, FSC-approved certifiers had certified over 200 different forest enterprises in roughly 20 countries, including tropical, temperate, and boreal forests covering over 10 million hectares (FSC 1998).

The Certification Process

Forestry operations may be certified at the forest management unit level (a specific block of forest, such as a logging concession), or certfication can be granted in the name of the forest resource manager who

TABLE 26-1 *Summary of Forest Stewardship Council (FSC) Principles and Criteria of Sustainable Forest Management (FSC 1996)*

Principle	Criteria
Principle #1: Compliance with laws and FSC Principles	Respect for all laws and administrative requirements Appropriate payment of fees and permits Compliance with CITES and other international agreements Conflicts between laws and regulations resolved Management areas protected from unauthorized uses Forest managers have long-term commitment to FSC
Principle #2: Long-term land tenure and use rights are clearly defined	Tenure or use rights are clearly documented Customary use rights are resolved and honored Conflicts are resolved legally and amicably
Principle #3: Legal and customary rights of indigenous people are protected	Indigenous people control use of their own land Indigenous uses not diminished or restricted Special cultural or religious sites are protected Indigenous people compensated for use of traditional knowledge and/or resources
Principle #4: Forest workers and local communities are supported	Communities given priority for training and employment Health and safety of workers and families protected Workers maintain right to organize Social impact assessment used in management planning Grievances resolved amicably, legally, and fairly
Principle #5: Optimal use of forest's multiple products is ensured	Operation is economically viable while protecting forest Processing and marketing emphasizes local production and highest and best use of raw materials Processing minimizes waste Operation emphasizes diversification and local economy Operation enhances fisheries, watersheds, and other values Rate of product harvest is permanently sustainable
Principle #6: Biological diversity is maintained and negative environmental impacts minimized	Environmental impact assessment used in planning Threatened and endangered species and habitats are protected with conservation zones and strict protection areas Ecological functions and values maintained in the forest Representative samples of ecosystems in their natural state are maintained at landscape and other levels Written guidelines available for protecting soil and other resources Chemical use minimized and restricted to approved chemicals Chemicals and waste stored or disposed of properly Use of biological control agents strictly monitored and use of genetically modified organisms prohibited Use of exotic species minimized, controlled, and monitored

TABLE 26-1 *Summary of Forest Stewardship Council (FSC) Principles and Criteria of Sustainable Forest Management (FSC 1996) (continued)*

Principle	Criteria
Principle #7: Formal management planning documents in place, with long-term objectives and process for updating	Management documents comprehensive in coverage Plan is regularly revised based on monitoring and new information Staff receives training and supervision to implement plan Forest managers make available to the public key aspects of the plan
Principle #8: Monitoring conducted on forest condition, product yields, and social and environmental impacts	Monitoring is consistent, frequent and technically sound Monitoring covers product yield, forest condition, changes in forest composition, environmental impacts, wildlife, and costs Documentation exists for chain of custody of products Monitoring results used in improving management Forest managers make publicly available monitoring results
Principle #9: Primary, old-growth, and other unique forests and sites are conserved	Tree planting supplements natural regeneration and does not replace natural forest Planting of trees can be used for natural forest regeneration, but carefully matched to ecosystem conservation requirements
Principle #10: Plantations do not supplant natural forest, and they help to reduce pressure on natural forests	Plantation management objectives stated in plan Plantation layout fosters natural forest conservation and restoration Structural diversity is emphasized in plantation management Species selection is carefully matched to site; native species emphasized over exotic species Designated forest areas in plantation are maintained or restored to natural forest Soil structure, fertility, and biological activity is protected Minimal use of chemicals; emphasis on integrated pest management Monitoring occurs to control negative ecological, environmental, and social impacts associated with plantations

assumes responsibility for consistently controlling the management quality in specific forest management units. In the latter case, operations may or may not be owned by the forest manager, but must be explicitly managed under his/her direct supervision. After an initial forest assessment, which is the basis for granting certification, the forest enterprises that qualify are certified for a five-year period. Independent forest audits are required each year of certified operations. Table 26-2 presents a simple step-by-step description of the certification process as typically implemented by FSC-accredited certifiers. The complete certification assessment and decision process typically lasts one to six months, depending on the size and complexity of the operation being evaluated and the issues encountered during the assessment.

TABLE 26-2 *FSC-Approved Certification Assessment Process*

Step #1	*Request for certification* The candidate operation submits a written application to the certifier, indicating which forest areas it wishes to submit to the certification process
Step #2	*Pre-assessment research* During this step a review is made of existing documentation (e.g., management plans, biological inventories, list of adjoining landowners or other affected stakeholders, etc.), and interviews conducted with key authorities, about the forestry situation in the operating area of the candidate operation
Step #3	*Regionalization of certification standards* Where regional standards do not yet exist, an assessment team develops a set of standards based upon general FSC guidelines and input from key stakeholders, including local communities, government agencies, scientific community, employees or contractors of the candidate operation, and other interested parties
Step #4	*Field assessment* The field assessment includes: a) review of written forest management plan, and b) field-checking and on-site verification of practices through field evaluations and interviews in the forest and nearby communities. This intensive field assessment step typically lasts a few days for a small operation (< 1,000 hectares), to two weeks for large operations (> 100,000 hectares), with multiple field or site visits to various parts of the operation
Step #5	*Development of draft assessment report* The draft assessment report includes a description of the precertification conditions, and recommendations for achieving certification. Also included is a recommendation of whether the operation is likely to achieve certification in the future
Step #6	*Review of draft assessment report* The candidate operation, and a minimum of two confidential independent peer reviewers, examine the draft assessment report. Peer reviewers are specifically asked whether they agree with the findings in the assessment report, including whether to certify or not, whether sampling or research was complete and systematic, and whether the findings of the assessment team seem well-founded, based on the experience of the peer reviewer
Step #7	*Certification decision* Based on the assessment report, peer review comments and comments from the candidate operation, the certifying agency decides whether or not to certify the candidate forest area. If approved, it is typical that there will be some (usually less than ten to fifteen) mandated improvements that will be required to be made by the certified operation over the initial 5 year certification contract period

Certification and Wildlife Conservation

Conserving wildlife and their habitat is one of the cornerstones of sustainable forest management and the certification process. A number of principles, criteria, indicators, and verifiers have been developed to maintain the structure, diversity, and ecological functions of forests under natural forestry management (Prabhu et al. 1996; Viana et al. 1996; Ervin and Pierce 1997). These encourage forest managers to protect rare and endangered species, maintain over-mature trees for habitat, minimize damage to residual trees and soils, and set aside areas of the harvest block as biological reserves and riparian corridors (see photos 26-1, 2a, b). Criteria, indicators, and verifiers that directly relate to wildlife conservation in managed tropical forests are outlined in table 26-3.

PHOTO 26-1 *Drastically overharvested dipterocarp forest in Malaysia, causing extensive changes in wildlife habitat and community structure. (R. Z. Donovan)*

PHOTO 26-2 *Unacceptable vs. acceptable logging practices: a) heavy rutting soil damage on forest road in Malaysia; b) minimal soil damage and low structure disturbance of residual forest. (R. Z. Donovan—both photos)*

TABLE 26-3 *Selected Criteria and Indicators Directly Related to Wildlife Conservation in Managed Tropical Forests*

Criteria	Indicators
Management activities, over time, sustain key biological components and ecological functions associated with maximum long-term biological productivity (Scientific Certification Systems, Inc., B.2)[a]	Length of managed rotations relative to ecological rotations management efforts designed to maintain the nutrient capital of managed areas
Wildlife and wildlife habitats are considered, protected and enhanced during the course of timber management operations and as a distinct element of overall management of the forest (Scientific Certification Systems, Inc., B.3)	Regular involvement of wildlife biology expertise extent of acquisition, analysis and utilization of data on wildlife, habitat and species degree of retention of desirable habitat, e.g., cavity trees, downed logs, horizontally and vertically diverse cover vegetation
Immigration, settlement, hunting, and timber extraction along logging roads is controlled (SmartWood 2.5)	Uncontrolled incursions into forest management areas are not occurring; monitoring system is implemented, i.e., forest and other resource management areas are visited by staff or contractors on a regular basis
No timber harvesting is taking place in high risk areas (e.g., highly erodable or wet) or within pre-designated buffer zones for rivers and streams (SmartWood 4.12)	Protection (or buffer) zones are equal to twice the width of perennial stream courses (e.g., if stream is 20 meters wide, buffer zone should be 20 meters on each side) and with minimum buffers on each side
Field assessments of biological resources and non-timber forest products are conducted in forest management areas (SmartWood 5.1)	Assessments conducted prior to, during, and after harvesting

[a] Information in parentheses refers to the certification organization and their specific criteria.
SOURCE Lammerts van Bueren et al. 1997

Certification standards also typically incorporate a number of criteria that indirectly affect the conservation of wildlife. SmartWood certification guidelines (SmartWood 1998), for example, require that all access to forest areas be controlled, and that such access be planned in consultation with local communities. This can provide an indirect control on both legal and illegal hunting activities, or protection of unique environmental features (habitat, species) that may be identified by local people. FSC's Principles and Criteria require all FSC-approved certifiers to incorporate similar criteria (T. Synnott, FSC Director, per-

sonal communication). Finally, careful control of the use of exotic species, chemicals, and biological control agents, and a prohibition on the
use of genetically-modified organisms (see table 26-1), help to preserve
the integrity of wildlife communities.

To date, the effect of certification on wildlife or habitat conservation have not been definitively documented or researched. Anecdotal
experience indicates that certified operations have been able to achieve
conservation of critical wetlands and other unique habitats in their
forest management areas. In Bolivia, certified operations have been
able to introduce anti-hunting measures (e.g., meat substitution food
programs for employees) that have reduced or eliminated hunting, and
it appears that all the certified operations are placing a higher priority
on habitat conservation than they might have without certification (R.
Z. Donovan, personal observation). Many unknowns remain, however, and certified forestry operations should be assessed to document
the effect of certification on wildlife conservation.

Challenges to Conserving Wildlife in Certified Forests

Certified forest enterprises are generally recognized as being more environmentally *friendly* than conventional timber-harvesting practices, however a number of obstacles remain before these operations can be truly
heralded as conserving the full suite of wildlife occurring within natural
tropical forests. The conservation challenges occur at both the forest management unit and global levels of the certification process.

Wildlife Conservation in the Certified Forest Management Unit

A major obstacle to conserving wildlife in certified forest operations
is that we have very little information about the impacts of timber harvesting operations on wild animals and their habitat (see chapters 4-14).
In addition, the certification process does not include sufficient time or
resources to conduct well-designed biological inventories or assessments, which further complicates this condition. Assessors must rely on
the data provided to them by forestry operations and relevant scientific
literature, and combine these with a set of indicators and verifiers that
often include vague and ambiguous measures of wildlife presence and
abundance (see tables 26-1 and 26-3). Determining whether sufficient
keystone resources exist within the residual landscape, for example, is a

difficult assessment to make by even the best ecologists. Certifiers can easily identify certain gross forest management impacts on wildlife communities and their habitat (*fatal flaws*)—such as hunting-threatened or endangered species by company employees or contractors (see table 26-4). The paucity of ecological information about most wildlife taxa hampers the capacity of certifiers and other stakeholders to definitively verify the short- and long-term impacts of timber harvests on most wildlife species (Putz and Viana 1996).

TABLE 26-4 *Listing of Fatal Flaws That Relate to the Conservation of Wildlife and Their Habitat[1]*

---

1. Threatened or endangered species (as listed under CITES or national laws) are being hunted or harvested.

2. Forest manager makes no attempt to control hunting or third party access to forest lands or facilities.

3. Timber harvesting causes consistent (i.e., multiple locations, multiple occasions) negative impacts to rivers and streams in forest management area through erosion, inappropriate stream crossings, pushing of debris into streams, etc.

4. Illegal chemicals are used either in the forest or at wood processing facilities.

5. Toxic chemicals are stored in open areas with no security, near sensitive biological areas (e.g., streams, delicate habitat).

6. Silvicultural operations are designed and conducted with sole focus on production of wood fiber and blatant disregard for biological or ecological integrity.

---

[1] This table was developed using FSC Principles and Criteria and various FSC-approved certifier standards as background. Neither FSC- nor FSC-approved certifiers formally use the term *fatal flaw* in their documentation.

Hunting often has greater environmental impacts on wildlife than many of the direct habitat disturbance impacts associated with timber harvesting operations (see chapters 15–17). It is very difficult for many forest enterprises to control hunting by their employees, and even harder for certification assessors to determine their success in these endeavors.

Wildlife Conservation in the Global Certification Context

Several impediments exist to the widespread adoption of certification practices in tropical forests, and the wildlife conservation measures they promote. To begin, the economics of certification remain unclear. While positive gains to forest operators adopting reduced-impact logging measures has been documented in areas (Cabarle et al. 1995; Uhl et al. 1997; see chapter 28), no thorough studies of the benefits to forestry firms or society from applying certification level management have been undertaken. The substantial *up-front* costs and unclear benefits associated with certification have proven unattractive to many operators, keeping them on the *sidelines* until the wisdom or profitability of these ventures can be more definitively determined. Finally, markets for certified wood have been unstable, due in part to the absence of clear public demand or customer awareness in consuming countries, and in part to the relative unpredictability of certified wood product supplies. For many, the potential for ecological certification as a market-based incentive to reduce tropical forest exploitation remains questionable (Merry and Carter 1997).

Future Directions

For certification to close the gap between theoretical and applied sustainable forest management—which includes the conservation of wildlife populations and their habitat (International Tropical Timber Organization 1993)—efforts must be advanced to:

- Quantify and qualify the response of wildlife populations to timber-harvesting operations, in the near and long term
- Develop rapid and cost-efficient indicators or verifiers of these responses
- Identify the optimal size and distribution of reserves within both the forest management unit and the managed forest matrix
- Assess the role of keystone species and resources to wildlife
- Separate direct (e.g., habitat disturbance) from indirect (e.g., hunting) impacts of harvesting on wildlife

Equipped with this information, resource managers will be much better positioned to design silvicultural interventions that balance economic gains and ecological impacts. Projects such as BOLFOR in Bolivia are helping to define these applied guidelines (Townsend 1996b; see chapter 29), which rely heavily upon an integration of foresters and wildlife managers. Detailed experimental designs for evaluating these variables are reviewed in chapters 18 and 19.

The economics of certification have also been an obstacle to the widespread adoption of these programs. New models of certification (e.g., *resource manager* model introduced by SmartWood in 1995) are being developed, which show promise for reducing the costs and complexity of certification for small landowners (SmartWood Program 1997). The resource manager structure allows for certification of forest and wildlife managers that manage other people's forestlands, creating an economy of scale which facilitates lowering of certification costs and increased technical assistance to small landowners on wildlife management and other forest management requirements. Though such structures have been implemented in North America, there are currently no examples of certified resource managers in tropical regions.

Finally, the Forest Stewardship Council is implementing a series of national and region-specific certification standards development initiatives in nine different countries—and in some countries, multiple regions within those countries (e.g., Canada, U.S.) (Simpson 1998). Such initiatives incorporate what might be referred to as *state of the art* expectations in terms of managing wildlife and conserving wildlife habitat in each region. These initiatives will have concrete results in the form of region- or country-specific certification standards, which are to be updated approximately every three to five years by FSC-sponsored regional or national working groups.

Conclusions

Certification is a tool that can improve commercial forest management from a wildlife conservation perspective. Certification standards incorporate the need for proactive management of wildlife resources, helping to protect species and their habitat at both the community and landscape levels. This is achieved by requiring forest managers to adopt reduced-impact logging measures (see chapters 21, 23, and 24), thereby lowering the negative environmental impacts associated with timber harvesting. Certification is not a panacea, however. In many regions, the organizations working to develop certification standards have a very limited technical base of wildlife management experience or research material on which to base certification requirements. In most tropical countries, the absence of solid, sustainable wildlife and forest management experience means that those commercial operations and researchers that are trying to improve will have to undertake demanding, but pioneering, work to address these issues. This requires well-designed, applied research programs to identify wildlife requirements and silvicultural practices that minimize disturbance to native fauna communities (see chapters 18 and 19). Ultimately, certification will prove most useful as a wildlife conservation tool where it is used to develop:

- Practical alternatives for biodiversity conservation that concessionaires, communities, loggers, legislators, foresters, environmental groups (*watchdogs*), and local wildlife specialists (public or private) can apply
- Working partnerships between wildlife specialists and commercial forestry operations
- Biodiversity management research that effectively bridges the needs, capabilities, and resources of commercial forest managers and wildlife specialists

ACKNOWLEDGMENTS

The author would like to acknowledge the assistance of participants in the 1996 Bolivia workshop entitled *Effects of Logging on Wildlife in the Tropics* (13-15 November, Santa Cruz), Robert Fimbel, and several anonymous reviewers for their constructive comments on earlier drafts of this chapter.

Chapter 27

CAN FORESTRY CARBON-OFFSET PROJECTS PLAY A SIGNIFICANT ROLE IN CONSERVING FOREST WILDLIFE AND THEIR HABITATS?

Elizabeth Losos

Among the most prominent global environmental challenges facing policy makers and scientists today are tropical de-forestation and global warming. The two are not unrelated. Forest systems respond to and play a role in regulating the global climate. In fact, tropical deforestation and related land use changes contribute approximately one fifth of the carbon released into the atmosphere annually—the burning of fossil fuels being responsible for most of the rest (IGBP Terrestrial Carbon Working Group 1998). It is from this connection that *forestry carbon-offset projects*—tools aimed at reducing carbon dioxide in the atmosphere—were developed (see box 27-1). Within the conservation community, interest has been building over whether forestry carbon-offset projects hold promise for improving the management and conservation of forests. In this chapter, the background of tropical forestry carbon-offset projects is explained, as well as some of the benefits and shortfalls of these instruments for conserving tropical forest wildlife and their habitat.

Carbon-Offset Projects

Carbon-offset projects typically take one of two forms: carbon emissions reduction or sequestration. Emissions reduction entails decreasing the amount of carbon dioxide (CO_2) produced or released through measures such as increasing efficiency in power plants, extending consumer conservation measures, avoiding deforestation through forest preservation, and improving forest management. Carbon sequestration, or *sink enhancement,* is achieved by removing CO_2 that has already been released into the atmosphere—through planting trees, increasing growth rates of existing trees, and changing agricultural practices to increase soil carbon uptake.

Box 27-1 *Forestry carbon-offsets and climate change.*

From a global perspective, are forestry carbon-offset projects large enough to play a significant role in stabilizing the level of carbon dioxide (CO_2) in the atmosphere? In theory, the answer is yes. Forests have an enormous capability to store carbon—currently retaining more than 92 percent of the world's terrestrial carbon pool—with tropical moist forests averaging between 155 and 187 ton C/ha and tropical dry forests between 27 and 63 ton C/ha (Cairns and Meganck 1994). Completely halting tropical deforestation could conserve roughly 1.2 to 3.0×10^6 ton C/yr—a sizable counterbalance to the approximate 6.0×10^6 ton C/yr released into the atmosphere through fossil fuel burning (Houghton 1990; Brown et al. 1992; Dixon et al. 1993; ICGP Terrestrial Carbon Working Group 1998).

The carbon sequestration potential of reforestation is also vast. Substantial areas that once supported forests could do so again in both the tropic and temperate zones (Trexler 1995). Reforesting the 8.7×10^8 ha of potentially available deforested or degraded land—areas forested in the past that now are unused for cropland or settlements—would offset about half of the carbon emitted annually from deforestation (Houghton 1990; Houghton 1991).

A working group of the Intergovernmental Panel on Climate Change estimated that by 2050, slowing deforestation and promoting natural forest regeneration from all regions, including temperate and boreal zones, could reduce or sequester about 12 to 15 percent of cumulative fossil fuel emissions over the same period (Watson et al. 1996).

It must be noted, however, that the great potential for offsetting worldwide carbon emissions through either deforestation abatement or reforestation is just that—a potential. There are two problems with assuming that forestry projects can be carried out on the scale described above. First, the vast areas of technically available land may not actually be available for reforestation (Sathaye et al. 1995). According to a study by Trexler et al., only 69 percent of technically available land was socioeconomically available. Second, to attain the levels of carbon mitigation described above, forestry projects would have to be carried out on a massive scale well beyond our current capabilities. Deforestation trends in the tropics would have to be almost completely (and unrealistically) reversed to make significant inroads on carbon emissions. To counter carbon emissions, reforestation would also have to occur over an enormous area (Houghton 1990; Marland and Marland 1994). Houghton (1990), for example, notes that offsetting every 1.0×10^6 ton C requires 1.5 to 2.0×10^8 ha of reforestation, or almost 1 percent of the earth's landmass. The reality is that, on a global scale, forestry is likely to offer only a limited contribution towards solving the problem of rising CO_2 levels. Because many energy-related projects have the capability of offsetting much greater quantities of carbon, smaller numbers of these projects are likely to have a greater impact on the global concentration of CO_2.

A voluntary program, called the Clean Development Mechanism (CDM), has been designed to take advantage of win-win opportunities created by carbon-offset projects. The Clean Development Mechanism, first introduced in the Kyoto Protocol of 1997, will likely be modeled after Joint Implementation activities that were first established by the United Nations Framework Convention on Climate Change. Clean Development Mechanism projects, however, would be restricted to greenhouse gas trading between countries included in Annex I of the Climate Change Convention (industrialized countries) and non-Annex I countries (developing world nations). If Clean Development Mechanism projects become an accepted means of greenhouse gas emissions reduction, units of carbon and other greenhouse gases that are sequestered or whose release is avoided in developing countries could be traded to offset units released in industrialized nations. These projects would have one or more donors from the industrialized world, such as an electric power utility or government agency, and one or more recipient organizations within developing countries. The donors would be able to achieve their carbon emissions targets at a reduced cost, because atmospheric carbon reduction is often cheapest in the developing world, and the recipients would benefit through the transfer of technology or shared economic or environmental benefits, as well as complete compensation for project costs (Dixon et al. 1993; Vellinga and Heintz 1995; Trexler 1995).

The future of carbon-offset activities is unknown. The Kyoto Protocol, which was the product of the third Conference of the Parties to the Climate Change Convention in December 1997, sets binding targets for greenhouse gas emissions by industrialized countries and provides for flexible methods such as Clean Development Mechanism projects to achieve these goals. Yet this protocol has not been ratified by enough nations to bring it into effect, nor is it likely to be in the current political environment. Activities aimed at reducing greenhouse gases, consequently, are still voluntary. Even if the Kyoto Protocol were ratified by 2000, the year in which the Clean Development Mechanism is scheduled to come into effect, the details remain sketchy as to how industrialized countries could use tradable carbon-offsets to help achieve their reductions. Details concerning how the Clean Development Mechanism would operate and which projects would be eligible will probably be worked out during the sixth Conference of the Parties to the Climate Change Convention in 2000—with the guidance of a special report currently being developed by the Intergovernmental

Panel on Climate Change detailing land-use change and carbon storage and sequestration activities.

Strong proponents and opponents exist to the concept of carbon-offsets, even within the environmental community and within developing nations (Maya 1995; Trexler 1995; World Wildlife Fund 1995). Of the wide range of arguments that have been levied against offset projects, most relate to environmental imperialism and national sovereignty. In this chapter, only those arguments that are unique to forestry carbon-offset projects will be considered.

The Attraction of Forestry for Carbon-Offset Projects

During a 1995 to 1999 pilot phase for experimenting with carbon-offset projects (USIJI Secretariat 1998a), forestry activities emerged as one of the most popular types of projects. As of November 1998, 12 of the 30 projects approved by the United States Initiative on Joint Implementation agency (USIJI)—an inter-agency body that oversees U.S. carbon crediting—were forestry carbon-offset projects, and 10 of these were in the tropics (see table 27-1). If all 12 of these forestry projects were fully funded and implemented, they would be responsible for roughly 136 million metric tons of CO_2 offset—representing about two-thirds of the total carbon-offset by all 30 projects (USIJI Secretariat 1998a; USIJI Secretariat 1998b). Six of the USIJI-approved forestry projects are already in progress. The remaining projects have not yet initiated on-site activities because they have had difficulties in obtaining funds and/or overcoming logistical and technical problems, which have delayed the projects (USIJI Secretariat 1998a).

Re-forestation (carbon sequestration), forest protection (carbon emissions reduction; see photo 27-1), and natural forest management (carbon emissions reduction; see photo 27-2) are the basis for most forestry carbon-offset projects. In each of these cases, the net carbon-offset is calculated as the amount of stored carbon, in both the forest and in long-lived wood end products, exceeding what would have been found under the non-project or *reference case* conditions. Reforestation is fairly straightforward: trees replace a lower biomass system. reforestation is typically carried out in plantations rather than through natural regeneration. For forest protection, carbon sinks are maintained through the abatement of otherwise imminent deforestation. Natural forest management reduces the release of carbon into the

TABLE 27-1 List of Forestry Carbon-Offset Projects Accepted Into the U.S. Initiative on Joint Implementation as of September 1998[a] (USIJI Secretariat 1998a; USIJI Secretariat 1998b)

Project Title	Country	Forestry Activities	Project Duration (yr)	Estimated Cumulative GHG Emission Reduction (metric tons CO_2)
Rio Bravo Carbon Sequestration Pilot Project	Belize	Natural forest management and protection	40	10,000,000
ECOLAND: Esquinas National Park	Costa Rica	Protection	16	1,343,000
Klinki Forestry Project	Costa Rica	Reforestation	46	7,216,000
Noel Kmepff M. Climate Action Forestry Project	Bolivia	Protection and natural forest management	30	53,190,000
Reforestation of Chiriqui Province	Panama	Reforestation	25	58,000
RUSAFOR: Saratov Afforestation Project	Russia	Afforestation and reforestation	60	293,000
Reforestation in Vologda	Russia	Reforestation	60	878,000
Forest Conservation-Bilsa Reserve	Ecuador	Protection	30	1,170,000
Scolel Te—Sustainable Land Management and Carbon Sequestration, Chiapas	Mexico	Agro-forestry and reforestation	30	1,210,000
Carbon Sequestration through Reduced Impact Logging	Indonesia	Natural forest management	40	134,000
Carbon Sequestration through Costa Rican Territorial and Financial Consolidation of Biological Reserves Project	Costa Rica	Protection and reforestation	20	57,000,000
Community Silviculture in Sierra Norte, Oaxaca	Mexico	natural forest management	30	3,065,000

[a] not all projects are funded.

atmosphere through two means: a) providing forest owners with a long-term income source that discourages conversion of the forest to other, less biomass-intense uses, and b) including a reduced-impact logging component that minimizes unnecessary biomass death (and subsequent carbon loss) over the short-term (see chapter 21). The sustainable management of timber, therefore, inadvertently leads to the sustainable retention of carbon (Ismail 1995).

The prevalence of forestry carbon-offset projects may appear surprising given that the energy industry would be expected to invest more in energy-related projects rather than forestry activities. Energy

PHOTO 27-1 *Old-growth tropical forests are one of the largest carbon sinks in the world. Their ease of monitoring, and biodiversity conservation value, make them an attractive option for carbon emission-reduction forest protection projects. (R. Fimbel)*

PHOTO 27-2 *Natural forest management emissions reduction projects, while difficult to quantify, have attracted some interest by carbon producers given their low capital costs. (R.A. Fimbel)*

companies have traditionally tried to comply with environmental requirements using *technological fixes* within their plants. The recent popularity of forestry projects may be related to two attractive features of forestry carbon-offset projects: their cost-effectiveness and secondary benefits.

Cost Effectiveness

Forestry is often considered the cheapest type of carbon-offset project per unit of carbon. This is largely due to the low capital costs of forestry projects. Forestry in the tropics is especially appealing because the opportunity costs of labor and land tend to be relatively low there, while forest growth and productivity are typically high (Trexler et al. 1989; Houghton 1990). A wide range of studies have suggested that carbon sequestration through tree planting usually costs less than U.S. $10/ton C and rarely over $30/ton C (Sedjo 1995). The cost of almost all non-forestry USIJI-approved carbon-offset projects by comparison are greater than $35/ton C, with one wind power project costing more than $4,000/ton C. The one notable exception is a fugitive gas capture project in Russia, whose costs are less than $0.05/ton C (USIJI Secretariat 1996a).

Ancillary Benefits

Forestry carbon-offset projects are also attractive to both industrialized country donors and developing countries because of their ancillary benefits. They tend to follow a *no regrets* policy such that the project makes sense for various ecological, social, and economic reasons in addition to the benefit of reducing carbon dioxide (Faeth et al. 1990; Godoy 1990; Winjum and Lewis 1993; Cairns and Meganck 1994; Trexler 1995). From the perspective of developing countries, forestry projects must generate significant secondary benefits, such as the protection of wildlife and its habitat, production of fuelwood, restoration of degraded land, and long-term income, in order to be attractive. If carbon-offset projects are not viable in their own right—aside from their ability to mitigate global warming—then host governments would likely not accept them (Godoy 1990; Brown 1998).

From the perspective of the industrialized country sponsors, particularly those from the private sector, the ancillary benefits generated by forestry projects are especially appealing because they can create highly visible, positive public relations (Pearce 1995; Petricone and Vetleseter 1995; USIJI Secretariat 1996b). Planting a tree or preserving wildlife within its forest habitat is an image that can be easily conveyed to the public. Decreasing CO_2 emissions through operational or technological changes within a power plant, even if the carbon emissions reduction is enormous, is largely invisible to the public (Trexler et al. 1989; Pearce 1995). Because the U.S. energy sector is currently undergoing deregulation and energy companies will soon be able to sell their products nationally, the highly visible and positive public image conveyed by forestry projects, with their environmental and carbon benefits, makes these carbon-offsets especially attractive (Martin 1997).

Unique Problems with Forestry Offset Projects

While forestry carbon-offset projects have some uniquely attractive features, such as their cost-effectiveness and ancillary benefits, they also have their own particular set of shortcomings. In general, the energy industry and some governmental agencies perceive that investment risks are higher for forestry offsets than energy-related offsets, such as fuel switching from oil to natural gas, or increased production efficiency. This apprehension is due in part to a lack of familiarity by carbon producers with identifying, financing, and monitoring forestry projects (Petricone and Vetleseter 1995). Some of these concerns, though, are real. They relate primarily to the fact that carbon sequestering and storage by forestry activities, in contrast to most energy projects, occur over a dispersed area and long timeframe, and are not permanent.

Measuring and Monitoring

One aspect of forestry carbon-offset projects that sets them apart from most energy-related projects is the inherent complexity of measuring, monitoring, and verification of carbon sequestered and stored in forests. Whereas carbon dioxide emitted from a power plant smokestack can be fairly easily and convincingly calculated, this is not the case for re-forestation, natural forest management, or protection projects (Dixon et al. 1993; Trexler 1995). Above-ground carbon storage is typically approximated from data on tree girth and height, wood density, and carbon content of a species. Measurements should ideally also be made for roots, soils, and standing litter crop, which may represent as much as two-thirds of the carbon stored in the project area (ICBP Terrestrial Carbon Working Group 1998). Carbon should also be monitored in the end-wood products, though this is virtually never done (Andrasko 1997). For sequestration measurements in re-forestation projects, additional dynamic data are needed on tree growth, mortality, and recruitment, at a minimum. Even in the rare circumstances where all these data are available, estimates of biomass and carbon can still vary almost twofold (Brown et al. 1989). Several guidelines have recently been published (e.g., MacDicken 1997), responding to the need to establish standardized, accurate, and cost-effective methods for calculating and monitoring carbon (Petricone and Vetleseter 1995). Further research on these on-the-ground techniques and complementary remote sensing are still needed, however, to develop precise measurements that are cost-effective (S. Brown, personal communication). In addition to monitoring carbon storage at the project level, there is also a need to develop protocols for monitoring at the national and regional scales, which could help avoid some of the leakage problems discussed below (ICGP Terrestrial Carbon Working Group 1998).

Protection and natural forest management projects also require information on de-forestation rates and local trends to monitor changes in carbon sinks. Accurate predictions require socio-economic information as well as bio-physical data, making this the most difficult baseline condition to accurately estimate (Andrasko 1997). This problem is exacerbated by the fact that host countries have the perverse incentive to exaggerate the baseline condition. A government might claim, for example, that it has no intention of protecting (or re-foresting) an area so as to attract money for a carbon-offset project (Pearce 1995; Brown 1998). Partly due to the uncertainty of carbon monitoring in natural forest management and protection projects, investors have tended to favor re-forestation projects (Brown et al. 1997).

Due to methodological uncertainty, imprecision, and the high costs of measurement and monitoring, some environmental groups have argued that forestry projects in general are impractical and should be excluded as internationally accepted carbon-offsets for fear that they cannot insure CO_2 will be sequestered at the levels promised (Hare and Stevens 1995; Pearce 1995; World Wildlife Fund 1995). Others, however, believe that such problems are not inherently worse than those posed by energy-related offsets (Trexler and Associates, Inc. 1997).

Leakage

For a carbon-offset project to be approved by United States Initiative on Joint Implementation Office, the project must document that its activities will not simply shift the release of carbon emissions outside the project boundaries, in space or time, thus creating *leakage* of additional carbon dioxide. In the case of forest protection and natural forest management, offset projects must demonstrate that:

- A forest is under imminent threat of destruction
- Without the project the forest would not otherwise be protected or sustainably managed
- The project does not displace deforestation pressures to another site
- For natural forest management and protection projects that result in decreased timber production, timber demand would not be displaced to

other sources—thus encouraging increased logging elsewhere (USIJI Secretariat 1996b; Brown et al. 1997)

This latter scenario occurred in southeast Asia in 1989, when a logging ban in Thailand appears to have increased deforestation in neighboring countries. The adoption of reduced-impact logging practices helps avoid this *displacement* problem because careful harvesting techniques can decrease the biomass (and carbon) waste without significant timber reduction (Pinnard and Putz 1996; see chapters 21 and 28; Brown et al. 1997).

Leakage can also be a serious concern for plantation projects, where the production of carbon-subsidized wood could discourage the establishment of non-subsidized commercial plantations.

Longevity and Permanence

Because forests are part of an active biological cycle, they are best viewed as important, but temporary carbon sinks that can buy time while working to reduce fossil fuel emissions (IGBP Terrestrial Carbon Working Group 1998). Carbon gains from forestry activities, unlike those from most energy-related projects, can be reversed virtually overnight. Forestry projects, however, are typically given no latitude for set-backs, such as when a protected forest is regazetted for other uses or infringed upon illegally—when trees in a plantation are harvested before the time specified in the project, or even when a fire or hurricane devastates a project site. These risks are exacerbated by the longevity of forestry projects, which are normally planned for up to 40 years or more. Such guarantees would be difficult to ensure over several decades anywhere in the world (Dixon et al. 1993; Andrasko 1995). In developing nations, where governments may be unstable, interest rates high, and land tenure ill defined, the issue of uncertainty is paramount.

The problems resulting from longevity and lack of permanence, while fairly unique to forestry carbon-offsets, can be partially countered by bundling together a wide variety of projects, thus hedging against localized natural or human-caused problems (Brown 1998; Frumhoff et al. 1998). This approach has been pioneered by Costa Rica through the sale of national carbon bonds that represent an amalgamation of many forestry carbon-offset projects throughout the country (USIJI Secretariat 1998b). Purchasers of carbon-offsets can also diversify their portfolios across several countries to reduce their risks.

Advantages and Disdvantages of Forestry Offsets for Biodiversity Conservation

To offset even modest carbon emissions, forestry projects must be large-scale—potentially translating into significant habitat preservation or restoration and wildlife protection (see box 27-1). A handful of companies and private foundations in the United States, including Applied Energy Services, New England Electric Company, Wisconsin Electric Power Company, and Edison Electric Institute, have already initiated forestry carbon-offset projects with positive biodiversity conservation components. In the Rio Bravo Conservation and Management Area of northwestern Belize, for example, a carbon-offset forestry project purchased over 137,000 ha of forest for protection and is promoting sustainable forest management techniques on another 44,000 ha. This subtropical humid forest, savanna, and marsh, which would have otherwise been converted into agricultural pasture, is home to jaguar (*Panthera onca*), white lipped peccary (*Tayassu tajacu*), ocelot (*Felis pardalis*), Baird's tapir (*Tapirus bairdii*), Great Curassow (*Crax rubra*), and other top carnivores and large frugivores (USIJI Secretariat 1998b).

Does this mean that all forestry offset projects represent a win-win scenario for biodiversity conservation? Not necessarily. While many features of forestry carbon-offset projects bring about positive contributions to biodiversity conservation, others may be less conducive—some even detrimental. Some of the strengths and weaknesses of forestry carbon-offset projects used as a tool for conserving forest wildlife and their habitat are described below.

Compatibility of Wildlife Conservation and Carbon Sequestration Priorities

The U.S. Initiative on Joint Implementation's approval process requires the identification of biodiversity conservation measures (USIJI Secretariat 1996b), however there are no incentives for their inclusion in offset projects. It is likely that the approval process will be similar for future Clean Development Mechanism projects. While there may be other reasons to incorporate a conservation component, it should not be assumed that this will automatically happen. The problem lies with the fact that priorities used to select carbon-offset projects do not necessarily parallel those for wildlife conservation.

The appeal of plantation forestry illustrates this apparent conundrum. Plantations are generally considered a more credible form of forestry carbon-offsets than protection or natural forest management projects because the latter two are plagued by problems such as carbon leakage. Because young forests sequester carbon at a higher rate than mature forests, plantations are also seen to have the greatest potential for absorbing carbon from the atmosphere (Brown et al. 1997; e.g., Asia: Brown 1996). The international climate change community consequently favors reforestation and afforestation over the protection and natural forest management of existing forests, despite the fact that the latter provide much higher wildlife conservation benefits. Because of the bias favoring reforestation and afforestation carbon-offset projects, the Kyoto Protocol (Articles 3, 6, and 12) is ambiguous as to whether protection and natural forest management offset projects will even be included in an international carbon-trading program.

The plantation-based carbon sequestration approach ignores the fact that old-growth tropical forests are among the world's largest reservoirs of carbon (see box 27-2), and that these forests (especially those in Amazonia) may even be sequestering significant amounts of carbon from the atmosphere (Grace et al. 1995; Phillips et al. 1998). Though their rate of carbon sequestration may be generally low, the value of old-growth forests as a carbon sink can be enormous. Natural forests tend to be much more effective at storing carbon than plantations, even when carbon storage in wood end products are considered (Harmon et al. 1990; Houghton 1991; Brown et al. 1997).

Box 27-2 *Carbon storage and old-growth forests.*

Old-growth tropical forest, with its large trees, is an extraordinary storehouse for carbon. Woody stems larger than 10 cm dbh in a 50ha Forest Dynamics Plot in lowland forest of Pasoh Forest Reserve, Peninsular Malaysia, for example, contain approximately 222 ton C/ha. Trees larger than 30 cm dbh, account for approximately 40 percent of the above-ground biomass (M. Boscolo, J. LaFrankie, P. Ashton, and N. Manokaran, unpublished data; wood volume equations from Forestry Department of Peninsular Malaysia; conversion factor for stem wood biomass to above-ground biomass from Brown et al. [1989]). A similar plot on Barro Colorado Island, Panama contains approximately 142 ton C/ha. Within this plot, only 0.6 percent of the trees are larger than 50 cm dbh, however 53 percent of the above-ground biomass is stored in these large stems (S. Lao, S. Bohlman, T. Guynup, R. Condit, S. Hubbell, and R. Foster, unpublished data; wood density equations from T. Guynup and literature; tree height equations from S. Bohlman [unpublished data]; conversion factor for stem wood biomass to above-ground biomass from Brown et al. 1989). Carbon storage is optimized for either protection or forest management projects by selecting mature forests and retaining the largest trees (Harmon et al. 1990). Biodiversity conservation is also optimized under this strategy, because many forest wildlife species are dependent on the habitat conditions created by old-growth forest (see chapters 2-14).

Thus, carbon-offset and wildlife conservation priorities may overlap more often than is generally acknowledged. The most effective carbon strategy may be to:

- Protect forests with large standing biomass and low productivity
- Re-forest lands with little standing biomass and low productivity
- Practice natural forest management for forests with high productivity, harvesting trees for use as long-lived products or as substitutes for fossil fuels (Marland and Marland 1994)

Such a strategy should also prove effective for conserving wildlife and their habitats. The habitats most critical to wildlife tend to be the ones with the greatest biomass—old-growth forests (Harmon et al. 1990). Late seral forest also would probably be most attractive for natural forest management, and techniques such as reduced-impact logging can have beneficial impacts for wildlife (see chapter 21). Finally, reforestation of degraded lands, especially through natural forest regeneration, can provide new habitats for wildlife and corridors connecting existing habitats.

Carbon and biodiversity interests can also diverge when selecting a project site. Units of carbon, sequestered anywhere in the world, are essentially equivalent. Biodiversity is not. Site selection using carbon mitigation criteria may, therefore, lead to biologically perverse results. Fernside (1995) noted that because of the strict requirements of the United States Initiative on Joint Implementation's approval process, the most attractive locations to establish a protected area or natural forest management site for carbon credit are near human populations where encroachment pressure is high. Yet these forests are often already extensively degraded. Forests in remote locations would be much easier and cheaper to protect, yet may not be eligible for carbon credits because of lack of an imminent threat. Nevertheless, during the pilot phase, strong cases have been made (and accepted by the USIJI) for an imminent threat of deforestation in remote areas such as Noel Kempff in Bolivia and Rio Bravo in Belize (see table 27-1).

On a broader scale, the countries with the greatest opportunities for carbon sequestration tend to overlap with the countries that are considered *biodiversity hot spots*, such as Brazil, Colombia, and Indonesia (Brown 1998). In a practical sense, however, carbon-offset projects will not necessarily be sited in countries that have the greatest carbon-offset potential. The private sector typically looks for politically and economically stable regions in which to invest, especially for long-term projects. This is one reason that Costa Rica, with its long-established democracy, has been so successful in promoting its carbon-offset projects (see table 27-1). The highest richness in biodiversity, however, is often found in politically volatile countries such as the Democratic Republic of Congo (formerly Zaire) and Colombia. As a result, the countries that are most appealing for investment in carbon-offset projects may not always be the places with the greatest importance for wildlife conservation.

New Funding Opportunities

The most frequently mentioned attraction of carbon-offset projects for biodiversity conservation is access to new sources of funding, particularly long-term sources. It is widely acknowledged that, depending on the outcome of international negotiations, carbon-offset funding has the potential to mobilize large transfers of funding and technology from the developed to the developing world for offset projects—including forestry projects with biodiversity conservation components (Trexler 1995; Vellinga and Heintz 1995; Frumhoff et al. 1998).

Funding for forestry projects through the Clean Development Mechanism, therefore, may provide the incremental funds needed to make wildlife conservation efforts financially viable. Wangwacharakul and Bowonwiwat (1995), for example, have shown that in Thailand, modest payments for carbon storage in protected areas could provide the incremental income necessary to preserve forests. The case of natural forest management is especially compelling because financial considerations tend to weigh even more heavily on logging decisions than park management decisions. Without the additional financial incentives, conventional logging or forest conversion is inevitably more financially attractive than natural forest management, as illustrated by a study of Costa Rican forestry by Kishor and Constantino (1993). According to the theoretical research of Boscolo et al. (1997), relatively inexpensive payments for carbon storage may reverse the economic bottom line so that natural forest management is more financially profitable than conventional logging. Their computer simulations, driven by a matrix growth model and calibrated with data from a 50-ha forest demographic plot in peninsular Malaysia, found that reduced-impact logging techniques that were estimated to reduce logging damage of non-commercial trees from 40 percent to 15 percent (Pinnard and Putz 1996) could preserve about 27 ton C/ha, at an additional cost of only $150/ha, or $5.50 per ton of carbon.

While Clean Development Mechanism projects may attract additional funding, they also impose additional costs. The project development, administration, and monitoring costs of forestry carbon-offset projects, in excess of what would normally be encountered with traditional forestry or biodiversity conservation projects, are exceedingly high. Consider, first, the costs of preparing a proposal. To submit a proposal to the U.S. Initiative on Joint Implementation for approved carbon credits, a rigorous, lengthy, and costly (tens of thousands of dollars or more) proposal must be prepared (Petricone and Vetleseter 1995). Although carbon emissions reduction targets are still voluntary, the U.S. Initiative on Joint Implementation gives official voluntary credits for carbon-offset projects that it approves. These voluntary credits may become real if and when the United States ratifies the Kyoto Protocol. For forestry projects, project development can be particularly demanding because of the need to document current and future carbon storage, rate of sequestration, and leakage. Once a project has been approved and initiated, substantial effort must then be allocated to fulfill initial measurement, monitoring, verification, and reporting requirements. Carbon monitoring in several existing forestry projects, which followed monitoring guidelines developed by Winrock International (MacDicken 1997), have been estimated to account for about 20 percent of project costs. Given the extent of these demands, small, nonprofit conservation organizations may not be able to divert staff energies to oversee such responsibilities, even with sufficient funding.

From a biodiversity conservation perspective, whether these funds will be new and additional rather than a reshuffling of current funds from the same pools, is somewhat in question. Will the same conservation and sustainable development projects be funded, but now under the name of climate change? For carbon-offset projects, the U.S. Initiative on Joint Implementation technically only approves projects that can show that the funds are additional to existing sources (USIJI Secretariat 1996b). In actuality, it is very hard to prove or disprove that the funds are additional.

Whether or not these funds are additional to current spending on conservation and sustainable development projects may depend on their origin. Many of the original carbon-offset projects have been developed and funded by the public sector, primarily through bilateral agreements between governments or international organizations such as the Global Environment Facility. These projects typically aim to link carbon mitigation with biodiversity conservation and other sustainable development activities. One of the most innovative public approaches has been the recent sale of national carbon credits from Costa Rica to the government of Norway for $10/ton C—totaling $2,000,000. Given the current budget-constrained climate, however, it appears likely that government funding for carbon-offset activities will represent a re-allocation of existing public funds.

This should not belittle the importance to biodiversity conservation of carbon mitigation funding by the public sector. Carbon-offset projects may prove critical in providing a rationale to maintain or even slightly increase existing funding sources, at a time when public support for international aid has been waning. While international sustainable development projects may appear too altruistic in times of domestic economic hardship, carbon-offset projects can be promoted as fulfilling the self-interests of industrialized nations because they link international assistance with the energy used by the developed world (Trexler 1995; Vellinga and Heintz 1995).

Over the medium term, the private sector may offer greater potential for generating new funds for offset projects (Trexler 1995). Some have estimated that while public sources like the Global Environment Facility are likely to invest only a few billion dollars for carbon-offset activities, private sources could input more than $150 billion (Maya 1995). Because industries have profit maximization as their highest priority, however, they are unlikely to be as concerned about sustainable development or biodiversity conservation as the public sector.

The magnitude of the investment in carbon-offset projects, especially by the private sector, will clearly depend on whether there is an internationally ratified treaty with binding targets for greenhouse gas emissions, and whether tradable carbon-offset could be used to reach these targets. Yet from the point of view of biodiversity conservation, the currently ambiguous nature of treaty negotiations may be an asset. As long as a treaty has not been ratified, businesses that invest in carbon-offset projects are interested in all of the benefits that accrue from carbon-offset projects rather than focusing solely on maximizing carbon reduction and minimizing cost. One of the most compelling non-carbon benefits of forestry projects is, as noted above, positive corporate public relations. If and when CO_2 emission standards become mandated, businesses may shift to projects that can sequester the greatest quantity of carbon, which are not typically forestry projects.

Political Facilitation or Liability

Packaging conservation activities with carbon-offset projects may provide the added political impetus necessary to initiate conservation efforts that would otherwise be blocked by lack of political will or competing interests. Costa Rica, for example, has mobilized a myriad of local partners in the government, private sector, and non-governmental community to support far-reaching forestry and conservation projects that, if fully funded, should offset over 57,000,000 metric ton of CO_2, and protect and regenerate secondary forests on 530,000 ha of wildlife habitat (USIJI Secretariat 1998a; USIJI Secretariat 1998b). It is possible, however, that linking wildlife conservation projects with the carbon-offsets can result in negative associations—such as environmental colonialism—for conservation activities. An otherwise acceptable forestry project may be rejected only due to its connection with the Clean Development Mechanism. The forestry carbon-offset project between the Innoprise Corporation and New England Power Company in Sabah, Malaysia exemplifies this scenario: The Government of Malaysia still has not approved this reduced-impact logging project because of the international political implications of emissions trading. The government is holding the project hostage while it negotiates legal international treaty issues related to the transfer of funds and technology from the developed to developing world. Although carbon benefits should add to the attractiveness of a wildlife conservation project, and vice versa, there could be instances in which combining the two makes a project too politically sensitive for local or national approval.

From the conservation perspective, long-term commitments are ideal for ensuring the success of protection and sustainable forest management projects. The fifteen- to sixty-year project lifespan of forestry carbon-offset projects is far longer than most conservation projects, which are often wedded to a two to three year funding cycle. This long time frame and the commitment from sponsors and developing country governments is a highly attractive feature of forestry carbon-offset projects for biodiversity conservation.

Is Forestry Carbon-Offset Likely to be a Serious Tool for Biodiversity Conservation?

Just as, in the minds of some, debt-for-nature swaps and the marketing of nontimber forest products have not lived up to their promise as the great panacea for wildlife and forest conservation, forestry carbon-offset projects will likely share the same fate. If and when funding for carbon-offset projects does increase because CO_2 emissions targets become mandatory, industry sponsors will likely focus primarily on energy-related projects (Frumhoff 1998). Of those forestry projects that they do support, most will continue to be in a handful of nations, often in areas close to population centers. Of the money invested in reforestation, protection, and sustainable management to offset CO_2 emissions, only a limited amount is likely to be funneled into the areas of greatest conservation needs.

This is not to say that carbon-offset projects have no role in the conservation of biodiversity. Even a relatively small carbon-offset project can translate into a sizable, long-term wildlife conservation project. Where the needs of carbon-offsets and biodiversity conservation overlap, the carbon component can attract badly needed funds. For those conservation organizations that have the capability to develop and carry out these large-scale projects, Clean Development Mechanism projects may represent another significant international funding source. Carbon-offset projects should be considered as just one more tool in the conservation biologists' chest.

ACKNOWLEDGMENTS

I would like to thank P. Brown, R. Fimbel, P. Frumhoff, and A. Grajal for helpful discussions and their review of previous versions of this manuscript. Financial support for this work and related activities of the Center for Tropical Forest Science have been made generously available by the John D. and Catherine T. MacArthur Foundation, Southern Company, and Smithsonian Tropical Research Institute.

Chapter 28

THE ECONOMICS OF SUSTAINABLE FOREST MANAGEMENT AND WILDLIFE CONSERVATION IN TROPICAL FORESTS

Neil Byron

Measures to conserve wildlife in tropical forests, or to improve the ecological sustainability and long-term timber productivity of these forests, are seldom greeted with enthusiasm by commercial logging operations. Industry (and in some instances, government) has opposed such measures, even when there are very strong grounds to believe they would be in the wider public interest (Ruitenbeck, 1998). Most forest operations follow management practices that prioritize their financial returns, since the bottom line rarely improves from providing nonmonetary benefits to the wider society or to future unborn generations. Wildlife is no exception, as forest animals are often viewed as just another resource to be exploited in the short term (parts II and III), or a nuisance to be avoided or eradicated. Quite simply, there is a serious divergence between what is in the long-term public interest and what is in the short-term commercial interest of loggers, concessionaires, and public officials subject to limited accountability for their actions (see chapter 18 for a discussion of the principle problems preventing good forest management in the tropics).

Economists and policy analysts are beginning to examine options to change the institutional, legislative and economic conditions that promote forest exploitation over sound management. There is growing evidence that *no regrets* or *win-win* changes in forest management technology are emerging, offering opportunities to comply with national and international conservation priorities, without making logging commercially nonviable. Some of these activities have the potential to substantially reduce the negative environmental impacts associated with logging and silvicultural practices, without significantly increasing operational costs (see case studies presented later in the chapter). In fact, efforts towards sustainable forest management (which can greatly enhance wildlife conservation) may actually save logging companies and governments money!

In this chapter, I examine the opportunities and limitations of economics to contribute to sustainable forest management and wildlife conservation. I begin with a brief overview of the constraints facing natural resource economists, ending with a discussion of where these individuals can contribute to the sustainable forest management debate. This is followed by a discussion of current efforts to bring about change in the timber harvest *status quo*. Two case studies reviewing the costs and benefits associated with conventional and reduced-impact logging practices are presented. The chapter closes with a discussion of how economics can contribute to the advancement of sustainable forest management practices in the near- and long-term.

Economic Constraints on and Contributions to Sustainable Forest Management

Constraints

Economics studies the production, distribution, and consumption of goods and services that are limited in society. As such, it is concerned with efficiency (not wasting scarce resources), equity (fairness, both within and between generations), and making decisions based on trade-offs between feasible options. While these principles are useful in guiding forest management and conservation policy, they cannot define an optimal state of management due to uncertain and imperfect knowledge about the goods and services involved.

The economics of sustainable forest management need to be understood in the context of incomplete knowledge, uncertainty and change (Loomis 1995; Kengen 1997; Bennett and Byron 1998), and a market system that has proven inefficient at conserving forests and at ensuring equitable access to forest benefits (IUCN 1980). This condition is a function of financial analyses that examine monetary benefits and costs to the decision-maker (while ignoring all or most of the other effects on society) being the norm in tropical forestry decision-making (Ruitenbeck, 1998). This restricted approach is understandable, because environmental and conservation values are mostly non-monetary and, therefore, difficult to quantify in financial terms (some values have been estimated for wildlife products however; for examples see box 28-1 and chapter 16). This leaves corporations and governments limited opportunities and incentives to include these other values in a broader economic analysis of management options (one that includes all costs and benefits to the whole of society). Compounding this problem is the fact that both products and services from the same area are inter-related, making value calculations very complicated. Furthermore, the way in which society-at-large values the environmental consequences associated with management activities in tropical forests has yet to be adequately determined. Finally, even if the potential benefits of forests were known and quantified, some forest-user groups would disagree with the results of a comprehensive analysis because an alternative use was more lucrative to them. In the face of imperfect knowledge about ecosystems, continuing disagreement over what constitutes *significant damage*, and uncertainty over what mix of environmental benefits future societies will most appreciate and value, comprehensive economic analyses of sustainable forest management initiatives are all but impossible.

> **Box 28-1** *The economic value of wildlife as incentives for wildlife conservation in production forests. (Mohamed Zakaria bin Hussin)*
>
> Malaysia's tropical forests are a home to approximately 300 species of mammals, 750 species of birds, 300 species of reptiles, 165 species of amphibians, 300 species of freshwater fish, and millions of invertebrates. If one could quantify the economic value of the goods and services these organisms provide (e.g., protein, medicinal products, and the pet trade, not to mention their role in pollination, seed dispersal, and nutrient cycling), it is almost certain that the economic values of wildlife conservation would far exceed any costs associated with protecting and managing this resource. Unfortunately, it is difficult to measure these values, as many of these goods and services are seldom or never traded in the marketplace. Nonetheless, some efforts have been made to assess the economic value of wildlife.
>
> Stuebing et al. (1993) and Caldecott and Nyaoi (1984) conducted studies on the economics of wildlife hunting in Sabah and Sarawak, respectively. The values obtained were expressed either as per animal, per kilogram of meat, or the total value of meat per year. They found that the value of milky stork (*Ibis cinereus*) was estimated to be RM246,000 per bird and RM10 per bird for spotted dove (*Streptopelia chinensis*). The value of Sambar deer (*Cervus unicolor*) meat to the economy of Sabah was estimated at RM530,400 per year, while the total value of the meat of barking deer (*Muntiacus muntjac*) and wild pig (*Sus barbatus*) in Sarawak were estimated at RM24,852,555 and RM3,335,332 per year, respectively. Wildlife products such as swiftlet nests bring in RM6.75 million per year in government revenue for Sarawak and RM6.0 million per year in government revenue for Sabah. In peninsular Malaysia, the value of elephant

was estimated at RM186,000 per animal at a 4 percent discount rate and RM244,000 per animal at a 10 percent discount rate (Vincent et al. 1993).

While it is relatively simple to determine the economic value of logs, assessing the economic value of wildlife conservation or other ecological and environmental functions is still in its infancy. Since the economic value of these resources is difficult to determine, their real potential as income generators has not been fully explored or realized. There is a strong need for studies to be carried out to quantify to the fullest extent the economic value of all forest goods and services. Only then will we have a complete view on the costs and benefits of comparing alternative land-use options (Abd. Ghani et al. 1998).

Contributions

Given the lack of clear information on the full range of values/benefits that a forest can provide, what exactly can economists contribute to maximizing the sustainable level of overall benefits from the resource (including wildlife)? Two broad roles for economists in natural resource planning and management can be identified:

– Providing conceptual frameworks for developing changes in policies and practices that can help to define directions for change (Ruitenbeck 1998). This begins with an assessment of how and why the *status quo* evolved, through an understanding of income/wealth distribution, and previous/present land-use policies and practices. This approach is relatively imprecise, and best suited to broad planning efforts at the national and regional level

– Assisting comparisons between two or more steady state forest management or wildlife conservation conditions (e.g., the *status quo* compared with the conditions if particular policy/practice changes were adapted). This approach has been very effective in comparing a finite set of easily quantifiable values, such as dollars spent/m^3 of wood harvested, in reduced-impact logging and conventional timber harvesting operations (see case studies later in this chapter)

The central principle underlying the above approaches is to identify conditions where the private benefits and costs to the decision-maker closely resemble the broader social benefits and costs. It is via this route that economics can help to promote conservation, through the presentation of alternative policy and practices for sustainable forest management that will be practical, politically acceptable, and adopted.

The Feasibility of Reducing the Environmental Impacts Associated with Timber Harvests

Logging operations are motivated by profits and dominated by commercial imperatives, and therefore tend to avoid costs that threaten to lower the bottom line (e.g., watershed protection and wildlife conservation measures). In economic jargon, there is a wide divergence between the private benefits and the social costs. Recognizing this, and the importance of logging to many tropical nations, some conservation organizations have sought to develop techniques that minimize the broader (environmental) costs without unduly increasing the direct costs of logging. This is one area where economists have been instrumental in helping to inform the debate surrounding the contributions of reduced-impact logging, because labor, equipment, and product inputs/outputs are easily quantifiable. Two studies that compare the costs and benefits associated with conventional and reduced-impact logging practices are reviewed below.

Case Study I—Costs and Benefits Associated with Forest Management Practices in Eastern Amazonia (adapted from Uhl et al. 1997 and Barreto et al. 1998)

Selective rain forest logging in eastern Amazonia is often done carelessly, resulting in much unnecessary forest damage. In an effort to raise the quality of timber-harvesting practices, the technical feasibility, efficiency, and profitability of reduced-impact logging (planned) practices were compared to conventional (unplanned) logging operations in side-by-side plots. Financial assessments of the two approaches, based on the net present value for 20- and 30-year cutting cycles, are reported here (see figure 28-1).

The cost to plan logging operations was estimated at U.S.$72/ha or approximately U.S.$1.80–2.05/m^3 for harvests of 35 to 40 m^3/ha (typical for the study region). More than 90 percent of this cost was in tree mapping, vine cutting, and planning logging maneuvers (e.g., skid trail layout, felling angle determinations, etc.). The careful planning of tree felling operations resulted in a 15 percent increase in productivity (m^3 felled/work hour) over the unplanned operation. The machine time (minutes/m^3 harvested) necessary to open logging roads and log landings was 37 percent less in the planned operation than in the unplanned operation, and the productivity of skidding logs to landings

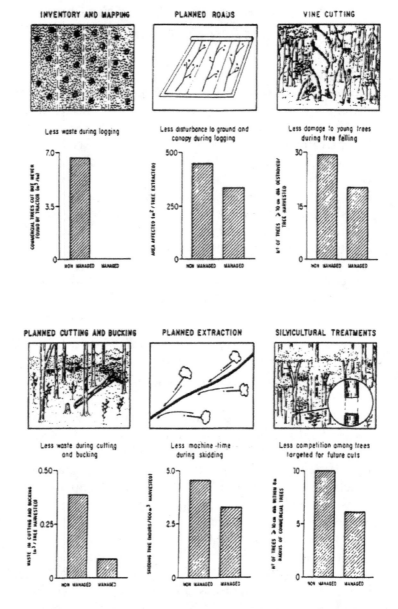

FIGURE 28-1 Six logging management steps that bring measurable benefits in eastern Amazonia terra firme forests. The benefits include: less waste during logging, less disturbance to the ground and canopy during road building and felling, less damage to young trees during felling, less waste during cutting and bucking, less machine use during skidding, and better growing conditions (less competition) for trees targeted for future cuts (Uhl et al. 1997).

(m^3 hauled/hour) was 27 percent greater in the planned operation (using wheeled skidders compared to standard bulldozer-based operations in unplanned extraction). In the absence of planning, 26 percent of the volume of timber that was felled was wasted: seven percent lost to poor felling techniques and 19 percent lost to felled trees that were never found by the tractor operators (see photos 28-1a, b). In the

PHOTO 28-1 *In the absence of planning and supervision, profits can be lost due to wasted wood, and habitat unnecessarily degraded: a) cut made too high on stem, wasting wood in base of butt-log; b) poor felling leading to log damage and placement in inaccessible area. (R. Fimbel)*

planned area, only one percent of the felled timber was wasted. Overall, increased work productivity and reduced waste in the planned logging operation resulted in financial benefits of U.S.$3.70/m³, which was about two times the cost of planning.

In addition to short-term economic benefits, carefully planned logging operations reduce damage to the forest, leaving a well-stocked stand (see figure 28-2). Good stocking, combined with the application of silvicultural treatments following logging, should result in greater timber production in managed forests. With planning (reduced-impact logging), it is estimated that 68 percent more timber volume could be extracted over a 30-year period than with conventional harvesting practices. Using discount rates ranging from 6 to 20 percent, estimates of net present value of timber extraction with planned forest management were 38 to 45 percent higher than without planning.

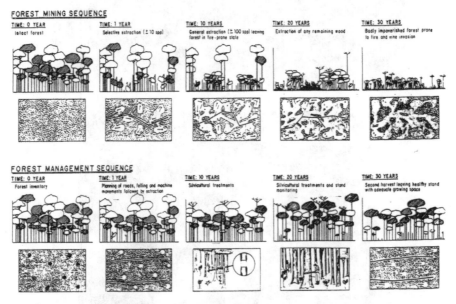

FIGURE 28-2 *Typical terra firme logging ("mining") practices that lead to forest degradation in eastern Amazonia (top sequence). The alternative is forest management and includes conducting forest inventory, planning of extraction activities, and silvicultural treatments. With this approach, sustainable cutting cycles might be reduced from 70 to 100 years to 30 to 40 years (bottom sequence). For each step in these sequences, the upper panel shows a side view of the forest, and the lower panel shows a view from above or close-up side view (from Uhl et al. 1997).*

Though forest planning (reduced-impact logging measures) appears to be economically attractive, there are various barriers to its widespread application in the region. Information on the technical aspects of these practices has not been transferred to landowners. Forestry regulations are not enforced. Additionally, any investment in forested land is perceived as risky because of frequent disputes over land ownership. Overcoming these barriers will require, at the very least, the implementation of a regional forest-use policy that is sensitive to both local economic conditions and forest ecology.

Case Study II—The Costs and Benefits of Low-Impact Logging Relative to Conventional Logging Practices in the Brazilian Amazon (Adapted from Holmes et al. 1999)

The Tropical Forest Foundation, its Brazilian subsidiary Funda Floresta Tropical, the United States Forest Service, and other collaborators, recently completed a cost-benefit assessment of conventional and low-impact logging practices (Holmes et al. 1999). This study was conducted in several 100ha cutting blocks at the Fazenda Cauaxi near Paragominas, Brazil. The sites were selectively logged, with an average removal of three to four trees/ha (approximately 25 m^3/ha). Preliminary findings indicate that low-impact measures cost less than conventional logging practices, while helping to preserve the composition and structure of the residual forest.

The total costs associated with conventional logging are greater than those incurred in low-impact logging operations (see table 28-1). This difference is a function of efficiency gains during skidding and log deck operations ($2.51/m^3 during low-impact logging vs. $4.12/m^3 in conventional logging) that offset increases in planning costs ($1.91/m^3 during low-impact logging vs. $0.91/m^3 in conventional logging).

Conventional logging operations also have a greater impact on forest structure and composition compared to low-impact logging practices. Ground area disturbed by heavy equipment was reduced approximately 50 percent in the low-impact logging sites (see table 28-2, figure 28-3), while residual commercial stock mortality was one-third the level found in conventionally logged forest (see table 28-3). Reductions in damage to future crop trees will increase the profitability of the next harvest, while helping to conserve habitat for wildlife.

Finally, a number of *disincentives* impede the widespread adoption of low-impact logging practices by the private sector, including the need for better trained practitioners (loggers and foresters) and lower

TABLE 28-1 *Costs and Returns of Conventional Logging Versus Low-Impact Logging Operations*

	Cost (U.S.$/m^3)	
Activity	Conventional Logging	Low-impact Logging
Preharvest		
Block layout	0.00	0.25
Inventory[a]	0.14	0.48
Vine cutting	0.00	0.13
Data processing	0.00	0.09
Mapmaking	0.00	0.19
	0.14	**1.14**
Harvest Planning		
Tree marking	0.00	0.14
Road planning	0.00	0.03
Road construction	0.28	0.14
Log deck planning	0.00	0.01
Log deck construction	0.29	0.14
Skid trail layout	0.00	0.31
	0.57	**0.77**
Harvest		
Felling and bucking	0.50	0.63
Skidding	2.12	1.24
Log deck operations	2.00	1.27
	4.62	**3.14**
Support, Logistics, Supervision	0.41	0.32
Total cost/m^3 of marketable wood	5.74	5.37

[a]Conventional logging operations do not have a planned inventory per se, however a tree prospector "cruises" for merchantable trees at the logging face, providing information directly to the loggers. In contrast, all trees ≥ 35 cm diameter at 1.3 m were located, measured, and mapped in the low-impact logging operation, prior to harvesting activities.

tariffs on equipment that directly reduces logging damage (e.g., winches and logging arches). There are also institutional impediments. If the logger has no expectation of returning for future harvests, for example, the benefits of lower costs and higher yields are then negligible in his calculations. With more secure and longer-term tenure, these future benefits might be sufficient to change his current behavior. The Tropical Forest Foundation and its partners are currently undertaking a study of these disincentives, and options for overcoming them.

The mentioned financial analyses are examples of no regrets or win-win logging, where the application of new techniques leads to both enhanced environmental outcomes and increased profits/reduced costs. One must bear in mind, however, that these studies represent only one component of a comprehensive forest management economic analysis—failing to consider the many non-timber benefits associated with healthy and productive forests. Maintaining natural forest structure and composition in lieu of conversion to pastures, agriculture, or plantations, provides a multitude of direct (e.g., water, timber and nontimber products) and indirect (ecosystem functions, conservation of wild genetic strains, undiscovered medicines and products) benefits to society (see chapters 1 and 30). At present, economists are at a loss

TABLE 28-2 *Percentage of Ground Area Disturbed by Conventional and Low-Impact Logging Practices*

Activity	Percent Disturbance (%)	
	Conventional Logging	Low-Impact Logging
Secondary roads	1.35	0.65
Log decks	1.05	0.63
Skid trails	7.66	3.90
Total Area Disturbed (%)	**10.06**	**5.18**

TABLE 28-3 *Percentage of Residual Commercial Stock Disturbed by Conventional and Low-Impact Logging Practices*

Health Class	Percent Disturbance (%)	
	Conventional Logging	Low-Impact Logging
Dying or dead	13.8	4.8
Damaged (dying?, recovering?)	6.6	6.5
Damaged (recovering)	8.3	10.0
Undamaged	71.3	78.7

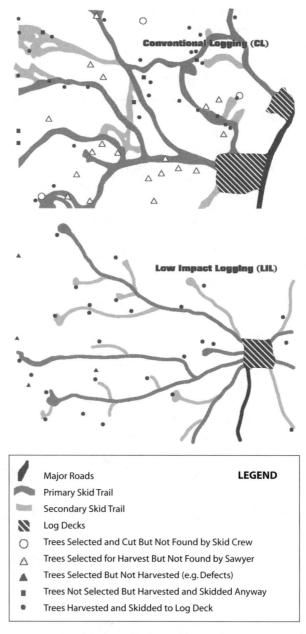

FIGURE 28-3 *A comparison of area disturbed and trees harvested using conventional logging and low-impact logging techniques in Fazenda Cauaxi, Brazil (from Tropical Forest Foundation, Alexandria, Virginia, USA).*

to quantify many of these attributes with any degree of certainty. The bottom line is that any cost-benefit analysis may underestimate the total benefits of SFM to society as a whole.

Economics and Future Forest Conservation Measures

There are considerable, nonquantifiable benefits related to retaining forests and conserving wildlife, but economists cannot say conclusively that they always outweigh the cost of their protection. The financial analyses outlined are a first step towards clarifying the costs associated with reducing the impacts of timber-harvesting practices on the forest environment. These approaches are quite limited in the variables that they can quantify, however. The development of a more complete picture of the economic conditions associated with sustainable forest management will require long-term, detailed costing of silvicultural options (see chapters 1 and 30). This long-term approach should help to identify prescriptions capable of increasing an operator's profits (through improved technology and avoidance of penalties), compared to current practices, while improving environmental outcomes (some can be specified, others will continue to be too difficult to value). Industry should endeavor to adopt these measures voluntarily, if for no other reason than profits are increased. There will also be little need for monitoring and enforcement; industry, society and wildlife gain from a healthier, more productive forest, and these practices will create a new *best practices* benchmark against which future refinements in management activities can be measured.

The above scenario holds promise for helping to conserve tropical forests in the future, however actions beyond current financial cost-benefit analyses of reduced-impact logging and planned, long-term economic studies of forestry practices are needed to promote a shift from conventional forestry practices to a sustainable standard. Economists must broaden their examinations to include issues such as insufficient re-investment in the resource, ineffective timber pricing systems, and inequality of profit sharing (see chapter 18). Reforms to current timber taxation/fee structures (e.g., flat taxes, competitive bidding based on estimated volumes, etc.), built upon thorough evaluations of potential gains and costs to society, are a crucial step towards resolving these issues and facilitating the transition from conventional to sustainable forest management. Through such reforms, forest departments and society as a whole should capture more of the resource value (thereby providing funds for training, supervision, and local stakeholder compensations), while bringing the price of commodities into closer balance with social and environmental costs (through the process of open, competitive bidding). These steps, combined with the promotion of long-term contracts between governments and forest operators (encourages investment); tax incentives to protect sensitive habitat within the harvest landscape (see chapters 15 and 29), and institutional and regulatory reforms (make industry more accountable for the environmental consequences of their forest practices), are all measures that economists can undertake to help governmental and nongovernmental groups in their efforts to establish sustainable standards for logging practices in tropical forests.

ACKNOWLEDGMENTS

The author gratefully acknowledges the generous advice and assistance of the editors and the anonymous reviewers.

Chapter 29

RAIN FOREST LOGGING AND WILDLIFE USE IN BOLIVIA

Management and Conservation in Transition

Damián I. Rumiz and Fernando Aguilar

Bolivia is a land-locked South American country encompassing a total area of 1,098,581 km^2, 48 percent of which is covered by forest (Robbins et al. 1995). Most of the forests occur in the lowlands east and northeast of the Andes, and range from humid evergreen forests in the north (Departments of Pando, north of La Paz, Beni and a fraction of Cochabamba), to dry deciduous forests in the south (east of Chuquisaca and Tarija). In between lies Santa Cruz, where a gradient from humid evergreen to semideciduous and dry forests occurs (see figure 29-1).

These diverse forest types harbor a significant portion of the country's rich flora, which is estimated at about 20,000 species (Moraes and Beck 1992), as well as much of the country's animal diversity. The biodiversity of Bolivia is one of the world's richest, with a vertebrate fauna containing at least 319 species of mammals, 1,358 of birds, more than 220 of reptiles, 123 of amphibians, and about 500 of fish (Ergueta and Morales 1996). Local communities rely on this biological diversity as a natural resource using, in one case, 498 plant and 29 animal species in northern Santa Cruz (FAN-WCS 1994). Timber companies have focused until recently on the extraction and export of a few valuable species, such as mahogany (*Swietenia macrophylla*), oak (*Amburana cearensis*) and cedar (*Cedrela odorata*) (López 1993). Other uses of the forested landscape include the generalized extraction of timber and nontimber products, hunting, and conversion to pastures and agriculture.

Forest management activities impact this natural landscape, with 5.7 million ha under forest concessions, and other forestlands in private properties and indigenous territories. Historically, no logging concession has been managed for sustainability, and invasion of illegal loggers, hunters, and campesinos (local peasant farmers) in these areas have damaged

FIGURE 29-1 *Map of Bolivia showing area with lowland forest cover and production forest sites mentioned in text.*

the forest (Mancilla 1996). Inconsistent legislation and lack of law enforcement are the main causes of forest mismanagement in Bolivia, along with numerous (and as yet poorly understood) socioeconomic issues. This chapter describes the legal framework for timber and wildlife management in Bolivia, the history and current situation of these sectors, proposed changes from new laws and policies, and their implications for wildlife conservation.

History of Wildlife Use, Logging, and Unsuccessful Laws

A Brief History of Wildlife and Forest Resource Exploitation

The history of heavy use of forest wildlife by colonists in the lowlands of Bolivia started with the rubber (*Hevea brasiliensis*) boom at the end of the 1800s, when thousands of adventurers moved along the rivers of the Amazon basin (La Paz, Beni, Pando, and later, Santa Cruz) and established temporary or permanent towns for which the main source of protein was wild meat and fish. When the rubber business declined, hunting of black caimans (*Melanosuchus niger*) in the 1930s began to supply the skin market and continued through the 1960s and 1970s with the harvesting of smaller caimans (*Caiman yacare*), otters, cats, and peccaries (Ribera 1996a). The opening of roads between Cochabamba and Santa Cruz in the 1950s, and between Santa Cruz and Trinidad in the 1970s, had an important impact on the exploitation of forest resources. Road building allowed the growth of the timber industry and colonization, while the northern lowlands of Pando and Beni continued with the seasonal collection of rubber and Brazil nuts (*Bertholettia excelsa*) in all areas accessible by river. The trade of live animals, mainly parrots and macaws for the pet market and monkeys for biomedical research, flourished in the 1970s and 1980s. These businesses created more incentives to attract or keep people who depended on other wildlife resources for food (mainly ungulates or birds) or specialty uses (e.g., large monkeys to bait cat traps) in the forest. Populations of several animal species declined or became extinct locally. Black caiman, giant and common river otters (*Pteronura brasiliensis* and *Lutra longicde-forestaudis*), jaguar (*Panthera onca*), ocelot (*Leopardus pardalis*), margay cat (*Leopardus wiedii*), several macaws and monkeys were severely impacted (Ribera 1996a).

The pressure from the international skin and pet markets declined after 1980, but other activities continued this process of faunal overexploitation (Marconi and Hanagarth 1985; Ribera 1996a). Commercial meat hunting for local markets, and forest activities such as logging, palm heart extraction, Brazil nut collection, oil exploration, and mining are currently having an impact on populations of large mammals, birds, and reptiles—although not in such an irreversible way as deforestation (deforestation in Bolivia was estimated at 160,000 to 200,000 ha/year in the early 1990s [Pacheco 1998], creating extensive agricultural fields in the lowlands of Santa Cruz). Finally, native communities dependent on wildlife resources for their subsistence have recently begun to experience increasing competition from loggers and commercial hunters for limited game. In areas, subsistence hunting is no longer sustainable because of reductions in hunting grounds and lack of areas functioning as sources of wild meat (Townsend 1996a). In spite of these hunting pressures, much land under forest concessions and indigenous territories still harbor rich faunal assemblages.

Forestry and Wildlife Laws Through the 1970s

The origin of the Bolivian legislation regarding logging and wildlife dates back to 1825, when the Congress of the new Republic enacted its first Presidential Decree to regulate the use of the forests. Close to 100 regulations directly related to wildlife have been passed in Bolivia since 1832 (Marconi 1991). Prohibitions alternated with authorizations and taxes for hunting and trade with wildlife species, all of which did little to protect the affected species (Ribera 1996a). Chinchillas went extinct in the Andes, while vicunas and many lowland tropical forest species were hunted for skins, pelts, meat, and as pets.

The first forestry legal framework was not completed until the mid-1970s, when the General Forestry Law of the Nation was passed as Law Decree (DL 11686) in 1974, and its complementary regulations as a Presidential Decree (DS 14459) in 1977. This law considered all forestlands as public property and included specific norms, mandates, and procedures to promote, regulate and oversee the felling, processing, and marketing of timber and nontimber forest products, while promoting forest protection and conservation. Among its main provisions, this law created the Forestry Development Center and established a system of licenses and revenues for the use of public timber and non-timber resources.

The Forestry Development Center was assigned to perform a wide range of functions and activities, which included formulating forest policies, administering forest resources, conducting inventories, classifying forestlands, granting logging permits, monitoring logging activities, controlling log and lumber flow, conducting research, and providing technical assistance. This office was also responsible for preventing any commercial logging or extraction activities in forested areas that would threaten biodiversity, watershed functions, and fragile areas.

In 1975, the Bolivian Congress passed the Wildlife, National Parks, Game and Fish Law (DL 12301), which gave the Forestry Development Center the responsibility of managing and conserving wildlife through granting and monitoring hunting permits. Even when it clearly defined different types of hunting and established strict prohibitions, its complementary regulations were never passed, so it lacked provisions such as fines and penalties for offenders.

The enforcement of these laws faced several problems. The Forestry Development Center could not comply with many of its mandates because of the shortage of financial and human resources, and excessive political influence by individuals more interested in unrestricted logging (PAF 1995). Although each logging company was required to submit a forest inventory and a management plan to obtain logging permits, neither the accuracy nor the compliance of these documents were adequately evaluated or monitored. Some key tasks such as field supervision and controlling records of harvested volumes were poorly performed (see photo 29-1). Lack of enforcement resulted in the growth of illicit operations and loss of revenue for the forestry office. Small local logging enterprises flourished (often with the investment of just a chainsaw), and increasingly invaded concessions and protected areas.

PHOTO 29-1 *Stump of a mahogany tree cut under the minimum legal diameter (60 cm dbh) in a forest concession of Santa Cruz. (D. Rumiz)*

In the case of the 1975 Wildlife Law, the government collected some revenue from hunting, fishing, and trade permits, but could not penalize transgressors because of the lack of enforcement regulations. Although Bolivia endorsed the Convention on International Trade in Endangered Species (CITES) in 1974, the use of CITES permits was often fraudulent. According to Pacheco (1992), legal wildlife exports (with CITES certificates) between 1983 and 1990 included at least 27 species and 58,751 individuals of macaws, parrots, and parakeets; 3,754 live squirrel monkeys (*Saimiri sciureus*); and the skins of two peccaries (*Tayassu pecari* and *T. tajacu*, 32,477 skins), four species of cats (63,570), caimans (491,641), and tegu lizards (*Tupinambis* spp., 455,472). These numbers were concentrated in a few years (mainly 1983, 1984 and 1985), with no exports in other years, suggesting that the actual trade was much higher and mostly illegal—similar to the previous decade for which there were no available CITES records.

The "Historical Ecological Pause"

The increasing conflicts between timber firms and growing peasant settlements, indigenous groups, and other forest users, together with the generalized non-compliance with the Forestry and Wildlife laws, induced the government to declare the so-called "Historical Ecological Pause" (DS 22407) in 1990. The granting of new logging permits in production forests was banned between 1990 and 1995. Firms with short-term (3 years) and medium-term contracts (10 years) were required to convert them into long-term (20 years) contracts.

A set of regulations for the Ecological Pause was promulgated in 1991 (DS 22884), including norms, mandates, and provisions to regulate forestry activities and conserve forest resources. The Forestry Development Center was required to strengthen its system of forest guards and to assign 20 percent of stumpage fees to protect production forests from illegal invasion and logging. Timber companies were mandated to forbid hunting in their logging areas and to feed their crews with beef, rather than wild meat. Exporters were given a five-year deadline to bring their logging into compliance with sound forest management practices under the monitoring of the Ministry of Agriculture and Peasant Affairs and the General Environment Secretary.

The application of the law was less successful. As a consequence of the Ecological Pause, all timber firms did shift to 20-year contracts. Exports of timber products increased from 97,559 tons in 1990 to 112,655 tons in 1995, with mahogany (*Swietenia macrophylla*) dominating by volume the national production, but decreasing from 142,946 cubic meters in 1990 to 71,508 in 1995 (CNF 1995). No practices promoting sustainability were adopted and production figures showed decreasing mahogany stocks in the forests of Santa Cruz and Beni (López 1993)—particularly evident in the Chimanes region (Ross 1992) (see figure 29-1). Extraction and saw milling techniques were mostly wasteful and inefficient, while the forest industry concentrated in high-grading the forests and exporting rough sawn wood. Hunting to feed logging crews at field camps and mills—or for sale at local markets—continued as before, further impoverishing wildlife populations in production forests (see chapter 15). As with the 1974 Forestry Law, most of the tasks assigned to the forestry office and other institutions by the Ecological Pause were not accomplished (Andaluz et al. 1996), and its application did not mean any real change for the forestry regime.

The "Permanent and General Ban" Law and Environmental Law

In addition to the specific norms included in the Ecological Pause regarding wildlife protection, the government of Bolivia enacted the Permanent and General Ban Law (DS 22641) in 1990, forbidding all harassment, capture, and gathering of wild animals and their products. Only the scientific collection of specimens was allowed, while subsistence hunting was assumed to be legal for indigenous communities because of their dependence on the wildlife resource. Under this law, which is still in effect, commercial hunting of any species (or harvesting of wildlife products) must be specifically authorized by a presidential decree on a case-by-case basis.

Enforcement of this law was nonexistent and violations were open and widespread. Hunting by logging crews, activities of sport hunting clubs, sale of bush meat in restaurants, and sale of pets in the streets continued unabated (see photos 29-2, 29-3). The organized trade of skins and live animals from the forest decreased, however, either due to the new policy or to the general decline of international markets (Ribera 1996a).

Finally, the Environmental Law (L 1333) was enacted in 1992. It established a general framework and delineated responsibilities for the government, universities, and society in general to protect the environment and improve the quality of life through sustainable development. Because of its generalized character, however, it is of limited direct use in wildlife conservation (Ribera 1996a, b). Perhaps the most significant contribution of this law regarding wildlife conservation has been

PHOTO 29-2 *Tapir, deer, and bird bones found at a logging camp in Bajo Paraguá, northern Santa Cruz. (D. Rumiz)*

the groundwork it laid for two other related laws: the Law of Conservation of Biological Diversity (pending, it will replace the Wildlife Law of 1975) and the Law of Genetic Resources (enacted in 1997).

The Legal Framework of Forest Use and Wildlife Conservation in the 1990s

The Forestry Law of 1996

The institutional framework started changing dramatically in 1993. The Forestry Development Center disappeared in 1995, and there was an institutional gap in the administration of forest resources while the new Forestry Law was debated in Congress. The Forestry Law 1700 was eventually passed on July 12, 1996. This law and its regulations (DS 24453), enacted in late 1996, constitute a modern legal statute designed to address a set of persisting problems in the forestry sector and to enforce the sustainable management of Bolivian forests. These objectives are part of the structural and institutional changes that have been evolving in Bolivia in recent years, which included the creation of the Ministry of Sustainable Development and the Environment (re-

PHOTO 29-3 *Red brocket deer skins in a logging concession camp, Beni. (D. Rumiz)*

named in 1997 to Ministry of Sustainable Development and Planning) and the administrative de-centralization of the government at Departmental (Prefectures) and Municipal (community) levels in 1994.

One important feature of the new Forestry Law was the creation of the Forest Superintendency as a key institution in the implementation of the new forestry regime in coordination with the Ministry, Prefectures, Municipalities and National Fund for Forest Development. The Superintendency is in charge of granting concessions and permits for the use of public and private forests, approving management plans, conducting audits, collecting and distributing fees, and other activities related to law enforcement. In order to reduce political influence and institutional instability, the Superintendent is chosen by the President from a list of three candidates proposed by the Senate, stays in office for a period of six years (two years longer than the Presidency), and may be removed only after due process by the Supreme Court of Justice.

Forest concessions are awarded through a competitive process and for periods of 40 years, during which concessionaires must apply sound management plans and satisfy audits every five years in order to keep their logging rights. Local communities have priority for access to timber and non-timber concessions in areas open to contracts. Terms of reference for management plans also differentiate large-scale timber firms from local communities and indigenous groups, recognizing the need for specific guidelines and approaches to vary with the forest user. An annual fee per hectare, based on the productive area of each concession, replaces previous stumpage fees and royalties based on volume of timber harvested. Private land owners, local communities, and indigenous groups only pay the fee for the area to be harvested yearly according to their specific management plans.

Under the Forestry Law of 1996, regulations to elaborate and implement management plans require new technical information and analyses, forcing forest administrators to change traditional and wasteful forest exploitation practices, improve their technical capacity, and start focusing on long-term sustainability and environmental issues. Concessionaires cannot log riverine strips (from 10 m at each bank in non-erosive streams, up to 100 m in rivers with erosion-prone or flooded banks), wetlands, or steep ($\geq$ 45 percent) slopes, and may optionally set apart up to 30 percent of the concession for protection purposes without paying the per hectare fee for these areas. Land owners must also protect riverbanks, slopes, and wetlands, which are considered permanent *ecological easements* and must be delimited in each *property-ordering plan*. Nature reserves voluntarily established for a minimum of ten years on private property are tax-exempt. In the case of land-clearing permits for agriculture and cattle ranching, the minimum charge on the felling area is 15 times the minimum charge for concessions, plus 15 percent of the current market price of the felled trees as sawn wood. A summary of these regulations, mostly related to timber products, is listed in box 29-1.

> **Box 29-1** *Summary of Technical Regulations for Forest Management in Bolivia (according to the Forestry Law 1700, its complementary regulations, and technical norms).*
>
> Professional Responsibility:
>
> Foresters accredited by the Forestry Superintendency must sign reports of forest inventories, management plans, maps, and operative plans, thereby providing a legal accounting of the information reported.
>
> Vegetation Maps:
>
> Based on aerial photographs or satellite images and ground truthing, landscape and forest types must be classified according to characteristics of the terrain and relative height and density of trees, respectively. Other maps should show concession borders, administrative divisions, etc.
>
> > Map scale should be 1:100,000 for areas > 40,000 ha; 1:50,000 for areas 10,000 to 40,000 ha, and 1:20,000 for areas < 10,000 ha.
> >
> > Map should show the location of the sampling units for the inventory described below.
>
> Forest inventory:
>
> A general forest inventory must be conducted throughout the production area, to estimate tree species composition by category of use, abundance, size distribution, estimated volume, and re-generation by species.
>
> > Requested sampling intensities range from 0.1 percent ($\geq$ 200,000 ha) to 8 percent ($\leq$ 100 ha) for all trees $\geq$ 40 cm dbh (diameter at breast height), and half this intensity for trees between 20 and 40 cm dbh.
> >
> > Number of plots should be at least 100. For larger trees, plots should range between 2.0 and 0.1 ha, with larger plots in larger concessions.
> >
> > Increasingly smaller plots, included in each large one, should account for natural re-generation of trees in size classes between 10 to 20 cm dbh, 5 to 10 cm dbh, and smaller than 5 cm dbh, but taller than 1.30 m. The sampling protocols for regeneration are currently under revision.
> >
> > Palms should also be identified and counted (in fewer size classes and at lower sampling intensities than non-palm trees) to assess food sources for frugivores and potential non timber products. The sampling protocols for palms are currently under revision.

Minimum cutting diameters:

These are preliminary established sizes for some species and life zones until further studies can suggest better silvicultural practices. They range from 70 cm dbh for mahogany to 40 cm dbh for more common species.

Management plan :

It must establish a clear and sound strategy for the long-term sustainable production of forest resources, including:

— The list of harvestable timber species, annual volumes, and cutting cycles which would guarantee a sustainable production according to the proposed silvicultural treatments and the natural re-generation and growth of the forest estimated with the best information available

— Concrete actions to avoid the extinction of plant or animal species listed as endangered or vulnerable, to minimize negative impacts on species known as keystone resource for wildlife, and to prevent the degradation of soils and aquatic environments

— A list of large- and medium-sized vertebrates occurring in the area, based on interviews and surveys of animal signs, and noting their conservation status according to the *Red Data Book of Bolivian Vertebrates*

— A set of ecological reserves covering areas of restricted use such as riverine strips or steep slopes, including habitats and critical areas for wildlife refuge, feeding or reproduction

— Maps showing vegetation cover, administrative divisions, annual cutting areas and ecological reserves

— A monitoring plan to evaluate growth, yield and forest response to silvicultural treatments

— An annual operation plan defining areas, species, and volumes to harvest based on a 100 percent commercial inventory, and identifying marked trees, skid trails, extraction roads, log yards and other features on detailed maps of the censused area

The Superintendency implements these regulations through review and approval of the plans, while the control of field operations is handled jointly with the Municipalities. Management plans are monitored by periodic audits, remote sensing, and field inspections. Operations obtaining sustainable forestry certification through internationally recognized organizations (see chapter 26) will be relieved from audits and other controls made by the Superintendency. The Superintendency is growing in technical capacity and is receiving assistance from internationally-funded projects like BOLFOR (a seven-year, USAID-funded project in sustainable forest management). Many aspects of the regulations and their technical background were developed based on the experiences and research of this and other projects. A series of publications and handbooks are available to guide practices such as timber inventories (Dauber 1995), computerized mapping of logging compartments (Camacho 1998), wildlife conservation in forest management (Rumiz and Herrera 1998), and how local social groups may establish forest management areas (Andaluz et al. 1998).

Although it is too early to assess the impact of this new law on the conservation of forest wildlife, the influence on the forestry sector is clear. First, the total area under concessions has decreased because of the non-renewal of contracts and the reduction in area of individual concessions of companies not ready to pay the per-hectare fee. Companies remaining in business are changing their traditional logging and marketing practices, and improving their technical capacity to conduct forest management, product development, and marketing (Guzmán, 1998). Several companies are voluntarily seeking international evaluation of their operations and are going through the process of being certified as *sustainably managed*. Local social groups and indigenous communities are in the process of securing forest areas and seeking technical assistance to manage them. Hunting seems to have decreased in some concessions (Rumiz, personal observation) due to pressure from the certification process, growing control, and the field presence of technical assistance provided by international projects. Stronger action in control and training involving municipalities, however, is still needed.

An important body of information is being generated from forest inventories and the establishment of permanent plots in production forests. Data on vulnerable or endangered species of wildlife and their resources will also be collected, and ecological reserves will be established in managed forests. It is not clear, however, how and by whom the impacts of forest management on vegetation and wildlife will be monitored in the long-term, and how these results are going to be incorporated into new, more sustainable practices.

A National Proposal for Conservation and Sustainable Management of Wildlife

Apart from the expected changes in wildlife conservation derived from the Forestry Law, a series of institutional developments occurred in this recent period that will affect the administration of wildlife issues. Since the Law of Reform to the Executive Power was put into effect in 1993 (DS 23792), the National Directorate of Biodiversity Conservation (in the Ministry of Sustainable Development and Environment) has been in charge of wildlife and protected areas. This office has promoted major philosophical and structural changes for the management of wildlife, protected areas, and genetic resources. The National Program for Wildlife Conservation, proposed by the

Biodiversity Conservation office, has the objective of promoting the long-term permanence of wild plants and animals through the involvement of local communities in the application of conservation measures and the generation of benefits from the use of these species. This strategy includes developing pilot case studies of sustainable management of species with economic importance (such as vicuñas [*Vicugna vicugna*] for wool in the Andes, or caimans [*Caiman yacare*] and capybaras [*Hydrochaeris hydrochaeris*], for skin in the lowlands), by protecting and monitoring species that are threatened or the subject of harvesting, and by controlling the use, transport, and trade of wildlife. This office was renamed the General Directorate of Biodiversity Conservation (in the Ministry of Sustainable Development and Planning) in 1997.

A critical point on biodiversity to be addressed by the new legislation is the right of indigenous communities to hunt for subsistence purposes, and also to legally sell wildlife products such as skins. Still unresolved are the rights of non-indigenous local communities and the mixed situation of communities involved in the exploitation of timber or non-timber forest products such as Brazil nuts—but who rely on wildlife for food.

The National System of Protected Areas is also being developed under the administration of the same Directorate. Several evaluations and analyses (Marconi 1992; Mérida 1995; Ribera 1995; 1996b)— from a total of almost fifty areas mentioned in different sources as having an assortment of objectives, legal status, and conservation values—contributed to a protected area management strategy. A program for coordinated management of eight protected areas is currently in progress with special funding, while other areas are awaiting funds to be surveyed and included under management. Two, new, large, protected areas—Madidi and Kaa-Iya del Chaco National Parks—with their adjacent Natural Areas of Integrated Management, were declared during 1996. They include more than 5 million ha for protection and the sustainable use of wildlife. Management of the Kaa-Iya del Chaco area is also being implemented with an innovative joint administration between a conservation NGO and an indigenous group (Taber et al. 1997). All plans regarding conservation and sustainable use of wildlife, however, as well as management of protected areas, will be hindered if new legislation does not establish efficient control and enforcement. The long-term value of protected areas must be defended against the imminent threat from mining or habitat conversion by farmers or ranchers.

The Future of Forests and Wildlife in Bolivia

Previous analyses of environmental laws in Bolivia before 1992 have described them as highly dispersed, disorganized, out-of-date, incomplete, vague, or incorrect (Marconi 1991), and as "probably the most disobeyed and disregarded of the whole legislation" (Ribera 1996a). Reasons why these laws failed to accomplish their objectives include their lack of reality and their juridical inefficiency (lack of complementary regulations, procedures, and enforcement mechanisms), the marginality of environmental issues in the economic development of the country, and an over-estimation of the command and control power of the state (Andaluz 1996a).

The new forestry legislation takes these problems into consideration and provides some mechanisms to control them (Andaluz 1996b; Mancilla and Andaluz 1996). Among these, *social control* is exerted through the participation of any concerned citizen (or organization) willing to verify inventories, forest operations and management plans, and through forestry professionals being held accountable for their documents and actions. Local municipal governments are also in charge of selecting 20 percent of production forest for local concessions, supporting local concessionaires in the implementation of management plans, and helping the Forest Superintendency in monitoring local forest activities. Similar mechanisms should be built into biodiversity legislation, to promote the involvement of local inhabitants and the public in general in decisions about the environment. The long-term value of public protected areas should be defended against pollution from mining and unsustainable use of resources.

For social control to work, education and training are necessary at different levels. The public must understand the importance of the ecological services provided by the forest and the damage that de-forestation and uncontrolled hunting have on biodiversity. For professionals and politicians there is the need to change culturally towards a more sustainable use of resources. Technical assistance for developing and monitoring management plans, training officials to control field activities in concessions, (compliance with annual harvesting plan, reserves, no hunting, cutting diameters), and forming road control posts (origin, permits, wildlife products), are priorities.

The future of the Bolivian forest and its wildlife depends upon the passing of pending biodiversity legislation and further implementation and enforcement of the Forestry Law. Proper coordination and compatibility between the Forestry Law and the regulations of the recently-passed Agrarian Reform Law must be taken into account to avoid dangerous confusion, conflicts of interests, and overlap of attributions between both legal bodies affecting the forest-land tenancy and use.

ACKNOWLEDGMENTS

We wish to thank John Nittler, Antonio Andaluz, and Andrew Taber for reviewing early versions of the manuscript, and Robert Fimbel for his major editorial contribution during this process.

Part VII

SYNOPSIS

Chapter 30

LOGGING AND WILDLIFE IN THE TROPICS

Impacts and Options for Conservation

Robert A. Fimbel, Alejandro Grajal, and John G. Robinson,
with input from all contributors to the volume

Large expanses of tropical forests are subject to logging, with only a fraction of these forests managed sustainably for timber production (Poore et al. 1989; Johnson and Cabarle 1993). Technological, social, political, and economic constraints promote the rapid, unsustainable harvest of forest resources. One consequence of harvesting practices is that many wildlife species in production forests are experiencing population declines and local extinctions (Grieser Johns 1997; Laurance and Bierregaard 1997; see chapters 4–14, 24).

In recent years, foresters and conservationists have begun promoting sustainable forest management as a means to achieve long-term yields of important timber and non-timber forest products, while minimizing the environmental impacts associated with logging. Their efforts at sustainable forest management remain stronger in theory than application, however, with most efforts to date focused on reduced-impact logging measures and economic assessments of these practices. Other components of sustainable forest management, such as wildlife conservation, have received less attention. In this concluding chapter, we briefly recount:

- ⌐ The impact of logging on wildlife
- ⌐ How degraded wildlife populations might influence forest functions
- ⌐ Practices that reduce logging impacts on wildlife
- ⌐ Approaches to integrating wildlife conservation into forest management practices

Impacts of Logging on Wildlife and Their Habitat

Most extant tropical rain forests have been logged or will be in the near future (Bruenig 1996; Whitmore 1997). Logging intensities range from one tree every few hectares (e.g., *Swietenia macrophylla* in South America; see chapter 8), to 15 or more trees/hectare in the lowland dipterocarp forests of southeast Asia (Bertault and Kadir 1998; see chapters 9 and 18). These practices affect wildlife by altering their habitat (food, water, cover, and reproduction sites) and interactions with other organisms (competition, mutualism, pollination, dispersal, etc.). Even under the best of circumstances, commercial logging has a negative impact on some wildlife species because the spatial scales and intensities of these operations are usually larger and more intense than the natural disturbance events to which those species are adapted. *Under conditions of moderate-light selective logging, ample recovery periods, and accessible refugia sites within the managed landscape, most species appear capable of persisting in production forests* (Grieser Johns 1997; see chapters 4–14). *Indirect activities associated with logging, such as hunting* (see chapters 15–17), *burning, or land conversion, may have greater impacts on wildlife populations than the direct habitat disturbances caused by cutting* (see chapters 3–14, 18, and 24).

Habitat Modification

A habitat is a place where an organism (plant or animal), or community of organisms, lives and thrives (Meffe and Carroll 1997). Fire and conversion of forestlands to agriculture, pastures, or plantations drastically disturbs habitat, leading to declines and local extinctions in forest wildlife due to fragmentation and habitat loss (see chapters 8, 9, and 20). Repeat logging in an area, without adequate recovery periods for restoring structural and composition conditions characterizing the initial forest, also leads to a cascading degradation of the habitat (see chapters 9 and 29). In general, logging modifies habitat in several ways:

1. Logging causes a *shift in the diameter distribution* of a forest, with post-harvested stands often containing lower structural diversity (high densities of small- to medium-sized stems), and fewer cavity/den sites (see chapters 2, 17, 23, and 24). Enrichment plantings compound this situ-

ation, decreasing understory structure (through brushing) and disrupting microhabitat conditions (see chapters 7 and 8)

2. Logging causes a *shift towards earlier successional vegetation,* which may simplify the floral composition (especially where vines come to dominate disturbed areas; see chapter 2). In monodominant forest-types, however, logging can increase tree diversity and the availability of fruit resources and leaf protein contents, thereby improving habitat conditions for many wildlife species (see chapter 4).

3. Logging causes *changes in floral and fruit composition* that alter the availability of certain food resources—with large-seeded, shade-tolerant species often replaced by pioneer species (see chapters 2, 7, 10, 12, 17, 21, and 24). Grasses and browse frequently increase in logged sites, while fruit abundance tends to decline (except in monodominant forest types; see chapters 4 and 5).

4. Diameter limit cutting often leads to *high grading* or *creaming* of the forest, reducing re-generation of commercial species where cutting diameters include most seed-bearing individuals (see chapter 21). In some areas, selection cutting of mahogany (*Swietenia macrophylla*) has eliminated nearly 100 percent of the seed-bearing individuals from the forest (L. Snook, personal communication). The implications of such actions on the wildlife community and the ecological processes they per-form remain unknown.

5. Logging *fragments the forest canopy.*

 ⌐Gaps created by logging are usually larger and more frequent than natural treefalls, allowing more light into the understory (see chapters 2 and 8). These openings tend to be connected and expand over time with the mortality of injured trees, leading to fragmentation and increases in edge (see chapters 2, 6, and 8). Group selection and clear cutting lead to the largest gaps and drastic changes in the understory microhabitat and microclimate, often leading to vine tangles (see chapters 2 and 10). Alteration of the microhabitat and microclimatic conditions found in the forest understory cause the decline of interior forest wildlife, while populations of edge (generalist) species often rise (see chapters 7, 11, 12, 13, and 24).

 ⌐Vine cutting prior to logging diminishes intercrown connectivity, impacting the movements of non-volant arboreal animals (see chapter 2). On a positive note, this practice can reduce post-harvest gap size (lower residual crown damage).

— Roads act as physical barriers to some canopy and ground species (e.g., non-volant arboreal mammals and understory birds). They also introduce noise, increase soil temperatures, compaction, erosion and sedimentation, and allow access for illegal activities such as land clearing and hunting (see chapters 8, 21, and 24).

— Thinnings between harvests isolate trees (reduces connectivity), lower tree diversity, and expose the understory to higher insolation (see chapters 2, 8, 12, and 20).

6. Logging *alters the soil environment* by reducing porosity, increasing specific gravity, altering litter quality and quantity, decreasing pH, broadening extremes in temperature and humidity regimes, and increasing runoff and erosion. These conditions impact numerous wildlife species and disrupt a number of ecological processes in the soil and forest litter layer, including predator/prey relationships and nutrient cycling (see chapters 11, 12, 13, and 14). In general, soil disturbance is more severe under logging than natural treefalls, with microhabitat degradation increasing as the intensity of logging increases (see chapter 2).

7. Logging *alters aquatic environments*.

— Stream crossings and harvesting in riparian areas lead to increases in stream temperatures and sedimentation loads (the latter reduces the interstitial spaces between cobbles and pool depths), fluctuations in water nutrient and pH levels, and the invasion of exotic aquatic species (see chapters 11 and 14).

— Artificial ponds, created by road impoundments, serve as conduits for competitive generalist amphibians and diseases such as malaria (see chapter 11). These can benefit some forest amphibians however by providing breeding sites.

Hunting Impacts

Hunting is commonly associated with forestry activities, often in adjacent unlogged parcels and in many areas—in flagrant violation of laws protecting wildlife (see chapters 16–17). These illegal operations lead to revenue losses and long-term degradation of the faunal resource (see chapters 8 and 29). It is unclear how declines in the populations of game animals affect ecological processes such as pollination, seed dispersal, and soil nutrient cycles, but they may have cascading effects on the future composition and productivity of production forests (see chapters 3, 5, and 24). Finally, local populations often suffer loss of game due to increased hunting associated with influxes of sportsmen, commercial hunters, and/or hunting by company employees (see chapters 15 –16).

Activities That Promote Hunting in Logged Forest

⁻ Concession operations facilitate hunting activities via *easy access, available transport, and worker demand* for wildlife products (see chapters 17, 21, and 28)

⁻ Hunters in concessions (loggers, commercial, or sport) usually have *better hunting equipment* than local subsistence hunters (guns, vehicles, spotlights), allowing for rapid depletion of wildlife in an area (especially large game; see chapter 16)

⁻ Concession *workers tend to consume more wild meat* per meal than local subsistence hunters, and are often more wasteful in the usage of the animals they hunt (killing for recreational purposes in some cases; see chapters 16-17)

Hunting Impacts on Fauna

⁻ *Rare, threatened, and endangered species are commonly hunted* without regard to their protection status (see chapters 15-16)

⁻ *Large, slow-reproducing species are at the greatest risk* of over-hunting (see chapters 4 and 6). Hunting can have a greater negative impact on these species than the direct impacts (habitat disturbance) associated with silvicultural interventions (see chapters 15 and 20)

⁻ Hunting over time can have a *knock-down effect* on the wildlife population, where the elimination of large, easily-hunted animals leads to the hunting of smaller and less preferred species (see chapters 16-17)

Wildlife Responses to Logging Activities

Most wildlife species appear capable of surviving within logged landscapes, although their numbers may drastically decline (see chapters 4–14 and 24). There is no evidence to suggest that a species has gone extinct in direct response to timber-harvesting practices (Grieser Johns 1997:186). Historical studies of logging impacts on wildlife are often difficult to interpret, however, given different study designs, logging intensities, and environmental conditions (see chapters 4 and 18). In general, a species' response to logging is a function of habitat quality (food resources, competition, predation, cover, and breeding sites) and the availability of these resources during periods of forest recovery (see chapter 12).

Wildlife Groups or Guilds That Tend to Decline Following Light-Moderate Logging Intensities

The more intensive the logging, and the shorter the recovery period, the greater the impact of logging on interior-forest dependent wildlife becomes (see chapters 8, 14, 21, and 25). *Animals at particular risk are those that are rare, have low rates of reproduction, or have restricted home ranges (see chapters 5–6).*

The response of individual species to logging is dependent upon the intensity of logging (in time and space), the availability of refugia sites, hunting pressures, and the ecology of the species in question (behavior, habitat requirements). Wildlife species studied to date are reviewed in chapters 2 to 17, 24, and 29. In general, *groups that demonstrate negative responses to light-moderate logging intensities include:*

1. *Primates:* Competition for depleted resources and hunting pressure appears to have the greatest impact on large-bodied, slow-moving species, and those that consume a high percentage of seeds in their diets (see chapter 4).

2. *Large-to-medium terrestrial mammals:* Species at risk are those that participate in mass migrations such as pigs and peccaries (due to food loss and hunting pressure), selective browsers dependent on closed-forest conditions (e.g., Saola in Indochina), roadside feeders (easily hunted), and animals with restricted ranges (e.g., rhinoceros species in southeast Asia; see chapter 5).

3. *Small terrestrial and arboreal mammals* (rodents and marsupials): Members of the arboreal frugivore-granivore group and species with limited range distributions often decline due to competition for scarce food resources from generalist *edge* species (see chapters 6 and 24).

4. *Bats:* Specialized insectivores and vertebrate predators appear to decline following the loss of prey species and roosting sites after logging, while

some very specialized frugivores also decline due to elimination of understory roosts and/or the loss of critical food resources (see chapter 7).

5. *Birds:* Selection logging has its greatest impacts on terrestrial and understory birds, including mixed-flock assemblages, terrestrial insectivores, sallying insectivores, large-bodied species (including raptors in some areas), and, in some instances, arboreal frugivores. Habitat disruption leading to the loss of food resources, suitable nesting sites, and overhunting, contributes to the decline of these groups (see chapters 8, 9, 10, and 24).

6. *Herpetofauna* (amphibians and reptiles): Most forest-dependent species decline due to changes in microclimatic conditions (the forest floor is warmer and drier following canopy fragmentation, degrading cover sites and prey populations of litter arthropods), destruction of wetland habitats (especially important for larval development), and competition/ predation from invasive generalists (see chapter 11).

7. *Invertebrates:* Very little is known about the impacts of logging on invertebrates. In general, species sensitivity appears to be a function of population size, life-cycle stage, mobility, and impacts on food and cover resources. Groups with specialized habitat requirements such as moths, termites, ants, and soil litter earthworms (food resources and microclimatic conditions) may be at greatest risk (see chapters 12–13).

8. *Aquatic vertebrates:* Floodplain fish, species dependent on interstitial cobble spaces and deep pools for reproduction, and fishes requiring cool, oxygen-rich waters, are at greatest risk of decline following logging due to habitat degradation (especially siltation and changes in the temperature and nutrient regimes) and competition from invasive species (competitors and algal blooms; see chapter 14).

Groups or Guilds That Tend to Increase Following Light-Moderate Logging Intensities

Generalist species that use edge conditions and early successional resources tend to increase after logging. Their increase can offset the loss of interior-forest species, leading to an overall increase in wildlife diversity (Hunter 1990; see chapters 4–10, and 24). Examples of generalist species that tend to increase in response to logging include herbivores (browsing and grazing mammals where hunting is low), dispersers of pioneer plants, and exotic species (see chapters 5, 6, 8, 9, 12, and 16).

Species or groups of species that respond to logging in a predictable and consistent pattern, either decreasing or increasing following a harvest event, may serve as *indicators* of interior-habitat conditions within the residual forest. Over time, their responses should help to identify the degree of forest recovery. The potential of indicators to track changes in forest habitat conditions and ecological processes (their correlation to other plants, animals, and functions) remains in question, however (Lawton et al. 1998).

Declining Forest Wildlife Populations and Ecological Functions

Wildlife regulate numerous ecological functions that help to maintain the health and productivity of tropical forests through their actions as pollinators, dispersers, consumers, predators, and decomposers (Meffe and Carroll 1994; see chapters 3–14). They also serve *keystone* roles, creating environmental conditions that support the existence of other plant and animal species (e.g., elephant modifications of forest regeneration; Struhsaker 1997).

Changes in ecological processes and forest productivity due to reductions in select wildlife populations following logging are poorly understood (see chapters 11–12), especially as they relate to regeneration (see chapter 3, 8, and 12). Many wildlife species that are vulnerable to logging disturbances (see chapters 4–11, 14, and 24) are considered high-quality dispersers, even though their actual seed dispersal roles have seldom been measured (see chapter 3). This is especially true for large-bodied birds and mammals that are the primary dispersal agents of large-seeded, shade tolerant plants (see chapter 3). Some commercial tree species are even dependent on a very small group of animals for their dispersal, such as *Cola lizae* in Gabon, whose only known disperser is the lowland gorilla (*Gorilla gorilla gorilla*; Tutin et al. 1991), and *Hymenea courbaril* (Caesalpiniaceae) in Costa Rica that is principally dispersed by agoutis (Hallwachs 1993). Both dispersers are hunted for wild meat in logging camps (see chapters 15 and 17). To a limited extent, ecological redundancy within tropical forest systems may help to compensate for disruptions in ecosystem processes (see chapters 3 and 12), however questions of overall quality and quantity in these processes remain. Maintaining the long-term health and productivity of production forests requires the conservation of wildlife populations in production forest landscapes.

Forest Management Practices That Mitigate the Impacts of Logging on Wildlife and Their Habitat

A number of technical options exist for conserving wildlife within the production forest landscape. For these activities to succeed, several conditions must first be met.

Conditions for Wildlife Conservation in the Production Forest Landscape

The success of forest management *practices designed to mitigate the impacts of logging on wildlife and their habitat rely on several conditions:*

1. Timber production goals are secondary to *conservation goals* for areas of the permanent forest estate that contain globally or regionally outstanding biological diversity (see chapter 20).

2. Forestry departments assume the lead in *planning, implementing, and supervising* management practices on the permanent forest estate, which accounts for most of the production forest in tropical countries (see chapters 21 and 24).

3. Conservation measures can be undertaken using *existing technologies* and in a *cost-effective* manner (see chapter 28).

4. Planning and implementation processes are *transparent*, allowing for input from forest stakeholders (see chapters 21 and 26).

5. Management strategies are adapted to the specific requirements of each site. There is no *one size fits all* solution to timber production and wildlife conservation, with different areas of the permanent forest estate subject to different management strategies based on ecological and socioeconomic conditions (see chapter 20).

Wildlife Conservation in the Management Unit (or Concession)

The permanent forest estate is a matrix of forest-use zones, including production areas, community forests, and reserves (see chapters 20, 23, and 25). Production areas (commonly referred to as management units or concessions) are slated for timber-harvesting and other silvicultural interventions, and typically range in size from a few hundred hectares to well over 100,000 hectares. *The conservation of interior forest species is maximized where harvesting practices are distributed in a shifting-mosaic* (Bormann and Likens 1979; Harris 1984; see chapter 23) *that emphasizes connectivity between areas of similar habitat types* (especially those in a mature condition; see chapter 12). *Complementing this mosaic of harvest sites with a protected area network of reserves and corridors* (usually riparian) *that conserve critical habitats and late successional conditions provides refugia for many rare and logging-sensitive forest animals* (see chapter 23). All stages of forest succession, however—from recently logged to near-full recovery—conserve more wildlife habitat than forest converted to other land uses (e.g., plantations, agriculture, or pasture; see chapters 10 and 21).

The management unit is a miniature version of the production forest landscape, containing a shifting mosaic of production sites (annual cutting blocks—see below) and reserve areas. Forest wildlife is best conserved in management units when:

1. *Management contracts encourage investment* in forest management practices (e.g., integration of reduced-impact logging measures). To achieve this, the size of the management unit and length of the contract period must be compatible with the silvicultural prescriptions (adequate volume and area to allow for economic harvests and recovery periods between harvests, based upon the ecology of the exploited species; see chapters 8, 10, 19, and 29). Contractual periods of 30 to 50+ years, covering an appropriate land area, should be considered (see chapter 8-11, and 26). This approach helps to ensure a stable yield of wood products, which in turn encourages investment in forest management practices. These investments provide for the long-term health and productivity of the forest—an essential step toward minimizing the risk of forest conversion and habitat loss.

2. Management planning leads to the *delineation of production and reserve areas*. This process requires information drawn from forest

inventories, topographic maps, air photos, and in some instances, ground reconnaissance surveys (see chapters 19 and 23).

3. *Reserves comprise at least 10 percent of the land area*, in addition to those sites removed from harvesting due to low stocking or inaccessibility (chapters 16 and 21). To the extent possible, reserve areas should be interconnected to facilitate wildlife movements.

4. *Silvicultural prescriptions meet ecological, social and economic criteria* (see chapter 26). These treatments must be grounded in the ecology of the commercial species, natural disturbance patterns, and efforts to conserve the plant and animal communities comprising these forestlands (especially rare, threatened, or endangered species; see chapters 20-22). The challenge is to mimic natural disturbance events in the gentlest ways, and on the smallest possible scale (see chapters 21 and 22).

5. *Access is controlled.* Controlling access is probably the most cost-effective conservation measure available to limit illegal hunting, cutting, and land conversion (see chapters 8, 9, 11, 14, 16, 17, 21, and 26).

Conservation measures in community forests are similar to the management unit (or concession) noted throughout this section, with the exception that the scale of logging activities is likely to be smaller and less intensive than in commercial operations, and the emphasis on managing non-timber forest products is higher (see chapter 25). Agreements on exactly how the forest is to be managed, and how to monitor activities once they begin, are key to launching a successful community-based conservation management program.

Reserve Areas

Within management unit reserve areas are designed to conserve plant and animal species dependent on interior forest conditions; the genetic diversity in these populations; the ecological processes they mediate, and to serve as re-colonization centers during forest recovery (see chapters 6, 9, 12, 16, 23, 28, and box 30-1).

> **Box 30-1** *A quick checklist of habitats to maintain and protect in managed tropical forests.*
>
> Here is a quick checklist for forest managers of wildlife habitat categories that could be maintained or protected in managed tropical forests (categories pertain to PAN elements 1 and 2 [see chapter 23 text for explanation]).

PAN Element 1—Key Habitat Elements in
The Managed Forest Matrix

Snags and down logs
Trees prone to hollowing
Epiphyte patches
Large seed- and fruit-bearing trees
Large overstory/emergent trees
Sites and habitat elements for species critical to forest health and productivity
- Habitat for essential plant pollinators
- Habitat for essential seed dispersers
- Habitat for fungal spore dispersers
- Habitat for predators of seeds, seedlings, and animals
- Habitat for nutrient cyclers and insectivore prey
Individual trees or forest groves supporting colonies of birds

PAN Element 2—Zones of Protected Forest Within the
Managed Forest Matrix

Habitat for priority wildlife species
Buffers for streams and roads
Scarce and declining habitats
Unique wildlife habitats
- Riverine gallery forests
- Salt licks and mineral soil
- Caves and rock outcrops
- Natural forest openings
- Palm groves
- Lagoons and other water sources
- Low hills within seasonally flooded forests
Other considerations and criteria for delineating protection zones
- Habitat corridors and connections
- Natural disturbances
- Logging effects and general suitability of the logged forest matrix
- Sites of cultural or religious interest
- Sites of other use to humans

When considering *areas of wildlife habitat to reserve from production*, priority should be given to:

1. Sites known to support *rare or unique species and their habitat* (see discussion under "Stocking and topographic surveys").

2. *Representative areas of all forest types* should be maintained within the production landscape (see chapters 7 and 9). While biological diversity tends to be highest in lowland rain forests, all forest types support distinct communities. This includes the productive areas of the management unit, where reserve areas of this forest type help to conserve both wildlife and the genetic diversity of commercial timber species (especially those species having low natural stocking levels such as *Shorea leprosula*; Wickneswari et al. 1997). Areas of unique or high biodiversity that overlap with areas slated for silvicultural interventions are a logical starting point for reserve consideration; see chapters 20 and 23).

3. Larger *reserves* (100–500+km^2) conserve more species than small reserves (1–20km^2; Meffe and Carroll 1994). Large reserves are best suited for delineation at the landscape level prior to the allocation of management units. At the management unit level, small reserves serve to protect species of special interest, especially those that have small home ranges (see chapters 6, 16, and 21).

4. *Primary vegetation corridors* (especially evergreen riparian and floodplain sites), with minimum widths of 200 to 400m, where possible (Laurance and Bierregaard 1997; see chapters 2, 8, 9, 12, 14, 16, and 21), to facilitate the movement of interior-forest dependent species across the disturbed landscape. Narrower corridors (10–50 m wide) along streams and rivers also serve to conserve aquatic communities from siltation and environmental changes, while providing refugia for some terrestrial forest wildlife species (see chapters 14 and 21).

Controlling Hunting and Access

Logging operators must accept responsibility for enforcing existing hunting regulations and prohibiting commercial hunting within the concessions they manage (see chapters 10, 16, 21, and photo 30-1). Penalties or conservation bonds may be necessary to encourage enforcement of game regulations (see chapters 17 and 21). In general, a program *to limit hunting within a concession* should include measures to:

1. *Control access* within the harvest area. Individuals within concessions should be limited to logging personnel and regulatory staff. All roads should be closed at the conclusion of management activities to discourage trespassing (see chapters 8, 16, 17, 21, and 26).

PHOTO 30-1 Bongo (*Tragelaphus euryceros*) are an important game species to both subsistence and sport hunters. Logging roads greatly increase accessibility to interior forest and species such as bongo can experience rapid declines in their populations without measures to limit hunting pressures. (R. Carroll)

2. *Ban firearms* in the concession (except for protection purposes), and disperse workers throughout the concession to discourage social-oriented hunting (see chapter 16).

3. *Prohibit hunting of rare/threatened/endangered species,* as well as important seed dispersers (see chapters 3 and 15).

4. *Provide domestic sources of protein to workers* in logging camps as an alternative to wild meat (see chapters 15, 16, and 21).

5. *Link pay to productivity* in an effort to reduce time available to employees for hunting (see chapters 15–16).

6. *Assist indigenous peoples* with sustainable harvesting of their game resources. Document current use and population levels, help to set catch limits, discourage taking species that do not tolerate hunting pressure such as primates, and develop alternative sources of protein (see chapters 15 and 21).

7. *Increase conservation education programs* for forest employees and local communities, outlining the importance of wildlife to forest health and society as a whole (see chapters 16–17).

Wildlife Conservation in the Annual Cutting Block

The annual cutting block is that portion of the management unit or concession to be harvested within a one-year period. Conservation measures include the application of reduced-impact logging practices and the reservation of small protected areas or individual habitat elements (large snags, keystone trees; see box 30-1).

For each annual cutting block, a tactical plan outlining technical procedures and planning details for the harvesting operations must be developed (Sist et al. 1998), along with adequate training and manuals for supervisors and technicians so that they can institute the prescriptions developed for that forest (see chapters 18–19). The silvicultural procedures to be applied during the harvest must be based on the following information if disturbances to habitat structure and composition are to be minimized.

Stocking and Topographic Surveys

Stocking and topographic surveys assess the abundance, distribution, and accessibility of commercial species. These surveys should also identify potential areas to reserve from management activities (e.g., steep slopes, riparian areas, critical wildlife habitat), with these sites clearly delineated on maps and in the forest (see chapters 19 and 21). *Sites in the annual cutting block that contain one or more of the following properties should be considered for protection:*

1. *Rare, threatened, and endangered species* (see chapters 20 and 23). Provide buffer of one tree-length as a minimum to protect the habitat from external disturbance. Buffers of 100 to 200 m help to minimize disturbance to the forest microclimate (Laurance and Bierregaard 1997:508).

2. *Keystone plant species* (e.g., figs; see photo 30-2) and fruiting species in general (large-seeded, shade-tolerant species appear to be at greatest risk). Protect either through avoidance or by refraining from poisons and girdling during intermediate practices (see chapters 3, 7, 9, 21, and 23).

3. Probable *habitat of sensitive species* (e.g., caves), *unusual habitat types (e.g., rock-outcrops), and sites used by animals performing important ecological functions (*e.g., roosting sites used by pollinators or dispersers; see chapters 3, 7, and 23). Buffer as noted above.

4. *Nesting sites*, including trees with large cavities and emergants (see chapters 2, 9, 10, 21, 23, and 26).

PHOTO 30-2 Three important components of a healthy tropical forest: a) strangler fig (keystone resource); b) red-knobbed hornbill (*Phyticeros cassidix*) disperses the seeds of numerous tree species; and c) cavities in trees that serve as nest site for birds, mammals, and many invertebrate species. (M. F. Kinnaird)

5. *Standing and down coarse woody debris*. These individual habitat elements often support large faunal communities and should be left as undisturbed as possible (see chapters 12 and 23).

6. *Stream corridors* (buffer strips 10–40+ m wide) to prevent patch isolation while reducing nutrient, water, and sediment loads to streams (see chapters 8, 14, 21, and 26). Riparian areas also tend to be evergreen year-round, providing scarce habitat during dry periods (see chapters 2, 8, 12, 16, and 21).

7. Production forest between areas reserved for topographic, biological, and/or cultural reasons (to serve as *corridors to connect the reserve sites*; see chapter 16).

Reduced-Impact Logging Practices

Reduced-impact logging techniques (see box 30-2) help to protect residual vegetation (including advanced re-generation), minimize soil compaction and litter disturbance (70 percent of animals spend some or all of their life cycles in the soil or leaf litter; see chapters 11–13), and maintain overall site productivity. Conserving habitat minimizes the influence of logging on interior forest microclimate and microhabitat conditions that control populations of many vertebrates (especially birds, amphibians, and reptiles) and most invertebrate species (see chapters 7, 9, 11, and 12–14). To maximize their effectiveness, reduced-impact logging techniques must be refined to accommodate the silvicultural, stocking and topographic characteristics of the forest.

Box 30-2 Reduced-impact logging guidelines for conserving wildlife habitat.

Reduced-impact logging practices help minimize the impacts of logging on forest structure, composition, wildlife, and their habitat. Below is a generic overview of the three main components in a reduced-impact logging operation: planning, roads, and harvesting practices. To minimize logging impacts on the environment, these guidelines must be adapted to local site conditions. This requires careful planning, training, and supervision at all stages of the operation. For more detailed examples of reduced-impact logging guidelines, the reader is encouraged to review ITTO (1990b), FAO (1996), and Sist et al. (1998a, b).

Planning

Planning all activities in advance, and then training/supervising the cutting crews during logging implementation, are essential components for the success of reduced-impact logging measures (see chapter 21). Planning components are as follows:

—Conducting stock and topographic surveys in advance of cutting to identify the commercial species, areas of the annual cutting block that should be considered for cutting, and sites to reserve from logging (steep slopes, wetlands, wildlife habitat, cultural values; Sist et al. 1998).

—Selecting a silvicultural harvesting system based on the re-generation ecology of the timber species. Openings should be designed to promote natural re-generation or release advanced re-generation. These systems negate the need for enrichment plantings, which are both costly and destructive to understory wildlife habitat (see chapter 22).

—Selecting an extraction system compatible with the topography of the site to minimize excess damage to the residual stand. Where slopes regularly exceed 30 percent, consider a skyline cable system instead of ground-based skidders and tractors. Where they exceed 60–70 percent, reserve from cutting or cut sparingly using helicopters (Sist et al. 1998b).

—Avoiding the harvest on sensitive sites (e.g., wetlands and steep slopes), and during the rainy season when soils tend to compact (see chapters 14 and 16). When harvesting in the dry season, keep disturbances to a minimum as many invertebrates, reptiles and amphibians use the soil as a refuge against the heat (see chapters 11–13).

Road Design

Stock and topographic surveys are essential to the proper design of a logging road network. Engineering practices that minimize the impacts of roads on interior forest habitat include:

—Minimizing the total road network length by keeping roads as straight as possible

—Minimizing the total width of the roadway (7.5 m for major haul roads, 5 m for minor roads)

—Keeping the size of landing areas to a minimum (< 0.5 ha)

—Avoiding grades over 10 to 20 percent except for short distances (<500 m)

—Minimizing stream crossing. Where crossings are necessary, bridges should be constructed. In those limited instances where streams or rivers must be forged, crossings should only occur where the riverbed is firm

—Applying cross-drains and contouring the road surface to move water off the road. This can reduce erosion by 90 percent and dry roads quickly

—Keeping vehicles off roads during wet weather (decreases erosion and maintenance costs)

—Controlling access during and after silvicultural interventions (restricts hunting and forest conversion)

—Closing and stabilizing roads using cross-drains following cutting

For further explanation of these practices, see chapters 8, 9, 12, 14, 16, 17, 21, and 26.

Harvesting Practices

The felling and removal of every tree in a selection harvest should be carefully planned and executed to minimize damage to the felled tree, residual forest and its habitat.

- When selecting cut trees, reserve some individuals for post-harvest seed production (distribution is a function of advanced re-generation, seed production, and dispersal characteristics) and those supporting known nesting sites of important seed dispersers (e.g., hornbills; see chapters 2 and 21)

- Cut vines interconnecting target trees with surrounding canopy approximately one year in advance of the harvest to minimize crown damage to surrounding trees (see chapters 2, 8, and 21)

- Apply directional felling techniques to minimize damage during felling and to prepare stems for removal (see chapters 14, 21, and 24). Felling considerations include:

  - Flagging skid trails in advance of felling

  - Felling trees at a 30° angle to the proposed skid trail to facilitate removal

  - Felling into existing gaps where possible

  - Avoiding cutting trees that will fall down slope or across riparian areas

  - Buffering riparian areas with no cutting or reduced-intensity cutting (see box table 30-1)

  - Winching logs to skidder to minimize understory disturbance

- Create gaps consistent with the ecological requirements of the harvested species, keeping gaps to a minimum wherever possible (maintains connectivity for arboreal animals, while helping to maintain closed-forest conditions in the understory; see chapters 2, 6, 10, and 21). No species of tree appears to require gaps greater than 0.5 ha to regenerate. Specific considerations include:

  - Gaps less than 0.3 to 0.4 ha appear to minimize the development of vine tangles (Struhsaker 1997)

  - Edge effects from gaps penetrate 10 to 250 m into undisturbed forest, depending upon the variable or lifeform in question (Laurance and Bierregaard 1997:508)

BOX TABLE 30-1 *Buffer Zone Widths for Permanent Flowing Streams (modified from Sist et al. 1998)*

Stream Width (between banks in meters)	Buffer Width (to each side of stream in meters)
1–10	10
11–20	20
21–40	40
> 40	100

Postharvest Conservation Activities

The length of recovery periods between harvests is dependent upon the growth characteristics of the target species and the intensity of the previous harvest, and may range from 30 to 80+ years (see chapters 8, 9, 10, 11, and 19). *During the recovery period, post-harvest monitoring should be conducted to assess logging impacts on select fauna, their habitat, and ecological processes, in an effort to refine future silviculture practices.* Candidates for biological monitoring are a) rare, threatened, and endangered species (see chapter 15); *b)* species of commercial interest (see chapter 21), and c) animals sensitive to logging activities (especially earthworms and other decomposers, herpetofauna, and understory birds; see chapters 8-13). Guilds are generally easier to monitor and more informative indicators than individual species. Ecological processes (phenology, re-generation, and decomposition; see chapters 12, 13, 19, and 21), and structural habitat attributes (canopy closure, vertical structure, soil disturbance, stem mortality; see chapters 2 and 19) also merit monitoring. Partnerships with conservation organizations and other research institutions can help operators with the implementation and interpretation of monitoring programs (see chapters 19 and 26). Finally, protecting harvest sites from fire, conversion, and illegal interventions during their recovery phases is critical to maintaining a productive forest and wildlife habitat (see chapters 8, 9, 11, and 12).

Current Mechanisms for Integrating Wildlife Conservation Into Forest Management Practices

The vast majority of forest wildlife and their habitat can be conserved in managed production forests following the general forest management guidelines outlined above. While there is still a great need to conduct further research on technical forestry/wildlife practices so that silvicultural systems maximize the conservation of interior forest species (see box 30-3), the major stumbling blocks to wildlife conservation efforts in production forests lie in the socio-political arenas. *Public policy, economic pressures, and social conditions are the principal factors determining how timber harvests are conducted in most tropical nations. Breaking the status quo and positioning forest managers to adopt well-designed and implemented timber harvesting prac-*

tices is key to conserving wildlife and the forest resource in the long term (see box 30-3).

> **Box 30-3** *Research to advance the conservation of wildlife and their habitat in production forests.*
>
> Contributors to the volume identified several main areas where future research would help to: a) refine silvicultural treatments to reduce the impacts of logging on wildlife and their habitat, especially species of concern, and b) bring about change in the socio-political sector that drives these initiatives. Research priorities focus on study design, assessments of logging impacts on habitat and species ecology, and socio-political issues. A selection of the contributors' research recommendations are outlined below (see individual chapters for a detailed discussion of research needs).
>
> ## Experimental Designs
>
> The results presented in this book were derived from a wide range of studies assessing the response of select wildlife to forest management practices (usually selection harvesting systems with moderate- to low-supervision, although the impacts associated with other disturbance regimes, including forest conversion, were occasionally discussed). The methodologies employed in these studies were equally diverse, making it difficult to draw direct comparisons between sites, species, or different time periods. Contributors to the volume noted high priorities for future logging wildlife related research as follows:
>
> - Identifying where improved forest management practices are likely to be adopted in the near-term, and then concentrating research efforts in these areas (see chapters 7, 14, 18, 19, 20, and 24); demonstration forests at the major forest-type level, and *within concession* monitoring and adaptive management programs, is one possible approach for concentrating limited research funds and personnel (see chapter 19)
> - Identifying regional and local questions of ecological and economic interest to foresters, operators, and other forest stakeholders (see chapters 8 and 19)
> - Standardizing research methodologies to the greatest extent possible to facilitate comparisons between studies at the same site and between sites (see chapters 7 and 11).
> - Applying the standardized methodologies in well-replicated, long-term, pre-/post-studies of silvicultural interventions and their impacts on forest productivity, wildlife, and ecological processes (see chapters 4, 6, 7, 9–12, 15, 18, and 19).
> - Developing rapid assessment techniques and indicators capable of tracking the impacts of forest management practices on biodiversity (see chapters 19, 23, and 26)
> - Incorporating training and educational activities into all research efforts

Pre- Versus Postlogging Comparisons

Pre- versus postlogging comparisons seek to separate logging impacts on wildlife and their habitat from normal background variation due to biotic, edaphic, and topographic conditions in heterogeneous tropical forests.

Habitat characteristics. Basic information is lacking about the structure, composition, and functions of most tropical wildlife habitats, and how they respond to logging. Contributors identified the following as priority research areas to address these topics:

1. Structure

   - Changes in gap distribution, size, and quantity over time (see chapter 2)
   - The influence of forest management practices on resource availability (food, cover, mobility; see chapter 2)
   - The maximum distance between usable habitat patches in a disturbed landscape (for species of interest; see chapter 2)
   - The penetration of edge effects in the wake of logging, and its influence on habitat quality (see chapter 6)

2. Composition

   - The role of vines within logged and unlogged forests (food, cover, connectivity; see chapter 2)
   - re-generation dynamics before and after silvicultural interventions, including:

       - The reproductive biology of tropical timber species (pollinators, primary and secondary dispersers; see chapters 3 and 21)

       - Seed dispersers and dispersal mechanisms (quantity and quality, including commercial species; see chapters 3, 9, and 21)

       - The availability of recruitment sites and seedbed requirements (see chapter 3)

       - The correlation between seedtrees and re-generation establishment (see chapters 3 and 21)

3. General habitat

   - Wildlife habitat requirements (food, cover, breeding site requirements of individual species and guilds; see chapters 3, 5, 8-11, 18, 19, and 23)
   - The influence of logging on the microhabitat (structure, composition, climate) of species considered at risk from logging disturbance (see chapters 8 and 9)
   - Comparisons of logging and natural disturbance events, including rates of recovery (see chapters 2, 13, 14, 19, and 22)

4. Protected area networks

   - The role of reserves and corridors as refugia, breeding and re-colonization sites, and areas that maintain ecological processes (see chapters 4, 6, 9, and 23)

⎯The optimum size and spatial distribution of reserves within production landscapes (see chapters 2, 8, 19, 21, 23, and 26)

⎯The magnitude of edge effects and their influence on reserve habitat quality (see chapter 6)

⎯The role of protected area networks (PANs) in contributing to timber production (see chapter 23)

Species ecology and ecological processes. Basic information is lacking about the ecology of most tropical wildlife species, the ecological functions that they perform, and how these respond to logging. Contributors identified the following as priority research areas to address these topics:

1. Basic ecology

⎯Taxonomic clarification for certain wildlife groups (e.g., bats and invertebrates; see chapters 7 and 12)

⎯The basic ecology of forest dwelling organisms (habitat requirements, role in ecological processes, reproductive biology), and their response to disturbance events (see chapters 5, 6, 7, 11, 12, and 23); priority species are those that are rare, threatened, or endangered; have commercial or cultural value; serve as game, and perform important pollination, dispersal, or keystone functions

2. Ecological processes

⎯For species of interest, clarify the ecological functions they perform in logged and unlogged environments (see chapters 3, 6, 7, 12, and 23), especially those related to foodwebs and decomposition processes (see chapters 11, 12, and 14), and predator-prey relationships (see chapter 8)

⎯Keystone resources and their role in natural and managed systems (see chapters 3, 8, and 26)

⎯The capacity of compensatory mechanisms to maintain ecological processes following logging (see chapter 12)

⎯Cascading effects on animal communities and their resources following disturbance (see chapter 12)

Sociopolitical Issues

In many developing nations, logging practices are driven more by the social, political, and economic sectors than by the science of forestry. Contributors identified the following as priority research areas to bring greater balance between timber management and wildlife conservation in production forests.

Commitment and enforcement.

⎯How to develop long-term commitment to forest management goals (including supporting foresters in their efforts) in the face of high interest rates, political instability, and questions of land tenure rights (see chapters 25 and 27)

⎯How to ensure effective control and enforcement at the management unit (concession) level (see chapter 29)

Stakeholder involvement.

— How to adequately involve stakeholders in the forest management process, and to increase the transparency of these operations (see chapters 18, 19, and 25)

— The question of commercial hunting/trading in wildlife products as an inalienable right of indigenous peoples (see chapter 29)

Training/education.

— How to develop education and training programs to advance conservation via social participation in the management of natural resources (see chapters 25 and 29)

— How to best advance technical training for scientists (including foresters) in developing nations (basic taxonomy, conservation biology, ecosystem management, biometrics; see chapters 6, 7, 18, and 19), including the development of regional and local manuals that outline best management guidelines for the implementation of silvicultural practices (see chapter 26)

Economics and financial support.

— How to develop funding mechanisms to support applied research efforts for achieving sustainable forest management and wildlife conservation in production forests (see chapters 6 and 25)

— How to best market conservation values associated with improvements in forest management practices to both forest operators and society as a whole (see chapters 26–28)

— How to improve the profile and marketability of products from well-managed forests (see chapters 25 and 26)

There are presently several incentives available to promote logging practices that help conserve wildlife and their habitat. Incentives imply benefits, and are likely to be most successful where they are sympathetic to and mutually reinforcing at all levels, from loggers and forest operators to foresters and society as a whole (see chapter 28). Incentives promoting the conservation of wildlife and their habitat in production forests include considerations of multiple-use management, reforms to concession contracts and taxation schedules, increased profitability through applications of reduced-impact logging, and direct support (technical, financial, etc.) from donors and conservation NGOs.

Multiple-Use Management

Forests provide a multitude of products and services in addition to timber (see chapters 20, 21, 23, 27, and 28): biodiversity prospecting (economics) and protection (ecological functions, genetic diversity), sequestration of carbon, non-timber forest products (including wildlife products), and watershed protection (to name but a few). Considering these factors in forest management planning, with the goal of minimizing the impacts of logging on them, can pay dividends to landowners and society. Watershed planning and protection, for example, reduces risks of flooding, clogging of navigable waterways, and disruption of hydropower production, while conserving aquatic communities (see chapters 14 and 21). Studies of wildlife are helping to identify sustainable hunting quotas, which can then be incorporated into management plans to permit controlled harvest of this resource (see chapters 15 and 17). Permanent sample plots are beginning to provide useful information to forest managers (and hence to the refinement of the management process) on commercial species' growth-and-yields, rare/threatened/endangered species' responses to forest management practices, and changes in ecological processes in response to these activities (see chapter 3). Finally, natural forests are more effective at storing carbon than plantations, with well-managed and old growth forests among the world's largest reservoirs for carbon (see chapter 27). *It is thus prudent to conserve forest wildlife and other forest resources, given our limited knowledge of their roles in maintaining the health and productivity of our environment* (pollination, seed dispersal, pest control, decomposition, and nutrient cycling), *and their potential to yield direct and indirect economic benefits to forest operators and society at large* (see chapters 3, 7, 8, 16, 17, 21, and 23).

Concession Contracts and Taxation Schedules

Wildlife conservation in tropical forests is dependent upon a stable and profitable forest products industry. *Reforms are needed for concession contracts and taxation schedules* in an effort to promote further investment in the forestry sector. Recommendations include:

1. Concessions should be awarded through a *competitive process* that maximizes revenue generation while rewarding good management practices (see chapter 29).

2. Concessions should be allocated for *long periods of time* (30–50+ years), as this provides a stable resource base for investment by the operator in infrastructure (including local communities) and forest management activities (regeneration and recruitment become important for future cuts). The operator becomes vested in forest management through this process. This approach necessitates large concessions to allow adequate recovery periods between harvests (see chapter 21).

3. Concession operators should pay an *annual fee (tax) based on the area* under management, its productivity, and ease of access. The fee could initially be low to encourage infrastructure development (roads, mills), and gradually rise as a company's profits increase (M. Fay, personal communication). This easy-to-implement-and-enforce process also provides incentives for establishing small, protected areas within the management unit (wetlands, grottos and steep slopes, riparian areas, unique habitats), as non-production areas could be removed from the fee schedule (see chapters 20, 21, 23, and 29).

4. Further *tax incentives* might be provided to concession holders willing to reserve areas of productive forest from logging for periods of 10 or more years, while strong penalties would be imposed for areas of the forest converted to other land use (see chapter 29).

5. *Conservation bonds* posted by the concession operator would help to enforce existing environmental regulations (return of bond based on pre-/postevaluations of forest condition; see chapter 17).

6. A *wild meat tax* would help to provide incentives for consumption of domestic meat sources, with some of the revenues generated going back to rural communities (see chapter 17).

Reduced-Impact Logging Practices

Many forest management activities aimed at reducing environmental impacts are among the most profitable to forest operators (see chapter 21). The *planning and supervision costs associated with reduced-impact logging practices can be offset by:*

- Creating more efficient road networks
- Reducing machine time and wear
- Reducing labor costs
- Recovering higher volumes of material
- Streamlining skidding and yarding operations
- Reducing residual stand damage (future crop tree protection), and
- Reducing the need for expensive and problematic (and ecologically degrading) artificial regeneration programs (see chapters 2, 8, 21, 23, and 28)

Perhaps the greatest profitability however is the opportunity to reharvest a site multiple times, cutting more volume and higher quality trees than in secondary forests degraded by poorly-conducted logging practices (many of which will be converted to other land-uses, given their low timber values).

Donor and Conservation NGO Support

Donor organizations and conservation NGOs can help to directly support wildlife conservation in production forests by:

1. *Training* foresters, research scientists, policy makers, and local stakeholders to identify areas of high biological diversity or high sensitivity to disturbance, techniques to inventory and monitor wildlife resources to ensure their inclusion in the management process, and the production of manuals to serve as guidelines for the implementation of improved forestry practices (see chapters 17–19 and 29)

2. *Leveraging policy reforms* with donations to build national capacity and implement a nationwide biodiversity conservation strategy (see chapter 19, 21, and 29)

3. *Pressuring expatriate firms* to comply with existing conservation regulations within the countries in which they are working (see chapter 17)

4. *Supporting carbon-offset initiatives* devoted to natural forest conservation. Low capital costs associated with forestry practices may help to attract donors interested in carbon-offset projects designed to advance sustainable forest management principles (see chapters 20, 21, 26, and 27). The high public-relations value associated with wildlife conservation has attracted attention in the carbon-sequestering arena to assist in improving forest management efforts (see chapter 27)

5. *Promoting transparency* in the forestry sector. Certification raises the quality and transparency of forest management practices and efforts at wildlife conservation (see chapters 15, 17, 20, 21, 23, 26, 28, and 29). This process can be of direct benefit to concessionaires by alleviating the need for national audits by government foresters and increasing the value of wood products (see chapters 26 and 29). Future plans to certify national-level resource managers should help to lower the costs associated with certification processes (see chapter 26).

Future Prospects for Integrating Wildlife Conservation Into Forest Management Practices

Contributions to this volume highlight the impacts of logging on forest wildlife in the tropics, and to a lesser extent how these impacts may influence ecological processes, concluding that wildlife can be conserved in production forest landscapes using a variety of planning and technical approaches ranging from reserves to reduced-impact logging. There are also biological, financial, social, and political incentives for promoting these conservation efforts. So why is there not a spontaneous adoption of reduced-impact logging practices and other measures designed to move forestry closer to a state of sustainable forest management?

Limiting Factors

There are a number of inter-dependent factors that serve as *obstacles to improvements in the forestry sector.* Some of the more prominent limiting factors include:

- Large upfront financial investments necessary to implement reduced-impact logging
- Too few well-trained professional managers
- Ineffective taxation systems
- Poor timber pricing
- High discount rates
- Unstable political bureaucracies built on corruption and perceptions that forests are both limitless and best converted to other land uses (Myers 1996; see chapter 18 and box 18-1)

Because of these conditions, foresters have often been unable to directly control forest management activities, and governments have not invested in the training, infrastructure, and research necessary to implement scientific forest management practices. Finally, in the international conservation arena, a polarized controversy continues over sustainable forest management vs. long-term forest preservation (Putz 1994; Rice et al. 1997; Bowles et al. 1998; Cabarle 1998). This leaves both donor agencies and governments sitting on the fence waiting for the scientific community to come to some agreement as to how tropical forests should best be managed.

Breaking the "Log Jam"

Current trends in population growth and natural resource consumption indicate that the majority of tropical forests will be targeted for some form of forest management in the near future. For this reason, most contributors believe that the choice will rarely be between "to cut or not to cut," but rather to identify approaches where the forest landscape is managed at different levels of intensities—preservation in select areas of high biodiversity, conversion in areas of high human pressure and low biodiversity, and most areas placed under some level of sustainable forest management (see chapter 20). To implement this scenario, some options for promoting wildlife conservation and sustainable forestry are being explored (as outlined above and throughout the volume), however these piecemeal initiatives are too few and too small to bring about the rapid changes that are needed in the forestry sector. Continuing along the current path will eventually bring sustainability to a sub-set of the current tropical forests, with many areas irreversibly degraded through species and habitat loss. To help catalyze widespread change, governments, international forestry agencies, conservation NGOs, development agencies (European Union, USAID, JICA), and large donors (World Bank, IMF, regional reconstruction banks) must immediately adopt unified standards as to what are and are not acceptable logging practices in tropical forests. These minimum technological and social standards (perhaps drawing upon the principals and criteria set forth by organizations such as the Forest Stewardship Council, ITTO, and FAO; see chapters 21 and 26) would then become an integral component of all loans and donations. They could also become a requirement for admission into global trade organizations such as the World Trade Organization, Commonwealth of Nations, and similar institutions (bringing greater transparency to the process). Tropical forests are of local, national, and global concern, and the time is right for the social-political-economic sectors to apply their influence to the conservation of wildlife and other forest resources. It is not an overstatement to say that, in the near future, the economic and biological health of the planet and many nations will depend on the way that tropical forests are used.

Literature Cited

Abd. Ghani, A. N., R. Mohd, S. Mohamed, F. R. Ibrahim, and M. Z. Hussin. 1998. Economic valuation of forest goods and services in Malaysia and its implications on Sustainable Forest Management. Paper presented at the Workshop on Forestry Economics and Policy: R&D Towards Achieving Sustainable Forest Management, 10–11 November 1998. Serdang, Selangor, Malaysia: Universiti Putra Malaysia.

Adams, P. W. and J. O. Ringer. 1994. The effects of timber harvesting and forest roads on water quantity and quality in the Pacific Northwest: Summary and annotated bibliography. Special publication of the Forest Engineering Department. Corvallis, Ore.: Oregon State University.

Adams, P. W., R. L. Beschta, and H.A. Forelich. 1988. Mountain logging near streams: Opportunities and challenges. In Oregon State University and the International Union of Forestry Research Organizations, eds., *Proceedings of the International Mountain Logging and Pacific Northwest Skyline Symposium*, December 1998, pp. 153–162. Portland, Ore.: Oregon State University.

Adler, G. H. and D. W. Kestell. 1998. Fates of Neotropical tree seeds influenced by spiny rats (*Proechimys semispinosus*). *Biotropica* 30:677–681.

ADB (Asian Development Bank). 1994. Yunnan-Simao forestation and sustainable wood utilization project (People's Republic of China). Project document. Manila: Asian Development Bank.

Aeschlimann, A. 1963. Observations sur *Philantomba maxwelli* (Hamilton-Smith), une antilope de la forêt éburnéenne. *Acta Tropica* 20:341–368.

Ahmad, A. H. 1994. The ecology of mousedeer (*Tragulus* species) in a Bornean rainforest, Sabah, Malaysia. MPhil thesis, University of Aberdeen.

Aide, T. M. and J. K. Zimmerman. 1990. Patterns of insect herbivory, growth, and survivorship in juveniles of a neotropical liana. *Ecology* 71:1412–1421.

Alcorn, J. B. 1993. Indigenous peoples and conservation. *Conservation Biology* 7:424–426.

Alexander, B. 1994. People of the Amazon fight to save the flooded forest. *Science* 264:606–607.

Alexandre, D.Y. 1978. Le role disseminateur des elephants en foret de Tai, Cote d'Ivoire. *Terre et Vie* 32:47–72.

Alho, C. J. R. 1982. Brazilian rodents: their habitats and habits. In M. A. Mares and H. H. Genoways, eds., *Mammalian Biology in South America*, Special Publication Series Volume 6, pp. 143–166. Pittsburgh, Pa.: University of Pittsburgh Pymatuning Laboratory of Ecology.

Alho, C. J. R., T. E. Lacher Jr., and H. C. Goncalves. 1988. Environmental degradation in the Pantanal ecosystem. *BioScience* 38:164–171.

Allan, D. G., J. A. Harrison, R. A. Navarro, B.W. van Wilgen, and M. W. Thompson. 1997. The impact of commercial afforestation on bird populations in Mpumalanga province, South Africa—insights from bird-atlas data. *Biological Conservation* 79:173–185.

Allmon, W. D. 1991. A plot study of forest floor litter frogs, Central Amazon, Brazil. *Journal of Tropical Ecology* 7:503–522.

Allport, G., M. Ausden, P. V. Hayman, P. Robertson, and P. Wood. 1989. The conservation of the birds of Gola Forest, Sierra Leone. Cambridge, United Kingdom: ICBP (International Council for Bird Preservation) study report no. 38.

Alvard, M. S. 1994. Conservation by native peoples: Prey choice in a depleted habitat. *Human Nature* 5:127–154.

Alvarez, J. and M. R. Willig. 1993. Effects of treefall gaps on the density of land snails in the Luquillo Experimental Forest of Puerto-Rico. *Biotropica* 25:100–110.

Ammann, H. 1985. Contributions to the ecology and sociology of the Javan rhinoceros (*Rhinoceros sondiacus Desm.*). Ph.D. diss., University of Basel.

Ananthakrishnan, T. N. 1996. *Forest Litter Insect Communities: Biology and Chemical Ecology.* Lebanon, N.H.: Science Publishers, Inc.

Andaluz, A. 1996a. Marco legal e institucional para la conservación y el uso sostenible de los bosques y tierras forestales. In BOLFOR, eds., *Hacia el Manejo Forestal Sostenible,* 139–175. Santa Cruz, Bolivia: BOLFOR (Bolivia Sustainable Forestry Management Project), MDSMA (Ministerio de Desarrollo Sostenible y Medio Ambiente).

——. 1996b. Mecanismos de control social en la nueva ley forestal. *Boletín BOLFOR (Bolivia Sustainable Forestry Management Project)* 7:8–9.

Andaluz, A., W. Cordero, and R. Saucedo. 1998. Manual práctico sobre la Resolución Ministerial #133/97 sobre las Agrupaciones Sociales del Lugar (ASL). Miscellaneous report. Santa Cruz, Bolivia: BOLFOR (Bolivia Sustainable Forestry Management Project).

Andaluz, A., E. Vásquez, and V. C. Aguilar. 1996. Evaluación de la aplicación del decreto de reglamentación de la Pausa Ecológica Histórica. Sector Forestal (D.S. 22884). Documento técnico 11/15. Santa Cruz, Bolivia: BOLFOR (Bolivia Sustainable Forestry Management Project).

Andrasko, K. 1995. Program notes. USIJI (United States Initiative on Joint Implementation) Program Conference. Washington, D.C.: USIJI Secretariat.

——. 1997. Forest management for greenhouse gas benefits: Resolving monitoring and issues across project and national boundries. *Mitigation and Adaptation Strategies for Global Change* Special Issue 2:117–132.

Andren, H. 1994. Effects of habitat fragmentation on birds and mammals in landscapes with different proportions of suitable habitats: A review. *Oikos* 71:355–366.

Andrén, O., J. Bengston, and M. Clarholm. 1995. Biodiversity and species redundancy among litter decomposers. In H. P. Collins, G. P. Robertson and M. J. Kluge, eds., *The Significance and Regulation of Soil Biodiversity,* pp. 141–151. The Netherlands: Kluwer Academic Publishers.

Anonymous. 1983. Guidelines for the selective logging of rainforest areas in north Queensland state forests and timber reserves. Brisbane, Australia: Queensland Forest Service.

——. 1991. Harvesting, marketing and resources management manual. Brisbane, Australia: Queensland Forest Service.

——. 1995. *Information Malaysia.* Kuala Lumpur, Malaysia: Berita Publishing Sdn. Bhd.

Ansell, W. F. H. 1950. The Yellow-backed duiker—elusive and secretive. *African Wild Life* 4:151–153.

Anstey, S. 1991. *Wildlife Utilization in Liberia.* Gland, Switzerland: WWF (World Wildlife Fund) International.

Anthony, H. E. 1929. Two genera of rodents from South America. *American Museum Novitates* 383:1–6.

API Dimako. 1995. Généralités sur l'a ménagment des forêts de production de la province de l'Est. Yaounde, Cameroon: Projet API de Dimako, Ministere de l'Environnement et des Forêts.

Appanah, S. and F. E. Putz. 1984. Climber abundance in virgin dipterocarp forest and the effect of pre-felling climber cutting on logging damage. *Malaysian Forester* 47:335–342.

Araujo-Lima, C. and M. Goulding. 1997. *So Fruitful a Fish.* New York, N.Y.: Columbia University Press.

Arcese, P. and A. R. E. Sinclair. 1997. The role of protected areas as ecological baselines. *Journal of Wildlife Management* 61:587–602.

Arentz, F. 1992. Low impact logging-helicopter logging in Papua New Guinea. *ITTO (International Tropical Timber Organization) Tropical Forest Management Update* 6:6.

——. 1994. Papua New Guinea forests on the line. *ITTO (International Tropical Timber Organization) Tropical Forest Update* 4:25.

Arita, H. T., J. G. Robinson, and K. H. Redford. 1990. Rarity in neotropical forest mammals and its ecological correlates. *Conservation Biology* 4:181–192.

Armitage, I. and M. Kuswanda. 1989. Forest management for sustainable production and conservation in Indonesia. UTF/INS/065: INS Forestry Studies. Field Document no. 1–2. Jakarta, Indonesia: Ministry of Forestry, Directorate General of Forest Utilization.

Ascorra, C. F., D. L. Gorchov, and F. Cornejo. 1993. The bats from Jenaro Herrera, Loreto, Peru. *Mammalia* 4:533–552.

Ascorra, C. F. and D. E. Wilson. 1992. Bat frugivory and seed dispersal in the Amazon, Loreto, Perú. Publicaciones del Museo de Historia Natural, Universidad Nacional Mayor de San Marcos, *Serie A Zoología* 43:1–6.

Ascorra, C. F., D. E. Wilson, and M. Romo. 1991. Lista anotada de los quirópteros del Parque Nacional Manú, Perú. Publicaciones del Museo de Historia Natural, Universidad Nacional Mayor de San Marcos, *Serie A Zoología* 42:1–14.

Ashton, P. S. 1988. Dipterocarp biology as a window to the understanding of tropical forest structure. *Annual Review of Ecology and Systematics* 19:347–370.

Asia-Pacific Forestry Commission. 1997. DRAFT: Code of practice for forest harvesting in Asia Pacific. Rome, Italy: FAO (Food and Agricultural Organization).

Asibey, E. O. A. 1974. The grasscutter, *Thyronomys swinderianus,* Temmick, in Ghana. *Symposium of the Zoological Society of London* 34:161–170.

——. 1978. Primate conservation in Ghana. In D. J. Chivers and W. Lane-Petter, eds., *Recent Advances in Primatology,* Volume 2, pp. 55–59. London, United Kingdom: Academic Press.

Asquith, N. M., J. Terborgh, A. E. Arnold, and C. M. Riveros. 1999. The fruits the agouti ate: *Hymenaea courbaril* seed fate when its disperser is absent. *Journal of Tropical Ecology* 15:229–235.

Asquith, N. M., S. J. Wright, and M. J. Clauss. 1997. Does mammal community composition control recruitment in neotropical forests? Evidence from Panama. *Ecology* 78:941–946.

Assitou, N. and J. G. Sidle. 1995. Congo game conservation: limits of a legal approach. In J. A. Bissonette, and P. R. Krausman, eds., *Integrating People and Wildlife for a Sustainable Future*, pp. 509–511. Bethesda, Md.: The Wildlife Society.

Augspurger, C. K. and S. E. Franson. 1988. Input of wind-dispersed seeds into light gaps and forest sites in a neotropical forest. *Journal of Tropical Ecology* 4: 239–252.

Aulerich, E. 1995. Applying skylines to partial cuts in the tropics. *Asian Timber* (November): 41–44.

Auzel, P. 1996a. Agriculture/extractivisme et exploitation forestière. Etude de la dynamique des modes d'exploitation du milieu dans le nord de lUFA de Pokola, nord Congo. WCS (Wildlife Conservation Society)/GEF (Global Environment Fund) Congo. Bomassa: Republic of Congo.

——. 1996b. Evaluation de l'impact de la chasse sur la faune des forêts dAfrique Centrale, nord Congo. Mise au point de méthodes basées sur l'analyse des pratiques et les résultats des chasseurs locaux. WCS (Wildlife Conservation Society)/GEF (Global Environment Fund) Congo. Bomassa: Republic of Congo.

Avila-Pires, T. C. S. 1995. Lizards of Brazilian Amazonia (Reptilia: Squamata). *Zoologische Verhandelingen Leiden* 299:1–706.

Ayres, J. M. 1986. Uakaris and Amazonian flooded forest. Ph.D. diss., University of Cambridge.

Ayres, J. M., D. De Magalhães Lima, E. De Souza Martins, and J. L. K. Barreiros. 1990. On the track of the road: Changes in subsistence hunting in a Brazilian Amazonian Village. In J.G. Robinson and K.H. Redford, eds., *Neotropical Wildlife Use and Conservation,* pp. 82–92. Chicago, Ill.: University of Chicago Press.

Babweteera, F. 1998. Influence of gap size and age on the diversity and abundance of climbers in the Budongo Forest Reserve, Uganda. Unpublished MSc thesis, Makerere University.

Baillie, J. and J. Groombridge. 1996. *1996 IUCN (International Union for Conservation of Nature and Natural Resources) List of Threatened Mammals* Gland and Cambridge: IUCN (International Union for Conservation of Nature and Natural Resources).

Baillie, S. R. 1991. Monitoring terrestrial breeding bird populations. In F. B. Goldsmith ed., *Monitoring for Conservation and Ecology*, pp. 112–132. London, United Kingdom: Chapman and Hall.

Baker, W. L. 1989. Landscape ecology and nature reserve design in the Boundary Waters Canoe Area, Minnesota. *Ecology* 70:23–35.

Balasubramanian, P. and P. V. Bole. 1993. Seed dispersal by mammals at Point Calimere Wildlife Sanctuary, Tamil Nadu. *Journal of the Bombay Natural History Society* 90:33–44

Balmford, A. 1996. Extinction filters and current resilience: The significance of past selection pressures for conservation biology. *Trends in Ecology and Evolution* 11:193–196.

Balun, L., P. Emrik, and L. Orsak. 1996. A study on plant species diversity and spatial patterns in rain forest communities from Sulka Area in New Britian Island, Papua New Guinea. Rabul, Papua New Guinea: Pacific Heritage Foundation.

Barbier, E. B. 1993. Economic aspects of tropical deforestation in South-East Asia. *Global Ecology and Biogeography Letters* 3:1–20.

Barbier, E. B., J. C. Burgess, J. Bishop, and B. Aylward. 1994. *The Economics of the Tropical Timber Trade*. London, United Kingdom: Earthspan Publications.

Barnes, B. V., S. Spurr, D. Zak, and S. Denton. 1998. *Forest Ecology.* 4th Edition. New York, N.Y.: John Wiley and Sons.

Barnes, R. F. W. 1990. Deforestation trends in tropical Africa. *African Journal of Ecology* 28:161–173.

Barnes, R. F. W, K. L. Barnes, M. P. T. Alers, and A. Blom. 1991. Man determines the distribution of elephants in the rain forests of northeastern Gabon. *African Journal of Ecology* 29:54–63.

Barr, K., H. Moller, E. Christmas, P. Lyver, and J. Beggs. 1996. Impacts of introduced common wasps (*Vespula vulgaris*) on experimentally placed mealworms in a New Zealand beech forest. *Oecologia* 105:266–270.

Barreto, P., P. Amaral, E. Vidal, and C. Uhl. 1998. Costs and benefits of forest management for timber production in eastern Amazonia. *Forest Ecology and Management* 108:9–26.

Barrette, C. 1977. Some aspects of the behaviour of muntjacs in Wilpattu National Park. *Mammalia* 41:1–34.

Barros, A. C. and C. Uhl. 1995. Logging along the Amazon River and estuary: Patterns, problems and potential. *Forest Ecology and Management* 77:87–105.

Barthem, R. and M. Goulding. 1997. *The Catfish Connection*. New York, N.Y.: Columbia University Press.

Bastable, H. G., W. J. Shuttleworth, R. L. G. Dallarosa, and C. A. Nobre. 1993. Observations of climate, albedo, and surface radiation over cleared and undisturbed Amazonian forest. *International Journal of Climatology* 13:783–796.

Basu, P., E. Blanchart, and M. Lepage. 1996. Termite (Isoptera) community in the Western Ghats, South India: Influence of anthropogenic disturbance of natural vegetation. *European Journal of Soil Biology* 32:113–121.

Baudenon, P. 1958. Écologie des petites et moyennes antilopes dans le Togo meridional. *Mammalia* 22:285–293.

Baur, G. N. 1964.*The Ecological Basis of Rainforest Management*. Rome, Italy: FAO (Food and Agricultural Organization).

Bawa, K. S. 1990. Plant-pollinator interactions in tropical rain forests. *Annual Review of Ecology and Systematics* 21:399–422.

Bawa, K. S. and S. L. Krugman. 1991. Reproductive biology and genetics of tropical trees in relation to conservation and management. In A. Gómez-Pompa, T. C. Whitmore, and M. Hadley, eds., *Rain Forest Regeneration and Management*, pp. 119–136. M. A. B. (Man and the Biosphere) series Vol. 6, Parthenon Publishing Group, Paris, France: UNESCO (United Nations Educational, Scientific, and Cultural Organization).

Bawa, K. S. and R. Seidler. 1998. Natural forest management and conservation of biodiversity in tropical forests. *Conservation Biology* 12:46–55.

Bawa, K. S., D. R. Perry, and J. H. Beach. 1985. Reproductive biology of tropical lowland rain forest trees. 1. Sexual systems and incompatibility mechanisms. *American Journal Botany* 72:331–345.

Bazzaz, F. A. 1990. Regeneration of tropical forests: Physiological responses of pioneer and secondary species. In A. Gomez-Pompa, T. C. Whitmore, and M. Hadley, eds., *Rainforest Regeneration and Management*, pp. 91–118. M. A. B. (Man and the Biosphere) Series Vol. 6 Paris, France: Parthenon Publishing Group. Paris, France: UNESCO (United Nations Educational, Scientific, and Cultural Organization).

BCN (Biodiversity Conservation Network). 1997a. Annual report: Getting down to business. Washington D.C.: Biodiversity Conservation Network, Biodiversity Support Program.

——. 1997b. Biodiversity Conservation Network: Evaluating issues of business, the environment, and local communities. A website at: www.BCNet.org.

Beaman, R. S., J. H. Beaman, C. W. Marsh, and P. V. Woods. 1985. Drought and forest fires in Sabah in 1983. *Sabah Society Journal* 8:10–30.

Beattie, A. J. and I. Oliver. 1994. Taxonomic minimalism. *Trends in Ecology and Evolution* 9:488–490.

Beccaloni, G. W. 1997. Vertical stratification of the ithomiine butterfly (Nymphalidae: Ithomiinae) mimicry complexes: The relationship between adult flight height and larval host-plant height. *Biological Journal of the Linnean Society* 62:313–341.

Bedward, M., R. L. Pressey, and D. A. Keith. 1992. A new approach for selecting fully representative reserve networks: Addressing efficiency, reserve design and land suitability with an iterative analysis. *Biological Conservation* 62:115–125.

Beehler, B. M. 1983. Frugivory and polygamy in birds of paradise. *Auk* 100:1–12.

Beehler, B. M., K. S. R. Krishna Raju, and S. Ali. 1987. Avian use of man-disturbed forest habitats in the Eastern Ghats, India. *Ibis* 129:197–211.

Beir, Beier, P., and R. F. Noss. 1998. Do habitat corridors provide connectivity? *Conservation Biology* 12(6):1241–1252.

Belshaw, R. and B. Bolton. 1993. The effect of forest disturbance on the leaf litter ant fauna in Ghana. *Biodiversity and Conservation* 2:656–666.

Benavides, M. and M. Pariona. 1995. *The Yanesha Forest Cooperative and Community-based Management in the Central Peruvian Forest*. Madison, WI: Land Tenure Center, University of Wisconsin.

Bengston, J., H. Setälä, and D. W. Zheng. 1996. Food webs and nutrient cycling in soils: interactions and positive feedbacks. In G. A. Polis and K. O. Winemiller, eds., *Food Webs: Integration of Patterns and Dynamics*, pp. 30–38. New York, N.Y.: Chapman and Hall.

Bennett, Hennessey, A. 1995. A study of the meat trade in Ouesso, Republic of Congo. Miscellaneous report. Bronx, N.Y.: WCS (Wildlife Conservation Society).

Bennett, A. F. 1987. Conservation of mammals within a fragmented forest environment: the contributions of insular bio-geography and autecology. In S. A. Saunders, G. W. Arnold, A. A. Burbidge, and A. J. M. Hopkins, eds., *Nature Conservation: The Role of Remnants of Native Vegetation*, pp. 41–52. SCIRO and CALM, Australia: Surrey Beatty and Sons Pty, Limited.

Bennett, C. P. A. and R. N. Byron. 1998. Valuing resource valuation: Exploring the role of quantitative valuation of Indonesia's forest resources, CIFOR Occasional Paper. Bogor, Indonesia: CIFOR (Centre for International Forestry Research).

Bennett, E. L. and Z. Dahaban. 1995. Wildlife response to disturbances in Sarawak and their implications for forest management. In R. B. Primack and T. E. Lovejoy, eds., *Ecology, Conservation and Management of Southeast Asian Rain Forests*, pp. 66–86. New Haven, Conn.: Yale University Press.

Bennett, E. L., A. J. Nyaoi, and J. Sompud. 1995. A conservation management study of wildlife hunting in Sabah and Sarawak. Miscellaneous report. Bronx, N.Y.: WCS (Wildlife Conservation Society).

——. 1999. Saving Borneo's bacon: The sustainability of hunting in Sarawak and Sabah. In J. G. Robinson and E. L. Bennett, eds., *Hunting for Sustainability in Tropical Forests*, pp. 305–324. New York, N.Y.: Columbia University Press.

Bennett, L., A. Edwards, A. Rabinowitz, L. White, and C. Fairgrieve. In press. Observing animals. In L. J. T. White and A. Edwards, eds., A Methods Manual for Ecological and Socioeconomic Surveys in the African Rain Forest. New York, N.Y.: WCS (Wildlife Conservation Society).

Benstead, J. P., M. L. J. Stiassny, P. V. Loiselle, K. J. Riseng, and N. Raminosoa. 2000. River ronservation in Madagascar. In P. J. Boon, B. R. Davies, and G. E. Petts, eds. *Global Perspectives on River Conservation: Science Policy and Practice*, pp. 205–231. New York, N.Y.: John Wiley and Sons.

Bertault, J. G. and K. Kadir, eds. 1998. *Silvicultural research in a lowland mixed dipterocarpt forest of East Kalimantan: The contribution of STREK (Silvicultural treatment for the regeneration of logged-over forest in East Kalimantan) project.* Montpellier, France: CIRAD-Forêt (Centre de coopération internationale en recherche agronomique pour le développement).

Bertault, J. G. and P. Sist. 1997. An experimental comparison of different harvesting intensities with reduced-impact and conventional logging in East Kalimantan, Indonesia. *Forest Ecology and Management* 94:209–218.

Beschta, R. L. 1991. Stream habitat management for fish in the northwestern United States: The role of riparian vegetation. *American Fisheries Society Symposium* 10:53–58.

Bierregaard, R. O. and T. E. Lovejoy. 1989. Effects of forest fragmentation on Amazonian understory bird communities. *Acta Amazonica* 19:215–241.

Bierregaard, R .O., T. E. Lovejoy, V. Kapos, A. A. dos Santos, and R. W. Hutchings. 1992. The biological dynamics of tropical forest fragments. *Bioscience* 42:859–866.

Bilby, R. E. and G. E. Likens. 1980. Importance of organic debris dams in the structure and function of stream ecosystems. *Ecology* 61:1107–1113.

Bisbal, E., F. J. 1994. Consumo de fauna silvestre en la zona de Imataca, Estado Bolivar, Venezuela. *Interciencia* 19:1–6.

Blair, J. M., R. W. Parmelee, and P. Lavelle. 1995. Influences of earthworms on biogeochemistry. In P. F. Hendrix, ed., *Earthworm Ecology and Biogeography in North America,* pp. 127–158. Chelsea: Lewis Scientific Publishers.

Blake, J. G. and W. G. Hoppes. 1986. Influence of resource abundance on use of treefall gaps by birds in an isolated woodlot. *Auk* 103:328–340.

Blake, S. 1994. A reconnaissance survey of the Kabo logging concession south of the Nouabale-Ndoki national park northern Congo. Miscellaneous report. Bronx, N.Y.: (WCS) Wildlife Conservation Society.

——. 1995. A reconnaissance survey in the Likouala swamps of northern Congo and its implications for conservation. MSc thesis, University of Edinburgh.

Blake, S., E. Rogers, J. M. Fay, and M. Ngangoue. 1995. Swamp gorillas in northern Congo. *African Journal of Ecology* 33:285–290.

Blanchart, E. and J. M. Julka. 1997. Influence of forest disturbance on earthworm (Oligochaeta) communities in the Western Ghats (South India). *Soil Biology and Biochemistry* 29:303–306.

Blau, W. S. 1980. The effect of environmental disturbance on a tropical butterfly population. *Ecology* 61:1005–1012.

Blockhus, J. M., M. Dillenbeck, J. A. Sayer, and P. Wegge. 1992. *Conserving Biological Diversity in Managed Tropical Forests.* Gland, Switzerland and Cambridge, United Kingdom: IUCN (International Union for Conservation of Nature and Natural Resources) Forest Conservation Programme/ ITTO(International Tropical Timber Organization).

Bloxam, Q. M. C., J. L. Behler, E. R. Rakotovao, H. J. A. R. Randriamahazo, K. T. Hayes, S. J. Tonge, and J. U. Ganzhorn. 1996. Effects of logging on the reptile fauna of the Kirindy forest with special emphasis on the flat-tailed tortoise (*Pyxis planicauda*). *Primate Report* 46:189–203.

Bodmer, R. E. 1989. Frugivory in Amazonian artiodactyla: Evidence for the evolution of the ruminant stomach. *Journal of Zoology* 219:457–467.

——. 1990a. Ungulate frugivores and the browser-grazer continuum. *Oikos* 57:319–325.

——. 1990b. Fruit patch size and frugivory in the lowland tapir (*Tapirus terrestris*). *Journal of Zoology* 222:121–128.

——. 1990c. Responses of ungulates to seasonal inundations in the Amazon floodplain. *Journal of Tropical Ecology* 6:191–201.

——. 1994. Managing wildlife with local communities in the Peruvian Amazon: The case of the Reserva Comunal Tamshiyacu-Tahuayo. In D.Western and R. M. Wright, eds., *Natural Connections: Perspectives in Community-based Conservation,* pp. 113–134. Washington, D.C.: Island Press.

Bodmer, R. E., T. G. Fang, and L. Moya. 1988. Estudio y manejo de los pecaries (*Tayassu tajacu* y *Tayassu pecari*) en la Amazonia Peruana. *Notas Cientificas/Matero* 1(2):18–25.

Bodmer, R. E., T. G. Fang, I. Moya, and R. Gill. 1994. Managing wildlife to conserve Amazonian forests: Population biology and economic considerations of game hunting. *Biological Conservation* 67:29–35.

Bodmer, R. E., P. Puertas, J. Garcia, D. Dias, and C. Reyes. 1999. Game animals, palms and people of the flooded forests: Management considerations for the Pacaya-Samiria National Reserve, Peru. In C. Padoch, J. M. Ayres, M. Pinedo-Vásquez, and A. Henderson (eds.) *Várzea Diversity, Development, and Conservation of Amazonia's Whitewater Floodplains,* pp. 217–231. Advances in Economic Botany. Volume 13. New York, N.Y.: The New York Botanical Garden Press.

Bolger, D. T., T. A. Scott, and J. T. Rotenberry. 1997. Breeding bird abundance in an urbanizing landscape in coastal southern California. *Conservation Biology* 11(2):406–421.

Bolton, B. 1994. *Identification Guide to Ant Genera of the World.* Cambridge, Mass.: Harvard University Press.

Bonaccorso, F. J. 1979. Foraging and reproductive ecology in a Panamanian bat community. Bulletin of the Florida State Museum. *Biological Sciences* 24:359–408.

Bond, W. J. 1994. Do mutualisms matter? Assessing the impact of pollinator and disperser disruption on plant extinction. *Philosophical Transactions of the Royal Society of London, Series B* 344:83–90.

——. 1995. Assessing the risk of plant extinction due to pollinator and disperser failure. In J.H. Lawton and R.M. May, eds., *Extinction Rates,* pp. 131–146. Oxford, United Kingdom: Oxford University Press.

Bongers, F. 1998. Manipulation of light in tropical rain forest. In Tropenbos Foundation, ed., *Research in Tropical Rain Forests: Its Challenges for the Future. Seminar Proceedings, 25–26 November 1997,* pp. 169–184. The Netherlands: The Tropenbos Foundation, Wageningen.

Boose, E. R., D. R. Foster, and M. Fluet. 1994. Hurricane impacts to tropical and temperate forest landscapes. *Ecological Monographs* 64:369–400.

Bormann, F. H. and G. E. Likens. 1979. *Pattern and Process in a Forested Ecosystem.* New York, N.Y.: Springer-Verlag.

Borner, M. 1979. A field study of the Sumatran rhinoceros: ecology, behaviour and conservation situation in Sumatra. Ph.D. diss., University of Basel.

Bosch, J. M. and J. D. Hewlett. 1982. A review of catchment experiments to determine the effect of vegetation changes on water yield and evapotranspiration. *Journal of Hydrology* 55:3–23.

Boscolo, M., J. Bonjourno, and T. Panayotou 1997. Simulating options for carbon sequestration through improved management of a lowland tropical rainforest. *Environment and Development Economics* 2:214–263.

Bossel, H. and H. Krieger. 1991. Simulation model of natural tropical forest dynamics. *Ecological Modelling* 59:37–72

Bouché, M. B. 1977. Stratégies lombriciennes. *Ecological Bulletin* 23:122–132.

Boucher, D. H. 1990. Growing back after hurricanes: catastrophes may be critical to rain forest dynamics. *BioScience* 40:163–166.

Bowen, S. H. 1983. Detritovory in neotropical fish communities. *Environmental Biology of Fishes* 9:137–144.

Bowles, I. A., A. B. Rosenfeld, C. A. Sugal, and R. A. Mittermeier. 1998. *Natural Resource Extraction in the Latin American Tropics: A Recent Wave of Investment Poses New Challenges for Biodiversity Conservation.* Washington, D.C.: Conservation International.

Bowles, I. A., R. E. Rice, R. A. Mittermeier, and G. A. B. da Fonseca. 1998. Logging and tropical forest conservation. *Science* 280:1899–1900.

Bowman, D. M. J. S., J. C. Z. Woinarski, D. P. A. Sands, A. Wells, and V. J. McShane. 1990. Slash-and-burn agriculture in the wet coastal lowlands of Papua New Guinea: Response of birds, butterflies and reptiles. *Journal of Biogeography* 17:227–239.

Boyle, T. J. B., M. Lawes, N. Manokaran, R. Prabhu, J. Ghazoul, S. Sastrapradja, H. C. Thang, V. Dale, H. Eeley, B. Finegan, J. Soberon, and N. E. Stork. 1998. Criteria and indicators for assessing the sustainability of forest management: A practical approach to assessment of biodiversity. CIFOR Working Paper. Bogor, Indonesia: CIFOR (Centre for International Forestry Research).

Braby, M. F. 1995. Seasonal-changes in relative abundance and spatial-distribution of Australian lowland tropical satyrine butterflies. *Australian Journal of Zoology* 43:209–229.

Braithwaite, L. W., J. Turner, and J. Kelly. 1984. Studies on the arboreal marsupial fauna of eucalypt forest being harvested for woodpulp at Eden, NSW. III. Relationships between faunal densities, eucalypt occurrence and foliage nutrients, and soil parent materials. *Australian Wildlife Research* 11:41–48.

Branan, W. V., M. C. M. Werkhoven, and R. L. Marchington. 1985. Food habits of the brocket and white-tailed deer in Suriname. *Journal of Wildlife Management* 49:972–976

Brawn, J. D. and S. K. Robinson. 1996. Source-sink population dynamics may complicate the interpretation of long-term census data. *Ecology* 77:3–12.

Bray, D.B. 1991. The struggle for the forest: Conservation and development in the Sierra Juárez. *Grassroots Development* 15:13–25.

Bray, D.B., M. Carreón, L. Merino, and V. Santos. 1993. On the road to sustainble forestry. *Cultural Survival Quarterly* 17(1):38–41.

Brokaw, N.V. L. 1983. Groundlayer dominance and apparent inhibition of tree regeneration by *Aechmea magdalenae* (Bromeliaceae) in a tropical forest. *Tropical Eco-logy* 24:194–200.

——. 1985. Gap-phase regeneration in a tropical forest. *Ecology* 66:682–687.

——. 1987. Gap-phase regeneration of three pioneer tree species in a tropical forest. *Journal of Ecology* 75:9–19.

Brokaw, N. V. L. and E. P. Mallory. 1993. Vegetation of the Rio Bravo Conservation and Management Area, Belize. Miscellaneous report. Manomet, Mass.: Manomet Bird Observatory.

Brosius, J. P. 1986. Rivers, forest and mountain: the Penan Gang landscape. *Sarawak Museum Journal* 39(57)(new series):173–184.

Brosset, A. and P. Charles-Dominique. 1990. The bats from French Guiana: A taxonomic, faunistic and ecological approach. *Mammalia* 54:509–560.

Brosset, A., P. Charles-Dominique, A. Cockie, J. F. Cosson, and D. Mason. 1996. Bat communities and deforestation in French Guiana. *Canadian Journal of Zoology* 74:1974–1982.

Brown, J. H., D. W. Mehlman, and G. C. Stevens. 1995. Spatial variation in abundance. *Ecology* 76(7):2028–2043.

Brown, N. and M. Press. 1992. Logging rainforests the natural way? *New Scientist* 1812:25–29.

Brown, N. D. 1996. A gradient of seedling growth from the centre of a tropical rain forest gap. *Forest Ecology and Management* 8:239–244.

Brown, N. D. and T. C. Whitmore. 1992. Do dipterocarp seedlings really partition tropical rain forest gaps? *Philosophical Transactions of the Royal Society of London, Series B* 335:369–378.

Brown, P. 1998. *Climate, Biodiversity, and Forests: Issues and Opportunities for the Kyoto Protocol*. Washington, D.C.: WRI (World Resources Institute) and IUCN (International Union for Conservation of Nature and Natural Resources).

Brown, P., B. Cabarle, and R. Livernash. 1997. *Carbon Counts: Estimating Climate Mitigation in Forestry Projects*. Washington, D.C.: WRI (World Resources Institute).

Brown, S. 1996. Mitigation potential of carbon dioxide emissions by management of forests in Asia. *Ambio* 25:273–278.

Brown, S., A. Gillespie, and A. Lugo. 1989. Biomass estimation methods for tropical forests with applications to forest inventory data. *Forest Science* 35:881–902.

Brown, S., A. E. Lugo, and L. R. Iverson. 1992. Processes and lands for sequestering carbon in the tropical forest landscape. *Water, Air, and Soil Pollution* 64:139–155.

Brown, S., A. Lugo, S. Silander, and L. Liegel. 1983. Research history and opportunities in the Luquillo experimental forest. New Orleans, USA: USDA (United States Department of Agriculture) Forest Service, Southern Forest Experiment Station, General Technical Report SO-44.

Brown, W. C. and A. C. Alcala. 1986. Comparison of the herpetofaunal species richness of Negros and Cebu Islands, Philippines. *Silliman Journal* 33:74–86.

Bruenig, E. F. 1993. The ITTO (International Tropical Timber Organization) guidelines for the sustainable management of natural and planted tropical forests. In H. Lieth and M. Lohmann, eds., *Restoration of Tropical Forest Ecosystems*, pp. 137–143. Dordrecht, Netherlands: Kluwer Academic Publishers.

——. 1996. *Conservation and Management of Tropical Rainforests: An integrated Approach to Sustainability*. Wallingford, United Kingdom: CAB (Centre for Agriculture and Bioscience).

Bruijnzeel, L. A. 1990. *Hydrology of Moist Tropical Forests and Effects of Conversion: A State of Knowledge Review*. Paris: UNESCO (United Nations Educational, Scientific, and Cultural Organization) IHP, Humid Tropics Programme.

Bruijnzeel, L. A. and W. R. S. Critchley. 1994. Environmental impacts of logging moist tropical forests. IHP Humid Tropics Programme Series No. 7. Paris, France: UNESCO (United Nations Educational, Scientific, and Cultural Organization).

Bryant, D., D. Nielsen, and L. Tangley. 1997. *The Last Frontier Forests: Ecosystems and Economies on the Edge*. Washington, D. C.: World Resources Institute.

Bryant, R. C. 1914. *Logging*. New York, N.Y.: John Wiley and Sons.

Buchmann, S. L. and G. P. Nabhan. 1996. *The Forgotten Pollinators*. Washington, D.C.: Island Press.

Buckland, S. T., D. R. Anderson, K. P. Burnham, and J. L. Laake. 1993. *Distance Sampling: Estimating Abundance of Biological Populations*. London, United Kingdom: Chapman and Hall.

Bugo, H. 1995. The significance of the timber industry in the economic and social development of Sarawak. In R. B. Primack and T. Lovejoy, eds., *Ecology, Conservation, and Management of Southeast Asian Rainforests*, pp. 221–240. New Haven, Conn. and London, United Kingdom: Yale University Press.

Bull, J. J. 1980. Sex determination in reptiles. *Quarterly Review of Biology* 55:3–21.

Burgess, P. F. 1971. Effect of hill logging on hill dipterocarp forest. *Malayan Nature Journal* 24:231–237.

——. 1975. Silviculture in the hill forests of the Malay Peninsula. Kuala Lumpur, Malaysia: Malaysian Forest Department Research Pamphlet 66. Jalan Swettenham.

Burghouts, T. B. A., G. Ernsting, G. W. Korthals, and T. H. De Vries. 1992. Litterfall, leaf-litter decomposition and litter invertebrates in primary and selectively logged dipterocarp forest in Sabah, East Malaysia. *Philosophical Transactions of the Royal Society of London, Series B*. 335:407–416.

Burke, I. C., W., K. Laurenroth, and D. P. Coffin. 1995. Soil organic matter recovery in semiarid grasslands: Implications for the conservation reserve program. *Ecological Applications* 5(3):793–801.

Burnett, S. E. 1992. Effects of a rainforest road on movements of small mammals: mechanisms and implications. *Wildlife Research* 19:95–104.

Burney, D.J. 1993. Recent animal extinctions: recipes for disaster. *American Scientist* 81:531–541.

Burns, R. 1966. *A Choice of Burns's Poems and Songs*. London, United Kingdom: Faber and Faber.

Buschbacker, R., C. Uhl, and E. A. S. Serrao. 1992. Reforestation of degraded Amazon pasture lands. In M.K. Wali, ed., *Ecosystem Rehabilitation*, pp. 257–274. The Hague, Netherlands: Academic Publishing.

Bush, M. B. and P. A. Colinvaux. 1994. Tropical forest disturbance: Paleoecological records from Darien, Panama. *Ecology* 75:1761–1768.

Bush, M. B., D. R. Piperno, P. A. Colinvaux, P. E. de Oliviera, L. A. Krissek, M. C. Miller, and W. E. Rowe. 1992. A 14,300-yr paleoecological profile of a lowland tropical lake in Panama. *Ecological Monographs* 62:251–275.

Butynski, T. M. 1990. Comparative ecology of blue monkeys (*Cercopithecus mitis*) in high- and low-density subpopulations. *Ecological Monographs* 60:1–26.

Butz Huryn, V. M. 1997. Ecological impacts of introduced honeybees. *Quarterly Review of Biology* 72:275–297.

Caballé, G. 1977. Multiplication végétative en forêt dense du Gabon de la liane *Entada sclerata* (Mimosoideae). *Adansonia, Series 2*: 17:215–220.

Cabarle, B. 1998. Logging in the rain forests. *Science* 281:1453–1454.

Cabarle, B., R. J. Hrubes, C. Elliot, and T. Synnott. 1995. Certification accreditation: The need for credible claims. *Journal of Forestry* 93(4):12–16.

Cairns, M. A. and R. A. Meganck. 1994. Carbon sequestration, biological divresity, and sustainable development: Integrated forest management. *Environmental Management* 18:13–22.

Caldecott, J. 1987. *Hunting and Wildlife Management in Sarawak*. Washington, D.C.: WWF (World Wildlife Fund).

——. 1991a. Monographie des Bartschweines. *Bongo* 18:54–68.

——. 1991b. Eruptions and migrations of bearded pig populations. *Bongo* 18:233–243.

Caldecott, J. O. 1988. *Hunting and Wildlife Management in Sarawak*. Gland, Switzerland and Cambridge, United Kingdom: IUCN (International Union for the Conservation of Nature).

——. 1996. *Designing Conservation Projects*. Cambridge, United Kingdom: Cambridge University Press.

Caldecott, J. O., and A. Nyaoi. 1984. A conservation management study for hunted wildlife in Sarawak. Progress Reports 1, 2, and 3 on file with NPWO (unpublished).

Caldwell, J. P. 1989. Structure and behavior of Hyla geographica tadpole schools, with comments on classification of group behavior in tadpoles. *Copeia* 1989:938–950.

——. 1993. Brazil nut fruit capsules as phytotelmata: Interactions among anuran and insect larvae. *Canadian Journal of Zoology* 71:1193–1201.

——. 1994. Natural history and survival of eggs and early larval stages of *Agalychnis calcarifer* (Anura: Hylidae). *Herpetological Natural History* 2:57–66.

——. 1996a. Diversity of Amazonian anurans: The role of systematics and phylogeny in identifying macroecological and evolutionary patterns. In A. C. Gibson, ed., *Neotropical Biodiversity and Conservation*, pp. 73–88. Los Angeles, Calif.: Mildred E. Mathias Botanic Garden Miscellaneous Publication Number 1.

——. 1996b. The evolution of myrmecophagy and its correlates in poison frogs (Family Dendrobatidae). *Journal of Zoology, London* 240:75–101.

——. 1997. Pair bonding in spotted poison frogs. *Nature* 385:211.

Caldwell, J. P. and M. C. Araújo. 1998. Cannibalistic interactions resulting from indiscriminate predatory behavior in tadpoles of poison frogs (*Anura: Dendrobatidae*). *Biotropica* 30:92–103.

Caldwell, J. P. and C. W. Myers. 1990. A new poison frog from Amazonian Brazil, with further revision of the quinquevittatus group of Dendrobates. *American Museum Novitates* 2988:1–21.

Caldwell, J. P. and L. J. Vitt. 1999. Dietary asymmetry in leaf litter frogs and lizards in a transitional northern Amazonian rain forest. *Oikos* 84:383–397.

Calvo-Irabien, L. M. and Islas-Luna, A. 1999. Predispersal predation of an understory rainforest herb *Aphelandra aurantiaca* (Acanthaceae) in gaps and mature forest. *American Journal of Botany* 86:1108–1113.

Camacho, O. 1996. Analisis del impacto del aprovechamiento forestal sobre la vegetacion natural de un bosque seco subtropical en Lomerìo, Santa Cruz, Bolivia. Thesis, Facultad de Ciencias Agricolas, Universidad Autonoma Gabriel Rene Moreno.

——. 1998. Planificación computarizada del aprovechamiento de los bosques. *Boletín BOLFOR (Bolivia Sustainable Forestry Management Project)* 12:5

Campbell, E. J. F. and D. McC. Newberry. 1993. Ecological relationships between lianas and trees in lowland rain forest in Sabah, East Malaysia. *Journal of Tropical Ecology* 9:469–490.

Canaday, C. 1997. Loss of insectivorous birds along a gradient of human impact in Amazonia. *Biological Conservation* 77:63–77.

Canham, C. D., J. S. Denslow, W. J. Platt, J. R. Runkle, T. A. Spies, and P. S. White. 1990. Light regimes beneath closed canopies and treefall gaps in temperate and tropical forests. *Canadian Journal of Forest Research* 20:620–631.

Cannon, C. H. and M. Leighton. 1994. Comparative locomotor ecology of gibbons and macaques: selection of canopy elements for crossing gaps. *American Journal of Physical Anthropology* 93:505–524.

Cannon, C. H., D. R. Peart, M. Leighton, and K. Kartawinata. 1994. The structure of lowland rainforest after selective logging in West Kalimantan, Indonesia. *Forest Ecology and Management* 67:49–68.

Cant, J. G. H. 1988. Positional behavior of long-tailed macaques (*Macacca fasicularis*) in Northern Sumatra. *American Journal of Physical Anthropology* 76:29–37.

———. 1992. Positional behavior and body size of arboreal primates: A theoretical framework for field studies and an illustration of its application. *American Journal Physical Anthropology* 88:273–283.

Carlson, J. Y., C. W. Andrus, and H. A. Froehlich. 1990. Woody debris, channel features, and macroinvertebrates of streams with logged and undisturbed riparian timber in northeastern Oregon, U.S.A. *Canadian Journal of Fisheries and Aquatic Sciences* 47:1103–1111.

Carranza, J., L. M. Arias-De-Reyna, and C. Ibáñez. 1982. Uso del espacio y movimientos en una comunidad de quirópteros neotropicales. *Historia Natural* 2:177–190.

Carroll, R.W. 1996. Relative density range extension and conservation potential of the lowland gorilla (*Gorilla-gorilla-gorilla*) in the Dzanga-Sangha region of southwestern Central African Republic. *Mammalia* 52:309–324.

Carson, R. T. 1998. Valuation of tropical rainforests: Philosophical and practical issues in the use of contingent valuation. *Ecological Economics* 24:15–29.

Cartmill, M and K. Milton. 1977. The lorisiform wrist joint and the evolution of "brachiating" adaptations in the Hominoidea. *American Journal Physical Anthropology* 47:249–272.

Cassells, D. S., D. A. Gilmour, and M. Bonnell. 1985. Catchment response and watershed management in the tropical rainforest in north-eastern Australia. *Forest Ecology and Management* 10:155–175.

Caswell, H. 1989. *Matrix Population Models: Construction, Analysis, and Interpretation.* Sunderland, Mass.: Sinauer Associates.

Catalán, A. 1993. El proceso de deforestación en Venezuela entre 1975–1988. MARNR-DGSISAV. *Revista Ambiente.* 49:1–12.

Cates, R. G. and G. H. Orians. 1975. Successional status and the palatability of plants to generalized herbivores. *Ecology* 56:410–418.

Catinot, R. 1997. *The Sustainable Management of Tropical Rainforests.* Paris, France: Scytale Publishing.

Centre Technique Forestier Tropical (CTFT). 1985. *Inventaire des ressources forestières du sud Cameroun.* Nogent-sur-Marne, France: CTFT, Deprtement Forestier du CIRAD.

———. 1989. *Mémento du Forestier: Techniques rurales en Afrique.* Paris, France: Ministère de la Coopération et du Développement.

Chambers, J. C. and J. A. MacMahon. 1994. A day in the life of a seed: Movements and fates of seeds and their implications for natural and managed systems. *Annual Review of Ecology and Systematics* 25:263–292.

Chaplin, G. E. 1985. An integrated silvicultural solution to weedy climber problems in the Solomon Islands. *Commonwealth Forestry Review* 64:133–139.

Chapman, C. A. 1995. Primate seed dispersal: Coevolution and conservation implications. *Evolutionary Anthropology* 4:74–82.

Chapman, C. A. and L. J. Chapman. 1995. Survival without dispersal? Seedling recruitment under parents. *Conservation Biology* 9:675–678.

———. 1996. Frugivory and the fate of dispersed and non-dispersed seeds of six African tree species. *Journal of Tropical Ecology* 12:491–504.

———. 1997. Forest regeneration in logged and unlogged areas of Kibale National Park, Uganda. *Biotropica* 29:396–412.

Chapman, C. A., L. J. Chapman, and K.E. Glander. 1989. Primate populations in northwestern Costa Rica: Potential for recovery. *Primate Conservation* 10:37–44.

Chapman, D. W. 1988. Critical review of variables used to define effects of fines in redds of large salmonids. *Transaction of the American Fisheries Society* 117:1–21.

Chapman, L. J., C. A. Chapman, and R. W. Wrangham. 1992. Balanites wilsoniana: Elephant dependent dispersal? *Journal of Tropical Ecology* 8:275–283

Chardonnet, P., H. Fritz, N. Zorzi, and E. Feron. 1995. Current importance of traditional hunting and major contrasts in wild meat consumption in sub-saharan Africa. In J.A. Bissonette, and P.R. Krausman, eds., *Integrating People and Wildlife for a Sustainable Future*, pp. 304–307. Bethesda, Md.: The Wildlife Society.

Charles-Dominique, P. 1986. Inter-relation between frugivorous vertebrates and pioneer plants: Cecropia, birds and bats in French Guyana. In A. Estrada, and T. H. Fleming, eds., *Frugivores and Seed Dispersal*, pp. 119–135. Dordrecht, Netherlands: Dr. W. Junk Publisher.

Charles-Dominique, P., M. Atramentowicz, M. Charles-Dominique, H. Gerard, A. Hladik, C. M. Hladik and M. F. Prevost. 1981. Les mamíferes frugivores arboricoles nocturnes d'une foret guyanaise: inter-relations plantes-animaux. *Review Ecologue*. (Terre et Vie) 35:341–435.

Chesser, R. T. and S. J. Hackett. 1992. Mammalian diversity in South America. *Science* 256:1502–1504.

Cheung, P. S. C. 1992. Climate Change in Xishuangbanna, S. China and its implications. Paper presented at the IVth World Congress on National Parks and Protected Areas, Caracas, Venezuela.

Chiariello, N. 1984. Leaf energy balance in the wet lowland tropics. In E. Medina, H. A. Mooney, and C. Vazquez-Yanes, eds., *Physiological Ecology of Plants in the Wet Tropics*, pp. 85–98. Dordrecht, Netherlands: Dr. W. Junk Publisher.

Chin, S.C. 1985. Agriculture and resource utilization in a lowland rainforest Kenyah community. *Sarawak Museum Journal 25(56)* (new series): Special Monograph No. 4.

Chung, A. Y. C. 1993. A study of ant fauna (*Hymenoptera: Formicidae*) of the tropical rainforest (primary and secondary) in Danum Valley, Sabah. MSc thesis, Universiti Kebangsaan.

Clark D. A. and D. B. Clark. 1984. Spacing dynamics of a tropical rainforest tree: Evaluation of the Janzen-Connell model. *The American Naturalist* 124:769–788.

Clark, D. B., D. A. Clark, and P. M. Rich. 1993. Comparative analysis of microhabitat utilization by saplings of nine tree species in neotropical rain forest. *Biotropica* 25:297–407.

Coleman, D. C. 1996. Energetics of detritivory and microbivory in soil in theory and practice. In G. A. Polis and K. O. Winemiller, eds., *Food Webs: Integrations of Patterns and Dynamics*, pp. 39–50. New York, N.Y.: Chapman and Hall.

Collar, N. J., M. J. Crosby, and A. J. Stattersfield. 1994. *Birds to watch 2: The World List of Threatened Birds*. Birdlife Conservation Series no. 4. Cambridge: Birdlife International.

Collins, N. M. 1980. The distribution of soil macrofauna on the west ridge of Gunung (Mount) Mulu, Sarawak. *Oecologia* 44:263–275.

Collins, N. M. and M. G. Morris. 1985. *Threatened Swallowtail Butterflies of the World*. Gland, Switzerland: IUCN (International Union for Conservation of Nature and Natural Resources).

Collins, N. M., J. A. Sayer, and T. C. Whitmore. 1991. *The Conservation Atlas of Tropical Forests—Asia and the Pacific*. New York, N.Y.: Simon and Schuster.

Colwell, R. K. and J. A. Coddington. 1995. Estimating terrestrial biodiversity through extrapolation. In D.L. Hawksworth, ed., *Biodiversity Measurement and Estimation*, pp. 101–118. London, United Kingdom: Chapman and Hall Publishers.

Connell, J. H. 1971. On the role of natural enemies in preventing competitive exclusion in some marine animals and in forest trees. In P. J. den Boer and G. Gradwell, eds., *Dynamics of Populations*, pp. 298–312. Wageningen, The Netherlands: Pudoc.

——. 1978. Diversity in tropical rainforests and coral reefs. *Science* 199:1302–1309.

Connell, J. H. and M. D. Lowman. 1989. Low diversity tropical rainforests: Some possible mechanisms for their existence. *American Naturalist* 134:88–119.

Connell, J. H., M. D. Lowman, and I. R. Noble. 1997. Subcanopy gaps in temperate and tropical forests. *Australian Journal of Ecology* 22:163–168.

Conner, R. H. 1978. Snag management for cavity nesting birds. In R. M. DeGraaf, ed., *Proceedings of the workshop Management of Southern Forests for Nongame Birds*, pp. 120–128. U.S. Forest Service General Technical Report SE-14.

Conway, S. 1982. *Logging Practices*. San Francisco, Calif.: Miller Freeman.

Coomes, D. A. and P. J. Grubb. 1996. Amazonian caatinga and related communities at La Esmeralda, Venezuela: Forest structure, physiognomy and floristics, and control of soil factors. *Vegatatio* 122:167–191.

Cooper, D. S. and C. M. Francis. 1998. Nest predation in a Malaysian lowland rain forest. *Biological Conservation* 85:199–202.

Corbet, G. B. and J. E. Hill. 1980. *A World List of Mammal Species*. London, United Kingdom: British Museum (Natural History).

Cordero, W. 1995. Uso de bueyes en operaciones de aprovechamiento forestal en áreas rurales de Costa Rica. Estudio Monografico de Explotacion Forestal 3. Rome, Italy: FAO (Food and Agricultural Organization).

Cowles, R. B. and C. M. Bogert. 1944. A preliminary study of the thermal requirements of desert lizards. *Bulletin of the American Museum of Natural History* 83:265–276.

Cox, P. A., T. Elmqvist, E. D. Pierson, and W. E. Rainey. 1991. Flying foxes as strong interactors in South Pacific island ecosystems: A conservation hypothesis. *Conservation Biology* 5:448–454.

Coyle, F. A. 1981. Effects of clearcutting on the spider community of a southern Appalachian forest. *Journal of Arachnology* 9:285–298.

Crampton, L. H. and M. R. Barclay. 1996. Habitat use by bats in fragmented and unfragmented aspen mixedwood stands of different ages. In M. R. Barclay and R. M. Brigham, eds., *Bats and Forests Symposium, 1995 Victoria, British Columbia, Canada*, pp. 238–259. British Columbia: Ministry of Forests Research Program.

Crawford, R. J. M. 1984. Activity, group structure and lambing of blue duikers C. monticola in the Tsitsikamma National Parks, South Africa. *South African Journal of Wildlife Research* 14:65–68.

Crome, F. H. J. 1975. The ecology of fruit pigeons in tropical northern Queensland. *Australian Wildlife Reserach* 2:155–185.

——. 1991. Wildlife conservation and rain forest management - Examples from north east Queensland. In A. Gómez-Pompa, T. C. Whitmore, and M. Hadley eds., *Rain Forest Regeneration and Management*, pp. 407–418. UNESCO (United Nations Educational, Scientific, and Cultural Organization) Man and the Biosphere Series Volume 6. Park Ridge, N.J.: Parthenon Publishing Group.

Crome, F. H. J. and L. A. Moore. 1989. Display site constancy of bowerbirds and the effects of logging on Mt. Windsor Tableland, North Queensland. *Emu* 89:47–52.

Crome, F. H. J. and G. C. Richards. 1988. Bats and gaps: Microchiropteran community structure in a Queensland rain forest. *Ecology* 69:1960–1969.

Crome, F. H. J., L. A. Moore, and G. C. Richards. 1992. A study of logging damage in upland rainforest in north Queensland. *Forest Ecology and Management* 49:1–29.

Crome, F. H. J., M. R. Thomas, and L. A. Moore. 1996. A novel Bayesian approach for assessing impacts of rain forest logging. *Ecological Applications* 6:1104–1121.

Crow, T. R. and E. J. Gustafson. 1997. Ecosystem management: Managing natural resources in time and space. In K.A. Kohm and J.F. Franklin, eds., *Creating a Forestry for the 21st Century. The Science of Ecosystem Management*, pp. 215–228. Washington, D.C: Island Press.

Crump, M. L. 1971. Quantitative analysis of the ecological distribution of a tropical herpetofauna. *Occasional Papers of the Museum of Natural History, University of Kansas* 3:1–62.

Crump, M. L. and N. J. Scott Jr. 1994. Visual encounter surveys. In W. R. Heyer, M. A. Donnelly, R. W. McDiarmid, L. C. Heyek, and M. S. Foster, eds., *Measuring and Monitoring Biological Diversity: Standard Methods for Amphibians*, pp. 84–92. Washington, D.C.: Smithsonian Institution Press.

Culotta, E. 1995. Fifty shades of rain-forest green. *Science* 269:31.

Cummins, K. W., C. A. Tryon, Jr., and R. T. Hartman. 1966. Organism-substrate relationships in streams. Special Publication 4. Pittsburgh, Pa.: University of Pittsburgh, Pymatuning Laboratory of Ecology.

Cunha, C. C. J. 1986. O desmatamento em Mato Grosso do Sul. In *Manual de educacao ambiental para professores de 1 e 2 graus de MS*. Campo Grande, Brazil: INAMB.

Curnutt, J., J. Lockwood, H. K. Luh, P. Nott, and G. Russell. 1994. Hotspots and species diversity. *Nature* 367:326–327.

Currie, D. J. 1991. Energy and large-scale patterns of animal and plant-species richness. *American Naturalist* 137:27–49.

Cuvelier A. 1996. Problems and ways of improving forest exploitation in Madagascar. *Primate Report* 46–1:133–148

d'Huart, J-P. 1978. *Ecologie de l'hylochere (Hylochoerus meinerzhageni Thomas) au Parc National des Virungas*. 2e serie fasc. 25. Brussels, Belgium: Foundation pour Favoriser, les Recherches Scientifique en Afrique.

——. 1993. The Forest Hog. In W. L .R. Oliver, ed., *Pigs, Peccaries and Hippos Status Survey and Conservation Action Plan*, pp. 84–93. Gland, Switzerland and Cambridge, United Kingdom: IUCN/SSC (Species Survival Commission).

Dagasto, M. D. 1992. Effect of habitat structure on positional behavior and substrate use in Malagasy lemurs. *American Journal of Physical Anthropology* 14:678.

Dahaban, Z. 1996. Effects of selective logging on wildlife in a hill dipterocarp forest of Sarawak. Ph.D. diss., Universiti Kebangsaan Malaysia.

Dangerfield, J. M. 1990. Abundance, biomass and the diversity of soil macrofauna in savanna woodland and associated managed habitats. *Pedobiologia* 34:141–150.

Daniels, R. J. R. 1996. Landscape ecology and conservation in the western Ghats, South India. *Ibis* 138:64–69.

Danielsen, F. and M. Heegaard. 1995. Impact of logging and plantation development on species diversity: A case study from Sumatra. In O. Sandbukt, ed., *Management of Tropical Forests: Towards an Integrated Perspective*, pp. 73–92. Oslo: University of Oslo.

Darwin, C. 1881. *The Formation of Vegetable Mould Through the Action of Worms, with Observations of Their Habits*. London, United Kingdom: Murray.

Datta, A. and S. P. Goyal. 1996. Comparison of forest strusture and use by the Insiian giant squirrel (*Ratufa indica*) in two riverine forests if central India. *Biotropica* 28:394–399.

Dauber. E. 1995. Guía práctica y teórica para el diseño de un inventario forestal de reconocimiento. Miscellaneous report. Santa Cruz, Bolivia: BOLFOR (Bolivia Sustainable Forestry Management Project).

Davies, A. G. 1987. *The Gola Forest Reserves, Sierra Leone: Wildlife Conservation and Management*. Gland, Switzerland: IUCN (International Union for Conservation of Nature and Natural Resources) Tropical Forest Programme.

——. 1991. Survey methods employed for tracking duikers in Gola. *Gnuseletter*: 10(1):9–12

Davies, A. G. and B. Birkenhager. 1990. Jentink's duiker in Sierra Leone—Evidence from the Freetown Peninsula. *Oryx* 24:143–146.

Davies, A. G. and J. B. Payne. 1982. A faunal survey of Sabah. Malaysia Kuala Lumpur, Malaysia: WWF (World Wildlife Fund).

Davies, A. G. and P. Richards. 1991. Rain forest in Mende life: Resource and subsistence strategies in rural communitites around Gola North Forest Reserve (Sierra Leone). Unpublished report. London, United Kingdom: ESCOR (Economic and Social Committee for Overseas Research)/ODA (Overseas Development Administration).

Davies, G. 1986. The orang-utan in Sabah. *Oryx* 20:40–45.

Davies, K. F. and C. R. Margules. 1998. Effects of habitat fragmentation on carabid beetles: experimental evidence. *Journal of Animal Ecology* 67:460–471.

Davies, R. G., P. Eggleton, L. Dibog, J. H. Lawton, D. E. Bignell, A. Brauman, C. Hartmann, L. Nunes, J. Holt, and C. Rouland. 1999. Successional response of a tropical forest termite assemblage to experimental habitat perturbation. *Journal of Applied Ecology* 36:946–962.

Davis, D. D. 1962. Mammals of the lowland rainforest of North Borneo. *Bulletin 31 of the Natural Museum, State of Singapore.*

Dawkins, H. C. and M. S. Philip. 1998. *Tropical Moist Forest Silviculture and Management: A history of success and failure.* Wallingford, United Kingdom: CAB (Centre for Agriculture and Bioscience).

De Camino, R. 1998. Research needs for a low impact logging operation in the Brazilian Amazon: the case of precious woods / Mil Madereira Itacaotiara. In Tropenbos, ed., *Research in Tropical Rain Forests: Its Challenges for the Future Seminar Proceedings,* pp. 43–56. Wageningen, The Netherlands: TROPENBOS Foundation.

De Graaf, N. R. 1986. *A Silvicultural System for Natural Regeneration of Tropical Rain Forest in Suriname.* Wageningen, The Netherlands: Wageningen Agricultural University.

De Graaf, N. R. and R. L. H. Poels. 1990. The Celos management system: A polycyclic method for sustained timber production in South American rainforest. In A. B. Anderson, ed., *Alternatives to Deforestation: Steps Towards Sustainable Use of the Amazon Rainforest,* pp. 116–127. New York, N.Y.: Columbia University Press.

De Graaf, N. R., R. L. H. Poels, and R. S. A. R. van Rompaey. 1999. Effect of silvicultural treatment on growth and mortality of rainforest in Surinam over long periods. *Forest Ecology and Management* 124:123–135.

De Kroon, H., A. Plaisier, J. Van Groenendael, and H. Caswell. 1986. Elasticity: The relative contribution of demographic parameters to population growth rate. *Ecology* 67:1427–1431.

Dean, W. R. J. and S. J. Milton. 1995. Plant and invertebrate assemblages on old fields in the arid southern Karoo, South Africa. *African Journal of Ecology* 33:1–13.

DeAngelis, D. L. 1992. *Dynamics of Nutrient Cycling and Food Webs.* New York, N.Y.: Chapman and Hall.

Decher, J. 1997. Conservation, small mammals, and the future of sacred groves in West Africa. *Biodiversity and Conservation* 6:1007–1026.

Dekeyser, P. L. and A. Villiers. 1955. Cephalophe a dos jaune et cephalophe de Jentink. *Notes Africaines* 66:54–57.

DellaSala, D. A., J. R. Strittholt, R. F. Noss, and D. M. Olson. 1996. A critical role for core reserves in managing inland northwest landscapes for natural resources and biodiversity. *Wildlife Society Bulletin* 24:209–221.

Demment, M. W. and P. J. van Soest. 1985. A nutritional explanation for body size patterns of ruminant and non-ruminanant herbivores, *American Naturalist* 125:641–672.

Denslow, J. S. 1980. Gap partitioning among tropical rain forest trees. *Biotropica* 12:47–55.

———. 1987. Tropical rainforest gaps and tree species diversity. *Annual Review of Ecology and Systematics* 18:431–451.

Devoe, N. N. 1989. Differential seeding and regeneration in openings and beneath closed canopy in subtropical wet forest. Ph.D. diss., Yale University.

Dewberry, T. 1992. Protecting the biodiversity of rivers and riparian ecosystems: The National River Public Land Policy Development Project. In R .E. McCabe, ed.,

Transactions of the 57th North American Wildlife and Natural Resources Conference, Charlotte, North Carolina, pp. 424–432. Washington, D.C.: Wildlife Management Institute.

DeVries, P. J. 1988. Stratification of fruit-feeding nmphalid butterflies in a Costa-Rican rainforest. *Journal of Research on the Lepidoptera*, 26:98–108.

Dial, R. and J. Roughgarden. 1995. Experimental removal of insectivores from rain forest canopy: Direct and indirect effects. *Ecology* 76:1821–34.

Diamond, J. 1989. Quaternary megafaunal extinctions: Variations on a theme by Paganini. *Journal of Archeological Science* 16:167–175.

Diamond, J. M., K. D. Bishop, and S. van Balen. 1987. Bird survival in an isolated Javan woodland: island or mirror? *Conservation Biology* 1:132–142.

Dickinson, M. B., J. C. Dickinson, and F. E. Putz. 1996. Natural forest management as a conservation tool in the tropics: Divergent views on possibilities and alternatives. *Commonwealth Forestry Review* 75:309–315.

Didden, W. A. M. 1990. Involvement of Enchytraeidae (Oligochaeta) in soil structure evolution in agricultural fields. *Biology and Fertility of Soils* 9:152–158.

Didham, R. K., J. Ghazoul, N. E. Stork, and A. J. Davis. 1996. Insects in fragmented forests: A functional approach. *Trends in Ecology and Evolution* 11:255–260.

Dinerstein, E. and C. M. Wemmer. 1988. Fruits Rhinoceros eat: Dispersal of *Trewia nudiflora* (Euphorbiaceae) in lowland Nepal. *Ecology* 69:1768–1774.

Dinerstein, E., D. M. Olson, D. J. Graham, A. L. Webster, S. A. Primm, M. P. Bookbinder, and G. Ledec. 1995. *A Conservation Assessment of the Terrestrial Ecoregions of Latin America and the Caribbean*. Washington, D.C.: WWF (World Wildlife Fund) and The World Bank.

Dittrich, L. 1972. Beobachtungen bei der Haltung von *Cephalophus*-Arten sowie zur Fortpflanzung und Jugendentwicklung von *Cephalophus dorsalis* und *C.rufilatus* in Gefangenschaft. *Zoologischer Garten* 42(1/2):1–16.

Dixon, J. R. and P. Soini. 1975. The reptiles of the upper Amazon basin, Iquitos region, Peru I. Lizards and Amphisbaenians. *Milwaukee Public Museum Contributions in Biology and Geology* 4:1–58.

———. 1977. The reptiles of the upper Amazon basin, Iquitos region, Peru II. Crocodilians, turtles and snakes. *Milwaukee Public Museum Contributions in Biology and Geology*. 12:1–91.

Dixon, R. K. K .J. Andrasko, F. G. Sussman, M. A. Lavinson, M. C. Trexler, and T. S. Vinson. 1993. Forest sector carbon offset projects: Near-term opportunities to mitigate greenhouse gas emissions. *Water, Air, and Soil Pollution* 70:561–577.

Dixon, R. K., J. A. Perry, E. L. Vanderklein, and F. H. Hiol. 1996. Vulnerability of forest resources to global climate change: Case study of Cameroon and Ghana. *Climate Research* 6:127–133.

Domenici, P. V. 1996. Editorial: The reality of science funding. *Science* 273:1319.

Doube, B. M. and O. Schmidt. 1997. Can the abundance or activity of soil macrofauna be used to indicate the biological health of soils? In C. E. Pankhurst, B. M. Doube, and V. V. S. R. Gupta, eds., *Biological Indicators of Soil Health*, 265–296. Wallingford, United Kingdom: CAB (Centre for Agriculture and Bioscience).

Douglas, I., T. Greer, K. Bidin, and M. Spilsbury. 1993. Impacts of rainforest logging on river systems and communities in Malaysia and Kalimantan. *Global Ecology and Biogeography Letters* 3:245–252.

Douglas I., T. Greer, W. W. Meng, T. Spences, and W. Sinun. 1990. The impact of commercial logging on a small rainforest catchment in Ulu Segama, Sabah, Malaysia. In R.R. Ziemer, C. L O'Loughlin and L. S. Hamilton, eds., *Research Needs and Applications to Reduce Erosion and Sedimentation in Tropical Steeplands*, pp. 165–173. Walingford, United Kingdom: International Association of Hydrological Sciences Press.

Douglas, I., T. Spencer, T. Greer, K. Bidin, W. Sinun, and W. W. Meng. 1992. The impact of selective commercial logging on stream hydrology, chemistry, and sediment loads in the Ulu Segama rain forest, Sabah, Malaysia. *Philosophical Transactions of the Royal Society London, Series B* 335:397–406.

Dounias, E., A. Hladik, and C.M. Hladik. 1996. De la resource disponible à la ressource exploitée: méthode de quantification des ressources alimentaires dans les régions forestières et les savanes du Cameroun. In A. Froment, I. de Garine, C. Binam Bikoi, and J.F. Loung, eds., *Bien Manger et Bien Vivre: Anthropologie alimentaire et développement en Afrique intertropicale: du biologique au social*, pp. 55–66. Paris, France: Orstom.

Downer, C. C. 1996. The mountain tapir, endangered flagship of the high Andes. *Oryx* 30:45–58.

Dranzoa, C. 1995. Bird populations of primary and logged forests in the Kibale Forest National Park, Uganda. Unpublished Ph.D. diss., Makerere University.

Driscoll, P. V. 1985. The effects of logging on bird populations in lowland New Guinea rainforest. Ph.D. diss., Brisbane, Australia: University of Queensland.

——. 1986. Activity of understory birds in relation to natural treefalls and intensive logging in lowland New Guinea rain forest. In R. Hermann, ed., *18th IUFRO (International Union of Forestry Research Organizations) World Congress in Ljubljana, Yugoslavia. Division 1.* Volume 2, pp. 620–632. Vienna, Austria: IUFRO.

Driscoll, P. V. and J. Kikkawa. 1989. Bird species diversity of lowland tropical rainforests of New Guinea and Northern Australia. In M. L. Harmelin-Vivien and F. Bourliere, eds., *Vertebrates in Complex Tropical Systems*, pp. 123–152. New York, N.Y.: Springer-Verlag.

Dubost, G. 1979. The size of African forest artiodactyls as determined by the vegetation structure. *African Journal of Ecology* 17:1–17.

——. 1980. L'écologie et la vie sociale du céphalophe bleu (*Cephalophus monticola*), petit ruminant forestier africain. *Zeitschrift für Tierpsychologie.* 54:205–266.

——. 1983. Le comportement de *Cephalophus monticola* Thunberg et *C. dorsalis* Gray, et la place des céphalophes au sein des ruminants. (1 partie). *Mammalia* 47:141–177.

——. 1984. Comparison of the diets of frugivorous ruminants of Gabon. *Journal of Mammology* 65:298–316.

Dubost, G. and F. Feer. 1988. Variabilite comportementale a l'interieur du genre *Cephalophus*, par l'exemple de *C. rufilatus*. *Zeitschrift für Säugetierkunde* 53:31–47.

Dudgeon, D. 1994. The need for multi-scale approaches to the conservation and management of tropical inland waters. In D. Dudgeon and P. K. S. Lam. Mitteilungen, eds., *Inland Waters of Tropical Asia and Australia: Conservation and Management.* Internationale Vereinigung fuer Theoretische und Angewandte Limnologie 24:11–16.

Dudley, J. P., A. Y. Mensah-Ntiamoah, and D. G. Kpelle. 1992. Forest elephants in a rainforest fragment: preliminary findings from a wildlife conservation project in Ghana. *African Journal of Ecology* 30:116–126.

Dudley, N., J. P. Jeanrenaud, and F. Sullivan. 1995. *Bad Harvests? The Timber Trade and Degradation of the World's Forests.* London, United Kingdom: Earthscan Publications.

Duellman, W. E. 1978. The biology of an equatorial herpetofauna in Amazonian Ecuador. *Occasional Papers of the Museum of Natural History, the University of Kansas* 65:1–352.

——. 1979. The South American Herpetofauna: A panoramic view. *University of Kansas Museum of Natural History Monographs* 7:1–28.

——. 1990. Herpetofaunas in neotropical rainforests: Comparative composition, history, and resource use. In A. H. Gentry, ed., *Four Neotropical Rainforests*, pp. 455–505. New Haven, Conn.: Yale University Press.

——. 1993. Amphibians in Africa and South America: Evolutionary history and ecological comparisons. In P. Goldblatt, ed., *Biological Relationships between Africa and South America*, pp. 200–243. New Haven, Conn.: Yale University Press.

——. 1995. Temporal fluctuations in abundances of anuran amphibians in a seasonal Amazonian rainforest. *Journal of Herpetology* 29:13–21.

Duellman, W. E. and M. S. Hoogmoed. 1992. Some hylid frogs from the Guiana highlands, northeastern South America: New species, distributional records, and a generic reallocation. *Occasional Papers of the Museum of Natural History, University of Kansas* 147:1–21.

Duellman, W. E. and M. Lizana. 1994. Biology of a sit-and-wait predator, the leptodactylid frog *Ceratophrys cornuta*. *Herpetologica* 50:51–64.

Duellman, W. E. and J. D. Lynch. 1988. Anuran amphibians from the Cordillera de Cutucu, Ecuador. *Proceedings of the Academy of Natural Sciences of Philadelphia* 140:125–142.

——. 1997. Frogs of the genus Eleutherodactylus in western Ecuador: systematics, ecology, and biogeography. *University of Kansas Special Publication* 23:1–236.

Duellman, W. E. and J. R. Mendelson. 1995. Amphibians and reptiles from northern Departamento Loreto, Peru: Taxonomy and biogeography. *University of Kansas Science Bulletin* 55:329–376.

Duellman, W. E. and A. W. Salas. 1991. Annotated checklist of the amphibians and reptiles of Cuzco Amazonico, Peru. *Occasional Papers of the Museum of Natural History, the University of Kansas* 143:1–13.

Duellman, W. E. and R. Schulte. 1992. Description of a new species of Bufo from Northern Peru with comments of phenetic groups of South American toads (*Anura Bufonidae*). *Copeia* 1992:162–172.

Duellman, W. E. and L. Trueb. 1986. *Biology of Amphibians*. New York, N.Y.: McGraw and Hill.

Duff, A. B., R. A. Hall, and C. W. Marsh. 1984. A survey of wildlife in and around a commercial tree plantation in Sabah. *The Malaysian Forester* 47:197–213.

Dung, V. V., P. M. Giao, N. N. Chinh, and J. MacKinnon. 1994. Discovery and conservation of the Vu Quang ox in Vietnam. *Oryx* 28:16–21.

Dunne, T. 1979. Sediment yield and land use in tropical catchments. *Journal of Hydrology* 42:281–300.

Dyer, R. C. 1988. Remote sensing identification of tornado tracks in Argentina, Brazil, and Paraguay. *Photogrammetric Engineering and Remote Sensing* 54:1429–1435.

Dykstra, D. and R. Heinrich. 1992. Sustaining tropical forests through environmentally sound harvesting practices. *Unasylva* 43:9–15.

—— (eds.). 1994. *Forest Codes of Practice: Contributing to Environmentally Sound Forest Operations*. FAO Forestry Paper 133, FAO, Rome, Italy.

——. 1996. *FAO (Food and Agricultral Organization) Model Code of Forest Harvesting*. Rome, Italy: FAO (Food and Agricultral Organization).

Edwards, C. A. and P. L. Bohlen. 1996. *Earthworm Ecology and Biology*. London, United Kingdom: Chapman and Hall.

Edwards, R. L. 1992. Can the species richness of spiders be determined? *Psyche* 100:185–208.

Eggeling, W. J. 1947. Observations on the ecology of the Budongo rainforest, Uganda. *Journal of Ecology* 34:20–87.

Eggleton, P. D., D. E. Bignell, W. A. Sands, N. A. Mawdsley, J. H. Lawton, T. G. Wood, and N. C. Bignell. 1996. The diversity, abundance and biomass of termites under differing levels of disturbance in the Mbalmayo Forest Reserve, southern Cameroon. *Philosophical Transactions of the Royal Society of London, Series B* 351:51–688.

Eggleton, P. D., D. E. Bignell, W. A. Sands, B. Waite, T. G. Wood, and J. H. Lawton. 1995. The species richness of termites (Isoptera) under differing levels of forest disturbance in the Mbalmayo Forest Reserve, southern Cameroon. *Journal of Tropical Ecology* 11:85–98.

Ehrlen, J. and J. Van Groenendael. 1998. Direct perturbation analysis for better conservation. *Conservation Biology* 12:470–474.

Ehrlich, P. 1980. The strategy of conservation 1980–2000. In M. Soulé, and B. Wilcox, eds., *Conservation Biology: An Evolutionary Perspective*, pp. 329–344. Sunderland, Mass.: Sinauer Assoc. Inc.

Ehrlich, P. R. and A. H. Ehrlich. 1981. *Extinction: The Causes and Consequences of the Disappearance of Species*. New York, N.Y.: Random House.

Ek, R. C. 1997. *Botanical Diversity in the Tropical Rain Forest of Guyana. Tropenbos–Guyana Series 4*. Wageningen, The Netherlands: The Tropenbos Foundation.

Elliot, C. and R. Z. Donovan. 1996. Introduction. In V. M. Viana, J. Erwin, R. Z. Donovan, C. Elliot, and H. Gholz, eds., *Certification of Forest Products: Issues and Perspectives*, pp. 1–10. Washington, D.C.: Island Press.

Elliot, C. and V. M. Viana. 1996. Potential inequalities and unintended effects of certification. In V. M. Viana, J. Erwin, R.Z. Donovan, C. Elliot, and H. Gholz, eds., *Certification of Forest Products. Issues and Perspectives*, pp. 137–144. Washington, D.C.: Island Press.

El-Swaify, S. A. and E. W. Dangler. 1982. *Rainfall Erosion in the Tropics: A State-of-the-art. Soil Erosion and Conservation in the Tropics*. Madison, WI: American Society of Agronomy Special Publication 43.

Emmons, L. H. 1984. Geographic variation in densities and diversities of non-flying mammals in Amazonia. *Biotropica* 16:210–222.

——. 1995. Mammals of rain forest canopies. In M. D. Lowman and N. M. Nadkarni, eds., *Forest Canopies*, pp. 199–223. San Diego, Calif.: Academic Press.

Emmons, L. H. and F. Feer. 1990. *Neotropical Rainforest Mammals: A Field Guide*. Chicago, Ill.: The University of Chicago Press.

Emmons, L. H., A. Gautier-Hion, and G. Dubost. 1983. Community structure of the frigivorous-folivorous forest mammals of Gabon. *Journal of the Zoological Society of London* 199:209–222.

Engstrom, M. D., T. E. Lee, and D. E. Wilson. 1987. *Bauerus dubiaquercus. Mammalian Species* 282:1–3.

Ergueta, P. and C. de Morales., eds. 1996. *Libro Rojo de los Vertebrados de Bolivia*. La Paz, Bolivia: Centro de Datos para la Conservación.

Ervin, J. and A. Pierce. 1997. *Sample Certification Standards for the FSC's (Forest Stewardship Council) Principles and Criteria. Draft 2*. Oaxaca, Mexico: FSC (Forest Stewardship Council).

Estrada, A. and R. Coates-Estrada. 1984. Fruit eating and seed dispersal by Howling Monkeys (Alouatta palliata) in the tropical rain forest of Los Tuxtlas, Mexico. *American Journal of Primatology* 6:77–91.

Estrada, A., R. Coates-Estrada, D. J. Meritt, S. Montiel, and D. Curiel. 1993. Patterns in frugivore species richness and abundance in forest islands and in agricultural hab itats in Los Tuxtlas, Mexico. *Vegetatio* 107/108:245–257.

Evans, J. 1982. *Plantation Forestry in the Tropics*. Oxford, United Kingdom: Oxford University Press.

Everett, M. J., 1977. *A Natural History of Owls*. London, United Kingdom: Hamlyn.

Everett, R. and K. Robson. 1991. Rare cliff-dwelling plant species as biological monitors of climate change. *Northwest Environmental Journal* 7:352–353.

Everham, E. M. III and N. V. L. Brokaw. 1996. Forest damage and recovery from catastrophic wind. *The Botanical Review* 62:113–185.

Eves, H.E. 1995. Socioeconomics of natural resource utilization in the Kabo logging concession northern Congo. Report to Wildlife Conservation Society, New York.

Ewel, J. J. 1980. Tropical succession: manifold routes to maturity. *Biotropica* 12:2–7.

Ewel, J. and L. Conde. 1978. *Environmental Implication of Any-Species Utilization of the Moist Tropics. Proceedings of Conference on Improved Utilization of Tropical Forest. Forest Products Laboratory.* Madison, Wisc.: U.S. Forest Service.

——. 1980. Potential ecological impact of increased intensity of tropical forest utilization. *Biotropica Special Publication* 11:1–70.

Fa, J .E., J. Juste, J. Perez del Val, and J. Castroviejo. 1995. Impact of market hunting on mammal species in Equatorial Guinea. *Conservation Biology* 9:1107–1115.

Faeth, P., C. Cort, and R. Livernah. 1994. Evaluating the carbon sequestration benefits of forestry projects in developing countries. Washington, D.C.: World Resources Institute.

Faeth, P., M. C. Trexler, and D. Page. 1990. *Sustainable Forestry as a Response to Global Warming: A Central American Perspective.* Washington, D.C.: WRI (World Resources Institute).

Fairgrieve, C. 1995. The comparative ecology of blue monkeys (*C.mitis stuhlmannii*) in logged and unlogged forest, Budongo Forest Reserve, Uganda. Unpublished Ph.D. diss., University of Edinburgh.

Falkner, M. B. and T. J. Stohlgren. 1997. Evaluating the contribution of small national park areas to regional biodiversity. *Natural Areas Journal* 17:324–330.

FAN (Fundación Amigos de la Naturaleza)–WCS (Wildlife Conservation Society). 1994. Plan de manejo, Reserva de Vida Silvestre Ríos Blanco y Negro. Fundación Amigos de la Naturaleza, Wildlife Conservation Society, Santa Cruz, Bolivia

Fanshawe, J. H. 1995. The effects of selective logging on the bird community of Arabuko-Sokoke Forest, Kenya. Unpublished Ph.D. diss., University of Oxford.

FAO (Food and Agriculture Organization). 1979. Way Kambas Management Plan 1980–1985. Field Report 5 of FAO (Food and Agriculture Organization) Project INS/78/061. Rome, Italy: FAO of the United Nations.

——. 1982. *Management and Utilization of Mangroves in Asia and the Pacific.* FAO Environment Paper 3, Rome Italy.

——. 1985. *Mangrove Management in Thailand, Malaysia and Indonesia.* FAO Environment Paper 4, Rome Italy.

——. 1993. *Forest Resources Assessment 1990: Tropical Countries.* Rome, Italy: FAO of the United Nations.

——. 1997. *State of the World's Forests.* Rome, Italy: FAO of the United Nations.

Faurie, A. S. and M. R. Perrin. 1993. Diet selection and utilisation in blue duikers (*Cephalophus monticola*) and red duikers (*C. natalensis*). *Journal of African Zoology* 107:287–299.

Fay, J. M. 1991. An elephant (*Loxodonta africana*) survey using dung counts in the forests of the Central African Republic. *Journal of Tropical Ecology* 7:25–36.

——. 1997. The ecology, social organization, populations, habitat and history of the western lowland gorilla (*Gorilla gorilla gorilla*). Ph.D. diss., Washington University.

Fay, J. M. and M. Agnagna. 1989. A population survey of forest elephants (*Loxodonta africana cyclotis*) in northern Congo. Report to WCS (Wildlife Conservation Society) /EEC (European Economic Community)/WWF (World Wildlife Fund) African elephant program, New York, N.Y.

Fay, J. M., M. Agnagna, and J.M. Moutsambote. 1990. A survey of the proposed Nouabale conservation area in northern Congo. Report to WCS (Wildlife Conservation Society), New York, N.Y.

Fearnside, P. M. 1989. Extractive reserves in Brazilian Amazonia. *BioScience* 39:387–293.

Feder, M. E. and W. W. Burggren. 1992. *Environmental Physiology of the Amphibians.* Chicago, Ill.: University of Chicago Press.

Feener, D. H. and E. W. Schupp. 1998. Effect of treefall gaps on the patchiness and species richness of Neotropical ant assemblages. *Oecologia* 116:191–201.

Feer, F. 1979. Observations ecologiques sur le neotrague de Bates (*Neotragus batesi* de Winton, 1903, Artiodactyle, Ruminant, Bovide) du Nord-Est du Gabon. *Revue d'Ecologie (Terre et Vie)* 33:159–239.

———. 1988. Strategies Ecologiques de deux Especes de Bovidés Sympatriques de la Forêt Sempervirente Africaine (*Cephalophus callipygus* et C. *dorsalis*): Influence du rythme d'activité. DSc thesis, Université Pierre et Marie Curie.

———. 1989. Comparaison des regimes alimentaires de *Cephalophus callipygus* et C. *dorsalis*, bovides sympatriques de la foret sempervirente africaine. *Mammalia* 53(4):563–604.

———. 1993. The potential for sustainable hunting and rearing of game in tropical forests. In C. M. Hladik, A. Hladik, O. F. Linares, H. Pagezy, A. Semple, and M. Hadley, eds., *Tropical Forests, People and Food: Biocultural Interactions and Applications to Development*, 691–708. Paris, France: UNESCO (United Nations Educational, Scientific, and Cultural Organization).

———. 1995. Seed dispersal in African forest ruminants. *Journal of Tropical Ecology* 11:683–689.

———. 1999. Effects of dung beetles (Scarabaeidae) on seeds dispersed by howler monkeys (*Alouatta seniculus*) in the French Guianan rain forest. *Journal of Tropical Ecology* 15:129–142

Feinsinger, P., W. H. Busby, K. G. Murray, J. M. Beach, W. Z. Pounds, and Y. B. Linhart. 1988. Mixed support for spatial heterogeneity in species interactions: hummingbirds in a tropical disturbance mosaic. *The American Naturalist* 131:33–57.

Fenton, M. B. 1972. The structure of aerial-feeding bat faunas as indicated by ears and wing elements. *Canadian Journal of Zoology* 50:287–296.

———. 1990. The foraging behaviour and ecology of animal-eating bats. *Canadian Journal of Zoology* 68:411–422.

Fenton, M. B., L. Acharya, D. Audet, M. B. C. Hickey, C. Merriman, M. K. Obrist, D. M. Syme, and B. Adkins. 1992. Phyllostomid bats (Chiroptera: Phyllostomidae) as indicators of habitat disruption in the neotropics. *Biotropica* 24:440–446.

Fernside, P. M. 1995. Global warming response options in Brazil's forest sector: Comparison of project-level costs and benefits. *Biomass and Bioenergy* 8:309–322.

Fetcher, N., S. F. Orberbauer, and B. R. Strain. 1985. Vegetation effects on microclimate in lowland tropical forests in Costa Rica. *International Journal of Biometeorology* 29:145–155.

Fimbel, C. 1994a. Ecological correlates of species success in modified habitats may be disturbance- and site-specific: The primates of Tiwai Island. *Conservation Biology* 8:106–113.

———. 1994b. The relative use of abandoned farm clearings and old forest habitats by primates and a forest antelope at Tiwai, Sierra Leone, West Africa. *Biological Conservation* 70:277–286.

Fimbel, C. C., R. A. Fimbel, and B. K. Curran. 1996. *The Lobéké Forest, Southeast Cameroon. Summary of Activities, Period 1988–1995*. Miscellaneous report. New York, N.Y.: Wildlife Conservation Society.

Fimbel, R. A. and F. Runge. Submitted. Semi-deciduous forest in the Lobeke Reserve, SE Cameroon: Recent history, present composition, and future prospects. *African Journal of Ecology*.

Fimbel, R. A., E. L. Bennett, R. Z. Donovan, P. C. Frumhoff, A. Grajal, R. E. Gullison, D. J. Mason, F. E. Putz, J. G. Robinson, and D. I. Rumiz. 1998. The potential for sustainable forest management to conserve wildlife within tropical forest landscapes. Issues and Policy Paper no. 4., Bronx, N.Y.: Wildlife Conservation Society.

Fimbel, R. A., T. G. O'Brien, A. A. Dwiyahreni, B. van Balen, J. Ghazoul, S. Hedges, P. Hidayat, K. Liston, E. Widodo, and N. Winarni. 1999. Faunal surveys in unlogged forest of the Inhutani II Malinau timber concession. Bulungan Research Report Series. CIFOR (Centre for International Forestry Research), Bogor, Indonesia.

Findely, J. S. 1993. *Bats: A Community Perspective*. New York, N.Y.: Cambridge University Press.

Fischermann, B. 1996. Contexto social y cultural. In Centurión T. R and I. Kraljevic, eds., *Las plantas útiles de Lomerío*, 25–53. Santa Cruz, Bolivia: BOLFOR (Bolivia Sustainable Forestry Management Project)-Herbario USZ (Universidad Autónoma Gabriel Rene Moreno)-CICOL (Central Intercomunal Campesina del Oriente de Lomerió).

Fittkau, E. J. and H. Klinge. 1973. On biomass and trophic structure of the Central Amazonian rain forest ecosystem. *Biotropica* 5:2–14.

Fitzgerald, C. H. and C. W. Selden. 1975. Herbaceous weed control accelerates growth in a young poplar plantation. *Journal of Forestry* 77:21–22.

Fitzgibbon, C. D., H. Mogaka and J. H. Fanshawe. 1995. Subsistence hunting in Arabuko-Sokoke forest, Kenya, and its effects on mammal populations. *Conservation Biology* 9:1116–1126.

Fleagle, J. G. 1980. Locomotion and posture. In D.J. Chivers, ed., *Malayan Forest Primates*, 191–207. New York, N.Y.: Plenum Press.

——. 1985. Size and adaptation in primates. In W. L. Jungers, ed., *Size and Scaling in Primate Biology*, 1–19. New York, N.Y.: Plenum Press

Fleagle, J. G. and R. A. Mittermeier. 1980. Locomotor behavior, body size, and comparative ecology of seven Surinam monkeys. *American Journal of Physical Anthropology* 52:301–314.

Fleming, T. H. 1986. Opportunism versus specialization: The evolution of feeding strategies in frugivorous bats. In A. Estrada and T. H. Fleming, eds., *Frugivores and Seed Dispersal, Tasks for Vegetation Science*. Volume 15, 105–118. Dordrecht, Netherlands: Dr. W. Junk Publishers.

——. 1988. *The Short-tailed Fruit Bat*. Chicago, Ill.: University of Chicago Press.

Fleming, T. H. and E. R. Heithaus. 1981. Frugivorous bats, seed shadows and the structure of tropical forests. *Biotropica* 13(supplement):45–53.

Fleming, T. H., R. Breithwisch, and G. H. Whitesides. 1987. Patterns of tropical vertebrate frugivore diversity. *Annual Review of Ecology and Systematic* 18:91–109.

Flynn, R. 1978. The Sumatran rhinoceros in the Endau-Rompin National Park of Peninsula Malayasia. *Malay Naturalist* 4:5–12.

Flynn, R. W. and M. T. Abdullah. 1984. Distribution and status of the Sumatran rhinoceros in Peninsular Malaysia. *Biological Conservation* 28:253–273.

Fogden, M. P. L. 1972. The seasonality and population dynamics of equatorial forest birds in Sarawak. *Ibis* 114:307–343.

Folgarait, P. J. 1998. Ant biodiversity and its relationship to ecosystem functioning: A review. *Biodiversity and Conservation* 7:1221–1244.

Fonseca, G. A. B. 1989. Small mammal species diversity in brazilian tropical primary and secondary forests of different sizes. *Review Brasilian Biology* 6:381–422.

Fonseca, G. A. B. and J. G. Robinson. 1990. Forest size and structure: competitive and predatory effects on small mammal communities. *Biological Conservation* 53:265–294.

Foose, T. J. and N. van Strien, eds. 1997. *Asian Rhinos: Status Survey and Conservation Action Plan*. Gland, Switzerland: IUCN (International Union for Conservation of Nature and Natural Resources).

Ford, H. A. and G. W. H. Davison. 1995. Forest avifauna of Universiti Kebangsaan Malaysia and some other forest remnants in Selangor, peninsular Malaysia. *Malayan Nature Journal* 49:117–138.

Fore, L. S., J. R. Karr, and R. W. Wisseman. 1996. Assessing invertebrate responses to human activities: Evaluating alternative approaches. *Journal of the North American Benthological Society* 15:212–231.

Forget, P-M. 1990. Seed-dispersal of *Vouacapoua americana* (Caesalpiniaceae) by cavimorph rodents in French Guiana. *Journal of Tropical Ecology* 6:459–468.

Forget, P-M. and D. S. Hammond. In press. Plant-vertebrate relationships in Guianan rainforests. In H. ter Steege and D. S. Hammond, eds., *Tropical Forests of the Guianas*. Dordrecht, Netherlands: Kluwer Academic.

Forman, R. T. T. 1995. *Land Mosaics: The Ecology of Landscapes and Regions*. Cambridge, United Kingdom: Cambridge University Press.

Foster, D. R., M. Fluet, and E. R. Boose. 1999. Human or natural disturbance: Landscape-scale dynamics of the tropical forests of Puerto Rico. *Ecological Applications* 9:555–572.

Foster, R. B. 1980. Heterogeneity and disturbance in tropical vegetation. In M. E. Soulé, and B. A. Wilcox, eds., *Conservation Biology: An Evolutionary-ecological Perspective*, 75–92. Sunderland, Mass.: Sinauer Associates, Inc.

——. 1982. Famine on Barro Colorado Island. In E. G. Leigh Jr., A. S. Rand, and D. S. Windsor, eds., *The Ecology of a Tropical Forest: Seasonal Rhythms and Long-term Changes*, 201–212. Washington, D.C.: Smithsonian Institution Press.

Foster, R. B., J. Arce, and T. S. Wachter. 1986. Dispersal and the sequential plant communities in Amazonian Peru floodplain. In A. Estrada, and T.H. Fleming, eds., *Frugivores and Seed Dispersal*, 357–370. Sunderland, Mass.: Sinauer.

Fox, J. E. D. 1968a. Exploitation of the Gola Forest. *Journal of the West African Science Association* 13:185–201

——. 1968b. Logging damage and the influence of climber cutting prior to logging in the lowland dipterocarp forest of Sabah. *Malayan Forester* 31:326–347.

——. 1976. Constraints on the natural regeneration of tropical moist forest. *Forest Ecology and Management* 1:37–65.

Fragoso, C. and P. Lavelle. 1972. Earthworm communities of tropical rain forests. *Soil Biology and Biochemistry* 24:1397–1408.

Fragoso, J. M. V. 1991a. The effect of hunting on tapirs in Belize. In Robinson, J. G. and K. E. Redford, eds., *Neotropical Wildlife Use and Conservation*, 154–162. Chicago, Ill.: The University of Chicago Press.

——. 1991b. The effects of selctive logging on Baird's tapir. In M. A. Mares and D. J. Schimdly, eds., *Latin American Mammalogy: History, Biodiversity and Conservation*, pp. 295–304. Norman, Okla.: University of Oklahoma Press.

——. 1997. Tapir-generated seed shadows: Scale-dependent patchiness in the Amazon rain forest. *Journal of Ecology* 85:519–529

Francis, C. M. 1994. Survival rates and movements of babblers in Malaysian lowland rainforest (abstract). *Journal für Ornithologie* 135:497.

Frankie, G. W., H. G. Baker, and P. A. Opler. 1974. Comparative phenological studies of tropical wet and dry forests in the lowlands of Costa Rica. *Journal of Ecology* 62:881–919.

Fredericksen, N. J., B. Flores, T. S. Fredericksen, and D. Rumiz. 1999. Wildlife use of different size logging gaps in a Bolivia tropical dry forest. BOLFOR (Bolivia Sustainable Forestry Management Project) Technical Document, Santa Cruz, Bolivia.

Fredericksen, T. S. and B. Mostacedo. 2000. Regeneration of timber species following selection logging in a Bolivian tropical dry forest. *Forest Ecology and Management* 131:47–55.

Fredericksen, T. S. and N. J. Fredericksen. 1998. Preferential use of forest tree species by bark gleaning birds in a dry forest in eastern Bolivia. Encuentro Boliviano para la Conservacion de las Aves, Estacion Biologico Beni, Bolivia.

Fredericksen, T. S., D. Rumiz, M. J. Justiniano Bravo, and R. Agaupe Abacay. 1999b. Harvesting free-standing fig trees for timber in Bolivia: Potential implications for forest management. *Forest Ecology and Management* 116:151–161.

Freemark, K. E. and H. G. Merriam. 1986. Importance of area and habitat heterogeneity to bird assemblages in temperate forest fragments. *Biological Conservation* 36:115–141.

Freese, C.H. 1996. *The Commercial, Consumptive Use of Wild Species: Managing it for the Benefits of Biodiversity.* Washington, D.C.: WWF (World Wildlife Fund),

Friend, J. A. 1987. Local decline, extinction and recovery: relevance to mammal populations in vegetation remnants. In S. A. Saunders, G. W. Arnold, A. A. Burbidge and A. J. M. Hopkins, eds., *Nature Conservation: The Role of Remnants of Native Vegetation*, 53–64. SCIRO and CALM, Australia: Surrey Beatty and Sons Pty Limited.

Friends of the Earth. 1991. *Life after logging? The Role of Tropical Timber Extraction in Species Extinction.* London, United Kingdom: Friends of the Earth.

Frumhoff, P. 1995. Conserving wildlife in tropical forests managed for timber. *Bioscience* 45:456–464.

Frumhoff, P. C. and E. C. Losos. 1998. Setting priorities for conserving biological diversity in tropical timber production forests. Policy report. The Union of Concerned Scientists and the Center for Tropical Forest Science, Smithsonian Institution. Cambridge, Mass. and Washington, D.C.

Frumhoff, P. C., D. C. Goetze, and J. J. Hardner. 1998. Linking solutions to climate change and biodiversity loss through the Kyoto Protocol's Clean Development Mechanism. UCS Reports, Union of Concerned Scientists, Cambridge, Mass..

FSC (Forest Stewardship Council). 1996. *Principles and Criteria for Forest Management.* Oaxaca, Mexico: FSC.

——. 1998. Report on January 1998 Meeting of FSC Board of Directors. FSC, Oaxaca, Mexico: FSC.

Fuentes, M. 1995. How specialized are fruit-bird interactions? Overlap of frugivore assemblages within and between plant species. *Oikos* 74:324–330.

Fujita, M. S. and M. D. Tuttle. 1991. Flying foxes (Chiroptera: Pteropodidae): Threatened animals of key ecological and economical importance. *Conservation Biology* 5:455–463

Furness, R. W., J. J. D. Greenwood, and P. J. Jarvis. 1993. Can birds be used to monitor the environment? In R.W. Furness, and J. J. D. Greenwood, eds., *Birds as Monitors of Environmental Change*, 1–41. London, United Kingdom: Chapman and Hall.

Futuyma, D. 1973. Community structure and stability in constant environments. *American Naturalist* 107:443–446.

Gaceta Oficial de Bolivia. 1996. Ley forestal #1700 Articulo 29 deg III f). Ano XXXVI #1944 (12 julio). La Paz, Bolivia.

Gadgil, M. and F. Berkes. 1993. Indigenous knowledge for biodiversity conservation. *Ambio* 22:151–156.

Galletti, M. and F. Pedroni. 1994. Seasonal diet of capuchin monkey (*Cebus apella*) in a forest in south-east Brazil. *Journal of Tropical Ecology* 10:27–39.

Galletti, M., F. Pedroni, and L. P. C. Morellato. 1994. Diet of the brown howler monkey (*Alouatta fusca*) in a forest fragment in southeastern Brazil. *Mammalia* 48:111–118.

Gally, M. and P. Jeanmart. 1996. Etude de la chasse villageoise en forêt dense humide d'Afrique centrale. Travail de fin d'études. Faculté Universitaire des Sciences Agronomiques de Gembloux, Gembloux, France.

Ganzhorn, J. U. 1995a. Low-level forest disturbance effects on primary production, leaf chemistry and lemur populations. *Ecology* 76:2084–2096.

——. 1995b. Cyclones over Madagascar: Fate or fortune? *Ambio* 24:124–125.

Ganzhorn, J. U. and J-P Sorg, eds. 1996. Ecology and Economy of a tropical dry forest in Madagascar. *Primate Report* Special Issue 46-1.

Ganzhorn, J. U., A. W. Ganzhorn, J-P. Abraham, L. Andriamanarivo, and A. Ramananjatovo. 1990. The impact of selective logging on forest structure and tenrec populations in western Madagascar. *Oecologia* 84:126–133.

Garber, P.A. and J. D. Pruetz. 1995. Positional behavior in moustached tamarin monkeys: Effects of habitat on locomotor variability and locomotor stability. *Journal of Human Evolution* 28:411–428.

García-Montial, D. C. and F. N. Scatena. 1994. The effects of human activity on the structure and composition of a tropical forest in Puerto Rico. *Forest Ecology and Management* 63:57–78.

Gardner, A. L. 1977. Feeding habits. In R. J. Baker, J. K. Jones, and D. C. Carter, eds., *Biology of Bats of the New World Family Phyllostomatidae Part II*, 293–350. Special publications No. 13. Lubbock,Tex.: The Museum of Texas Technical University.

Gardener, A. L., R. K. LaVal, and D. E. Wilson. 1970. The distributional status of some Costa Rican bats. *Journal of Mammalogy* 51:712–729.

Gardner, S. M., M. R. Cabido, G. R. Valladares, and S. Diaz. 1995. The influence of habitat structure on arthropod diversity in Argentine semi-arid Chaco forest. *Journal Vegetation Science* 6:349–356.

Garman, G. C. and J. R. Moring. 1993. Diet and annual production of two boreal river fishes following clearcut logging. *Environmental Biology of Fishes* 36:301–311.

Garrison, R. W. and M. R. Willig. 1996. Arboreal invertebrates. In D. P. Reagan and R. B. Waide, eds., *The Food Web of a Tropical Rain Forest*, 183–246 Chicago, Ill.: University of Chicago Press.

Gartshore, M. E., P. D. Taylor, and I. S. Francis. 1995. Forest birds in Côte d'Ivoire. A survey of Tai National Park and other forests and forestry plantations, 1989–91. *Birdlife International Study Report* 58:1–81.

Garwood, N. C., D. P. Janos, and N. Brokaw. 1979. Earthquake-caused landslides: A major disturbance to tropical forests. *Science* 205:997–999.

Gascon, C. 1991. Population- and community-level analyses of species occurrences of central Amazonian rainforest tadpoles. *Ecology* 72:1731–1746.

Gaston, K. J. 1991. The magnitude of global species richness. *Conservation Biology* 5:282–296.

——. 1994. *Rarity: Population and Community Biology.* Series 13. London, United Kingdom: Chapman and Hall.

Gautier-Hion, A. 1984. Seed dispersal by African forest cercopithecines. *La Terre et la Vie: revue d'ecologie appliquee.* 39:159–166.

Gautier-Hion, A. and G. Michaloud. 1989. Are figs always keystone resources for tropical frugivorous vertebrates? A test in Gabon. *Ecology* 70:1826–1833.

Gautier-Hion, A., J-M. Duplantier, R. Quris, F. Feer, C. Sourd, J-P. Decoux, G. Dubost, L. Emmons, C. Erard, P. Hecketweiler, A. Moungazi, C. Roussilhon, and J-M. Thiollay. 1985. Fruit characters as a basis of fruit choice and seed dispersal in a tropical forest vertebrate community. *Oecologia* 65:324–337.

Gautier-Hion, A., L. H. Emmon, and G. Dubost. 1982. A comparison of the diets of three major groups of primary consumers of Gabon (primates, squirrels, and ruminants). *Oecologia.* 45:182–189.

Gebo, D. L. and C. A. Chapman. 1995a. Habitat, annual, and seasonal effects on positional behavior in red colobus monkeys. *American Journal of Physical Anthropology* 96:72–82.

——. 1995b. Positional behaviour in five species of old world monkeys. *American Journal of Physical Anthropology* 97:49–76.

Geist, V. 1988. How markets for wildlife meat and parts, and the sale of hunting privileges, jeopardize wildlife conservation. *Conservation Biology* 2:15–26.

Gemeinschaft fur Technische Zusammenarbeitung (GTZ). 1979. Etat actuel des parcs nationaux de la Comoe et de Tai ainsi que de la reserve d'Azagny et propositions visant a leur conservation et a leur developpement aux fins de promotion du tourisme. Tome III: Parc National de Tai. Technical Co-operation Report GTZ-PN 73.2085.6–01.100.

Gentry A. H. 1982. Patterns of neotropical plant species diversity. *Evolutionary Biology* 15: 1–84

——. 1991. The distribution and abundance of climbing plants. In F. E. Putz and H. A. Mooney, eds., *The Biology of Vines*, 3–52. Cambridge, United Kingdom: Cambridge University Press.

Gerhardt, K. 1993. Tree seedling development in tropical dry abandoned pasture and secondary forest in Costa Rica. *Journal Vegetation Science* 4:95–102.

Ghazoul, J. 1997. The pollination and breeding system of Dipterocarpus obtusifolius (Dipterocarpaceae) in dry deciduous forests of Thailand. *Journal of Natural History* 31:901–916.

——. 2000. Indirect effects of logging on the reproductive ecology of a butterfly-pollinated tree. *Journal of Ecology.* In press.

Ghazoul, J., K. A. Liston, and T. J. B. Boyle. 1998. Disturbance induced density dependent seed set in *Shorea siamensis* (Dipterocarpaceae), a tropical forest tree. *Journal of Ecology* 86:462–473.

Giao, P. M., D. Tuoc, V. V. Dung, E. D. Wikramanayake, G. Amato, P. Arctander, and J. MacKinnon. 1998. Description of *Muntiacus truongsonensis*, a new species of muntjac (Artiodactyla: Muntiacidae) from Central Vietnam, and implications for conservation. *Animal Conservation* 1:61–68.

Gibbons, D. R. and E. O. Salo. 1973. An annotated bibliography of the effects of logging on fish of the western United States and Canada. USDA Forest Service General Technical Report PNW–10.

Gibbs, J. P., M. L. Hunter, and S. M. Melvin. 1993. Snag availability and communities of cavity nesting birds in tropical versus temperate forests. *Biotropica* 25:236–241.

Gilbert, G. S., S. P. Hubbell, and R. B. Foster. 1994. Density and distance-to-adult effects of a canker disease of trees in a moist tropical forest. *Oecologia* 98:100–108

Gilbert, L. E. 1980. Food web organization and the conservation of neotropical diversity. In M. E. Soulé and B. A. Wilcox, eds., *Conservation Biology: An Evolutionary-Ecological Perspective*, 11–33. Sunderland, Mass.: Sinauer Associates.

Giller, P. S. 1984. *Community Structure and the Niche.* London, United Kingdom: Chapman and Hall.

Gillman, G. F., D. F. Sinclair, R. Knowlton, and M. G. Keys. 1985. The effect on some soil chemical properties of the selective logging of a north Queensland rainforest. *Forest Ecology and Management* 12:195–214.

Gilman, E. L. 1997. Community based and multipurpose protected areas: A model to select and manage protected areas with lessons from the Pacific Islands. *Coastal Management* 25:59–91.

Gilmour, D. A. 1971. The effects of logging on streamflow and sedimentation in a north Queensland rainforest catchment. *Commonwealth Forestry Review* 50:38–48.

——. 1977. Logging and the environment with particular reference to soil and stream protection in tropical rainforest situations. In J. L. Davis, ed., *Guidelines for Watershed Management*, 223–235. FAO (Food and Agriculture Organization) Conservation Guide 1, Rome, Italy: FAO.

Glanz, W. E. 1991. Mammalian densities at protected versus hunted sites in central Panama. In J. G. Robinson and K. H. Redford, eds., *Neotropical Wildlife Use and Conservation*, 163–173. Chicago, Ill.: University of Chicago Press.

Godoy, H. C. 1990. The greenhouse effect and protected areas in Central America: A contribution to stop the changes in global climate. In P. Faeth, M. C. Trexler, and D. Page, eds., *Sustainable Forestry as a Response to Global Warming: A Central American perspective*, 25–39. Washington, D.C.: WRI (World Resources Institute).

Goldingay, R., G. Daly, and F. Lemckert. 1996. Assessing the impacts of logging on reptiles and frogs in the montane forests of southern New South Wales. *Wildlife Research* 23:495–510.

Golley, F. B. 1983. Decomposition. In F. B. Golley, ed., *Tropical Rain Forest Ecosystems-Structure and Function*, 101–114. Amsterdam: Elsevier, Ecosystems of the World 14a.

Gómez-Pompa, A. and Burley, F. W. 1991. The management of natural tropical forests. In A. Gómez-Pompa, T. C. Whitmore and M. Hadley, eds., *Rain Forest Regeneration and Management,* 3–18. Paris, France: M. A. B.-series Vol. 6, UNESCO (United Nations Educational, Scientific, and Cultural Organization)/Parthenon, Press.

Gonzalez, G. and X. Zou. 1999a. Earthworm influence on N availability and the growth of *Cecropia schreberiana* in tropical pasture and forest soils. *Pedobiologia* 43(6):824–829.

———. 1999b. Plant and litter influences on earthworm abundance and community structure in a tropical wet forest. *Biotropica* 31:486–493.

González, G., X. Zou and S. Borges. 1996. Earthworm abundance and species composition in abandoned tropical croplands: comparisons of tree plantations and secondary forests. *Pedobiologia* 40:385–391.

Gonzalez, G., X. Zou, A. Sabat, and N. Fetcher. 1999. Earthworm abundance and distribution pattern in contrasting plant communities within a tropical wet forest. *Caribbean Journal of Science* 35:93–100.

González-M., A. and M. Alberico. 1993. Selección de hábitat en una comunidad de mámiferos pequeños en la Costa Pacífica de Colombia. *Caldasia* 17:313–324.

Gorchov, D. L., F. Cornejo, C. Ascorra, and M. Jaramillo. 1993. The role of seed dispersal in the natural regeneration of rain forest after strip-cutting in the Peruvian Amazon. *Vegetation* 107/108:339–349.

Gotmark, F. and C. Nilsson. 1992. Criteria used for protection of natural areas in Sweden, 1909–1986. *Conservation Biology* 6(2):220–231.

Gottsberger, G. 1978. Seed dispersal by fish in the inundated regions of Humaitá, Amazonia. *Biotropica* 10:170–183.

Goulding, M. 1980. *The Fishes and the Forest: Explorations in Amazonian Natural History.* Berkeley, Calif.: The University of California Press.

Goulding, M., N. J. H. Smith, and D. J. Mahar. 1996. *Floods of Fortune: Ecology and Economy Along the Amazon.* New York, N.Y.: Columbia University Press.

Grace, J., J. Lloyd, J. McIntyre, A. C. Miranda, P. Meir, H. S. Miranda, C. Nobre, J. Moncrieff, J. Massheder, Y. Malhi, I. Wright, and J Gash. 1995. Carbon Dioxide uptake by an undisturbed tropical rain forest in Southwest Amazonia, 1992 to 1993. *Science* 270:778–80.

Graham, C. H., T. C. Moermond, K.A. Kristensen and J. Mvukiyumwami. 1995. Seed dispersal effectiveness by two bulbuls on *Maesa lanceolata,* an African montane forest tree. *Biotropica* 27:479–486.

Grainger, A. 1987. The future environment for forest management in Latin America. In *Management of the Forests of Tropical America: Prospects and Technologies,* pp. 1–9. Washington, D.C.: Institute of Tropical Forestry/USDA Forest Service.

Gram, W. K., and J. Faaborg. 1997. The distribution of Neotropical migrant birds wintering in the El Cielo Biosphere Reserve, Tamaulipas, Mexico. *Condor* 99(3):658–670.

Grand, T. I. 1984. Motion economy within the canopy: Four strategies for mobility. In P.S. Rodman and J. G. H. Cant, eds., *Adaptations for Foraging in Nonhuman Primates,* pp. 54–72. New York, N.Y.: Columbia University Press.

Grant, J. D. and W. F. Laurance. 1991. Court size and maintenance in the tooth-billed catbird. *Sunbird* 21:90–92.

Green, M. J. B., M. G. Murray, G. C. Bunting, and J. R. Paine. 1996. *Priorities for Biodiversity Conservation in the Tropics.* Cambridge, United Kingdom: World Conservation Monitoring Centre.

Greenburg, R. 1981. The abundance and seasonality of forest canopy birds on Barro Colorado Island, Panama. *Biotropica* 13:241–251.

Gregory, K. J. 1992. Vegetation and river channel process interactions. In P. J. Boon and G. E. Petts, eds., *River Conservation and Management,* pp. 255–270. New York, N.Y.: John Wiley and Sons.

Gribel, R. 1988. Visits of *Caluromys lanatus* (Didelphidae) to flowers of *Pseudobombax tomentosum* (Bombacaceae): A probable case of pollination by marsupials in Central Brazil. *Biotropica* 20:344–347.

Grieser Johns, A. 1996. Bird population persistence in Sabahan logging concessions. *Biological Conservation* 75:3–10.

——. 1997. *Timber Production and Biodiversity Conservation in Tropical Rain Forests*. Cambridge, United Kingdom: Cambridge University Press.

Grieser Johns, A. and B. Grieser Johns. 1995. Tropical forest primates and logging: Long-term co-existence? *Oryx* 29:205–211.

Grime, J. P. 1977. Evidence for the existence of three primary strategies in plants and its relevance to ecological and evolutionary theory. *American Naturalist* 111:1169–1194.

Grindal, S. D. 1996. Habitat use by bats in fragmented forests. In Barclay, R. M. R. and R. M. Brigham, eds., *Bats and Forests Symposium; 1995; Victoria, British Columbia, Canada*, pp. 260–272. British Columbia: Ministry of Forests Research Program.

Groves, C. P. 1967. On the rhinoceroses of South East Asia. *Säugetierkundliche Mitteilungen*. 15:221–237.

Grubb, P. J. 1995. Mineral nutrition and soil fertility in tropical rain forests. In A. E. Lugo, and C. Lowe, eds., *Tropical Forests: Management and Ecology*, pp. 308–330. New York, N.Y.: Springer-Verlag.

Guamán, A, and J. Montaño. 1988. *Mapa de isoyetas e isotermas del Departamento de Santa Cruz*. Santa Cruz, Bolivia: CORDECRUZ-SENAMHI.

Guariguata, M. R., and M. A. Pinard. 1998. Ecological knowledge of regeneration from seed in neotropical forest trees: Implications for natural forest management. *Forest Ecology and Management* 112:87–99.

Guariguata, M. R., J. J. R. Adame, and B. Finegan, 2000. Seed removal and fate in two selectively logged lowland forests with constrasting protection levels. *Conservation Biology* 14:1046–1054.

Guinart, D. 1997. Los mamíferos del bosque semideciduo neotropical de Lomerío (Bolivia). Interacción indígena. Ph.D. diss., University of Barcelona.

Gullison, R. E. and J. J. Hardner. 1993. The effects of road design and harvest intensity on forest damage caused by selective logging: Empirical results and a simulation model from Bosque Chimanes, Bolivia. *Forest Ecology and Management* 59:1–14.

Gullison, R. E., S. N. Panfil, J. J. Strouse, and S. P. Hubbell. 1996. Ecology and management of mahogany (Swietenia macrophylla King) in the Chimanes Forest, Beni, Bolivia. *Botanical Journal of the Linnean Society* 122:9–34.

Gumal, M. T. and E. L. Bennett. 1995. *Report on a Field Trip to Penan Communities of Belaga District*, 11th–14th December 1995. Kuching, Sarawak: Sarawak Forest Department.

Guzmán, R. 1998. Planes de manejo en el nuevo regimen forestal. *Boletín BOLFOR (Bolivia Sustainable Forestry Management Project)* 14:7–8

Hails, C. J. and F. Jarvis. 1987. *Birds of Singapore*. Singapore: Times Editions.

Hall, J. B. and M. D. Swaine. 1981. *Distribution and Ecology of Vascular Plants in a Tropical Rain Forest. Forest vegetation of Ghana*. Geobotany series Vol. 1. Dordrecht, Netherlands: Dr. W. Junk Publishers.

Hallwachs, W. 1993. Agoutis (Dasyprocta punctata): The inheritors of guapinol (*Hymenea courbaril*: Leguminosae). In A. Estrada and T. H. Fleming, eds., *Frugivores and Seed Dispersal*. Volume 15, *Tasks for Vegetation Science*, pp. 285–304. Dordrecht, Netherlands: Dr. W. Junk Publishers.

Halpin, P. N. 1997. Global climate change and natural-area protection: Management responses and research directions. *Ecological Applications*. 7(3):828–843.

Hamer, K. C., J. K. Hill, L. A. Lace, and A. M. Langhan. 1997. Ecological and biogeographical effects of forest disturbance on tropical butterflies of Sumba, Indonesia. *Journal of Biogeography* 24:67–75.

Hamilton, A. C. 1982. *Environmental History of East Africa*. New York, N.Y.: Academic Press.

Hamilton, L. S. and P. N. King. 1983. *Tropical Forested Watersheds*. Boulder, Colo.: Westview Press.

Hamilton, L. S. and S. C. Snedaker, eds. 1984. *Handbook for Mangrove Area Management*. Gland, Switzerland: UNEP.

Hamilton, W. D. and R. May. 1977. Dispersal in stable habitats. *Nature* 269:578–581

Hammond, D. S. 1996. A compilation of known Guianian timber trees and the significance of their dispersal mode, seed size, and taxonomic affinity to tropical rain forest management. *Forest Ecology and Management* 83:99–116.

Hammond, D. S. and V. K. Brown. 1995. Seed size of woody plants in relation to disturbance, dispersal, soil type in wet Neotropical forests. *Ecology* 76:2544–2561.

Hammond, D. S., V. K. Brown, and R. Zagt. 1999. Spatial and temporal patterns of seed attack and germination in a large-seeded neotropical tree species. *Oecologia* 119:208–218.

Hammond, D. S., S. Gourlet-Fleury, P. van der Hout, H. ter Steege, and V. K. Brown. 1996. A compilation of known Guianan timber trees and the significance of their dispersal mode, seed size and taxonomic affinity to tropical rain forest management. *Forest Ecology and Management* 83:99–116.

Hammond, D. S., A. Schouten, L. van Tienen, M. Weijerman, and V. K. Brown. 1992. The importance of being a forest animal: Implications for Guyana's timber trees. In *NARI/CARDI* (National Agricultural Research Institute in Guyana / the Caribbean Agricultural Research and Development Institute) eds., *Proceedings of the 6th Annual NARI/CARDI Review Conference*, pp. 144–151. Mon Repos, Guiana: NARI

——. 1998. Disturbance, phenology and life-history characteristics: Factors influencing frequency-dependent invertebrate attack on tropical seeds and seedlings. In D. Newberry, N. Brown and H. Prins, eds., *Population and Community Dynamics in the Tropics*. Oxford, United Kingdom: Blackwell.

Handley, C. O., Jr. 1967. Bats of the canopy of an Amazonian forest. *Atas do Simpósio sôbre a Biota Amazônica* 5:211–215.

——. 1976. Mammals of the Smithsonian Venezuelan Project. *Brigham Young University Science Bulletin, Biological Series* 20:1–91.

Hanski, I. 1987. Dung beetles. In H. Lieth and M. J. A. Werger, eds., *Tropical Rainforest Ecosystems, Biogeography, and Ecological Studies*, pp. 489–511. Elsevier, Amsterdam: Ecosystems of the World, 14b.

——. 1991. Single-species metapopulation dynamics: Concepts, models and observations. *Biological Journal of the Linnean Society* 42:17–38.

Hanski, I., A. Moilanen, and M. Gyllenberg. 1996. Minimum viable metapopulation size. *American Naturalist* 147:527–541.

Happold, D. C. D. 1973. *Large Mammals of West Africa*. Harlow, United Kingdom: Longman Scientific and Technical.

Harcourt, C. and J. Thornback. 1990. *Lemurs of Madagascar and the Comoros: The IUCN (International Union for Conservation of Nature and Natural Resources) Red Data Book*. Gland, Switzerland, and Cambridge, United Kingdom: IUCN (International Union for Conservation of Nature and Natural Resources).

Hardner, J. J. and R. Rice. 1999. Rethinking forest concession policies. In K. Keipi, ed., *Forest Resource Policy in Latin America*, pp. 161–194. Baltimore, Md.: Johns Hopkins University Press.

Hare, J. and A. Stevens. 1995. Joint implementation: A critical approach. In C. J. Jepma, ed., *The Feasibility of Joint Implementation*, pp. 79–85. The Netherlands: Kluwer Academic Publishers.

Harmon, M. E., W. K. Ferrell, and J. F. Franklin. 1990. Effects on carbon storage of conversion of old-growth forests to young forests. *Science* 247:699–701.

Harper, J. 1977. *Population Biology of Plants*. New York, N.Y.: Academic Press.

Harris, L. D. 1984. *The Fragmented Forest: Island Biogeography and the Preservation of Biotic Diversity*. Chicago, Ill.: University of Chicago Press.

Harrison, R. D. 2000. Repercussions of El Nino: drought causes extinction and the breakdown of mutualism in Borneo. *Proceedings of the Royal Society of London, Series B* 267:911–915.

Harrison, R. L. 1992. Toward a theory of inter-refuge corridor design. *Conservation Biology* 6:293–295.

Hart, J. A. 1985. Comparative dietary ecology of a community of frugivorous forest ungulates in Zaire. Ph.D. diss., University of Michigan.

Hart, J. A. and T. B. Hart. 1989. Ranging and feeding behaviour of okapi (*Okapia johnstoni*) in the Ituri Forest of Zaire: Food limitation in a rain-forest herbivore? In P. A. Jewell and G. M. O. Maloiy, eds., *The Biology of Large African Mammals in their Environment*, pp. 31–50. Oxford, United Kingdom: Clarendon Press.

Hart, J. A. and S. Kiyengo. 1989. Rapport d'une mission de prospection au Parc National del la Maiko, Zaire. Institut Zairois pour la Conservation del la Nature, Kinshasa, Zaire.

Hart, T. B. 1985. The ecology of a single-species-dominant forest and a mixed forest in Zaire, Africa. Unpublished Ph.D. diss., Michigan State University.

Hart, T. B., J. A. Hart, and P. G. Murphy. 1989. Monodominant and species rich forests of the humid tropics: causes for their co-occurrence. *American Naturalist* 133:613–633.

Hart, T. B., J. A. Hart, R. Dechamps, M. Fournier, and M. Ataholo. 1997. Changes in forest composition over the last 4,000 years in the Ituri basin, Zaire. In L. J. G. van der Maesen, X.M. van der Burgt, and J.M. van Medenbach de Rooy, eds., *The Biodiversity of African Plants*, pp. 545–563. Dordrecht, Netherlands: Kluwer Academic Publishers.

Hartshorn, G. S. 1978. Tree falls and tropical forest dynamics. In P. B. Tomlinson and M. H. Zimmerman, eds., *Tropical Trees as Living Systems*, pp. 617–638. Cambridge, United Kingdom: Cambridge University Press.

——. 1980. Neotropical forest dynamics. *Biotropica* 12:23–30.

——. 1989. Application of gap theory to tropical forest management: Natural regeneration in strip clear-cuts in the Peruvian Amazon. *Ecology* 70:567–569.

——. 1990a. Natural forest management by the Yanesha Forestry Cooperative in Peruvian Amazonia. In A. B. Anderson, ed., *Alternatives to Deforestation: Steps Toward Sustainable Use of the Amazon Rain Forest*, pp. 128–138. New York, N.Y.: Columbia University Press.

——. 1990b. An overview of Neotropical forest dynamics. In A. H. Gentry, ed., *Four Neotropical Rainforests*, pp. 585–599. New Haven, Conn.: Yale University Press.

——. 1995. Ecological basis for sustainable development in tropical forests. *Annual Review of Ecology and Systematics* 26:155–175.

Hartshorn, G. S. and W. Pariona. 1993. Ecologically sustainable forest management in the Peruvian Amazon. In C. Potter and J. Cohen, eds. *Perspectives on Biodiversity: Case Studies of Genetic Resource Conservation and Development*, pp. 151–166 Washington, D.C: AAAS Press.

Hashimoto, C. 1995. Population census of the chimpanzees in the Kalinzu forest, Uganda: comparison between methods with nest counts. *Primates* 36:477–488.

Hatch, U. and M. E. Swisher. eds. *Tropical Managed Ecosystems: New Perspectives on Sustainability*. Oxford, United Kingdom: Oxford University Press.

Hawkins, A. F. A. and L. Wilme. 1996. Effects of logging on forest birds. *Primate Report* 46–1:203–213.

Hawkins, A. F. A., P. Chapman, J. U. Ganzhorn, Q. M. C. Bloxham, S. C. Barlow, and S. J.Tonge. 1990. Vertebrate conservation in Ankarana and Analamera Special Reserves, northern Madagascar. *Biological Conservation* 54:83–110.

Hawthorne, W. D. 1993. Forest regeneration after logging. Findings of a study in the Bia South Game production Reserve, Ghana. ODA Forestry Series No. 3.

———. 1995. Ecological profiles of Ghanaian forest trees. Tropical Forestry papers No. 29, University of Oxford, United Kingdom: Oxford Forestry Institute.

Heck, A. L., G. Vanbelle, and D. S. Simberloff. 1975. Explicit calculation of the rarification diversity measurement and the determination of sufficient sample size. *Ecology* 56:1459–1461.

Heideman, P. D. 1989. Temporal and spatial variation in the phenology of flowering and fruiting in a tropical rainforest. *Journal of Ecology* 77:1059–1079

Heinen, J. T. 1992. Comparisons of the leaf litter herpetofauna in abandoned Cacao plantations and primary rain forest in Costa Rica: Some implications for faunal restoration. *Biotropica* 24:431–439.

Heinichen, I. G. 1972. Preliminary notes on the suni, *Nesotragus moschatus* and red duiker, *Cephalophus natalensis. Zoologica Africana* 7:157–165.

Heliovaara, K. and R. Vaisanen. 1984. Effects of modern forestry on northwestern European forest invertebrates: A synthesis. *Acta Forestalia Fennica* 189:1–32.

Heltshe, J. F. and N. E. Forrester. 1983. Estimating species richness using the jackknife procedure. *Biometrics* 39:1–11.

Henderson, M. 1997. Forest futures for Papua New Guinea: Logging or community forestry? In B. Burt and C. Clerk., eds., *Environment and Development in the Pacific Islands*, pp. 45–68. Australia and Port Moresby, PNG National Centre for Development Studies and the University of Papua New Guinea, Canberra.

Hendrison, J. 1990. Damage-controlled logging in managed tropical forest in Surinam. Wageningen Agricultural University, Wageningen, The Netherlands.

Henle, K. and R. Apfelbach. 1985. Auswilderung und Beobachtungen zur Biologie des Schwarzruckenducker (*Cephalophus dorsalis* Gray, 1846). *Säugetierkundliche Mitteilungen* 32:75–82.

Herrera, C. M. 1984. A study of avian frugivores, bird-dispersed plants, and their interaction in mediterranean scrubland. *Ecological Monographs* 54:1–23.

Herrera, R. A., R. P. Capote, L. Menendez, and M. E. Rodrigues. 1990. Silvigenesis stages and the role of mycorrhiza in natural regeneration in Sierra Del Rosario, Cuba. In A. Gomez-Pompa, T. C. Whitmore, and M. Hadley, eds., *Rainforest Regeneration and Management*, pp. 211–222. Paris, France: Parthenon Publishing Group.

Hershkovitz, P. 1972. The recent mammals of the Neotropical Region: a zoogeographic and ecological review. In A. Keast, F. C. Erk and B. Glass, eds., *Evolution, Mammals, and Southern Continents*, pp. 311–431. Albany, N.Y.: New York Press.

Heydon, M. J. 1994. The Ecology and Management of Rain Forest Ungulates in Sabah, Malaysia: Implications of Forest Distrubance. Final Report. Unpublished report. Institute of Tropical Biology, Dept. of Zoology, University of Aberdeen, Aberdeen, Scotland.

Heydon, M. J. and P. Bulloh. 1996. The impact of selective logging on sympatric civet species. *Oryx* 30:31–36.

———. 1997. Mousedeer densities in a tropical rainforest: The impact of selective logging. *Journal of Applied Ecology* 34:484–496.

Heyer, W. R. 1988. On frog distribution patterns east of the Andes. In P. E. Vanzolini, and W. R. Heyer, eds., *Proceedings of a Workshop on Neotropical Distribution Patterns*, pp. 245–273. Rio de Janeiro, Brasil: Academia Brasileira de Ciências.

Heyer, W. R., M. A. Donnelly, R. W. McDiarmid, L. C. Heyek, and M.S . Foster. 1994. *Measuring and Monitoring Biological Diversity: Standard Methods for Amphibians*. Washington, D.C.: Smithsonian Institution Press.

Higgs, A. J. 1981. Island biogeography theory and nature reserve design. *Journal of Biogeography* 8:117–124.

Hill, J. K. 1999. Butterfly spatial distribution and habitat requirements in a tropical forest: impacts of selective logging. *Journal of Applied Ecology* 36:564–572.

Hill, J. K., K. C. Hamer, L. A. Lace, and W. M. T. Banham. 1995. Effects of selective logging on tropical butterflies on Buru, Indonesia. *Journal of Applied Ecology* 32:454–460.

Hill, J. E., and J. D. Smith. 1984. *Bats: A Natural History*. Austin, Tex.: University of Texas Press.

Hill, K., J. Padwe, C. Bejyvagi, A. Bepurangi, F. Jakugi, R. Tykuarangi, and T. Tykuarangi. 1997. Impact of hunting on large vertebrates in the Mbaracayu Researve, Paraguay. *Conservation Biology*. 11(6)1339–1353.

Hilty, S. L. 1980. Flowering and fruiting periodicity in a premontane rainforest in Pacific Colombia. *Biotropica* 12:292–306.

Hladik, A. and S. Miquel. 1990. Seedling types and plant establishment in an African rain forest. In K.S. Bawa and M. Hadley, eds., *Reproductive Ecology of Tropical Forest Plants*, M.A.B. Series Vol. 7, pp. 261–282. Paris, France: UNESCO (United Nations Educational, Scientific, and Cultural Organization)/Parthenon.

Hladik, C. M., S. Bahuchet, and I. de Garine. 1990. Food and nutrition in the African rain forest. Paris, France: UNESCO (United Nations Educational, Scientific, and Cultural Organization)-MAB (Man and the Biosphere Series).

Hobbs, H. H., Jr. and C. W. Harte, Jr. 1982. The shrimp genus Atya (Decapoda: Atyidae). *Smithsonian Contributions to Zoology* No. 364.

Hoch, G. A. and G. H. Adler. 1997. Removal of black palm (*Astrocaryum standleyanum*) seeds by spiny rats (*Proechimys semispinosus*). *Journal of Tropical Ecology* 13:51–58.

Hoffmann, B. D., A. N. Andersen, and G. J. E. Hill. 1999. Impact of an introduced ant on native rain forest invertebrates: *Pheidole megacephala* in monsoonal Australia. *Oecologia* 120:595–604.

Hölldobler, B. and E. O. Wilson. 1990. *The Ants*. Cambridge, Mass.: Belknap Press of Harvard University Press.

Holling, C. S., ed. 1978. *Adaptive Environmental Assessment and Management*. Volume 3. *International series on applied systems analysis*. New York, N.Y.: John Wiley and Sons.

——. 1992. Cross-scale morphology, geometry, and dynamics of ecosystems. *Ecological Monographs* 62(4):447–502.

Holloway, J. D., A. H. Kirk-Spriggs, and V. K. Chey. 1992. The response of some rain forest insect groups to logging and conversion to plantation. *Philosophical Transactions of the Royal Society of London, Series B* 335:425–436.

Holmes, T. P., G. M. Blate, J. C. Zweede, R. Pereira Jr., P. Barreto, F. Boltz, and R. Bauch. 1999. Financial costs and benefits of reduced-impact logging relative to conventional logging in the Eastern Amazon. Phase 1 Final Report. Miscellaneous report. Washington, D.C.: Tropical Forest Foundation.

Holt, J. A. and R. J. Coventry. 1988. The effects of tree clearing and pasture establishment on a population of mound-building termites (Isoptera) in North Queensland. *Australian Journal of Ecology* 13: 21–325.

Holthuijzen, A. M. A. and J. H. A. Boerboom. 1982. The Cecropia seedbank in the Surinam lowland rain forest. *Biotropica* 14:62–68.

Hommel, P. W. F. M. 1987. *Landscape-ecology of Ujung Kulon (West Java, Indonesia)*. Utrecht, The Netherlands: Hommel Publishing.

Hoogerwerf, A. 1970. *Udjung Kulon: The Land of the Last Javan Rhinoceros*. R.J. Brill, Leiden.

Hoppe-Dominik, B. 1989. Habitatspräferenz und Nahrungsansprüche des Waldbüffels, *Syncerus caffer nanus*, im Regenwald des Elfenbeinküste. Ph.D. diss., Braunschweig, Germany.

Horn, M. H. 1997. Evidence for dispersal of fig seeds by the fruit-eating characid fish Brycon guatemalensis Regan in a Costa Rican tropical rain forest. *Oecologia* 109:259–264.

Horne, R. and J. Hickey. 1991. Ecological sensitivity of Australian rainforests to selective logging. *Australian Journal of Ecology* 16:119–129.

Houghton, R. A. 1990. The future role of tropical forests in affecting the carbon dioxide concentration of the atmosphere. *Ambio* 19:204–209.

——. 1991. Tropical deforestation and atmospheric carbon dioxide. *Climatic Change* 19:99–118.

Howard, A. F. and J. Valerio. 1996. Financial returns from sustainable forest management and selected agricultural land-use options in Costa Rica. *Forest Ecology and Management* 81:35–49.

Howard, P. 1991. *Nature Conservation in Uganda's Tropical Forest Reserves*. Gland, Switzerland: IUCN (International Union for Conservation of Nature and Natural Resources).

Howard, P., T. Davenport, and R. Matthews. 1996a. *Kibale National Park Biodiversity Report*. Kempala: Uganda Forest Department.

——. 1996b. *Budongo Forest Reserve Biodiversity Report*. Kempala: Uganda Forest Department.

Howden, H. F. and V. G. Nealis. 1975. Effects of clearing in a tropical rain forest on the composition of the coprophagous scarab beetle fauna (Coleoptera). *Biotropica* 7:77–83.

Howe, H. F. 1977. Bird activity and seed dispersal of a wet tropical forest tree. *Ecology* 58:539–550.

——. 1982. Fruit production and animal activity in two tropical trees. In E. G. Leigh, Jr, A. S. Rand, and D. M. Windsor, eds., *The Ecology of a Tropical Forest: Seasonal Rhythms and Long-term Changes*, pp. 189–199. Washington, D.C.: Smithsonian Institution Press.

——. 1983. Annual variation in a neotropical seed-dispersal system. In S. L. Sutton and T. C. Whitmore, eds., *Tropical Rain Forest Ecology and Management*, pp. 211–227. Oxford, United Kingdom: Blackwell.

——. 1984. Implications of seed dispersal by animals for tropical reserve management. *Biological Conservation* 30:261–281.

——. 1986. Seed dispersal by fruit-eating birds and mammals. In D. R. Murray, ed., *Seed Dispersal*, pp. 123–189. Sydney, Australia: Academic Press.

——. 1989. Scatter- and clump-dispersal and seedling demography: hypothesis and implications. *Oecologia* 79:417–426.

——. 1993. Aspects of variation in a neotropical seed dispersal system. *Vegetatio* 107/108:149–162

Howe, H. F., W. W. Schupp, and L. C. Westley. 1985. Early consequences of seed dispersal for a Neotropical tree (*Virola surinamensis*). *Ecology* 66:781–791.

Howe, H. F. and J. Smallwood. 1982. Ecology of seed dispersal. *Annual Review of Ecology and Systematics* 13:201–228.

Howe, H. F. and G. A. Vande Kerckhove. 1981. Removal of wild nutmeg (*Virola surinamensis*) crops by birds. *Ecology* 62:1093–1106.

HPI (Heifer Project International). 1996a. Boyo rural integrated farmer's alliance, Cameroon: Project summary. Little Rock, Ark.: Heifer Project International.

——. 1996b. Bui north/Donga Mantung small-holder integrated agricultural projects, Cameroon: Project summary. Little Rock, Ark.: Heifer Project International.

Huber, M. E. 1994. An assessment of the status of the coral reefs of Papua New Guinea. *Marine Pollution Bulletin* 29:69–73.

Huhta, V. 1976. Effects of clear-cutting on numbers, biomass and community respiration of soil invertebrates. *Annals Zoologica Fennici* 13:63–80.

Hume, R., and T. Boyer. 1991. *Owls of the World*. Philadelphia, Pa.: Running Press Book Publishers.

Humphrey, S. R. and F. J. Bonaccorso. 1979. Population and community ecology. In R. J. Baker, J. K. Jones, and D. C. Carter, eds., *Biology of Bats*, pp. 409–441. Special Publications of The Museum Texas Tech University, No 16. Lubbock, Tex.: Texas Tech Press.

Humphrey, S. R., F. J. Bonaccorso, and T. L. Zinn. 1983. Gild structure of surface-gleaning bats in Panama. *Ecology* 64:284–294.

Hunte, W. 1978. The distribution of freshwater shrimps (*Atyidae and Palaeo-monidae*) in Jamaica. *Zoological Journal of the Linnean Society* 64:135–150.

Hunter, M. L. Jr. 1990. *Wildlife, Forests, and Forestry: Principles of Managing Forests for Biological Diversity.* Englewood Cliffs, N.J.: Prentice Hall.

Hutchings, P., C. Payri, and C. Gabre. 1994. The current status of coral reef management in French Polynesia. *Marine Pollution Bulletin* 29 26–33.

Hutchinson, I. D. 1988. Points of departure for silviculture in humid tropical forests. *Commonwealth Forestry Review* 67:223–230.

Hutchison, V. H. and R. K. Dupré. 1992. Thermoregulation. In M. E. Feder, and W. W. Burggren, eds., *Environmental Physiology of the Amphibians*. pp. 206–249. Chicago, Ill.: University of Chicago Press.

Hvidberg-Hansen, H. 1970. Utilization of the white-lipped peccary (*Tayassu albirostris ILLINGER*) in Peru. Unpublished report to FAO Forestry Research and Training Project UNDP / SF No 116. Lima, Peru: La Molina University.

Hynes, H. B. N. 1975. The stream and its valley. Verhandlungen der Internationalen Vereinigung fur Theoretische und Angewandte *Limnologie* 19:1–15.

Ibáñez, C. J. 1981. Biología y ecología de los murciélagos del Hato "El Frío", Apure, Venezuela. Doñana, *Acta Vertebrata* 8:1–271.

IGBP (International Geosphere-Biosphere Programme) Terrestrial Carbon Working Group. 1998. The terrestrial carbon cycle: Implications for the Kyoto Protocol. *Science* 280:1393–1394.

Inger, R. F. 1980a. Densities of floor-dwelling frogs and lizards in lowland forests of southeast Asia and Central America. *American Naturalist* 115:761–770.

——. 1980b. Relative abundances of frogs and lizards in forests of Southeast Asia. *Biotropica* 12:14–22.

Inger, R. F. and R. K. Colwell. 1977. Organization of contiguous communities of amphibians and reptiles in Thailand. *Ecological Monographs* 47:229–253.

Inger, R. F., H. B. Shaffer, M. Koshy, and R. Bakde. 1987. Ecological structure of a herpetological assemblage in south India. *Amphibia-Reptilia* 8:189–202.

Inger, R. F. and H. K. Voris. 1993. A comparison of amphibian communities through time and from place to place in Bornean forests. *Journal of Tropical Ecology* 9:409–433.

Isabirye-Basuta, G. and J. M. Kasenene. 1987. Small rodent populations in selectively felled and mature tracts of Kibale forest, Uganda. *Biotropica* 19:260–266.

Ismail, R. 1995. An economic evaluation of carbon emission and carbon sequestration for the forestry sector in Malaysia. *Biomass and Bioenergy* 8: 281–292.

ITTO (International Tropical Timber Organization). 1990a. *The Promotion of Sustainable Forest Management: A Case Study in Sarawak, Malaysia.* Yokohama, Japan: International Tropical Timber Organization.

——. 1990b. *Guidelines for Sustainable Management of Natural Tropical Forests.* Yokohama, Japan: International Tropical Timber Organization.

——. 1991. ITTO guidelines for the sustainable management of natural tropical forests. *ITTO Technical Series* 5:1–18.

——. 1993. *ITTO Guidelines on the Conservation of Biological Diversity in Tropical Production Forests.* ITTO Policy Development Series Number 5, Yokohama, Japan.

——. 1996. *Annual Review and Assessment of the World Tropical Timber Situation 1995.* Yokohama, Japan: International Tropical Timber Organization.

IUCN (International Union for Conservation of Nature and Natural Resources). 1980. *World Conservation Strategy.* International Union for the Conservation of Nature and Natural Resources. Gland, Switzerland.

——. 1991. *The Conservation Atlas of Tropical Forests: Asia and the Pacific.* London and Basingstoke: MacMillan Press.

——. 1994. *Guidelines for Protected Area Management Categories.* CNPPA with the assistance of WCMC. International Union for the Conservation of Nature and Natural Resources. Gland, Switzerland.

Jacobs, M. R. 1955. Growth habits of the eucalypts. Canberra, Australia: Forestry and Timber Bureau, Commonwealth Government Printer.

Jacobson, S. K. 1995. Introduction: Wildlife conservation through education. In S. K. Jacobson, ed., *Conserving Wildlife—International Education and Conservation Approaches,* pp. xxiii–xxxi. New York, N.Y.: Columbia University Press.

Janis, C. M. 1976. The evolutionary strategy of the Equidae and the origins of rumen and caecal fermentation. *Evolution* 30:757–774.

Janos, D. P., C. T. Sahley, and L. H. Emmons. 1995. Rodent dispersal of vesicular-arbuscular mycorrhizal fungi in amazonian Peru. *Ecology* 76:1852–1858.

Jansen, P. A. and P. M. Forget. In press. Scatterhoarding rodents and tree regeneration in French Guiana. In F. Bongers, P. Charles-Dominique, P. M. Forget, and M. Théry, (eds)., *Nouragues: Dynamics and Plant-Animal Interactions in a Neotropical Rainforest.* Kluwer, Dordrecht, the Netherlands

Janson, C. H. 1983. Adaptation of fruit morphology to dispersal agents in a Neotropical forest. *Science* 219:187–189.

Janson, C. H. and L. H. Emmons. 1990. Ecological structure of the nonflying mammal community at Cocha Cashu Biological Station, Manu National Park, Peru. In A. H. Gentry, ed., *Four Neotropical Rainforests*, pp. 314–338. New Haven, Conn. and London, United Kingdom: Yale University Press.

Janzen, D. H. 1970. Herbivores and the number of tree species in tropical forests. *American Naturalist* 104:501–528.

——. 1975. *The Ecology of Plants in the Tropics.* London, United Kingdom: Edward Arnold.

——. 1979. How to be a fig. *Annual Review of Ecological Systematics* 10:13–51.

——. 1983. The dispersal of seeds by vertebrate guts. In D. J. Futuyma and M. Slatkin, eds., *Coevolution*, pp. 232–262. Sunderland, Mass.: Sinauer Associates.

——. 1982a. Removal of seeds from horse dung by tropical rodents: Influence of habitat and amount of dung *Ecology* 63:1887–1900.

——. 1982b. Seeds in tapir dung in Santa Rosa National Park, Costa Rica. *Brenesia* 19/20:129–135.

——. 1987. Insect diversity of a Costa Rican dry forest: Why keep it, and how? *Biological Journal of the Linnean Society* 30:343–356.

——. 1988. Guanacaste National Park: Tropical ecological and biocultural restoration. In J. Cairns, ed., *Rehabilitating Damaged Ecosystems,* pp.143–192. Boca Raton, Fla.: CRS Press, Inc.

Janzen, D. H. and P. S. Martin. 1982. Neotropical anachronisms: The fruit the Gomphoteresate. *Science* 215:19–27.

Janzen, D. H. and C. Vázquez-Yanes. 1991. Aspects of tropical seed ecology of relevance to management of tropical forested wildlands. In A. Gomez-Pompa, T. C. Whitmore, and M. Hadley, eds., *Rain Forest Regeneration and Management*, pp.137–157 M.A.B. Series Vol. 6 Paris, France: UNESCO (United Nations Educational, Scientific, and Cultural Organization)/Parthenon.

Jayasekera, P. 1995. Elephants in logging operations in Sri Lanka. FAO Forest Harvesting Case-Study 5, FAO, Rome, Italy.

Jha, L. K. 1997. *Shifting Cultivation.* New Delhi, India: APH Publishing Corporation.

Joanen, T., L. McNease, R. Elsey, and M. A. Staton. 1994. The commercial consumptive use of the American alligator (*Alligator mississippiensis*) in Louisiana: Its effect on conservation–a case study. Rockefeller Wildlife Refuge, Grand Chenier, La.

Johns, A. D. 1983a. Tropical forest primates and logging—Can they co-exist? *Oryx* 17:114–118.

——. 1983b. Ecological effects of selective logging in a West Malaysian rain forest. Ph.D. diss., University of Cambridge.

——. 1985. Selective logging and wildlife conservation in tropical rain forest: problems and recommendations. *Biological Consevation* 31:355–375.

——. 1986a. Effects of selective logging on the behavioral ecology of West Malaysian primates. *Ecology* 67:684–694.

——. 1986b. Effects of habitat disturbance on rain forest wildlife in Brazilian Amazonia. Final report to project US-302, World Wildlife Fund US, Washington DC.

——. 1986c. Effects of selective logging on the ecological organization of a peninsular Malaysian rain forest avifauna. *Forktail* 1:65–79.

——. 1987. The use of primary and selectively logged rain forest by Malaysian Hornbills (Bucerotidae) and implications for their conservation. *Biological Conservation* 40:179–190.

——. 1988. Effects of "selective" timber extraction on rain forest structure and composition and some consequences for frugivores and folivores. *Biotropica* 20:31–37.

——. 1989a. Recovery of a peninsular Malaysian rain forest avifauna following selective timber logging: The first twelve years. *Forktail* 4: 89–105.

——. 1989b. Timber, the environment and wildlife in Malaysian rain forests. Report to ODA (Overseas Development Administration)/NERC (Natural Environment Resource Council) project F3CR26/G1/05, Institute of South-east Asian Biology, Aberdeen, Scotland: University of Aberdeen.

——. 1991a. Responses of Amazonian rain forest birds to habitat modification. *Journal of Tropical Ecology* 7:417–437.

——. 1991b. Forest disturbance and Amazonian primates. In H. O. Box, ed., *Primate Responses to Environmental Change*, pp. 115–135. London, United Kingdom: Chapman and Hall.

——. 1992a. Species conservation in managed tropical forests. In T. C. Whitmore and J. A. Sayer, eds., *Tropical Deforestation and Species Extinction*, pp. 15–53. London, United Kingdom: Chapman and Hall.

——. 1992b. Vertebrate responses to selective logging: Implications for the design of logging systems. *Philosophical Transactions of the Royal Society, London (Series B)* 335:437–442.

Johns, A. D. and J. M. Ayres. 1987. Southern bearded sakis beyond the brink. *Oryx* 21:164–167.

Johns, A. D., R. H. Pine, and D. E. Wilson. 1985. Rain forest bats—An uncertain future. *Bat News* 5:2–3.

Johns, A. D. and J. P. Skorupa. 1987. Responses of rain forest primates to habitat disturbance: A review. *International Journal of Primatology* 8:157–191.

Johns, J. S. 1996. Logging damage during planned and unplanned logging in eastern Amazon. *Forest Ecology and Management* 89:59–77.

Johns, J. S., P. Barreto, and C. Uhl. 1996. Logging damage in planned and unplanned logging operations in the eastern Amazon. *Forest Ecology and Management.* 89:59–77.

Johns, R. J. 1986. The instability of the tropical ecosystem in New Guinea. *Blumea* 31:341–364.

———. 1992. The influence of deforestation and selective logging operations on plant diversity in Papua New Guinea. In T. C. Whitmore, and J. A. Sayer, eds., *Tropical Deforestation and Species Extinction,* pp. 143–147. London, United Kingdom: Chapman and Hall.

Johnson, C. N. 1996. Interactions between mammals and ectomycorrhizal fungi. *Tree* 11:503–507.

Johnson, N. and B. Cabarle. 1993. *Surviving the Cut: Natural Forest Management in the Humid Tropics.* Washington, D.C.: World Resources Institute.

Jones, J. K., J. Arroyo-Cabrales, and R. D. Owen. 1988. Revised checklist of bats (Chiroptera) of Mexico and Central America. Occasional Papers, The Museum, Texas Tech University No. 120:1–34.

Jonkers, W. B. J. 1987. *Vegetation Structure, Logging Damage, and Silviculture in a Tropical Rain Forest in Suriname.* Wageningen, The Netherlands: Agricultural University.

Jordan, C. F. 1987. Shifting cultivation. In C. F. Jordan, ed., *Amazonian Rain Forests: Ecosystem Disturbance and Recovery,* pp. 9–23. Berlin, Germany: Springer.

———. 1990. Nutrient cycling processes and tropical forest management. In A. Gomez-Pompa, T. C. Whitmore, and M. Hadley, eds., *Rainforest Regeneration and Management,* pp. 159–180. Paris, France: Parthenon Publishing Group.

Jordano, P. 1987. Patterns of mutualistic interactions in pollination and seed dispersal: connectance, dependence asymmetries, and coevolution. *The American Naturalist* 129:657–677.

———. 1992. Fruits and frugivory. In M. Fenner, ed., *Seeds: The Ecology of Regeneration in Plant Communities,* pp. 105–156 Wallingford, United Kingdom: CAB (Centre for Agriculture and Bioscience).

———. 1993. Geographical ecology and variation of plant-seed disperser interactions: southern Spanish junipers and frugivorous thrushes. *Vegetatio* 107/108:85–104.

———. 1994. Spatial and temporal variation in the avian-frugivore assemblage of *Prunus mahaleb*: Patterns and consequences. *Oikos* 71:479–491.

Jori, F. and J. M. Noel. 1996. Guide pratigue d'élevage d'aulacodes au Gabon. Vétérinaires Sans Frontières, Berthelot.

Julien-Laferriere, D. and M. Atramentowicz. 1990. Feeding and reproduction of three didelphid marsupials in two neotropical forests (French Guiana). *Biotropica* 22:404–415.

Jullien, M. and J. M. Thiollay. 1996. Effects of rain forest disturbance and fragmentation: Comparative changes of the raptor community along natural and man-made gradients in French Guiana. *Journal of Biogeography* 23:7–25.

Julliot, C. 1996. Seed dispersal by Red howling monkeys (*Alouatta seniculus*) in the tropical rain forest of French Guiana. *International Journal of Primatology* 17:239–258.

Jusoff, K. 1991. A survey of soil disturbance from tractor logging in a hill forest of peninsular Malaysia. In S. Appanah, F. S. Ng, and R. Ismail, eds., *Malaysian Forestry and Forest Products Research*, pp. 16–21. Kepong, Malaysia: Forest Research Institute Malaysia.

Kalina, J. 1988. Ecology and behaviour of the black and white casqued hornbill (*B. cylindricus subsquamata*) in the Kibale Forest Reserve. Unpublished Ph.D. diss., Michigan State University.

Kalko, E. K. V. 1995. Echolocation signal design, foraging habitats and guild structure in six Neotropical sheath-tailed bats (*Emballonuridae*).In P.A. Racey and S. M. Swift, eds., *Ecology, Evolution and Behaviour of Bats,* pp. 259–273. Oxford, United Kingdom: The Zoological Society of London.

———. 1997. Diversity in tropical bats. In Ulrich, H., ed., *Proceedings of the International Symposium on Biodiversity and Systematics in Tropical Ecosystems,* pp.13–43. Bonn, Germany: Zoologisches Forschungsinstitut und Museum Alexander Koenig.

Kalko, E. K. V., C. O. Handley, and D. Handley. 1996. Organization, Diversity, and Long-term Dynamics of a Neotropical Bat Community. In Cody, M. and Smallwood J., eds., *Long-term Studies of Vertebrate Communities*, pp. 503–553, New York, N.Y.: Academic Press.

Kapos, V. 1989. Effects of isolation on the water status of forest patches in the Brazilian Amazon. *Journal of Tropical Ecology* 5:173–185.

Karr, J. R. 1982. Avian extinction on Barro Colorado Island, Panama: A reassessment. *American Naturalist* 119:220–239.

Karr, J. R., J. D. Nichols, M. K. Klimkiewicz, and J. D. Brawn. 1990. Survival rates of birds of tropical and temperate forest: Will the dogma survive? *American Naturalist*. 136:277–291.

Karr, J. R. and D. R. Dudley. 1981. Ecological perspective on water quality goals. *Environmental Management* 5:55–68.

Kasenene, J. M. 1987. The influence of mechanized selective logging, felling intensity and gap-size on the regeneration of a tropical moist forest in the Kibale Forest Reserve, Uganda. Ph.D. Diss., Michigan State University.

Kasenene, J. M. and P. G. Murphy. 1991. Post-logging tree mortality and major branch losses in Kibole Forest, Uganda. *Forest Ecology and Management* 46:295–307.

Kasran, B. 1988. Effect of logging on sediment yield in a hill dipterocarp forest in peninsular Malaysia. *Journal of Tropical Forest Science* 1:56–66.

Kasran, B. and A. R. Nik. 1994. Suspended sediment yield resulting from selective logging practices in a small watershed in peninsular Malaysia. *Journal of Tropical Forest Science* 7:286–295.

Kauffman, J. B. and C. Uhl. 1990. Interactions of anthropogenic activities, fire, and rain forests in the Amazon Basin. In J. G. Goldammer, ed., *Fire in the Tropical Biota: Ecosystem Processes and Global Challenges*, pp. 117–133. Berlin, Germany: Springer-Verlag.

Kauffman, J. B., C. Uhl, D. L. Cummings. 1988. Fire in the Venezuelan Amazon 1: Fuel biomass and fire chemistry in the evergreen rainforest of Venezuela. *Oikos* 53:167–175.

Kavanagh, M., A. A. Rahim and C. J. Hails. 1989. Rainforest conservation in Sarawak: An International Policy for WWF (World Wildlife Fund). Kuala Lumpur, Malaysia and Gland, Switzerland: WWF (World Wildlife Fund) Malaysia.

Kavanagh, R. P. and Bamkin, K. L. 1995. Distribution of nocturnal forest birds and mammals in relation to the logging mosaic in South-eastern New South Wales, Australia. *Biological Conservation* 71:41–53.

Kay, R. N. B. 1987. The comparative anatomy and physiology of digestion in Tragulids and Cervids and its relation to food intake. In C. M. Wemmer, ed., *Biology and Management of the Cervidae*, pp. 214–222. Washington, D.C.: Smithsonian Institution Press.

Kay, R. N. B and A. G. Davies. 1994. Digestive physiology. In A. G. Davies and J. F. Oates, eds., *Colobine Monkeys: Their Ecology, Behaviour and Evolution*, pp. 229–249. Cambridge, United Kingdom: Cambridge University Press.

Keay, R. W. J. 1989. *Trees of Nigeria*. Oxford, United Kingdom: Claredon Press.

Keller, G, G. P. Bauer, and M. Aldana. 1995. *Caminos Rurales con Impactos Mínimos: Un Manual de Capacitación con Enfasis Sobre Planificación Ambiental, Drenajes, Establización de Taludes, y Control de Erosión.* Guatemala: U. S. Agency for International Development.

Kemp, A. C. and M. I. Kemp. 1975. Report on a study of hornbills in Sarawak, with comments on their conservation. Final report to project 2/74, World Wildlife Fund Malaysia. Miscellaneous report. Kuala Lumpur, Malaysia.

Kemp, N., L. M. Chan, and M. Dilger. 1995. Site description and conservation evaluation: Phu Mat Nature Reserve, Con Cuong District, Nghe An Province, Vietnam. Frontier-Vietnam Scientific Report No. 5. London, United Kingdom: The Society for Environmental Exploration, .

Kemper, C. and D. T. Bell. 1985. Small mammals and habitat structure in lowland rain forest of Peninsular Malaysia. *Journal Tropical Ecology* 1:5–22.

Kengen, S. 1997. *Forest Valuation for Decision-making: Lessons of Experience and Proposals for Improvement.* Rome, Italy: Food and Agricultural Organisation of the United Nations.

Kerans, B. L. and J. R. Karr. 1994. Benthic index of biotic integrity (B-IBI) for rivers of the Tennessee Valley. *Ecological Applications* 4:768–785.

Kershaw, M., G. M. Mace, and P. H. Williams. 1995. Threatened status, rarity, and diversity as alternative selection measures for protected areas: A test using Afrotropical antelopes. *Conservation Biology* 9(2):324–334.

Keskimies, P. and R. A. Visnen. 1986. *Monitoring Bird Populations: A Manual of Methods Applied in Finland.* Finnish Museum of Natural History. Helsinki, Finland: University of Helsinki.

Kikkawa, J. and P. D. Dwyer. 1992. Use of scattered resources in rain forest of humid tropical lowlands. *Biotropica* 24:293–308.

Killeen, T. J., B. T. Louman, and T. Grimwood. 1990. La ecología paisajística de la región de Concepción y Lomerío en la provincia de Ñuflo de Chavez, Santa Cruz, Bolivia. *Ecología en Bolivia* 16:1–45.

Kiltie, R. A. 1981. Stomach contents of rain forest peccaries (*Tayassu tajacu* and *T. pecari*). *Biotropica* 13:234–236.

——. 1982. Bite force as a basis for niche differentiation between rain forest peccaries (*Tayassu tajacu* and *T. pecari*). *Biotropica* 14:188–195.

Kiltie, R. A. and J. Terborgh. 1984. Ecology and Behaviour of Rain Forest Peccaries in Southeastern Peru. *National Geographic Research Reports* 17:873–882.

——. 1983. Observations on the behavior of rain forest peccaries in Peru: why do white-lipped peccaries form herds? *Zeitschrift für Tierpsychologie.* 62:241–255.

Kingdon, J. 1982. *East African Mammals, Volume 3*, parts C and D (Bovids). London, United Kingdom: Academic Press.

——. *Island Africa:The Evolution of Africa's Rare Animals and Plants.* London, United Kingdom: Collins.

Kishor, N. M. and L. F. Constantino. 1993. *Forest Management and Competing Land Uses: An Economic Analysis for Costa Rica.* Washington, D.C.: The World Bank, Latin America Technical Department, Environment Division.

Klasson, B. and J. Cedergen. 1996. Felling the right way: Some hints on the art and science of directional felling. *ITTO Tropical Forest Update* 6:5–7.

Klein, B. C. 1989. Effects of forest fragmentation on dung and carrion beetle communities in central Amazonia. *Ecology* 70:1715–1725.

Klein, M. and J. Heuveldop. 1993. A management planning concept for sustained yield of tropical forest in Sabah, Malaysia. *Forest Ecology and Management* 61:277–297.

Knight, D. H. 1975. A phytoscoiological analysis of species rich tropical forests on Barro Colorado Island, Panama. *Ecological Monographs* 45:259–284.

Kohm, K. A. and J. F. Franklin. 1997. *Creating a Forestry for the 21st Century. The Science of Ecosystem Management.* Washington, D.C.: Island Press.

Koppert, G. 1996. Méthodologie de l'enquète alimentaire. In A. Froment, I. de Garine, C. Binam Bikoi, and J.F. Loung, eds., *Bien Manger et Bien Vivre: Anthropologie Alimentaire dans les Régions Forestières et les Savanes du Cameroun,* pp. 89–98. Paris, France: Orstom.

Koppert, G. and C.M. Hladik. 1990. Measuring food consumption. In C. M. Hladik, S. Bahuchet, and I. de Garine, eds., *Food and Nutrition in the African Rain Forest,* pp. 59–61. Paris, France: UNESCO (United Nations Educational, Scientific, and Cultural Organization)-MAB.

Korsgaard, S. 1992. *An Analysis of Growth Parameters and Timber Yield Prediction.* Copenhagen, Denmark: The Council for Development Research.

Krebs, C. J. 1989. *Ecological Methodology.* New York, N.Y.: Harper Collins.

Kremen, C. 1992. Assessing the indicator properties of species assemblages for natural areas monitoring. *Ecological Applications.* 2:203–217.

Kremen, C., R. K. Colwell, T. L. Erwin, D. D. Murphy, R. F. Noss, and M. A. Sanjayan. 1993. Terrestrial arthropod assemblages: Their use in conservation planning. *Conservation Biology* 7:796–808.

Kremen, C., K. Lance and I. Raymond. 1998. Interdisciplinary tools for monitoring conservation impacts in Madagascar. *Conservation Biology* 12:549–563

Kremen C., K. Lance, I. Raymond, and A. Weiss. 1998. Monitoring natural resource use on the Masoala Peninsula, Madagascar: A tool for managing integrated conservation and development projects. In K. Saterson, R. Margoluis, and N. Salafsky, eds., *Measuring Conservation Impact: An Interdisciplinary Approach to Project Monitoring and Evaluation*, pp. 63–82. Washington, D.C.: Biodiversity Support Program.

Kuhn, H-J. 1968. Der Jentink-Ducker. *Natur und Museum* 98:17–23.

———. 1966. Der Zebraducker, *Cephalophus doria. Z. fur Saugetierkunde* 31:282–293.

Laake, J. L., S. T. Buckland, D. R. Anderson, and K. P. Burnham. 1993. *Distance Users Guide Version 2.0.* Fort Collins, Colo.: Colorado Cooperative Fish and Wildlife Research Unit, Colorado State University,

Lahm, S. 1994a. Hunting and wildlife in northeastern Gabon: Why conservation should extend beyond protected areas. *Institut de Recherche en Ecologie Tropicale.* Makokou, Gabon.

———. 1994b. Ecology and economics of human/wildlife interaction in northeastern Gabon. Ph.D. diss., New York University.

Laidlaw, R. K. 1994. The Virgin Jungle Reserves of peninsular Malaysia: The ecology and dynamics of small protected areas in managed forest. Ph.D. diss., University of Cambridge.

Lamb, D. 1990. *Exploiting the Tropical Rain Forest: An Account of Pulpwood Logging in Papua New Guinea.* Paris, France: UNESCO (United Nations Educational, Scientific, and Cultural Organization).

Lamb, F. B. 1966. *Mahogany of Tropical America: Its Ecology and Management.* Ann Arbor, Michigan: University of Michigan Press.

Lambert, F. R. 1990. Avifaunal changes following selective logging of a North Bornean rainforest. Unpublished report to the Danum Valley Management Committee, Sabah Foundation. Institute of Tropical Biology, Aberdeen, Scotland: University of Aberdeen.

———. 1991. The conservation of fig-eating birds in Malaysia. *Biological Conservation* 58:31–40.

———. 1992. The consequences of selective logging for Bornean lowland forest birds. *Philosophical Transactions of the Royal Society, London, Series B* 335:443–457.

Lambert, F. R. and A. G. Marshall. 1991. Keystone characteristics of bird-dispersed Ficus in a Malaysian lowland rain forest. *Journal of Ecology* 79:793–809.

Lamprecht, H. 1989. *Silviculture in the Tropics.* Eschborn, Federal Republic of Germany: Deutsche Gesellschaft für Technische Zusammenarbeit.

Langdale-Brown, H. A., H. A. Osmaston, and J. G. Wilson. 1964. *The Vegetation of Uganda and its Bearing on Land-use.* Entebbe, Uganda: Uganda Government Printer.

Langholz, J. 1996. Economics, objectives, and success of private nature reserves in sub-Saharan Africa and Latin America. *Conservation Biology* 10:271–280.

Lanly, J. P., K. D. Singh, and K. Janz. 1991. FAO's 1990 reassessment of tropical forest cover. *Nature and Resources* 27:21–26.

LaSalle, J. and I. D. Gauld. 1993. Hymenoptera: Their diversity, and their impact on the diversity of other organisms. In J. LaSalle and I. D. Gauld eds., *Hymenoptera and*

Biodiversity, pp. 1–25 Wallingford, United Kingdom: CAB (Centre for Agriculture and Bioscience).

Lasebikan, B. A. 1975. The effect of clearing on soil arthropods of a Nigerian rain forest. *Biotropica* 7:84–89.

Laurance, W. F. 1990. Comparative responses of five arboreal marsupials to tropical forest fragmentation. *Journal of Mammalogy* 71:641–653.

——. 1994. Rainforest fragmentation and the structure of small mammal communities in tropical Queensland. *Biological Conservation* 69:23–32.

——. 1998. A crisis in the making: Responses of Amazonian forests to land use and climate change. *Trends in Ecology and Evolution* 13:411–415.

Laurance, W. F. and R. O. Bierregaard, Jr., eds., 1997. *Tropical Forest Remnants: Ecology, Management and Conservation of Fragmented Communities*. Chicago, Ill.: University of Chicago Press.

Laurance, W. F. and S. G. W. Laurance. 1996. Responses of five arboreal marsupials to recent selective logging in tropical Australia. *Biotropica* 28:310–322.

Laurent, R. B. 1973. A parallel survey of equatorial amphibians and reptiles in Africa and South America. In B. J. Meggers, E. S. Ayensu, and D. W. Duckworth, eds., *Tropical Forest Ecosystems in Africa and South America: A Comparative Review*, pp. 259–266 Washington, D.C.: Smithsonian Institute.

LaVal, R. K. 1977. Notes on Some Costa Rican Bats. *Brenesia* 10/11:77–83.

LaVal, R. K. and H. S. Fitch. 1977. Structure, movements and reproduction in three Costa Rican bat communities. *Occasional Papers of the Museum of Natural History the University of Kansas* 69:1–28.

Lavelle, P. and B. Pashanasi. 1989. Soil macrofauna and land management in Peruvian Amazonia (Yurimaguas, Loreto). *Pedobiologia* 33:283–291.

Lavelle, P. and A. Martin. 1992. Small-scale and large-scale effects of endogeic earthworms on soil organic matter dynamics in soils of the humid tropics. *Soil Biology and Biochemistry* 24:1491–1498.

Lawton, J. H. 1994. What do species do in ecosystems? *Oikos* 71:367–374.

Lawton, J. H., D. E. Bignell, B. Bolton, G. F. Bloemers, P. Eggelton, P. M. Hammond, M. Hodda, R. D. Holt, T. B. Larsen, N. A. Mawdsley, N. E. Stork, D. S. Srivastava, and A. D. Watts. 1998. Biodiversity inventories, indicator taxa and effects of habitat modification in tropical forests. *Nature* 391:72–76.

Lawton, J. H. and V. K. Brown. 1993. Redundancy in ecosystems. In E. D. Schulze and H. A. Mooney, eds., *Biodiversity and Ecosystem Function*, pp. 255–270. New York, N.Y.: Springer.

Lawton, R. O. 1990. Canopy gaps and light penetration into a wind-exposed tropical lower montane rain forest. *Canadian Journal of Forest Research* 20:659–667.

Lawton, R. O. and F. E. Putz. 1988. Natural disturbance and gap-phase regeneration in a wind-exposed tropical cloud forest. *Ecology* 69:764–777.

Leakey, R. J. G. and J. Proctor. 1987. Invertebrates in the litter and soil at a range of altitudes on Gunung Silam, a small ultrabasic mountain in Sabah. *Journal Tropical Ecology* 3:119–129.

Lee, K. E. 1985. *Earthworms: Their Ecology and Relationship with Soils and Land Use*. Sydney, Australia: Academic Press.

Lee, K. N. 1993. *Compass and Gyroscope: Integrating Science and Politics for the Environment*. Washington, D.C: Island Press.

Lee, S. S., Y. M. Dan, I. D. Gauld, and J. Bishop. eds. 1998. *Conservation, Management and Development of Forest Resources*. Kepong, Malaysia: Forest Research Institute.

Lefkovitch, L. P. 1965. The study of population growth in organisms grouped by stages. *Biometrics* 21:1–18.

Leigh, E. G., Jr, S. J. Wright, and E. A. Herre. 1993. The decline of tree diversity on newly isolated tropical islands: A test of the null hypothesis and some implications. *Evolutionary Ecology* 7:76–102

Leighton, M. and D. R. Leighton. 1983. Vertebrate responses to fruiting seasonality within a Bornean rain forest. In S. L. Sutton, T. C. Whitmore and A. C. Chadwick, eds., *Tropical Rain Forest Ecology and Management*, pp. 181–196. Oxford, United Kingdom: Blackwell.

Leighton, M. and N. Wirawan. 1986. Catastrophic drought and fire in Borneo tropical rain forest associated with the 1982–1983 El Nino southern oscillation event. In G. T. Prance, ed., *Tropical Forests and the World Atmosphere*, pp. 75–102. Washington, D.C.: American Academy for the Advancement of Science.

Lekagul, B. and P. Round. 1986. *A Field Guide to the Birds of Thailand*. Bangkok, Thailand: Saha Karn Bhaet.

Lemmens, R. H. M. J., I. Soerianegara and W. C. Wong, eds., 1995. *Plant Resources of South-east Asia* Vol. 5(2): *Timber Trees: Minor Commercial Timbers*. Wageningen, The Netherlands: Pudoc Scientific Publishers.

Lenski, R. E. 1982. Effects of forest cutting on two Carabus species: Evidence for competition for food. *Ecology* 63:1211–1217.

Leopold, A. 1949. *A Sand County Almanac and Sketches Here and There*. New York, N.Y.: Oxford University Press.

Lepage, M. 1984. Distribution, density and evolution of Macrotermes bellicosus nests (Isoptera: Macrotermitinae) in the North-east of Ivory Coast. *Journal Animal Ecology* 53:107–117.

Lerdau, M. 1996. Insects and ecosystem function. *Trends in Ecology and Evolution* 11:151.

LeSica, P. 1992. Autecology of the endagered plant Howellia aquatilis; implications for management and reserve design. *Ecological Applications* 2(4):171–180.

Letouzey, R. 1968. *Etude Phytogéographique du Cameroun*. Paris, France: Paul Lechevalier Editions.

———. 1985. Notice de la carte phytogéographique du Cameroun au 1:500,000. Toulouse, France: Institut de la Carte Internationale de la Végétation.

Leuthold, W. 1977. *African Ungulates: A Comparative Review of Their Ethology and Behavioural Ecology*. Berlin, Germany: Springer-Verlag.

Levey, D. J. 1986. Methods of seed processing by birds and seed deposition patterns. in A. Estrada, and T. H. Fleming, eds., *Frugivores and Seed Dispersal: Tasks for Vegetation Science*. Volume 15, pp. 147–158. Dordrecht, Netherlands: Dr. W. Junk Publishers.

Levey, D. J. 1988a. Tropical wet forest treefall gaps and distributions of understory birds and plants. *Ecology* 69:1076–1089.

———. 1988b. Spatial and temporal variation in Costa Rican fruit and fruit-eating bird abundance. *Ecological Monographs* 58:251–269.

———. 1990. Habitat-dependent fruiting behaviour of an understory tree, Miconia centrodesma, and tropical treefall gaps as keystone habitats for frugivores in Costa Rica. *Journal of Tropical Ecology* 6:409–420.

Levings, S. C. and D. M. Windsor. 1982. Seasonal and annual variation in litter arthropod populations. In E. G. Leigh, A. S. R. and D. W. Windsor, eds., *Ecology of a Tropical Forest: Seasonal Rhythms and Long-Term Changes*, pp. 355–387. Oxford, United Kingdom: Oxford University Press.

Lewis, D., G. B. Kaweche, and A. Mwenya. 1990. Wildlife conservation outside protected areas: Lessons from an experiment in Zambia. *Conservation Biology* 4:171–180.

Lieberman, D. 1996. Demography of tropical tree seedlings: A review. In M.D. Swaine, ed., *The Ecology of Tropical Forest Seedlings*. M.A.B. series Volume 17, pp. 131–138. Paris, France: ESCO/Parthenon.

Lieberman, D. M. Lieberman, and C. Martin. 1987. Notes on seeds in elephant dung from Bia National Park, Ghana. *Biotropica* 19:365–369.

Lieberman, D., J. B. Hall, M. D. Swaine and M. Lieberman. 1979. Seed dispersal by baboons in the Shai hills, Ghana. *Ecology* 60:65–75.

Lieberman, M. and D. Lieberman. 1986. An experimental study of seed ingestion and germination in a plant-animal assemblage in Ghana. *Journal of Tropical Ecology* 2:113–126.

Lieberman, M., D. Lieberman, and R. Peralta. 1989. Forests are not just Swiss cheese: canopy stereogeometry of non-gaps in tropical forests. *Ecology* 70:550–552.

Lieberman, S. S. 1986. Ecology of the leaf litter herpetofauna of a neotropical rain forest: La Selva, Costa Rica. *Acta Zoologica Mexicana* 15:1–72.

Lim, B. H. and A. Sasekumar. 1979. A preliminary study on the feeding biology of mangrove forest primates, Kuala Selangor, Malaysia. *Malaysian Nature Journal* 33:105–112.

Lindemalm, F. 1997. Forest certification and community forestry as a means of preserving biodiversity in a natural tropical production forest. Minor Field Study No. 24. Masters thesis, Swedish University of Agricultural Sciences.

Lindenmayer, D. B., R. B. Cunningham, and C. F. Donnelly. 1993. The conservation of arboreal marsupials in the montane ash forests of the Central Highlands of Victoria, South-east Australia, IV: The presence and abundance of arboreal marsupials in retained linear habitats (wildlife corridors) within logged forest. *Biological Conservation,* 66: 207–221.

Lindenmayer, D. B., R. B. Cunningham, M. T. Tanton, A. P. Smith, and H. A. Nix. 1990. The conservation of arboreal marsupials in the montane ash forests of the central highlands of Victoria, south-east Australia: I. Factors influencing the occupancy of trees with hollows. *Biological Conservation* 54:111–131.

Lizarraga, I. P., and A. B. Helbingen. 1998. El Proceso Social de Formulacion de la Ley Forestal de Bolivia de 1996. La Paz, Bolivia: Center for International Forestry Research, Centro de Estudios para el Desarrollo Laboral y Agrario, Taller de Iniciativas en Estudios Rurales y Reforma Agraria, and Programa Manejo de Bosques de la Amazonia Boliviana.

Lodge, D. J. 1996. Microorganisms. In D. P. Reagan and R. B. Waide, eds, *The Food Web of a Tropical Rain Forest,* pp. 53–108 Chicago, Ill.: University of Chicago Press.

Lodge, D. J., W. H. McDowell, and C. P. McSwiney. 1994. The importance of nutrient pulses in tropical forests. *Trends in Ecology and Evolution* 9:384–387.

Lodge, D. J., F. N. Scatena, C. E. Asbury, and M. J. Sanchez. 1991. Fine litterfall and related nutrients inputs resulting from Hurricane Hugo in subtropical wet and lower montane rain forests of Puerto Rico. *Biotropica* 23:336–342.

Loiselle, B. A. and J. G. Blake, 1999. Dispersal of melastome seeds by fruit-eating birds of tropical forest understory. *Ecology* 80:330–336.

Loomis, J. B. 1993. *Integrated Public Lands Management: Principles and Applications to National Forests, Parks, Wildlife Refuges, and BLM Lands.* New York, N.Y.: Columbia University Press.

——. 1995. Review of: The economic value of biodiversity. *Journal of Economic Literature.* 33:2026–2027.

López, J. 1993. *Recursos forestales de Bolivia y su aprovechamiento.* La Paz, Bolivia: Cooperación Técnica Holandesa.

Louman, B. 1996. The use of small portable sawmills in forest management in Papua New Guinea. *Rural Development Forestry Network, Network Paper* 19d:16–25.

Lovejoy, T. E. 1974. Bird diversity and abundance in Amazon forest communities. *Living Bird* 13:127–191.

Lovejoy, T. E. and R. O. Bierregaard Jr. 1990. Central amazonian forests and the minimun critical size of ecosystems project. In A. H. Gentry, ed., *Four Neotropical Rainforests,* pp. 60–71. New Haven, Conn. and London, United Kingdom: Yale University Press.

Lovejoy, T. E., R. O. Bierregaard, Jr., A. B. Rylands, J. R. Malcolm, C. E. Quintela, L. H. Harper, K. S. Brown, Jr., A. H. Powell, G. V. N. Powell, H. O. R. Schubart, and

M. B. Hays. 1986. Edge and other effects of isolation on Amazon forest fragments. In M. E. Soulé, ed., *Conservation Biology: The Science of Scarcity and Diversity*, pp. 257–285. Sunderland, Mass.: Sinauer.

Lowe-McConnell, R. H. 1987. *Ecological Studies in Tropical Fish Communities*. New York, N.Y.: Cambridge University Press.

Lowman, M. D. 1984. An assessment of techniques for measuring herbivory: Is rainforest defoliation more intense than we thought? *Biotropica* 16(4): 264–268.

——. 1991. The impact of herbivorous insects on Australian rainforest tree canopies. In G .L. Werren and A. P. Kershaw, eds., *The Rainforest Legacy*. Volume III: Australian National Rainforest Study Report, pp. 177–190. Washington, D.C.: World Wildlife Fund.

Lowman, M. D. 1992a. Herbivory in Australia rainforest, with particular reference to the canopies. *Biotropica* 24:263–272.

——. 1992b. Leaf growth dynamics and herbivory in five species of Australian rain-forest canopy trees. *Journal of Ecology* 80:433–447.

Ludwig, D., R. Hilborn, and C. J. Walters. 1993. Uncertainty, resource exploitation and conservation: Lessons from history. *Science* 260:17–36.

Lugo, A. E. 1992. Comparison of tropical tree plantations with secondary forests of similar age. *Ecological Monographs* 62:1–41.

Lugo, A. E., M. Brinson, C. Ceranevivas, C. Gist, R. Inger, C. Jordan, H. Liegh, W. Milstead, P. Murphy, N. Smythe, and S. Snedaker. 1974. Tropical ecosystem structure and function. in E. G. Farnworth, and F. B. Golley, ed., *Fragile Ecosystems*, pp. 67–111. New York, N.Y.: Springer-Verlag.

Lumer, C. and R. D. Schoer. 1986. Pollination of *Blakea austin-smithii* and *B. penduliflora* (Melastomataceae) by small rodents in Costa Rica. *Biotropica* 18:363–364.

Lumpkin, S. and K. Kranz. 1984. *Cephalophus sylvicultor*. *Mammal species* 225:1–7.

Mabberley, D. J. 1992. *Tropical Rain Forest Ecology*. 2nd Edition. New York, N.Y.: Blackie Academic and Professional.

MacArthur, R. H. and E. Pianka. 1966. On optimal use of a patchy environment. *American Naturalist* 100:603–609.

MacDicken, D. G. 1997. *A Guide to Monitoring Carbon Storage in Forestry and Agroforestry Projects*. Arlington, Va.: Winrock International.

Macedo, D. S. and A. B. Anderson. 1993. Early ecological changes associated with logging in an Amazon floodplain. *Biotropica* 25:151–163.

Machtans, C. S., M. Villard, and S. J. Hannon. 1996. Use of riparian buffer strips as movement corridors by forest birds. *Conservation Biology* 10:1366–1379.

MacKinnon, J., 1982. Nature Conservation in Sabah. Field Report 26 of FAO project INS/78/061, Bogor, Indonesia.

MacKinnon, J., C. MacKinnon, G. Child, and J. Thorsell. 1986. *Managing Protected Areas in the Tropics*. Gland, Switzerland: International Union for the Conservation of Nature and Natural Resources.

MacKinnon, J. and I. Warsito. 1982. *Gunung Palung Reserve, Kalimantan Barat: Preliminary Management Plan*. Bogor, Indonesia: UNDP/FAO.

Maisels, F. and A. Gautier-Hion. 1994. Why are caesalpinoideae so important for monkeys in hydromorphic rainforests of the Zaire basin. In J. I. Sprent and D. McKey, eds., *Advances in Legume Systematics 5: the Nitrogen Factor*, pp. 189–204. Kew: Royal Botanical Gardens.

Maitre, H. F. and M. Hermeline. 1985. Dispositifs d'etude de l'evolution de la forêt dense Ivorienne suivant differentes modalites d'intervention sylvicole. SODEFOR, Republique de Cote d'Ivoire.

Majer, J. D. and G. Beeston. 1996. The biodiversity integrity index: an illustration using ants in western Australia. *Conservation Biology* 10:65–73.

Malcolm, J. R. 1990. Estimation of mammalian densities in continuous forest North of Manaus. In A. Gentry, ed., *Four Neotropical Rainforests*, pp. 339–357. New Haven, Conn.: Yale University Press.

——. 1991. The small mammals of Amazonian forest fragments: Pattern and process. Ph.D. diss., University of Florida.

——. 1994. Edge effects in central Amazonian forest fragments. *Ecology* 75:2438–2445.

——. 1995. Forest structure and the abundance and diversity of neotropical small mammals. In M. D. Lowman and N. M. Nadkarni, eds., *Forest Canopies*. pp. 179–197. New York, N.Y.: Academic Press.

——. 1997. Biomass and diversity of small mammals in Amazonian forest fragments. In W. F. Laurance and R. O. Bierregaard, Jr., eds., *Tropical Forest Remnants: Ecology, Management and Conservation of Fragmented Communities*, pp. 207–221 Chicago, Ill.: University of Chicago Press.

——. 1998. A model of conductive heat flow in forest edges and fragmented landscapes. *Climatic Change* 39:487–502.

——. 2001. Extending models of edge effects to diverse landscape configurations. In R. O. Bierregaard, Jr., C. Gascon, T. E. Lovejoy, and R. C. Mesquita, eds., *Lessons from Amazonia: The Ecology and Management of a Fragmented Forest*. New Haven, Conn.: Yale University Press.

Malcolm, J. R. and J. C. Ray. 2000. Influence of timber extraction routes on central African small mammal communities, forest structure, and tree diversity. *Conservation Biology* 14(6): 1623–1638.

Malmer, A. and H. Grip. 1990. Soil disturbance and loss of infiltrability caused by mechanized and manual extraction of tropical rainforest in Sabah, Malaysia. *Forest Ecology and Management*, pp. 1–12.

Mancilla, R. and A. Andaluz. 1996. Cambios Sustanciales en la Legislación Forestal Nacional. *Boletín BOLFOR (Bolivia Sustainable Forestry Management Project)* 7:6–8.

Manomet Observatory for Conservation Sciences. 1996a. Impacts of silvicultural trials on birds and tree regeneration in the Chiquibul Forest Reserve. A report prepared for the Forest Planning and Management Project, Ministry of Natural Resources, Belmopan, Belize. Manomet, Mass.: Manomet Observatory for Conservation Sciences.

Manomet Observatory for Conservation Sciences. 1996b. *The Shifting Mosaic Model*. Manomet, Mass.: Manomet Observatory for Conservation Sciences.

March, J. G., J. P. Benstead, C. M. Pringle, and F. N. Scatena. 1998. Migratory drift of larval freshwater shrimps into two tropical streams, Puerto Rico. *Freshwater Biology* 40: 261–274.

Marconi, M. 1991. Catálogo de legislación ambiental en Bolivia. Unpublished report. La Paz, Bolivia: Centro de Datos para la Conservación.

——. 1992. El Sistema Nacional de Areas Protegidas y las Areas Bajo Manejo Especial. In M. Marconi, ed., *Conservación de la Diversidad Biológica en Bolivia*, pp. 321–370. La Paz, Bolivia: Centro de Datos para la Conservación.

Marconi, M, and W. Hanagarth. 1985. Lista de especies de vertebrados protegidas y sujetas a comercio (unpublished). La Paz, Bolivia: Instituto de Ecología-UMSA.

Margoluis, R. and N. Salafsky. 1998. *Measures of Success: Designing, Managing, and Monitoring Conservation and Development Projects*. Washington, D.C.: Island Press.

Marinissen, J. C. Y. and P. C. de Ruiter. 1993. Contribution of earthworms to carbon and nitrogen cycling in agroecosytems. *Agriculture, Ecosystems and Environment* 47:59–74.

Marks, S. A. 1996. Local hunters and wildlife surveys: An assessment and comparison of counts from 1989, 1990, and 1993. *African Journal of Ecology* 34:237–257.

Marland, G. and S. Marland. 1994. Should we store carbon in trees? *Water, Air, and Soil Pollution* 64:181–195.

Marn, H. M. and W. J. B. Jonkers. 1981. Logging damage in tropical high forest. UNDP/FAO. Working Paper No. 5, Fo:MAL/76/008. Kuching, Malaysia: Forest Department.

Marn, H. M., E. Vel, and D. (K. H.) Chua. 1981. Planning and cost studies in harvesting in the mixed dipterocarp forest of Sarawak Technical Report for FAO FO:MAL/76/008. Kuching, Sarawak, Malaysia: Forest Department.

Marquis, R. J. and C. J. Whelan. 1994. Insectivorous birds increase growth of white oak through consumption of leaf-chewing insects. *Ecology* 75:2007–14.

Marsh, C. W. and A. G. Greer. 1992. Forest land-use in Sabah: An introduction to Danum Valley. *Philosophic Transactions of the Royal Society of London, Series B* 335:331–339.

Marsh, C. W. and W. L. Wilson. 1981. *A Survey of Primates in Peninsular Malaysian Forests.* Kuala Lumpur: University Kebangsaan, Malaysia.

Marsh, D. M. and P. B. Pearman. 1997. Effects of habitat fragmentation on the abundance of two species of leptodactylid frogs in an Andean montane forest. *Conservation Biology* 11:1323–1328.

Marshall, A. G. and M. D. Swaine. ed. 1992. *Tropical Rain Forest: Disturbance and recovery.* London, United Kingdom: The Royal Society.

Marshall, E. 1995. Homely fish draws attention to Amazon deforestation. *Science* 267:814.

Marshall, G. 1991. *The Political Economy of Logging: the Barnett Inquiry into corruption of the Papua New Guinea Timber Industry.* Port Moresby, Papua New Guinea: University of Papua New Guinea.

Martikainen, P., J. Siitonen, L. Kaila, P. Punttila, and J. Rauh. 1999. Bark beetles (*Coleoptera, Scolytidae*) and associated beetle species in mature managed and old-growth boreal forests in southern Finland. *Forest Ecology and Management* 116 233–245.

Martikainen, P., J. Siitonen, P. Punttila, L. Kaila, and J. Rauh. 2000. Species richness of Coleoptera in mature managed and old-growth boreal forests in southern Finland. *Biological Conservation* 94:199–209.

Martin, N. L. 1997. Preparing for the new power generation. *Corporate Philanthropy Report* 12:1–10.

Martin-Smith, K. M. 1998. Effects of disturbance caused by selective timber extraction on fish communities in Sabah, Malaysia. *Environmental Biology of Fishes* 53:155–167.

Martini, A. M. Z., N. de A. Rosa, and C. Uhl. 1994. An attempt to predict which Amazonian tree species may be threatened by logging activities. *Environmental Conservation* 21:152–162.

Maser, C. and J. R. Sedell. 1994. *From the Forest to the Sea: The Ecology of Wood in Streams, Rivers, Estuaries, and Oceans.* Delray Beach, Fla.: St. Lucie Press.

Mason, D. J. 1995. Forestry and the conservation of Venezuela's forest birds. Ph.D. diss., University of Wisconsin, Madison.

——. 1996. Responses of Venezuelan understory birds to selective logging, enrichment strips, and vine cutting. *Biotropica* 28:296–309.

Matola, S. 1974. Food of a tapir, Tapirus indicus. *Malayan Naturalist Journal* 28:90–93.

May, R. M. 1988. How many species are there on earth? *Science* 241:18–24.

Maya, R. S. 1995. Joint implementation: Cautions and options for the South. In C.J. Jepma, ed., *The Feasibility of Joint Implementation*, pp. 209–217. Netherlands: Kluwer Academic Publishers.

Mayer, J. J. and R. M. Wetzel. 1987. *Tayassu pecari. Mammalian Species* 293:1–7.

Mbalele, M. 1978. Part of African culture. *Unasylva* 29:16–17.

McCallum, R. and N. Sekhran. 1996. *Lessons Learned Through ICAD Experimentation in Papua New Guinea.* Port Vila, Vanuatu: United Nations South Pacific Forestry Development Programme.

McClure, H. E. and H.B. Othman. 1965. Avian biomass of Malaya. 2. The effect of forest destruction on a local population. *Bird-banding*, 36:242–269.

McCoy, J. 1995. Responses of blue and red duikers to logging in the Kibale forest of western Uganda. MS Thesis, University of Florida, Gainesville.

McFarland-Symington, M. 1988. Environmental determinants of population densities in Ateles. *Primate Conservation* 9:74–79.

McFoy, J. 1995. Responses of blue and red duikers to logging in the Kibale forest of Western Uganda. Master's thesis, Gainesville, Fla.: University of Florida.

McGeoch, M. A. 1998. The selection, testing and application of terrestrial insects as bio-indicators. *Biological Reviews of the Cambridge Philosophical Society* 73:181–201.

McKey, D. 1975. The ecology of coevolved seed dispersal systems. In E. Lawrence, and P.H. Raven, eds., *Coevolution of Animals and Plants*, pp. 159–191. Austin, Tex.: University of Texas Press.

McKey, D. B. 1978. Soils, vegetation, and seed-eating by black colobus monkeys. In G.G. Montgomery ed., *The Ecology of Arboreal Folivores*, pp. 423–438. Washington, D.C.: Smithsonian Institute Press.

McNab, B. K. 1971. The structure of tropical bat faunas. *Ecology* 52:352–358.

Medellín, R. A. 1993. Estructura y diversidad de una comunidad de murciélagos en el trópico húmedo mexicano. In R. A. Medellín, and G. Ceballos, eds., *Avances en el Estudio de los Mamíferos de México*, pp. 333–354. México, D.F.: Publicaciones especiales, Asociación Mexicana de Mastozoología.

——. 1994. Mammal diversity and conservation in the Selva Lacandona, Chiapas, México. *Conservation Biology* 8:780–799.

Medina, E. 1995. Physiological ecology of trees and application to forest management. In A.E. Lugo, and C. Lowe, eds., *Tropical Forests: Management and Ecology*, pp. 289–307. New York, N.Y.: Springer-Verlag.

Meehan, W. R., 1991. Influences of forest and rangeland management on salmonid fishes and their habitats. *American Fisheries Society Special Publication 19*.

Meffe, G. K. and C. R. Carroll. 1994. *Principles of Conservation Biology*. Sunderland, Mass.: Sinauer Associates, Inc.

Meijerink, A. M. J., W. van Wijngaarden, S. Amier Asrun, and B. H. Maathuis. 1988. Downstream damage caused by upstream land degradation in the Komering River Basin. *ITC Journal* 1988:96–108.

Menon, V., R. Sukumar, and A. Kumar. 1997. A god in distress: Threats of poaching and the ivory trade to the Asian elephant in India. Bangalore, India: Asian Elephant Conservation Centre.

Merenlender, A. M., C. Kremen, M. Rakotondratsima and A. Weiss. 1998. Monitoring impacts of natural resource extraction on lemurs of the Masoala Peninsula, Madagascar. *Conservation Ecology* 2: Article 5. *http://www.consecol.org/Journal/vol2/iss2/art5/*.

Mérida, G. 1995. Políticas del Sistema Nacional de Areas Protegidas. Unpublished report. Dirección Nacional de Conservación de la Biodiversidad - Ministerio de Desarrollo Sostenible y Medio Ambiente.

Merry, F. D. and D. R. Carter. 1997. Certified wood markets in the US: Implications for tropical deforestation. *Forest Ecology and Management* 92:221–228.

Merz, G. 1981. Reserches sur la biologie de nutrition et les habitats preferences de l'elephant de foret, Loxodonta africana cyclotis Matschie, 1990. *Mammalia* 45:299–305.

——. 1986. Movement patterns and group size of the African forest elephant Loxodonta africana cyclotis in the Tai National Park, Ivory Coast. *African Journal of Ecology* 24:61–68.

Miller, B. W. 1998. Community ecology of non-phyllostomid bats in northwestern Belize. Ph.D. diss., University of Kent, Durrell Institute of Conservation and Ecology.

Ministerio das Minas e Energia, Secretaria-Geral, Projecto RADAMBRASIL. 1982. *Levantamento de Recursos Naturais, 27*. Brasilia, Brazil: MME/Projecto Radambrasil.

Ministerio do Interior. 1974. Estudos Hidrologicos da Bacia do Alto Paraguai. Relatorio tecnico, Vol. 1. Ministerio do Interior, Brasilia, Brazil: Departamento Nacional de Obras e Saneamento.

Ministry of Forestry of the People's Republic of China/WWF (World Wildlife Fund), 1989. *National Conservation Management Plan for the Giant Panda and its Habitat.* Hong Kong: WWF (World Wildlife Fund).

Miranda, M., A. B. Uribe Q., L. Hernadez, J. Ochoa G. and E. Yerena. 1998. *All That Glitters is Not Gold: Balancing Conservation and Development in Venezuela's Frontier Forests.* Washington, D. C.: WRI (World Resources Institute.)

Mitani, M. 1990. A note on the present situation of the primate fauna found from southeastern Cameroon to northern Congo. *Primates* 31:625–634.

Mitchell, A. 1990. *The Fragile South Pacific: An Ecological Odyssey.* Austin, Tex.: University of Texas Press.

Mitchell, J. C., J. A. Wicknick, and C. D. Anthony. 1996. Effects of timber harvesting practices on peaks of otter salamander (*Plethodon hubrichti*) populations. *Amphibian & Reptile Conservation* 1:15–19.

Mitra, S. S. and F. H. Sheldon. 1993. Use of an exotic tree plantation by Bornean lowland forest birds. *Auk* 110:529–540.

Mittermeier, R. A. 1990. Hunting and its effect on wild primate populations in Suriname. In J. G. Robinson and K. H. Redford, eds., *Neotropical Wildlife Use and Conservation,* pp. 93–107. Chicago, Ill.: University of Chicago Press.

Moad, A. S. and J. L. Whitmore. 1994. Tropical forest management in the Asia-Pacific region. *Journal of Sustainable Forestry* 1:25–63.

Moermond, T. C. and J. S. Denslow. 1985. Neotropical frugivores: Patterns of behavior, morphology and nutrition with consequences for fruit selection. In P. A. Buckley, M. S. Foster, E. S. Morton, R. S. Ridgely, and N. G. Buckley, eds., *Neotropical Ornithology,* pp. 865–897. Lawrence, Kans.: Allen Press.

Mogollon, L. F. and J. Comerma. 1995. Suelos de Venezuela. Palmaven (PDVSA). Caracas, Venezuela: Edit. Ex Libris C. A.

Moguel, M., Oguel, P., and V. M. Toledo. 1999. Biodiversity conservation in traditional coffee systems of Mexico. *Conservation Biology* 13(1):11–21.

Mohd, Zakaria-Ismail. 1994. Zoogeography and biodiversity of the freshwater fishes of Southeast Asia. *Hydrobiologia* 285:41–48.

Mok, W. Y., D. E. Wilson, L. A. Lacey, and R. C. C. Luizao. 1982. Lista atualizada de quirópteros da Amazônia Brasileira. *Acta Amazonica* 12:817–823.

Moll, D. and K. P. Jansen. 1995. Evidence for a role in seed dispersal by two tropical herbivorous turtles. *Biotropica* 27:121–127

Mondolfi, E. 1976. Fauna silvestre de los bosques húmedos de Venezuela. In E. Mondolfi, and P. Rambach, eds., *Conservación de los Bosques Húmedos de Venezuela,* pp. 113–181. Consejo de Bienestar Rural, Caracas: Sierra Club.

Montgomery, G. G. and M. Sunquist. 1978. Habitat selection and use by two-toed and three-toed sloths. In G.G. Montgomery, ed., *The Ecology of Arboreal Foliovores,* pp. 329–359. Washington, D.C.: Smithsonian Institution Press.

Mooney, H. A., J. H. Cushman, E. Medina, O. E. Sala, and E. D. Schultz, eds. 1996. *Functional Role of Biodiversity: A Global Perspective.* New York, N.Y.: Wiley and Sons.

Moore, J. C. and P. C. De Ruiter. 1991. Temporal and spatial heterogeneity of trophic interactions within below-ground food webs. *Agriculture, Ecosystems and Environment* 34:371–397.

Moore, J. C., P. C. De Ruiter, and W. H. Hunt. 1993. Influence of productivity on stability of real and model ecosystems. *Science* 261:906–908.

Moore, J. C., D. Walter, and W. H. Hunt. 1988. Arthropod regulation of microbiota and mesobiota in below ground food webs. *Annual Review of Entomology* 33:419–439.

Moraes, M. and S. Beck. 1992. Diversidad florística de Boliviain M. Marconi, ed., *Conservación de la Diversidad Biológica en Bolivia*, pp. 73–111. La Paz, Bolivia: Centro de Datos para la Conservación.

Morellato, P. C. and H. F. Leitao-Filho. 1996. Reproductive phenology of climbers in a southeastern Brazilian forest. *Biotropica* 28:180–191.

Morrison, D. W. 1979. Apparent male defense of tree hollows in the fruit bat *Artibeus jamaicensis*. *Journal of Mammalogy* 60:11–15.

Moulds, F. R.1988. Forest management in Australia. *Commonwealth Forestry Review* 67:65–70.

Moura-Costa, P. 1996. Tropical forestry practices for carbon sequestration. In A. Schulte and D. Schöne, eds., *Dipterocarp Forest Ecosystems: Toward Sustainable Management*, pp. 308–334. Singapore: World Scientific.

Murray K. G. 1988. Avian seed dispersal of three neotropical gap-dependent plants. *Ecological Monographs* 58:271–298.

Myers, N. 1996. The world's forests: Problems and potentials. *Environmental Conservation* 23:156–168.

Myers, N. 1980. *Conversion of Tropical Moist Forests*. Washington, D.C.: National Academy of Sciences.

Myster, R. W., J. R. Thomlinson, and M. C. Larsen. 1997. Predicting landslide vegetation in patches on landscape gradients in Puerto Rico. *Landscape Ecology* 12:309–320.

Naeem, S., L. J. Thompson, S. P. Lawler, J. H. Lawton, and R. M. Woodfin. 1994. Declining biodiversity can alter the performance of ecosystems. *Nature* 368:734–737.

Naiman, R. J., ed. 1992. *Watershed Management: Balancing Sustainability and Environmental Change*. New York, N.Y.: Springer-Verlag.

Naiman, R. J., H. Decamps, and M. Pollock. 1993. The role of riparian corridors in maintaining regional biodiversity. *Ecological Applications* 3:209–212.

Naranjo, E. J. 1995. Habitos de alimentacion del tapir (*Tapir bairdii*) en el bosque lluvioso tropical de Costa Rica. *Vida Silvestris Neotropical* 4:108–113.

National Audubon Society, Procter and Gamble Company, and Pennsylvania State University. 1996. Guidelines to maintain biodiversity in Pennsylvania forests: 1996 study plan. Working Document. New York, N.Y.: National Audubon Society.

Naughton-Treves, L. 1996. Uneasy neighbors: Wildlife and farmers around Kibale National Park, Uganda. Ph.D. diss., University of Florida.

Navarro, G. 1992. Sectorización ecológica previa de la Reserva Ríos Blanco y Negro. Unpublished report to FAN. Santa Cruz, Bolivia.

——. 1995. La vegetación de la región de Lomerío. In Centurión T. R. and I. Kraljevic, eds., *Las Plantas Útiles de Lomerío*, pp 54–88. BOLFOR (Bolivia Sustainable Forestry Management Project)-Herbario USZ-CICOL (Central Intercomunal Campesina del Oriente de Lomerió, Santa Cruz, Bolivia.

Nectoux, F. and Y. Kuroda. 1989. *Timber from the South Seas*. Gland, Switzerland: WWF (World Wildlife Fund) International.

Nelson, B. W., V. Kapos, J. B. Adams, W. J. Oliveira, O. P. G. Braun, and I. L. do Amaral. 1994. Forest disturbance by large blowdowns in the Brazilian Amazon. *Ecology* 75: 853–858.

Nepstad, D.C., C. Uhl, and E. A. S. Serrao. 1991. Recuperation of a degraded Amazonian landscape: Forest recovery and agricultural restoration. *Ambio* 20:248–255.

New, T. R. 1999. Entomology and nature conservation. *European Journal of Entomology* 96:11–17.

Newbery, D. M. and H. de Foresta. 1985. Herbivory and defense in pioneer gap and understory trees in tropical rain forest in French Guiana. *Biotropica* 17:283–244.

Newing, H. 1994. Behavioural Ecology of Duikers (*Cephalophus* spp.) in Forest and Secondary Growth, Taï, Côte d'Ivoire. Ph.D. diss., University of Stirling, Kent.

Newmark, W. D. 1991. Tropical forest fragmentation and the local extinction of under-storey birds in the eastern Usambara Mountains, Tanzania. *Conservation Biology* 5:67–78.

Ngui, S.K. 1991. National parks and wildlife sanctuaries in Sarawak. In R. Kiew. *The State of Nature Conservation in Malaysia*, pp. 90–198. Kuala Lumpur, Malaysia: Malaysian Nature Society.

Nicholson, D. I., N. B. Henry, and J. Rudder. 1988. Stand changes in north Queensland rainforests. *Proceedings of the Ecological Society of Australia* 15:61–80.

Nieminen, M. 1996. Migration of moth species in a network of small islands. *Oecologia* 108:643–651.

Norberg, U. M. and J. M. V. Rayner. 1987. Ecological morphology and flight in bats (Mammalia; Chiroptera): Wing adaptations, flight performance, foraging strategy and echolocation. *Philosophic Transactions of the Royal Society of London, Series B Biological Sciences* 316:335–427.

Nordin, M. 1978. Voluntary Food Intake and Digestion by the Lesser Mousedeer. *Journal of Wildlife Management* 42:185–187.

Nordin, M. and M. Zakaria. 1997. Some effects of logging in mixed lowland diptero-carp forests on birds. In Ong Beng Gaik, ed., *State of the Environment in Malaysia*, pp. 161–166. Malaysia: Consumers' Association of Penang Press.

Noss, A.J. 1995. Duikers, cables and nets: A cultural ecology of hunting in a central African forest. Ph.D. diss., University of Florida.

——. 1997. Challenges to nature conservation with community development in central African forests. *Oryx* 31:180–188

Nowak, R. M. 1991. *Walker's Mammals of the World*. 5th Edition. 2 volumes. Baltimore, Md. and London, United Kingdom: The Johns Hopkins University Press.

Nummelin, M. 1989. Seasonality and effects of forestry practices on forest floor arthropods in the Kibale Forest, Uganda. *Fauna Norvegica Series B* 36:17–25.

——. 1990. Relative habitat use of duikers, bush pigs and elephants in virgin and selectively logged areas of the Kibale forest, Uganda. *Tropical Zoology* 3:111–120.

——. 1992. Invertebrate herbivory in virgin and managed sites in Kibale Forest, Western Uganda. *African Journal of Ecology* 30:213–222.

——. 1996. Community structure of invertebrates in primary and managed forest sites in the Kibale Forest, western Uganda. *Tropical Ecology* 23:201–213.

Nummelin, M. and L. Borowiec. 1991. *Cassidinae* beetles in primary and managed forest in Kibale Forest, Western Uganda. *African Journal of Ecology* 30:10–17.

Nummelin, M. and H. Fürsch. 1992. Coccinellids of the Kibale Forest, western Uganda: Comparison between virgin and managed sites. *Tropical Zoology* 5:155–166.

Nummelin, M. and I. Hanski. 1989. Dung beetles in primary and managed forests in Kibale Forest, Western Uganda. *Journal of Tropical Ecology* 5:349–352.

Nussbaum, R., J. Anderson, and T. Spencer. 1995. Factors limiting the growth of indigenous tree seedlings planted on degraded rainforest soils in Sabah, Malaysia. *Forest Ecology and Management* 74:149–159.

Nwoboshi, L. C. 1982. *Tropical Silviculture: Principles and Techniques*. Nigeria: Ibadan University Press.

Nyland, R. D. 1996. *Silviculture: Concepts and Applications*. New York, N.Y.: McGraw-Hill Companies, Inc.

Oates, J. F. 1996. Habitat alteration, hunting and the conservation of folivorous primates in African forests. *Australian Journal of Ecology* 21:1–9.

——. 1977. The social life of a black-and white Colobus monkey (*Colobus guereza*). *Zeitschrift fur Tierpsychologie* 45:1–60.

Oates, J. F., G. H. Whitesides, A. G. Davies, P. G. Waterman, S. M. Green, G. L. Dasilva, and S. Mole. 1990. Determinants of variation in tropical forest primates: New evidence from West Africa. *Ecology* 71:328–43

Oboussier, H. 1966. Zur Kenntnis der Cephalophinae. *Zeitschrift fur Morphologie und Okologie Tiere* 57:259–273.

Obua, J. O. 1992a. Frugivore abundance along a logging gradient in the Kibale Forest, western Uganda. In R.T. Wilson, ed., *Birds and the African Environment: Proceedings of the 8th Pan African Ornithological Congress. Annales Musee Royal de l'Afrique Centrale* (Zoologie) 268:327–333

———. 1992b. The influence of fruit profiles on avian feeding strategies of the Kibale Forest, Uganda. Unpublished MSc thesis, Makerere University.

Ochoa, J. 1988. Inventario de los mamiferos de la reserva florestal de Ticoporo y la Serrania de los Pijiguaos,Venezuela. *Acta Cientifica Venezolana* 39: 269–280.

———. 1992. Venezuela's bats: A case for conservation. *Bats* 10:10–13

———. 1993. Diseño de un sistema de corredores de vida silvestre en bosques productores de maderas de la Guayana venezolana. In A. Vega, ed., *Conservation Corridors in the Central American Region,* pp. 366–382. Gainesville, Fla.: Tropical Research and Development, Inc.

———. 1995. Los mamiferos de la Región de Imataca. *Acta Cientifica Venezolana* 46:1–14.

———. 1997a. Sensibilidades potenciales de una comunidad de mamiferos en un bosque productor de maderas de la Guayana Venezolana. *Interciencia* 22:1–15.

———. 1997b. El aprovechamiento forestal en la Guayana Venezolana: Evaluación ecológica e implicaciones para la conservación de los mamiferos de la region. Ph.D. diss., Universidad de los Andes.

———. 1998. Análisis preliminar de los efectos del aprovechamiento de maderas sobre la composición y estructura de bosques de la Guayana venezolana. *Interciencia* 23:197–207.

———. 2000. Efectos de la extracción de maderas sobre la diversidad de pequeños mamiferos en bosques de tierras bajas de la Guayana venezolana. *Biotropica* 32(1):146–164.

Ochoa, J., M. Aguilera, and P. Soriano. 1995. Los mamiferos del Parque Nacional Guatopo: lista actualizada y estudio comunitario. *Acta Cientifica Venezolana* 46:174–187.

Ochoa, J., C. Molina, and S. Giner. 1993. Inventario y estudio comunitario de los mamiferos del Parque Nacional Canaima, con una lista de las especies registradas para la Guayana venezolana. *Acta Cientifica Venezolana* 44:245–262.

Ochoa, J., J. Sánchez, M. Bevilacqua, and R. Rivero. 1988. Inventario de los mamiferos de la Reserva Forestal de Ticoporo y la Serranía de Los Pijiguaos, Venezuela. *Acta Cientifica Venezolana* 39:269–280.

ODA (Overseas Development Administration). 1995. Malaysia-United Kingdom: Conservation, management and development. Programme document. London, United Kingdom: South-east Asia Development Division, Overseas Development Administration.

ODA/FRR (Overseas Development Administration/Fountain Renewable Resources). 1996. Inception report for the Ghana Forest Sector Development Project. ODA Contract no. CNTR 955528A, London and Brackley: ODA (Overseas Development Administration) and FRR (Fountain Renewable Resources Limited).

Odum, H. T. 1970. Summary: An emerging view of the ecological system at El Verde. In H.T. Odum, and R. F. Pigeon, eds., *A Tropical Rain Forest: A Study of Irradiation and Ecology at El Verde, Puerto Rico,* pp. 1–283. Springfield, Va.: S.S. Atomic Energy Commission, National Technical Information Service.

Odum, H. T. and R. F. Pigeon. 1970. *A Tropical Rain Forest: A Study of Irradiation and Ecology at El Verde, Puerto Rico.* Springfield, Va.: S. S. Atomic Energy Commission, National Technical Information Service.

O'Farrell, M. J. 1997. Use of echolocation calls for the identification of free-flying bats. *Transactions of the Western Section of the Wildlife Society* 33:1–8.

O'Farrell, M. J., C. Corben, W. L. Gannon, and B. W. Miller. 1999. Confronting the dogma: A reply. *Journal of Mammalogy.* 80:297–302.

O'Farrell, M. J. and W. L. Gannon. 1999. A comparison of acoustic versus capture techniques for the inventory of bats. *Journal of Mammalogy* 80:24–30.

O'Farrell, M. J. and B. W. Miller. 1997. A new examination of echolocation calls of some neotropical bats (Emballonuridae and Mormoopidae). *Journal of Mammalogy* 87:954–963.

——. 1999. Use of vocal signatures for the inventory of free-flying Neotropical bats. *Biotropica* 31:507–516.

O'Farrell, M. J., B. W. Miller, and W. L. Gannon. 1999b. Qualitative identification of free-flying bats using the Anabat Detector. *Journal of Mammalogy* 80:11–23.

Oliver, C. D. and B. C. Larson. 1990. *Forest Stand Dynamics.* New York, N.Y.: McGraw-Hill, Inc.

Oliver, I. and A. J. Beattie. 1996a. Designing a cost-effective invertebrate survey: a test of methods for rapid assessment of biodiversity. *Ecological Applications* 6:594–607.

——. 1996b. Invertebrate morphospecies as surrogates for species: A case study. *Conservation Biology* 10:99–109.

Olivier, R. C. D. 1978. On the ecology of the Asian elephant *Elaphas maximus* Linn., with particular reference to Malaya and Sri Lanka. Ph.D. diss., University of Cambridge.

Olmsted, I. and E. R. Alvarez-Buylla. 1995. Sustainable harvesting of tropical trees: Demography and matrix models of two palm species in Mexico. *Ecological Applications* 5:484–500

Olson, D. M. and E. Dinerstein. 1998. *The Global 200: A Representation Approach to Conserving the Earth's Distinctive Ecoregions.* Washington, D.C.: WWF (World Wildlife Fund).

Ong, J. E. 1995. The ecology of mangrove conservation and management. *Hydrobiologia* 295:343–351.

Ongley, E. D. 1993. Water quality data programmes for developing land use and resource policies: Latin America. In FAO ed. *Prevention of Water Pollution by Agriculture and Related Activities. Proceedings of the Food and Agriculture Organization (FAO) of the United Nations, Santiago, Chile 1992*, pp. 263–272. Rome, Italy: FAO.

Orsak, L. 1996. *An Important Lesson Learnt from the ICAD (Integrated Conservation and Development) Project in Lak, Southern New Ireland Province: The Importance of Motivation and Conviction.* Madang, Papua New Guinea: Christensen Research Institute.

Ostertag, R. 1998. Below ground effects of canopy gaps in a tropical wet forest. *Ecology* 79:1294–1304.

Ouellet, M., J. Bonin, J. Rodrigue, J-L. DesGranges, and S. Lair. 1997. Hindlimb deformities (ectromelia, ectrodactyly) in free-living anurans from agricultural habitats. *Journal Wildlife Diseases* 33:95–104.

Owen-Smith, R. N. 1988. *Megaherbivores.* Cambridge, United Kingdom: Cambridge University Press.

Owiunji, I. 1996. The long-term effects of selective logging on the bird communities in the Budongo Forest Reserve, Uganda. Unpublished MSc thesis, Makerere University.

Owiunji, I. and A. J. Plumptre. 1998. Bird communities in logged and unlogged compartments in Budongo Forest, Uganda. *Forest Ecology and Management* 108:115–126.

Pacheco, L. F. 1992. El valor de nuestra fauna silvestre. *Instituto de Ecología, La Paz, Documentos Serie Zoología* 2:1–14.

Pacheco, P. 1998. *Estilos de Desarrollo, Deforestacion y Degradacion de los Bosques en las Tierras Bajas de Bolivia.* La Paz, Bolivia: CIFOR (Centre for International Forestry Research), CEDLA (Centro de Estudios para el Desarrollo Laboral Agrario) Taller de Iniciativas en Estudios Rurales y Reforma Agraria (TIERRA).

Pacheco, V., B. D. Patterson, J. L. Patton, L. H. Emmons, S. Solari, and C. F. Ascorra. 1993. List of mammal species known to occur in Manu Biosphere Reserve, Perú. Publicaciones del Museo de Historia Natural, *Universidad Nacional Mayor de San Marcos, Serie A Zoología* 1–12.

PAF (Plan de Acción Forestal). 1995. Situación del sector forestal del Departamento del Beni. Ministerio de Desarrollo Sostenible y Medio Ambiente, Secretaría de Recursos Naturales y Medio Ambiente, Subsecretaria de Recursos Naturales Renovables, Plan de Acción Forestal para Bolivia, La Paz.

Pahl, L. I., J. W. Winter and G. Heinsohn. 1988. Variation in responses of arboreal marsupials to fragmentation of tropical rainforests in North Eastern Australia. *Biological Conservation* 46:71–82.

Palmer, J. and T. Synnott. 1992. The management of natural forests. In N. P. Sharma, ed. *Managing the World's Forests*, pp. 337–373. Dubuque, Iowa: Kendall/Hunt Publishing Company.

Palmer, J. R. 1975. Towards more reasonable objectives in tropical high forest management for timber production. *Commonwealth Forestry Review* 54:273–289.

Pancel, L. 1993. *Tropical Forestry Handbook*. New York, N.Y.: Springer-Verlag.

Panayotou, T. and P. J. Ashton. 1992. *Not By Timber Alone: Economics and Ecology for Sustainable Tropical Forests*. Washington, D. C.: Island Press.

Pankhurst, C. E. 1997. Biodiversity of soil organisms as an indicator of soil health. In C. E. Pankhurst, B. M. Doube, and V. V. S. R. Gupta, eds., *Biological Indicators of Soil Health*. pp. 297–324. Wallingford, United Kingdom: CAB (Centre for Agriculture and Bioscience).

Pannell, C. M. 1989. The role of animals in natural regeneration and the management of equatorial rain forests for conservation and timber production. *Commonwealth Forestry Review* 68:309–313.

Parmelee, R. W., M. H. Beare, W. Cheng, P. F. Hendrix, S. J. Rider, D. A. Crossley, Jr., and D. C. Coleman. 1990. Earthworms and enchytraeids in conventional and no-tillage agroecosystems: A biocide approach to assess their role in organic matter breakdown. *Biology and Fertility of Soils* 10:1–10.

Parren, M. P. E. and N. R. De Graaf. 1995. The quest for natural forest management in Ghana, Côte d'Ivoire and Liberia. Tropenbos Series Volume 13. Wageningen, The Netherlands: The Tropenbos Foundation.

Paton, D. C. 1993. Honeybees in the Australian environment: Does *Apis mellifera* disrupt or benefit the native biota? *Bioscience* 43:95–103.

Pattemore, V. and J. Kikkawa. 1974. Comparison of bird populations in logged and unlogged rain forest in Wiangarie State Forest, N.S.W. *Australian Forestry* 37:188–198.

Patterson, B. D. 1991. The integral role of biogeographic theory in the conservation of tropical forest diversity. In M. A. Mares and D. J. Schmidly, eds., *Latin American Mammalogy: History, Biodiversity and Conservation*, pp. 124–149. Norman, Okla. and London, United Kingdom: University of Oklahoma Press.

Pavez L. and A. Bojanic H. 1998. El proceso social de formulación de la Ley Forestal de Bolivia de 1996. Unpublished report. CIFOR (Centre for International Forestry Research., Bogor, Indonesia.

Payne, J. 1988. Orang-utan conservation in Sabah. Kuala Lumpur, Malaysia: WWF (World Wildlife Fund) Malaysia.

Payne, J. C. 1992. A field study of techniques for estimating densities of duikers in Korup National Park, Cameroon. MS thesis, University of Florida.

Pearce, D. 1995. Joint implementation: A general overview. In C. J. Jepma, ed., *The Feasibility of Joint Implementation*, pp. 15–31. Netherlands: Kluwer Academic Publishers.

Pearce, J. and K. Ammann. 1995. *Slaughter of the Apes: How the Tropical Timber Industry is Devouring Africa's Great Apes*. London, United Kingdom: World Society for the Protection of Animals.

Pearman, P. B. 1997. Correlates of amphibian diversity in an altered landscape of Amazonian Ecuador. *Conservation Biology* 11:1211–1225.

Pearson, D. L. 1971. Vertical stratification of birds in a tropical dry forest. *Condor* 73:46–55.

Pechmann, J. H. K., D. E. Scott, R. D. Semlitsch, J. P. Caldwell, L. J. Vitt, and J. W. Gibbons. 1991. Declining amphibian populations: The problem of separating human impacts from natural fluctuations. *Science* 253:892–895.

Pelusa, N. L. 1992. *Rich Forests, Poor People*. Berkeley, Calif.: University of California Press.

Peña, F. 1996. Determinación de rendimientos y costos de operaciones iniciales de manejo forestal en la zona de Lomerío. Undergraduate thesis, UAGRM (Universidad Autónoma Gabriel Rene Moreno).

Perdue, M. and J. D. Steventon. 1996. Partial cutting and bats: A pilot study. In Barclay, R. M. R. and R. M. Brigham, eds., *Bats and Forests Symposium; 1995*, pp. 273–276. Victoria, British Columbia: Ministry of Forests Research Program.

Peres, C. A. 1993. Structure and spatial organisation of an Amazonian terra firme forest primate community. *Journal of Tropical Ecology* 9:259–276.

———. 1994. Indigenous reserves and nature conservation in Amazonian forests. *Conservation Biology* 8(2):586–588.

———. 2000a. Identifying keystone plant resources in tropical forests: The case of gums from Parkia pods. *Journal of Tropical Ecology* 16:287–317.

———. 2000b. Effect of subsistence hunting on vertebrate community structure in Amazonian forests. *Conservation Biology.*14:240–253.

Peres, C. A, L. C. Schiesari and L. C. Dias-Leme. 1997. Vertebrate predation of Brazilnuts (*Bertholletia excelsa*, Lecythidaceae), an agouti-dispersed Amazonian seed crop: A test of the escape hypothesis. *Journal of Tropical Ecology* 13:69–79.

Pérez-Salicrup, D. R. 1998. Effects of liana cutting on trees and tree seedlings in a tropical forest in Bolivia. Ph.D. diss., Department of Biology, University of Missouri-St. Louis.

Perrin, M. R., A. E. Bowland, and A. S. Faurie. 1992. Niche separation between the blue duiker *Philantomba monticola* and the Red duiker *Cephalophus natalensis*. *Ungulates* 91:201–204

Perry, D. A. and M. P. Amaranthus. 1997. Disturbance, recovery and stability. In K.A. Kohm and J.F. Franklin eds., *Creating a Forestry for the 21st Century: The Science of Ecosystem Management,* pp. 31–56. Washington, D.C.: Island Press.

Peters, C. M. 1994. Sustainable harvest of non-timber plant resources in tropical moist forest: An ecological primer. Washington, D.C.: Biodiversity Support Program, USAID (U.S. Agency for International Development).

Peterson, G., C. R. Allen, and C. S. Holling. 1998. Ecological resilience, biodiversity, and scale. *Ecosystems* 1:6–18

Petricone, S. and T. Vetleseter. 1995. U.S. participation in Joint Implementation Projects: Benefits, obstacles and policy responses. DE-AP01-95P030245.A000. Washington, D.C.: Center for Sustainable Development in the Americas, U.S. Department of Energy.

Pfeffer, P. and J. O. Caldecott. 1986. The bearded pig (*Sus barbatus*) in East Kalimantan and Sarawak. *Journal of the Malaysian Branch of the Royal Asiatic Society* 59:81–100

Pfiffer, W. J. 1996. Litter invertebrates. In D. P. Reagan and R. B. Waide, eds., *The Food Web of a Tropical Rain Forest*, pp. 137–182. Chicago, Ill.: University of Chicago Press.

Philip, M. S. 1965. Working Plan for the Budongo Central Forest Reserve. Third Revision 1964–1974. Entebbe, Uganda: Government Printer.

Phillips, J. D. 1989. An evaluation of factors determining the effectiveness of water quality buffer zones. *Journal of Hydrology* 107:133–145.

Phillips, O. L. and A. H. Gentry. 1994. Increasing turnover through time in tropical forests. *Science* 263:954–958.

Phillips, O. L., Y. Malhi, N. Higuchi, W. F. Laurance, P. V. Nunez, R. M. Vasquez, S. G. Laurance, L. V. Ferreira, M. Stern, S. Brown, and J. Grace. 1998. Change in the carbon balance of tropical forests: Evidence from long-term plots. *Science* 282:439–442.

Pianka, E. R. 1986. *Ecology and Natural History of Desert Lizards*. Princeton, N.J.: Princeton University Press.

Pickett, S. T. A. and M. J. McDonnell. 1989. Changing perspectives in community dynamics: A theory of successional forces. *Trends in Ecology and Evolution* 4:241–245.

Pickett, S. T. A. and P. S. White. 1985. *The Ecology of Natural Disturbance and Patch Dynamics*. New York, N.Y.: Academic Press, Inc.

Pimm, S. L. 1982. *Food Webs*. New York, N.Y.: Chapman and Hall.

——. 1991. *The Balance of Nature*. Chicago, Ill.: University of Chicago Press.

Pimm, S. L., J. H. Lawton, and J. E. Cohen. 1991. Food web patterns and their consequences. *Nature* 350:669–674.

Pinard, M. A. 1994. The reduced-impact logging project. *ITTO (International Tropical Timber Organization) Tropical Forest Update* 4(3):11–12.

Pinard, M. A, B. Howlett, and D. Davidson. 1996. Site conditions limit pioneer tree recruitment after logging of dipterocarp forests in Sabah, Malaysia. *Biotropica* 28:2–12.

Pinard, M. A. and F. E. Putz. 1996. Retaining forest biomass by reducing logging damage. *Biotropica* 28:278–295.

——. 1997. Monitoring carbon sequestration benefits associated with a reduced-impact logging project in Malaysia. *Mitigation and Adaptation Strategies for Global Change* 2:203–215.

Pinard, M. A., F. E. Putz, and J. C. Licona. 1999. Tree mortality and vine proliferation following a wildfire in a sub-humid tropical forest in eastern Bolivia. *Forest Ecology and Management* 116:247–252.

Pinard, M. A., F. E. Putz, D. Rumiz, R. Guzman, and A. Jardim. 1999. Ecological characterization of tree species to guiding forest management decisions in seasonally-dry forests in Lomerìo, Bolivia. *Forest Ecology and Management* 113:201–213.

Pinard M. A., F. E. Putz, J. Tay, and T. Sullivan. 1995. Creating timber harvest guidelines for a reduced-impact logging project in Malaysia. *Journal of Forestry* 93:41–45.

Pinedo-Vasquez, M., D. Zarin, P. Jipp, and J. Chota-Inuma. 1990. Use-values of tree species in a communal forest reserve in northeast Peru. *Conservation Biology* 4:405–416.

Plotkin, M. and L. Famolare, eds. 1992. *Sustainable Harvest and Marketing of Rain Forest Products*. Washington, D.C.: Island Press.

Plumptre, A. J. 1990. Forest Antelope Research: in Rwanda. *Gnusletter* 9(1):3.

——. 1994. The effects of long-term selective logging on blue duikers in Budongo Forest Reserve. *Gnusletter* 13(1 and 2):15–16

——. 1995. The Budongo Forest Reserve: The effect of 60 years of selective logging on tree distributions and forest structure. *Commonwealth Forestry Review* 74:253–258.

——. 1996a. Changes following sixty years of selective timber harvesting in the Budongo Forest Reserve, Uganda. *Forest Ecology and Management* 89:101–113

——. 1996b. Two nests of Nahan's Francolin in the Budongo Forest Reserve, Uganda. *Bulletin of the African Bird Club* 3:37–38.

——. 1997. Shifting cultivation along the trans-African highway and its impact on the understory bird community in the Ituri Forest, Zaire. *Bird Conservation International* 7:317–329

Plumptre A. J. and V. Reynolds. 1994. The effect of selective logging on the primate populations in the Budongo Forest Reserve. *Journal of Applied Ecology* 31:631–41.

——. 1996. Censusing chimpanzees in the Budongo forest. *International Journal of Primatology* 17:85–99.

——. 1997. Nesting behavior of chimpanzees: Implications for censuses. *International Journal of Primatology* 18:475–485.

Plumptre, A. J., V. Reynolds, and C. Bakuneeta. 1994. The contribution of fruit eating primates to seed dispersal and natural regeneration after selective logging. Final report of ODA project R4738. Oxford, United Kingdom: Biological Anthropology, Oxford University.

——. 1997. The effects of selective logging in monodominant forests on biodiversity. Report to ODA. Oxford, United Kingdom: Biological Anthropology, Oxford University.

Poffenberger, M. 1994. The resurgence of community forest management in Eastern India. In D. Western and R. M. Wright, eds., *Natural Connections: Perspectives in Community-based Conservation*, pp. 53–79. Washington, D.C.: Island Press.

Polis, G. A. and D. R. Strong. 1996. Food web complexity and community dynamics. *American Naturalis* 147:813–846.

Pollard, E. 1977. A method for assessing changes in abundance of butterflies. *Biological Conservation* 12:115–131.

Pomeroy, D. 1992. *Counting Birds*. Nairobi, Kenya: AWF (African Wildlife Federation) technical handbook No. 6.

Poore, D., P. Burgess, J. Palmer, S. Rictbergen, and T. Synnott. 989. *No Timber Without Trees. Sustainability in the Tropical Forest.* London, United Kingdom: Earthscan Publications.

Poore, D. and J. Sayer. 1987. *The Management of Tropical Moist Forest Lands: Ecological Guidelines.* Gland, Switzerland: IUCN (International Union for Conservation of Nature and Natural Resources).

Popma, J., F. Bongers, and J. Meave del Castillo. 1988. Patterns in the vertical structure of the tropical lowland rain forest of Los Tuxtlas, Mexico. *Vegetatio* 74:81–91.

Poulsen, B. O. 1994. Movements of single birds and mixed-species flocks between isolated fragments of cloud forest in Ecuador. *Studies on Neotropical Fauna and Environment* 3:149–160.

Powell, R. A., J. W. Zimmerman, D. E. Seaman, and J. F. Gilliam. 1996. Demographic analyses of a hunted black bear population with access to a refuge. *Conservation Biology* 10:224–234.

Prabhu, R., C. J. P. Colfer, P. Venkateswarlu, L.C. Tan, R. Soekmadi, and E. Wollenberg. 1996. Testing criteria and indicators for the sustainable management of forests: Phase 1 final report. CIFOR Special Publication. Bogor, Indonesia: Center for International Forestry Research.

Prance, G. T. 1977. The phytogeographic subdivisions of Amazonia and their influence on the selection of biological reserves. In G. T. Prance and T. Elias, eds. *Extinction is Forever*, pp. 195–213. New York, N.Y.: New York Botanical Garden.

Preen, A. R. 1981. The effects of selective logging on the vertebrate fauna of a tropical rainforest in northeast Queensland. BSc Hons. thesis, James Cook University.

Prendergast, J. R., R. M. Quinn, J. H. Lawton, B. C. Eversham, and D. W. Gibbons. 1993. Rare species, the coincidence of biodiversity hotspots and conservation strategies. *Nature* 365:335–337.

Pressey, R. L., C. J. Humphries, C. R. Margules, R. I. Vane-Wright, and P. H. Williams. 1993. Beyond opportunism: Key principles for systematic reserve selection. *Trends in Ecology and Evolution* 8:124–128.

Primack, R. B. and P. Hall. 1992. Biodiversity and forest change in Malaysian Borneo. *Bioscience* 42:829–837.

Pringle, C.M. and F. N. Scatena. 1999a. Factors affecting aquatic ecosystem deterioration in Latin America and the Caribbean. In U. Hatch and M. E. Swisher, eds., *Tropical Managed Ecosystems: New Perspectives on Sustainability*, pp. 104–113. Oxford, United Kingdom: Oxford University Press.

——. 1999b. Freshwater resource development: Case studies from Puerto Rico and Costa Rica. In U. Hatch and M. E. Swisher, eds., *Tropical Managed Ecosystems: New Perspectives on Sustainability*, pp. 114–121. Oxford, United Kingdom: Oxford University Press.

Prins, H. H. T. and J. M. Reitsama. 1989. Mammalian biomass in an African equatorial rain forest. *Journal of Animal Ecology* 58:851–61

Putz, F. E. 1984a. Liana biomass and leaf area of a "terra firme" forest in the Rio Negro Basin. *Biotropica* 15: 185–189.

———. 1984b. The natural history of lianas on Barro Colorado Island, Panama. *Ecology* 65: 1713–1724.

———. 1992. Silvicultural effects of lianas. In F. E. Putz, and H.A. Mooney eds., *The Biology of Vines*, pp. 493–501. Cambridge, United Kingdom: Cambridge University Press.

———. 1993. Consideration of the ecological foundations of natural forest management in the America tropics. Durham, North Carolina: The Center for Tropical Conservation.

———. 1994. Towards a sustainable forest. *ITTO* (International Tropical Timber Organization) *Tropical Forest Update* 4(3):7–10.

Putz, F. E. and N. V. L. Brokaw. 1989. Sprouting of broken trees on Barro Colorado Island, Panama. *Ecology* 70:508–512.

Putz, F. E. and H. T. Chan. 1986. Tree growth, dynamics, and productivity in a mature mangrove forest in Malaysia. *Forest Ecology and Management* 17:211–230.

Putz, F. E. and M. A. Pinard. 1993. Reduced-impact logging as a carbon-offset method. *Conservation Biology* 7:755–757.

Putz, F. E. and V. Viana. 1996. Biological challenges for certification of tropical timber. *Biotropica* 28:323–330.

Rabe, F. W., and N. L. Savage. 1979. A methodology for the selection of aquatic natural areas. *Biological Conservation* 15:291–300.

Rabinowitz, A. 1993. *Wildlife Field Research and Conservation Training Manual*. New York, N.Y.: Paul-Art Press Inc.

Ráez-Luna, E. F. 1995. Hunting large primates and conservation of the Neotropical rain forests. *Oryx* 29:43–48.

Raguso, R. A. and J. Llorente-Bousquets. 1990. The butterflies (Lepidoptera) of the Tuxtlas Mts., Veracruz, Mexico, revisited: Species-richness and habitat disturbance. *Journal of Research on the Lepidoptera* 29:105–133.

Rahm, U. 1960. Territoriumsmarkierung mit der Voraugendruse beim Maxwell-Ducker (*Philantomba maxwelli*). *Säugetierkundliche Mitteilungen* 8:140–141.

Rainey, W. E., E. D. Pierson, T. Elmqvist, and P.A.Cox. 1995. The role of flying foxes (Petropodidae) in oceanic island ecosystems of the Pacific. In P. A. Racey and S. M. Swift, eds., *Ecology, Evolution and Behaviour of Bats*, pp. 47–62. Oxford, United Kingdom: Clarendon Press.

Ralls, K. 1973. *Cephalophus maxwelli. Mammalian Species* 31:1–4.

Ralph, C. J., G. R. Geupel, P. Pyle, T. E. Martin, and D. F. DeSante. 1993. *Handbook of Field Methods for Monitoring Land Birds*. General Technical Report PSW-GTR-144. Albany, Calif.: Pacific Southwest Research Station, Forest Service, U.S. Department of Agriculture.

Reagan, D. P. 1996. Anoline lizards. In D. P. Reagan and R. B. Waide, eds., *The Food Web of a Tropical Rain Forest*, pp. 321–345. Chicago, Ill.: University of Chicago Press.

Reagan, D. P., G. R. Camilo, and R. B. Waide. 1996. The community food web: Major properties and patterns of organization. In D. P. Reagan and R. B. Waide, eds., *The Food Web of a Tropical Rain Forest*, pp. 461–510. Chicago, Ill.: University of Chicago Press.

Reagan, D. P. and R. B. Waide. 1996. *The Food Web of a Tropical Rain Forest*. Chicago, Ill.: University of Chicago Press.

Rebelo, A. G., and W. R. Siegfried. 1992. Where should nature reserves be located in the Cape Floristic Region, South Africa? Models for the spatial configuration of a reserve network aimed at maximizing the protection of floral diversity. *Conservation Biology* 6(2):243–252.

Reddy, M. V., V. R. K. Kumar, V. R. Reddy, P. Balshouri, D. F. Yule, A-L. Cogle, and L. S. Jangawad. 1995. Earthworm biomass response to soil management in semi-arid tropical Alfisol agroecosystems. *Biology and Fertility of Soils* 19:317–321.

Redford, K. H. 1990. The ecologically noble savage. *Orion* 9:24–29.

——. 1992. The empty forest. *Bioscience* 42:412–422.

Redford, K. H. and J. G. Robinson. 1987. The game of choice: Patterns of indian and colonist hunting in the Neotropics. *American Anthropologist* 89:650–667.

——. 1991. Subsistence and commercial uses of wildlife in Latin America. In J. G. Robinson, and K. H. Redford, eds., *Neotropical Wildlife Use and Conservation*, pp. 6–23. Chicago, Ill.: University of Chicago Press.

Regal, P. J. 1982. Pollination by wind and animals: Ecology of geographic patterns. *Annual Review of Ecology and Systematics* 13:497–524.

Reid, J. W. and R. E. Rice. 1997. Assessing natural forest management as a tool for tropical forest conservation. *Ambio* 26:382–386.

Reilly, J., H. Spedding, and G. Apriawan. 1997. Preliminary observations on the Sumatran rhino in Way Kambas National Park, Indonesia. *Oryx* 31:143–150

Reinthal, P. N. and M. L. J. Stiassny. 1991. The freshwater fishes of Madagascar: A study of an endangered fauna with recommendations for a conservation strategy. *Conservation Biology* 5:231–243.

Remsen, J. V., Jr. 1985. Community organization and ecology of birds of high elevation humid forests of the Bolivian Andes. *Ornithological Monographs* 36:733–756.

Restrepo, C. and N. Gómez. 1998. Responses of understory birds to anthropogenic edges in a neotropical montane forest. *Ecological Applications* 8:170–183.

Restrepo, C., S. Heredia, and N. Gomez. In review. Understory fruit abundance in a neotropical montane forest: The influence of edges and tree gaps. *Ecology*.

Ribera, M. O. 1995. Información resumida en matrices y documentos cortos sobre el sistema Nacional de areas Protegidas de Bolivia. Unpublished report. La Paz, Bolivia: Dirección Nacional de Conservación de la Biodiversidad - Ministerio de Desarrollo Sostenible y Medio Ambiente.

Ribera, M. O. 1996a. *Guía para la Categorización de Especies Aamenazadas de Vertebrados e Implementación de Acciones para su Conservación*. La Paz, Bolivia: Centro de Datos para la Conservación.

——. 1996b. Análisis sobre categorías de manejo, dependencias jurisdiccionales, declaratoria de areas protegidas y zonificación de manejo. *Revista Boliviana de Ecología y Conservación Ambiental* 1:71–76

Rice, R. E., R. E. Gullison, and J. W. Reid. 1997. Can sustainable management save tropical forests? *Scientific American* 276:44–49.

Richards, G. C. 1995. A review of ecological interactions of fruit bats in Australian ecosystems. In P. A. Racey and S. M. Swift, eds., *Ecology, Evolution and Behaviour of Bats*, pp. 79–96. Oxford, United Kingdom: Clarendon Press.

Richards, P. W. 1996. *The Tropical Rain Forest: An Ecological Study*. 2nd Edition. Cambridge, United Kingdom: Cambridge University Press.

Rick, C. M. and R. I. Bowman. 1961. Galapagos tomatoes and tortoises. *Evolution* 15:407–417.

Ridgely, R. S. and G. Tudor. 1989. *The Birds of South America: The Oscine Passerines*. Austin, Tex.: University of Texas Press.

Ridley, H. N. 1930. *The Dispersal of Plants Throughout the World*. Ashford, Kent, United Kingdom: L. Reeve and Co.

Riera, B. and D. Y. Alexandre. 1988. Surface des chablis et temps de renouvellement en forêt dense tropicale. *Acta Oecologia* 57:773–782.

Ristau, T., S. B. Horsley, and D. S. deCalesta. 1995. Establishment report for impact of glyphosate and sulfometureon on diversity of plants and wildlife in Allegheny hardwoods. Warren, Pa.: Northeast Forest Experiment Station, USDA Forest Service Report 4110-FS-NE-4152-163.

Rith Najarian, J. C. 1998. The influence of forest vegetation variables on the distribution and diversity of dragonflies in a northern Minnesota forest landscape: A preliminary study (Anisoptera). *Odonatologica* 27:335–351.

Rivera, V. S., P. M. Cordero, I. A. Cruz, and J. J. C. Domingo. 1996. *Situación de la lapa verde (Ara ambigua) en Costa Rica.* San Jose, Costa Rica: Ministerio de Ambiente y Energia; Union Mundial de la Naturaleza.

Robbins, C. S. 1979. Effect of forest of fragmentation on bird populations. In R. M. DeGraaf and K. E. Evans, eds., *Management of North Central and Northeastern Forests for Nongame Birds*, pp. 198–212. U.S. Forest Service General Technical Report NC-51.

Robbins, E., R. Kenny, and W. F. Hyde. 1995. Concesiones forestales y la Política Industrial Forestal en Bolivia. Documento Técnico 11/95. Santa Cruz, Bolivia: BOLFOR (Bolivia Sustainable Forestry Management Project).

Roberts, K. A. 1991. Monitoring terrestrial breeding bird populations. In F. B. Goldsmith, ed., *Monitoring for Conservation and Ecology*, pp. 179–212. London, United Kingdom: Chapman and Hall.

Robinson, G. R., R. D. Holt, M. S. Gaines, S. P. Hamburg, M. L. Johnson, H. S. Fitch, and E. A. Martinko. 1992. Diverse and contrasting effects of habitat fragmentation. *Science* 257:524–526.

Robinson, J. G. and K. H. Redford. 1986. Body size, diet, and population density of neotropical forest mammals. *The American Naturalist* 128:665–680.

———. 1989. Body size, diet, and population variation in Neotropical forest mammal species: Predictors of local extinction? In K. H. Redford and J. F. Eisenberg, eds., *Mammals of the Americas: Essays in Honor of Ralph M. Wetzel*, pp. 567–594. Gainesville, Fla.: Sandhill Crane Press.

———. 1991. Sustainable harvest of neotropical forest mammals. In J. G. Robinson and K. H. Redford eds., *Neotropical Wildlife Use and Conservation*, pp. 415–429. Chicago, Ill.: University of Chicago Press.

Robinson, S. K., J. Terborgh, and C. A. Munn. 1990. Lowland tropical forest bird communities of a site in Western Amazonia. In A. Keast, ed., *Biogeography and Ecology of Forest Bird Communities*, pp.229–258. The Hague, The Netherlands: SPB Academic Publishing.

Robinson, S. K., F. R. Thompson II, T. M. Donovan, D. R. Whitehead, and J. Faaborg. 1995. Regional forest fragmentation and the nesting success of migratory birds. *Science* 267:1987–1989.

Roderick, G. K. 1996. Geographic structure insect populations: Gene flow, phylogeography, and their uses. *Annual Review of Entomology* 41:325–352.

Rodriques, M., F. Olmos, and M. Galetti. 1993. Seed dispersal by tapir in south-eastern Brazil. *Mammalia* 57:460–461.

Rodríguez, L. O. and J. E. Cadle. 1990. A preliminary overview of the herpetofauna of Cocha Cashu, Manu National Park, Peru. In A. H. Gentry, ed., *Four Neotropical Rainforests*, pp. 410–425. New Haven, Conn.: Yale University Press.

Rodríguez, L. O. and W. E. Duellman. 1994. Guide to the frogs of the Iquitos Region, Amazonian Peru. *University of Kansas Museum of Natural History, Special Publication* 22:1–80.

Ross, D. 1992. Administración futura del Subproyecto Forestal Chimanes. Unpublished report. Santa Cruz, Bolivia: Sweed Forest Consulting.

Ross, S. M., J. B. Thornes, and S. Nortcliff. 1990. Soil hydrology, nutrient and erosional response to the clearance of terra firme forest. Maraca Island, Roraima, northern Brazil. *The Geographical Journal* 156 267–282.

Roubik, D. W. 1996. African honeybees as exotic pollinators in French Guiana. In A. Matheson, S. L. Buchmann, C. O'Toole, P. Westrich, and I. H. Williams, eds., *The Conservation of Bees*, pp. 173–182. London, United Kingdom: Academic Press.

Rouw, A. de, H. C. Vellema and W. A. Blokhuis. 1990. Land unit survey of the Tai region, south-west Cote d'Ivoire. Tropenbos Technical Series No. 7. Wageningen, The Netherlands: Tropenbos Foundation.

Rudel, T. K. and B. Horowitz. 1993. *Tropical Deforestation: Small Farmers and Land Clearing in the Ecuadorian Amazon*. New York, N.Y.: Columbia University Press.

Ruitenbeck, H. J. 1998. Rational Exploitations: Economic Criteria and Indicators for Sustainable Forest Management of Tropical Forests. CIFOR(Centre for International Forestry Research) Occasional paper. Bogor, Indonesia: CIFOR.

Rukundo, T. N. 1996. The long-term effect of canopy treatment on tree diversity in Budongo Forest: An evaluation of a 25-hectare permanent plot study. Unpublished MSc thesis, Makerere University.

Rumiz, D. I. and J. C. Herrera F. 1988. La evaluación de la fauna silvestre y su conservación en los bosques de producción de Bolivia. Unpublished report. Santa Cruz, Bolivia: BOLFOR (Bolivia Sustainable Forestry Management Project).

Rumiz, D. I. and A. B. Taber. 1994. Un relevamiento de mamíferos y algunas aves grandes de la Reserva de Vida Silvestre Ríos Blanco y Negro, Bolivia: Situación actual y recomendaciones. Working Papers #3. New York, N.Y.: Wildlife Conservation Society.

Rumiz, D. I. and L. Solar R. 1996. La caza, su impacto y aporte económico en una concesión forestal del Bajo Paraguá. Unpublished report. Santa Cruz, Bolivia: BOLFOR (Bolivia Sustainable Forestry Management Project).

Rykken, J. J., D. E. Capen, and S. P. Mahabir. 1997. Ground beetles as indicators of land type diversity in the Green Mountains of Vermont. *Conservation Biology* 11:522–530.

Ryti, R. T. 1992. Effect of the focal taxon on the selection of nature reserves. *Ecological Applications* 2(4):404–410.

Sabah Forest Department. 1989. *Forestry in Sabah*. Sandakan, Sabah, Malaysia: Sabah Forestry Department.

Sabatier, D. 1985. Saisonnalité et déterminisme du pic de fructification en forêt guyanaise. *Revue d'Ecologie* (Terre et Vie) 40:289–320

Sader, S. A. and A. T. Joyce. 1988. Deforestation rates and trends in Costa Rica, 1940 to 1983. *Biotropica* 20:11–19.

Saenger, P., E. J. Hegerl, and J. D. S. Davie. 1983. Global status of mangrove ecosystems. *The Environmentalist* 3(Supplement 3):1–88.

Saetersdal, M., J. M. Line, and H. J. B. Birks. 1993. How to maximize biological diversity in nature reserve selection: Vascular plants and breeding birds in deciduous woodlands, western Norway. *Biological Conservation* 66:131–138.

Saffirio, G. and R. Scaglion. 1982. Hunting efficiency in acculturated and unacculturated Yanomama villages. *Journal of Anthropological Research* 38:315–327.

Salafsky, N. 1994. Forest gardens in the Gunung Palung region of West Kalimantan, Indonesia. *Agroforestry Systems* 28:237–268.

——. 1997. *Eleven Steps for Setting up Community-based Timber Harvesting Enterprises: An Overview of IRECDP Experience in New Britain, Papua New Guinea*. Washington, D.C.: Biodiversity Support Program.

Salafsky, N., B. Cordes, M. Leighton, M. Henderson, W. Watt, and R. Cherry. 1998. Chainsaws as a tool for conservation: A comparison of community-based timber production enterprises in Papua New Guinea and Indonesia. London, United Kingdom: *Rural Forestry Network Development Paper* 22b.

Salafsky, N., B. L. Dugelby, and J. W. Terborgh. 1993. Can extractive reserves save the rain forest? An ecological and socioeconomic comparison of nontimber forest product extraction systems in Peten, Guatemala, and West Kalimantan, Indonesia. *Conservation Biology* 7:39–52.

Saldarriaga, J. G., D. C. West, M. L. Tharp, and C. Uhl. 1988. Long-term chronosequence of forest succession in the upper Rio Negro of Colombia and Venezuela. *Journal of Ecology* 76:938–958.

Salick, J., A. Mejia, and T. Anderson. 1995. Non-timber forest products integrated with natural forest management, Rio San Juan, Nicaragua. *Ecological Applications* 5:878–895.

Salo, E. O. and T. W. Cundy, eds. 1986. *Streamside Management: Forestry and Fishery Interactions*. Contribution No. 57. Proceedings of a symposium held at the University of Washington, February 12–14, 1986. Seattle, Washington: University of Washington, College of Forest Resources.

Salo, J., R. Kalliola, I. Hakkinen, Y. Makinen, P. Niemela, M. Puhakka, and P. D. Coley. 1986. River dynamics and the diversity of Amazon lowland forest. *Nature* 322: 254–258.

Salo, J. S. and R. J. Kalliola. 1990. River dynamics and natural forest regeneration in the Peruvian Amazon. In A. B. Anderson, ed., *Alternatives to Deforestation: Steps Towards Sustainable Use of the Amazon Rain Forest*, pp. 245–256. New York, N.Y.: Columbia University Press.

Samways, M. J. 1990. Species temporal variability: Epigaeic ant assemblages and management for abundance and scarcity. *Oecologia* 84 482–490.

——. 1993a. *Insect Conservation Biology*. London, United Kingdom: Chapman and Hall.

——. M. J. 1993b. Insects in biodiversity conservation: Some perspectives and directives. *Biodiversity and Conservation* 2:258–282.

Sanchez Pego, M. A. 1995. The forest enterprise of the indigenous community of Nuevo San Juan Parangaricutiro, Michoacan, Mexico. Madison, Wisconsin: Institute for Envrionmental Studies and the Land Tenure Center.

Sandstrom, U. 1992. Cavities in trees: Their occurrence, formation and importance for hole-nesting birds in relation to silvicultural practise. Ph.D. diss., Swedish University of Agricultural Sciences.

Sanford, R. L., J. Saldarriaga, K. L. Clark, C. Uhl, and R. Herrera. 1985. Amazon rainforest fires. *Science* 227: 53–55.

Santiapillai, C. and P. Jackson. 1990. *The Asian Elephant —An Action Plan for its Conservation*. Gland, Switzerland: IUCN (International Union for Conservation of Nature and Natural Resources).

Santillán, H. 1996. Efectos del aprovechamiento en dos sitios de estudio, Oquiriquia Bajo Paraguá. Unpublished report. Santa Cruz, Bolivia: ETSFOR (Escuela Técnica Superior Forestal)-BOLFOR (Bolivia Sustainable Forestry Management Project).

Sarawak Forest Department. 1994. *Forestry in Sarawak, Malaysia*. Kuching, Sarawak: Sarawak Forest Department.

Sargent, C., T. Husain, N. A. Kotey, J. Mayers, E. Prah, M. Richards, and T. Treve. 1994. Incentives for the sustainable management of the tropical high forest in Ghana. *Commonwealth Forestry Review* 73:155–163.

Sathaye, J., Makundi, W., and Andrasko, D. 1995. A comprehensive mitigation assessment process (COMAP) for the evaluation of forestry mitigation options. *Biomass and Bioenergy.* 8:345–356.

Saunders, D. A., G. W. Arnold, A. A. Burbridge, and A. J. M. Hopkins. 1987. *Nature Conservation: The Role of Remnants of Native Vegetation*. Chipping Norton, NSW, Australia: Surrey Beatty and Sons, PTY Limited.

Saxon, E. C. 1990. Disturbance regimes in north Queensland rainforests: A re-evaluation of their relationship to species richness and diversity. *Australian Journal of Ecology* 15 241–244.

Sayer, J. A. and P. Wegge. 1992. Biological conservation issues in forest management. In J.M. Blockhus , M. Dillenbeck, J. A. Sayer, and P. Wegge eds., *Conserving Biolog-*

ical Diversity in Managed Tropical Forests, pp. 1–12. Proceedings of an IUCN workshop, Perth, Australia, 30 November–1 December, 1990. Gland, Switzerland: IUCN (International Union for Conservation of Nature and Natural Resources)/ ITTO (International Tropical Timber Organization).

Sayer, J. A., P. A. Zuidema, and M. H. Rijks. 1995. Managing for biodiversity in humid tropical forests. *Commonwealth Forestry Review* 74:282–287.

Sazima, O. 1989. Peach-fronted parakeet feeding on winged termites. *Wilson Bulletin* 101(4):656–657.

Scatena, F. N. 1990. Selection of riparian buffer zones in humid tropical steeplands. In. R.R. Ziemer, C. L O'Loughlin and L.S. Hamilton, eds. *Research Needs and Applications to Reduce Erosion and Sedimentation in Tropical Steeplands*, pp. 328–337. Publication Number 192. Walingford, United Kingdom: International Association of Hydrological Sciences Press.

Schaller, G. B. 1967. *The Deer and the Tiger: A Study of Wildlife in India*. Chicago, Ill.: Chicago University Press.

Schaller, G. B. and Crawshaw, P. G. 1980. Movement patterns of jaguar. *Biotropica* 12:161–168.

Schaller, G. B and Rabinowitz, A. 1995. The saola or spindlehorn bovid (*Pseudoryx nghetinhensis*) in Laos. *Oryx* 29:107–114

Schaller, G. B., N. X. Dang, L. D. Thuy, and V. T. Son. 1990. Javan rhinoceros in Vietnam. *Oryx* 24:77–80.

Schaller, G. B., H. Jinchu, P. Wenshi, and Z. Jing. 1989. *The Giant Pandas of Wolong*. Chicago, Ill.: University of Chicago Press.

Schemske, D. W. and N. Brokaw. 1981. Treefalls, and the distribution of understory birds in a tropical forest. *Ecology* 62:938–945.

Schenkel, R. and L. Schenkel-Hulliger. 1969. The Javan rhinoceros in Udjung-Kulon Nature Reserve: Its ecology and behaviour. *Acta Tropica* 26:97–135.

Schep, J. 1996. *International Trade for Local Development: The Case of Solomon Western Islands Fair Trade (SWIFT)*. Rabaul, Papua New Guinea: Pacific Heritage Foundation.

Schmidt, R. C. 1991. Tropical rain forest management:a status report. In A. Gómez-Pompa, T.C. Whitmore, and M. Hadley, eds., *Rain Forest Regeneration and Management. M.A.B. (Man and the Biosphere) series Volume 6*, pp. 181–207. Paris, France: UNESCO (United Nations Educational, Scientific, and Cultural Organization)/Parthenon.

Schnitzler, H. U. and E. K. V. Kalko. 1998. How echolocating bats search for food. In: T. H. Kunz, P. A. Racey, eds. Bats: phylogeny, morphology, echolocation, and conservation biology. pp. 183–196. Washington, D.C.: Smithsonian Institution Press.

Schoener, T. W. 1989. Food webs from the small to the large. *Ecology* 70:1559–1589.

Schoener, T. W. and D. A. Spiller. 1987. Effect of lizards on spider populations: manipulative reconstruction of a natural experiment. *Science* 236:949–952.

Schoenly, K., R. A. Beaver, and T. A. Heumier. 1991. On the trophic relations of insects: A food web approach. *American Naturalist* 137:597–638.

Schowalter, T. D., J. W. Webb, and D. A. Crossley, Jr. 1981. Community structure and nutrient content of canopy arthropods in clearcut and uncut forest ecosystems. *Ecology* 62:1010–1019.

Schreiber, A., R. Wirth, M.Riffel, and H. van Rompaey. 1989. *Weasels, Civets, Mongooses, and their Relatives: An Action Plan for Conservation of Mustelids and Viverrids*. Gland, Switzerland: IUCN (International Union for Conservation of Nature and Natural Resources).

Schultz, A. H. 1956. The occurrence and frequency of pathological and teratological conditions and of twinning among non-human primates. *Primatologia* 1:965–1014.

Schultze, E-D. and H. A. Mooney, eds. 1994. *Biodiversity and Ecosystem Function*. Berlin, Germany: Springer-Verlag.

Schupp E. W. 1993. Quantity, quality and the effectiveness of seed dispersal by animals. *Vegetation* 107/108:15–29

Scott, N. J., Jr. 1976. The abundance and diversity of the herpetofaunas of tropical forest litter. *Biotropica* 8:41–58.

Sedjo, R. A. 1995. The economics of managing carbon via forestry: An assessment of existing studies. In C.J. Jepma, ed., *The Feasibility of Joint Implementation*, pp.315–327. Netherlands: Kluwer Academic Publishers.

Senn, R., L. D. Ford, J. Grant, L. L. Lind, D. Schumann, and T. Thake. 1993. *Sustainable forestry program for Papua New Guinea, the Solomon Islands, and Vanuatu.* Washington, D.C.: U.S. Agency for International Development.

Sessions, J. and R. Heinrich. 1993a. Forest roads in the tropics. In L. Pancel, ed., pp. 1269–1324. *Tropical Forestry Handbook.* New York, N.Y.: Springer-Verlag.

——. 1993b. Harvesting. In L. Pancel, ed., *Tropical Forestry Handbook*, pp. 1325–1423. New York, N.Y.: Springer-Verlag.

Sheil, D. 1996. The ecology of long term change in a Ugandan rainforest. Unpublished Ph.D. diss., University of Oxford.

Shenon, P. 1994. Logging firms descend on Papua New Guinea. *New York Times News Service*: June 5.

Short, J. 1981. Diet and feeding behaviour of the forest elephant. *Mammalia* 45:177–186.

Shugart, H. H., M. S. Hopkins, I. P. Burgess, and A. T. Mortlock. 1980. The development of a succession model for subtropical rainforest and its application to assess the effects of timber harvest at Wiangarie state forest, New South Wales. *Journal of Environmental Management* 11:243–265.

Silva, J. L. and S. D. Strahl. 1990. Human impact on populations of chachalacas, guans, and curassows (Galliformes: Cracidae) in Venezuela. In J.G. Robinson and K.H. Redford, eds., *Neotropical Wildlife Use and Conservation*, pp. 36–52. Chicago, Ill.: University of Chicago Press.

Silver, W. L., S. Brown, A. E. Lugo. 1996b. Biodiversity and biogeochemistry in tropical forests. In G. Orians, R. Dirzo, and H. Cushman, eds., *Biodiversity and Ecosystem Processes in Tropical Forests*, pp. 49–67. New York, N.Y.: Springer-Verlag.

——. 1996a. Effects of changes in biodiversity on ecosystem function in tropical forests. *Conservation Biology* 10:17–24.

Silvertown, J., M. Franco, I. Pisanty, and A. Mendoza. 1993. Comparative plant demography: Relative importance of life-cycle components to the finite rate of increase in woody and herbaceous perennials. *Journal of Ecology* 81:465–476.

Simons, L. 1998. Indonesia: Plague of fires. *National Geographic Magazine* 194:100–119.

Simpson, J. 1998. Regions with standards in progress or endorsed (September 1998). Unpublished report. Oaxaca, Mexico: Forest Stewardship Council.

Sist, P., D. P. Dykstra, and R. A. Fimbel. 1998. Reduced impact logging guidelines for hill dipterocarp forests in Indonesia. *CIFOR (Center for International Forestry Research)Occasional Paper 15.* Bogor, Indonesia: CIFOR.

Sist, P., T. Nolan, J-G. Bertault, and D. Dykstra. 1998. Harvesting intensity versus sustainability in Indonesia. *Forest Ecology and Management* 108:251–260.

Skorupa, J. P. 1986. Responses of rain forest primates to selective logging in the Kibale Forest, Uganda: a summary report. In K. Benirschke, ed., *Primates: The Road to Self-sustaining Populations*, pp. 57–70. New York, N.Y.: Springer-Verlag.

——. 1988. The effects of selective timber harvesting on rain forest primates in Kibale Forest, Uganda. Ph.D. diss., University of California.

Skorupa, J. P. and J. M. Kasenene. 1984. Tropical forest management: Can rates of natural treefalls help guide us? *Oryx* 18:96–101.

Slough, B. G. and G. Mowat. 1996. Lynx population dynamics in an untrapped refugium. *Journal of Wildlife Management* 60:946–961.

SmartWood Program. 1997. *SmartWood Policy for Resource Manager Certification.* New York, N.Y.: Rainforest Alliance.

——. 1998. *SmartWood Generic Guidelines for Assessing Forest Management.* New York, N.Y.: Rainforest Alliance.

Smith, A. P. 1987. Repuestas de hierbas del sotobosque tropical a claros ocasionados por la caida de arboles. *Revista Biologia Tropical* 35:111–118.

Smith, D. M. 1986. *The Practice of Silviculture.* New York, N.Y.: John Wiley and Sons.

Smith, D. M., B. C. Larson, M. J. Kelty, and P. M. S. Ashton. 1997. *The Practice of Silviculture: Applied Forest Ecology.* 9th Edition. New York, N.Y.: John Wiley and Sons.

Smith, J. D. and H. H. Genoways. 1974. Bats of Margarita Island, Venezuela, with zoogeographic comments. *Bulletin of the Southern California Academy of Science* 73:64–79.

Smyth, N. 1987. The importance of mammals in neotropical forest management. In J. C. Figueroa, F. H. Wadsworth and S. Branham, eds., *Management of the Forests of Tropical America: Prospects and Technologies*, pp. 79–98. San Juan: Institute of Tropical Forestry, in U.S.D.A. Forest Service and University of Puerto Rico.

Snook, L. K. 1996. Catastrophic disturbance, logging and the ecology of mahogany (*Swietenia macrophylla King*): grounds for listing a major tropical timber species in CITES. *Botanical Journal of the Linnean Society* 122:35–46.

Snow, D. W. 1981. Tropical frugivorous birds and their food plants: A world survey. *Biotropica* 13:1–14

——. 1982. *The Cotingas: Bellbirds, Umbrellabirds and Other Species.* Ithaca, N.Y.: Cornell University Press.

Sobrevila, C. and P. Bath. 1992. Evaluacion ecologica rapida: Un manual para usuarios de América Latina y el Caribe. Unpublished report. Arlington, Va.: The Nature Conservancy.

Soerianegara, I. and R. H. M. J. Lemmens, eds., 1993. *Plant Resources of South-East Asia Vol. 5(1): Timber trees: Major Commercial Timbers.* Wageningen, The Netherlands: Pudoc Scientific Publishers.

Sohmer, S. H., and R. Gustafson. 1987. *Plants and Flowers of Hawaii.* Honolulu, Hawaii: University of Hawaii Press.

Solar, L. 1996. Aprovechamiento de la fauna silvestre y actividades de búsqueda maderera en el Bajo Paraguá, Santa Cruz, Bolivia. Undergraduate thesis, UAGRM (Universidad Autónoma Gabriel Rene Moreno).

Sonke, B. 1998. Les forêts de la réserve de faune du Dja. *Canopée* 12:22–23.

Soriano, P. J. 1983. La comunidad de quirópteros de las selvas nubladas en los Andes de Mérida. Patrón reproductivo de los frugívoros y las estrategias fenológicas de las plantas. Magister Scientiae thesis, Universidad de Los Andes.

——. 1985. Ecología de comunidades. In M. Aguilera, ed., *El Estudio de los Mamíferos en Venezuela, Evaluación y Perspectivas*, pp. 105–111. Asociación Venezolana para el Estudio de los Mamíferos, Fondo Editorial Acta Científica, Caracas, Venezuela: Venezolana.

Soriano, P. and C. Lulow. 1988. Efectos de las inundaciones estacionales sobre poblaciones de pequeños mamíferos en los Llanos Occidentales de Venezuela. *Ecotropicos* 1:3–10.

Sosef, M. S. M., L. T. Hong, and S. Prawirohatmodjo, eds. 1998. *Plant Resources of South-East Asia. Volume 5(3). Timber Trees: Lesser-known Timbers.* Wageningen, The Netherlands: Pudoc Scientific Publishers.

Soule, M. E. 1994. The onslaught of alien species, and other challenges of the coming decades. *Natural Areas Report* 6(2):1, 8.

Sparrow, H. R., T. D. Sisk, P. R. Ehrlich, and D. D. Murphy. 1994. Techniques and guidelines for monitoring neotropical butterflies. *Conservation Biology* 8:800–809.

Spiller, D. A. and T. W. Schoener. 1990a. Lizards reduce food consumption by spiders: Mechanisms and consequences. *Oecologia* 83:150–161.

———. 1990b. A terrestrial field experiment showing the impact of eliminating top predators on foliage damage. *Nature* 347:469–472.

———. 1994. Effects of top and intermediate predators in a terrestrial food web. *Ecology* 75:182–196.

Spitzer, K., J. Jaros, J. Havelka, and J. Leps. 1997. Effect of small-scale disturbance on butterfly communities of an Indochinese montane rainforest. *Biological Conservation* 80:9–15.

Spitzer, K., V. Novotny, M. Tonner, and J. Leps. 1993. Habitat preferences, distribution and seasonality of the butterflies in a montane tropical rain forest, Vietnam. *Journal of Biogeography* 20:109–121.

Steege, H., R. G. A. Boot, L. C. Brouwer, J. C. Caesar, R. C. Ek, D. S. Hammond, P. P. Haripersaud, P. van der Hout, V. G. Jetten, A. J. van Kekem, M. A. Kellman, Z. Khan, A. M. Polak, T. L. Pons, J. Pulles, D. Raaimakers, S. A. Rose, J. J. van der Sanden, and R. J. Zagt. 1996. *Ecology and Logging in a Tropical Rain Forest in Guyana: With Recommendations for Forest Management.* Wageningen, The Netherlands: The Tropenbos Foundation.

Steel, E. A. 1994. Study of the value and volume of bushmeat commerce in Gabon. Unpublished report to WWF (World Wildlife Fund). Libreville, Gabon: World Wildlife Fund.

Stevens, S. 1997. *Conservation Through Cultural Survival: Indigenous People and Protected Areas.* Washington, D.C.: Island Press.

Stewart, M. M. and L. L. Woolbright. 1996. Amphibians. In D. P. Reagan and R. B. Waide, eds., *The Food Web of a Tropical Rain Forest,* pp. 273–320. Chicago, Ill.: University of Chicago Press.

Stiles, E. W. 1992. Animals as seed dispersers. In M. Fenner, ed., *Seeds: The Ecology of Regeneration in Plant Communities,* pp. 87–104. Wallingford, United Kingdom: CAB (Centre for Agriculture and Bioscience).

Stirzaker, R. J., J. B. Passioura, and Y. Wilms. 1996. Soil structure and plant growth: Impact of bulk density and biopores. *Plant and Soil* 185:151–162.

Stocker, G. C. 1985. Aspects of gap regeneration theory and the management of tropical rainforests. In K. R. Shepherd, and H. V. Richter, eds., *Managing the Tropical Forest,* pp. 225–228. Canabera, Australia: Australian National University.

Stocker, G. C. and A. K. Irvine. 1983. Seed dispersal by cassowaries (*Casuarius casuarius*) in north Queensland's rainforests. *Biotropica* 15:170–176

Stocks, A. and G. Hartshorn. 1993. The Palcazú Project: Forest management and native Yánesha communities. *Journal of Sustainable Forestry* 1:111–135.

Stokes, B. J. and J. F. McNeel. 1990. Wood damage from mechanical felling. U. S. Forest Service, Southern Forest Experiment Station Research Paper SO-258.

Stokland, J. N. 1997. Representativeness and efficiency in bird and insect conservation in Norwegian boreal forest reserves. *Conservation Biology* 11:101–111.

Stoms, D. M., M. I. Bochert, M. A. Moritz, F. W. Davis, and R. L. Church. 1998. A systematic process for selecting representative research natural areas. *Natural Areas Journal* 18:338–349.

Stork, N. E. 1988. Insect diversity: Facts, fiction and speculation. *Biological Journal of the Linnean Society* 35:321–337.

Stork, N. E., T. J. B. Boyle, V. Dale, H. Eeley, B. Finegan, M. Lawes, N. Manokaran, P. Prabhu, and J. Soberon. 1997. Criteria and indicators for assessing the sustainability of forest management: Conservation of biodiversity. CIFOR Working Paper No. 17. Bogor, Indonesia: CIFOR (Centre for International Forest Research).

Stouffer, P. C. and R. O. Bierregaard. 1995a. Use of Amazonian forest fragments by understory insectivorous birds. *Ecology* 76:2429–2445.

———. 1995b. Effects of forest fragmentation on understory hummingbirds in Amazonian Brazil. *Conservation Biology* 9:1085–1094.

Strickland, D. 1967. Ecology of the rhinoceros in Malaya. *Malay Naturalists Journal* 20:1–17.

Strickland, S. S. 1986. Long-term development of Kejamaan subsistence: An ecological study. *Sarawak Museum Journal 36(57)*(new series):117–172.

Stromayer, K. A. K. and A. Ekobo. 1991a. Biological survey of the lake Lobéké region southeastern Cameroon. Bronx, N.Y.: Wildlife Conservation Society.

——. 1991b. Biological survey of the proposed Boumba Bek forest reserve, east province Cameroon. New York, N.Y. and Washington, D.C.: Report to Wildlife Conservation Society and World Wildlife Fund.

——. 1991c. Biological surveys of southeastern Cameroon. Bronx, N.Y.: Report to Wildlife Conservation Society.

Struhsaker, T. T. 1997. *Ecology of an African Rain Forest: Logging in Kibale and the Conflict Between Conservation and Exploitation*. Gainesville, Fla.: University Press of Florida.

Struhsaker, T. T. and L. Leland. 1979. Socioecology of five sympatric monkey species in Kibale Forest, Uganda. *Advances in the Study of Behaviour* 9:159–227.

Struhsaker, T. T., J. S. Lwanga, and J. M. Kasenene. 1996. Elephants, selective logging and forest regeneration in the Kibale Forest, Uganda. *Journal of Tropical Ecology* 12:45–64.

Struhsaker, T. T. and J. F. Oates. 1975. Comparison of the behavior and ecology of red colobus and black-and-white colobus monkeys in Uganda: A summary. In R. H. Tuttle, ed., *Socioecology and Psychology of Primates*, pp. 103–123. The Hague, The Netherlands: Mouton.

Stuebing, R. 1995. Wildlife diversity, abundance and management. ITTO (International Tropical Timber Organization)/Model Forest Management Area Phase 1, Field Report No. 1. Kuching, Sarawak: Sarawak Forest Department.

Stuebing, R. B., J. Gasis, and B. H. Lee. 1993. The economics of wildlife in Sabah: An ecological perspective. *Sabah Museum Journal* 1:73–88.

Suarez, E., J. Stallings, and L. Suarez. 1995. Small-mammal hunting by two ethnic groups in Northwestern Ecuador. *Oryx* 29:35–42.

Sukumar, R. 1989. *The Asian Elephant: Ecology and Management*. Cambridge, United Kingdom: Cambridge University Press.

Sun, C., A. R. Ives, H. J. Kraeuter, and T. C. Moermond. 1997. Effectiveness of three turacos as seed dispersers in a tropical montane forest. *Oecologia* 112:94–103.

Sutton, S. L. and N. M. Collins. 1991. Insects and tropical forest conservation. In N. M. Collins and J. A. Thomas, eds., *The Conservation of Insects and their Habitats*, pp.405–424. London, United Kingdom: Academic Press.

Swaine, M. D. and T. C. Whitmore. 1988. On the definition of ecological species groups in tropical rain forests. *Vegetatio* 75:81–86.

Swift, L. W., Jr. 1983. Duration of stream temperature increases following forest cutting in the southern Appalachian mountains. International Symposium on Hydrometereology. In A.I. Johnson and R.A. Clark, eds., pp. 273–275. Bethesda, Md.: American Water Resources Association.

Swift, L. W., Jr. and J. B. Messer. 1971. Forest cuttings raise temperatures of small streams in the southern Appalachians. *Journal of Soil and Water Conservation* 26:111–116.

Swift, M. J. and J. M. Anderson. 1993. Biodiversity and ecosystem function. In E. D. Schultze and H. A. Mooney, eds., *Biodiversity and Ecosystem Function*, pp. 15–41. Heidelberg, Germany: Springer-Verlag.

Symstad, A. J., D. Tilman, J. Wilson, and J. M. H. Knops. 1998. Species loss and ecosystem functioning: Effects of species identity and community composition. *Oikos* 81:389–397.

Synnott, T. J. 1985. A checklist of the flora of the Budongo Forest Reserve, Uganda, with notes on ecology and phenology. CFI Occasional Paper No. 27. Oxford, United Kingdom: Oxford Forestry Institute.

Taber, A. B., G. Navarro S., and M. A. Arribas. 1997. A new park in the Bolivian Grand Chaco—An advance in tropical dry forest conservation and community based management. *Oryx* 31:189–198.

Tangley, L. 1992. *Mapping Biodiversity: Lessons from the field I*. Unpublished report. Washington, D.C.: Conservation International.

Tanner, E. V. J., V. Kapos, and J. R. Healey. 1991. Hurricane effects on forest ecosystems in the Caribbean. *Biotropica* 23:513–521.

Tanner, H. 1996. *Tala Dirigida con Motosiera en Bosques Tropicales*. Turrialba, Costa Rica: Centro Agronómico Tropical de Investigacion y Enseñaza Serie Técnica No. 23.

Temple, S. A. 1977. Plant-animal mutualism: Coevolution with dodo leads to near-extinction of plant. *Science* 197:885–886.

Terborgh, J. 1980. Causes of tropical species diversity. In R. Nohring, ed., *Acta XVII Congressus Internationalis Ornithologici*, pp. 955–961. Berlin, Germany: Deutschen Ornithologen-Gesellschaft.

——. 1986a. Keystone plant resources in the tropical forest. In M. Soule, ed., *Conservation Biology: The Science of Scarcity and Diversity*, pp. 330–344. Sunderland, Mass.: Sinauer Publishers.

——. 1986b. Community aspects of frugivory in tropical forests. In A. Estrada and T. H. Fleming, eds,. *Frugivores and Seed Dispersal. Tasks for Vegetation Science* Vol. 15, pp. 371–384. Dordrecht, Netherlands: Dr W. Junk Publishers.

——. 1992a. Maintenance of diversity in tropical forests. *Biotropica* 24:283–292.

——. 1992b. *Diversity and the Tropical Rainforest*. New York, N.Y.: Scientific American Library.

Terborgh, J. and B. Winter. 1980. Some causes of extinction. In M. E. Soulé and B. A. Wilcox, eds., *Conservation Biology: An Evolutionary-ecological Perspective*, pp. 119–133. Sunderland, Mass.: Sinauer Publishers.

Terwilliger, V. J. 1978. Natural history of Baird's tapir on Barro Colorado Island, Panama Canal Zone. *Biotropica* 27:96–105.

Tewe, G. O. and S. S. Ajaji. 1982. Performance and nutritional utilization by the African giant rat (Cricetomys gambianus, W.) on household waste of local foodstuffs. *African Journal of Ecology* 20:37–41.

TFF (Tropical Forestry Foundation). 1995. Sustainable forest management programs in Brazil: Low-impact logging (LIL) models. Alexandria, Va.: TFF (Tropical Forestry Foundation) Publication No. 1.

Thiollay, J. M. 1985a. Composition of Falconiforms communities along successional gradients from primary rain forest to secondary habitats. In I. Newton and R. Chancellor, eds., *Conservation Studies of Raptors*, pp. 181–190. Cambridge, United Kingdom: International Council for Bird Preservation, ICBP Technical Publication no 5.

——. 1985b. Raptor community structure of a primary rain forest in French Guiana and effect of human hunting pressure. *Raptor Research* 18:117–122.

——. 1986. Structure comparée du peuplement avien dans trois sites de forêt primaire en Guyane. *Revue Française d'Ecologie (Terre et Vie)* 41:59–105.

——. 1989. Area requirements for the conservation of rain forest raptors and game birds in French Guiana. *Conservation Biology* 3:128–137.

——. 1990. Comparative diversity of temperate and tropical forest bird communities: The influence of habitat heterogeneity. *Acta Oecologica* 11:887–911.

——. 1992. Influence of selective logging on bird species diversity in a Guianan rain forest. *Conservation Biology* 6:47–63

——. 1994. Structure, density and rarity in an Amazonian rain forest bird community. *Journal of Tropical Ecology* 10:449–481.

——. 1995. The role of traditional agroforests in the conservation of rain forest bird diversity in Sumatra. *Conservation Biology* 9:335–353.

——. 1997. Disturbance, selective logging and bird diversity: A neotropical forest study. *Biodiversity and Conservation* 6:1155–1173.

———. 1998. Current status and conservation of Falconiformes in tropical Asia. *Journal of Raptor Research* 32:40–55.

———. 1999. Responses of an avian community to rain forest degradation. *Biodiversity and Conservation* 8:513–534.

Thomas, C. D. 1991. Habitat use and geographic ranges of butterflies from the wet lowlands of Costa Rica. *Biological Conservation* 55:269–281.

Thomas, C. D., M. C. Singer, and D. A. Boughton. 1996. Catastrophic extinction of population sources in a butterfly metapopulation. *American Naturalist* 148:957–975.

Thomas, J. A. 1983. A quick method of estimating butterfly numbers during surveys. *Biological Conservation* 27:195–211.

Thomas, R. and A. Gaa Kessler. 1996. Nonanoline reptiles. In D. P. Reagan and R. B. Waide, eds., *The Food Web of a Tropical Rain Forest*, pp. 347–362. Chicago, Ill.: University of Chicago Press.

Thomas, S. C. 1991. Population densities and patterns of habitat use among anthropoid primates of the Ituri Forest, Zaire. *Biotropica* 23:68–83.

Tian, G. 1998. Effect of soil degradation on leaf decomposition and nutrient release under humid tropical conditions. *Soil Science* 163:897–906.

Tian, G., L. Brussaard, and B. T. Kang. 1992. Biological effects of plant residues with contrasting chemical composition under humid tropical conditions: Decomposition and nutrient release. *Soil Biology and Biochemistry* 24:1051–1060.

———. 1995. Breakdown of plant residues with contrasting chemical composition under humid tropical conditions: Effects of earthworms and millipedes. Soil Biology and Biochemistry 27:277–280.

Timm, R. M., D. E. Wilson, B. L. Clauson, R. K. LaVal, and C. S. Vaughn. 1989. Mammals of the La Selva-Braulio Carrillo Complex, Costa Rica. *North American Fauna* 75:1–162.

Tocher, M. 1997. A comunidade de anfibios da Amazônia central: Diferenças na composição específica entra a mata primária e pastagens. In C. Gascon and P. Montinho, eds., *Floresta Amazônica: Dinâmica, Regeneração Manejo*, pp. 219–232. Manaus, Brazil: Instituto Nacional de Pesquisas da Amazônia.

Toft, C. A. 1980. Feeding ecology of Panamanian litter anurans: Patterns in diet and foraging mode. *Journal of Herpetology* 15:139–144.

———. 1981. Seasonal variation in populations of Panamanian litter frogs and their prey: A comparison of wetter and drier sites. *Oecologia* 47: 34–38.

Torres, J. A. 1994. Wood decomposition on Cyrilla racemiflora in a tropical montane forest. *Biotropica* 26:124–140.

Townsend, W. R. 1996a. *Nyao ito Caza y pesca de los Siriono en Ibiato*. Instituto de Ecologia. La Paz, Bolivia: FUNDECO (Fundación para el Desarrollo de la Ecología).

———. 1996b. Por qué Hacer un Programa de Manejo de la Fauna Silvestre de Lomerio? Miscellaneous report. Santa Cruz, Bolivia: BOLFOR (Bolivia Sustainable Forestry Management Project) Project.

Toy, R. J. 1988. The pre-dispersal insect fruit-predators of Dipterocarpaceae in Malaysian rain forest. Ph.D. diss., University of Aberdeen.

Traveset, A. 1998. Effect of seed passage through vertebrate frugivores' guts on germination: A review. *Perspectives in Plant Ecology, Evolution and Systematics* 1/2:151–190.

Trexler and Associates, Inc. 1997. Report of the Biotics Offsets Assessment Workshop. Washington, D.C.: Office of Policy, Planning and Evaluation, Environmental Protection.

Trexler, M. C. 1995. Carbon offset strategies: A private sector perspective. In C. J. Jepma, ed., *The Feasibility of Joint Implementation*, pp. 233–248. The Netherlands: Kluwer Academic Publishers.

Trexler, M. C., P. E. Faeth, and J. M. Kramer. 1989. *Forestry as a Response to Global Warming: An Analysis of the Guatemala Agroforestry and Carbon Sequestration Project*. Washington, D.C.: World Resources Institute.

Trexler, M. C. and C. Haugen. 1995. *Keeping it Green: Tropical Forestry Opportunities for Mitigating Climate Change*. Washington, D.C.: World Resources Institute.

Tropenbos. 1994. Annual Report 1994. Wageningen, The Netherlands: The Tropenbos Foundation.

Troup, R. S. 1928. *Silvicultural Systems*. Oxford, United Kingdom: Oxford University Press.

Tuomisto, T., K. Ruokolainen, R. Kalliola, A. Linna, W. Danjoy, and Z. Rodriguez. 1995. Dissecting Amazonian biodiversity. *Science* 269:63–66.

Turkalo, A. and M. Fay. 1995. Studying elephants by direct observation: Results from the Dzanga clearing, Central African Republic. *Pachyderm* 20:45–51.

Turner, I.M. and A.C. Newton. 1990. The initial responses of some tropical rain forest tree seedlings to a large gap environment. *Journal of Applied Ecology* 27:605–608.

Tutin, C. E. G. and M. Fernandez. 1984. Nationwide census of Gorilla (*Gorilla g. gorilla*) and Chimpanzee (*Pan t. troglodytes*) populations in Gabon. *American Journal of Primatology* 6: 313–336.

——. 1993. Relationships between minimum temperature and fruit production in some tropical forest trees in Gabon. *Journal of Tropical Ecology* 9:241–248.

Tutin, C. E. G., L. J. T. White, and A. Mackanga-Missandzou. 1996. Lightning strike burns large forest tree in the Lope Reserve, Gabon. *Global Ecology and Biogeography Letters* 5:36–41.

——. 1997. The use by rain forest mammals of natural forest fragments in an equatorial African savanna. *Conservation Biology* 11:1190–1203.

Tutin, C. E. G., E. A. Williamson, M. E. Rogers and M. Fernandez. 1991. A case study of a plant-animal relationship: *Cola lizae* and lowland gorillas in the Lope Reserve, Gabon. *Journal of Tropical Ecology* 7:181–199.

Twilley, R. R., S. C. Snedaker, A. Yanez-Arancibia, and E. Medina. 1995. *Mangrove Systems*. In V. H. Heywood ed., *Global Biodiversity Assessment*, pp. 387–393. Cambridge, United Kingdom: Cambridge University Press.

U.S. Department of Commerce. 1988. Marketing in the Congo. Overseas business report 88–05. Washington, D.C: U.S. Department of Commerce.

Uhl, C. 1987. Factors controlling succession following slash-and-burn agriculture in Amazonia. *Journal of Ecology* 75:377–407.

Uhl, C. and R. Buschbacher. 1985. A disturbing synergism between cattle ranch burning practices and selective tree harvesting in the eastern Amazon. *Biotropica* 17:265–268.

Uhl, C. and I. Guimaraes. 1989. Ecological impacts of selective logging in the Brazilian Amazon: A case of study from the Paragominas Region of the State of Para. *Biotropica* 21:98–106.

Uhl, C. and J. B. Kauffman. 1990. Deforestation, fire susceptibility, and potential tree responses to fire in the eastern Amazon. *Ecology* 71:437–449.

Uhl, C. and I. C. G. Vieira. 1989. Ecological impacts of selective logging in the Brazilian Amazon: A case study from the Paragominas Region of the State of Para. *Biotropica* 21:98–106.

Uhl, C., P. Barreto, A. Veríssimo, E. Vidal, P. Amaral, A. C. Barros, C. Souza Jr., J. Johns, and J. Gerwing. 1997. Natural resource management in the Brazilian Amazon: An integrated research approach. *BioScience* 47:160–168.

Uhl, C., R. Buschbacher, and E. A. S. Serrao. 1988. Abandoned pastures in eastern Amazonia. I. Patterns of plant succession. *Journal of Ecology* 76:663–681.

Uhl, C., K. Clark, H. Clark, and P. Murphy. 1981. Early plant succession after cutting and burning in the Upper Rio Negro region of the Amazon Basin. *Journal of Ecology* 69:631–649.

Uhl, C., K. Clark, N. Dezzeo, and P. Maquirino. 1988. Vegetation dynamics in Amazonian treefall gaps. *Ecology* 69:751–763.

Uhl, C., C. Jordan, K. Clark, H. Clark, and R. Herrera. 1982. Ecosystem recovery in Amazon catinga forest after cutting and burning, and bulldozer clearing treatments. Oikos 38:313–320.

Uhl, C., J. B. Kauffman, and D.L. Cummings. 1988. Fire in the Venezuelan Amazon 2: Environmental conditions necessary for forest fires in the evergreen rainforest of Venezuela. *Oikos* 53:176–184.

Uhl, C., D. Nepstad, R. Buschbacher, K. Clark, J. B. Kauffman, and S. Subler. 1990. Studies of ecosystem response to natural and anthropogenic disturbance provided guidelines for designing sustainable land-use systems in Amazonia. In A. B. Anderson, ed., *Alternatives to Deforestation: Steps Towards Sustainable Use of the Amazon Rain Forest*, pp. 24–42. New York, N.Y.: Columbia University Press.

Uhl, C., A. Verissimo, M. Maria Mattos, Z. Brandino, and I. Guimaraes Vieira. 1991. Social, economic, and ecological consequences of selective logging in an Amazon frontier: The case of Tailandia. *Forest Ecology and Management* 46:243–273.

UK Government. 1994. *Sustainable Forestry: The UK Programme*. London, United Kingdom: HMSO (Natural Environment Resource Council).

Ulrich, K. E., T. M. Burton, and M. P. Oemke. 1993. Effects of whole-tree harvest on epilithic algal communities in headwater stream. *Journal of Freshwater Ecology* 8:83–92.

U.S. Initiative on Joint Implementation. 1997. Activities implemented jointly: 2nd report to the U.N. Framework Convention on Climate Change. Volume 2. Washington, D.C.: Environmental Protection Agency, Office of Policy, Planning and Evaluation.

USIJI (United States Initiative on Joint Implementation) Secretariat. 1996a. Joint implementation projects selected: U.S. companies employ innovative solutions to global climate change. Washington, D.C.: U.S. Initiative on Joint Implementation.

——. 1996b. Activities implemented jointly: First report to the Secretariat of the United Nations Framework Convention on Climate Change. Washington, D.C.: US Goverment.

——. 1998a. Activities implemented jointly: Second report to the Secretariat of the United Nations Framework Convention on Climate Change: Accomplishmnets and descriptions of projects accepted under the U.S. Initiative on Joint Implementation. Vol. 1. EPA 236 –R- Washington, D.C.: Environmental Protection Agency.

——. 1998b. USIJI Project Fact Sheet. Washington, D.C.: U.S. Initiative on Joint Implementation.

Utrera, L. A. 1996. Influencia de la actividad forestal sobre las comunidades de mamíferos en la Reserva Forestal de Caparo. MS diss., Universidad de Los Andes.

Van Bueren, E. M. L. and J.F. Duivenvoorden. 1996. Towards priorities of biodiversity research in support of management of tropical rain forests. Wageningen, The Netherlands: The Tropenbos Foundation.

Van Bueren, L., M. Erik, and E. M. Blom. 1997. *Principles, Criteria, Indicators: Hierarchical Framework for the Formulation of Sustainable Forest Management Standards*. The Netherlands: The Tropenbos Foundation.

Van der Meer, P. J., F. J. Sterck, and F. Bongers. 1998. Tree seedling performance in canopy gaps in a tropical rain forest at Nouragues, French Guiana. *Journal of Tropical Ecology* 2:119–137

Van der Pijl, L. 1982. *Principles of Dispersal in Higher Plants*. 3rd Edition. Berlin, Germany: Springer.

Van Helden, F. 1996. Issues in the production and marketing of sustainable timber from community based projects in Papua New Guinea and the Solomons Islands. Consultative Meeting on Sustainable Timber, Rabaul, Papua New Guinea, 1996. Rabaul, Papua New Guinea: Pacific Heritage Foundation.

Van Roosmalen, M. G. M. 1985. *Fruits of the Guianan Flora*. Utrecht, The Netherlands: Utrecht University.

Van Schaik, C. P., J. W. Terborgh, and S. J. Wright. 1993. The phenology of tropical forests: Adaptive significance and importance for primary consumers. *Annual Review of Ecology and Systematics* 24:353–377.

Van Straalen, N. M. 1997. Community structure of soil arthropods as a bioindicator of soil health. In C. E. Pankhurst, B. M. Doube, and V. V. S. R. Gupta, eds., *Biological Indicators of Soil Health*, pp. 235–264. Wallingford, United Kingdom: CAB (Centre for Agriculture and Bioscience)

Van Strien, N. J. 1985. The Sumatran Rhinoceros *Dicerorhinos sumatrensis* Fischer 1814 in the Gunung Lueser Reserve National Park, Sumatra, Indonesia: Its distribution, ecology and conservation. Unpublished report. Doorn, Holland.

Vanclay, J. K. 1989. Modelling selection harvesting in tropical rainforests. *Journal of Tropical Forest Science* 1:290–294.

Vander Wall, S. B. 1990. *Food Hoarding in Animals*. Chicago, Ill.: University of Chicago Press.

——. S. B. 1994. Removal of wind-dispersed pine seeds by ground-foraging vertebrates. *Oikos* 69:125–132.

Vane-Wright, R. I, C. J. Humphries, and P. H. Williams. 1991. What to protect? Systematics and the agony of choice. *Biological Conservation* 55:235–254.

Vannote, R. L., G. W. Minshall, K. W. Cummins, J. R. Sedell, and C. E. Cushing. 1980. The river continuum concept. *Canadian Journal of Fisheries and Aquatic Sciences* 37:130–137.

Veenendaal, E., M. Swaine, V. K. Agyeman, D. Blay, I. K. Abebrese, and C. E. Mullins. 1996. Differences in plant and soil water relations in and around a forest gap in West Africa during the dry season may influence seedling establishment and survival. *Journal of Ecology* 84:83–90

Veillon, J. P. 1976. Las deforestaciones en los Llanos Occidentales de Venezuela desde 1950 hasta 1975. In E. Mondolfi, and P. Rambach, eds., *Conservación de los Bosques Húmedos de Venezuela*, pp. 97–110. Caracas, Venezuela: Sierra Club, Consejo de Bienestar Rural.

Vellinga, P. and R. Heintz. 1995. Joint implementation: A cost-benefit analysis. In C. J. Jepma, ed., *The Feasibility of Joint Implementation*, pp. 69–77. The Netherlands: Kluwer Academic Publishers.

Venable, D. L. and J. S. Brown. 1993. The population-dynamic functions of seed dispersal. *Vegetatio* 107/108:31–55.

Veríssimo, A., P. Barreto, R. Tarifa, and C. Uhl. 1995. Extraction of a high-value natural resource in Amazonia: The case of mahogany. *Forest Ecology and Management* 72:39–60.

Verschuren, J. 1989. Habitats mammals and conservation in the Congo. *Bulletin de l'Institut Royal des Sciences Naturelles de Belgique.Biologie.* 59:169–180.

Viana, V. M. 1990. Seed and seedling availability as a basis for management of natural forest regeneration. In A.B. Anderson, ed., *Alternatives to Deforestation: Steps Towards Sustainable Use of the Amazon Rain Forest*, pp. 99–115. New York, N.Y.: Columbia University Press.

Viana, V. M., J. Ervin, R. Z. Donovan, C. Elliot, and H. Gholz, eds. 1996. *Certification of Forest Products: Issues and Perspectives*. Washington, D.C.: Island Press.

Vickers, W. T. 1990. Hunting yields and game composition over ten years in an Amazonian Indian Territory. In J. G. Robinson and K. H. Redford, eds,. *Neotropical Wildlife Use and Conservation*, pp. 53–81. Chicago, Ill.: University of Chicago Press.

Vidal, E., J. Johns, J. J. Gerwing, P. Barreto, and C. Uhl. 1997. Vine management for reduced-impact logging in eastern Amazonia. *Forest Ecology and Management* 98:105–114.

Vincent, J. R. 1995. Timber trade, economics, and tropical forest management. In R. B. Primack and T. E. Lovejoy, eds., *Ecology, Conservation, and Management of Southeast Asian Rainforests*. pp. 241–262. New Haven, Conn.: Yale University Press.

Vincent, J., D. Brooks, A. K. Gandapur, and K. Alamgir. 1991. Substitution between tropical and temperate logs. *Forest Science* 37:1484–1491.

Vincent, J. R., L. F. Wan, Y. T. Chang, M. Nooriha, and G. W. H. Davison. 1993. Malaysian national Conservation strategy towards sustainable development. Volume 4: Natural Resource Accounting. Kuala Lumpur, Malaysia: Economic Planning Unit, Prime Minister's Department.

Vinson, S. B., G. W. Frankie, and J. Barthell. 1993. Threats to solitary bees in a neotropical dry forest. In J. LaSalle and I. D. Gauld, eds., *Hymenoptera and Biodiversity*, pp. 53–81. Wallingford, United Kingdom: CAB (Centre for Agriculture and Bioscience).

Virani, M. Z. A. 1994. Ecology of the endangered Sokoke scops owl (Otus ireneae). MS thesis, University of Leicester.

Vitousek, P. M. 1990. Biological invasions and ecosystem processes: Towards an integration of population biology and ecosystem studies. *Oikos* 57:7–13.

Vitt, L. J. 1996. Biodiversity of Amazonian lizards. In A. C. Gibson, ed., *Neotropical Biodiversity and Conservation*. Los Angeles, Calif.: Mildred E. Mathias Botanic Garden Miscellaneous Publication Number 1.

Vitt, L. J. and J. P. Caldwell. 1994. Resource utilization and guild structure of small vertebrates in the Amazon forest leaf litter. *Journal of Zoology, London* 234:463–476.

Vitt, L. J. and G. R. Colli. 1994. Geographical ecology of a neotropical lizard: *Ameiva ameiva* (Teiidae) in Brazil. *Canadian Journal of Zoology* 72:1986–2008.

Vitt, L. J. and P. A. Zani. 1996a. Organization of a taxonomically diverse lizard assemblage in Amazonian Ecuador. *Canadian Journal of Zoology* 74:1313–1335.

——. 1996b. Ecology of the elusive tropical lizard *Tropidurus* [Uracentron] *flaviceps* (Tropiduridae) in lowland rain forest of Ecuador. *Herpetologica* 52:121–132.

Vitt, L. J., T. C. S. Avila-Pires, J. P. Caldwell, and V. Oliveira. 1998. The impact of individual tree harvesting on thermal environments of lizards in Amazonian rain forest. *Journal of Conservation Biology* 12:654–664.

Vitt, L. J., P. A. Zani, and M. C. Espósito. 1999. Historical Ecology of Amazonian lizards: Implications for community ecology. *Oikos* 87(2):286–294.

Vitt, L. J., P. A. Zani, and C. M. Lima. 1997. Heliotherms in tropical rain forest: The ecology of *Kentropyx calcarata* (Teiidae) and *Mabuya nigropunctata* (Scincidae) in the Curuá-Una of Brazil. *Journal of Tropical Ecology* 13:199–220.

Vivien, J. and J. J. Faure. 1985. *Arbres des Forêts Dense d'Afrique Centrale*. Paris, France: Ministere de Relations Exterieures, Cooperation et Developpment. Agence de Cooperation Culturrelle et Technique.

Vogel, S. 1988. *Life's Devices: The Physical World of Animals and Plants*. Princeton, N.J.: Princeton University Press.

Vos, E. C. and M. J. Kooistra. 1994. The effect of soil structure differences in a silt loam soil under various farm management systems: soil physical properties and simulated land qualities. *Agriculture Ecosystems and Environment* 51:227–238.

Voss, R. S. and L. H. Emmons. 1996. Mammalian diversity in neotropical lowland rainforests: a preliminary assessment. *American Museum of Natural History* 230:3–115.

Wachtel, P. S. 1993. Asia's sacred groves. *International Wildlife* 23:24–27.

Wadsworth, F. H. 1987. A time for secondary forests in tropical America. In J. C. Figueroa Colón , F. H. Wadsworth, and S. Branham, eds., *Management of the Forests of Tropical America: Prospects and Technologies*, pp. 189–197 Washington, D.C.: U.S.D.A. Forest Service.

——. 1997. Forest production for tropical America. USDA Forest Service Agricultural Handbook 710, Washington, D.C.: USDA Forest Service.

Waide, R. B. And A. E. Lugo. 992. A research perspective on disturbance and recovery of a tropical montane forest. In J. G. Goldammer, ed., *Tropical Forests in Transition*, pp. 173–190. Basel, Switzerland: Berkhauser-Verlag.

Walker, B. H. 1992. Biodiversity and ecological redundancy. *Conservation Biology* 6:18–23.

Walker, L. R., N. V. L. Brokaw, D. J. Lodge, and R. B. Waide. 1991. Ecosystem, plant, and animal responses to hurricanes in the Caribbean. *Biotropica* 23(4A).

Walker, L. R., D. J. Lodge, N. V. L. Brokaw, and R. B. Waide. 1991. An introduction to hurricanes in the Caribbean. *Biotropica* 23:313–316.

Wallace, J. B. and M. E. Gurtz. 1986. Response of Baetis mayflies (*Ephemeroptera*) to catchment logging. *American Midland Naturalist* 115:25–41.

Wallace, R. B., R. L. E. Painter, A. B. Taber, and J. M. Ayres. 1996. Notes on a distributional river boundary and southern range extension for two species of Amazonian primates. *Neotropical Primates* 4:149–151.

Walters, C. 1986. *Adaptive Management of Renewable Resources*. New York, N.Y.: MacMillan Publishing Company.

Wangwacharakul, V. and R. Bowonwiwat. 1995. Economic evaluation of CO_2 response optons in the forestry sector: The case of Thailand. *Biomass and Bioenergy* 8:293–307.

Ward, J. P. and P. J. Kanowski. 1983. Implementing control of harvesting operations in north Queensland rainforests. In K. R. Shepherd, and H. V. Richter, eds., *Managing the Tropical Forest*, pp. 165–186. Canabera, Australia: Australian National University.

Watson, R. T., M.C. Zinyowera, R. H. Moss, eds. 1996. Impacts, adaptations and mitigation of climate change: Scientific-technical analyses. IPCC Second Assessment Report: Climate change in 1995. New York, N.Y.: Cambridge University Press.

Waterman, P. G. and K. Kool. 1994. Colobine food selection and plant chemistry. In A.G. Davies and J.F. Oates, eds., *Colobine Monkeys: Their Ecology, Behaviour and Evolution*, pp. 251–284. Cambridge,United Kingdom: Cambridge University Press.

Waters, T. F. 1995. Sediment in streams: Sources, biological effects, and control. *American Fisheries Society Monograph* 7, Bethesda, Md..

WCS (Wildlife Conservation Society). 1998a. Program to inventory and monitor select plants and animals within production forest areas of communities participating within SUBIR Phase III. Miscellaneous Report. Bronx, N.Y.: Wildlife Conservation Society.

———. 1998b. The effects of management and natural disturbance on the biodiversity of mixed hardwood-spruce stands in the Adirondacks. Year one progress report. Miscellaneous Report. Bronx, N.Y.: Wildlife Conservation Society.

WCS (Wildlife Conservation Society) and Sarawak Forest Department. 1996. *A Master Plan for Wildlife in Sarawak*. Bronx, N.Y.: Wildlife Conservation Society and Kuching, Sarawak: Sarawak Forest Department.

Weaver, P. L. 1987. Enrichment planting in tropical America. In Figueroa, J., F. Wadsworth, and S. Branham, eds., *Management of the Forest of Tropical America: Prospects and Technologies*, pp. 259–277. Rio Piedras, Puerto Rico: Institute of Tropical Forestry.

Webb, E. L. 1997. Canopy removal and residual stand damage during controlled selective logging in lowland swamp forest of northeast Costa Rica. *Forest Ecology and Management* 95:117–129.

Webb, L. J., M. S. Hopkins, P. A. R. Young, J. Kikkawa, and T. E. Lovejoy. 1985. Conservation of tropical rain forest isolates. In J. Davidson, Tho Yow Pong, and M. Bijleveld, eds., *The Future of Tropical Rain Forests in South East Asia*, pp. 55–68. Gland, Switzerland: International Union for Conservation of Nature and Natural Resources.

Webster, D. W. and J. K. Jones Jr. 1984. Notes on a collection of bats from Amazonian Ecuador. *Mammalia* 48:247–252.

Webster, J. R., S. W. Golladay, E. F. Benfield, J. L. Meyer, W. T. Swank, and J. B. Wallace. 1992. Catchment disturbance and stream response: An overview of stream research at Coweeta Hydrologic Laboratory. In. P. J. Boon and G. E. Petts, eds.,

River Conservation and Management. pp. 231–253. New York, N.Y.: John Wiley and Sons.

Weisenseel, K., C. A. Chapman, and L. J. Chapman. 1993. Nocturnal primates of Kibale Forest: Effects of selective logging on prosimian densities. *Primates* 34:445–450.

Welcomme, R. L. 1985. River fisheries. Food and Agriculture Organization (FAO), Technical Paper 262. Rome, Italy: Food and Agriculture Organization (FAO).

Welcomme, R. L. and D. Hagborg. 1977. Towards a model of a floodplain fish population and its fishery. *Environmental Biology of Fishes* 2:7–24.

Welden, C. W., S. W. Hewett, S. P. Hubbell, and R. B. Foster. 1991. Sapling survival, growth and recruitment: Relationship to canopy height in a neotropical forest. *Ecology* 72:35–50.

Wells, D. R. 1988. The bird fauna of Pasoh Forest Reserve, Negeri Sembilan. *Enggang* 1:7–17.

Wells, M. and K. Brandon. 1992. *People and Parks: Linking Protected Area Management with Local Communities.* Washington, D.C.: The World Bank.

Welsch, D. 1991. *Riparian Forest Buffers: Function and Design for Protection and Enhancement of Water Resources.* Publication no. NA-PR-07091. U.S. Department of Agriculture, Forest Service (Northeastern Area State and Private Forestry).

Western, D. and R. M. Wright, eds., 1994. *Natural Connections: Perspectives in Community-based Conservation.* Covelo, Calif.: Island Press.

Wharton, C. H. 1968. Man, fires and wild cattle in southeast Asia. *Proceedings of the Annual Tall Timbers Fire Ecology Conference* 8:107–167.

Wheelwright, N. T. 1985. Fruit size, gape width, and the diets of fruit-eating birds. *Ecology* 66:808–818.

Wheelwright, N. T., W. A. Haber, K.G. Murray, and C. Guindon. 1984. Tropical fruit-eating birds and their food plants: A survey of a Costa Rican lower montane forest. *Biotropica* 16:173–192

Whigham D. F., M. B. Dickinson, and N. V. L. Brokaw. 1999. Background canopy gap and catastrophic wind disturbances in tropical forests. In L. R. Walker, ed., *Ecosystems of Disturbed Ground,* pp. 237–267. Amsterdam, The Netherlands: Elsevier.

Whigham, D. F., J. F. Lynch, and M.B. Dickinson. 1998. Dynamics and ecology of natural and managed forests in Quintana Rio, Mexico. In R. B. Primack, D. Bray, H. A. Galletti, and I. Ponciana, eds., *Timber, Tourists, and Temples: Conservation and Development in the Maya Forest of Belize, Guatemala, and Mexico,* pp. 267–281. Washington, D.C.: Island Press.

White, F. 1983. *The Vegetation of Africa: A Descriptive Memoir to Accompany the UNESCO/AETFAT/UNSO Vegetation Map of Africa.* Paris, France: UNESCO (United Nations Educational, Scientific, and Cultural Organization).

White, J. A. 1972. Forest Inventory of the Gola Forest Reserves. TA3155, Unpublished report. Rome, Italy: FAO

White, L. J. T. 1992. Vegetation history and logging disturbance: Effects on rain forest mammals in the Lopé Reserve, Gabon (with special emphasis on elephants and apes). Ph.D. diss., University of Edinburgh.

White, L. J. T. 1994a. Patterns of fruit-fall phenology in the Lopé Reserve, Gabon. *Journal of Tropical Ecology* 10:289–312.

——. 1994b. The effects of commercial mechanised selective logging on a transect in lowland rainforest in the Lopé Reserve, Gabon. *Journal of Tropical Ecology* 10:313–322.

——. 1994c. Biomass of rain forest mammals in the Lopé Reserve, Gabon. *Journal of Animal Ecology* 63:499–512.

White, L. J. T. and C. E. G. Tutin. 2001. Why chimpanzees and gorillas respond differently to logging: A case study from Gabon that cautions against generalizations. In. W. Weber., A. Veder, H. Simons-Morland, L. White, and T. Hart, eds., *African Rainforest Ecology and Conservation.* New Haven, Conn.: Yale University Press.

White, L. J. T., C.E.G. Tutin, and M. Fernandez. 1993. Group composition and diet of forest elephants, *Loxodonta africana cyclotis* Matschie 1900, in the Lopé Reserve, Gabon. *African Journal of Ecology* 31:181–199

White, P. S. and S. T. A. Pickett. 1985. Natural disturbance and patch dynamics: An introduction. In S. T. A. Pickett and P. S. White, eds., *The Ecology of Natural Disturbance and Patch Dynamics,* pp. 3–13. San Diego, Calif.: Academic Press.

Whitehead, G. K. 1972. *Deer of the World.* London, United Kingdom: Constable.

Whitesides, G. H., J. F. Oates, S. M. Green, and R. P. Kluberdanz.1988. Estimating primate densities from transects in a West African rain forest: A comparison of techniques. *Journal African of Ecology* 57:345–367.

Whitington, C. and U. Tresucon. 1991. Selection and treatment of food plants by white-handed gibbons (*Hylobates lar*) in Khao Yai National Park, Thailand. *Natural History Bulletin of the Siam Society* 39:111–122.

Whitman, A . A., J. M. Hagan III, and N. V. L. Brokaw. 1998. Effects of selective logging on tropical forest birds in northern Belize. *Biotropica* 30:449–457.

Whitmore, T. C. 1974. Chance with time and the role of cyclones in tropical rain forest on Kolmbangara, Solomon Islands. *Commonwealth Forestry Institute.* Paper 46.

——. 1980. Conservation in tropical rain forest. In M. Soulé, and B. Wilcox, eds., *Conservation Biology: An Evolutionary Perspective,* pp. 303–318. Sunderland, Mass.: Sinauer Associates Inc.

——. 1984. *Tropical Rain Forests of the Far East,* 2nd Edition. Oxford, United Kingdom: Clarendon Press.

——. 1990. *An Introduction To Tropical Rain Forests.* Oxford, United Kingdom: Clarendon Press.

——. 1991. Tropical rain forest dynamics and its implications for management. In A. Gomez-Pompa, T. C. Whitmore, and M. Hadley, eds., *Rain Forest Regeneration and Management, Man and the Biosphere Series.* Volume 6, pp. 67–89. Paris, France: UNESCO (United Nations Educational, Scientific, and Cultural Organization).

——. 1997. Tropical forest disturbance, disappearance, and species loss. In W. F. Laurance and R. O. Bierregaard Jr., eds.,*Tropical Forest Remnants. Ecology, Management, and Conservation of Fragmented Communities,* pp. 3–12. Chicago, Ill.: University of Chicago Press.

——. 1998. *An Introduction to Tropical Rain Forests,* 2nd Edition. Oxford, United Kingdom: Oxford University Press.

Whitmore, T. C. and D. Brown. 1996. Dipterocarp seedling growth in rain forest canopy gaps during six and a half years. *Philosophical Transaction of the Royal Society London* 351:1195–1203.

Whitmore, T. C. and J. A. Sayer. 1992. *Tropical Deforestation and Species Extinction.* IUCN (International Union for Conservation of Nature and Natural Resources) Forest Conservation Programe. London, United Kingdom: Chapman and Hall.

Whitney, K. D., M. K. Fogiel, A. M. Lamperti, K. M. Holbrook, D. J. Stauffer, B. D. Hardesty, V. T. Parker, and T. B. Smith. 1998. Seed dispersal by Ceratogymna hornbills in the Dja Reserve, Cameroon. *Journal of Tropical Ecology* 14:351–371

Wickneswari, R., C. T. Lee, M. Norwati, and T.J. Boyle. 1997. Effects of logging on the genetic diversity of six tropical rainforest species in a regenerated mixed dipterocarp lowland forest in Peninsular Malaysia. Paper presented at CIFOR (Centre for International Forestry Research) Wrap-Up Workshop on Impact of Disturbance, Bangalore, India, August 1997.

Wilcox, B. A. 1995. Tropical forest resources and biodiversity: The risks of forest loss and degradation. *Unasylva* 46:43–49.

Wilczynski, C. J. and S. T. A. Pickett. 1993. Fine root biomass within experimental canopy gaps: Evidence for a below-ground gap. *Journal of Vegetation Science* 4:571–574.

Wild, E. R. 1995. New genus and species of Amazonian microhylid frog with a phyloghenetic analysis of New World genera. *Copeia* 1995:837–849.

Wilkie, D. S. 1987. Impact of swidden agriculture and subsistence hunting on diversity and abundance of exploited fauna in the Ituri forest of northeastern Zaire. Ph.D. diss., University of Massachusetts.

——. 2001. Forest area and deforestation in central Africa: Current knowledge and future directions. In W. Weber, A. Veder, H. Simons Morland, L . J. T. White, and T. Hart, eds., *African Rain Forest Ecology and Conservation*. New Haven, Conn.: Yale University Press.

Wilkie, D. S. and B. Curran. 1991. Why do Mbuti hunters use nets? Ungulate hunting efficiency of bows and nets in the Ituri rain forest. *American Anthropologist* 93:680–689.

Wilkie, D. S. and J. T. Finn. 1990. Slash-burn cultivation and mammal abundance in the Ituri forest, Zaïre. *Biotropica* 22:90–99.

Wilkie, D. S., B. Curran, R. Tshombe, and G. A. Morelli. 1998. Modeling the sustainability of subsistence farming and hunting in the Ituri Forest of Zaire. *Conservation Biology* 12:137–147.

Wilkie, D. S., J. G. Sidle, and G. C. Boundzanga. 1992. Mechanized logging, market hunting and a bank loan in Congo. *Conservation Biology* 6:570–580.

Wilkie, D. S., A. Vedder, R. Oko, and B. Curran. 1994. Second year evaluation of the Congo Forest Conservation Project/Nouabalé-Ndoki Project. Miscellaneous report. New York, N.Y.: Wildlife Conservation Society.

Wilkinson, G. S. 1985. The social organization of the common vampire bat. I. Pattern and cause of association. *Behavioral Ecology and Sociobiology* 17:111–122.

Wilks, C. 1990. *La Conservation des EcosystemForistieres du Gabon*. Gland, Switzerland and Cambridge, United Kingdom: IUCN (International Union for Conservation of Nature and Natural Resources).

Williams, K. D. and G. W. Petrides. 1980. Browse use, feeding behaviour and management of the Malayan tapir. *Journal of Wildlife Management* 44:489–494.

Williamson, E. A. 1988. Behavioural ecology of western lowland gorillas in Gabon. Unpublished Ph.D. diss., University of Stirling.

Williamson, E. A., C. E. G. Tutin, C. E. G., M. E. Rogers, and M. Fernandez. 1990. Composition of the diet of lowland gorillas at Lopé in Gabon. *American Journal of Primatology* 21:256–77.

Willis, E. O. 1974. Populations and local extinctions of birds on Barro Colorado Island, Panamá. *Ecological Monographs* 44:153–169.

Willmer, P. G. and S. A. Corbet. 1981. Temporal and microclimatic partitioning of the floral resources of *Justicia aurea* amongst a concourse of pollen vectors and nectar robbers. *Oecologia* 51:67–78.

Willott, S. J. 1999. The effects of selective logging on the distribution of moths in a Bornean rainforest. *Philosophical Transactions of the Royal Society Of London Series B* 354:1783–1790.

Willson, M.F. 1992. The ecology of seed dispersal. In M. Fenner, ed., *Seeds: The Ecology of Regeneration in Plant Communities*, pp. 61–85. Wallingford, United Kingdom: CAB (Centre for Agriculture and Bioscience).

Willson, M. F., A. K J. Irvine and N. G. Walsh. 1989. Vertebrate dispersal syndromes in some Australian and New Zealand plant communities, with geographic comparisons. *Biotropica* 21:133–147

Wilson, C. C. and W.L. Wilson. 1975. The influence of selective logging on primates and some other animals in East Kalimantan. *Folia primatologica* 23:245–274.

Wilson, D. E. 1973. Bats faunas: A trophic comparison. *Systematic Zoology* 22:14–29.

——. 1990. Mammals of La Selva, Costa Rica. in A. Gentry, ed., *Four Neotropical Rainforests*, pp. 273–286. New Haven, Conn.:Yale University Press.

Wilson, D. E. and D. M. Reeder, eds. 1993. *Mammal Species of the World: a Taxonomic and Geographic Reference.* Washington, D.C. and London, United Kingdom: Smithsonian Institution Press.

Wilson, D. E., J. R. Cole, J. D. Nichols, R. Rudran, and M. S. Foster. 1996. *Measuring and Monitoring Biological Diversity: Standard Methods for Mammals.* Washington, D.C.: Smithsonian Institution Press.

Wilson, E. O. 1987. The little things that run the world (the importance and conservation of invertebrates). *Conservation Biology* 1:344–346.

Wilson, J. W. 1974. Analytical zoogeography of North American mammals. *Evolution* 28:124–140.

Wilson, W. L. and A. D. Johns. 1982. Diversity and abundance of selected animal species in undisturbed forest, selectively logged forest and plantations in East Kalimantan, Indonesia. *Biological Conservation* 24:205–218.

Winemiller, K. O. and G. A. Polis. 1996. *Food Webs: Integration of Patterns and Dynamics.* New York, N.Y.: Chapman and Hall.

Winjum, J. K. and D. K. Lewis. 1993. Forest management and the economics of carbon storage: The nonfinancial component. *Climate Research* 3:111–119.

Winter, J. W. 1979. The status of endangered Australian Phalangeridae, Petauridae, Burramyidae, Tarsipedidae, and the Koala. In M.J. Tyler, ed., *The Status of Endangered Australian Wildlife,* pp. 45–58. Adelaide, Australia: Royal Zoological Society of South Australia.

——. 1984. Conservation studies of tropical rainforest possums. In A. P. Smith and I. D. Hume, eds., *Possums and Gliders,* pp. 469–481. Sydney, Australia: Australian Mammal Society.

Wolda, H. 1979. Abundance and diversity of Homoptera in the canopy of a tropical forest. *Ecological Entomology* 14:181–190.

——. 1992. Trends in the abundance of tropical forest insects. *Oecologia* 89:47–52.

Wong, M. 1985. Understory birds as indicators of regeneration in a patch of selectively logged West Malaysian rainforest. In A. W. Diamond and T. E. Lovejoy, eds., *Conservation of Tropical Forest Birds.* International Council for Bird Preservation Technical Publication No. 4, pp 249–263. ICBP (International Council for Bird Preservation): Cambridge.

Wong, M. 1983. Understory phenology of the regenerating and virgin habitats in Pasoh Forest Reserve, Negri Sembilan, West Malaysia. *Malaysian Forester* 46:197–223.

——. 1986. Trophic organization of understory birds in a Malaysian dipterocarp forest. *Auk* 103:100–116.

Woodley, S., et al. 1998. North American Test of Criteria and Indicators of Sustainable Forestry: Volume 1 (Draft Report). Bogor, India: CIFOR (Centre for International Forestry Research).

Woodman, N., N. A. Slade, and R. M. Timm. 1995. Mammalian community structure in lowland tropical Peru, as determined by removal trapping. *Zoological Journal Linnean Society* 113:1–20.

Woodman, N., R. M. Timm, N. A. Slade, and T. J. Doonan. 1996. Comparison of traps and baits for censusing small mammals in Neotropical lowlands. *Journal of Mammology* 77: 274–281.

Woods, P. 1989. Effects of logging, drought, and fire on structure and composition of tropical forests in Sabah, Malaysia. *Biotropica* 21:290–298.

Woolbright, L. L. and M. M. Stewart. 1987. Foraging success of the tropical frog, *Eleutherodactylus coqui*: The cost of calling. *Copeia* 1987:69–75.

World Bank. 1991. *Forest Policy Paper.* Washington, D.C.: The World Bank.

Wrangham, R. W., C. A. Chapman, and L. J. Chapman. 1994. Seed dispersal by forest chimpanzees in Uganda. *Journal of Tropical Ecology* 10:355–368.

WRI (World Resources Institute). 1994. *World Resources, 1994–95: A Guide to the Global Environment*. New York, N.Y.: Oxford University Press.

WRI (World Resources Institute), PNUMA and UICN (International Union for Conservation of Nature and Natural Resources). 1995. Planificación nacional de la biodiversidad. Translation by FES, Colombia. Washington, D.C.: World Resources Institute.

WRI (World Resources Institute), UNEP (United Nations Environmental Programme) and UNDP (United Nations Development Programme). 1990. *World Resources, 1990–1991*. New York, N.Y.: Oxford University Press.

Wright, A. C. S., D. H. Romney, R. H. Arbuckle, and V. E. Vial. 1959. Land in British Honduras, report of the British Honduras Land Use Survey Team. London, United Kingdom: Her Majesty's Statiory Office,.

Wunderle, J. M., A. Diaz, I. Velasquez, and R. Sharron. 1987. Forest openings and the distribution of understory birds in a Puerto Rican rain forest. *The Wilson Bulletin* 99:22–37.

WWF (World Wildlife Fund). 1995. Joint Implementation: Position Statement. Gland, Switzerland: World Wildlife Fund.

Wyatt, S. 1996. Sustainable forestry and chainsaw mills in Vanuatu. Rural Development Forestry Network, Network Paper 19d:1–15.

Wyatt-Smith, J. 1987. Red meranti-keruing forest. Manual of Malayan silviculture for inland forest. Part III, Chapter 8. Kepong. Malaysia: Forest Research Institute.

Wydowski, R. S. 1978. Responses of trout populations to alterations in aquatic environments: A review. In J. R. Moring, ed., *Proceedings of the Wild Trout-catchable Trout Symposium*, pp. 57–92. Portland, Ore.: Oregon Department of Fish and Wildlife.

Yamakoshi, G. 1998. Dietary responses to fruit scarcity of wild chimpanzees at Bossou, Guinea: Possible implications for ecological importance of tool use. *American Journal of Physical Anthropology* 106:283–295.

Yap, S. W., C. V. Chak, L. Majuakim, A. Anuar, and F. E. Putz. 1995. Climbing bamboo (*Dinochloa* spp.) in Sabah: Biomechanical characteristics, mode of ascent, and abundance in a logged-over forest. *Journal of Tropical Forest Science* 8:196–202.

Yost, J. A. and P. M. Kelley. 1983. Shotguns, blowguns, and spears: The analysis of technological efficiency. In R. B. Hames, and W. T. Vickers, eds., *Adaptive responses of native Amazonians*, pp. 189–224. New York, N.Y.: Academic Press.

Young, T. P. and S. P. Hubbell. 1991. Crown asymmetry, tree falls, and repeat disturbance in a broad leaved forest. *Ecology* 72:1464–1471.

Yumoto, T., T. Maruhashi, J. Yamagiwa, and N. Mwanza. 1995. Seed-dispersal by elephants in a tropical rain forest in Kahuzi-Biega National Park, Zaire. *Biotropica* 27:526–530.

Yusop, Z. and A. Suki. 1994. Effects of selective logging methods on suspended solids concentration and turbidity level in streamwater. *Journal of Tropical Forest Science* 7:199–219.

Zagt, R. J. 1997. *Tree Demography in the Tropical Rain Forest of Guyana. Tropenbos – Guyana Series 3*. Wageningen, The Netherlands: The Tropenbos Foundation.

Zagt, R. J. and M. J. A. Werger. 1998. Community structure and the demography of seedlings and adults in tropical rain forest, with examples from the Neotropics. In M. D. Newberry, N. Brown, and H. Prins, eds., *Population and Community Dynamics in the Tropics*, pp. 193–219. Oxford, United Kingdom: Blackwell.

Zagt, R. J. and R. G. A. Boot. 1997. The response of tropical trees to logging: A cautious application of matrix models. In R. J. Zagt. *Tree Demography in the Tropical Rain Forest of Guyana*. Ph.D. thesis, Utrecht University Series Vol. 3, pp. 167–214. Wageningen, The Netherlands: The Tropenbos Foundation.

Zakaria b. H., M. 1994. Ecological effects of selective logging in a lowland dipterocarp forest on avifauna, with special reference to frugivorous birds. Ph.D. diss., Universiti Kebangsaan.

——. 1996. Can our logged forest fully regenerate naturally? *Proceedings of the National Symposium of Natural Resources Science and Technology*, pp. 124–128. Sabah, Malaysia: University Malaysia Sabah Press.

Zakaria b. H., M. and M. Nordin. 1998. Comparison of frugivory by birds in primary and logged lowland dipterocarp forests in Sabah, Malaysia. *Tropical Biodiversity* 5(1):1–9.

Zar, J. H. 1984. *Biostatistical Analysis*. Englewood Cliffs, N.J.: Prentice-Hill, Inc.

Zhang Keying. 1986. The influence of deforestation of tropical rainforest on local climate and disaster in Xishuangbanna region of China. *Climatology Notes* 35:223–236. Tokyo, Japan.

Zimmerman, B. L. and M. T. Rodrigues. 1990. Frogs, snakes, and lizards of the INPA-WWF (World Wildlife Fund) reserves near Manaus, Brazil. In A. H. Gentry, ed., *Four Neotropical Rainforests*, pp. 426–454. New Haven, Conn.: Yale University Press.

Zimmerman, B. L. and D. Simberloff. 1996. An historical interpretation of habitat use by frogs in a central Amazonian forest. *Journal of Biogeography* 23:27–46.

Zimmerman, J. K., T. M. Aide, M. Rosario, M. Serrano, and L. Herrera. 1995. Effects of land management and a recent hurricane on forest structure and composition in the Luquillo Experimental Forest, Puerto Rico. *Forest Ecology and Management* 78:65–76.

Zimmerman, J. K., W. M. Pulliam, D. J. Lodge, V. Quiñones-Orfila, N. Fetcher, S. Guzmán-Grajales, J. A. Parrota, C. E. Asbury, L. R. Walker, and R. B. Waide. 1995. Nitrogen immobilization by decomposing woody debris and the recovery of tropical wet forest from hurricane damage. *Oikos* 72:314–322.

Zimmerman, J. K, M. R. Willig, L. R. Walker, and W. L. Silver. 1996. Introduction: Disturbance and Caribbean ecosystems. *Biotropica* 28:414–423.

Zou, X. 1993. Species effects on earthworm density in tropical tree plantations in Hawaii. *Biology and Fertility of Soils* 15:35–38.

Zou, X. and G. Gonzalez. 1997. Changes in earthworm density and community structure during secondary succession in abandoned tropical pastures. *Soil Biology and Biochemistry* 29:627–629.

Zou, X., C. P. Zucca, R. B. Waide, and W. H. McDowell. 1995. Long-term influence of deforestation on tree species composition and litter dynamics of a tropical rain forest in Puerto Rico. *Forest Ecology and Management* 78:147–157.

Zuidema, P. A., J. A. Sayer, and W. Dijkman. 1997. Forest fragmentation and biodiversity: The case for intermediate-sized conservation areas. *Environmental Conservation* 23:290–297.

Index